# Physicochemical Hydrodynamics

PRENTICE-HALL INTERNATIONAL SERIES
IN THE PHYSICAL AND CHEMICAL ENGINEERING SCIENCES

NEAL R. AMUNDSON, EDITOR, *University of Minnesota*

ADVISORY EDITORS

ANDREAS ACRIVOS, *University of California*
MICHEL BOUDART, *University of California*
THOMAS J. HANRATTY, *University of Illinois*
DAVID E. LAMB, *University of Delaware*
JOHN M. PRAUSNITZ, *University of California*
L. E. SCRIVEN, *University of Minnesota*

ARIS *Vectors, Tensors, and the Basic Equations of Fluid Mechanics*
LEVICH *Physicochemical Hydrodynamics*
LYKOV AND MIKHAYLOV *Theory of Energy and Mass Transfer*

PRENTICE-HALL, INC.

PRENTICE-HALL INTERNATIONAL, INC., UNITED KINGDOM AND EIRE

PRENTICE-HALL OF CANADA LTD., CANADA

# Physicochemical Hydrodynamics

Veniamin G. Levich
*Corresponding Member of the Academy of Sciences, U.S.S.R.*

*Translated by*
Scripta Technica, Inc.

Prentice-Hall, Inc.
*Englewood Cliffs, N.J.*
1962

PRENTICE-HALL INTERNATIONAL, INC.

*London • Tokyo • Sydney • Paris*

PRENTICE-HALL OF CANADA, LTD.
PRENTICE-HALL DE MEXICO, S.A.



Library of Congress Catalog Card Number: 62-20437

*Printed in the United States of America*

67444-C

# Foreword from the Editor

To perceive so early the significance of a new subject, to conceive then of giving it such wide dominion, to contribute heavily to 20 years of its development, to consolidate that development with an authoritative monograph, all add up to an impressive accomplishment by V. G. Levich. His book in translation will be revealing to the reader who has been kept by a language barrier from intimacy with the considerable researches by Levich himself, his colleagues and compatriots. Brief encounters with his work have been tantalizing: a few articles in Acta Physicochimica URSS when English predominated in the wartime heyday of that journal; a paper communicated to the first Discussion of the Faraday Society; fragmentary secondhand accounts in Western electrochemical and, more recently, chemical engineering literature; some private translations of bits and pieces which have had limited circulation. The glimpses provided by L. D. Landau and E. M. Lifshitz's Fluid Mechanics prompted one reviewer of that translation to surmise that "problems involving small-scale viscous flow, perhaps with an interface between two fluids or with temperature variations, of the kind encountered in chemical engineering, have been treated extensively in U.S.S.R." The reader will find these kinds of problems among several equally important others falling in the purview of Physicochemical Hydrodynamics.

Professor Veniamin Grigorievich Levich, Corresponding Member of the Academy of Sciences of the U.S.S.R., is regarded there as a physical chemist and theoretical physicist; yet he is above all an engineering scientist. He is distinguished by broad scientific competence, unusual physical insight, and real concern for the application of his work. His modus operandi in the present monograph is characterized by quick progress of basic aspects, bold approximations, incisive mathematical analysis, and continual striving for lucid interpretations and useful generalizations. His blend of applied chemistry, applied physics and fluid mechanics is of particular interest to chemical engineers, while his frequent preoccupation with electrode processes holds special appeal for anyone interested in electrochemistry and electroanalytical chemistry. The central theme in this volume is the elucidation of mechanisms of transport phenomena and the conversion of understanding so gained into plain, useful tools for application. Higher mathematics is employed throughout the text, neither as an end in itself nor as an elegant dress for the subject matter, but as a potent implement for sharpening physical understanding.

The development of Physicochemical Hydrodynamics has taken place in what evidently are highly favorable circumstances for the growth of an engineering science. V. G. Levich in his role of principal has benefited from the active interest and support of Academicians A. N. Frumkin, the distinguished physical chemist, and L. D. Landau, the eminent physicist, as well as from the opportunity for creative collaboration with chemists, physicists, and engineers at the Academy of Sciences, Moscow State University, and Moscow Physical Engineering Institute. The present volume is indeed a testimonial to the fertility of interaction among related scientific disciplines.

It is altogether fitting that V. G. Levich's monograph, already a standard reference in the Soviet Union, should appear in translation as one of the first in an international series of volumes devoted to the physical and chemical engineering sciences.

The broad coverage and novel approaches of the author have complicated the work of translating and editing. Without the expert assistance of Prof. A. Acrivos and Mr. J. S. Newman of the University of California at Berkeley, Prof. T. J. Hanratty of the University of Illinois, and Prof. S. Bruckenstein and Prof. W. R. Ranz of the University of Minnesota, the labor of transforming the translation into a more accurate and readable state would have been prolonged and the outcome would have been less successful. Each of these colleagues at the final stage edited one or more chapters falling in his fields of specialization. Credit is due Mr. J. H. Davis of Prentice-Hall, Inc. for his stewardship of the entire undertaking. Every effort has been made to give the reader a rendition faithful to the author's intentions. A number of misprints and a few obvious mistakes in the second Soviet edition have been corrected, and citations of the originals have been inserted where reference is made to Soviet translations of Western texts.

L. E. Scriven

Minneapolis, July 1962

## FROM THE SOVIET PUBLISHER

Physicochemical hydrodynamics is a new area of research which has developed at the boundary zone between physics and chemistry. It is concerned with problems raised by the effect of fluid motion on chemical and physicochemical transformations and by the effect of physicochemical factors on the motion of fluids. A number of new theoretical problems concerning heat transfer and the motion and break-up of drops, bubbles, and jets have been reviewed in the second edition (the first edition appeared in 1952). The theory of a rotating disk electrode, which has recently assumed considerable importance in precise physicochemical measurements, has been developed further.

This graduate text is designed for research workers in physics, physical chemistry and chemistry, as well as for use by students taking advanced courses in theoretical physics and physical chemistry.

## PREFACE TO THE SECOND EDITION

The second edition (the first appeared in 1952), incorporates new sections on the theory of heat transfer in fluids and the theory of motion and break-up of drops, jets, bubbles, etc. A number of new aspects of convective diffusion and other problems formulated by the author and his collaborators are reviewed. Some of these new problems are published here for the first time anywhere. In addition, the book contains brief reviews of numerous experimental studies, conducted both in the Soviet Union and abroad, providing experimental verification of the author's theory of convective mass transfer in fluids.

The original work presented in this book was done at the Institute of Electrochemistry of the Academy of Sciences of the USSR and in the Department of Theoretical Physics of the Moscow Institute of Engineering Physics. The experimental work was conducted in the Department of Electrochemistry of the Moscow State University and at the Institute of Electrochemistry of the Academy of Sciences of the USSR.

Once again I should like to express my deep gratitude to Academician A. N. Frumkin for his numerous, and, to me, very valuable comments and advice, as well as for organizing the experimental studies in the above-mentioned institutions.

I wish to express my deep gratitude to Academician L. D. Landau for his very valuable advice.

I take great pleasure in thanking Dr. Z. Grabowski (Warsaw); Dr. J. Koutecky (Prague); Dr. J. Aygar (Cambridge); N. A. Bagotskaya, Cand. Chem. Sci.; Prof. B. N. Kabanov; Prof. S. S. Kutateladze; A. M. Rozen, Cand. Phys.-math. Sci.; I. L. Rozenfel'd, Dr. Chem. Sci.; G. L. Natanson, Dr. Chem. Sci.; and Prof. N. A. Fuks, for their valuable counsel and comments, which I have incorporated in the second edition of this book.

I wish to thank E. A. Aykazyan, Cand. Chem. Sci.; G. L. Vidovich; Yu. B. Pleskov; V. D. Yukhtanova; A. I. Fedorova, Cand. Chem. Sci.; Asst. Prof. V. P. Galushko (University of Dnepropetrovsk); Prof. Ya. V. Durdin; and Z. U. Dukhnyakova, Cand. Chem. Sci. (University of Leningrad); Prof. I. R. Krichevskiy; and Yu. V. Tsekhanskaya, Cand. Chem. Sci., whose experimental studies were of great assistance in the verification and development of our theoretical work.

The detailed discussions of the theory of polarographic maxima, held at a seminar of the institute of Physical Chemistry, Academy of Sciences of Czechoslovakia (Prague) have been of great value to me.

I should like to express my gratitude to Academician P. Brdicka, who conducted the seminar, and to all the participants, in particular to Dr. J. Koutecky and Dr. G. Koryta.

In the preparation of the second edition of this book, I have been assisted tremendously by Research Associates R. R. Dogonadze and Yu. A. Chizmadzhev. The participation of the following persons should also be acknowledged: Physicomathematical Sciences Yu. B. Ivanov, Cand. Phys.-math. Sci.; V. A. Kir'yanov; B. S. Grafov; Yu. I. Yalamov; E. B. Breshenkova; and V. Yu. Filinovskiy.

I take preat pleasure in offering them all my sincere gratitude.

V. Levich

## FROM THE FOREWORD TO THE FIRST EDITION

The development of physical chemistry in recent decades has been marked by the wide application of methods of investigation used in physics. In particular, the methods of theoretical physics are being used increasingly to solve various problems that lie on the borderline between physics and chemistry.

In the present book, which deals with problems in physicochemical hydrodynamics, methods of computation employed in theoretical physics have been applied to the study of some physicochemical problems in liquid-phase flow.

By physicochemical hydrodynamics we understand the aggregate of problems dealing with the effect of fluid flow on chemical or physicochemical transformations as well as the effect of physicochemical factors of fluid flow.

The chief purpose of this book is to systematize the concepts in physicochemical hydrodynamics that have been developed by us over the period of the last ten years in close cooperation with the scientific staff of the Electrochemistry Division of the Institute of Physical Chemistry, Academy of Sciences of the USSR.

I wish to express my deep gratitude to the Chief of the Division, Academician A. N. Frumkin, as well as to my teacher, Academician L. D. Landau, for the generous measure of extremely valuable suggestions and counsel given me both during the investigations and in the preparation of this book. I consider it a pleasure to acknowledge my indebtedness to Prof. B. N. Kabanov, Prof. I. N. Kuzminykh, and Senior Research Associates T. A. Kryukova, I. A. Bagotskaya, A. V. Gorodetskaya, and E. I. Popova, whose experimental studies contributed to the development of the theory presented here.

V. Levich

# Contents


Preface to the Second Edition. . . . . . . . . . . . . . . . . . . . . . .
From the Foreword to the First Edition. . . . . . . . . . . . . . . . .

Chapter I. Introduction . . . . . . . . . . . . . . . . . . . . . . . . . . 1
1. The Equations of Hydrodynamics . . . . . . . . . . . . . . . 1
2. Similarity of Hydrodynamic Phenomena . . . . . . . . . . 6
3. Fluid Flow at High Reynolds Numbers. Boundary Layer 9
4. Turbulent Flow . . . . . . . . . . . . . . . . . . . . . . . . . . . 20
5. Flow Past a Body With Appreciable Curvature . . . . . 34
Bibliography . . . . . . . . . . . . . . . . . . . . . . . . . . . . . . 36

Chapter II. Convective Diffusion in Liquids . . . . . . . . . . . . 39
6. Diffusion Kinetics . . . . . . . . . . . . . . . . . . . . . . . . 39
7. General Information on Diffusion Kinetics in the Liquid Phase . . . . . . . . . . . . . . . . . . . . . . . . . . . . . . . 40
8. Convective Diffusion in Liquids . . . . . . . . . . . . . . . 46
9. Boundary Conditions for the Equation of Convective Diffusion . . . . . . . . . . . . . . . . . . . . . . . . . . . . . . . 53
10. General Theory of Convective Diffusion in Liquids. . . 57
11. Solution of the Equation for Convective Diffusion to the Surface of a Rotating Disk . . . . . . . . . . . . . . . . . 60
12. Intermediate Kinetics on the Surface of a Rotating Disk. The Method of the Uniformly Accessible Surface . . 72
13. Reduction of the Equation of Convective Diffusion to the Form of the Heat Conduction Equation. . . . . . . . . . 78
14. Diffusion to a Free-Falling Solid Particle . . . . . . . . 80
15. Diffusional Flow to the Surface of A Plate in a Flowing Fluid . . . . . . . . . . . . . . . . . . . . . . . . . . . . . . . . 87
16. Analogy Between Convective Diffusion and Surface Drag 91
17. Solution of the Equation of Convective Diffusion for a Plate, Under Conditions of Intermediate Kinetics . . 93
18. Relaxation of the Diffusion Process. . . . . . . . . . . . . 102
19. Modeling Heterogeneous Chemical Reactions . . . . . . 109
20. An Interior Problem — Diffusion in Laminar Flow in a Tube . . . . . . . . . . . . . . . . . . . . . . . . . . . . . . . 112
21. Distribution of Matter Introduced Into a Fluid Stream. 116
22. Convective Diffusion in a Two-Component Liquid System Near the Critical Solution Point. . . . . . . . . . . . . . 120
23. Diffusional Flow in Natural Convection. The Case of a Vertical Plate . . . . . . . . . . . . . . . . . . . . . . . . . . 127
Bibliography . . . . . . . . . . . . . . . . . . . . . . . . . . . . . . 136

Chapter III. Diffusion Rates in Turbulent Flow . . . . . . . . . . . 139
24. General Laws Governing Mass Transfer in Turbulent Flow . . . . . . . . . . . . . . . . . . . . . . . . . . . . . . . 139
25. Diffusional Flow . . . . . . . . . . . . . . . . . . . . . . . . 144
26. Diffusional Flow to the Inside Surface of a Tube and to the Surface of a Plate . . . . . . . . . . . . . . . . . . . 154
27. Diffusional Flow to a Rotating Disk . . . . . . . . . . . . . 157
28. Diffusional Flow to the Surface of a Non-Streamlined Body . . . . . . . . . . . . . . . . . . . . . . . . . . . . . . . 160
29. Diffusional Flow to a Rough Surface in a Laminar Stream. Dissolution From the Corner of a Body . . . 162
30. Diffusional Flow at a Rough Surface From a Turbulent Stream Near the Surface . . . . . . . . . . . . . . . . . 166
31. Concerning the Analogy Between Convective Diffusion and Surface Friction in Turbulent Flow . . . . . . . . . . 171
32. Motion of Particles Suspended in a Turbulent Flow . . 174
33. Diffusion to Particles Suspended in a Turbulent Fluid. The Elementary Extraction Process . . . . . . . . . . . 176
Bibliography . . . . . . . . . . . . . . . . . . . . . . . . . . . 183

Chapter IV. Heat Transfer in Fluids . . . . . . . . . . . . . . . . 185
34. Heat Transfer in Fluids . . . . . . . . . . . . . . . . . . . . 185
35. The Simplest Problems of Convective Heat Transfer . 190
36. Heat Transfer in Turbulent Flow . . . . . . . . . . . . . . 195
37. Theory of Convective Heat Transfer in Liquid Metals . 197
38. General Interpolation Formula for Heat Flow to Fluids at Any Value of the Prandtl Number . . . . . . . . . . 204
Bibliography . . . . . . . . . . . . . . . . . . . . . . . . . . . 206

Chapter V. Certain Problems in the Theory of Coagulation of Dispersions Involving Liquids and Gases . . . . 207
39. Smoluchowski's Theory . . . . . . . . . . . . . . . . . . . . 207
40. Gradient Coagulation . . . . . . . . . . . . . . . . . . . . . 211
41. Theory of Coagulation of Colloids in Turbulent Flows. 213
42. A Second Mechanism of Coagulation of Aerosol Particles in Turbulent Flows . . . . . . . . . . . . . . . . . 219
43. Impaction of Aerosols and Colloids . . . . . . . . . . . . 221
Bibliography . . . . . . . . . . . . . . . . . . . . . . . . . . . 230

Chapter VI. Passage of Current Through Electrolytic Solutions 231
44. State of Quasi-Equilibrium in an Electrolytic Cell . . . 231
45. Current Flow in an Electrolytic Cell . . . . . . . . . . . . 237
46. Concentration Overpotential . . . . . . . . . . . . . . . . . 242
47. Chemical Overpotential. Hydrogen Overpotential . . . . 250
48. Comparison of Various Factors That Determine the Value of the Current in a Cell . . . . . . . . . . . . . . . 257
49. Current Distribution in An Electrolytic Cell in the Absence of Concentration Overpotential . . . . . . . . . . 260

50. Flow of Current Through an Agitated Electrolyte .... 278
51. Current in a Binary Electrolyte ................ 280
52. Theory of a Disk Electrode in a Binary Electrolyte .. 286
53. Current in the Presence of a Supporting Electrolyte .. 293
54. Diffusion Current at the Surfaces of Disk Electrodes and Plates in the Presence of a Supporting Electrolyte .. 297
55. Comparison of Theory and Experiment .......... 301
56. Quantitative Verification of the Theory. Laminar Flow Regime ........................ 307
57. Quantitative Verfication of the Theory. Turbulent Flow Regime ........................ 317
58. Application of the Theory of Convective Diffusion to the Solution of Electrochemical Problems .......... 323
59. Dissolution of Homogeneous Metals in Acids. Dissolution of Inclusions ........................ 336
60. Study of Kinetic and Catalytic Electrochemical Processes by Means of a Rotating Disk Electrode ..... 345
61. Nonstationary Convective Diffusion. Time Required to Establish a Stationary Regime ............... 357
62. The Case of a Given Surface Concentration ........ 358
63. Establishment of a Regime at a Given Current Density .. 364
Bibliography ........................ 367

Chapter VII. Motion Induced by Capillarity .............. 372
64. The Interfacial Region ........................ 372
65. Conditions of Equilibrium Between Two Fluid Phases at Rest ........................ 373
66. Motion Induced by Capillarity .................. 380
67. Rate of Capillary Rise ........................ 382
68. Thermocapillary Motion ........................ 384
69. Effect of Surface-Active Substances on Liquid Flow ... 390
Bibliography ........................ 394

Chapter VIII. Motion of Drops and Bubbles in Fluid Media .. 395
70. Motion of Liquid Drops in Fluid Media ........... 395
71. Comparison of the Rybczynski-Hadamard Formula with Experimental Data ........................ 402
72. Diffusional Flux to a Moving Drop .............. 404
73. Fall of Drops in Liquid Media in the Presence of Surface-Active Substances ........................ 409
74. Fall of Drops in the Presence of a Surface-Active Material, Adsorption Being the Rate Determining Step .. 413
75. Fall of Drops in the Presence of a Surface-Active Material, the Transfer Rate Being Determined by Diffusion in the Bulk of the Medium and by Surface Diffusion ........................ 416
76. Comparison of Different Theories ............... 420
77. Comparison of the Different Retarding Mechanisms .. 423

78. Limits of Applicability of the Theory . . . . . . . . . . 426
79. Motion of Large Drops . . . . . . . . . . . . . . . . . . . . 429
80. Motion and Dissolution of Gas Bubbles in Liquids . . . 432
81. Motion of Very Small Bubbles . . . . . . . . . . . . . . . 434
82. Motion of Medium-Size Bubbles . . . . . . . . . . . . . . 436
83. Experimental Verification of the Theoretical Expression for the Velocity of Bubble Motion in a Liquid Medium . . . . . . . . . . . . . . . . . . . . . . . . . . . 448
84. Motion of Large Bubbles . . . . . . . . . . . . . . . . . . 451
85. Breakup of Bubbles . . . . . . . . . . . . . . . . . . . . . 452
86. Breakup of Drops . . . . . . . . . . . . . . . . . . . . . . 454
87. Breakup of Drops in Turbulent Liquid Flow . . . . . . . 457
88. Breakup of Drops in the Turbulent Flow of a Liquid Past a Wall . . . . . . . . . . . . . . . . . . . . . . . . . 460
89. Breakup of Drops in Turbulent Gas Flow and Breakup of Bubbles . . . . . . . . . . . . . . . . . . . . . . . . . . 464
90. Dissolution of Gas Bubbles. Elementary Theory of the Bubbling Process . . . . . . . . . . . . . . . . . . . . . . 465
91. Calculation of the Rate of Gas Bubble Dissolution . . . 466
92. Dissolution of a Gas Bubble Suspended in a Turbulent Liquid Stream . . . . . . . . . . . . . . . . . . . . . . . . 469
Bibliography . . . . . . . . . . . . . . . . . . . . . . . . . . 470

Chapter IX. Motion of Particles in Electrolytic Solutions . . . 472
93. Electrokinetic Phenomena . . . . . . . . . . . . . . . . . 472
94. Electrophoretic Motion Near a Plane Surface (Electro-Osmosis) . . . . . . . . . . . . . . . . . . . . . . . . . . . 473
95. Electrophoresis of Solid Dielectric Particles . . . . . 475
96. Electrophoresis of Ideally Polarizable Metal Particles. Calculation of Forces for the Case of a Diffuse Double Layer . . . . . . . . . . . . . . . . . . . . 482
97. Electrophoresis of Ideally Polarizable Metal Particles. Calculation of Forces for the Case of a Helmholtz Double Layer . . . . . . . . . . . . . . . . . . 485
98. Electrocapillary Motion of Mercury Drops in an Electric Field . . . . . . . . . . . . . . . . . . . . . . . . . 490
99. Motion of Liquid Metal Drops in an Electric Field . . 493
100. Motion of Nonideally Polarizable Drops in an Electric Field . . . . . . . . . . . . . . . . . . . . . . . . . . . . . . 506
101. Comparison of Theory and Experiment . . . . . . . . . . 510
102. Fall of Mercury and Emulsified Drops in a Gravitational Field . . . . . . . . . . . . . . . . . . . . . . . . . . 512
103. Potentials of Falling Drops . . . . . . . . . . . . . . . . 519
104. Effect of a Magnetic Field on the Fall of Drops of Mercury . . . . . . . . . . . . . . . . . . . . . . . . . . . . 526
Bibliography . . . . . . . . . . . . . . . . . . . . . . . . . . 530

Chapter X. Theory of the Polarographic Method . . . . . . . . . 532
105. The Polarographic Method . . . . . . . . . . . . . . . . . 532

106. Liquid Motion in a Hanging Mercury Drop........ 534
107. Diffusion Current at a Dropping Mercury Electrode in the Presence of Supporting Electrolytes. The Case of Reversible Reactions .............. 537
108. Corrections to the Equation for the Diffusion Current at a Dropping Mercury Electrode Allowing for Surface Curvature and Non-Steady Efflux of the Liquid .. 544
109. Diffusion Current at a Dropping Mercury Electrode in the Presence of Supporting Electrolyte. Irreversible Reactions.......................... 549
110. Current at a Dropping Mercury Electrode in a Binary Salt Solution ......................... 556
111. Polarographic Maxima ..................... 561
112. Polarographic Maxima of the Second Kind........ 562
113. Polarographic Maxima of the First Kind......... 567
114. Maxima on the Positive and Negative Segments of the Electrocapillary Curve.................... 577
115. Suppression of Polarographic Maxima and Some Practical Applications ...................... 582
Bibliography ......................... 589

Chapter XI. Waves on a Liquid Surface .............. 591
116. Waves on the Surface of an Ideal Liquid ......... 591
117. Waves on the Surface of a Viscous Liquid ........ 599
118. Wave Motion on the Surface of a Low-Viscosity Liquid 603
119. Wave Motion on the Surface of a Highly Viscous Liquid 607
120. Effect of Surface-Active Substances on the Wave Motion of a Liquid ....................... 608
121. Damping of Capillary Waves by Surface-Active Substances ............................ 609
122. Damping of Waves by Soluble Surface-Active Agents .. 620
123. Break-up of a Liquid Jet at Low Flow Velocities. The Case of Symmetrical Deformations........... 626
124. Break-up of a Liquid Jet at Low Flow Velocities. The Case of Arbitrary Deformations............. 634
125. Break-up of a Jet at High Velocities. Atomization .. 639
126. Break-up of High Velocity Jets. The Case of Long Waves ............................. 646
127. Break-up of a Liquid Mass of Arbitrary Shape. Concluding Remarks ..................... 650
128. Capillary Waves on the Surface of a Drop ........ 653
129. Wave Generation on a Flat Liquid Surface........ 655
130. Generation by Wind of Large-Amplitude Deep-Water Waves and Their Damping by Turbulent Friction... 660
Bibliography ........................ 667

Chapter XII. Motion and Diffusion in Thin Liquid Films .... 669
131. Flow of Thin Liquid Films ................. 669
132. The Method of "Blowing-off" Thin Liquid Films ... 672

133. Thickness of Film Remaining of the Surface of a Solid Withdrawn From a Quiescent Liquid .......... 674
134. Wave Motion in Thin Liquid Layers ............. 683
135. Turbulent Motion in a Film ................ 689
136. Dissolution of Gas at the Liquid-Gas Interface of a Flowing Film: The Elementary Scrubbing Process .. 692
Bibliography ........................... 699

# Physicochemical
# Hydrodynamics

CHAPTER I

# Introduction

## 1. THE EQUATIONS OF HYDRODYNAMICS

This book deals with the application of hydrodynamic theory to selected physicochemical problems. Modern hydrodynamical concepts, such as the theories of viscous fluid flow, turbulence, and boundary layers, are used to a considerable extent.

Full treatment of all of these problems is not within the scope of this book. We have therefore assumed that the reader is familiar with the basic principles of hydrodynamics [1]. Certain more specialized aspects of hydrodynamics, however, are presented in appropriate sections of the book. In addition, a brief summary of fundamentals appears in this chapter.

The following discussion is limited to the motion of incompressible liquids, and thus we assume that the fluid density is constant in time and space.

The state of a moving incompressible fluid is fully described if for each point in space and for each instant the following four quantities can be defined: the three components of the fluid velocity $\mathbf{v}$ and the pressure $p$. In an incompressible fluid the velocity $\mathbf{v}$ satisfies the equation of continuity

$$\operatorname{div} \mathbf{v} = \frac{\partial v_x}{\partial x} + \frac{\partial v_y}{\partial y} + \frac{\partial v_z}{\partial z} = 0, \tag{1.1}$$

which expresses the law of conservation of matter.

Three further relations required to determine the desired functions are the equations of motion of an element of the fluid. In vectorial form the first is

$$\rho \frac{d\mathbf{v}}{dt} = -\operatorname{grad} p + \mu \Delta \mathbf{v} + \mathbf{f}. \tag{1.2}$$

This, based upon a unit of fluid volume, is known as the Navier-Stokes equation. The left side comprises the product of the mass of the unit and its acceleration and the right represents the sum of the external forces acting on the unit. The vector $\mathbf{f}$ represents the volume force exerted on the element of fluid; gravity is an example of a volume force. The negative of the pressure gradient is that volume force which acts on the fluid element when pressure changes from point to point. Thus, if we regard a certain volume of fluid as isolated, the net force acting on it is equal to

$$-\oint p\, d\mathbf{S},$$

where dS is an element of the surface enclosing that volume. Converting the surface integral to a volume integral we obtain

$$-\oint p\, d\mathbf{S} = -\int \operatorname{grad} p\, dv.$$

The latter integral represents the total force acting over the enclosed volume. It follows that - grad p, is the force per unit volume. Since it is not the pressure but only its gradient which is required in the equation of motion, the pressure itself may be reckoned from an arbitrary datum.

The term $\mu\Delta\mathbf{v}$, where $\mu$ is the viscosity of the fluid, accounts for the effect of the viscous forces. The viscous nature of the fluid, i.e., its internal friction, is manifested only when one region of the fluid moves relative to another. Faster moving layers of the fluid entrain slower moving ones, and momentum is transferred from the faster to the slower layers. The particular volume force $\mu\Delta\mathbf{v}$ arises in those fluids in which the transfer obeys Newton's law of friction, and the viscous properties are described by a single constant value of viscosity $\mu$. Such fluids are termed normal, or Newtonian.*

Normal fluids include water and aqueous solutions of inorganic and many organic substances. A number of organic liquids, alcohols, hydrocarbons, liquid metals, glycerine, certain resins, glasses, and gases are further examples.

The values of viscosity $\mu$ for different fluids show an unusually wide spread. Some examples are given in Table 1.

Despite this wide variation of viscosity, all of these fluids follow Newton's law of viscosity strictly. There is however a broad class of fluids for which Newton's law of friction does not apply. Such fluids are usually termed non-Newtonian, or anomalous (i.e., complex).

---

*It should be stressed that the viscous properties of a compressible Newtonian fluid are characterized by two constants: the viscosity $\zeta$ and a second viscosity $\mu$. The second enters the equation of motion as a coefficient of the term involving the divergence of velocity. Since div $\mathrm{v} = 0$ is an incompressible fluid, this term does not appear in the equation of motion above, and the second viscosity is not shown.

Table 1

| Substance | Viscosity at 20° C | Kinematic viscosity $\nu = \frac{\mu}{\rho}$ |
|---|---|---|
| Water | 0.010 | 0.010 |
| Air | $1.8 \cdot 10^{-4}$ | 0.150 |
| Mercury | 0.0156 | 0.0012 |
| Glycerine | 8.5 | 6.8 |

The properties of non-Newtonian fluids are considered in this book although their study undoubtedly constitutes one of the subjects of physicochemical hydrodynamics. Unfortunately no well-grounded concepts for a theory of flow of non-Newtonian fluids exist at present. Notwithstanding a very large number of theoretical studies in this field, no consistent quantitative theory with which to treat their hydrodynamic properties has been established.

If we break the acceleration into parts and assume the density of the fluid to be constant, we can rewrite equation (1.2) as

$$\frac{\partial \mathbf{v}}{\partial t} + (\mathbf{v}\,\mathrm{grad})\,\mathbf{v} = -\,\mathrm{grad}\,\frac{p}{\rho} + \nu\,\Delta\mathbf{v} + \frac{\mathbf{f}}{\rho} \tag{1.3}$$

or in component form as

$$\frac{\partial v_i}{\partial t} + v_k\frac{\partial v_i}{\partial x_k} = -\frac{1}{\rho}\frac{\partial p}{\partial x_i} + \nu\frac{\partial^2 v_i}{\partial x_k^2} + \frac{f_i}{\rho}. \tag{1.3'}$$

In equation (1.3′), as always hereafter, summation is to be taken over subscripts which appear twice. Thus, we sum over the subscript k which has the values 1, 2, 3.

The quantity $\nu$, where

$$\nu = \frac{\mu}{\rho} \quad (\mathrm{cm}^2/\mathrm{sec}),$$

is known as the kinematic viscosity of the fluid.

If $f_i$ is omitted, the Navier-Stokes equation can be considerably simplified. Using equation (1.1), we rewrite equation (1.3′) in the form

$$\frac{\partial \rho v_i}{\partial t} = \frac{\partial}{\partial x_k}\left[-p\delta_{ik} + \rho v_i v_k + \mu\left(\frac{\partial v_i}{\partial x_k} + \frac{\partial v_k}{\partial x_i}\right)\right], \tag{1.4}$$

where $\delta_{ik} = \begin{cases} 1 \text{ when } i = k \\ 0 \text{ when } i \neq k \end{cases}$

The equivalence of expressions (1.4) and (1.3′) is confirmed by the fact that

$$\frac{\partial}{\partial x_k}(v_i v_k) = v_k \frac{\partial v_i}{\partial x_k} + v_i \frac{\partial v_k}{\partial x_k} = v_k \frac{\partial v_i}{\partial x_k}$$

and

$$\frac{\partial^2 v_k}{\partial x_i \partial x_k} = \frac{\partial}{\partial x_i} \frac{\partial v_k}{\partial x_k} = 0,$$

since equation (1.1) written in component form is

$$\frac{\partial v_k}{\partial x_k} = 0.$$

If the bracketed expression in equation (1.4) is designated as $p_{ik}$

$$p_{ik} = -p\delta_{ik} + \rho v_i v_k + \mu\left(\frac{\partial v_i}{\partial x_k} + \frac{\partial v_k}{\partial x_i}\right), \tag{1.5}$$

we obtain

$$\frac{\partial \rho v_i}{\partial t} = \frac{\partial p_{ik}}{\partial x_k}. \tag{1.6}$$

The quantity $p_{ik}$ is known as the stress tensor, and since $p_{ik} = -p\delta_{ik} + \mu\left(\frac{\partial v_i}{\partial x_k} + \frac{\partial v_k}{\partial x_i}\right) + \rho v_i v_k$, it represents the total of the nine quantities $p_{xx}$, $p_{yy}$, $p_{xy}$, $p_{xz}$, etc.

From its very definition it is clear that the stress tensor is symmetrical in an isotropic medium, i.e.,

$$p_{ik} = p_{ki}.$$

Indeed, for example,

$$p_{xy} = \mu\left(\frac{\partial v_x}{\partial y} + \frac{\partial v_y}{\partial x}\right) + \rho v_x v_y = \mu\left(\frac{\partial v_y}{\partial x} + \frac{\partial v_x}{\partial y}\right) + \rho v_y v_x = p_{yx}. \tag{1.7}$$

Thus, only six of the nine quantities of $p_{ik}$ are independent.

To clarify the meaning of the tensor $p_{ik}$, let us integrate equation (1.6) over an arbitrary volume, and apply the Gauss-Ostrogradsky theorem to the right side of the equation. Because the summation is with respect to the subscript k, the right side of equation (1.6) is actually a divergence

$$\frac{\partial}{\partial t}\int(\rho v_i)\,dv=\int\frac{\partial p_{ik}}{\partial x_k}\,dv=\oint p_{ik}\,ds_k. \tag{1.8}$$

Consequently equation (1.8) describes the momentum change in an arbitrary fluid volume. The change in momentum within the volume is equal to the net momentum transfer across the surface enclosing that volume. Thus, $p_{ik}$ represents momentum transfer. For example, component $p_{xy}$ is the x-component of momentum transfer across a unit surface which is perpendicular to the y-axis.

$$p_{xy}=(\rho v_x)\,v_y+\mu\left(\frac{\partial v_x}{\partial y}+\frac{\partial v_y}{\partial x}\right).$$

The first term on the right is the x-component momentum transfer accompanying physical transfer (convection) of fluid volume across a surface perpendicular to the y-axis. The second term represents the momentum transfer caused by the fluid viscosity. The viscous properties of the fluid assure the transfer of a portion of the momentum from regions of greater velocity to regions of lesser velocity.

The system of equations of motion (1.1) and (1.2) must be supplemented by a set of boundary conditions.

Numerous experimental studies of the flow of Newtonian liquids past the surface of a solid body wetted by them have established that the layer of fluid immediately adjacent to the surface remains motionless, or, as often stated, the fluid adheres to the solid surface. Velocity measurements have shown that the thickness of this stationary layer is quite small — limited to several individual molecular layers (see Section 132). Nevertheless, the absence of slip past the surface is highly important to fluid flow in general. An analogous phenomenon takes place in gases, when their densities are sufficiently great.

Thus, we may assume that the boundary condition holds at all solid surfaces in contact with a moving liquid

$$\mathbf{v}=0. \tag{1.9}$$

On this situation the fluid exerts on each unit area of the solid a force which is numerically equal to the rate of momentum transfer across the surface.

At the interface between two flowing phases, e.g., fluids, or a liquid and a gas, the velocity does not vanish. Instead, the following boundary conditions must be satisfied:

1) The tangential components of velocity are equal

$$v_t^{(1)}=v_t^{(2)}; \tag{1.10}$$

2) The normal components of velocity vanish

$$v_n^{(1)}=v_n^{(2)}=0; \tag{1.11}$$

3) The forces acting between the fluids are equal and opposite, i.e.,

$$F_n^{(1)} = F_n^{(2)}, \tag{1.12}$$

$$F_t^{(1)} = F_t^{(2)}, \tag{1.13}$$

where the indices 1 and 2 refer to the different fluids.

At the free surface of a fluid the tangential force component vanishes:

$$F_t = 0. \tag{1.14}$$

During the motion of a viscous fluid, energy is dissipated within it. Analysis [2] shows that the energy dissipated in a unit volume is given by

$$\begin{aligned} -\frac{dE}{dt} &= \int \frac{\mu}{2}\left(\frac{\partial v_i}{\partial x_k} + \frac{\partial v_k}{\partial x_i}\right)^2 dV = \\ &= \mu\left[-\int (\mathrm{rot}\,\mathbf{v})^2\, dV + \int \frac{\partial \mathbf{v}^2}{\partial n}\, ds - 2\int [\mathbf{v}\ \mathrm{rot}\,\mathbf{v}]\,\mathbf{n}\, ds\right]. \end{aligned} \tag{1.15}$$

From a mathematical viewpoint, solution of the system of hydromechanical equations presents considerable difficulty, because nonlinear partial differential equations are involved. In practice their general solution is possible in but a few special cases. Thus, an effort is generally made to simplify the equations, and then to find approximate solutions for the simplified system.

## 2. SIMILARITY OF HYDRODYNAMIC PHENOMENA

Dimensional analysis and similarity theory have been applied widely to hydrodynamics and related subjects, especially to heat transfer theory.

Here the exposition is limited to the simplest concepts of the hydrodynamic theory of similarity.

It is to be noted that the methods of similarity theory and dimensional analysis, which represent a scientific basis for the modeling of physical phenomena, are used not only in theoretical studies, but in engineering practice as well. They have found especially wide acceptance in the USSR.

In this book we limit our treatment of similarity theory to a determination of the conditions for similarity in hydrodynamic flow. We make extensive use of this particular aspect of the theory later for background on the theory of similarity. The reader is referred to a number of original papers and monographs [3].

Certain special problems involving application of the theory of similarity to heterogeneous chemical reactions are analyzed in Section 19.

To establish necessary and sufficient conditions for the similarity of two flows, let us consider the flow of a viscous fluid. The equations for flow must be expressed in dimensionless form and therefore all dimensional variables that appear in the hydrodynamic equations must be expressed in terms of factors characteristic of these variables. For instance, let a fluid flow around a body whose characteristic dimension is $l$, or let the fluid flow inside a tube whose radius is $l$. Then the dimensions of the body or the radius of the tube are the characteristic length factors for the regions in which fluid motion occurs. All linear dimensions can therefore be dimensionless ratios of the form $X_i = \frac{x_i}{l}$. In like fashion let $U_0$ represent the velocity of the stream flowing past the body or entering the tube. $U_0$ represents a characteristic velocity of the motion and may be chosen as the velocity factor. All velocities can then be expressed as dimensionless ratios $V_i = \frac{v_i}{U_0}$. Employing these length and velocity factors, we can write the Navier-Stokes equations for the steady flow of an incompressible fluid as

$$\frac{U_0^2}{l} V_k \frac{\partial V_i}{\partial X_k} = -\frac{1}{\rho} \frac{\partial p}{\partial X_i} \frac{1}{l} + \nu \frac{U_0}{l^2} \frac{\partial^2 V_i}{\partial X_k^2}, \tag{2.1}$$

or

$$V_k \frac{\partial V_i}{\partial X_k} = -\frac{\partial P}{\partial X_i} + \frac{\nu}{U_0 l} \frac{\partial^2 V_i}{\partial X_k^2}, \tag{2.2}$$

where $P = \frac{p}{\rho U_0^2}$ is the dimensionless pressure.

Expressed in dimensionless form equation (1.2) contains a dimensionless parameter known as the Reynolds number:

$$\mathrm{Re} = \frac{U_0 l}{\nu}.$$

Equation (2.2) can now be rewritten

$$V_k \frac{\partial V_i}{\partial X_k} = -\frac{\partial P}{\partial X_i} + \frac{1}{\mathrm{Re}} \frac{\partial^2 V_i}{\partial X_k^2}. \tag{2.3}$$

Equation (2.3) must be supplemented by the continuity equation, which in dimensionless form is

$$\frac{\partial V_k}{\partial X_k} = 0. \tag{2.4}$$

Furthermore, if there is to be a single-valued solution of any hydrodynamic problem, boundary conditions must be formulated for the set of surfaces that enclose the space in which fluid motion occurs.

Let us consider two fluid streams moving in geometrically similar regions (i.e., in regions that become interchangeable merely by changing the scale of length).

If the boundary conditions for both flows are identical and if both have identical Reynolds numbers, the dimensionless equations of motion for the two flows will be identical and the flows will be similar both geometrically and dynamically. Thus, geometrical similarity, identity of boundary conditions and equality of Reynolds numbers are made up of sufficient conditions for the similarity of two flows. For example, consider two spheres of radii $R_1$ and $R_2$ with streams of the same fluid flowing past them at different velocities $U_1$ and $U_2$ such that $\frac{U_1}{U_2} = \frac{R_2}{R_1}$; compare this to two identical spheres in flows of different velocities such that $\frac{U_1}{\nu_1} = \frac{U_2}{\nu_2}$. The Reynolds number contains the arbitrary quantities $U_0$, $l$ and $\nu$, and viscosity is a physical property of the fluid. The characteristic velocity and dimension may have any values determined by the boundary conditions. Dimensionless parameters, similar to the Reynolds number and made up of arbitrarily chosen values, are known as the controlling dimensionless numbers. All other dimensionless quantities characteristic of the moving fluid are functions of these controlling numbers. Any hydrodynamic variable may, therefore be expressed as a function of the controlling numbers and the dimensionless coordinates. For example, the velocity of the fluid may be expressed as

$$V_i = \frac{v_i}{U_0} = f\left(\mathrm{Re}, \frac{x_i}{l}\right).$$

In steady flow of an incompressible fluid there is only one controlling number — the Reynolds number. Thus, the dimensionless shear force acting on one square centimeter of the surface past which the fluid streams is equal to

$$\tau = f(\mathrm{Re}).$$

In more complex cases, for unsteady flow or for flow in the presence of an external field of volume forces, etc., other controlling parameters enter along with the Reynolds number. In these cases the flows are similar if the geometrical conditions are similar, the initial and boundary conditions are identical, and all the controlling dimensionless numbers have the same respective numerical values.

The more complex cases are not considered in this book. Instead our discussion is confined to the steady flow of an incompressible fluid. For this case the flow regime is determined by the value of the Reynolds number.

## 3. FLUID FLOW AT HIGH REYNOLDS NUMBERS. BOUNDARY LAYER

In practice the most frequent case encountered is fluid motion at high Reynolds numbers.

When $\mathrm{Re} \gg 1$, the last term in equation (2.3) may be ignored, provided the derivative $\frac{\partial^2 V_i}{\partial X_k^2}$, for one reason or another, does not attain any exceedingly high values.

Omitting the last term in (2.3), we may write

$$V_k \frac{\partial V_i}{\partial X_k} = -\frac{\partial P}{\partial X_i},$$

or upon introducing dimensional quantities

$$v_k \frac{\partial v_i}{\partial x_k} = -\frac{\partial}{\partial x_i}\left(\frac{p}{\rho}\right).$$

In the general case of unsteady flow in the presence of external forces we have

$$\frac{\partial \mathbf{v}}{\partial t} + (\mathbf{v}\nabla)\mathbf{v} = -\operatorname{grad}\frac{p}{\rho} + \mathbf{f}\cdot\frac{1}{\rho}. \tag{3.1}$$

In equation (3.1) viscosity is disregarded. This indicates that viscous forces at high Reynolds numbers are small and play a secondary role.

A fluid with no viscosity usually is termed ideal. Equation (3.1), expressing the law of motion of an ideal fluid, is known as Euler's equation.

Elimination of the viscosity term, which converts the Navier-Stokes to the Euler equation, constitutes a very important simplification. Euler's equation is first order (unlike the Navier-Stokes equation which is second order) and thus can often be integrated in general form.

In the case of steady motion under the influence of conservative external forces, for which $\mathbf{f} = -\operatorname{grad} U$, Euler's equation may be rewritten in the form

$$(\mathbf{v}\operatorname{grad})\mathbf{v} = -\operatorname{grad}\left(\frac{p}{\rho} + \frac{U}{\rho}\right). \tag{3.2}$$

In order to integrate equation (3.2), we will introduce the concept of streamlines. A streamline is a curve whose tangent at every point has the direction of the velocity vector. In steady flow, the streamlines represent trajectories of the fluid particles. Using the identity

$$\operatorname{grad}\frac{v^2}{2} = (\mathbf{v}\operatorname{grad})\mathbf{v} + [\mathbf{v}\operatorname{rot}\mathbf{v}]$$

and noting that the vector $[\mathbf{v} \operatorname{rot} \mathbf{v}]$ is perpendicular to the velocity vector v, we can express the projection of (3.2) on an arbitrary streamline $l$,

$$\left(\operatorname{grad} \frac{v^2}{2}\right)_l = -\left[\operatorname{grad}\left(\frac{p}{\rho} + \frac{U}{\rho}\right)\right]_l,$$

or

$$\frac{\partial}{\partial l}\left(\frac{v^2}{2} + \frac{p}{\rho} + \frac{U}{\rho}\right) = 0,$$

where (for the given streamline)

$$\frac{v^2}{2} + \frac{p}{\rho} + \frac{U}{\rho} = \text{const.} \tag{3.3}$$

Equation (3.3) is known as Bernoulli's theorem. It represents the general integral of the equations of motion for an ideal incompressible fluid. This theorem is analogous, in some degree, to the energy principle of ordinary mechanics. Bernoulli's theorem shows that as we go from regions of higher flow velocity to regions of lower velocity, the pressure of the fluid changes in the opposite direction.

Since Euler's equations, unlike the Navier-Stokes equations, are first-order differential equations, the boundary conditions for an ideal fluid must be changed. For example, vanishing of all velocity components of a fluid at a solid surface is inconsistent with Euler's equations. In an ideal fluid, interaction with a solid body does not exist, since viscosity is absent. Thus, the tangential velocity component is not restricted, and only the normal velocity component reduces to zero at the surface of the solid body:

$$v_n = 0 \text{ (at the surface of a solid body)}$$

Another important conclusion may be drawn from Euler's equations: in an ideal fluid, velocity circulation is conserved in any closed circuit that moves with the fluid

$$\oint \mathbf{v}\, d\mathbf{l} = \int \operatorname{rot} \mathbf{v}\, d\mathbf{s} = \text{const.}$$

It follows from this that if initially $\operatorname{rot} \mathbf{v} = 0$ for a given streamline, the motion along this streamline will remain irrotational thereafter. In particular, any motion of an ideal fluid, started from rest, is irrotational. (This is valid for fluids for which $\rho = \Phi(p)$.)

The motion of a fluid whose velocity at every point equals zero is termed "potential motion".

In potential motion, the velocity of the fluid may always be expressed as

$$\mathbf{v} = \text{grad}\,\varphi, \tag{3.4}$$

where $\varphi$, a function of time ξ position, is termed the velocity potential. When velocity is so represented, the condition rot $\mathbf{v}=0$ is automatically satisfied.

If we apply equation (3.4) to the continuity equation (1.1), we see that the velocity potential must satisfy the Laplace equation

$$\Delta\varphi = 0. \tag{3.5}$$

At the surface of solid bodies which confine a region of fluid motion, the velocity potential must satisfy the boundary conditions

$$v_n = \frac{\partial\varphi}{\partial n} = 0$$

(at the surface of the solid body), where n is the normal to the surface.

In an ideal fluid the velocity potential and, consequently, the velocity distribution may be derived from the solution of a boundary value problem well-known in mathematical physics.

The pressure distribution may be determined from Euler's equation. Since in potential flow rot $\mathbf{v}=0$ everywhere in the fluid, it is possible to write

$$(\mathbf{v}\nabla)\,\mathbf{v} = \text{grad}\,\frac{v^2}{2}.$$

Then Euler's equation assumes the form

$$\text{grad}\left(\frac{\partial\varphi}{\partial t} + \frac{v^2}{2} + \frac{p}{\rho} + \frac{U}{\rho}\right) = 0;$$

hence, it follows that

$$\frac{U}{\rho} + \frac{\partial\varphi}{\partial t} + \frac{v^2}{2} + \frac{p}{\rho} = \text{const.} \tag{3.6}$$

Another important simplification may be included for the case of a steady flow regime, in which $\frac{\partial\varphi}{\partial t} = 0$ and

$$\frac{p}{\rho} + \frac{v^2}{2} + \frac{U}{\rho} = \text{const.} \tag{3.7}$$

Unlike (3.3), the constant in (3.7) has the same value for all streamlines in the fluid.

It follows from Bernoulli's equation (3.3) that when an ideal fluid flows across a solid body, the greatest pressure is attained at that point where the fluid velocity becomes zero (ignoring the influence of gravity). Such a point is known as the point of incidence of flow, or the stagnation point.

When the potential flow of a fluid is steady, the conditions under which that fluid may be considered incompressible are easily defined. The change in density of an ideal fluid is always both adiabatic and reversible (hence isentropic), since no dissipation of energy occurs:

$$\delta\rho = \left(\frac{\partial\rho}{\partial p}\right)_S \delta p,$$

where S is the entropy.

From the Bernoulli equation (3.3) we may conclude that the change in pressure of the fluid is related to its kinetic energy by the approximation

$$\delta p \sim \frac{\rho v^2}{2},$$

since for steady motion $\frac{\partial\varphi}{\partial t} = 0$.

Moreover, it is known that the velocity of sound in a fluid is equal to

$$c = \sqrt{\left(\frac{\partial p}{\partial \rho}\right)_S}.$$

Hence,

$$\delta\rho \sim \frac{\rho v^2}{c^2},$$

or

$$\frac{\delta\rho}{\rho} \sim \frac{v^2}{c^2}.$$

A fluid may be considered incompressible when $\frac{\delta\rho}{\rho} \ll 1$. This inequality is satisfied when the fluid velocity v is low in comparison with the speed of sound c in the fluid.

Thus, the steady flow of an ideal fluid may be described by equations of motion whose solution and analysis are relatively simple.

The approximations employed for an ideal fluid are inadequate, however, for the motion of real fluids even at very high Reynolds numbers. This point is well indicated by the so-called d'Alembert paradox. This paradox states that the drag on a solid body in steady motion through an ideal fluid is zero.

This may be seen clearly in the example of a sphere moving steadily through an ideal fluid. Reasoning based on the Bernoulli equation shows that the force acting on the front hemisphere is exactly matched by the force acting on the rear hemisphere.

The absurdity of this conclusion indicates that the laws of motion for an ideal fluid, and Bernoulli's equation in particular, have limited application. It has been found that viscosity exerts a very significant influence in the region immediately adjacent to the surface of a solid body. Furthermore, in this region the law of conservation of velocity circulation is not valid. It has also been shown that the equations of motion of an ideal fluid admit discontinuous solutions. These solutions, moreover, are not single-valued.

The ideal fluid approximation is unsatisfactory for describing fluid motion near a phase interface. The character of flow in the neighborhood of a solid surface is usually treated in classical hydrodynamics. In Section 80 we also deal with the character of flow in the vicinity of a liquid-gas interface.

The flow velocity must become zero in a real fluid at the surface of a solid body. By the same token, it follows from the equations of motion of an ideal fluid — equations inposing no restrictions on the tangential component of fluid velocity near the solid surface — that the fluid in this region moves with a velocity comparable to that of the flow at a significant distance from the solid surface.

Thus, near the surface of the solid body there must be a thin zone in which the tangential velocity component undergoes a very abrupt change from a high value at the outer border of the zone to zero at the solid surface.

The retardation of the fluid in the boundary layer is caused by viscous forces alone. Mathematically speaking, the velocity gradient in the boundary layer is very large in a direction normal to the wall. The viscosity terms in the Navier-Stokes equations, which depend on derivatives in the same direction accordingly are large even if the fluid viscosity is low.

Although the boundary layer occupies an extremely small volume, it exerts a significant influence on the motion of the fluid. The phenomena that take place in the boundary layer are the source of hydrodynamic resistance to the motion of solids through fluids. Thus, the boundary layer is highly important in many of the problems in physicochemical hydrodynamics.

For ease in manipulation the equations of fluid flow in the boundary layer may be simplified considerably. In a thin boundary layer, all quantities change rapidly in the direction perpendicular to the wall while their tangential rate of change is comparatively small.

Moreover, for a sufficiently short length of the body, the flow in the boundary layer may be regarded as laminar (provided the dimensions of the body are large compared to the thickness of the boundary layer).

Let us consider the steady laminar motion of a fluid, choosing the y-axis perpendicular to the surface of the body and the x-axis along the surface in the direction of flow.

The equations of motion, (1.1) and (1.2), for the components in steady motion, take the following forms:

$$v_x \frac{\partial v_x}{\partial x} + v_y \frac{\partial v_x}{\partial y} = -\frac{1}{\rho}\frac{\partial p}{\partial x} + \nu\left(\frac{\partial^2 v_x}{\partial x^2} + \frac{\partial^2 v_x}{\partial y^2}\right), \tag{3.8}$$

$$v_x \frac{\partial v_y}{\partial x} + v_y \frac{\partial v_y}{\partial y} = -\frac{1}{\rho}\frac{\partial p}{\partial y} + \nu\left(\frac{\partial^2 v_y}{\partial x^2} + \frac{\partial^2 v_y}{\partial y^2}\right), \tag{3.9}$$

$$\frac{\partial v_x}{\partial x} + \frac{\partial v_y}{\partial y} = 0. \tag{3.10}$$

If we designate the thickness of the boundary layer as $\delta_0$ and the dimensions of the body as $l$, it may be assumed that the change in velocity along the y-axis takes place over distances of the order of $\delta_0$, and along the x-axis over distances of the order of $l$.

The entire zone of motion may be roughly subdivided into two regions: a region of inviscid motion and a boundary layer region in which viscosity plays an important part.

In the first region, we can omit the viscosity terms in the Navier-Stokes equation and substitute Euler's equation.

To simplify the Navier-Stokes equations within the boundary layer, we can then utilize the fact that the thickness of this layer is very small compared to its length along the body. Let us first introduce dimensionless coordinates into equations (3.8) to (3.10) and define

$$x = lX, \qquad y = \delta_0 Y. \tag{3.11}$$

These coordinates range between the limits

$$0 \leqslant X \leqslant 1, \qquad 0 \leqslant Y \leqslant 1. \tag{3.12}$$

With the new variables, equations (3.8) to (3.10) take the following forms:

$$\frac{v_x}{l}\frac{\partial v_x}{\partial X} + \frac{v_y}{\delta_0}\frac{\partial v_x}{\partial Y} = -\frac{1}{\rho l}\frac{\partial p}{\partial X} + \frac{\nu}{l^2}\frac{\partial^2 v_x}{\partial X^2} + \frac{\nu}{\delta_0^2}\frac{\partial^2 v_x}{\partial Y^2}, \tag{3.8'}$$

$$\frac{v_x}{l}\frac{\partial v_y}{\partial X} + \frac{v_y}{\delta_0}\frac{\partial v_y}{\partial Y} = -\frac{1}{\rho\delta_0}\frac{\partial p}{\partial Y} + \frac{\nu}{l^2}\frac{\partial^2 v_y}{\partial X^2} + \frac{\nu}{\delta_0^2}\frac{\partial^2 v_y}{\partial Y^2}, \tag{3.9'}$$

$$\frac{1}{l}\frac{\partial v_x}{\partial X} + \frac{1}{\delta_0}\frac{\partial v_y}{\partial Y} = 0. \tag{3.10'}$$

To compare the magnitudes of the velocity components $v_x$ and $v_y$, we see from (3.10′) that

$$v_y = -\frac{\delta_0}{l}\int_0^1 \frac{\partial v_x}{\partial X}\,dY. \tag{3.13}$$

The derivative $\frac{\partial v_x}{\partial X}$ has no unusual features, and its integral, within the limits zero and 1, is of the order of $v_x$. Therefore, in the boundary layer,

$$v_y \sim \frac{\delta_0}{l} v_x \ll v_x. \tag{3.14}$$

Using equation (3.14) the different terms in equation (3.8) may be evaluated. Since Y varies within the limits prescribed by equation (3.12), the derivatives

$$\frac{\partial v_x}{\partial Y} \sim \frac{\partial^2 v_x}{\partial Y^2} \sim v_x.$$

In similar fashion

$$\frac{\partial v_x}{\partial X} \sim \frac{\partial^2 v_x}{\partial X^2} \sim v_x.$$

Therefore, we can disregard the term from the right side of equation (3.8′)

$$\frac{\nu}{l^2}\frac{\partial^2 v_x}{\partial X^2} \sim \frac{v_x}{l^2}\nu$$

as compared to the term

$$\frac{\nu}{\delta_0^2}\frac{\partial^2 v_x}{\partial Y^2} \sim \frac{v_x}{\delta_0^2}\nu.$$

On the left side of (3.8′), however, both terms have the same order of magnitude:

$$\frac{v_x}{l}\frac{\partial v_x}{\partial X} \sim \frac{v_x^2}{l};$$

In view of (3.14),

$$\frac{v_y}{\delta_0}\frac{\partial v_x}{\partial Y} \sim \frac{v_x v_y}{\delta_0} \sim \frac{v_x^2}{l}.$$

Equation (3.8) can therefore be written in its final form:

$$v_x \frac{\partial v_x}{\partial x} + v_y \frac{\partial v_x}{\partial y} = -\frac{1}{\rho}\frac{\partial p}{\partial x} + \nu \frac{\partial^2 v_x}{\partial y^2}. \qquad (3.15)$$

Knowing that all terms in equation (3.15) are of the same order of magnitude, we can estimate the thickness of the boundary layer

$$\frac{v_x}{l}\frac{\partial v_x}{\partial X} \simeq \frac{\nu}{\delta_0^2}\frac{\partial^2 v_x}{\partial Y^2}. \qquad (3.16)$$

If the velocity $v_x$ at the outer edge of the boundary layer attains the value $U_0$, then, from equation (3.16), we get

$$\frac{U_0}{l} \sim \frac{\nu}{\delta_0^2},$$

or

$$\delta_0 \sim \sqrt{\frac{\nu l}{U_0}} \sim \sqrt{\frac{l^2}{\mathrm{Re}}} \sim \frac{l}{\sqrt{\mathrm{Re}}}. \qquad (3.17)$$

Thus, $\delta_0$ is smaller than the length of the body by a factor approximately equal to the reciprocal of the square root of the Reynolds number. Of course, it is essential here that the inequality $\sqrt{\mathrm{Re}} \gg 1$ be fulfilled. This estimate may be confirmed by more rigorous calculations (see below).

However, it should be emphasized that the concept of a well defined thickness of the boundary layer requires qualification. The transition from viscous flow in the boundary layer to inviscid flow in the main stream is smooth and gradual. The quantity $\delta_0$ represents the thickness of the region across which the principal change in velocity from zero to $U_0$ takes place.

It also follows from equation (3.15) that the derivative

$$\frac{1}{\rho}\frac{\partial p}{\partial x} \sim \frac{U_0^2}{l}. \qquad (3.18)$$

Evaluating the terms of equation (3.9), we find their order of magnitude to be $\frac{U_0^2 \delta_0}{l^2}$, i.e., smaller than that of the terms of equation (3.8) by the factor $\frac{\delta_0}{l}$.

From this it follows that the pressure gradient normal to the surface, which appears in equation (3.9), equals

$$\frac{1}{\rho}\frac{\partial p}{\partial y} \sim \frac{\delta_0 U_0^2}{l^2}.$$

Comparing this with (3.18), we obtain the following approximation

$$\frac{\partial p}{\partial y} \sim \frac{\delta_0}{l}\frac{\partial p}{\partial x}.$$

This means that the pressure gradient in the direction of the normal is small compared to the change of pressure along the surface. Therefore, excluding second-order terms, we can write in place of equation (3.9)

$$\frac{\partial p}{\partial y}=0. \tag{3.19}$$

Equation (3.19) shows that the pressure does not change in the normal direction but remains equal to the pressure outside the boundary layer. Therefore the pressure variation in the x-direction within the boundary layer is determined by the change in pressure outside. The latter may be determined from Bernoulli's theorem.

The boundary layer equations admit an exact solution for the case of flow past a semi-infinite plate when its leading edge encounters a fluid moving at a velocity $U_0$ (Figure 1).

In order to find solutions to equations (3.15) and (3.19), for the boundary conditions

$$v_x=v_y=0 \quad \text{at} \quad y=0, \tag{3.20}$$

$$v_x \to U_0 \text{ as } y \to \infty \text{ (outside the boundary layer)}, \tag{3.21}$$

we note at the outset that the velocity in the outer region is constant. It follows from the Bernoulli equation that the pressure in the outer region is also constant. As a result, the term containing the pressure gradient in equation (3.15) can be omitted and the equation rewritten

$$v_x\frac{\partial v_x}{\partial x}+v_y\frac{\partial v_x}{\partial y}=\nu\frac{\partial^2 v_x}{\partial y^2}. \tag{3.22}$$

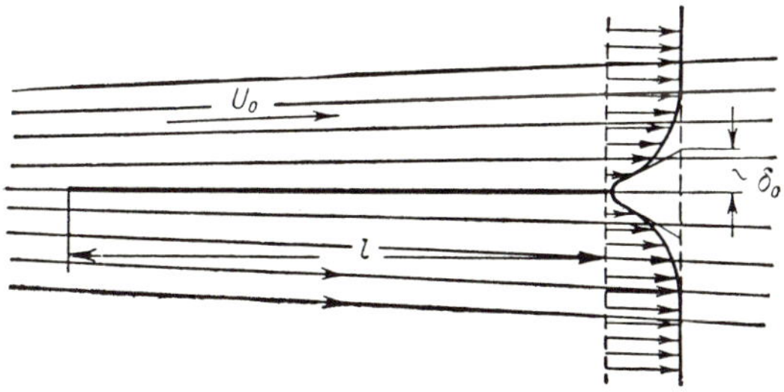


Figure 1. Flow past a plate.

To find the distribution of the boundary layer velocities $v_x$ and $v_y$ satisfying the continuity equation (3.10), we use the stream function $\psi$, defined by

$$v_x=\frac{\partial\psi}{\partial y}; \tag{3.23}$$

$$v_y = -\frac{\partial\psi}{\partial x}. \tag{3.24}$$

In this case, equation (3.10) becomes an identity.

Equations (3.10) and (3.22) and the boundary conditions contain no characteristic parameter having the dimension of length. It is therefore convenient in equation (3.22) to define a new dimensionless variable

$$\eta = \frac{1}{2}\sqrt{\frac{U_0}{\nu x}}\cdot y. \tag{3.25}$$

Let us put the stream function $\psi$ in the form

$$\psi = \sqrt{\nu U_0 x}\cdot f(\eta), \tag{3.26}$$

Here it is evident that $f(\eta)$ satisfies an ordinary differential equation. Thus, we have

$$v_x = \frac{\partial\psi}{\partial y} = \frac{1}{2}U_0 f'(\eta), \tag{3.27}$$

$$v_y = -\frac{\partial\psi}{\partial x} = \frac{1}{2}\sqrt{\frac{\nu U_0}{x}}(\eta f' - f), \tag{3.28}$$

$$\frac{\partial v_x}{\partial x} = -\frac{1}{4}\frac{U_0}{x}\eta f'', \tag{3.29}$$

$$\frac{\partial v_x}{\partial y} = \frac{U_0}{4}\sqrt{\frac{U_0}{\nu x}}f'', \tag{3.30}$$

$$\frac{\partial^2 v_x}{\partial y^2} = \frac{1}{8}U_0\frac{U_0}{\nu x}f'''. \tag{3.31}$$

On substituting the corresponding quantities into equation (3.22), we obtain

$$f''' + f''f = 0. \tag{3.32}$$

Substitution of $v_x$ and $v_y$ into the boundary conditions (3.20) and (3.21) gives

$$f = f' = 0 \quad \text{at} \quad \eta = 0, \tag{3.33}$$

$$f' = 2 \quad \text{at} \quad \eta \to \infty. \tag{3.34}$$

If we let $f''(0) = \alpha$, the solution of equation (3.32) can be written in the form suggested in [4]

$$f = \frac{\alpha\eta^2}{2!} - \frac{\alpha^2\eta^5}{5!} + \frac{11\alpha^3\eta^8}{8!} + \ldots \tag{3.35}$$

For large values of $\eta$, the series is inconvenient to use. In such cases the limiting value of f that satisfies condition (3.34) is obtained by means of numerical integration.

The following value has been determined in this manner [4]

$$\alpha = 1.33.$$

The drag force $F_x$ acting on one side of the plate is given by

$$F_x = \int_0^b \int_0^l \mu \left(\frac{\partial v_x}{\partial y}\right)_{y=0} dx\, dz = \frac{\alpha \rho U_0^2}{2} bl \sqrt{\frac{\nu}{U_0 l}}, \tag{3.36}$$

where b is the width of the plate.

Instead of the drag force, a so-called drag coefficient is often employed. It is defined as

$$K_f = \frac{F_x}{\rho \frac{U_0^2}{2} \cdot bl} = \frac{\alpha}{\sqrt{\text{Re}}}. \tag{3.37}$$

If the thickness of the boundary layer $\delta_0$ is defined as the distance from the surface of the plate to the point where $v_x$ attains a value equal to 90 percent of the velocity of the main stream, $U_0$, then, by numerical computation, we obtain the following value for $\delta_0$ (Figure 2):

$$\delta_0 = 5.2 \sqrt{\frac{\nu x}{U_0}}. \tag{3.38}$$

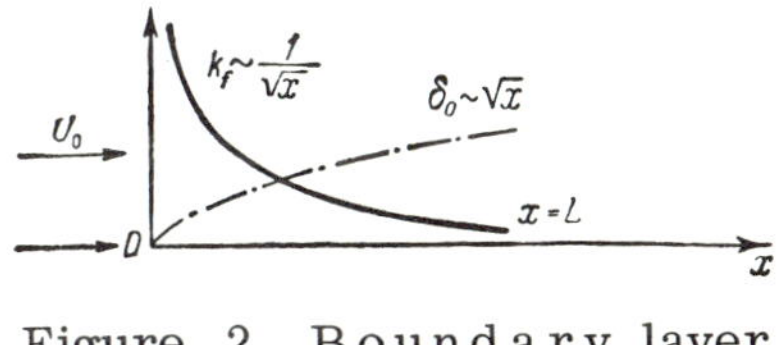


Figure 2. Boundary layer thickness and shear stress as functions of the coordinate x along a plate.

Qualitatively, the formulas derived for a plate are applicable to an arbitrary body with a small curvature.

Let us write the distribution of velocities for $\eta \ll 1$ in explicit form* (Figure 3):

---

*It would be incorrect to verify the inequality $\frac{\partial^2 v_x}{\partial y^2} \gg \frac{\partial^2 v_x}{\partial x^2}$ by substituting the expansion (3.29) in it. In the neighborhood of the point y = 0 the behavior of the continuous function $v_x$ is defined by the boundary condition (3.20).

$$v_x = \frac{U_0}{2}\left(\alpha\eta - \frac{\alpha^2\eta^4}{4!} + \dots\right) \approx \frac{1.33U_0}{4}\sqrt{\frac{U_0 y^2}{\nu x}} \approx \frac{U_0 y}{\delta_0}, \tag{3.39}$$

$$v_y \approx \frac{\alpha}{4}\sqrt{\frac{U_0\nu}{x}}\,\eta^2 \approx \frac{\alpha}{16}\frac{U_0^{3/2} y^2}{\nu^{1/2} x^{3/2}} \approx \frac{\nu y^2}{\delta_0^3}. \tag{3.40}$$

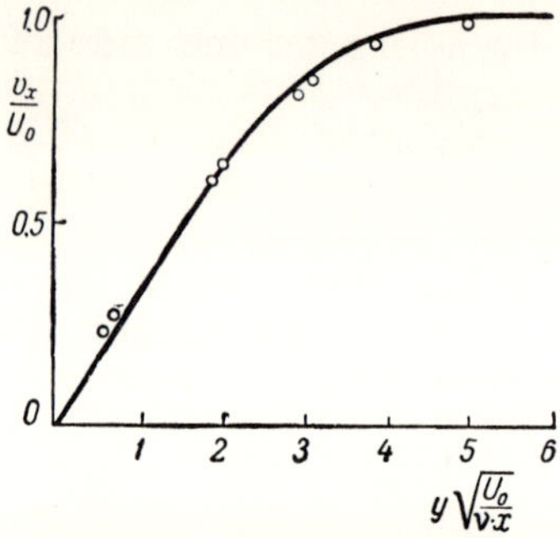


Figure 3. Distribution of the tangential component of velocity in the boundary layer on a plate.

Circles designate experimental values.

## 4. TURBULENT FLOW

At high Reynolds numbers, the nature of fluid flow usually differs from that examined in the preceding sections.

At a certain value of the Reynolds number, steady laminar flow gives way to distinctly unsteady, chaotic motion in which only on the average is there net flow in a particular direction.

The steady advance of a fluid in separate layers is known as laminar flow. The unsteady, chaotic motion in which the flow velocity fluctuates about some average value is known as turbulent flow.*

Studies of the transition to the turbulent regime have shown that it is related to the inherent instability of laminar flow at Reynolds numbers in excess of some critical value $Re_{cr}$.

---

*Very often in general literature, and especially in physicochemical literature, the discussion of turbulent motion emphasizes its rotational nature. While it is true that turbulent motion is usually rotational, it may sometimes be irrotational, or potential motion (although in such a case the motion is noticeably damped). Furthermore, laminar flow of a viscous fluid is also rotational. The distinguishing feature of the turbulent regime is the chaotic and distinctly unsteady nature of the motion of the fluid particles, and not the rotational character that this motion may have.

The gross over-all motion of a fluid is always subject to infinitesimal disturbances. At $Re < Re_{cr}$, disturbances that occur in the fluid are rapidly damped. At $Re > Re_{cr}$, disturbances are not damped, but rather reinforce each other. If these disturbances are periodic in nature and if their frequencies are incommensurable, the effect of their superposition will be the establishment of a chaotic regime. Random eddies are also superposed on the basic motion. In the region $Re \gg Re_{cr}$ the magnitudes of the eddy velocities are comparable to the average velocity.

If we observe the trajectory of a particular mass of fluid in turbulent flow, we find it highly complicated and involved; its tendency toward a systematic advance can only be described in terms of averages. The trajectory in this case is somewhat similar to the motion of a gas molecule in a stream of gas. This indicates that the theory of turbulent flow must be statistical in nature. A quantitative theory of turbulence has not been perfected. However, thanks to the work of Soviet scientists A. A. Fridman, L. V. Keller, A. N. Kolmogorov, L. D. Landau, A. M. Obukhova, L. G. Loitsyansky and M. D. Millionshchikov, as well as von Karman, Prandtl, Heisenberg, Lin, Taylor and others it has been advanced considerably. We can, therefore, state that a qualitative, or perhaps a semi-quantitative, theory of turbulent motion has evolved [5].

Let us first* examine qualitatively the general nature of turbulent motion at $Re \gg Re_{cr}$. Such motion is known as developed turbulence.

Eddy velocities of extremely varied magnitudes are superposed upon the average motion of a fluid having a velocity U. Turbulence eddies must be characterized by their velocities, and by the distances over which these velocities change significantly. These distances are known as the scale of motion. The most rapid eddy motion also has the largest scale of motion. The velocities $v'$ of the most rapid eddies are approximately

$$v' \approx \Delta U, \tag{4.1}$$

where $\Delta U$ is the change in the average velocity over a distance equal to the scale of an eddy $l$. Thus, for the example of turbulent motion in a tube, the largest scale $l$ of turbulence eddies is equal to the diameter of the tube, and the eddy velocities vary within the range of average velocity over that distance, i.e., they are of the order of the maximum value of the velocity at the center of the tube.

Such large scale eddies contain the main part of the kinetic energy of turbulent motion.

The Reynolds numbers of these motions, defined as $\frac{\Delta U \cdot l}{\nu}$, have values equal to the Reynolds number of the stream taken as a whole.

---

*What follows is in accordance with the L. D. Landau and M. M. Lifshitz presentation [5].

Together with these large scale eddies, turbulent flow also includes eddies of smaller scale $\lambda$, with lesser velocities $v_\lambda$. Although the number of such small eddies is very large, they represent only a small portion of the kinetic energy of the stream. Nevertheless, small eddies play an important part in turbulent flow.

To clarify their role, let us examine the Reynolds number that corresponds to an eddy of scale $\lambda$, i.e., $Re_\lambda = \frac{v_\lambda \lambda}{v}$. The smaller the value of $\lambda$ and of the corresponding velocity $v_\lambda$, the smaller is the value of $Re_\lambda$. For large scale eddies the Reynolds number is very large. Therefore, in fluid motion with a scale $\lambda \approx l$, viscous forces actually have no effect. Such motion takes place without energy dissipation. The superposition of large scale eddies on each other creates small scale eddies, for which the Reynolds numbers rapidly decrease with decreasing $\lambda$.

At a certain value of $\lambda = \lambda_0$ the Reynolds number for the corresponding motion $Re_{\lambda_0} = \frac{v_{\lambda_0}\lambda_0}{\nu}$ is approximately unity. This means that in the region of $\lambda_0$, viscous forces begin to have a noticeable effect on the motion of the fluid. Eddy motion of a scale $\lambda_0$ is accompanied by dissipation of energy.

With a large quantity of small scale motion, there is a considerable dissipation of energy, which is transformed to heat. This energy is continually drawn by small scale motions from the large scale motions, so that one may visualize the existence of a continuous transfer of energy from large scale eddies to eddies progressively smaller in scale, until in eddies with scale on the order of $\lambda_0$ it is converted to heat. Small scale motions serve as a "bridge" by means of which the kinetic energy of large scale motions may be converted into thermal energy. For steady state fluid flow the process of energy transfer is also steady in nature. Eddies of a given scale receive as much energy from larger scale eddies as they in turn pass on to smaller scale eddies. Thus, although turbulent motion occurs only at relatively high Reynolds numbers, it is accompanied by considerable dissipation of energy. From this standpoint it is possible to define a certain effective "eddy viscosity" $\mu_{turb}$ appropriate to turbulent flow. This "eddy viscosity" expresses energy losses occurring in the flow per second, per unit volume, by an equation analogous to (1.15):

$$\varepsilon = -\frac{dE}{dt} = \mu_{turb}\left(\frac{\Delta U}{l}\right)^2. \tag{4.2}$$

The order of magnitude of the eddy viscosity may be determined from considerations of similarity.

It is clear from the foregoing that the magnitude of $\epsilon = -\frac{dE}{dt}$, is not a function of the scale of the motion but is a characteristic constant for a given flow. In particular, for the largest scale motions, it equals the energy dissipated in the process of creating the smaller scale motions. This process occurs at high Reynolds numbers and can not be a function of the molecular viscosity $\mu$ of the fluid. Therefore, $\epsilon$ must be determined from quantities characteristic of large scale turbulent motion. These include the following: the velocity $\Delta U$, the scale of motion $l$ and the density of the fluid $\rho$ (no other quantities besides these and $\mu$ enter into the hydrodynamic equations). We can combine $\Delta U$, $l$ and $\rho$ into a single dimensional quantity, $[\epsilon]$ in whose units are erg/cm³ sec

$$\varepsilon \approx \rho \frac{(\Delta U)^3}{l}. \tag{4.3}$$

Comparing (4.2) and (4.3), we have

$$\mu_{\text{turb}} = \rho \,\Delta U l. \tag{4.4}$$

The corresponding kinematic viscosity may be expressed as

$$\nu_{\text{turb}} = \frac{\mu_{\text{turb}}}{\rho} = \Delta U l. \tag{4.5}$$

These last equations can also be derived on the basis of analogy already mentioned that exists between turbulent motion and random motion of gas molecules. If the analogy between these two types of motion is closely followed, the scale of motion $l$ may be considered as the analog of the length of the mean free path, and the eddy velocity as the analog of the average velocity of the gas molecules.

In this way equations (4.4) and (4.5) may be derived directly from the well-known equations of the kinetic theory of gases. Developing the analogy further, we can write the approximation for eddy velocity as is usually done in the kinetic theory of gases:

$$\Delta U \approx l \frac{\partial U}{\partial l}. \tag{4.6}$$

The effective eddy viscosity is very large in comparison with the ordinary viscosity. Indeed, having set up the relation $\nu \frac{\nu}{\text{turb}}$, we find

$$\frac{\nu}{\nu_{\text{turb}}} = \frac{\nu}{\Delta U l} \approx \frac{1}{\text{Re}} \ll 1. \tag{4.7}$$

With the aid of $\mu_{\text{turb}}$, we can express the drag acting on 1 cm² of a solid surface (shear stress) in the form

$$\tau = \frac{F}{S} \approx \mu_{\text{turb}} \left(\frac{\Delta U}{l}\right) \approx \mu_{\text{turb}} \frac{\partial U}{\partial l} \approx \rho l^2 \left(\frac{\partial U}{\partial l}\right)^2 = \alpha \rho l^2 \left(\frac{\partial U}{\partial l}\right)^2, \tag{4.8}$$

where $\alpha$ is a certain numerical factor.

In the following we consider two special cases of turbulent motion: motion having a scale $\lambda \ll l$, i.e., small scale turbulence at a distance from the solid walls, and turbulent motion close to the solid walls. In these cases it is possible to determine the characteristics of the turbulent flow on the basis of similarity.

Let us first examine small scale motion ($\lambda \ll l$) in a volume of fluid. Let us assume, however, that $\lambda \gg \lambda_0$, so that the type of motion is inviscid. Let us find the velocity $v_\lambda$ of turbulence eddies in scale $\lambda$ (or, in other words, the change in the motion velocity over a distance on the order of $\lambda$).

The quantity $v_\lambda$ can be a function only of $\rho$, $\lambda$ and the constant $\epsilon$, since these characterize motion on any scale. The motion of the fluid (at $\lambda \gg \lambda_0$) cannot be a function of the fluid viscosity $\nu$. Neither can it be a direct function of the scale $l$ or of the velocity of the flow U (since $\lambda \ll l$).

The only combination of the quantities $\rho$, $\lambda$ and $\epsilon$, having the dimensions of the velocity is $\left(\frac{\epsilon\lambda}{\rho}\right)^{1/3}$. Thus,

$$v_\lambda \approx \left(\frac{\varepsilon\lambda}{\rho}\right)^{1/3}. \qquad (4.9)$$

Using (4.3) to express $\epsilon$ in terms of $\Delta U$, we find

$$v_\lambda \approx \Delta U \left(\frac{\lambda}{l}\right)^{1/3}. \qquad (4.9')$$

Thus, the eddy velocities for motion of scale $\lambda$ are smaller than the velocity of the main flow by the factor $\left(\frac{\lambda}{l}\right)^{1/3}$.

The reductions in velocity and scale are matched by a corresponding reduction in the Reynolds number, according to the relation

$$\mathrm{Re}_\lambda = \frac{v_\lambda \lambda}{\nu} = \frac{\Delta U \lambda^{4/3}}{\nu l^{1/3}} = \mathrm{Re}\left(\frac{\lambda}{l}\right)^{4/3}.$$

At certain scale $\lambda_0$, known as the inner scale (microscale) of turbulence, the Reynolds number $\mathrm{Re}_{\lambda_0}$ is found to be approximately unity. Evidently,

$$\lambda_0 \approx \frac{l}{\mathrm{Re}^{3/4}} \approx \left(\frac{\nu^3\rho}{\varepsilon}\right)^{1/4}. \qquad (4.10)$$

Starting at this scale level, the motion of the fluid is viscous in nature. Turbulence eddies of a scale $\lambda \leq \lambda_0$ do not suddenly disappear, but are gradually damped due to the effects of viscosity.

Let us now examine the case of turbulent motion near a solid surface by first considering fluid flowing past a flat plate of infinite extent in the plane $y = 0$. Let the mean flow be in the x-direction and the average velocity be $v_x = U$. The average velocity is, in general, a function of the distance of the fluid layer from the surface of the solid body, and thus $U = U(y)$. Upon the average motion of the fluid in the x-direction, there is a superposed eddy motion in all directions.

Let us find the function $U(y)$. To do this we may employ equation (4.8) by rewriting it in the form

$$\frac{\partial U}{\partial y} = \sqrt{\frac{\tau}{\rho\alpha}} \cdot \frac{1}{l}. \tag{4.11}$$

Because all points of the infinite plane $y = 0$ along which there is flow are entirely equivalent, the shear stress $\tau$ is constant over all planes. This may be interpreted as follows. The quantity $\tau$ represents the momentum transferred from the flowing fluid to the wall. Within the fluid flowing along the wall there is a continuous transfer of momentum (a constant momentum flux equal to $\tau$) from rapidly moving, more distant fluid layers to layers adjacent to the wall.

Since the momentum transfer must satisfy the law of conservation of momentum, and, by assumption, is the same along the entire surface (i.e., it is not a function of the coordinate x), it must be constant and in a direction normal to the wall (i.e., it can not be a function of the coordinate y). This ignores changes in momentum transfer arising from energy dissipation caused by molecular viscosity.

Considering that $\tau$ is constant, we can now rewrite equation (4.11)

$$U = \sqrt{\frac{\tau}{\rho\alpha}} \int \frac{dy}{l} + \text{const} = \frac{v_0}{\sqrt{\alpha}} \int \frac{dy}{l(y)} + C_1, \tag{4.12}$$

where $C_1$ is a constant of integration and v designates the quantity

$$v_0 = \sqrt{\frac{\tau}{\rho}}. \tag{4.13}$$

In order to integrate the equation (4.12), it is necessary to determine the scale of the motion as a function of the distance $l(y)$ separating the fluid layer from the solid surface. The special feature of the flow situation we are examining is that the conditions determining the flow regime do not include the dimensions of the body or any other linear quantity which could be used to describe a characteristic scale of large turbulence eddies $l$. It is logical, therefore, to assume that

$$l(y) \approx y. \tag{4.14}$$

The approximation (4.14) shows that the eddy scale increases with distance from the solid wall. And the latter is the only linear quantity which can be used to describe the flow regime. Such an assumption appears to be quite natural: the solid wall retards the fluid moving nearby, so that the velocity gradually diminishes as the wall is approached.

Using the approximation (4.14), we can derive from equation (4.12)

$$U=\frac{v_0}{\sqrt{a}}\ln y+C_1. \tag{4.15}$$

The physical significance of the quantity $v_0$ is clarified by noting that the quantity $v'$, according to expression (4.1), is equal to

$$v'\approx\Delta U\approx U(y+l)-U(y)\approx v_0.$$

Thus $v_0$ is the velocity of the turbulence eddies that are characteristic of the flow. In order to determine the constant $C_1$ we must remember that the reduction in the scale of turbulence eddies as the wall is approached is matched by a corresponding reduction in the Reynolds number

$$\mathrm{Re}=\frac{v_0 l(y)}{\nu}.$$

At a certain $l=\delta_0$ it is approximately equal to unity.

In the region $y<\delta_0$, known as the viscous sublayer, the nature of the flow is viscous. The thickness of the viscous sublayer is given by the condition

$$\frac{v_0\delta_0}{\nu}\approx 1,$$

or

$$\delta_0=a\frac{\nu}{v_0}, \tag{4.16}$$

where a is a proportionality factor. The constant in (4.15) is chosen in such a manner so that at $y\sim\delta_0$ the average flow velocity becomes a small quantity on the order of the characteristic velocity of the turbulence eddies $v_0$.

Then, for the average velocity, the so-called logarithmic profile is obtained (Figure 4):

$$U=\frac{v_0}{\sqrt{a}}\ln\frac{v_0 y}{a\nu}. \tag{4.17}$$

Expressing $v_0$ in terms of $\tau$ by means of equation (4.12), we finally obtain

$$U=\sqrt{\frac{\tau}{\rho\alpha}}\ln\frac{y}{a\nu}\sqrt{\frac{\tau}{\rho}}. \tag{4.18}$$

The eddy viscosity may be written in the form

$$\nu_{turb}\sim v_0 l\sim v_0 y\sim\nu\frac{y}{\delta_0}. \tag{4.19}$$

The logarithmic profile of velocities (4.13) contains two unknown constants: $\alpha$ and a. Their values must be determined from experimental data for the velocity distribution in the vicinity of the solid surface. This may be done most conveniently by introducing the dimensionless factor

$$y_*=\frac{v_0 y}{\nu}\sim\frac{y}{\delta_0}.$$

The dimensionless ratio $\frac{U}{v_0}$ is presented as a function of log $y_*$ in Figure 4. The data for this plot were obtained from numerous measurements of velocity distribution in the vicinity of the solid wall.

It is evident that the velocity distribution can be represented by a simple logarithmic relationship only at $y_* \geq 30$. In that region $\alpha \approx 0.17$. The determination of a directly from the curve in the region $y_* \geq 30$ has no meaning, however, since by definition, a relates to the region in which $Re = y_* \sim 1$, i.e., to the viscous sublayer.

Contemporary hydrodynamics do not offer any singular point of view regarding the velocity distribution in the viscous sublayer. Two hypotheses have been suggested:

1) The Prandtl hypothesis [4], which has been accepted widely, states that in the region $y< \delta_0$, the fluid motion is entirely laminar. Prandtl himself named this region the "laminar sublayer", and based his hypothesis on the fact that at $y<\delta_0$ the Reynolds number is found to be smaller than unity.

The shear stress $\tau_0$ in the laminar sublayer evidently may be expressed by the equation

$$\tau_0=\rho\nu\cdot\frac{dU}{dy}. \tag{4.20}$$

The velocity distribution can, therefore, be expressed by the linear equation

$$U=\frac{\tau_0}{\rho\nu}y+C.$$

The integration constant must equal zero, since the velocity of the fluid at the solid surface is zero. Therefore, for $y < \delta_0$

$$U = \frac{\tau_0}{\rho\nu} y. \tag{4.21}$$

The linear and logarithmic velocity profiles over the respective ranges up to and beyond their intersection do not provide satisfactory agreement with the experimental distribution data shown in Figure 4.

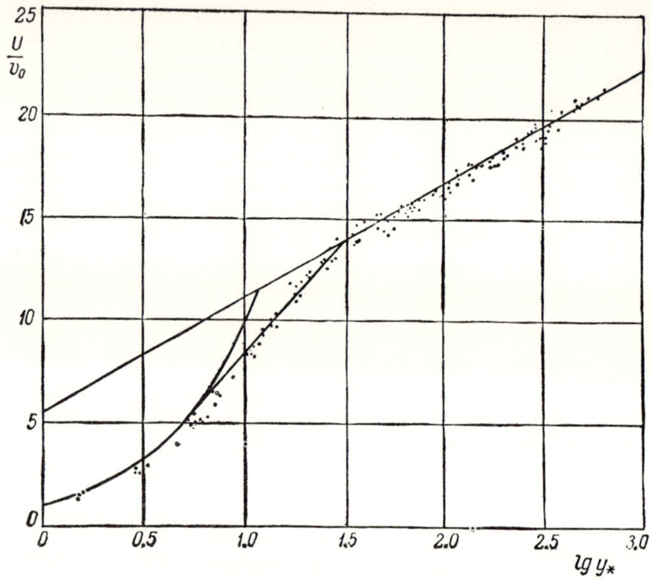


Figure 4. Distribution of average velocity for turbulent flow past a plate.

Von Karman, therefore, proposed a theory for the case of turbulent flow past a solid plane, in which the flow is divided into three regions: (a) a region of turbulent flow, (b) a "buffer" layer, and (c) a laminar sublayer.

According to von Karman [4], the turbulent flow in the buffer layer is damped as it approaches the solid wall, and the same law (4.14) as for the turbulent flow region is also applicable here. In the buffer layer, however, the effect of viscosity becomes significant. Consequently the values of the constants $\alpha$ and a in the logarithmic velocity distribution equation must differ from their values in the main turbulent zone.

2) The other hypothesis, presented jointly by L. D. Landau and the author of this book [1, 6], states that the turbulent motion in the viscous sublayer does not suddenly disappear, but is gradually damped as it approaches the wall. The equations for the damping of turbulence eddies in the viscous sublayer, i.e., the dependence of $l$ on y, can no longer be applied on the basis of dimensional

considerations, as is the case for the region of developed turbulence. All quantities in the viscous sublayer may be functions of viscosity, and the distance from the wall is no longer the sole quantity with a linear dimension. The equations for the damping of the turbulent flow in a viscous sublayer may be derived as follows. The distribution of the average velocity in this layer has the same form as in laminar flow, i.e.,

$$v_x \sim y.$$

Although turbulence eddies do not originate in the viscous sublayer, they enter it from the side $y > \delta_0$. The eddy velocities have the same magnitude as the average velocities in the sublayer. Therefore,

$$v'_x \sim y.$$

In view of the continuity equation

$$\frac{\partial v'_x}{\partial x} + \frac{\partial v'_y}{\partial y} = 0$$

the normal component of eddy velocity is

$$v'_y = -\int \frac{\partial v'_x}{\partial x} dy \sim y^2.$$

The proportionality coefficient in the expression for $v_y$ can be evaluated using the condition that, at $y \approx \delta_0$, the eddy velocity $v'_y$ at the boundary of the viscous layer is of the same order of magnitude as the characteristic velocity of the turbulent flow $v_0$. Therefore,

$$v'_y = v_0 \frac{y^2}{\delta_0^2}. \tag{4.22}$$

Thus, in a viscous sublayer the tangential and normal components of the average velocity and of the eddy velocities vary as a function of distance in the same way as the distribution of velocities in a laminar boundary layer. This, in essence, is the extent of the resemblance between a viscous sublayer and a laminar boundary layer.

In order to evaluate the coefficient of eddy viscosity in the viscous sublayer, the scale of the turbulent motion must be determined. This may be done as follows. In a viscous sublayer Re is less than unity, and the second-order terms in the Navier-Stokes equations are small compared to the first-order terms. The velocity distribution in a viscous sublayer can therefore be determined by linear equations only. If a certain spectrum of eddies penetrates a viscous sublayer, the interaction between separate eddies ceases. The flow then becomes a sum of independent periodic motions, whose periods T remain constant throughout the viscous sublayer.

Thus, it may be assumed that the periods of the turbulence eddies within a viscous sublayer are not a function of the distance y from the wall. The scale of the eddy motion in the y-direction is equal to

$$l \approx v_y' \cdot T, \tag{4.23}$$

or, since T is not a function of y,

$$l \sim y^2.$$

For $y \approx \delta_0$, the scale of the motion must equal that of the turbulent boundary layer, i.e., $l \approx \delta_0$. Therefore, the normalization factor is equal to $\frac{1}{\delta_0}$, and

$$l = \frac{y^2}{\delta_0}. \tag{4.24}$$

The scale of eddy motions in the viscous sublayer decreases with the distance y from the wall more rapidly than in the turbulent boundary layer.

By definition, the momentum transferred by eddies in the direction of the wall is given by the equation

$$\tau_{turb} = \rho \cdot \nu_{turb} \frac{dU}{dy} = \rho \cdot v_y' l \frac{dU}{dy} \sim \frac{\rho v_0 y^4}{\delta_0^3} \frac{dU}{dy}. \tag{4.25}$$

Therefore, the eddy viscosity is

$$\nu_{turb} \sim \frac{v_0 y^4}{\delta_0^3} \sim \nu \left(\frac{y}{\delta_0}\right)^4. \tag{4.26}$$

At $y > \delta_0$, the momentum transferred by eddies is less than the momentum transferred by molecular viscosity, and the eddy viscosity is less than the viscosity $\nu$. In view of this, at $y > \delta_0$, it may be assumed that the shear stress $\tau$ coincides approximately with the quantity $\tau_0$, and that the average velocity profile is defined by equation (4.20). Nevertheless, turbulence eddies exist up to the wall itself. It is shown later that the existence of turbulence eddies in the viscous sublayer is of significant inportance in the transfer of heat and mass toward a solid surface. In addition, knowledge of the existence of turbulence eddies in the viscous sublayer provides the basis for a theoretical approach in determining the velocity distribution in the boundary region.

It should be noted that another point of view also exists concerning the nature of damping of turbulence eddies in the viscous sublayer [7] — namely, that the scale of eddy motion is not expressed by formula (4.24) but by the same law (4.14) that applies to the region $y > \delta_0$. This leads to an eddy viscosity given by

$$\nu_{turb} \sim \nu \left( \frac{y}{\delta_0} \right)^3. \tag{4.27}$$

A choice between the two hypotheses, and between the two viewpoints of the second hypothesis concerning the damping of eddies, may be made only on the basis of experimental data. Detection of turbulence eddies and determination of equations describing their damping in the proximity of the wall present a highly complicated problem. Moreover, the existing measurement techniques do not provide the means for a definitive selection of either one of the hypotheses that have been advanced. It has been found that convincing data could be obtained in studying the diffusion of dissolved matter in turbulent flow.

As shown below (see Section 57), facts exist which support the hypothesis of the gradual reduction of turbulence in a viscous sublayer in accordance with equation (4.24). We therefore accept this equation for the purpose of subsequent discussion and utilize expression (4.26) for $\nu_{turb}$ in determining the velocity profile in the boundary zone. In deriving the velocity distribution in that zone, we assume that the transfer of momentum is accomplished by turbulence eddies which have already been affected by the viscosity. The significance of this is that the expression of the reduction of eddy viscosity (4.19) is not applicable in the boundary zone. Thus, in the boundary zone, it is convenient to choose an interpolation formula for the coefficient of eddy viscosity (a similar formula was given by G. P. Piterskikh). The formula is intermediate between (4.19) and (4.26)

$$\nu_{turb} = b\nu \left( \frac{y}{\delta_0} \right)^2, \tag{4.28}$$

and for shear stress

$$\tau = \rho(\nu + \nu_{turb}) \frac{dU}{dy} = \rho \left[ \nu + b\nu \left( \frac{y}{\delta_0} \right)^2 \right] \frac{dU}{dy}, \tag{4.29}$$

where b is an undetermined constant.

Integrating (4.29), we obtain the relation describing the distribution of the average velocity in the boundary zone,

$$U = \frac{v_0}{\sqrt{b}} \operatorname{arctg} \sqrt{b} \frac{y}{\delta_0} + c. \tag{4.30}$$

The constants b and c are determined by assuming that the velocity distribution corresponds to (4.21) at $y_* \approx 5$, and to (4.17) for $y_* = 30$.

In its final form the relations for the average velocity distribution may be expressed as

$$\frac{U}{v_0} = y_*, \qquad 0 \leqslant y_* \leqslant 5, \tag{4.31}$$

$$\frac{U}{v_0} = 10 \operatorname{arctg}(0.1 y_*) + 1.2, \qquad 5 \leqslant y_* \leqslant 30, \tag{4.32}$$

$$\frac{U}{v_0} = 5.5 + 2.5 \ln y_*, \qquad y_* > 30. \tag{4.33}$$

These show better agreement with measured velocity distributions (See Figure 4) than the previous expressions.

In reality we can never, of course, duplicate the condition of flow past an infinite plane for which these relations were derived. Nevertheless, they can be applied quite effectively to the boundary layer problems of flow pas a plate or within a tube.

Let us first examine the flow of a fluid past a semi-infinite plate. As noted in Section 3, the Reynolds number of the boundary zone increases with the distance from the leading edge. If it reaches the critical value $Re_\delta = Re_{cr}$, the flow in the boundary layer becomes turbulent (Figure 5). The critical value of the Reynolds number for a flat plate is approximately 1,500, but this value is substantially reduced by various types of disturbances in the boundary layer, such as are produced by wire probes or other small protuberances. In contrast to the case of the infinite surface examined above, the turbulent boundary layer formed in the stream flowing past a smooth plate not only varies with the coordinate x, but has a finite thickness d as well. (By convention we establish the origin at the leading edge of the plate.) Internally, the boundary layer has a velocity distribution with the logarithmic profile (4.17). At the border of the boundary layer, i.e., at y = d, the flow velocity is equal to the velocity of the main stream $U_0$:

$$U_0 = \frac{v_0}{\sqrt{\alpha}} \ln \frac{v_0 d}{10\nu}. \tag{4.34}$$

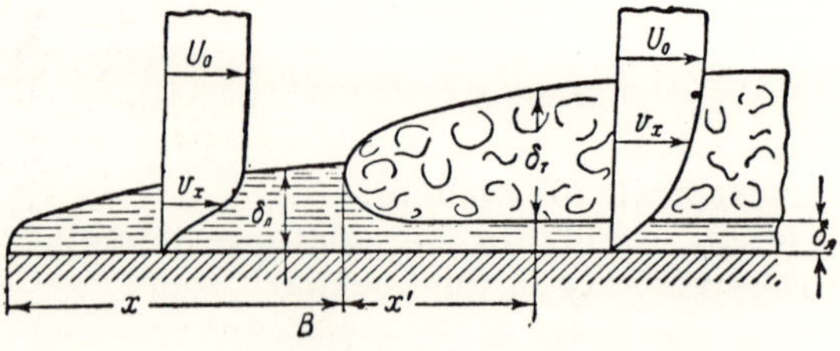


Figure 5. Formation of a turbulent boundary layer on a plate.

Up to point B the flow in the boundary layer layer is laminar, beyond that point it is turbulent.

The thickness of the boundary layer and the characteristic velocity $v_0$ vary with y in such a manner that this velocity has a constant value $U_0$ at the border of the boundary layer.

The second equation relating $v_0$ to d may be derived in the following manner. The slope of the curve d (x) is equal to the ratio of the velocities normal and tangential to the plate

$$\frac{\partial d(x)}{\partial x} \approx \frac{v_y}{v_x}.$$

However, at the boundary of the layer $v_x \approx U$ , and the magnitude of $v_y$ is on the order of the velocity $v_0$. Thus,

$$\frac{\partial d(x)}{\partial x} \approx \frac{v_0}{U_0}.$$

Since $v_0$ is a weak function of x, the thickness of the layer d can be expressed approximately as

$$d \approx \frac{v_0 x}{U_0}. \tag{4.35}$$

The thickness of the turbulent boundary layer is proportional to the distance from the leading edge of the plate. Therefore,

$$U_0 = \frac{v_0}{\sqrt{\alpha}} \ln \frac{v_0^2 x}{10 U_0 \nu}. \tag{4.36}$$

The drag on the plate is equal to

$$F = \int_0^b \int_0^l \tau \, dx \, dz = K_f \frac{\rho U_0^2}{2} bl,$$

where $K_f$ is the drag coefficient.

Simple computation shows that $K_f$ may be determined from the equation given in [8]

$$\frac{1}{\sqrt{K_f}} = 4.13 \lg (\mathrm{Re}\, K_f). \tag{4.37}$$

For rough calculation of the drag coefficient in turbulent flow past a plate, we can use the empirical relation of [8] as follows:

$$K_f = \frac{0.074}{\mathrm{Re}^{\frac{1}{5}}}. \qquad (4.37\ )$$

All the other relations derived above for flow past infinite plane may be directly applied to the case of a plate.

The velocity distribution curve shown in Figure 4 applies precisely to the case of flow past a plate. The logarithmic profile of the distribution of the average velocity may be applied to turbulent fluid flow in a tube. Because of the slow change of the logarithmic term in equation (4.17), this formula is applicable in calculating the average fluid velocity in a tube. Thus, for tubes,

$$U_R = \frac{v_0}{\sqrt{\alpha}} \ln \frac{R v_0}{a\nu},$$

where $U_R$ is the flow velocity in the center of the tube and R is the radius of the tube.

Figure 6 shows the velocity profiles for laminar and turbulent flow regimes in a tube.

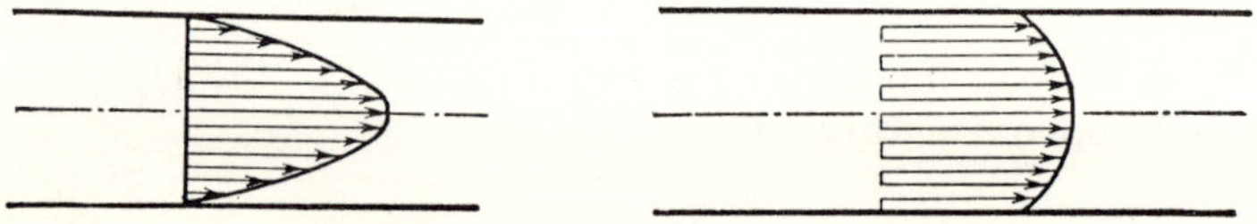

Figure 6. Velocity distribution in laminar (left) and turbulent (right) flow in a tube.

It is often important to relate the eddy velocity $v_0$ to the velocity $U_0$.

$$v_0 = \sqrt{\frac{\tau}{\rho}} = \sqrt{\frac{k_f}{2}}\, U_0 \approx \frac{\sqrt{k_f}}{1.41}\, U_0. \tag{4.38}$$

For flow past a plate and for intermediate values of Reynolds numbers ($\lesssim 10^5$) the approximation derived in [8] is applicable,

$$v_0 \approx \frac{0.27}{0.58} \frac{U_0}{\mathrm{Re}^{4.0}} = 0.47 \frac{U_0}{\mathrm{Re}^{4.0}}. \tag{4.39}$$

Similarly, for a tube the following may be written [8]

$$v_0 \approx \frac{0.16 U_R}{\mathrm{Re}^{\frac{1}{8}}} = \frac{0.2\bar{U}}{\mathrm{Re}^{\frac{1}{8}}}. \tag{4.40}$$

## 5. FLOW PAST A BODY WITH APPRECIABLE CURVATURE

The flow of a fluid past the surface of a solid plate was discussed above. In practice it is often necessary to study fluid motion past a body having appreciable curvature, such as a cylinder, a sphere, etc. Such bodies are usually described as having a non-streamlined shape.

The configuration of the flow past a non-streamlined surface is more complex than in the case of a plate. By the same token, the transfer of mass toward such a surface is also more complicated.

As a typical example, let us examine flow past a surface of a cylinder whose axis is perpendicular to the direction of the fluid stream.

Direct hydrodynamic measurements have shown that the fluid flows smoothly past the upstream portion of the cylinder forming a boundary layer which differs in nature from the boundary layer formed on a plate (Figure 7).

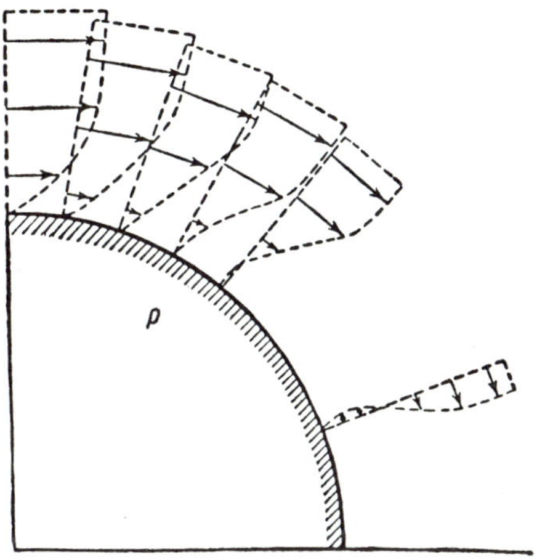


Figure 7. The appearance of separation in flow past a cylinder.

The essential difference between the two boundary layers, i.e., that on the surface of a cylinder and that on a plate, lies in the fact that the velocity and the pressure in the fluid passing around the cylinder vary from one point to the next. Beyond the boundary layer, pressure and flow velocity are determined by the Bernoulli equation (3.3). Therefore, as already noted in Section 3, the velocity of the fluid is least at the stagnation point on the cylinder, and increases uniformly up to the midpoint, beyond which it begins to decline. On the other hand, pressure is greatest at the stagnation point, decreases up to the midpoint, and increases over the downstream portion of the cylinder.

Over the upstream portion of the cylinder the fluid in the boundary layer moves in the direction of the pressure gradient; in the downstream part, against the pressure gradient. It is clear that this gradient initially will retard the slowly moving fluid layers adjacent to the surface of the body. The velocity profile will change as shown in Figure 7. At some point P, pressure in the opposite direction (pressure drag) will completely stop the motion of the fluid layers next to the surface. Beyond that point, there will be flow reversal: fluid will move in the opposite direction at the surface. Here the fluid layers within the boundary layer are forced out into the mainstream flow, so that, effectively, the fluid layers formerly within the boundary layer, are separated from the surface of the solid body. The separated boundary layer forms a jet that flow into the main fluid stream. The surface of the body in Figure 7 is shown by a solid curve; the straight dotted lines indicate the streamlines; and the dotted curved lines show the velocity profile.

It has been experimentally established (in agreement with theory) that at point P eddies begin to separate and are carried away by the stream. The separation of eddies starts at Reynolds numbers around 20.

At higher Reynolds numbers (100 to 200) the motion of the jet is unstable, and the fluid motion becomes turbulent beyond the point of separation (Figure 8). This effect leads to a considerable increase in the resistance of a non-streamlined body immersed in a moving fluid, as compared to a plate. This resistance is known as the drag and is associated with the energy dissipated in the zone of turbulence, which is termed the wake. This resistance is considerably

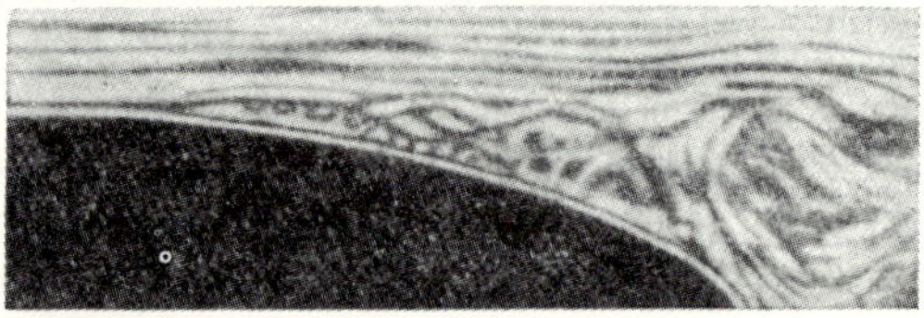

Figure 8. Appearance of turbulence in the downstream region beyond the point of separation.

greater than the usual viscous resistance. The preceding discussion points to the role played by the boundary layer on the surface of a body having appreciable curvature and on the regime of the flow past that body. It has been found that the drag F experienced by the body, may be expressed by the formula

$$F = K_f \frac{\rho U_0^2}{2} S, \tag{5.1}$$

where $K_f$ is a constant coefficient (for a body of a given configuration).

## BIBLIOGRAPHY

1. Landau, L. D., and E. M. Lifshitz, Mekhanika sploshuykh sred (Mechanics of Continuous Media), Gostekhizdat, 1944; L. D. Landau and E. M. Lifshitz, Fluid Mechanics, Pergamon

Press, London, 1959. L. G. Loitsyanskii, Mekhanika zhidkostei i gazov (Fluid Mechanics), Gostekhizdat, 1950; N. E. Kochin, I. A. Kibel and N. V. Roze, Teoreticheskaya gidrodinamika (Theoretical Hydrodynamics), Vols. I and II, Gostekhizdat, 1948; N. J. Kotschin, I. A. Kibel, and N. W. Rose, Theoretische Hydromechanik, Bd I und II, Akademie-Verlag, Berlin, 1954. L. Prandl, Gidroaerodinamika (Hydroaerodynamics), IL, 1951. Cf. L. Prandtl, Essentials of Fluid Dynamics, Blackie, London, 1952, and Prandtl-Tietjens, Hydro- and Aerodynamics, Dover reprint, New York, 1957.

2. Landau, L. D., and E. M. Lifshits, Mekhanika sploshnykh sred (Mechanics of Continuous Media), Gostekhizdat, 1944, p. 53; Lamb, Gidrodinamika (Hydrodynamics) Gostekhizdat, 1947, p. 726. H. Lamb, Hydrodynamics, 6th ed., Dover reprint, New York, 1945.

3. Sedov, L. I., Metody podobiya i razmernosti v mekhanike (Methods of Similarity and Dimensions in Mechanics), Gostekhizdat, 1951; L. I. Sedov, Similarity and Dimensional Methods in Mechanics, New York, 1957. M. V. Kirpichev and P. K. Konakov, Matematicheskiya osnovy teorii podobiya (Mathematical Bases for Similarity Theory), AN SSSR Press, 1949; Sb. Teoriya podobiya i modelirovaniye (Symposium, Similarity Theory and Similitude), AN SSSR Press, 1951; A. A. Gukhman, Fizicheskie osnovy teploperedachi (Physical Bases of Heat Transfer), Gosenergoizdat, 1934.

4. Sb. Sovremennoe sostoyanie gidroaerodinamiki vyazkoi zhidkosti (Collection: Current Aspects of the Hydroaerodynamics of a Viscous Fluid), Edited by S. Goldstein, Vol. I, IL 1949, p. 149. S. Goldstein, Modern Developments in Fluid Dynamics, Vols. I and II, Clarendon, Oxford, 1938.

5. Landau, L. D., and E. M. Lifshitz, Chap. II Mikrostruktura turbulentnogo potoka (Microstructure of Turbulent Flow); A. M. Obukhov and A. M. Yaglom, Prikl. matem. i mekhan. (Applied Mathematics and Mechanics), 15, 4, (1951); A. N. Kolmogorov, Lokal'naya struktura turbulentnosti v neszhimaemoi zhidkosti pri ochen bol'shikh chislakh Reinoldza (Local Structure of Turbulence at Very High Reynolds Numbers), Dokl. AN USSR 30, No. 4 (1941); A. N. Kolmogorov, Rasseyanie energii pri lokal'no izotropnoi turbulentnosti, (Dispersion of Energy in Local Isotropic Turbulence), Dokl. AN SSSR 32, No. 1 (1941); A. M. Obukhov, O raspredelenii energii v spektre turbulentnogo potoka (Concerning Energy Distribution in the Spectrum of a Turbulent Flow), Izvest. AS USSR, series geography, and geophysics, Nos. 4 & 5, (1941); L. G. Loitsyanskii, Nekotorye zakonomernosti isotropnogo turbulentnogo potoka (Certain Laws Governing Isotropic Turbulent Flows), Transactions of the Central Aero-Hydrodynamical Institute (im. N. E. Zhukovskii), No. 440 (1939);

M. D. Millionshchikov, K teorii isotropnoi turbulentnosti, (On the Theory of Isotropic Turbulence), Izvest. AN SSSR, series geography and geophysics, Nos. 4 & 5, (1941); A. Townsend, Proc. Camb. Phil. Soc. 44, No. 4 (1948); L. Richardson, Proc. Roy. Soc. A. 110, No. 756 (1926); A. M. Obukhov, Dokl. AS USSR 67, No. 4, (1949); H. Stommel, Journ. Marin, Res. 8, No. 3 (1949); L. Prandtl, Gidroaerodinamika (Hydro-Aerodynamics) IL, 1951; G. Batchelor, Teoriya odnorodnoi isotropnoi turbulentnosti (Theory of Homogeneous Isotropic Turbulence), IL, 1955. G. K. Batchelor, The Theory of Homogeneous Turbulence, Cambridge, 1956.

6. Levich, V. G., Acta physicchimica URSS 19, 117 (1943); same author, Doctorate dissertation, Kazan, 1943.
7. Vielstich, Z. f. Elektrochem. 57, No. 8 (1953). C. Lin, R. Moulton and G. Putnam, Ind. Eng. Chem. 45, 636 (1953).
8. Sb. Sovremennoe sostoyanie gidroaerodinamiki vyazkoi zhidkosti (Current Aspects of the Hydroaerodynamics of a Viscous Fluid) Ed. by S. Goldstein, IL, 1948, Vol. II, Sections 153 to 155.

CHAPTER II

# Convective Diffusion in Liquids

## 6. DIFFUSION KINETICS

The nature of most chemical and physicochemical operations is, as we suggested in the introduction, largely determined by hydrodynamic factors. Among such operations, heterogeneous transformations in liquids and gases are of primary importance. By heterogeneous, we mean chemical and physicochemical reactions or transformations that take place on certain surfaces, such as phase boundaries and surfaces with catalytic properties. Under this broad definition we include: all catalytic reactions, adsorption, and desorption at solid and liquid surfaces; dissolution, and precipitation of crystals from solutions and melts; electrochemical reactions at the surfaces of electrodes immersed in electrolytes; evaporation, sublimation, and condensation; gas adsorption by liquid and solid surfaces, etc.

All heterogeneous reactions involve several steps. The first of these is the transfer of reacting particles to the surface on which the reaction occurs (the reaction surface). The second involves heterogeneous reaction itself (chemical transformation, adsorption or desorption, the discharge and the formation of ions, etc.). The third step is the transfer of the reaction products away from the reaction site.

In some cases, the heterogeneous reaction process is accompanied by secondary effects — such as the release of large amounts of heat or the appearance of bubbles due to the formation of gaseous reaction products — which complicate the course of the reaction.

The net rate of the heterogeneous process is determined by the combined effect of its separate stages. However, the rate of the entire process is governed by the slowest step if one step is significantly slower than the others. This is common for reactions occurring in successive stages. In those cases where the slow step involves the introduction or removal of reactants, the reaction is said to be diffusion controlled and is governed by the laws of diffusion kinetics. If, on the other hand, the chemical or physical transformations constitute the slow step, the rate of the reaction is determined by the kinetics of these processes. When neither of these limiting cases apply, we shall refer to the process as an intermediate heterogeneous reaction, for which the rates of transfer of the reactants and of the chemical reaction are comparable.

Of greatest interest from a chemical viewpoint are those reactions whose rates are determined by chemical reaction kinetics. However, the majority of heterogeneous reactions and especially those of industrial importance are diffusion controlled. Examples of such reactions are: adsorption processes for gases soluble in liquids; most cases of dissolution of solid materials in liquids; most electrochemical and fast catalytic reactions; and many others. The investigation of the kinetics of diffusion processes, therefore, has great practical significance, and, as pointed out below, may in some cases be of value to the hydrodynamic theories of turbulence and heat transfer. In addition, even when the kinetics of the heterogeneous transformations is of interest, factors relating to the diffusion step should be excluded from the experimentally determined over-all rate of the reaction. These factors to be excluded can be readily and accurately determined if the theoretical expression for the rate of diffusion is known.

There is a close relationship between the transport of matter and the transfer of heat in a fluid medium. The many analogies which exist between the two are examined in more detail in later sections of this book. At this point we only note that one of the most interesting applications of the theory of diffusion kinetics is its applicability to complex problems in heat transfer, especially to those involving large thermal stresses.

In accordance with the objectives of this book, we shall limit our discussion to the problems of diffusion kinetics in the liquid phase. Gases will be discussed only to illustrate the similarities and differences of mass transfer in the two media. The diffusion kinetics of reactions in the solid state (so-called topochemical reactions) will not be examined in this book.

The general theory of liquid-phase diffusion processes is developed in the following sections. The treatment includes applications to heterogeneous reactions at a liquid-solid interface and touches on some characteristics of the intermediate-type reactions.

The diffusion kinetics of electrochemical systems is examined in Chapter VI, and the diffusion kinetics at liquid-liquid and at gas-liquid interfaces is discussed in Chapters VIII, X, and XII.

## 7. GENERAL INFORMATION ON DIFFUSION KINETICS IN THE LIQUID PHASE

Studies on the rates of heterogeneous processes in liquids began with the simplest problem in diffusion kinetics — namely the dissolution of solids in liquids. Upon analysis, the copious experimental data collected by A. N. Shchukarev [1] yielded the following empirical equation for dissolution:

$$Q = k(c_S - c_0)S, \tag{7.1}$$

where Q is the amount of material dissolved per unit time, S is the surface area of the dissolving body, $c_S$ is the concentration of a saturated solution, $c_0$ is the actual concentration of the solution at a given instant, and k is a proportionality constant.

Further studies, primarily by Nernst [2, 3, 4], showed that k is proportional to D, the diffusion coefficient of the substance in the liquid. Thus, the expression for Q may be written in the form

$$Q = D\frac{(c_S - c_0)}{\delta}S, \tag{7.2}$$

where $\delta$ is a constant.

Equation (7.2) expresses the fundamental point of Nernst's theory of heterogeneous reactions — namely, that the liquid immediately adjacent to the surface of the dissolving solid is always saturated. The rate of the heterogeneous reaction, when governed by diffusion kinetics,* is determined by the rate of diffusion between the concentrated liquid (adjacent to the dissolving substance) and the bulk of the solution. If $\delta$ designates the width of the region through which particles diffuse, then equation (7.2) expresses the usual law for steady state diffusion in a static liquid between two infinite planes, a distance $\delta$ apart, with known concentrations $c_S$ and $c_0$. The concentration profile between the two planes can be determined by linear interpolation.

In general, however, the rate of diffusion-controlled reactions in a static medium is very low because the value of the diffusion coefficient D is very small in liquids. Therefore, reactions occurring in static media must be limited to those which proceed according to the

---

* In the first version of the Nernst theory, it was assumed that all heterogeneous processes proceed according to diffusion kinetics. Later, other heterogeneous reactions were discovered whose rates are controlled by the transformations themselves and not by diffusion. It should be borne in mind, however, that the main point of either theory is its assumption concerning the mechanism of the diffusional rather than the heterogeneous process. Because this error in the initial version is no longer questioned, we concentrate our present criticism on this assumption. A large number of heterogeneous processes are known which are not diffusion controlled. As examples we may cite the discharge of hydrogen and oxygen ions at the surfaces of metallic electrodes, many cases of crystallization, etc.

diffusion kinetics of the unsteady state. For example, the rate of an electrochemical reaction in a motionless medium is so low that there is a noticeable current between the electrodes only during the time before the establishment of the steady state. (Further details are given in Chapter VI.) In practice, in the great majority of cases, efforts are made to increase the rate of reactions by agitating the liquid with stirrers, rotating the electrodes, or by taking advantage of natural convection. Because small density changes in low viscosity liquids (e.g., water) lead to the incidence of natural convection, it is rarely possible to conduct a reaction in a completely static medium. The differences in concentration which are established during a reaction, or the release of heat, bring about changes in the liquid density which lead to natural convection and finally to the thorough mixing of the liquid. Only by the use of special devices, or by gelling the liquid, can natural convection be avoided in a reactor.

The mechanism of mass transport in a moving liquid should therefore be examined. The value of $\delta$ in the empirical equation (7.2) can be determined from the measured values of Q, the concentrations $c_s$ and $c_0$, and the diffusion coefficient D. It has been found that under ordinary mixing conditions, the quantity $\delta$ has a magnitude of $10^{-2}$ to $10^{-4}$ cm, which is extremely small compared to the dimensions of the usual reaction vessels. This led Nernst to assume that the $\delta$ in equation (7.2) represents the thickness of the layer across which the diffusion occurs in a moving liquid. Later, direct measurements by A. G. Samartsev [5] showed that a constant concentration of the solution does indeed exist down to the minimum measurable distance from a reaction surface. According to the Nernst theory, there is a thin layer of static liquid immediately adjacent to the surface of the solid body — a layer through which diffusion of the reacting molecules takes place. Beyond this layer, the concentration in the bulk of the solution is constant because of the liquid motion. $\delta$ has been termed the thickness of the Nernst diffusion layer. Since the liquid in the diffusion layer is assumed to be static, the concentration distribution across it is easily found by solving the diffusion equation in a static medium. This concentration distribution, as shown in Figure 12, is linear. Thus, according to the Nernst theory, equation (7.2) should be interpreted as relating the rate of transfer of a diffusing substance to the concentration difference $(c_s - c_0)$, for both a flowing and a stagnant liquid.

Neither the thickness of the diffusion layer $\delta$ nor its dependence on the fluid velocity can be calculated from the Nernst theory. Experimental determinations (using measured values of Q) have shown that $\delta$ as a function of velocity has the form

$$\delta \approx \frac{1}{v^n},$$

where v is the velocity of the moving liquid. The exponent n has been found to depend on the experimental conditions (for more details, see

Section 55), with different authors reporting different values for n, ranging from $n = \frac{1}{2}$ to $n = 1$.

The Nernst theory has been used extensively, especially in electrochemistry, and in its time played an important role in the development of the theory of heterogeneous reactions. The general idea of the Nernst theory — the concept that a thin liquid layer exists adjacent to an electrode where a change in concentration occurs — has been confirmed by A. G. Samartsev's data. Over the past 20 years, however, a number of papers have appeared which justly criticize this theory, so that today it is regarded as unsatisfactory. For example, experimental measurements [6] (see also Section 9) have shown that liquid motion may be observed* at distances of $\approx 10^{-5}$ cm from the solid wall. Thus, Nernst's assumption that the liquid is static within a layer 10 to 1000 times thicker contradicts empirical evidence and must be rejected. If it is accepted, however, that the liquid is not static inside the diffusion layer, the linear equation for concentration distribution no longer has a sound basis.

Equation (7.2), which represents a theoretically sound expression for diffusional flow in a static medium, should be regarded as purely empirical when applied to a medium in motion. The nontheoretical nature of this formula is especially apparent since the thickness of such a diffusion layer is found to be a function of the diffusion coefficient of the particles. This indicates that, under given experimental conditions, there is a diffusion layer thickness corresponding to each ion or atom diffusing in the solution. The Nernst theory is also unsatisfactory for another reason: it does not permit the value of the flow Q to be determined or even approximated on any theoretical grounds since the thickness $\delta$ cannot be computed a priori. It is, therefore, a qualitative theory of diffusion kinetics for heterogeneous reactions. It should be stressed, however, that the fact that the liquid retains its mobility down to very small distances from the solid wall, even though such distances are considerably smaller than $\delta$, does not imply the absence of a diffusion step in the process. Nevertheless, this strange conclusion was reached by some authors [7, 8], who attempted to remove this contradiction by proposing theories in which the diffusion step of the heterogeneous transformation was omitted.**

---

* In the experiments conducted by Fage and Townsend, the motion of a liquid near a solid surface was studied under turbulent conditions. The movement of suspended colloidal particles was observed by means of an ultramicroscope and it was found that the particles retained their mobility at distances from the wall on the order of $10^{-5}$ cm.

** The Moelwyn-Hughes assertion of complete disagreement between the Nernst theory and the observed facts (Moelwyn-Hughes'

The Nernst theory has been used primarily in studies of dissolution, in theories treating the passage of current through solutions, and in other cases of heterogeneous reactions at a liquid-solid interface.

Another theory of diffusion kinetics in heterogeneous reactions, basically similar to that of Nernst, has been generally accepted in chemical engineering technology. This is the so-called film theory proposed by Langmuir [9], according to which a thin layer exists near the interface between the moving phases. The properties of the layer, which has been termed "coating" or "film", differ from those in the bulk of the moving phases. It consists of a region lying between the stationary medium and the moving phase. For example, according to the film theory, the properties of the liquid film adjacent to a liquid-solid body interface differ, as far as mass or energy transfer are concerned, from those in the bulk of the liquid. Two films are formed at the interface of a moving liquid-gas system. A liquid film is established in the liquid phase side of the interface, and a gas film on the gas phase side of the interface. The properties of the liquid and gas films differ from those of the respective main streams. With so general a description, it is difficult to argue against the film theory. It will be shown later that regions of different flow regimes actually exist near phase interfaces (although the physicochemical properties in these regions are the same as those of the respective pure phases).

The film theory, however, allows one to draw additional conclusions concerning the properties of the interfacial films. It is assumed that there is no relative motion between the phases within the films. This means that the film at a solid-liquid interface is static just as in the Nernst theory [10]. Similarly, for films at a gas-liquid interface, this means that there is no relative motion between the liquid and gas films [11], and it is further postulated that the liquid and gas within the films are in strictly laminar flow, and that all velocity components are parallel to the interface. Thus, the film theory proposes that the motion within the film plays no part in the transport of matter and that the latter depends solely on molecular diffusion through the film. Film thicknesses can be determined experimentally by measuring the quantities diffusing through the film and relating them by means of equations similar to equation (7.2).

In discussing the film theory reference is usually made to the experiments of Langmuir [9], who discovered the existence of a fairly thick layer (of the order of several millimeters) adjacent to a

---

calculations indicate that the thickness of the diffusion layer must be approximately 1 cm) is the result of an arithmetic error, as pointed out by Yu. G. Siver. In computing the volume of the solution, expressed in cubic cm per liter, he omitted the factor $10^3$, which caused a corresponding increase in the reported values of $\delta$.

gas-solid interface. The major temperature drop was observed to occur in this layer. However, for a solid-liquid interface, the same remarks apply as in the case of the Nernst theory. Thus, for films near a liquid-solid interface, Langmuir's assumption that the tangential motion in the film has no effect on the transfer of matter is erroneous. This will be demonstrated in Section 18. A detailed examination of the phenomena which take place at this interface shows that the tangential motion is usually accompanied by a motion normal to the interface (see Section 15). We shall also see later that the assumption of no turbulent motion at the interface (and the accompanying mixing of the fluid) also contradicts experimental evidence.

As with the Nernst theory, the film theory provides only a qualitative description of the mass transfer in a moving fluid. It does not allow the a priori computation of the diffusional flux.

We have discussed these two theories in such detail because they have been widely used in various fields of chemistry and physical chemistry and also in studies of various engineering operations.

Despite their clarity and simplicity, the Nernst and film theories are largely incorrect. The limits of their usefulness in the qualitative study of mass transfer in liquids and gases are discussed below. At the same time we shall present evidence in support of an empirical expression of the type (7.2) for mass transfer in a moving fluid.

Among other studies on the theory of diffusional processes, we should first draw attention to the work of D. A. Frank-Kamenetskiy [12], tho was the first to utilize similarity methods in examining the phenomena of mass transfer. These methods have been widely accepted in the fields of hydrodynamics and theoretical heat transfer (similarity theory is treated in Section 19). Despite their great value, however, the methods of similarity do not elucidate the mechanisms of diffusional processes in moving media.

The inadequacy of the Nernst theory led E. Eucken [13] to abandon the concept of a static layer and to obtain a rigorous solution of the equations for convective diffusion. However, he made a computational error (see Section 15) which for practical cases invalidates his conclusions.

Phenomena which are quite similar to convective diffusion have been widely studied in the theory of heat transfer. The transport of matter in a moving fluid is considered in diffusion kinetics, while the transfer of energy under the same conditions is the subject of the heat transfer theory. Heat transfer problems have been attacked during recent decades by hydrodynamicists and heat transfer engineers from both theoretical and experimental points of view [14].

The hazy concepts of the Nernst and film theories in the field of diffusional phenomena contrast sharply with the quantitative theory of heat transfer in moving media that applies to various flow regimes and different geometric configurations. As demonstrated below, the essential principles of the heat transfer theory can be readily adapted to diffusional phenomena in gases. In the case of liquids, although the basic principles of the theory remain unchanged, a different quantitative picture will be obtained for mass transport. This is mainly due to some specific characteristics of the liquids which have to be accounted for in the diffusional processes.

## 8. CONVECTIVE DIFFUSION IN LIQUIDS

In the following discussion we examine the behavior of a liquid (or a gas) containing some added material. We term such a mixture a solution even though this term usually denotes only matter dissolved in liquids (when matter is dissolved in gases, the solutions are termed gaseous mixtures).

In general, we deal with solutions containing a single dissolved substance, although in some cases we shall discuss solutions containing several solutes. The composition of a solution is characterized by its concentration c, which is defined as the number of particles of dissolved matter per unit volume of liquid. In cases where it is more convenient to use molar concentrations, i.e., when the composition is described by the number of gram-moles per $cm^3$, the concentration will be denoted by C.

When a liquid containing dissolved matter is in a state of equilibrium, the following thermodynamic conditions are satisfied: there is no macroscopic motion, and the temperature T, the pressure p, and the partial free energy (chemical potential) $\mu$ (T, p, c) are constant.

Initially, and until we prove otherwise, it will be assumed that any displacement from equilibrium in the liquid must be caused by macroscopic motion (flow) and by differences in the solute concentration from one point to another. The temperature of the liquid will be assumed constant, and, while we do not assume that the pressure is strictly constant, we shall suppose that the pressure gradient is relatively small (see below).

The motion of the liquid is determined by the volume and surface forces acting on it and can be described by the equations of motion and continuity given in Section 1.

The transport of solute in a moving liquid is governed by two quite different mechanisms. First, there is molecular diffusion as a result of concentration differences; second, solute particles are entrained by the moving liquid and are transported with it. The combination of these two processes is called convective diffusion of solute in a liquid.

If some transformation involving solute particles takes place in the liquid or at its interface, the concentration may vary from point to point and as a function of time.

Let us write a differential equation satisfied by the function c (x, y, z, t) in the moving liquid. We shall assume that the liquid motion is steady and laminar and that the liquid is incompressible. The turbulent flow regime will be examined in the next chapter.

The variation in concentration from point to point destroys the equilibrium and leads to a diffusional transport of the solute. If the concentration gradients in the liquid are small, we can assume that the partial potential gradients are also small. In that case, the diffusional flux (the number of particles passing through 1 cm² of an imaginary surface in the solution in 1 sec) can be written as

$$\mathbf{j}_D = -\alpha \operatorname{grad} \mu, \tag{8.1}$$

where $\alpha$ is a proportionality constant, assumed to be positive. The significance of the minus sign will become clear later. Molecular diffusion is the mechanism of transport of particles causing the appearance of a diffusional flux in the liquid.

By expressing the gradient of $\mu$ in terms of concentration, we obtain

$$\mathbf{j}_D = -\alpha \left(\frac{\partial \mu}{\partial c}\right)_{T,\,p} \operatorname{grad} c.$$

In the foregoing we have assumed that no temperature gradient exists and that the pressure gradient is sufficiently small. Thus, the mass flow resulting from the pressure gradient is negligible in comparison with the flow resulting from concentration differences. The minus sign indicates that the diffusional flux proceeds from regions of higher to regions of lower concentration. By introducing the diffusion coefficient D, defined by the equation

$$D = \alpha \left(\frac{\partial \mu}{\partial c}\right)_{T,\,p},$$

we can represent the diffusion flux as

$$\mathbf{j}_D = -D \operatorname{grad} c.$$

The diffusion coefficient D is a function of both the concentration of the solution and of the temperature.

If the concentration of the solution is low, D can be considered constant and independent of composition. Then the diffusional flux is proportional to the concentration gradient and its direction is toward the decreasing concentrations. It is precisely the case which we shall consider in the following.

When the solute is in a moving liquid, it is entrained by the flowing stream. Then, in addition to the diffusional flux $\mathbf{j}_D$ passing in one second across 1 cm² of an imaginary plane constructed in the liquid a flow of matter

$$\mathbf{j}_{conv} = c\mathbf{v}$$

travels with the volume of liquid $\mathbf{v}$ passing through the plane in one second.

The total mass flux is the sum of the convective and the diffusional fluxes and is expressed by the vector

$$\mathbf{j} = c\mathbf{v} - D \operatorname{grad} c. \tag{8.2}$$

If the solution is not isothermal or if it is subjected to significant external forces (for example, a gravitational field, or in the case of an ionic solution an electric field), additional terms will enter into the equation. The motion of charged particles (ions) in a moving solution and in the presence of an electric field is examined in detail in Chapter VI.

The expression presented for the total mass flux with a constant value for the diffusion coefficient D is valid, as we have stressed above, only for sufficiently dilute solutions. It is also important to note that D is not defined directly as the proportionality coefficient in the flux j, but by the derivative $\left(\frac{\partial \mu}{\partial c}\right)_{T,p}$. This plays an important part in the examination of phenomena which occur at the critical point (see Section 22).

Let us consider an arbitrary volume V in a body, and let us express the net flow of particles entering and leaving it per unit time. The number of particles passing through the surface S in one second equals

$$Q = -\oint \mathbf{j}\, d\mathbf{S},$$

where the integral is taken over the surface S that surrounds the volume V. A direction outward from the surface is chosen as the positive direction for the normal vector.

If $\frac{\partial c}{\partial t}$ is the change in the number of particles per unit volume per second, then the change in the number of particles in the volume V is equal to $\int \frac{\partial c}{\partial t} dV$. Equating this to the number of particles entering the volume, we obtain

$$\int \frac{\partial c}{\partial t} dV = -\oint \mathbf{j}\, d\mathbf{S}.$$

The integral on the right side of the equation can be converted according to the Gauss-Ostrogradsky theorem, so that

$$\int \frac{\partial c}{\partial t} dV = -\int \operatorname{div} \mathbf{j}\, dV. \tag{8.3}$$

Since the volume was selected arbitrarily, it follows from equation (8.3) that

$$\frac{\partial c}{\partial t} = -\operatorname{div} \mathbf{j}, \tag{8.4}$$

or

$$\frac{\partial c}{\partial t} = \operatorname{div}(D \operatorname{grad} c) - \operatorname{div} c\mathbf{v}. \tag{8.5}$$

If we neglect the dependence of the diffusion coefficient on concentration and consider it constant, we can set

$$\operatorname{div}(D \operatorname{grad} c) = D \operatorname{div} \operatorname{grad} c = D\Delta c.$$

Furthermore,

$$\operatorname{div}(c\mathbf{v}) = (\mathbf{v} \operatorname{grad}) c + c \operatorname{div} \mathbf{v}.$$

Since the liquid is incompressible, div $\mathbf{v} = 0$. Therefore, equation (8.5) may finally be presented in the form

$$\frac{\partial c}{\partial t} + (\mathbf{v} \operatorname{grad}) c = D \Delta c. \tag{8.6}$$

Equation (8.6) is the general equation of convective diffusion. In terms of Cartesian coordinates it may be written more explicitly as follows:

$$\frac{\partial c}{\partial t} + v_x \frac{\partial c}{\partial x} + v_y \frac{\partial c}{\partial y} + v_z \frac{\partial c}{\partial z} = D\left(\frac{\partial^2 c}{\partial x^2} + \frac{\partial^2 c}{\partial y^2} + \frac{\partial^2 c}{\partial z^2}\right). \tag{8.7}$$

Since the liquid velocity is a function of the coordinate distances, the equation of convective diffusion is, mathematically, a second-order, partial differential equation with variable coefficients. The equation alone is insufficient for the determination of the concentration distribution $c(x, y, z, t)$. For a full solution of the problem, we must also know the boundary and the initial conditions to be satisfied by $c(x, y, z, t)$.

The problem of determining the function that satisfies the system consisting of an equation and boundary and initial conditions is known as a boundary value problem. For a certain class of boundary and initial conditions, a uniqueness theorem can be proved, according to which a given boundary value problem has one, and only one, solution.

The boundary and initial conditions depend upon the nature of the physicochemical processes in which the solute takes part. Various examples of boundary conditions are examined below, but for the time being we shall assume that all the necessary boundary and initial conditions are known.

The equation of convective diffusion is greatly simplified if the concentration distribution c is assumed to be independent of time. Whether such an assumption applies in a particular case will become clear when the physical conditions of the process are analyzed. If c is not a function of time, so that $\frac{\partial c}{\partial t}=0$, then, instead of (8.6), the equation for the concentration distribution becomes

$$(\mathbf{v}\,\mathrm{grad})\,c = D\,\Delta c. \tag{8.8}$$

Another important simplification of equation (8.5) occurs when the liquid medium is static. Then $\mathbf{v} = 0$, and equation (8.6) becomes identical to the usual molecular diffusion equation

$$\frac{\partial c}{\partial t} = D\,\Delta c. \tag{8.9}$$

As contrasted with the equation of convective diffusion, which contains coefficients that are functions of position (i.e., the velocity components $v_x$, $v_y$, and $v_z$), the equation for molecular diffusion is a second-order, partial differential equation with constant coefficients.

Solutions of this equation for various boundary and initial conditions have been the subject of detailed studies in mathematical physics in connection with the heat transfer theory. As is well known, the temperature distribution in a static medium satisfies the same type of equation as the concentration distribution, with the thermal diffusivity replacing the diffusion coefficient of the medium.

Several boundary value problems in a motionless medium are examined below, first to consider their physicochemical applications, and second to establish a basis of comparison between the solution of the diffusion problem in a static medium and its solution in a moving medium. In certain instances it is necessary to study the distribution of concentrations which change as a result of chemical or other processes (for example, recombination or ionization) occurring in the bulk of the liquid, or certain homogeneous chemical reactions involving the dissolved particles. In such instances, the equation for convective diffusion must be supplemented by a term which accounts for the appearance or disappearance of particles within the bulk of the liquid (sources or sinks). Let us assume that there is a mass source (or sink) in a volume element dV of the liquid. Let us designate by $Q_0(\mathbf{r})dV$ the number of particles which appear in the volume dV per second, where the volume dV surrounds the point $\mathbf{r}$. The quantity $Q_0$ is known as the source strength. If the particles disappear, then $Q_0$ is a negative quantity.

Adding $Q_0\,dV$ to the balance of particles in the volume V, we obtain, in place of (8.3),

$$\int \frac{\partial c}{\partial t}\, dV = -\oint \mathbf{j}\, d\mathbf{S} + \int Q_0\, dV,$$

whence

$$\frac{\partial c}{\partial t} = -\operatorname{div} \mathbf{j} + Q_0,$$

or

$$\frac{\partial c}{\partial t} + (\mathbf{v}\, \mathrm{grad})\, c = D\, \Delta c + Q_0. \tag{8.10}$$

It must be stressed that $Q_0$ is a volume source. Particles disappearing or appearing on a reaction surface are accounted for only by boundary conditions. The existence of sources or sinks on the boundary surface cannot be reflected directly in the diffusion equation since the latter represents the net balance of particles in the bulk of the solution. The convective diffusion equation, in meaning and in form, is similar to the Navier-Stokes hydrodynamic equation. The latter represents the net balance of momentum transported in the liquid, while the former is a material balance. In this case, therefore, it is possible to employ the same methods of solution which were applied to the Navier-Stokes equation.

Let us reduce the convective diffusion equation to a dimensionless form. This is accomplished by introducing a characteristic length L, along which the major change in concentration takes place, and a characteristic flow velocity $U_0$. Considering first a steady state process, we can write

$$\frac{v_x}{LU_0}\frac{\partial c}{\partial \frac{x}{L}} + \frac{v_y}{LU_0}\frac{\partial c}{\partial \frac{y}{L}} + \frac{v_z}{LU_0}\frac{\partial c}{\partial \frac{z}{L}} = \frac{D}{U_0 L^2}\left(\frac{\partial^2 c}{\partial \frac{x^2}{L^2}} + \frac{\partial^2 c}{\partial \frac{y^2}{L^2}} + \frac{\partial^2 c}{\partial \frac{z^2}{L^2}}\right).$$

Introducing the dimensionless velocity components $V_x = \frac{v_x}{U_0}$, $V_y = \frac{v_y}{U_0}$ and $V_z = \frac{v_z}{U_0}$, the dimensionless coordinates $X = \frac{x}{L}$, $Y = \frac{y}{L}$, and $Z = \frac{z}{L}$, and the dimensionless concentration $C = \frac{c}{c_0}$, where $c_0$ is the constant concentration in the bulk of the solution, we obtain

$$V_x \frac{\partial C}{\partial X} + V_y \frac{\partial C}{\partial Y} + V_z \frac{\partial C}{\partial Z} = \frac{D}{U_0 L}\left(\frac{\partial^2 C}{\partial X^2} + \frac{\partial^2 C}{\partial Y^2} + \frac{\partial^2 C}{\partial Z^2}\right). \tag{8.11}$$

The dimensionless ratio

$$\mathrm{Pe} = \frac{U_0 L}{D} \tag{8.12}$$

is known as the Peclet number.

Introducing the Peclet number into equation (8.11), we get

$$V_x \frac{\partial C}{\partial X} + V_y \frac{\partial C}{\partial Y} + V_z \frac{\partial C}{\partial Z} = \frac{1}{\mathrm{Pe}}\left(\frac{\partial^2 C}{\partial X^2} + \frac{\partial^2 C}{\partial Y^2} + \frac{\partial^2 C}{\partial Z^2}\right). \tag{8.13}$$

The left side of the convective diffusion equation represents the convective transfer of the matter in the liquid, while the right side describes molecular diffusion. It is clear that all the dimensionless terms in equation (8.13) are, in general, approximately unity. The relationship between the convective and the diffusional transfer of matter is, therefore, described by a single parameter — the dimensionless Peclet number. It plays the same role in convective diffusion as the Reynolds number does in fluid flow.

When the Peclet number is small, the terms on the left side of equation (8.13) can be neglected, and the right-hand terms can be set equal to zero. In this case the concentration distribution is determined largely by molecular diffusion; at sufficiently low Pe, mass transfer by convection is negligibly small. From the definition of the Peclet number (8.12), it follows that such a situation occurs (for a given D) at sufficiently low liquid velocities and in regions of small dimensions. Conversely, when the Pe number is large, the terms on the right side of equation (8.13) are small. In this case, the concentration distribution is determined essentially by convective transfer, and molecular diffusion can be neglected.

It should be stressed here that this argument does not apply to a region in the liquid where a sharp change in concentration occurs and where the concentration gradient along one of the coordinates has an especially high value. In these instances, the molecular diffusion terms must be retained even though their coefficient is small. We have already encountered such a case in our examination of the Prandtl boundary layer theory.

The dimensionless ratio of the Peclet number to the Reynolds number is known as the Prandtl number, Pr:

$$\mathrm{Pr} = \mathrm{Pe} : \mathrm{Re} = \frac{U_0 L}{D} : \frac{U_0 L}{\nu} = \frac{\nu}{D}. \tag{8.14}$$

The Prandtl number* is not a function of flow velocity or of the characteristic dimensions and is determined exclusively by the physical properties which characterize momentum transfer and the transport of matter by a purely molecular mechanism.

When the Prandtl number is equal to unity (i.e., when $\nu = D$), there is a similarity between momentum and mass transfer. In a region of the liquid where the molecular momentum transfer is controlling, the molecular mass transfer also predominates.

The obvious difference between the Prandtl numbers of liquids and gases is clearly illustrated if numerical values for the diffusion coefficients are substituted into equation (8.14). For gases, the diffusion coefficient and the kinematic viscosity are numerically of

---

* The Prandtl number for diffusion processes is sometimes known as the Schmidt number Sc.

the same order of magnitude, and hence $Pr \approx 1$. This is not true for liquids. The kinematic viscosity of common liquids, such as water, is approximately $\nu \approx 10^{-2}$ cm²/sec. The diffusion coefficients of molecules and ions in aqueous solutions are of the order of magnitude of $D \approx 10^{-5}$ cm²/sec; and for macromolecules, $D \approx 10^{-6}$ cm²/sec. Therefore, for water and comparable liquids, $Pr \approx 10^3$. As the kinematic viscosity of the liquid increases, the diffusion coefficient decreases in accordance with the relation*

$$D = \frac{\text{const}}{\nu}. \tag{8.15}$$

Thus, the Prandtl number increases as the square of the kinematic viscosity. In viscous fluids, the Prandtl number attains values of $10^6$ and higher. A high value of the diffusional Prandtl number signifies that, even at very low velocities, the transport of matter in a fluid by convection predominates over molecular diffusion. Molecular diffusion of matter in liquids is so slow that convective transfer of dissolved matter plays in general a more important role in liquids than in gases.

The relationship between heat and mass transfer in a moving liquid is a most important problem and will be examined in Chapter IV.

## 9. BOUNDARY CONDITIONS FOR THE EQUATION OF CONVECTIVE DIFFUSION

Since the equation of convective diffusion in liquids is a second-order, partial differential equation with variable coefficients, it can be solved exactly only for systems of simple configuration.

In order to solve the equations of convective diffusion, a set of boundary conditions must be established. As a rule these are specified at the reaction surface and at an appreciable distance from it in the bulk of the solution.

A specified concentration usually constitutes the boundary condition in the bulk of the solution. Since, as shown below, the change in concentration occurs in a thin liquid layer near the reaction surface, the condition far from the surface can be taken as the boundary condition at infinity, i.e.,

$$c = c_0 \quad \text{as} \quad y \to \infty, \tag{9.1}$$

---

* Equation (8.15) is only approximate. For more details on the interrelationship of diffusion and viscosity coefficients, see Section 11.

where y is the distance from the reaction surface. The significance of the conditions $y \rightarrow \infty$ will be clarified later. Frequently, $c_0$ can be set equal to zero, as in problems of dissolution.

The conditions at the reaction surface are more complex. Under steady-state conditions, the number of particles (molecules, atoms, or ions) which react on the surface per unit time equals the number of particles arriving at the surface. This mass flux to the surface equals

$$j = D\left(\frac{\partial c}{\partial n}\right)_{y=0}, \tag{9.2}$$

where $\left(\frac{\partial c}{\partial n}\right)_{y=0}$ is the derivative along the outward normal evaluated at the reaction* surface. The number of particles which react per unit time is generally given by the "reaction law"

$$q = k\,(c_1)^m, \tag{9.3}$$

where $c_1$ is the value of the concentration at the surface. The exponent m is known as the order of the reaction. It may be an integer or a fraction, since (9.3) may express the net result of several elementary reactions at the surface or may approximate a more complex function $q = f(c_1)$. The quantity k is a function only of temperature and not of the concentration of the reactants and is known as the rate constant of the reaction. It is usually taken to be the following function of temperature

$$k = k_0 e^{-\frac{U}{kT}}, \tag{9.4}$$

where U is the activation energy and T is the absolute temperature. The usual activation energy in heterogeneous reactions is 6 to 10 kilocalories per mole.

Under steady-state conditions, we have

$$j = q. \tag{9.5}$$

Substituting expressions (9.2) and (9.3) into (9.5), we obtain for a reaction of any given order

$$D\left(\frac{\partial c}{\partial n}\right) = kc_1^m \qquad \text{at the reaction surface}. \tag{9.6}$$

---

* Unlike (8.1), there is no minus sign in (9.2), because the direction of the outward normal is opposite to the vector (grad c).

In the general case of intermediate kinetics with $m \neq 1$, the boundary condition (9.6) is nonlinear, and the solution of the problem of convective diffusion becomes much more complex. If, however, the rate of one of the processes — either mass transfer or the reaction — is high compared to the other, the boundary condition (9.6) can be considerably simplified.

To analyze this simplification, let us rewrite (9.6) in dimensionless form, introducing the characteristic coordinate $N = \frac{n}{L}$ and the concentration $C_1 = \frac{c_1}{c_0}$.

Then we have

$$\frac{D}{kLc_0^{m-1}} \left(\frac{\partial C}{\partial N}\right)_{y=0} - C_1^m = 0. \tag{9.7}$$

The dimensionless quantity $K = \frac{D}{kLc_0^{m-1}}$ characterizes the relative significance of the rate of transfer of particles to the reaction surface and the rate of the reaction.

In the limiting case, $K \ll 1$, the first term in (9.7) is small compared to the second and can be neglected. Then, the boundary condition becomes

$$C_1 \approx 0 \qquad \text{at the reaction surface.} \tag{9.8}$$

Condition (9.8) has a simple meaning. From the definition of K, we see that small values of K correspond to relatively large values of the reaction rate constant k. In other words, the inequality $K \ll 1$ indicates that the reaction is relatively rapid compared to the rate of mass transfer. The over-all rate of the heterogeneous conversion is therefore determined by the slower rate of mass transfer.

Boundary condition (9.8) shows that when the reaction rate is sufficiently large all particles approaching the surface react instantaneously. The concentration of molecules in an infinitely thin layer of solution immediately adjacent to the reaction surface may become infinitesimal. This is the case where the difference between the concentrations in the bulk of the solution and at the reaction surface is greatest $\Delta c = c_0 - c_1 \approx c_0$. Since the flux of solute to the reaction surface is proportional to the difference in concentration, all other conditions being equal, the flux of solute, in the case under consideration, has a limiting high (maximum) value. We designate as the limiting diffusional flux, $j_{lim}$, the greatest possible flux of solute, which corresponds to the maximum difference in concentration. The diffusion-controlled regime established at $c = 0$ on the reaction surface is called the limiting (maximum) flux regime. It is very often met in practice and is examined later.

For the opposite extreme, $K \gg 1$, which corresponds to a small value of the reaction rate constant k and, consequently, to the case of a low reaction rate in comparison with the rate of mass transfer, the boundary condition (9.7) takes the form

$$\left(\frac{\partial c}{\partial n}\right) \approx 0 \quad \text{at the reaction surface.} \tag{9.9}$$

According to condition (9.9), the concentration in the entire solution is constant. The over-all rate of the process is determined by the reaction rate and is expressed by $q = kc_0^m$.

In the special case of first-order reactions ($m = 1$), the general boundary condition (9.6) has the form

$$D\left(\frac{\partial c}{\partial n}\right) = kc_1 \quad \text{at the reaction surface} \tag{9.10}$$

and represents a linear relationship. In this case, a solution of the intermediate kinetics problem is simplified.

When the removal of the reaction products from the surface is the slowest step, the conditions mentioned above require certain alterations. First of all, the concentration of the reaction products far from the reaction surface is very low compared to the concentration near the surface, so that expression (9.8) becomes

$$c \to 0 \quad \text{as} \quad y \to \infty. \tag{9.11}$$

Thus, condition (9.11) presupposes a very low total product formation. Furthermore, since the direction of the diffusion flux is reversed, the number of molecules leaving the reaction surface is determined not by equation (9.6) but by the following expression:

$$-D\left(\frac{\partial c}{\partial n}\right) = kc_1^m \quad \text{at the reaction surface.} \tag{9.12}$$

An example of such reactions is one involving autocatalysis, in which the reaction products are catalysts for the reaction itself. In such autocatalytic reactions, rapid removal of reaction products can reduce the reaction rate. On the other hand, when the removal of the products is slow, the diffusion equation with boundary conditions (9.11) and (9.12) may have no solution. This would mean that near the reaction surface there would be a continuous accumulation of reaction products which would catalyze and accelerate the process. Thus, a steady reaction rate could not be maintained.

Another important example involves the dissolution of a solid in a liquid, where the dissolution process itself is often much more rapid than the rate at which dissolved particles are removed. This corresponds to the boundary condition

$$c = c_{equil} \quad \text{(at the dissolving surface)}, \tag{9.13}$$

where $c_{equil}$ is the equilibrium concentration (the concentration of a saturated solution) in the liquid layer immediately adjacent to the dissolving surface.

## 10. GENERAL THEORY OF CONVECTIVE DIFFUSION IN LIQUIDS

The high value of the diffusion Prandtl number in liquids has considerable significance [15]. Since the Peclet number is equal to

$$\text{Pe} = \text{Re} \cdot \text{Pr}, \tag{10.1}$$

it is, in fact, greater than unity for Reynolds numbers as low as $10^{-2}$.

For the Reynolds numbers usually met in practice, the Peclet numbers are very high. Thus, even for relatively small Reynolds numbers, convective diffusion in a liquid is controlling (over molecular diffusion). In other words, the diffusion coefficient in liquids is so small that, even at low flow velocities, the mass transport by the moving liquid begins to predominate over molecular diffusion. The equation of convective diffusion (8.8) can be simplified accordingly; for then the term $D\Delta c$ is extremely small compared with the convective term and can be neglected:

$$v_x \frac{\partial c}{\partial x} + v_y \frac{\partial c}{\partial y} + v_z \frac{\partial c}{\partial z} = 0. \tag{10.2}$$

Equation (10.2) is satisfied by the solution

$$c = \text{const} \tag{10.3}$$

which obviously fulfills the condition of constant concentration far from the reaction surface.

Another possible solution of equation (10.2) is a concentration distribution such that the concentration gradient at every point in the fluid is perpendicular to the velocity vector. However, such a solution does not satisfy the boundary condition far from the reaction surface and has no physical significance.

The solution (10.3), however, does not apply near the reaction surface since it does not satisfy the boundary condition there. This is particularly evident for a system where the condition $c = 0$ must hold at the reaction surface. Near the surface, therefore, a thin liquid layer must exist in which the concentration changes rapidly. The derivatives of concentration with respect to distance are in this case very large, with the result that the right side of equation (8.7), representing molecular diffusion, becomes comparable to the left side, despite the small value of the diffusion coefficient.

Thus, at high Peclet numbers, the liquid may be nominally divided into two regions, the first, a region of constant concentration far from the reaction surface, and the second, a region of rapidly changing concentration in the immediate vicinity of the surface. This division corresponds to the boundary layer treatment as used in hydrodynamics for flows at high Reynolds numbers. The second region represents a very thin liquid layer which is analogous to Prandtl's boundary layer. Just as the effect of viscosity is significant in the momentum boundary layer, molecular diffusion must be taken into account in the thin layer adjacent to the reaction surface, which is therefore termed the diffusion boundary layer.

To find the order of magnitude for the thickness $\delta$ of the diffusion boundary layer, let us note that, according to equation (3.38), the thickness of the Prandtl boundary layer is proportional to the square root of the viscosity. In diffusion, the diffusion coefficient D is analogous to the kinematic viscosity, although it is numerically some one thousand times smaller. The thickness of the diffusion boundary layer, therefore, should also be considerably smaller than that of the Prandtl boundary layer.

This permits us to neglect all but the first term in (3.39), the expansion for the velocity components, since in the diffusion boundary layer $y < \delta < \delta_0$.

Let us evaluate the order of magnitude of the quantities in equation (8.7) at $y \approx \delta$, where $\delta$ is the thickness of the diffusion layer. We have, evidently (see Section 3),

$$\frac{\partial c}{\partial x} \approx \frac{c}{l},$$

$$\frac{\partial c}{\partial y} \approx \frac{c}{\delta},$$

$$\frac{\partial^2 c}{\partial x^2} \approx \frac{c}{l^2}, \tag{10.4}$$

$$\frac{\partial^2 c}{\partial y^2} \approx \frac{c}{\delta^2}, \tag{10.5}$$

hence

$$\frac{\partial^2 c}{\partial y^2} \gg \frac{\partial^2 c}{\partial x^2}, \tag{10.6}$$

and, therefore, the equation for convective diffusion in a diffusion boundary layer takes the form

$$v_x \frac{\partial c}{\partial x} + v_y \frac{\partial c}{\partial y} = D \frac{\partial^2 c}{\partial y^2}; \tag{10.7}$$

also,

$$v_y \frac{\partial c}{\partial y} \approx v_y \frac{c}{\delta}. \tag{10.8}$$

We may now replace $v_y$ by its value near the solid surface, in accordance with equation (3.40). We then obtain

$$v_y \frac{\partial c}{\partial y} \approx \frac{\nu \delta c}{\delta_0^3}. \tag{10.9}$$

In addition

$$v_x \frac{\partial c}{\partial x} \approx U_0 \frac{\delta}{\delta_0} \frac{c}{x} \approx v_y \frac{c}{\delta} \approx v_y \frac{\partial c}{\partial y}, \tag{10.10}$$

because, according to the continuity equation,

$$\frac{\partial v_x}{\partial x} \approx U_0 \frac{\delta}{\delta_0 x} \approx \frac{\partial v_y}{\partial y} \approx \frac{v_y}{\delta}. \tag{10.11}$$

Finally,

$$D \frac{\partial^2 c}{\partial y^2} \approx \frac{Dc}{\delta^2}. \tag{10.12}$$

For values of $y < \delta$, the diffusion term in equation (10.7) is negligible, while at $y = \delta$, it becomes comparable to the convective term.

Therefore, on comparing (10.9) with (10.12) at $y = \delta$, we find

$$\nu \delta \frac{c}{\delta_0^3} \approx \frac{Dc}{\delta^2}, \tag{10.13}$$

whence the thickness of the diffusion layer is

$$\delta \approx \left(\frac{D}{\nu}\right)^{1/3} \delta_0 = \frac{\delta_0}{\mathrm{Pr}^{1/3}}. \tag{10.14}$$

Thus, for $\mathrm{Pr} \approx 10^3$, the thickness of the diffusion boundary layer is about one tenth that of the Prandtl layer (Figure 9). Therefore, the tangential component of the liquid velocity at the outer edge of the diffusion boundary layer is about 10 percent of its value far from the solid surface.

From expressions (10.14) and (3.38) we may easily obtain

$$\delta \approx D^{1/3} \nu^{1/6} \sqrt{\frac{x}{U_0}}. \tag{10.15}$$

Thus, the diffusion boundary layer thickness is inversely proportional to the square root of $U_0$ the free stream velocity, and proportional to the square root of x, the distance from the leading edge of the body, and also is a function of the viscosity and of the diffusion coefficient of the particles.

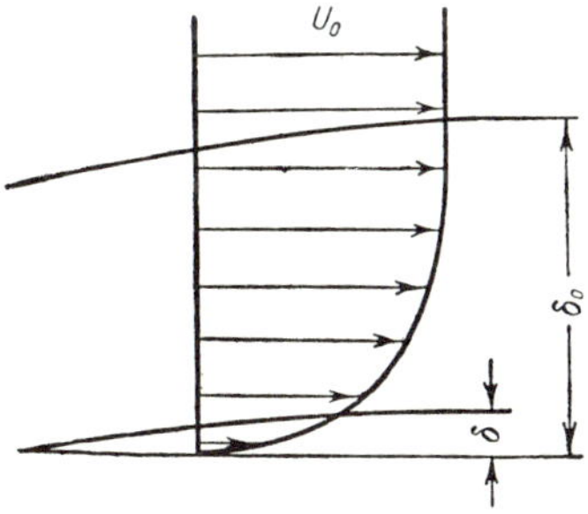

Figure 9. Thicknesses of the diffusion and hydrodynamic boundary layers.

Within the diffusion boundary layer, the calculated concentration of the solution shows a rapid change. Still, as a first approximation, this concentration change may be regarded as linear. Therefore, the expression for the diffusional flux may be approximated by

$$j_{\text{diff}} = \frac{Dc_0}{\delta},$$

i.e., by the same form as in the Nernst theory. In this case, however, $\delta$ is a known function of the physical properties of the liquid, of its velocity, as well as of the diffusion coefficient.

The concentration and diffusion flux distributions can be derived more rigorously by an exact solution of equation (10.7). However, such a solution can be obtained only when the configuration is extremely simple.

It is pertinent at this point to compare the notion of a diffusion boundary layer, defined by formula (10.15), with the diffusion layer concepts based upon the Nernst and film theories.

The boundary layer concept presented here differs in principle from the Nernst and film concepts. Specifically: 1) Our analysis accounts for the motion of the liquid and the convective mass transfer induced by it, while in the Nernst layer, the liquid is considered static. 2) In our treatment, convective and molecular diffusion both across and along the boundary layer are taken into account (in Section 17 we shall show that the tangential convective diffusion is important in the transport of solute). However, in the Nernst and film theories, it is assumed that the thickness of the layer does not change along the entire surface of the body, and diffusion in the tangential direction is entirely disregarded. Actually, the thickness of the thickness of the diffusion layer varies substantially along the surface, owing to tangential transport of particles. Therefore, the Nernst and the film theories may lead to appreciable errors. 3) The thickness of the Nernst, or of the film, layer was assumed to be constant for a given regime of motion. But in fact, the diffusion boundary layer has no clearly defined boundary. It is simply that region where the concentration gradient is at a maximum. Equation (10.15), therefore, shows that the thickness of the diffusion layer is a function not only of the physical properties and the velocity of the solution, but also of the diffusion coefficient. This indicates that the thickness of the boundary layer will depend on which species is diffusing, and that every material, with its own diffusion coefficient, will have its own corresponding boundary layer. When several materials are diffusing at the same time under given mixing conditions, several boundary layers can exist simultaneously.

## 11. SOLUTION OF THE EQUATION FOR CONVECTIVE DIFFUSION TO THE SURFACE OF A ROTATING DISK

The equations for convective diffusion acquire their simplest form when the surface of a rotating disk serves as the reaction site. The rotating disk is employed in electrochemistry and is convenient for studying chemical kinetics under laboratory conditions. Von Karman [15] and later Cochran [16] have solved the problem in which liquid is entrained by a rotating disk whose axis is perpendicular to its plane surface. Cochran's exact solution of the hydrodynamic equations provides the following picture of the fluid motion. Far from the rotating disk, the fluid moves toward the disk, and, in a thin layer immediately adjacent to its surface, the liquid acquires a rotating motion. The angular velocity of the fluid increases as the surface of the disk is approached, until the angular velocity of the rotating disk is finally attained. Furthermore, the fluid also acquires a radial velocity under the influence of the centrifugal force.

Let us present the principal considerations for determining the velocity distribution. This problem is of special interest in hydrodynamics, because it is one of the few examples where an exact solution of the hydrodynamic equations can be obtained. This exact solution gives the velocity distribution throughout the body of the viscous fluid.

The disk is assumed to be sufficiently large so that edge effects are negligible. In cylindrical coordinates. the Navier-Stokes and continuity equations take the form

$$\frac{v_\varphi}{r}\frac{\partial v_r}{\partial \varphi} + v_r\frac{\partial v_r}{\partial r} - \frac{v_\varphi^2}{r} + v_y\frac{\partial v_r}{\partial y} = -\frac{1}{\rho}\frac{\partial p}{\partial r} + \nu\left(\Delta v_r - \frac{v_r}{r^2} - \frac{2}{r^2}\frac{\partial v_\varphi}{\partial \varphi}\right) \quad (11.1)$$

$$\frac{v_\varphi}{r}\frac{\partial v_\varphi}{\partial \varphi} + v_r\frac{\partial v_\varphi}{\partial r} + \frac{v_r v_\varphi}{r} + v_y\frac{\partial v_\varphi}{\partial y} = -\frac{1}{\rho r}\frac{\partial p}{\partial \varphi} + \nu\left(\Delta v_\varphi + \frac{2}{r^2}\frac{\partial v_r}{\partial \varphi} - \frac{v_\varphi}{r^2}\right) \quad (11.2)$$

$$\frac{v_\varphi}{r}\frac{\partial v_y}{\partial \varphi} + v_r\frac{\partial v_y}{\partial r} + v_y\frac{\partial v_y}{\partial y} = -\frac{1}{\rho}\frac{\partial p}{\partial y} + \nu\Delta v_y, \quad (11.3)$$

$$\frac{1}{r}\frac{\partial v_\varphi}{\partial \varphi} + \frac{\partial v_r}{\partial r} + \frac{v_r}{r} + \frac{\partial v_y}{\partial y} = 0, \quad (11.4)$$

where $v_r$, $v_\varphi$, and $v_y$ are respectively, the radial, tangential, and axial velocity components, and

$$\Delta = \frac{\partial^2}{\partial r^2} + \frac{1}{r}\frac{\partial}{\partial r} + \frac{1}{r^2}\frac{\partial^2}{\partial \varphi^2} + \frac{\partial^2}{\partial y^2}.$$

At the surface of the disk the following boundary conditions must be satisfied:

$$v_r = 0, \quad v_\varphi = \omega r \quad \text{and} \quad v_y = 0 \quad \text{at} \quad y = 0, \quad (11.5)$$

where $\omega$ is the angular velocity of the rotating disk. The condition on the tangential velocity component $v_\varphi$ implies that the fluid near the disk rotates with it. Furthermore, the entrainment of fluid due to rotation of the disk leads to a significant radial velocity near its surface in a direction outward from the center toward the edge of the disk. In order then that fluid may be continuously supplied to the disk surface, an axial (vertical) flow must be maintained. Therefore, the boundary conditions at infinity are

$$v_r = 0, \quad v_\varphi = 0 \quad \text{and} \quad v_y = -U_0 \quad \text{as } y \to \infty. \tag{11.6}$$

The value of $U_0$ is determined from the solution of the problem itself. The minus sign indicates that the fluid velocity is toward the disk in the negative direction of the y axis. Because of axial symmetry, all derivatives with respect to the angle $\varphi$ vanish. The pressure in the fluid may be considered constant along the radius r.

Then, equations (11.1) to (11.3) may be rewritten as follows:

$$v_r \frac{\partial v_r}{\partial r} - \frac{v_\varphi^2}{r} + v_y \frac{\partial v_r}{\partial y} = \nu \left( \frac{\partial^2 v_r}{\partial y^2} + \frac{\partial^2 v_r}{\partial r^2} + \frac{1}{r} \frac{\partial v_r}{\partial r} - \frac{v_r}{r^2} \right), \tag{11.7}$$

$$v_r \frac{\partial v_\varphi}{\partial r} + \frac{v_r v_\varphi}{r} + v_y \frac{\partial v_\varphi}{\partial y} = \nu \left( \frac{\partial^2 v_\varphi}{\partial y^2} + \frac{\partial^2 v_\varphi}{\partial r^2} + \frac{1}{r} \frac{\partial v_\varphi}{\partial r} - \frac{v_\varphi}{r^2} \right), \tag{11.8}$$

$$v_r \frac{\partial v_y}{\partial r} + v_y \frac{\partial v_y}{\partial y} = -\frac{1}{\rho} \frac{\partial p}{\partial y} + \nu \left( \frac{\partial^2 v_y}{\partial y^2} + \frac{\partial^2 v_y}{\partial r^2} + \frac{1}{r} \frac{\partial v_y}{\partial r} \right). \tag{11.9}$$

Let us seek a solution satisfying the continuity equation and the boundary conditions (11.5) and (11.6) in the form

$$v_r = r\omega F(\xi), \quad v_\varphi = r\omega G(\xi), \quad v_y = \sqrt{\nu\omega} H(\xi),$$
$$p = -\rho\nu\omega P(\xi),$$

where the independent variable is the dimensionless quantity

$$\xi = \sqrt{\frac{\omega}{\nu}}\, y \tag{11.10}$$

and where F, G, H, and P are unknown functions which satisfy the equations

$$F^2 - G^2 + F'H = F'', \tag{11.11}$$

$$2FG + G'H = G'', \tag{11.12}$$

$$HH' = P' + H'', \tag{11.13}$$

$$2F + H' = 0 \tag{11.14}$$

and the boundary conditions

$$F=0, \quad G=1, \quad H=0 \qquad \text{at} \quad \xi=0, \tag{11.15}$$

$$F\to 0, \quad G\to 0, \quad H\to -\alpha \qquad \text{as} \quad \xi\to\infty, \tag{11.16}$$

all of which are obtained by substituting the definitions of $v_r$, $v_\varphi$, and $v_y$ into the Navier-Stokes and continuity equations, and into the boundary conditions. The constant $\alpha = \frac{U_0}{\sqrt{\nu\,\omega}}$ is to be determined.

The functions F, G and H may be made to satisfy the above equations and boundary conditions by formal series expansions. The nature of these expansions, for large values of $\xi$ (asymptotic expansion), is suggested by the boundary condition for H. Thus, since the asymptotic value of H($\xi$) is H $\to -\alpha$ as $\xi\to\infty$, and since the functions F and G become small as $\xi\to\infty$, it is possible to omit the quadratic terms in equation (11.11) and to rewrite it in the form

$$-F'\alpha \approx F'' \qquad \text{as} \quad \xi\to\infty.$$

Integration of this equation yields the asymptotic expression for F

$$F\sim e^{-\alpha\xi} \qquad \text{as} \quad \xi\to\infty.$$

Similarly, as $\xi\to\infty$ equation (11.12) acquires the form

$$-G'\alpha \approx G'',$$

from which the asymptotic expression for G is

$$G\sim e^{-\alpha\xi} \qquad \text{as} \quad \xi\to\infty.$$

It follows that the asymptotic expansions of functions F, G, and H must be exponentials of the form $e^{-\alpha\xi}$. The first terms of these expansions satisfying equations (11.11) to (11.14) and boundary conditions (11.15) and (11.16) are

$$F = Ae^{-\alpha\xi} - \frac{A^2+B^2}{2\alpha^2}e^{-2\alpha\xi} + \frac{A(A^2+B^2)}{4\alpha^4}e^{-3\alpha\xi} + \ldots,$$

$$G = Be^{-\alpha\xi} - \frac{B(A^2+B^2)}{12\alpha^4}e^{-3\alpha\xi} + \ldots,$$

$$H = -\alpha + \frac{2A}{\alpha}e^{-\alpha\xi} - \frac{A^2+B^2}{2\alpha^3}e^{-2\alpha\xi} + \frac{A(A^2+B^2)}{6\alpha^5}e^{-3\alpha\xi} + \ldots$$

In a similar manner, one can find expansions for small values of $\xi$ satisfying the equations and boundary conditions

$$F = a\xi - \frac{\xi^2}{2} - \frac{1}{3}b\xi^3 + \ldots,$$

$$G = 1 + b\xi + \frac{1}{3}a\xi^3 + \ldots,$$

$$H = -a\xi^2 + \frac{1}{3}\xi^3 + \ldots$$

The constants A, B, a, b, and $\alpha$ should be chosen so that the functions F, G, and H and the derivatives $F'$ and $G'$ remain continuous when one matches these two series. Then the higher-order derivatives will also be continuous in accordance with the equations of motion. Numerical integration leads to the following values:

$$a = 0.51023,^{*} \qquad b = -0.616, \qquad \alpha = 0.88447.^{*}$$
$$A = 0.934, \qquad B = 1.208.$$

The distribution of the pressure P follows automatically. We shall later utilize the value of $v_y$, which to a first approximation is

$$v_y \approx -0.89\sqrt{\nu\omega} \quad \text{as} \quad y \to \infty, \tag{11.17}$$

$$v_y \approx -0.51\sqrt{\frac{\omega^3}{\nu}}\,y^2 \quad \text{for} \quad y \ll \sqrt{\frac{\nu}{\omega}}. \tag{11.18}$$

The functions F, G, and H are shown graphically in Figure 10. From the form of these plots, which decrease exponentially

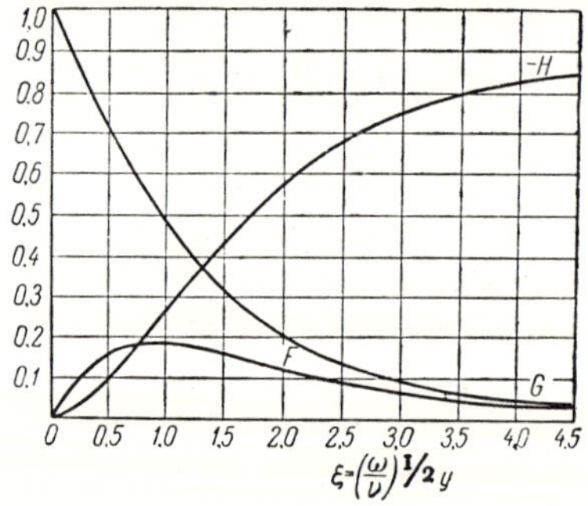


Figure 10. Graphs of the functions F, G, and H.

with increasing $\xi$, we obtain the following picture of the velocity distribution: at values of $\xi \approx 3.6$ (which corresponds to $y = \delta_0$), where

$$\delta_0 = 3.6\sqrt{\frac{\nu}{\omega}}, \tag{11.19}$$

$v_y$ attains 80 percent of its limiting value, and the velocity $v_\varphi$ decreases to 0.05 times its value at the surface of the disk. Thus, in

* From Sparrow and Gregg, J. Heat Transfer 81C, 249 (1959).

accordance with (11.19), we may regard $\delta_0$ as defining the thickness of the hydrodynamic boundary layer on the disk surface. Within the boundary layer, the radial and tangential velocity components are not zero, while beyond that layer, only axial motion exists (Figure 11.a).

The actual streamlines at the disk surface are shown in Figure 11.b (the method by which the photograph was obtained will be explained in Section 59). Here, we neglect the edge effect of the disk, which is permissible if the boundary layer thickness is small compared to the disk radius R.

We can now formulate the diffusion problem [17]. The convective diffusion equation in cylindrical coordinates takes the form

$$v_r \frac{\partial c}{\partial r} + \frac{v_\varphi}{r} \frac{\partial c}{\partial \varphi} + v_y \frac{\partial c}{\partial y} = D \left( \frac{\partial^2 c}{\partial y^2} + \frac{\partial^2 c}{\partial r^2} + \frac{1}{r} \frac{\partial c}{\partial r} + \frac{1}{r^2} \frac{\partial^2 c}{\partial \varphi^2} \right). \tag{11.20}$$

The boundary conditions are

$$c = c_0 \qquad \text{as} \quad y \to \infty, \tag{11.21}$$

where $c_0$ is the concentration in the bulk of the solution. For a maximum diffusional flux, the following condition prevails at the disk surface

$$c = 0 \qquad \text{at} \quad y = 0. \tag{11.22}$$

The case of intermediate kinetics is examined in the next section.

Let us seek a solution of equation (11.20) satisfying boundary conditions (11.21) and (11.22) in the form

$$c = c(y); \tag{11.23}$$

i.e., let us assume that the concentration is a function only of the distance from the disk surface and is not a function of either r or $\varphi$. Here we intentionally ignore effects at the edge of the disk where a solution in the form (11.23) cannot exist.* Assuming that (11.23) is applicable, we rewrite equation (11.20)

$$v_y(y) \frac{dc}{dy} = D \frac{d^2 c}{dy^2}. \tag{11.24}$$

---

* Let us stress that in the absence of convection, i.e., in the case of a stationary disk with an infinitely large surface, the diffusion problem has no steady-state solutions for an infinite fluid volume [i.e., for boundary condition (11.21)]. Also, in a gravitational field, there may be natural convection superimposed on the forced convection in the solution (see Section 23).

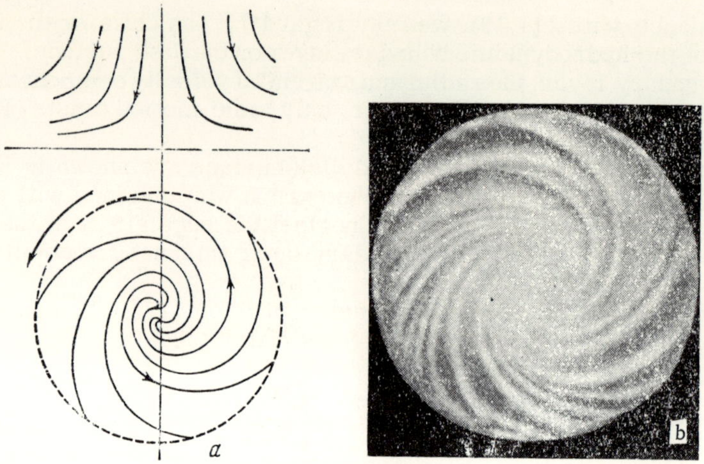


Figure 11. Distribution of streamlines at the surface of a rotating disk.
a — schematic; b — etched lines on the rotating disk.

By integrating (11.24), we obtain

$$\frac{dc}{dy} = a_1 \exp\left\{\frac{1}{D}\int_0^y v_y(z)\,dz\right\}, \tag{11.25}$$

and by a second integration

$$c = a_1 \int_0^y \exp\left\{\frac{1}{D}\int_0^t v_y(z)\,dz\right\} dt + a_2. \tag{11.26}$$

The constants $a_1$ and $a_2$ are found from boundary conditions (11.21) and (11.22). The latter immediately yields

$$a_2 = 0,$$

since, at $y=0$, the integral in (11.26) becomes zero. Also, because of boundary condition (11.21)

$$c_0 = a_1 \int_0^\infty \exp\left\{\frac{1}{D}\int_0^t v_y(z)\,dz\right\} dt. \tag{11.27}$$

In order to evaluate this integral, let us subdivide the interval of integration into two regions: from zero to $\delta_0$ and from $\delta_0$ to infinity, where $\delta_0$, as defined by formula (11.19), is the boundary layer thickness. Beyond the boundary layer, $v_y$ is given by formula (11.17). Within the boundary layer, $v_y$ varies in a complex fashion. As shown in Section 10, however, for high Prandtl numbers (i.e., for $D \ll \nu$), the major variation occurs in a region where the distance from the solid surface is small compared to $\delta_0$. For $y \ll \delta_0$, $v_y$ may be approximated by (11.18). Below, we shall substitute for $v_y$ its value from (11.18) and shall examine the justification for this step in the course of the calculations. Thus, we have

$$J = \int_0^\infty \exp\left\{\frac{1}{D}\int_0^t v_y(z)\,dz\right\} dt =$$

$$= \int_0^{\delta_0} \exp\left\{\frac{1}{D}\int_0^t v_y(z)\,dz\right\} dt + \int_{\delta_0}^\infty \exp\left\{\frac{1}{D}\int_0^t v_y(z)\,dz\right\} dt = J_1 + J_2.$$

Using (11.18) and neglecting the higher order terms in the expansion of $v_y$ we obtain for $J_1$

$$J_1 = \int_0^{\delta_0} \exp\left\{-\frac{\omega^{3/2} t^3}{5.88 D \nu^{1/2}}\right\} dt.$$

Introducing a new variable

$$u = \frac{\omega^{1/2} t}{\sqrt[3]{5.88}\, D^{1/3} \nu^{1/6}}, \qquad (11.27')$$

we find

$$J_1 \approx \frac{\sqrt[3]{6} D^{1/3} \nu^{1/6}}{\omega^{1/2}} \int_0^{\frac{\delta_0 \omega^{1/2}}{\sqrt[3]{6}\, D^{1/3} \nu^{1/6}}} e^{-u^3}\,du = \frac{1.81 D^{1/3} \nu^{1/6}}{\omega^{1/2}} \int_0^{2\left(\frac{\nu}{D}\right)^{1/3}} e^{-u^3}\,du.$$

For $\frac{\nu}{D} \gg 1$, the upper limit of the integral is significantly greater than unity, and, since the integrand decreases rapidly for values of the argument greater than unity, we can replace the upper limit by infinity.

Then

$$J_1 \approx \frac{1.81 D^{1/3} \nu^{1/6}}{\omega^{1/2}} \int_0^\infty e^{-u^3}\,du.$$

Thus, because of the rapid convergence of the integral for $\nu \gg D$, it is possible to use only the first term of the expansion of $v_y$.

The magnitude of $J_1$ is determined chiefly by the small values of u. The integral for $J_1$ may now be expressed in terms of the $\Gamma$ function,

$$\int_0^\infty e^{-u^3}du = \frac{1}{3}\int_0^\infty e^{-t}t^{-2/3}dt = \frac{1}{3}\Gamma\left(\frac{1}{3}\right) = \Gamma\left(1+\frac{1}{3}\right) = \Gamma\left(\frac{4}{3}\right) \approx 0.89.$$

Therefore, finally

$$J_1 = 1.61166 \frac{D^{1/3}\nu^{1/6}}{\omega^{1/2}} \; *$$

The integral $J_2$ is computed in a similar manner:

$$J_2 = \int_{\delta_0}^\infty \exp\left\{\frac{1}{D}\int_0^t v_y(z)\,dz\right\}dt = \int_{\delta_0}^\infty \exp\left\{-\frac{0.89\sqrt{\nu\omega}}{D}t\right\}dt =$$

$$= \frac{D}{0.89\sqrt{\nu\omega}}\exp\left\{-\frac{0.89\sqrt{\nu\omega}}{D}\delta_0\right\} \cong \frac{D}{0.89\sqrt{\nu\omega}}e^{-3\left(\frac{\nu}{D}\right)}.$$

For $\nu \gg D$, the value of the integral $J_2$ is very small compared to $J_1$ and can be omitted.

Thus, $J \approx J_1$, and from (11.27) we find that

$$a_1 = \frac{c_0}{J_1}.$$

Finally, for a concentration distribution that satisfied both the convective diffusion equation and the boundary conditions, we obtain

$$c = \frac{c_0}{J_1}\int_0^y \exp\left\{\frac{1}{D}\int_0^t v_y(z)\,dz\right\}dt =$$

$$= \frac{c_0}{1.61\left(\frac{D}{\nu}\right)^{1/3}\sqrt{\frac{\nu}{\omega}}}\int_0^y \exp\left\{\frac{1}{D}\int_0^t v_y(z)\,dz\right\}dt. \tag{11.28}$$

In order to determine c as a function of distance y from the surface of the disk (where y = 0) for small values of y (in comparison with $\delta_0$), we substitute $v_y$ in (11.28) by its definition from (11.18). Then the equation for the concentration distribution (for $y < \delta_0$) can be written in a symmetrical form

$$c = c_0\frac{\int_0^Y e^{-u^3}du}{\int_0^\infty e^{-u^3}du}, \tag{11.29}$$

---

* Numerical coefficient from Sparrow and Gregg, J. Heat Transfer 81C, 249 (1959).

where u is defined, as before, by formula (11.27′) and Y denotes the ratio

$$Y = y : \frac{\sqrt[3]{6}\, D^{1/3} \nu^{1/6}}{\omega^{1/2}} \approx \frac{2y}{\delta_0} \left(\frac{\nu}{D}\right)^{1/3}.$$

Equation (11.29) shows that, for small values of y, the concentration increases very rapidly with the distance y (more rapidly than exponentially).

In view of the rapid convergence of the integral for $Y > 1$, i.e., for $y > \frac{1}{2}\left(\frac{D}{\nu}\right)^{1/3} \delta_0$, Y can be replaced by infinity, and the value of concentration in equation (11.29) reduces to $c_0$ (i.e., to the concentration in the bulk of the solution).

The ratio $\frac{c}{c_0}$ from equation (11.29) is plotted in Figure 12.

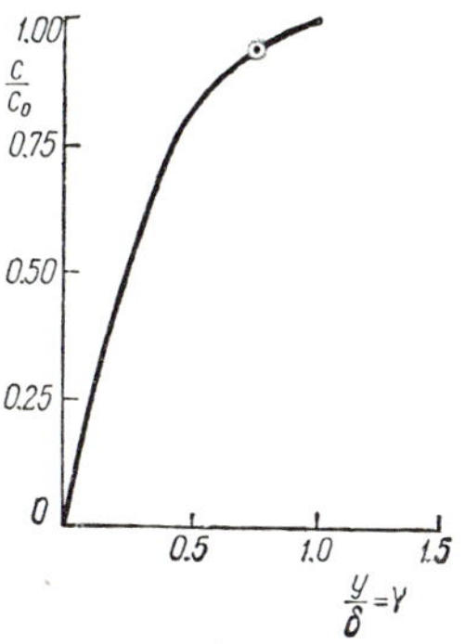


Figure 12. Concentration as a function of distance from the disk.

Differentiating (11.29) we obtain the mass flux to the disk surface

$$j = D \left(\frac{\partial c}{\partial y}\right)_{y=0} = \frac{Dc_0}{0.89} \frac{\omega^{1/2}}{\sqrt[3]{6}\, D^{1/3} \nu^{1/6}} = \frac{Dc_0}{1.61 \left(\frac{D}{\nu}\right)^{1/3} \sqrt{\frac{\nu}{\omega}}} = \tag{11.30}$$

$$= 0.62 D^{2/3} \nu^{-1/6} \omega^{1/2} c_0.$$

Expressions (11.29) and (11.30) are in full agreement with the general considerations given in Section 10. The thickness of the diffusion boundary layer, represented by δ, is given by

$$\delta = \frac{Dc_0}{j} = 1.61 \left(\frac{D}{\nu}\right)^{1/3} \sqrt{\frac{\nu}{\omega}} \approx 0.5 \left(\frac{D}{\nu}\right)^{1/3} \delta_0, \tag{11.31}$$

which agrees with expression (10.15), but, unlike (10.15), contains a numerical coefficient. The major change in concentration occurs in a region $\sim \delta$ in thickness. For the usual values of the diffusion coefficient $D \sim 10^{-5}\,cm^2/sec$ in water, and for $\nu \sim 10^{-2}\,cm^2/sec$, the thickness $\delta$ of the diffusion boundary layer constitutes only 5% of the thickness $\delta_0$ of the hydrodynamic boundary layer. This fully justifies the use of (11.18) for $v_y$ instead of the full expansion which holds for all values of y.

A significant characteristic of the diffusion boundary layer on a disk is the fact that its thickness is not a function of the distance from the axis of rotation, but is constant over the entire disk surface. The only exception is the zone at the edge, to which the foregoing analysis does not apply. The extent of this zone in which the edge of the disk does exert an effect apparently extends over a distance of the order of $\delta_0$.

Equation (11.30) shows that the diffusional flux to the surface of the disk is proportional to the 2/3 power of the diffusion coefficient D, to the (– 1/6) power of the kinematic viscosity, and to the 1/2 power of the angular velocity. Since the diffusion coefficient also varies in accordance with equation (8.15), the flux is in fact a function of kinetic viscosity according to the law $j \sim \nu^{-5/6}$. The total flow of matter to the disk surface is

$$I = \frac{\pi D c_0 R^2}{1.61\left(\frac{D}{\nu}\right)^{1/3}\sqrt{\frac{\nu}{\omega}}} \approx 1.9 D^{2/3} \nu^{-1/6} \sqrt{\omega}\, R^2 c_0. \tag{11.32}$$

The rotating disk apparatus used to measure the diffusional flux given by formula (11.32) is shown in Figure 13 (see also Section 56).

We now examine the effect of temperature on the diffusional flux to the disk. The diffusion coefficient D and the kinematic viscosity $\nu$ for the usual liquids are exponential functions of temperature:

$$D = D_0 e^{-\frac{U_D}{kT}}, \tag{11.33}$$

$$\nu = \nu_0 e^{+\frac{U_\nu}{kT}}, \tag{11.34}$$

where $U_D$ and $U_\nu$ are, respectively, the activation energies for diffusion and viscosity, and $D_0$ and $\nu_0$ relate to a reference temperature such as room temperature. The numerical values of $U_D$ in most cases are approximately 3000 cal/mole, and $U_\nu$, somewhat greater, is about 3600 cal/mole.

Substituting the expressions for D and $\nu$ into (11.30), we obtain

$$j \sim e^{-\frac{4U_D + U_\nu}{6kT}}. \tag{11.35}$$

The temperature dependence of j is simple, but the effective flux activation energy $U_j$ is complex:

Figure 13. Rotating disk electrode (exhibited at the Brussels World's Fair in 1958).

Let us stress that all equations in this section apply only to laminar flow past the disk. Laminar flow is ordinarily sustained up to Re $\sim 10^4$, but with very well polished and centered disks up to Re $\sim 10^5$. On the other hand, at lower rotation speeds (Re $\approx 10$), the thickness $\delta_0$ of the hydrodynamic layer becomes comparable to the radius of the disk.

Riddiford and Gregory [36] have computed the integral (11.27) employing the higher-order terms of the expansion of $v_y$. Use of a series for the function $H(\xi)$ makes it possible to compute the value of $v_y$ with any desired number of terms in the expansion in $\xi$.

Substituting the expression for $v_y$ into (11.27) and including the higher-order terms of the expansion in the function $H(\xi)$, we obtain, with an accuracy to terms of the third order,

$$J_1 \approx 1.61 \left(\frac{D}{\nu}\right)^{1/3} \sqrt{\frac{\nu}{\omega}} \left[1 + 0.35 \left(\frac{D}{\nu}\right)^{0.36}\right]. \tag{11.36}$$

The value of the integral $J_1$ is thus found to be somewhat greater than given by the zero approximation. Correspondingly, the total mass flow I is found to be somewhat smaller. The correction to the flux, given approximately by (11.32), is from three to five percent. As will be indicated in Chapter VI, the experimental values are given more closely by (11.32) than by (11.36). This means that there are additional corrections of the same magnitude but opposite in sign. One of them is for the edge effect. The maximum value of this correction apparently occurs for a static medium. The flowing fluid always carries matter away from the edge of the disk. If the disk has an outer rim, then the flux to the disk is given by the equation [35]

$$I' = 4RDc_0 \text{ and } \frac{I'}{I} = \frac{4}{\pi}\frac{\delta}{R} = 1.99\left(\frac{D}{\nu}\right)^{1/3}\sqrt{\frac{\nu}{\omega R^2}}.$$

Other corrections to the flux toward the disk, which increase its value beyond that given by (11.32) are associated with natural convection (see Section 23) and with the appearance of turbulence (see Chapter III). The first correction becomes significant at small Reynolds numbers, and the latter at high Reynolds numbers. A comparison of theoretically predicted values with experimental data, obtained in A. N. Frumkin's laboratory by precision measurements, shows that the corrections to the flux (11.32) listed above are compensated to a large degree by the use of the correction given in equation (11.36).

If a deposit forms on the disk surface, it grows for the most part at the edge, and the error of formula (11.32) increases.

## 12. INTERMEDIATE KINETICS ON THE SURFACE OF A ROTATING DISK. THE METHOD OF THE UNIFORMLY ACCESSIBLE SURFACE

We examine now the case of intermediate kinetics, when boundary condition (9.6) is fulfilled on the disk surface,

$$D\cdot\left(\frac{\partial c}{\partial y}\right)_{y=0} = kc_1^m, \tag{12.1}$$

where $c_1$ is the concentration at the disk surface, to be determined. The general solution (11.26) shows that

$$a_2 = c_1. \tag{12.2}$$

Substitution of (12.2) into (12.1) gives

$$Da_1 = ka_2^m. \tag{12.3}$$

Also, because of condition (11.21),

$$c_0 = a_1\int_0^\infty \exp\left\{\frac{1}{D}\int_0^t v_y(z)\,dz\right\}dt + a_2. \tag{12.4}$$

The integral of this equation has been computed above. Designating it by $\delta$, we find

$$c_0 = a_2 + a_1\delta,$$

whence

$$a_1 = \frac{c_0 - a_2}{\delta} = \frac{c_0 - c_1}{\delta}.$$

Then, from (12.3) we obtain the algebraic equation for the boundary concentration $c_1 = c_1(d, \delta, k, c_0)$

$$c_1^m + \frac{D}{k\delta} c_1 - \frac{Dc_0}{k\delta} = 0. \qquad (12.5)$$

The flux of matter to a unit of disk surface (equal to the rate of the reaction at the surface) has the form

$$j = Da_1 = D\frac{c_0 - c_1}{\delta}. \qquad (12.6)$$

Solving the algebraic equation (12.5) and using (12.6) we obtain the flux to the disk for the case of a reaction in the intermediate kinetics regime.

Entering into the boundary condition (12.1) are two quantities which characterize the reaction: the exponent m and the coefficient of proportionality k. They also enter into the theoretical equation of the reaction rate.

A rotating disk possesses an important feature that distinguishes it from other reaction surfaces in a moving liquid, that is, the thickness of the diffusion boundary layer has a constant value over the entire disk surface. This means that regardless of the distance from the axis of rotation, the conditions for transport of matter to any point on the surface of the disk are absolutely identical. Such a reaction surface, following the terminology proposed by D. A. Frank-Kamenetskiy [12], may be termed uniformly accessible (from the standpoint of diffusion).

In the case of such a uniformly accessible surface, the concentration $c_1$ of matter at the surface and the reaction rate $q = kc_1^m$ are constant over the entire surface. Equation (12.5) provides a single-valued relation between $c_1$ and $c_0$, provided that the mixing conditions determining the value $\delta$ are given. In this connection it is significant that the value of $\delta$ is not a function of the kinetics of the chemical conversion or of the distance along the surface.

Equation (12.5) may be conveniently solved in dimensionless form by a graphical method. By introducing the dimensionless concentration $\xi = \frac{c_1}{c_0}$ and the dimensionless parameter $\alpha = \frac{\delta k}{Dc_0^{1-m}} = \frac{1}{\mu}$ equation (12.5) may be rewritten as

$$\mu \xi^m = 1 - \xi . \tag{12.7}$$

Figure 14 presents a plot, reproduced from a book by Frank-Kamenetskiy [12], which provides an easy means for the solution of equation (12.7). The straight lines originating from the point $\xi = 1$ and having a slope $\frac{1}{\mu}$ correspond to the right side of equation (12.7). The curves originating from the point $\xi = 0$ represent the left side of

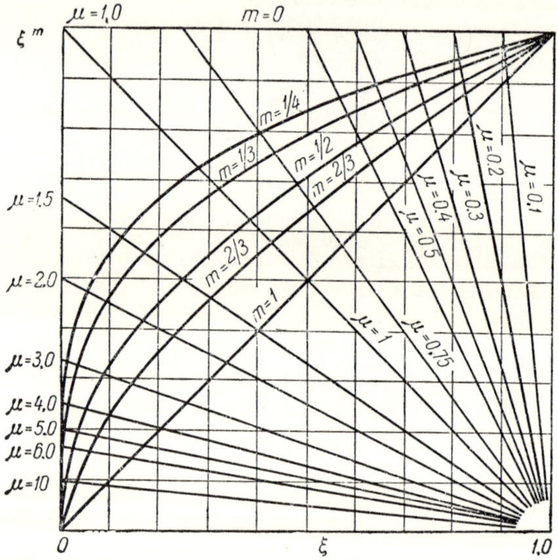


Figure 14. Diffusion kinetics of fractional order reactions. The dimensionless concentration $\xi$ is plotted along the abcissa. $\xi^m$ is plotted along the ordinate. Each curve is labeled with a value of the reaction order m. The straight lines originating from the point $\xi = 1$ represent the quantity $\frac{1-\xi}{\mu}$. Each is labeled with the value of $\mu$. The abcissa of the intersection of the curve $\xi^m$ and the straight line gives the value of the dimensionless concentration $\xi$ as a function of $\frac{1}{\mu}$ for a given order of reaction.

the equation for various values of m. The points of intersection of the curves and the straight lines correspond to solutions of the equation (12.7). At a given m, the different points of intersection correspond to the different values of the parameter $\mu$.

In practice, the exponent m is usually less than unity. Therefore, the curves shown on Figure 14 are given for only fractional values of m.

A first-order reaction, where m = 1, is the simplest case. Thus,

$$\xi = \frac{1}{1+\mu}, \quad \mu = \frac{\delta k}{D},$$

or

$$c_1 = \frac{Dc_0}{k\delta + D}$$

and

$$j = \frac{D(c_0 - c_1)}{\delta} = \frac{kc_0}{1 + \frac{\delta}{Dc_0} \cdot kc_0} = \frac{Dc_0}{\left[1.61\left(\frac{D}{\nu}\right)^{1/3}\sqrt{\frac{\nu}{\omega}} + \frac{D}{k}\right]} = \frac{j_k}{\left(1 + \frac{j_k}{j_d}\right)}, \quad (12.8)$$

where $j_k = kc_0$ is the limiting value of the reaction rate (for $c_1 = c_0$), while $j_d$ is the limiting diffusional flux (for $c_1 = 0$).

Formula (12.8) suggests that the diffusing substance should follow the diffusion path given by

$$\delta_1 = \delta + \frac{D}{k}.$$

The ratio $\frac{D}{k}$ represents an additional resistance to mass transfer.

Figure 15 shows (curve 1) the flux of particles to the disk as a function of the angular velocity $\omega$, according to formula (12.8). For low values of $\omega$, the rate of the process is diffusion controlled. At high $\omega$, there is a smooth transition to the reaction-controlled regime (kinetic region). As the exponent m decreases, the curves $j(\omega)$ change. The transition from the diffusion-controlled to the reaction-controlled regime takes place in an increasingly abrupt fashion. For values of $m \leqslant \frac{1}{2}$, we obtain curve 2. The experimental data illustrated in Figure 15 are discussed in Section 56.

It should also be noted that the methods of solution used above may easily be applied to studies of reaction rates whose order cannot be approximated by a power function.

In studying the reaction mechanism, the order (i.e., the value of m) and the energy of activation U should be determined. The values m and U may be determined from the theoretical equation (12.6) and from an experimental determination of the value of j.

Actually, to determine m and k, it is necessary to measure j at two different values of the mixing rate or concentration $c_0$. Rewriting expression (12.6)

$$c_1 = c_0 - \frac{j\delta}{D}$$

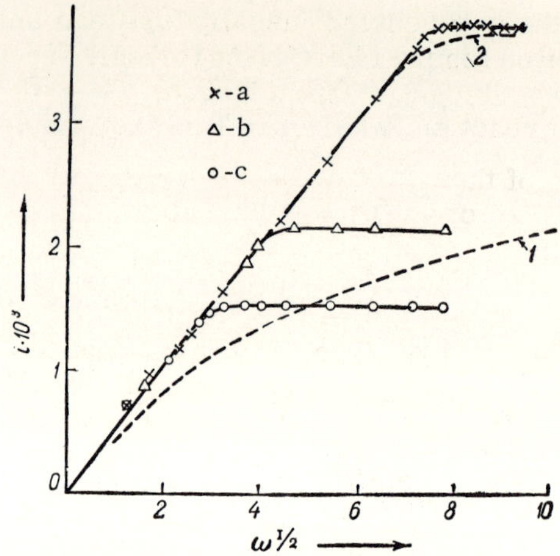


Figure 15. j as a function of the square root of the speed of rotation of the electrode at $\eta = 0.045$ v. a — 1N $H_2SO_4$, b — 1N HCl, c — 1N HBr; curve 1 is computed according to formula (12.8); curve 2 corresponds to the case of $m \leq \frac{1}{2}$.

and noting that

$$q = kc_1^m = k\left(c_0 - j\,\frac{\delta}{D}\right)^m = j,$$

we obtain

$$\left(1 - j\,\frac{\delta}{c_0 D}\right)^m = \frac{j}{kc_0^m} \tag{12.9}$$

and

$$k = \frac{j}{\left(c_0 - j\,\frac{\delta}{D}\right)^m}. \tag{12.10}$$

Knowing j for two different liquid velocities (different δ ), we can determine m from equation (12.9)

$$m \ln \frac{\left(1 - \frac{j_1 \delta_1}{c_0 D}\right)}{\left(1 - \frac{j_2 \delta_2}{c_0 D}\right)} = \ln \frac{j_1}{j_2}.$$

Knowing the values of m, we can find k by means of (12.10).

It would appear easy to determine the energy of activation U by means of equation (9.4) if k were known. In practice, however, this is not the case. Not only does the value of k change with temperature, but so do the values of the diffusion layer thickness $\delta$ and the diffusion coefficient D.

The dependence of these quantities on temperature can be easily observed using a first-order reaction as an example. Let us write equation (12.8) in the form

$$\frac{1}{k} = \frac{c_0}{j} - \frac{\delta}{D} = \frac{c_0}{j} - \text{const}\,\frac{\nu^{1/6}}{D^{2/3}}.$$

Introducing the activation energies for the kinematic viscosity and the diffusion coefficient, we find

$$e^{\frac{U}{kT}} = \frac{k_0 c_0}{j} - \text{const}\; e^{\frac{2U_D}{3kT} + \frac{U_\nu}{6kT}}. \tag{12.11}$$

The quantity $\dfrac{4U_D + U_\nu}{6}$ is usually in the range of 2 to 3 kilocalories. Inasmuch as the activation energy of a chemical reaction is considerably larger, of the order of several tens of kilocalories, we may assume

$$e^{\frac{U}{kT}} \approx \frac{k_0 c_0}{j} - \text{const}.$$

The measurement of j as a function of T makes it possible to find U immediately. However, as indicated by the foregoing discussion, the accuracy of this method is not very great. Furthermore, U may be found from equation (12.11).

The calculations become more involved for reactions of fractional order.

Let us once again consider the extent to which the method of a uniformly accessible surface applies.

It is clear from considerations developed in Section 10 that the thickness of the diffusion boundary layer generally changes from one point to another on the reaction surface. In general, therefore, the different points on the surface are not equivalent from the standpoint of diffusion, and thus the basic assumption of the method of a uniformly accessible surface is not valid.

It will be shown in Section 17 that the nonequivalence of points on the surface has a substantial effect on the magnitude of the mass flux; thus calculations based on the method of uniformly accessible surfaces give results substantially different from the correct values. The method of a uniformly accessible surface is applicable only in special cases.

At the same time, it is clear from the foregoing that this method is entirely appropriate for studying the mechanism of reactions occurring in the intermediate region, under the closely controlled conditions of mixing at the surface of the rotating disk. This is evident, for example, when the exponent m of the reaction is small. If the reaction occurs at the disk surface, then curve $j(\omega)$ has the form of a curve with a characteristic sharp break. On the other hand, if the reaction occurs on a surface that is not uniformly accessible, such as a plate, then the curve has another form. Because the thickness of the diffusion boundary layer is different at different sections of the surface, the observed rate of the process is represented by a curve without a definite break, approaching the form of curve 1 of Figure 15. Analysis of a curve obtained from a surface that is not uniformly accessible can often lead to faulty conclusions regarding the reaction order.

D. A. Frank-Kamenetskiy, in the monograph cited, examines the method of a uniformly accessible surface in detail and considers it to be a universal, if approximate, method for solving problems in diffusion kinetics.

Calculations made in the following sections of this book show that such a uniformly accessible surface, for example a rotating disk, is actually a special case. Large errors may result by treating a surface as uniformly accessible when it is not, since the diffusion layer thickness along the reaction surface often varies from ten to several hundred times its minimum value.

At the present time, the rotating disk is being used increasingly as a practical device for studying the kinetics of electrochemical reactions (see Sections 56-59).

Let us stress in conclusion that the foregoing examination refers to laminar liquid flow in the vicinity of a rotating disk. Experimental measurements have shown that flow around a disk is laminar up to Reynolds numbers $Re = \frac{\omega R^2}{\nu}$ of the order of $10^4$ to $10^5$.

## 13. REDUCTION OF THE EQUATION OF CONVECTIVE DIFFUSION TO THE FORM OF THE HEAT CONDUCTION EQUATION

Diffusion toward a rotating disk is a special case where an exact solution of the convective diffusion equation can be obtained. In other cases it is necessary to seek approximate solutions of the boundary layer equation. For this purpose we introduce a general procedure which permits us to reduce the boundary layer equation to the form of the heat conduction equation. The methods for integrating such equations have been worked out in detail.

We assume that the reaction surface has only a slight curvature, or, more exactly, that its radius of curvature is large compared to the thickness of the diffusion boundary layer. In real cases this condition is almost always fulfilled, since the thickness of the diffusion layer is very small. Then the equation of steady convective mass transfer in the diffusion boundary layer may be written similarly to equation (10.7)

$$v_x \frac{\partial c}{\partial x} + v_y \frac{\partial c}{\partial y} = D \frac{\partial^2 c}{\partial y^2}, \tag{13.1}$$

where $v_x$ and $v_y$ are known functions of the distances along the x and y axes which satisfy the continuity equation (1.1)

$$\frac{\partial v_x}{\partial x} + \frac{\partial v_y}{\partial y} = 0. \tag{13.2}$$

This equation leads to the following expressions for $v_x$ and $v_y$:

$$v_x = \frac{\partial \psi}{\partial y}; \quad v_y = -\frac{\partial \psi}{\partial x}, \tag{13.3}$$

where $\psi$ is the stream function.

We consider the stream function to be known from a solution of the hydrodynamic equations expressed in terms of the pertinent coordinates.

We rewrite equation (13.1) with new variables, replacing (x,y) by (x, $\psi$)

$$\left(\frac{\partial c(x, y)}{\partial x}\right)_y = \left(\frac{\partial c}{\partial x}\right)_\psi + \frac{\partial c}{\partial \psi}\frac{\partial \psi}{\partial x} = \left(\frac{\partial c}{\partial x}\right)_\psi - v_y \frac{\partial c}{\partial \psi}, \tag{13.4}$$

$$\frac{\partial c}{\partial y} = \frac{\partial c}{\partial \psi}\frac{\partial \psi}{\partial y} = v_x \frac{\partial c}{\partial \psi}, \tag{13.5}$$

$$\frac{\partial^2 c}{\partial y^2} = \frac{\partial \psi}{\partial y}\frac{\partial}{\partial \psi}\left(v_x \frac{\partial c}{\partial \psi}\right) = v_x \frac{\partial}{\partial \psi}\left(v_x \frac{\partial c}{\partial \psi}\right). \tag{13.6}$$

Substitution of equations (13.4) to (13.6) into equation (13.1) yields

$$\left(\frac{\partial c}{\partial x}\right)_\psi = D \frac{\partial}{\partial \psi}\left[v_x \frac{\partial c}{\partial \psi}\right]. \tag{13.7}$$

In equation (13.7) the concentration and the tangential velocity component are functions of x and $\psi$. As before, x is the distance along the surface and $\psi$ is the stream function entering through relations (13.3) and, for this reason, may be determined up to an arbitrary constant. Because $\psi$ enters through derivatives, we are at liberty to take $\psi = 0$ at the surface of the body where y = 0. Then the boundary conditions for equation (13.7) are

$$c = c_0 \quad \text{as} \quad \psi \to \infty \qquad (y \to \infty), \tag{13.8}$$

$$c = 0 \quad \text{as} \quad \psi \to 0 \quad \text{(at the surface of the body)}. \tag{13.9}$$

However, two obvious conditions are insufficient for a complete solution of equation (13.7), which contains second-order derivatives of $c(x, \psi)$ in $\psi$, and a first-order derivative in x.

One more condition must be given along with (13.8) and (13.9) to complete the formulation of a boundary value problem. For this last condition one may require that the expressions for the concentration have no singularity at the point of incidence of the flow on the body ($x = 0$, $y = 0$), i.e., shall remain finite and single valued. Thus, the missing condition may be stated as:

The concentration $c(x, \psi)$ has no singularity at the point of incidence of the flow on the body.

Equation (13.7) is a type of heat conduction equation, but with a thermal conductivity which is a function of distance and time,

$$\frac{\partial T}{\partial t} = \frac{\partial}{\partial x}\left[\varkappa(x, t)\frac{\partial T}{\partial x}\right]. \tag{13.10}$$

However, the integration of equation (13.10) is much simpler than the integration of equation (13.1).

A further simplification of equation (13.7) arises from the fact that we are interested in its solution only near the surface where $y = 0$, i.e., at small values of $\psi$. This permits us to retain only the first terms of a power series expansion of $v_x(\psi, x)$.

## 14. DIFFUSION TO A FREE-FALLING SOLID PARTICLE

Let us apply the method of integrating the convective diffusion equation developed in the preceding section, by examining diffusion to a solid particle descending freely in a solution. We assume the particle to be sufficiently small so that the corresponding Reynolds number $Re = \frac{Ua}{\nu}$ (where U is the rate of descent and a is the diameter of the particle) is small compared to unity. Then the liquid around the particle is in creeping motion. The velocity distribution near a spherical falling particle was solved by Stokes, and is well known as the Stokes problem.

The complete solution to this problem, in particular the determination of the velocity and the stream function, is given, for example, in the book by Kochin et al. [18]. Let us merely note that the velocity of the liquid flowing past the particle decreases gradually and smoothly with distance from the particle surface, and that there is no hydrodynamic boundary layer near the surface.

Nevertheless, a diffusion boundary layer does exist near the particle surface. Indeed, as pointed out above, the Peclet number

$Pe = \frac{Ua}{D} = Re \cdot Pr$ is approximately $10^3$ times greater than the Reynolds numbers $Re \approx \frac{1}{2}$; it is clear that the conditions $Pe \gg 1$ and $Re \ll 1$ are simultaneously fulfilled over a rather wide range of particle dimensions.

For $Pe \gg 1$, the change in concentration occurs in the thin boundary layer, in spite of the fact that the liquid velocity varies gradually.

The convective diffusion equation for the boundary layer in spherical coordinates r and $\theta$ (it is obvious that the concentration cannot be a function of the azimuthal angle $\varphi$) becomes

$$v_r \frac{\partial c}{\partial r} + \frac{v_\theta}{r} \cdot \frac{\partial c}{\partial \theta} = D\left(\frac{\partial^2 c}{\partial r^2} + \frac{2}{r}\frac{\partial c}{\partial r}\right). \tag{14.1}$$

Here we have omitted from the right side the angular portion of the Laplacian $\frac{1}{r^2 \sin\theta}\frac{\partial}{\partial\theta}\left(\sin\theta \frac{\partial c}{\partial \theta}\right)$, since the derivatives along the surface of a sphere are small compared to the derivatives along the radius-vector. We shall demonstrate later that a sharp change in concentration occurs in a layer whose thickness is small compared to the particle dimensions. In this case the diffusion layer is practically planar. This indicates that we are interested in solutions of the convective diffusion equation at values of r close to the radius a of the particle.

At small distances from the surface of the sphere, the stream function is

$$\psi = -\frac{U}{2}\sin^2\theta\left(r^2 - \frac{3}{2}ar + \frac{1}{2}\frac{a^3}{r}\right) \approx -\frac{3}{4}Uy^2 \sin^2\theta, \tag{14.2}$$

where U is the velocity of the falling particle.

Correspondingly, the tangential velocity component is

$$v_\theta = -\frac{1}{r\sin\theta}\frac{\partial\psi}{\partial y} \approx -\frac{1}{a\sin\theta}\frac{\partial\psi}{\partial y} = \frac{3}{2}U\frac{y}{a}\sin\theta. \tag{14.3}$$

Considering that at small values of y ($y \ll a$),

$$\frac{\partial^2 c}{\partial y^2} \gg \frac{2}{a}\frac{\partial c}{\partial y},$$

and using (14.3), we can rewrite the equation of convective diffusion in terms of the variables ( $\psi$, $\theta$ )

$$\frac{1}{a}\frac{\partial c}{\partial \theta} = Da\sin\theta \frac{\partial}{\partial\psi}\left[(a\sin\theta v_\theta)\frac{\partial c}{\partial\psi}\right]. \tag{14.4}$$

Expressing $v_\theta$ in terms of $\psi$, we find

$$\left(\frac{\partial c}{\partial\theta}\right)_\psi = Da^2\sin^2\theta\sqrt{3U}\frac{\partial}{\partial\psi}\left(\sqrt{-\psi}\frac{\partial c}{\partial\psi}\right). \tag{14.5}$$

The boundary conditions for equation (14.5) are

$$c = 0 \quad \text{at} \quad \psi = 0 \quad \text{(on the surface of the particle)}, \tag{14.6}$$

$$c = c_0 \quad \text{as} \quad \psi \to -\infty \quad \text{(far from the particle)}, \tag{14.7}$$

$$c = c_0 \quad \text{at} \quad \theta = 0, \ \psi = 0. \tag{14.8}$$

The meaning of the last condition is simple. The point $\theta = 0$, $\psi = 0$ represents the point of incidence of the liquid flow on the sphere. It is clear that at this point the incident flow has not yet been depleted by diffusion, and that its concentration is equal to the concentration in the bulk of the solution.

In order to integrate equation (14.5), we introduce a new variable

$$t = Da^2\sqrt{3U}\int \sin^2\theta\, d\theta = \frac{Da^2\sqrt{3U}}{2}\left(\theta - \frac{\sin 2\theta}{2}\right) + C_1. \tag{14.9}$$

We then obtain

$$\frac{\partial c}{\partial t} = \frac{\partial}{\partial \psi}\left(\sqrt{-\psi}\,\frac{\partial c}{\partial \psi}\right). \tag{14.10}$$

In solving partial differential equations of the type (14.10), we utilize a similarity method that is rarely presented in mathematical texts.* In view of this, we briefly describe here the basis of this method.

Equation (14.10) remains unchanged when the variables are transformed as follows:

$$\begin{aligned} \psi &\to \lambda\psi', \\ t &\to \mu t', \end{aligned} \tag{14.11}$$

provided the arbitrary changes in the scale of the coordinates are linked by the relationship

$$\mu = \lambda^{3/2}.$$

Transforming the variables (14.11) does not require a change in the boundary conditions (14.6) to (14.8). The solution of the boundary value problem, therefore, is invariant under the transformation (14.11), i.e.,

---

* An exception is the well-known textbook by A. N. Tikhanov and A. A. Samarskiy, Uravneniya matematicheskoy fiziki (Equations of Mathematical Physics), Gostekhizdat, 1951, page 252 (see also German translation: Differentialgleichungen der mathematischen Physik, Deutscher Verlag der Wissenschaften, 1959; cf. Birkhoff, Hydrodynamics, 2nd ed., Princeton University Press, 1960).

$$c(\psi, t) = c(\lambda\psi, \lambda^{3/2}t).$$

In view of this, a solution $c(\psi, t)$ may be sought such that a combination of variables $\psi$ and $t$ remains unchanged by transformations (14.11).

Such a combination is

$$\eta = \frac{-\psi}{t^{2/3}}.$$

Evidently, we have

$$\frac{\partial c}{\partial t} = -\frac{2\psi}{3t^{5/3}}\frac{dc(\eta)}{d\eta},$$

$$\frac{\partial c}{\partial \psi} = \frac{1}{t^{2/3}}\frac{dc(\eta)}{d\eta},$$

$$\frac{\partial}{\partial \psi}\left(\sqrt{-\psi}\,\frac{\partial c}{\partial \psi}\right) = \frac{1}{t}\frac{d}{d\eta}\left(\sqrt{\eta}\,\frac{dc(\eta)}{d\eta}\right).$$

Substituting these three equations into (14.10), we obtain the ordinary differential equation for c

$$-\frac{2}{3}\eta\frac{dc}{d\eta} = \frac{d}{d\eta}\left(\sqrt{\eta}\,\frac{dc}{d\eta}\right).$$

Introducing the following variable for convenience in integration

$$z = \sqrt{\eta},$$

we obtain finally

$$\frac{d^2c}{dz^2} + \frac{4}{3}z^2\frac{dc}{dz} = 0. \tag{14.12}$$

Integration of (14.12) is straightforward. Thus

$$c(z) = C_2\int_0^z \exp\left\{-\frac{4}{9}z^3\right\}dz + C_3, \tag{14.13}$$

where

$$z = \sqrt{\frac{3U}{4}}\,\frac{y\sin\theta}{\left[Da^2\sqrt{\frac{3U}{4}}\left(\theta - \frac{\sin 2\theta}{2}\right) + C_1\right]^{1/3}}. \tag{14.14}$$

The constants $C_1$, $C_2$, and $C_3$ are determined from the boundary conditions (14.6) to (14.8).

Condition (14.6) gives $C_3 = 0$. It follows, from condition (14.7) that

$$C_2 = \frac{c_0}{\int\limits_0^\infty \exp\left\{-\frac{4}{9} z^3\right\} dz}.$$

This integral can be expressed in terms of a gamma function. It is equal to

$$\int\limits_0^\infty e^{-\frac{4z^3}{9}} dz = \left(\frac{9}{4}\right)^{1/3} \frac{1}{3} \Gamma\left(\frac{1}{3}\right) = 1.15.$$

Finally, near the point of incidence of the flow on the particle where $\theta = 0$, the concentration c must be a single-valued, positive function of the angle $\theta$. At small values of $\theta$, we have,

$$z \approx \sqrt{\frac{3U}{4}} \frac{y\theta}{\left[C_1 + \frac{2\theta^3}{3} \cdot \frac{Da^2}{2} \sqrt{3U}\right]^{1/3}}.$$

In order for z to be a real, positive quantity at small $\theta$, $C_1$ must equal zero. Finally, we get

$$z = \sqrt[3]{\frac{3U}{4Da^2}} \frac{y \sin\theta}{\left(\theta - \frac{\sin 2\theta}{2}\right)^{1/3}},$$

$$c = \frac{c_0}{1.15} \int\limits_0^z \exp\left\{-\frac{4}{9} z^3\right\} dz. \qquad (14.15)$$

The diffusional flux to the surface of the particle is

$$j = D\left(\frac{\partial c}{\partial y}\right)_{y=0} = \frac{Dc_0}{1.15} \sqrt[3]{\frac{3U}{4Da^2}} \frac{\sin\theta}{\left(\theta - \frac{\sin 2\theta}{2}\right)^{1/3}}, \qquad (14.16)$$

which is proportional to the concentration $c_0$ of the diffusing matter, to the 1/3 power of U the velocity of the particle, to $D^{2/3}$, to $a^{-2/3}$, and to a function of the angle $\theta$. The latter is equal to $\sqrt[3]{\frac{2}{\pi}}$ for $\theta = \frac{\pi}{2}$, and to zero for $\theta = \pi$. For $\theta = 0$ equation (14.16) is indeterminate, but by resolving the indeterminacy we find that the function of the angle $\theta$ is unity at $\theta = 0$.

Thus, the mass flux is highest at the point of incidence $\theta = 0$. It decreases with increasing $\theta$ and becomes zero at the downstream side of the particle (Figure 16). The corresponding effective thickness of the diffusion layer is given by

$$\delta = \frac{1.15\left(\theta - \frac{\sin 2\theta}{2}\right)^{1/3}}{\sin\theta}\sqrt[3]{\frac{4Da^2}{3U}}. \tag{14.17}$$

This shows that $\delta$ increases with the angle $\theta$ and that it becomes infinite for $\theta = \pi$. Actually, of course, $\delta$ does not become infinite at $\theta = \pi$, and j is not zero. Let us note that, had we not set $C_1$ in equation (14.14) equal to zero, the value of $\delta$ at the point of incidence where $\theta = 0$ would be infinitely great, and the flux j would be equal to zero, which is obviously absurd.

In the course of the analysis, we assumed that the thickness $\delta$ of the diffusion layer is very small compared with the radius of the particle. It follows from equation (14.17) that, for a certain value of $\theta$ relatively close to $\pi$, the thickness of the diffusion layer attains a value comparable to that of the radius a of the particle. In this range of $\theta$, the theory presented above is not applicable. It is clear, however, that in the range $\theta \sim \pi$ there is no significant effect on the total mass flow I to the particle. The total flow is equal to

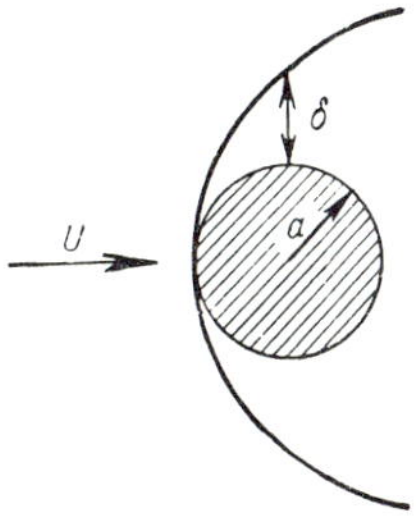


Figure 16. Diffusion boundary layer on the surface of a particle.

$$I = \int j\,ds = 2\pi a^2 \int_0^{\pi} j \sin\theta\, d\theta =$$

$$= \frac{Dc_0 a^{4/3}}{1.15}\sqrt[3]{\frac{3U}{4D}} \cdot 2\pi \int_0^{\pi} \frac{\sin^2\theta\, d\theta}{\left(\theta - \frac{\sin 2\theta}{2}\right)^{1/3}}. \tag{14.18}$$

Evaluation of the integral gives

$$I = 7.98 c_0 D^{2/3} U^{1/3} a^{4/3}. \tag{14.19}$$

The total rate of mass transfer to a solid particle moving at low velocities through a solution is thus shown to be proportional to $c_0$, $U^{1/3}$, and $a^{4/3}$.

It is interesting to note that the diffusional flux to the particle is not directly proportional to D, as is true in a static medium, but to a fractional power of D. This is not a circumstance unique to the particular problem examined here, but is general in nature. The change in concentration in a static medium occurs over a distance of the order of the dimensions of the body. In convective diffusion, the substance is carried by the moving liquid closer to the surface of the body, thus compensating in part for the small value of D. A similar analysis can be performed for diffusion to a rod-shaped particle (see Section 43).

Somewhat later, but independently of the author of this book, the problem of mass diffusion to the surface of a sphere was examined by G. A. Aksel'rud [19] who employed the approximate method of total mass balance in the boundary layer ("integral technique") using a polynomial for the concentration. His results, in general, agree well with ours. In his expression for the total rate of mass transfer to a sphere, he obtained the numerical coefficient 8.2, instead of the 7.98 of (14.19). It should be noted, however, that this agreement is somewhat fortuitous, and closer examination reveals significant inadequacies in the polynomial method. This method expresses the distribution of concentration as a polynomial in powers of $\frac{y}{\delta_A}$ with coefficients adjusted to satisfy the boundary conditions

$$c = c_0\left[\frac{2y}{\delta_A} - 2\left(\frac{y}{\delta_A}\right)^3 + \left(\frac{y}{\delta_A}\right)^4\right], \tag{14.20}$$

where $\delta_A$ is the thickness of the diffusion layer. At $y = \delta_A$, $c = c_0$.

For the thickness $\delta_A$, Aksel'rud obtains a value almost twice that of $\delta$. Owing to the selection of the coefficients in polynomial (14.20), however, the flux to the surface is

$$j = D\left(\frac{\partial c}{\partial y}\right)_{y=0} = D \cdot \frac{2c_0}{\delta_A},$$

so that Aksel'rud arrives at a value for j within a few percent of the corresponding solution (14.16).

Thus, polynomial approximations result in a substantial error for the concentration distribution, even though they yield at low values of y, a relatively accurate slope for the curve $\frac{\partial c}{\partial y}$.

When the dimension of the particle is so small that the Peclet number becomes less than unity, then convective diffusion plays no discernible part in the transport of matter. In this case, the convective term may be omitted in equation (8.8), which then takes the form

$$D\,\Delta c = 0.$$

A solution of the equation for mass diffusion to a sphere which is obviously a function only of the radial coordinate r, and which satisfies the boundary conditions $c = 0$ at the surface of the particle and $c = c_0$ at infinity, has the form

$$c = c_0\left(1 - \frac{a}{r}\right).$$

The diffusional flux to the surface of the sphere is

$$j = D\left(\frac{\partial c}{\partial r}\right)_{r=a} = \frac{Dc_0}{a}, \tag{14.21}$$

and the flow to the surface of the particle is correspondingly equal to

$$I = 4\pi\, D a\, c_0. \tag{14.22}$$

For Pe numbers less than unity, the following interpolation formula may be written:

$$I = 4\pi\, D c_0\, a \left(1 + 0.64\, \mathrm{Pe}^{1/3}\right), \tag{14.23}$$

which becomes (14.21) as Pe → 0, and (14.19) for Pe $\gg 1$.

## 15. DIFFUSIONAL FLOW TO THE SURFACE OF A PLATE IN A FLOWING FLUID

We now consider the case in which the reaction surface is a smooth plate with a fluid flowing around it [17].

We examine here the case of laminar flow in the boundary layer. We assume that the velocity of the fluid far from the plate equals U, and let h be the length and b the width of the plate. We also assume that the length and width of the plate are significantly larger than the thickness of the hydrodynamic boundary layer. The plate may then be regarded as semi-infinite, i.e., as occupying the right half-plane $y = 0$ $(x \geq 0)$, where x is the distance from the leading edge of the plate, and y is the distance along the normal. The flow past a semi-infinite plate has been briefly examined in Section 3.

Assuming the velocity distribution to be known, we can now consider the diffusion problem. The concentration distribution of the solute is a function of the coordinates x and y, but may be assumed independent of the coordinate z. The convective diffusion equation in a laminar boundary layer on the surface of the plate is

$$v_x \frac{\partial c}{\partial x} + v_y \frac{\partial c}{\partial y} = D\, \frac{\partial^2 c}{\partial y^2}. \tag{15.1}$$

On the right side of this equation, $\frac{\partial^2 c}{\partial x^2}$ has been omitted as being small in comparison with $\frac{\partial^2 c}{\partial y^2}$ [cf. (10.16)]. The velocity components $v_x$ and $v_y$ are determined by the expansions (3.39) and (3.40).

We note that both members on the left side of equation (15.1) are of the same order of magnitude (as was shown in Section 10). Although the velocity component $v_y$ is small compared to $v_x$, the derivative $\frac{\partial c}{\partial y}$ is considerably greater than $\frac{\partial c}{\partial x}$. In Eucken's analysis of convective diffusion to the surface of a plate [13], equation (15.1) was introduced. The term $v_y\, (\frac{\partial c}{\partial y})$ was, however, omitted from it without sufficient reason, and this led the author to faulty conclusions.

In this section we confine our study to a calculation of the maximum diffusional flux. Here, the reaction rate is infinitely large compared to the rate of mass transfer. The boundary conditions in this case are

$$\begin{aligned} c &= c_0 \quad \text{as} \quad y \to \infty, \\ c &= 0 \quad \text{at} \quad y = 0. \end{aligned} \tag{15.2}$$

In solving equation (15.1) we utilize the general method developed in Section 13.

Substituting for $v_x$ and $v_y$ the equations (3.39) and (3.40) with the stream function (3.26), we obtain, in place of (15.1),

$$x^{1/4}\frac{\partial c}{\partial x} = \beta \frac{\partial}{\partial \psi}\left(\sqrt{\psi}\frac{\partial c}{\partial \psi}\right), \tag{15.3}$$

where

$$\beta = \frac{D\left(\frac{\alpha}{2}\right)^{1/2} U^{3/4}}{\nu^{1/4}}. \tag{15.4}$$

Utilizing the same method of solution as described above, we introduce a new variable

$$\eta = \frac{\psi^{1/2}}{\left(\frac{\alpha}{2}\right)^{1/2}(\nu U)^{1/4} x^{1/4}} = \frac{1}{2}\sqrt{\frac{Uy^2}{\nu x}}, \tag{15.5}$$

and obtain

$$\frac{\partial c}{\partial x} = \frac{\partial c}{\partial \eta}\frac{\partial \eta}{\partial x} = -\frac{\psi^{1/2}}{4x^{5/4}} \cdot \frac{1}{\left(\frac{\alpha}{2}\right)^{1/2}(\nu U)^{1/4}} \cdot \frac{\partial c}{\partial \eta},$$

$$\frac{\partial c}{\partial \psi} = \frac{\partial c}{\partial \eta}\frac{\partial \eta}{\partial \psi} = \frac{1}{2}\frac{1}{x^{1/4}\psi^{1/2}} \cdot \frac{1}{\left(\frac{\alpha}{2}\right)^{1/2}(\nu U)^{1/4}} \cdot \frac{\partial c}{\partial \eta}.$$

With a simple transformation we get

$$\frac{d^2 c}{d\eta^2} + \left(\frac{\alpha}{2}\right)^{3/2}\frac{(\nu U)^{3/4}}{\beta} \cdot \frac{dc}{d\eta} \cdot \eta^2 = 0. \tag{15.6}$$

Substituting the value of $\beta$, we finally obtain

$$\frac{d^2 c}{d\eta^2} + \frac{\alpha}{2} \cdot \frac{\nu}{D}\eta^2 \frac{dc}{d\eta} = 0. \tag{15.7}$$

Integration of (15.7) gives

$$c = C_1 \int_0^\eta \exp\left\{-\frac{\alpha}{2}\frac{\nu}{D}\frac{z^3}{3}\right\} dz + C_2.$$

The boundary conditions (15.2) are used to find the integration constants $C_1$ and $C_2$. Finally, for a concentration distribution c(x, y) that satisfies equation (15.1) and boundary conditions (15.2), we obtain

$$c(x,\ y) = \frac{c_0 \int\limits_0^{\frac{1}{2}\sqrt{\frac{Uy^2}{\nu x}}} \exp\{-0.22\,\mathrm{Pr}\,z^3\}\,dz}{\int\limits_0^{\infty} \exp\{-0.22\,\mathrm{Pr}\,z^3\}\,dz}. \tag{15.8}$$

The integral in the denominator may be expressed in terms of a gamma function

$$\int\limits_0^{\infty} \exp\{-0.22\,\mathrm{Pr}\,z^3\}\,dz = \frac{\int\limits_0^{\infty} e^{-t^3}\,dt}{(0.22\,\mathrm{Pr})^{1/3}} = \frac{\frac{1}{3}\Gamma\left(\frac{1}{3}\right)}{(0.22\,\mathrm{Pr})^{1/3}} \approx \frac{0.89}{(0.22\,\mathrm{Pr})^{1/3}}.$$

Finally, therefore,

$$c(x,\ y) = \frac{(0.22\,\mathrm{Pr})^{1/3} c_0}{0.89} \int\limits_0^{\frac{1}{2}\sqrt{\frac{Uy^2}{\nu x}}} \exp\{-0.22\,\mathrm{Pr}\,z^3\}\,dz. \tag{15.9}$$

Graphical integration of this equation gives c as a function of $\frac{y}{\sqrt{x}}$, as shown in Figure 17.

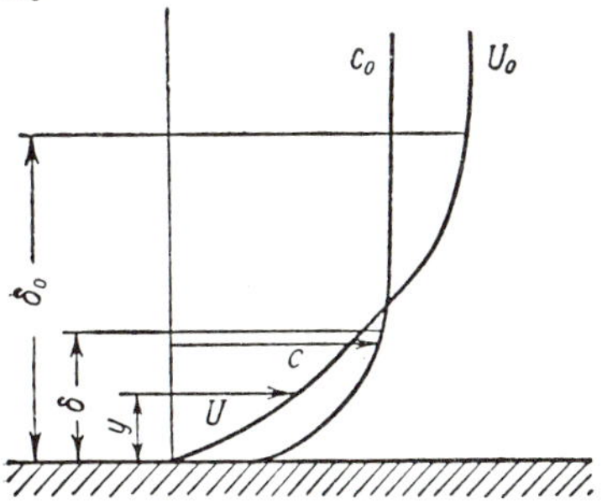


Figure 17. Concentration distribution and the tangential velocity at the surface of the plate.

It is clear from Figure 17 that for a given value of x, the concentration increases rapidly with increasing y, and attains, at a distance

$\delta$ from the plate, a value that is almost equal to the concentration $c_0$ in the bulk of the solution, where

$$\delta \approx 3\left(\frac{D}{\nu}\right)^{1/3}\sqrt{\frac{\nu x}{U}}. \tag{15.10}$$

The distance $\delta$ is, of course, the thickness of the diffusion boundary layer on the plate. Using equation (15.8), we can determine the diffusional flux to the plate as follows:

$$j = D\left(\frac{\partial c}{\partial y}\right)_{y=0} = \frac{Dc_0\,(0.22\mathrm{Pr})^{1/3}}{1.78}\sqrt{\frac{U}{\nu x}} = 0.34\,\frac{Dc_0\sqrt{U}}{\sqrt{\nu x}}\cdot\left(\frac{\nu}{D}\right)^{1/3}. \tag{15.11}$$

The thickness of the diffusion layer may now be given as

$$\delta = \frac{Dc_0}{j} \approx 3\left(\frac{D}{\nu}\right)^{1/3}\sqrt{\frac{\nu x}{U}} \approx 0.6\left(\frac{D}{\nu}\right)^{1/3}\delta_0\,, \tag{15.12}$$

where $\delta_0$ is the thickness of the hydrodynamic boundary layer.

The thickness of the diffusion layer at the surface of the plate, in contrast to the thickness of the diffusion layer on the surface of a rotating disk, is a function of the distance from the upstream edge of the plate and increases proportionally to $\sqrt{x}$. In other words, different points along the plate are not uniformly accessible from a diffusional point of view (Figure 18).

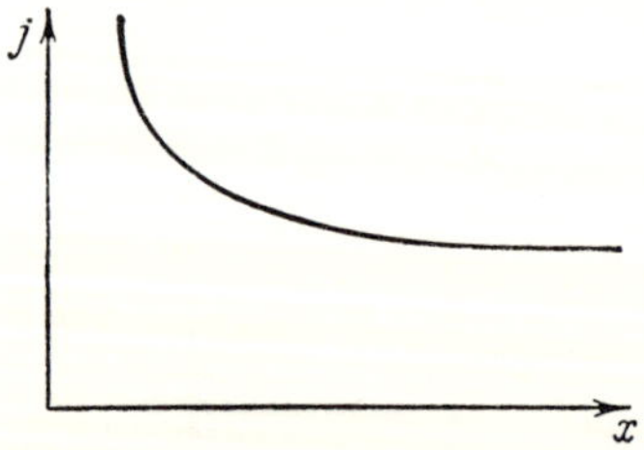


Figure 18. Diffusional flux as a function of distance from the upstream edge of a plate.

The tangential transport of solute by the moving fluid increases the effectiveness of the diffusional flux very substantially. The diffusional flux at the upstream edge of the plate is greater than at the downstream edge. Naturally, at the leading edge, at distances less than $\delta$ where the hydrodynamic boundary layer has not yet formed, equation (15.11) does not apply.

The maximum diffusional flux to the plate depends on D and $\nu$ in the same manner as for a rotating disk. The maximum flux to a plate is proportional to the square root of the fluid velocity U and is also inversely proportional to the square root of the distance from the edge of the plate.

In analyzing mass transport to a rotating disk, we have established the similarity between the diffusion and the hydrodynamic boundary layers. This similarity also exists for a plate for which the following approximation is valid:

$$\delta = 3\left(\frac{D}{\nu}\right)^{1/3}\sqrt{\frac{\nu x}{U}} \approx 0.6\left(\frac{D}{\nu}\right)^{1/3}\delta_0.$$

The total diffusion flow $I_{max}$ to the surface of a plate can be obtained by integrating $j_{max}$ along the surface

$$I_{max} = \int j_{max}\, dx\, dz = 0.68\, Dc_0 b\left(\frac{\nu}{D}\right)^{1/3}\sqrt{\frac{Uh}{\nu}}. \qquad (15.13)$$

Let us note that this relation was obtained with a somewhat different numerical coefficient by Kruzhilin [20] in connection with the analogous heat transfer problem for a plate. In Kruzhilin's work, however, a rough method of calculation was used in which the velocity and temperature distributions in the boundary layer were approximated by polynomials. Since such approximations are completely arbitrary, they introduce indeterminate errors into the computation. This led Kruzhilin to such incorrect conclusions as the belief that his expression, similar to expression (15.13), is valid for all $Pr \geqslant 0.5$. Actually, it is valid only for $Pr \ll 1$. The similarity of the formulas is for this reason somewhat fortuitous.

## 16. ANALOGY BETWEEN CONVECTIVE DIFFUSION AND SURFACE DRAG

We noted in Section 10 that the velocity distribution v in a flowing fluid and the concentration c of the solute are described by similar equations. According to Section 1, the hydrodynamic equations may be interpreted as equations of conservation of momentum. Similarly, the equations of convective diffusion may be considered as equations of conservation of mass. Here the transport of matter occurs both by molecular diffusion and by convection.

If these equations are written as boundary layer equations

$$v_x\frac{\partial c}{\partial x} + v_y\frac{\partial c}{\partial y} = \frac{\partial}{\partial y}\left(D\frac{\partial c}{\partial y}\right),$$

$$v_x\frac{\partial v_x}{\partial x} + v_y\frac{\partial v_x}{\partial y} = \frac{1}{\rho}\frac{\partial}{\partial y}\left(\mu\frac{\partial v_x}{\partial y}\right),$$

then the right side of both equations will contain derivatives of two fluxes: the molecular diffusional flux

$$j_{diff} = D\frac{\partial c}{\partial y}$$

and the momentum flux transferred by molecular friction, equal to the surface drag on 1 $cm^2$ of surface area,

$$\tau = \mu \frac{\partial v_x}{\partial y}.$$

The left sides of the equations contain expressions representing, respectively, the divergence of convective flow c and of convective momentum flux v $(\rho v_x)$.

When the values of the quantities $\nu = \frac{\mu}{\rho}$ and D are comparable, a complete analogy exists between the velocity field and the concentration field in the boundary layer. Here we must bear in mind that what has been said refers to a boundary layer in which the x component of velocity is large compared to $v_y$, so that "velocity field" means the $v_x$ component. In other words, in a boundary layer where $\nu = D$, the quantities $v_x$ and c are identically distributed. Actually, however, in a liquid, $\nu \gg D$; hence there can be no complete analogy between the distribution of $v_x$ and c. Nevertheless, a certain similarity between them exists. The velocity component $v_x$ changes from zero at the surface of the solid body to the velocity U of the irrotational flow at $y \approx \delta_0$, where $\delta_0$ is the thickness of the hydrodynamic (Prandtl) boundary layer. In the same way, c changes from a value $c = 0$ at the reaction surface to $c = c_0$ at $y \approx \delta$, where $\delta$ is the thickness of the diffusion boundary layer. Here, as shown in Section 10, the thicknesses are proportional to one another.

The shear stress (drag acting on 1 $cm^2$ of the surface area of a solid body) is of the order of magnitude

$$\tau \approx \mu \frac{U_0}{\delta_0},$$

and the diffusional flux is of the order of magnitude

$$j_{\text{diff}} \approx D \frac{c_0}{\delta}.$$

Since $\delta$ is proportional to $\delta_0$, it is always true that

$$j_{\text{diff}} \approx \tau,$$

and, in the extension to the entire plate,

$$I \approx F,$$

where $F = \int \tau dS$, the integral being taken over the plane of the reaction surface.

As an example of the analogy between diffusional flux and shear stress, let us examine these quantities in the case of flow past a plate. Calculations [see (3.36)] lead to the following expression for total drag (referred to one side of the plate):

$$F = 0.664\rho b\sqrt{\nu U^3 h}.$$

Comparing this with the value of $I_{max}$ found above, we have

$$I_{max} \approx \frac{1}{\rho}\frac{c_0}{U}\left(\frac{D}{\nu}\right)^{2/3} F. \tag{16.1}$$

Substituting the drag coefficient $K_f$ for the drag F, we find

$$I_{max} \approx \frac{K_f c_0 U S}{2\mathrm{Pr}^{2/3}}. \tag{16.2}$$

Thus, $I_{max}$ is proportional to the drag coefficient $K_f$, to the rate of convective transport by the irrotational flow $c_0$ U, and to the area of the plates. The proportionality coefficient is the quantity $\frac{1}{2\mathrm{Pr}^{2/3}}$.

Exactly the same expression is also obtained for the diffusional flow to a rotating disk. Thus, if the drag of a body of any shape is designated by $K_f$, then expression (16.2) determines the diffusion flux to that body.

At Pr = 1 there is full similarity between the transfer of matter and of momentum, and

$$I_{max} = \frac{K_f}{2} c_0 U S = \frac{F c_0}{\rho U}. \tag{16.3}$$

It should be remembered that formula (16.3) is valid only for streamlined shapes. When the shape of the body is not streamlined and, at a certain distance from the point of incidence, separation occurs, then formula (16.3) represents the diffusional flux only over that portion of the surface which extends to the line of separation, i.e., that portion around which there is laminar flow.

## 17. SOLUTION OF THE EQUATION OF CONVECTIVE DIFFUSION FOR A PLATE, UNDER CONDITIONS OF INTERMEDIATE KINETICS*

For the case of a first-order reaction proceeding at a rate comparable to that at which reagents flow to the plate surface, the boundary condition, in terms of the concentration at y = 0, must be

$$D\left(\frac{\partial c}{\partial y}\right)_{y=0} = kc_1. \tag{17.1}$$

---

* The contents of this and those of the next section are based on the studies reported in [21].

The boundary condition at infinity remains the same as in the diffusion controlled case, i.e., it takes the form of expression (9.1),

$$c \to c_0 \quad \text{as} \quad y \to \infty. \tag{17.2}$$

However, in order to solve this problem we require an additional condition which is equivalent to the initial condition in the diffusion equation. Thus, in the plane x = 0, where the point of incidence of the flow is located, we require that

$$c = c_0 \quad \text{at} \quad x = 0, \; y = 0. \tag{17.3}$$

The solution (15.8), previously obtained, automatically satisfies this condition.

To solve for the concentration distribution under conditions of intermediate kinetics, it is necessary and sufficient to find a solution to equation (15.1) that satisfies the boundary conditions noted above [21]. For this purpose, let us introduce the variables

$$\xi = \frac{1}{3} \beta x^{3/4}; \qquad \varphi = \sqrt{\psi}.$$

Then equation (15.1) takes the form

$$\frac{\partial c}{\partial \xi} = \frac{1}{\varphi} \frac{\partial^2 c}{\partial \varphi^2}. \tag{17.4}$$

This is an expression similar to the heat conduction equation, except that its diffusion coefficient is a function of the variable $\varphi$. It is possible now to rearrange equation (17.4) into a more convenient form by introducing the new variable

$$z = \frac{2}{3} \varphi^{3/2}.$$

Then equation (17.4) becomes

$$\frac{\partial^2 c}{\partial z^2} + \frac{1}{3z} \frac{\partial c}{\partial z} = \frac{\partial c}{\partial \xi}. \tag{17.5}$$

Equation (17.5) is a special case of the somewhat more general equation

$$\frac{\partial^2 c}{\partial z^2} + \frac{1 - 2p}{z} \frac{\partial c}{\partial z} = \frac{\partial c}{\partial \xi}, \tag{17.6}$$

which was examined in detail by Sutton [22]. For $p = 1/3$, (17.6) becomes equivalent to (17.5), whereas for $p = 1/2$, we have the usual heat conduction equation.

To integrate equation (17.6), Sutton developed a general method similar to the one proposed by Goursat for the integration of the heat conduction equation.

Using equation (17.5) as the example, we confine our treatment here to a general description of the method, and refer the readers to the original works of Goursat and Sutton for details.

We present the function sought as

$$c(\xi, z) = z^{1/3} S(\xi, z).$$

Then $S(\xi, z)$ must satisfy the equation

$$\frac{\partial^2 S}{\partial z^2} + \frac{1}{z}\frac{\partial S}{\partial z} - \frac{1}{9z^2} S = \frac{\partial S}{\partial \xi}, \tag{17.7}$$

in which the variables are separable.

Its solutions are, for example, the functions

$$e^{-\frac{\lambda^2}{4}(\xi-\rho)} J_{1/3}\left(\frac{\lambda z}{2}\right); \qquad e^{-\frac{\lambda^2}{4}(\xi-\rho)} J_{-1/3}\left(\frac{\lambda z}{2}\right),$$

where $\lambda$ and $\rho$ are arbitrary constants, and $J_{\pm 1/3}$ represents Bessel functions of order $\pm 1/3$ *. From these functions it is possible to construct two fundamental solutions of equation (17.7).

In selecting the basic solutions, it is convenient to take advantage of the invariance of equation (17.7) with respect to the transformation

$$\xi' \to -\frac{1}{\xi-\rho}; \qquad z' \to \frac{z}{\xi-\rho};$$

$$S' \to \frac{1}{\xi-\rho} e^{-\frac{z^2}{4(\xi-\rho)}} S\left(-\frac{1}{\xi-\rho}, \frac{z}{\xi-\rho}\right).$$

This means that if $S(\xi, z)$ is a solution of equation (17.7), then the function $S' = \frac{1}{\xi-\rho} e^{-\frac{z^2}{4(\xi-\rho)}} \cdot S\left(-\frac{1}{\xi-\rho}, \frac{z}{\xi-\rho}\right)$ is also a solution of that equation.

Let us select one of the following as the function S:

$$S_1 = \exp\left(\frac{\lambda^2\xi}{4}\right) I_{1/3}\left(\frac{\lambda z}{2}\right),$$

$$S_2 = \exp\left(\frac{\lambda^2\xi}{4}\right) I_{-1/3}\left(\frac{\lambda z}{2}\right),$$

where $\lambda$ is an arbitrary number and $I_{\pm 1/3}$ is the modified Bessel function of order $\pm 1/3$. Then we have two solutions for equation (17.5), which are henceforth termed the fundamental solutions of that equation:

---

* $I_\nu(x) = i^{-\nu} J_\nu(ix)$.

$$\chi_{1,2}(\xi, z; \lambda, \rho) = \frac{\lambda^{1/3} z^{1/3}}{2(\xi-\rho)} e^{-\frac{z^2+\lambda^2}{4(\xi-\rho)}} I_{\pm 1/3}\left\{\frac{\lambda z}{2(\xi-\rho)}\right\},$$

where the subscript 1 corresponds to the upper sign of the Bessel function and subscript 2 corresponds to the lower sign. This formula for the fundamental solutions $\chi_{1,2}$ is applicable only when $\xi > \rho$. Sutton showed that at $\xi \leq \rho$, $z > 0$, both of these functions are equal to zero.

The fundamental functions $\chi_{1,2}(\xi, z; \lambda, \rho)$ satisfy equation (17.5) for any values of the parameters $\lambda$ and $\rho$ and in that respect may be regarded as functions of four variables — the variables $(\xi, z)$ and the parameters $(\lambda, \rho)$. The construction of the fundamental functions $\chi_{1,2}$ for equation (17.5) is not unique and may be done in a similar manner for the general equation (17.6). Furthermore, for $p = 1/2$, it is possible to build a solution for the heat conduction equation in the same manner, by utilizing the relation between the Bessel functions $I_{\pm 1/2}$ and the hyperbolic sines and cosines. In that case we have

$$\chi_1 + \chi_2 = \frac{e^{-\frac{(z-\lambda)^2}{4(\xi-\rho)}}}{\sqrt{\pi}(\xi-\rho)^{1/2}}, \quad \xi > \rho.$$

With the help of the fundamental functions $\chi_{1,2}$ it is now possible to construct the solution of the boundary value problem associated with equation (17.5), i.e., to find a solution which satisfies the given boundary conditions

Indeed, let us construct the function

$$c(\xi, z) = \int_L [A(\lambda, \rho)\chi_1 + B(\lambda, \rho)\chi_2][P(\lambda, \rho)\,d\lambda + Q(\lambda, \rho)\,d\rho], \quad (17.8)$$

where $\chi_1$ and $\chi_2$ are fundamental solutions of equation (17.5) and A, B, P, and Q are functions of the parameters $\lambda, \rho$. Integration is along a certain contour in the half-plane $\lambda \geq 0$. By appropriate selection of the integration contour and of the functions A, B, P, and Q, we obtain a solution of the boundary value problem which satisfies equation (17.5) and the given set of boundary conditions.

Let us now apply this general method of solution to the system of boundary conditions mentioned earlier. Another example of the application of this method will be given later. A similar solution of boundary value problems in the theory of heat conduction may be found in Goursat's work (cf. Sutton [22]).

We select the straight line $\rho = 0$ as the path of integration, and set

$$A(\lambda, 0)P(\lambda, 0) = \lambda^{1/3} f_1(\lambda)$$

and

$$B(\lambda, 0)P(\lambda, 0) = \lambda^{1/3} f_2(\lambda).$$

For the sake of simplicity, we take the functions $f_1(\lambda)$ and $f_2(\lambda)$ to be continuous and bounded as $\lambda \to \infty$. We find that

$$c(\xi, z) = c_1(\xi, z) + c_2(\xi, z) =$$

$$= \int_0^\infty \frac{f_1(\lambda)\,\lambda^{2/3} z^{1/3}}{2\xi} \exp\left\{-\frac{z^2+\lambda^2}{4\xi}\right\} \cdot I_{1/3}\left(\frac{\lambda z}{2\xi}\right) d\lambda +$$

$$+ \int_0^\infty \frac{f_2(\lambda)\,\lambda^{2/3} z^{1/3}}{2\xi} \exp\left\{-\frac{z^2+\lambda^2}{4\xi}\right\} I_{-1/3}\left(\frac{\lambda z}{2\xi}\right) d\lambda, \tag{17.9}$$

where $c_1(\xi, z)$ and $c_2(\xi, z)$ are, respectively, equal to the first and the second terms.

We shall show now that the unknown functions $f_1(\lambda)$ and $f_2(\lambda)$ can be determined so as to satisfy the boundary and initial conditions (17.1) to (17.3). This will then complete the solution of the problem.

First of all, we note that in the variables $(z, \xi)$ we have

$$\lim_{y\to 0} D\left(\frac{\partial c}{\partial y}\right) = \lim_{z\to +0} D\left(\frac{\partial c(\xi, z)}{\partial z}\right)\frac{dz}{d\varphi}\frac{\partial \varphi}{\partial y} = 2^{-4/3}\beta^{4/3}\xi^{-1/3}\lim_{z\to +0}\left(z^{1/3}\frac{\partial c(\xi, z)}{\partial z}\right).$$

The system of boundary and initial conditions can therefore be rewritten as

$$2^{-4/3}\beta^{4/3}\xi^{-1/3}\lim_{z\to +0}\left(z^{1/3}\frac{\partial c(\xi, z)}{\partial z}\right) = k \lim_{z\to +0} c(\xi, z), \tag{17.1'}$$

$$\lim_{z\to\infty} c(\xi, z) = c_0, \tag{17.2'}$$

$$\lim_{\xi\to +0} c(\xi, z) = c_0. \tag{17.3'}$$

As shown by Sutton, the functions $c_1(\xi, z)$ and $c_2(\xi, z)$ have the following properties:

I. $$\lim_{\xi\to +0} c_{1,2}(\xi, z) = f(z).$$

II. $$\lim_{z\to\infty} c_{1,2}(\xi, z) = \lim_{z\to\infty} f(z) \qquad (\xi > 0).$$

III. $$\lim_{z\to\infty,\,\xi\to 0}\left\{z^{1/3}\frac{\partial c_{1,2}(\xi, z)}{\partial z}\right\} = \lim_{z\to\infty}\left\{z^{1/3}\frac{df}{dz}\right\} \qquad (\xi > 0).$$

IV. $$\lim_{z\to 0} c_1(\xi, z) = 0 \qquad (\xi > 0).$$

V. $$\lim_{z\to 0}\left\{z^{1/3}\frac{\partial c_1(\xi, z)}{\partial z}\right\} = \frac{2^{1/3}}{\Gamma\left(\frac{1}{3}\right)}\xi^{-1/3}\int_0^\infty f_1(2\sqrt{\xi\rho})\,e^{-\rho}\,d\rho \qquad (\xi > 0).$$

VI. $$\lim_{z\to 0} c_2(\xi, z) = \frac{1}{\Gamma\left(\frac{2}{3}\right)}\int_0^\infty f_2(2\sqrt{\xi\rho})\,e^{-\rho}\rho^{-1/3}\,d\rho \qquad (\xi > 0).$$

VII. $$\lim_{z\to 0}\left\{z^{1/3}\frac{\partial c_2(\xi, z)}{\partial z}\right\} = 0 \qquad (\xi > 0).$$

The last four properties are obtained by a series expansion of $I_{\pm 1/3}\left(\frac{\lambda z}{2\xi}\right)$ and the substitution in the integral of the expressions for $c_1(\xi, z)$ and $c_2(\xi, z)$. Properties I and II show that

$$\lim_{\xi \to 0} c(\xi, z) = f_1(z) + f_2(z) = c_0,$$

$$\lim_{z \to \infty} c(\xi, z) = \lim_{z \to \infty} f_1(z) + \lim_{z \to \infty} f_2(z) = c_0.$$

Therefore, conditions (17.2′) and (17.3′) will be satisfied if

$$f_2(\lambda) = c_0 - f_1(\lambda).$$

Finally we select $f_1(\lambda)$ in such a way that the remaining condition (17.1′) is satisfied.

Substituting the appropriate values into (17.1′) and noting the properties IV to VII and (17.2′), we obtain

$$\frac{\beta^{4/3}}{2\Gamma\left(\frac{1}{3}\right)} \xi^{-2/3} \int_0^\infty f_1(2\sqrt{\xi\rho})\, e^{-\rho}\, d\rho = \frac{k}{\Gamma\left(\frac{2}{3}\right)} \int_0^\infty [c_0 - f_1(2\sqrt{\xi\rho})]\, e^{-\rho} \rho^{-1/3} d\rho. \quad (17.10)$$

The unknown function $f_1(\lambda)$ may easily be determined from this relation. Let us set

$$\rho\xi = t; \quad \xi = \frac{1}{u}; \qquad f_1(2\sqrt{t}) = \varphi(t);$$

then

$$\frac{\beta^{4/3}\Gamma(2/3)}{2k\Gamma\left(\frac{1}{3}\right)} \int_0^\infty u\varphi(t)\, e^{-ut}\, dt = \int_0^\infty [c_0 - \varphi(t)]\, e^{-ut} \frac{dt}{t^{1/3}}.$$

By integrating the left-hand side by parts and noting that $\varphi(0) = 0$, we can show that

$$\frac{\beta^{4/3}\Gamma\left(\frac{2}{3}\right)}{2k\Gamma\left(\frac{1}{3}\right)} \int_0^\infty \varphi'(t)\, e^{-ut}\, dt = \int_0^\infty [c_0 - \varphi(t)]\, e^{-ut} t^{-1/3} dt,$$

from which it follows that $\varphi(t)$ is the solution of the differential equation

$$\frac{\beta^{4/3}\Gamma\left(\frac{2}{3}\right)}{2k\Gamma\left(\frac{1}{3}\right)} \varphi'(t) = [c_0 - \varphi(t)]\, t^{-1/3},$$

which satisfies the initial condition $\varphi(0) = 0$. We readily see that

$$\varphi(t) = c_0\left(1 - e^{-\frac{3}{2\eta}t^{2/3}}\right),$$

where

$$\eta = \frac{\Gamma\left(\frac{2}{3}\right)\beta^{4/3}}{2k\Gamma\left(\frac{1}{3}\right)}.$$

The required function is therefore

$$f_1(t) = \varphi\left(\frac{t^2}{4}\right) = c_0\left[1 - e^{-\frac{3}{2\eta}\cdot\frac{t^{4/3}}{4}}\right].$$

Substituting this into (17.9), we obtain the function $c(\xi, z)$ that satisfies equation (15.3) and the initial and boundary conditions (17.1) and (17.2). The properties of $f_1$ and $f_2$ are such that

$$\lim_{z\to 0} c(\xi, z) = \frac{c_0}{\Gamma\left(\frac{2}{3}\right)}\xi^{-2/3}\int_0^\infty e^{-\frac{3t^{2/3}}{2\eta}} e^{-t/\xi} t^{-1/3}\,dt,$$

or

$$\lim_{z\to 0} c(\xi, z) = \frac{c_0}{\Gamma\left(\frac{2}{3}\right)}\int_0^\infty e^{-\frac{3}{2\eta}(\xi\rho)^{2/3}} e^{-\rho}\rho^{-1/3}\,d\rho.$$

Thus the flux j onto the plate, expressed in terms of coordinates (x, y), is

$$j(x, 0) = \frac{kc_0}{\Gamma\left(\frac{2}{3}\right)}\int_0^\infty e^{-\gamma t^{2/3}} e^{-t} t^{-1/3}\,dt, \tag{17.11}$$

where, for the sake of brevity, we introduce the dimensionless number $\gamma$

$$\gamma = \frac{3^{1/3}\Gamma\left(\frac{1}{3}\right)kx^{1/2}}{\beta^{2/3}\Gamma\left(\frac{2}{3}\right)}. \tag{17.12}$$

Expanding the term $e^{-\gamma t^{2/3}}$ in a series, we obtain an expression for $j(x, 0)$ in powers of $\gamma$ which is convergent for all values of $\gamma$.

This series may, however, be used only in the region of small $\gamma$, yielding j in the form

$$j(x,\ 0)=\frac{kc_0}{\Gamma\left(\frac{2}{3}\right)}\left[\Gamma\left(\frac{2}{3}\right)-\Gamma\left(\frac{4}{3}\right)\frac{\gamma}{1!}+\Gamma\left(\frac{6}{3}\right)\frac{\gamma^2}{2!}-\frac{\gamma^3}{3!}R_3\right], \tag{17.13}$$

where

$$R_3=\int_0^{\infty} e^{-t}t^{5/3}\left[1-\frac{\gamma t^{2/3}}{4}+\frac{(\gamma t^{2/3})^2}{4\cdot 5}-\dots\right]dt.$$

For large values of $\gamma$, the asymptotic expansion of (17.11) should be used. Let us make the following substitution

$$\gamma t^{2/3}=u.$$

Expressed in terms of this new variable, (17.11) becomes

$$j(x,\ 0)=\frac{3}{2}\frac{kc_0}{\gamma}\frac{1}{\Gamma\left(\frac{2}{3}\right)}\int_0^{\infty} e^{-u}e^{-\left(\frac{u}{\gamma}\right)^{3/2}}du. \tag{17.14}$$

We now employ the series expansion of $e^{-\left(\frac{u}{\gamma}\right)^{3/2}}$, using the first three terms only

$$j(x,\ 0)=\frac{3^{2/3}}{2}\frac{c_0 3^{2/3}}{\Gamma\left(\frac{1}{3}\right)}\frac{1}{\sqrt{x}}\left[1-\gamma^{-3/2}\,\Gamma\left(\frac{5}{2}\right)+\right. \tag{17.15}$$

$$\left.+\gamma^{-3}\cdot\frac{1}{2}\Gamma\left(\frac{8}{2}\right)-\gamma^{-9/2}\frac{1}{3!}S_3\right],$$

where

$$S_3=\int_0^{\infty} e^{-u}u^{9/2}\left[1-\frac{1}{4}\left(\frac{u}{\gamma}\right)^{3/2}+\frac{1}{20}\left(\frac{u}{\gamma}\right)^{3/2\cdot 2}-\dots\right]du.$$

The physical significance of these solutions is clear from the following: j(x, 0) expresses the flux of solute toward 1 cm$^2$ of the plate. Near the edge of the plate the flux is determined only by the rate of the chemical reaction. On the other hand the convective mass transfer in the liquid acquires a more significant role as the distance from the edge of the plate increases. It is clear, however, that an expression for j(x) of the type of (17.13) becomes impractical after a certain value of x. At a large distance from the edge of the plate the convective transfer rate must also decrease, owing to depletion of the reacting substance in the solution. If it decreases to values that are small in comparison with the reaction rate, then convection becomes the factor controlling the rate of the over-all process. In this case the mass flux density must coincide with the maximum diffusional flux, as calculated in Section 15. However, before this limiting case is attained, a certain intermediate region exists. The transition from one region to the other occurs when $\gamma = 1$, for which the series in (17.13) converges very slowly.

Also of interest is the magnitude of

$$\delta = \frac{D\,[c_0 - c\,(x,\,0)]}{j},$$

which might be called the effective thickness of the diffusion boundary layer, and which corresponds to the definition of this quantity given earlier. This thickness also varies according to different laws at different values of x. Indeed, it follows from (17.1) and from the expression for j(x, 0) in formula (17.13) that, for $\gamma \ll 1$, the thickness of the boundary layer is

$$\delta \approx \frac{\Gamma\left(\frac{4}{3}\right)\gamma D}{\Gamma\left(\frac{2}{3}\right)k}, \tag{17.16}$$

whereas at large values of $\gamma$ it follows from (17.1) and (17.15) that

$$\delta \approx \frac{2}{3}\Gamma\left(\frac{2}{3}\right)\frac{\gamma}{k}D, \tag{17.17}$$

which is equivalent to the expression for δ given in Section 15 for the case of maximum flow (i.e., as $k \to \infty$).

Comparing the relations (17.16) and (17.17), we see that the ratio of the coefficients for $\frac{\gamma}{k}$ is approximately 0.7. In other words, the thickness of the diffusion boundary layer obeys different relations at different sections of the plate, and is a function of the ratio $\frac{\gamma}{k}$. The difference in the relations for δ is associated with the difference in the role of convective delivery of the solute parallel to and transverse to the plate at various locations. The diffusion boundary layer in the elementary Nernst theory was viewed as a characteristic quantity, which, in the limiting case, is a function of the fluid velocity. As shown above, Nernst's elementary theory has been found to be invalid. In the same way we also demonstrated previously that the thickness of the diffusion layer is a function of the physical properties of the diffusing particles (the diffusion coefficient).

The Nernst theory asserted that diffusion took place only in a direction normal to the reaction surface. In Section 15 the problem of the normal and the tangential transport of matter to the reaction surface could not be clarified, since we confined ourselves to the limiting case of a chemical reaction at a given rate. It follows from equations (17.16) and (17.17) that the thickness of the diffusion layer is a function of the diffusion coefficient, of the fluid properties (through the quantities $\beta$), and of the distance from the leading edge of the plate. Near the edge ($\gamma \ll 1$) the normal and tangential transport of solute are both substantial. Here, the magnitude of $\delta$ is determined by formula (17.16). Far from the leading edge ($\gamma \gg 1$), the solution near the surface of the plate is depleted by diffusion to the preceding sections. Here, the transport of solute normal to the surface is predominant, leading to the difference in the values of $\delta$ at small and large values of $\gamma$.

## 18. RELAXATION OF THE DIFFUSION PROCESS

The mechanism of convective diffusion in fluids can be clarified by considering reaction surfaces on which different parts have different properties. As pointed out below, such problems are not only of basic theoretical interest, but they may also have practical significance.

Let us imagine a liquid streaming past a semi-infinite plate. The solute reacts and disappears on the surface of this plate. We also assume that the reaction takes place only on a portion of the surface area of the plate, and that in other places the reaction is artificially inhibited. To be specific, let us assume that on an upstream portion $x < h_0$ of the surface the reaction rate is zero so that the mass flux $j = 0$ (where $h_0$ is the distance from the leading edge). This portion of the surface will be termed the coated portion. We assume that a limiting (maximum) regime occurs on the surface of the plate ($c_1 = 0$) starting from $x = h_0$ and continuing into larger values of x.

It is clear that at values of x considerably larger than $h_0$, the presence of the coated portion will not affect the rate of solute transfer and that the flux on the plate surface will be equal to the limiting diffusional flux calculated earlier for the case $h_0 = 0$.

On the other hand, at values of x close to $h_0$, the solution approaching the plate has not been depleted by diffusion over the upstream sections of the plate and, therefore, over a certain portion of the plate the diffusional flux will be noticeably higher than the maximum flux in the absence of the coated section.

The portion of the plate over which the limiting flux region is being established is called the relaxation area.

We begin the study of the relaxation diffusion process with the case of a flow past a semi-infinite plate. Later, several other instances of relaxation processes will be examined.

Our objective is to find the concentration distribution in the relaxation area and to determine its extent.

The distinguishing feature of the relaxation area is the fact that the transport of solute by the tangential velocity component contributes more to the total diffusional flow than in fully developed, limiting regimes. A detailed study of the relaxation process will, therefore, result in a better understanding of the role played by the tangential transfer of solute in convective diffusion.

We shall now consider the extent of the relaxation area from the point of view of the Nernst theory. According to that theory, the fluid within the diffusion layer is assumed to be static, and the transport of solute is accomplished by molecular diffusion. The presence of a coated portion of the plate, where the reaction is inhibited, changes the diffusion conditions over a distance which is of the order of the thickness of the diffusion layer. Here the solute diffuses from a non-depleted solution in the normal as well as in the tangential direction.

In accordance with the Nernst theory, a limiting regime must be established at a distance $x > h_0 + \delta$ from the leading edge of the plate.

Interpreted in the light of the diffusion boundary layer theory, the relaxation process appears differently. In order to determine the concentration distribution and diffusional flux, equation (15.3) for convective diffusion to the surface of a plate must be solved:

$$x^{1/4}\frac{\partial c}{\partial x} = \beta \frac{\partial}{\partial \psi}\left(\sqrt{\psi}\,\frac{\partial c}{\partial \psi}\right) \qquad (18.1)$$

with the following boundary conditions:

1) the flow of solute to the coated portion of the plate is equal to zero; hence,

$$\frac{\partial c}{\partial y} = 0 \quad \text{at} \quad y = 0 \qquad (x < h_0); \qquad (18.2)$$

2) a limiting regime exists on the uncoated portion of the plate surface

$$c = 0 \quad \text{at} \quad y = 0 \qquad (x > h_0); \qquad (18.3)$$

3) far from the plate, the solute concentration is

$$c = c_0 \quad \text{as} \quad y \to \infty; \qquad (18.4)$$

4) at the leading edge, where $x = 0$, $y = 0$

$$c = c_0. \qquad (18.5)$$

To solve this boundary value problem, we introduce into equation (18.1) new variables $\varphi$ and $\xi$, related to x and $\psi$ by the following:

$$\xi = \frac{1}{3}\beta x^{3/4}, \qquad \varphi = \sqrt{\psi}. \qquad (18.6)$$

Thus equation (18.1) takes the form

$$\frac{\partial c}{\partial \xi} = \frac{1}{\varphi} \frac{\partial^2 c}{\partial \varphi^2}. \tag{18.7}$$

In these variables, the boundary conditions are

$$c \to c_0 \quad \text{as} \quad \varphi \to \infty, \tag{18.8}$$

$$\frac{\partial c}{\partial \varphi} = 0 \quad \text{for} \quad \xi < \frac{1}{3} \beta h_0^{3/4}, \quad \varphi = 0, \tag{18.9}$$

$$c = 0 \quad \text{for} \quad \xi \geqslant \frac{1}{3} \beta h_0^{3/4}, \quad \varphi = 0. \tag{18.10}$$

It can be demonstrated that the above boundary value problem has a unique solution. If this can be proved, then the solution itself can be obtained by employing the solution of the analogous problem for a limiting regime over the entire plate.*

In solving the boundary value problem, we find, first of all, the solution for the limiting case in terms of the variables $\varphi$ and $\xi$, i.e., for the problem of diffusion to a semi-infinite plate with limiting regime over the entire surface. This solution can be obtained by substitution of the desired variables into equation (15.9). It is more convenient, however, to solve the limiting case of the boundary value problem in $\varphi$ and $\xi$ directly.

The boundary conditions are

$$c \to c_0 \quad \text{as} \quad \varphi \to \infty, \tag{18.11}$$

$$c = 0 \quad \text{at} \quad \varphi = 0, \tag{18.12}$$

$$c = c_0 \quad \text{at} \quad \xi = 0, \ \varphi = 0. \tag{18.13}$$

To find a solution for equation (18.7) with boundary conditions (18.11) to (18.13) we use the variable

$$\tau = \frac{\varphi}{\sqrt[3]{9}\, \xi^{1/3}}.$$

Making the appropriate substitution we replace (18.7) with

$$\frac{d^2 c}{d\tau^2} + 3\tau^2 \frac{dc}{d\tau} = 0. \tag{18.14}$$

The solution of (18.14) satisfying boundary conditions (18.11) to (18.13) is

* The results given below for relaxaticn on a plate were obtained by N. N. Meyman in 1947.

$$c_{\lim} = \frac{c_0 \int\limits_0^{\frac{\varphi}{\sqrt[3]{9\xi}}} e^{-\lambda^3}\, d\lambda}{\int\limits_0^{\infty} e^{-\lambda^3} d\lambda} = \frac{c_0}{\frac{1}{3}\Gamma\left(\frac{1}{3}\right)} \int\limits_0^{\frac{\varphi}{\sqrt[3]{9\xi}}} e^{-\lambda^3}\, d\lambda, \tag{18.15}$$

where the subscript lim indicates that the solution applies to the limiting regime. Since equation (18.7) is invariant with respect to the substitution of $\xi$ by $\xi$ + const, then, together with (18.15), the function

$$c = \frac{c_0}{\frac{1}{3}\Gamma\left(\frac{1}{3}\right)} \int\limits_0^{\frac{\varphi}{\sqrt[3]{9}\,(\xi + \text{const})^{1/3}}} e^{-\lambda^3}\, d\lambda$$

is also a solution of equation (18.14), but one that satisfies other boundary conditions.

Choosing the constant equal to

$$\text{const} = -\frac{1}{3}\beta h_0^{3/4}$$

we obtain the function

$$c = \frac{c_0}{\frac{1}{3}\Gamma\left(\frac{1}{3}\right)} \int\limits_0^{\frac{\varphi}{\sqrt[3]{9}\,\left(\xi - \frac{1}{3}\beta h_0^{3/4}\right)^{1/3}}} \exp\{-\lambda^3\}\, d\lambda,$$

which, as will be easily seen, is the solution of equation (18.7) that satisfies boundary conditions (18.11) and (18.12).

Reverting from the variables $\varphi$ and $\xi$ to the variables y and x, we have the complete solution for the mixed boundary value problem, equation (15.1) with boundary conditions (18.2) to (18.5),

$$c(x,\ y) = \begin{cases} c_0 & \text{for } x < h_0, \\ \dfrac{c_0}{\Gamma\left(\frac{1}{3}\right)} \displaystyle\int\limits_0^{\frac{1}{12}\frac{y^3}{x^{3/2}}\cdot\frac{1}{1-\left(\frac{h_0}{x}\right)^{3/4}}\cdot\frac{\beta^2}{2D^3}} \exp(-t)\, t^{-2/3}\, dt & \text{for } x > h_0. \end{cases} \tag{18.16}$$

It is simple to verify that the function thus determined and all its derivatives are continuous at any values of x and y, except for the critical point ($x = h_0$, $y = 0$).

Differentiating c(x, y) with respect to y and allowing $y \to 0$, we obtain the diffusional flux for $x > h_0$:

$$j(x, 0)=D\left(\frac{\partial c}{\partial y}\right)_{y=0}=\frac{3}{12^{1/3}}\frac{c_0}{\Gamma\left(\frac{1}{3}\right)}\frac{1}{\sqrt{x}}\frac{1\cdot\beta^{2/3}}{\left[1-\left(\frac{h_0}{x}\right)^{3/4}\right]^{1/3}2^{1/3}}. \tag{18.17}$$

Consequently, the ratio of the true flux to the limiting diffusional flux in the absence of the coated section is given by

$$\frac{j(x, 0)}{j_{\lim}}=\frac{1}{\left[1-\left(\frac{h_0}{x}\right)^{3/4}\right]^{1/3}}. \tag{18.18}$$

Thus, the ratio $j/j_{\lim}$ is a function only of the ratio $\frac{h_0}{x}$. For values of x of the order of $h_0$ — the length of the coated portion — the diffusional flux is substantially greater than the limiting flux. The quantity j is still perceptibly different from $j_{\lim}$ over a distance two to three times the length of the coated portion.

The dimensions of the relaxation area are found to be of the order of $h_0$. Thus, in contrast to the conclusion obtained on the basis of the Nernst theory, the effect of the coated section prevails over a very considerably distance, which could exceed the diffusion layer thickness $\delta$ by several orders of magnitude. The significance of this result is simple: at sections adjacent to the coated portion, the fluid streaming past the plate is not depleted (Figure 19). Because the tangential transfer of solute plays an essential part, the diffusional flux here greatly exceeds that of the limiting flow, i.e., the maximum possible flow of material transported in a direction normal to the surface. Only at a great distance from the coated portion,

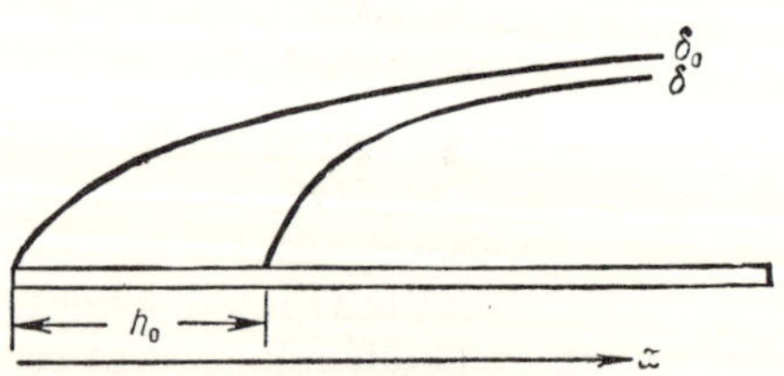


Figure 19. Hydrodynamic and diffusion boundary layers on a "coated" plate.

exceeding the thickness of the diffusion layer by several orders of magnitude, does the solution become so depleted that the basic transport is in a direction normal to the surface and the flux decreases to $j_{\lim}$. Thus, it is possible to reach the general conclusion that the tangential transfer plays an important part in the over-all transfer of matter to the surface. It is clear that the surface may be considered uniformly accessible from a diffusional standpoint only in certain special cases. In all instances wherein the tangential transfer of matter contributes to the total diffusional flow, the surface is not uniformly accessible. Neglecting the tangential transport and replacing real conditions with the assumption of uniform accessibility may lead to substantial errors.

Let us now examine another case of the relaxation process, a case involving a rotating disk with an inner portion which has been coated. *

Let the radius of the "coated" portion equal $R_0$.

The convective diffusion equation has the form (11.20). Transforming from the variables (r, y) to the new variables ($\psi$, r), where $\psi$ is the stream function, we obtain

$$\frac{\partial c}{\partial r}=Dr\frac{\partial}{\partial\psi}\left[rv_r\frac{\partial c}{\partial\psi}\right].$$

Eliminating $v_r$ in terms of $\psi$, we have

$$\frac{\partial c}{\partial r^3}=\gamma\frac{\partial}{\partial\psi}\left(\sqrt{\psi}\frac{\partial c}{\partial\psi}\right). \tag{18.19}$$

In the new variables, the boundary conditions have the form

$$c\to c_0 \quad \text{as} \quad \psi\to\infty, \tag{18.20}$$

$$c=0 \quad \text{at} \quad \psi=0,\ r\geqslant R_0, \tag{18.21}$$

$$\frac{\partial c}{\partial\psi}=0 \quad \text{at} \quad \psi=0,\ r<R_0. \tag{18.22}$$

Let us attempt to find a solution to the boundary value problem assuming that the concentration distribution is a function of the variable

$$\tau=\sqrt[3]{\frac{4}{9\gamma}}\frac{\sqrt{\psi}}{r}. \tag{18.23}$$

Then equation (18.19) takes the form

$$\frac{1}{3}\frac{d^2c}{d\tau^2}+\tau^2\frac{dc}{d\tau}=0.$$

Its solution, satisfying the boundary conditions of a limiting regime (18.20) to (18.22), in accordance with formula (15.9) is

$$c=\frac{3c_0}{\Gamma\left(\frac{1}{3}\right)}\int\limits_0^{\sqrt[3]{\frac{4}{9\gamma}}\frac{\sqrt{\psi}}{r}}e^{-\lambda^3}\,d\lambda=\frac{3c_0}{\Gamma\left(\frac{1}{3}\right)}\int\limits_0^{\tau}e^{-\lambda^3}\,d\lambda.$$

Since equation (18.19) is invariant with respect to the transformation $r^3\to r^3+\text{const}$, the solution of (18.19) that satisfies boundary conditions (18.20) to (18.22) is

---

* At our request, this problem was solved by R. M. Zaidel using the N. N. Meyman method.

$$c(r, y) = \frac{3c_0}{\Gamma\left(\frac{1}{3}\right)} \int_0^{\frac{1}{\sqrt[3]{6}}\left(\frac{\nu}{D}\right)^{1/3}\left(\frac{\omega}{\nu}\right)^{1/2}\frac{y}{\left(1-\frac{R_0^3}{r^3}\right)^{1/3}}} e^{-\lambda^3}\, d\lambda.$$

The corresponding diffusional flux to the surface of the disk is

$$j = D\left(\frac{\partial c}{\partial y}\right)_{y=0} = \begin{cases} \dfrac{Dc_0}{1.61\left(\frac{D}{\nu}\right)^{1/3}\sqrt{\frac{\nu}{\omega}}\left(1-\frac{R_0^3}{r^3}\right)^{1/3}}, & r > R_0, \\ 0, & r \leqslant R_0. \end{cases} \tag{18.24}$$

The general conclusions to be drawn from these equations correspond to those given above for the semi-infinite plate. The relaxation area has a magnitude approaching that of the radius of the "coated" area.

It is interesting to examine the case of a disk on which the reaction surface is an annulus with a radius $R_0$ and a width $\Delta R$ (where $\Delta R \ll R_0$).

The flux to the surface of the annulus is given by (18.24), and the total flow to the annulus, correspondingly, equals

$$I = j_0 \int_{R_0}^{R_0+\Delta R} \frac{2\pi r\, dr}{\left(1-\frac{R_0^3}{r^3}\right)^{1/3}} = \pi j_0 \left(r^3 - R_0^3\right)^{2/3}\Bigg|_{R_0}^{R_0+\Delta R} \approx 3^{2/3}\pi j_0 R_0^{4/3}(\Delta R)^{2/3}.$$

We see that the total flow I is proportional to the 4/3 power of the annulus radius $R_0$ and to the 2/3 power of its width $\Delta R$. This makes an annular surface highly advantageous for obtaining large diffusional flows. Indeed, if the entire surface of the disk were a working (reaction) surface, then an annulus of the same area would receive a diffusional flow

$$I_0 = 2\pi j_0 R_0 \Delta R.$$

Forming the ratio

$$\frac{I}{I_0} = \frac{3^{2/3}}{2}\left(\frac{R_0}{\Delta R}\right)^{1/3}, \tag{18.25}$$

we see that for $R_0 \gg \Delta R$, $I \gg I_0$.

The experimental verification of formulas (18.24) and (18.25) is discussed in Section 56.

## 19. MODELING HETEROGENEOUS CHEMICAL REACTIONS

The problem of modeling chemical conversions has been often discussed in the chemical literature [23 to 25], but has not yet been fully resolved.

On the basis of the theory of similarity of reactions, we first examine reactions governed solely by diffusion kinetics. Hence, the problem becomes that of modeling the convective diffusion process.

In keeping with previous work, we limit our attention to steady-state regimes of flow and diffusion. As a scale for measuring the concentration of material in the fluid, let us take the concentration far from the reaction surface. Then the hydrodynamic and diffusion boundary layer equations can be written in the dimensionless form

$$V_x \frac{\partial V_x}{\partial X} + V_y \frac{\partial V_x}{\partial Y} = -\frac{\partial P}{\partial X} + \frac{1}{\mathrm{Re}} \frac{\partial^2 V_x}{\partial Y^2}, \tag{19.1}$$

$$V_x \frac{\partial C}{\partial X} + V_y \frac{\partial C}{\partial Y} = \frac{1}{\mathrm{Pe}} \frac{\partial^2 C}{\partial Y^2}, \tag{19.2}$$

where

$$C = \frac{c}{c_0}, \quad \mathrm{Re} = \frac{Ul}{\nu} \text{ and } \mathrm{Pe} = \frac{Ul}{D}.$$

The boundary conditions have the form

$$V_x \to 1 \quad \text{as} \quad Y \to \infty;$$
$$V_x = V_y = 0 \quad \text{at} \quad Y = 0; \tag{19.3}$$
$$C \to 1 \quad \text{as} \quad Y \to \infty; \tag{19.4}$$
$$C = 0 \quad \text{at} \quad Y = 0. \tag{19.5}$$

Equations (19.1) and (19.2) include only two dimensionless numbers, Re and Pe. All other dimensionless quantities are functions of these numbers.

The Prandtl number may be introduced instead of Pe by setting

$$\mathrm{Pr} = \frac{\mathrm{Pe}}{\mathrm{Re}} = \frac{\nu}{D}.$$

The advantage of using the Prandtl number lies in the fact that it is a function only of the physical properties of the material, and not of the flow regime.

The diffusional flux per unit of surface area is of basic interest in diffusion problems. This may be conveniently expressed in the form of a dimensionless quantity by defining the diffusional flux or the Nusselt number by

$$j_{\lim} = \frac{Dc_0}{l} \mathrm{Nu},$$

i.e.,

$$\mathrm{Nu} = \frac{l}{Dc_0} j_{\mathrm{lim}} \tag{19.6}$$

The Nusselt number is related to the thickness of the diffusion boundary layer through the ratio

$$\mathrm{Nu} = \frac{l}{\delta}. \tag{19.7}$$

For a given geometry, the Nusselt number is a function of either Re and Pe, or of Re and Pr:

$$\mathrm{Nu} = f(\mathrm{Re}, \mathrm{Pe}), \tag{19.8}$$

$$\mathrm{Nu} = \varphi(\mathrm{Re}, \mathrm{Pr}). \tag{19.9}$$

The latter expression is generally used.

The local and average values of the Nusselt number must not be confused, since $j_{\mathrm{lim}}$ is a point function, applicable to a given point on the reaction surface. The value of the local Nu, determined by formula (19.6), may vary at different points along the surface.

Instead of the local Nusselt number, the average value may be used; it is given by

$$\overline{\mathrm{Nu}} = \frac{I_{\mathrm{lim}} l}{Dc_0 S}, \tag{19.10}$$

where $I_{\mathrm{lim}} = \int j_{\mathrm{lim}} dS$ and S is the area of the reaction surface;

$$\overline{\mathrm{Nu}} = \psi(\mathrm{Re}, \mathrm{Pr}). \tag{19.11}$$

We have computed the function $\psi$ (Re, Pr) for certain cases in preceding sections.

Let us now determine the dimensionless parameters which characterize a heterogeneous chemical reaction taking place in the intermediate regime. In this case, the boundary condition (19.5) must be replaced by condition (9.6).

Let us first examine the case of a first-order reaction. Instead of condition (19.5), we then have

$$D\left(\frac{\partial c}{\partial y}\right) = kc_1 \quad \text{at} \quad y = 0. \tag{19.12}$$

The quantities entering into condition (19.12) are not independent and their values are determined by integrating the equation of convective mass transport.

In converting to dimensionless quantities, we should bear in mind that

$$\frac{\partial c}{\partial y} \approx \frac{c_0}{\delta}, \text{ but not } \frac{c_0}{l}.$$

In order to employ dimensionless quantities, therefore, condition (19.12) should be taken as

$$D\left(\frac{\partial c}{\partial y}\right)_{y=0}=\frac{D}{l}\frac{l}{\delta}\left(\frac{\partial C}{\partial Y}\right)_{Y=0}c_0.$$

Then

$$\frac{\partial C}{\partial Y}=\lambda C \quad \text{at} \quad Y=0, \tag{19.13}$$

where the dimensionless parameter $\lambda$ equals

$$\lambda=\frac{kl}{D}\cdot\frac{\delta}{l}=\frac{kl}{D}\cdot f(\mathrm{Re},\ \mathrm{Pr}). \tag{19.14}$$

The function f(Re, Pr), in the form $\mathrm{Re}^{-1/2}\mathrm{Pr}^{-1/3}$, corresponds to that calculated in Section 17, and is determined from a solution of the convective diffusion problem (in a purely diffusional regime). The parameter $\lambda$ is the determining dimensionless criterion, insofar as it is made up of the arbitrary parameters k, $l$, D, and a function of Re and Pr. The values of $\lambda$ may vary for different points on the surface.

The parameter $\lambda$ thus defined has already appeared in Section 17, where its physical significance was made clear: it represents the ratio between the net rate of the chemical conversion process (the number of particles reacting per second at $c_0 = 1$) and the rate of diffusion across the boundary layer $\delta$ (the number of particles approaching 1 cm$^2$ of the surface per second, when the concentration of the solution $c_0 = 1$).

The dimensionless rate of the heterogeneous process on a unit of the surface area is

$$\mathrm{Nu}=\Phi(\mathrm{Re},\ \mathrm{Pr},\ \lambda). \tag{19.15}$$

Let us note that the function $\Phi$ in no way corresponds to $\varphi$ (Re, Pr) which enters into the expression for the Nusselt number under conditions where diffusion kinetics is controlling.

For given values of Re and Pr, $\Phi$ may vary if the criterion $\lambda$ also varies. An example of this case was examined in Section 17. It was shown that $\lambda$ is not only a function of k, but may change from point to point on the surface.

Boundary condition (19.5) may be expressed as follows:

$$\left(\frac{\partial C}{\partial Y}\right)_{Y=0}=\gamma C^m, \tag{19.16}$$

where the dimensionless group $\lambda$ is equal to

$$\gamma=\frac{kl}{D}\frac{\delta}{l}c_0^{m-1}=\frac{klc_0^{m-1}}{D}f(\mathrm{Re},\ \mathrm{Pr}). \tag{19.17}$$

The dimensionless reaction rate is

$$\mathrm{Nu} = \Phi(\mathrm{Re}, \mathrm{Pr}, \gamma). \tag{19.18}$$

The physical significance of $\lambda$ is as clear in this case as it is for a first-order reaction.

It should be noted that groups similar to $\lambda$ were proposed by D'yakonov, who chose the ratio of diffusion time as the similarity criterion. The meaning of the reaction time was not clearly defined by this author. Furthermore, D'yakonov ignored the fact that $\lambda$ may be a function of the distance x and, consequently, also of the motion of the fluid.

## 20. AN INTERIOR PROBLEM — DIFFUSION IN LAMINAR FLOW IN A TUBE

So far we have been examining only convective diffusion directed from without to the surface of a body where reaction occurs. It is of interest, however, to examine the case of a fluid moving inside a tube. In practice the flow inside tubes is usually turbulent.

The laws governing diffusion in laminar flow, however, are also worthy of attention. A solution of this problem presents no difficulties. Let us consider a laminar flow regime inside a tube of circular cross section. Under usual conditions, a laminar regime is maintained for Reynolds numbers not exceeding 2500. The entire flow in the tube may be viewed as consisting of two sections: the so-called inlet region and the Poiseuille flow region. In the inlet region, the fluid entering the tube is subject to the retarding action of the walls. Under the influence of this action, the initially flat velocity profile gradually assumes a Poiseuille profile (Figure 20). The length of the inlet region may be estimated as follows. The fluid entering the

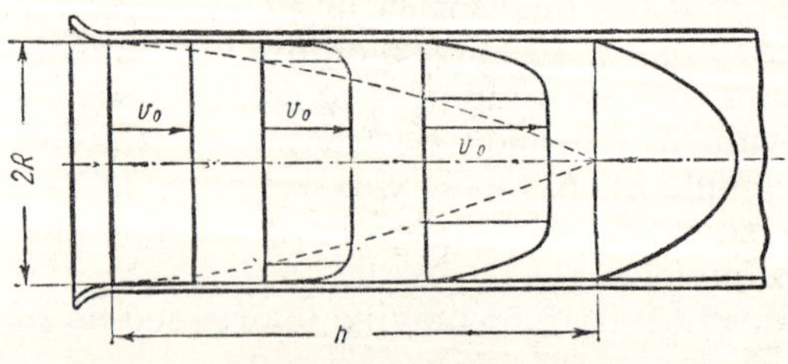


Figure 20. Establishment of a Poiseuille profile for laminar flow in a tube.

tube forms a boundary layer on the tube walls. As long as the thickness of the boundary layer is small compared to the radius R of the tube, it may be considered planar and equivalent to the boundary layer on a plate. It grows according to the relation (3.39) with increasing distance x from the tube inlet. When the thickness of the boundary layer becomes comparable to the radius of the tube, the entire interior of the tube is filled with a laminar flow regime that is subject to the effects of viscosity. The thickness of the central core of the flow, where the profile is flat, decreases gradually and approaches zero at $\delta_0 \sim R$.

The decrease in radius of the core is accompanied by an increase in the fluid velocity in that core, since the total fluid flow must remain constant. Thus, the combination of a gradually thickening boundary layer and of the accelerating core of the flow leads to the formation of a Poiseuille velocity profile at $x \sim h$.

Substituting the value of $\delta_0$ from (3.38), we find that the length of the inlet region is given by the following proportionality:

$$h \sim R \cdot \mathrm{Re}. \tag{20.1}$$

This expression also contains an important numerical coefficient, such that the length of the inlet region can be approximated by

$$h \sim 0.1\, R \cdot \mathrm{Re}. \tag{20.2}$$

Beyond this point, the formation of the Poiseuille velocity profile is completed. The flow thereafter is in the axial direction with a parabolic velocity profile given by

$$v = v_0\left(1 - \frac{r^2}{R^2}\right), \tag{20.3}$$

where $v_0 = \dfrac{R^2 \Delta p}{4 \mu L}$ is the maximum velocity at the axis of the tube $(r = 0)$ and $\Delta p$ is the pressure drop over the distance L.

On the basis of general considerations of the similarity between diffusion and surface drag, we can expect a somewhat similar situation in the case of diffusion toward a tube wall which acts as a reaction surface. In particular, we should expect the presence of an inlet region and a region of fully established concentration distribution. Since the diffusion coefficient in a liquid is small compared to its kinematic viscosity, the region over which the distribution of diffusing substance becomes established should be considerably longer than the inlet region required for establishing the Poiseuille velocity profile. Thus, the entire tube may be subdivided into three sections: the hydrodynamic inlet section of length h, the diffusion inlet section H, and the zone of developed regions.

In the first section, the diffusion of solute toward the wall of the tube occurs in a manner similar to that toward a plate.

In the second section, the velocity distribution is given by equation (20.3). The thickness of the diffusion boundary layer $\delta$, however, may still be regarded to be considerably smaller than the radius of the tube.

Let us examine the distribution of the solute and the diffusional flux in this section.

Because of the extremely small thickness of the diffusion layer, the diffusion occurs over small distances from the wall, where $r \approx R$. Therefore, by introducing a new variable

$$y = R - r,$$

we can confine the analysis to small values of the variable $y$ ($y \ll R$). Then

$$v \approx v_0 \frac{2y}{R}. \tag{20.4}$$

The equation of convective diffusion takes the form

$$\frac{2v_0}{R} y \frac{\partial c}{\partial x} = D \frac{\partial^2 c}{\partial y^2}, \tag{20.5}$$

where $x$ is the distance along the tube.

We shall determine only the maximum diffusional flux, for which the boundary conditions are

$$\begin{aligned} c &\to c_0 \quad \text{as} \quad y \to \infty, \\ c &= 0 \quad \text{at} \quad y = 0. \end{aligned} \tag{20.6}$$

Equation (20.5) is solved by introducing the dimensionless quantity

$$\eta = \left(\frac{v_0}{DR}\right)^{1/3} \frac{y}{x^{1/3}}.$$

Then, for $c(\eta)$, we obtain

$$\frac{d^2 c}{d\eta^2} + \frac{2}{3} \eta^2 \frac{dc}{d\eta} = 0. \tag{20.7}$$

The following expression is the solution of this equation which satisfies boundary conditions (20.6):

$$c = \frac{c_0 \int\limits_0^{\frac{y}{x^{1/3}}\left(\frac{v_0}{DR}\right)^{1/3}} \exp\left\{-\frac{2}{9}\eta^3\right\} d\eta}{\int\limits_0^{\infty} \exp\left\{-\frac{2}{9}\eta^3\right\} d\eta}. \tag{20.8}$$

The diffusional flux to the wall of the tube is given by

$$j_{\text{diff}} = D\left(\frac{\partial c}{\partial y}\right)_{y=0} = \frac{Dc_0}{x^{1/3}}\left(\frac{v_0}{DR}\right)^{1/3} \frac{1}{\int\limits_0^{\infty} \exp\left\{-\frac{2}{9}\eta^3\right\} d\eta} = \tag{20.9}$$

$$= \frac{Dc_0}{x^{1/3}}\left(\frac{v_0}{DR}\right)^{1/3} \cdot \frac{\sqrt[3]{0.23}}{0.89} = 0.67 c_0 D \left(\frac{v_0}{DRx}\right)^{1/3} = \frac{Dc_0}{\delta},$$

where the thickness of the diffusion layer $\delta$ is determined by

$$\delta = \frac{1}{0.67}\left(\frac{D}{\nu}\right)^{1/3} \sqrt[3]{\frac{\nu}{v_0 R}} \sqrt[3]{R^2 x}. \qquad (20.10)$$

The diffusional flow to the inner surface of the tube (in the region where the diffusional flow is becoming established) may be described by

$$I_{\text{diff}} = 2\pi R \int j_{\text{diff}}\, dx = 2.01\pi c_0 D \left(\frac{v_0 x^2}{DR}\right)^{1/3} R. \qquad (20.11)$$

An interesting feature of equations (20.9) to (20.11) is the fact that they do not contain the viscosity term (only $v_0$, at a given $\frac{\Delta p}{L}$, is a function of viscosity). It would be erroneous, however, to conclude that these equations are valid at any value of the ratio $\nu/D$. The assumption that $\nu \gg D$ was made in the process of deriving the above, inasmuch as the thickness of the diffusion boundary layer was supposed to be extremely small compared to the thickness of the hydrodynamic boundary layer.

Let us now find the order of magnitude of the length H, i.e., of the region where the diffusional regime is forming.

This length is determined by the relation

$$\delta \sim R,$$

or, if for $\delta$ we substitute its expression from (20.10), we obtain

$$H \sim \frac{v_0 R^2}{D} \sim \mathrm{Re} \cdot \mathrm{Pr} \cdot R.$$

Since $\mathrm{Pr} \gg 1$, then $H \gg h$, as was previously assumed.

In practice the quantity H/h is usually so large that the entire length of the tube does not suffice for the establishment of a steady diffusion regime. The entire length of the tube is taken up by the inlet section, for which equations (20.10) and (20.11) are valid.

Expressions for the maximum diffusional flow were computed by us in 1943. Prior to that, in 1928, they were obtained by Leveque [27], but Leveque did not define the zones in which his expressions were to be applied. Also, he arrived at somewhat different numerical coefficients [(2.16) instead of our (2.01), see (20.11)].

Let us next examine the phenomenon of relaxation diffusion, under the conditions of an interior problem. In particular, let us assume that there is dissolution of the inner surface of the tube and that the amount of dissolving matter is limited. After a certain time, the substance being dissolved will have been completely removed from the beginning of the tube. Thereafter the solution front will move gradually along the tube. We confine ourselves to the case of an established velocity profile and a non established concentration distribution, as examined in Section 18. The following, however, will now serve as boundary conditions:

$$c = 0 \quad \text{for} \quad x \leqslant h, \tag{20.12}$$

$$c = c_0 \quad \text{for} \quad x > h. \tag{20.13}$$

As was done above, let us take advantage of the invariance of the diffusion equation with respect to the transformation x → x + const, and let us write the concentration distribution that satisfies boundary conditions (20.12) and (20.13) in the form

$$c = c_0 - \frac{c_0 \int\limits_0^{y\left(\frac{v_0}{DR}\right)^{1/3}\frac{1}{(x-h)^{1/3}}} \exp\left\{-\frac{2}{9}\eta^3\right\} d\eta}{\int\limits_0^{\infty} \exp\left\{-\frac{2}{9}\eta^3\right\} d\eta}.$$

The flux of the dissolving substance is equal to

$$j = -D\left(\frac{\partial c}{\partial y}\right)_{y=0} = 0.67 c_0 D \left(\frac{v_0}{DR}\right)^{1/3} \frac{1}{(x-h)^{1/3}} \quad \text{for} \quad x \geqslant h.$$

We have assumed here that the concentration in the bulk of the solution is always equal to zero and that the solution is not enriched by the solute that dissolved near the tube inlet.

This assumption may not be valid for a very slowly moving fluid.

## 21. DISTRIBUTION OF MATTER INTRODUCED INTO A FLUID STREAM

The measurement of the flow velocity by introducing a radioactive substance or an electrolyte at a certain cross section, and subsequently measuring radioactivity or electrical conductivity at another cross section located downstream, has been utilized in many branches of engineering and especially in physiology (the measurement of the velocity of blood flow in arteries).

The complication encountered in this method arises because the substance introduced into the stream diffuses both downstream and upstream. If a certain amount of substance is introduced into the stream, as illustrated in Figure 21, then, due to convection and diffusion, the contours of the zone filled with substance become deformed and disappear. In measuring the electrical conductivity (or the activity) at a point downstream in the flow, the reading begins to increase at a certain point, reaches a maximum, and then again decreases to zero.

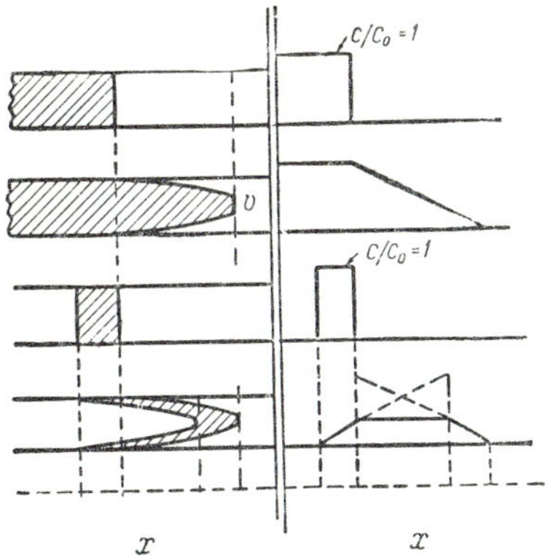


Figure 21. Distribution of the average concentration in the absence of molecular diffusion. The different graphs refer to different time intervals.

In a series of papers, G. I. Taylor [28] developed a theory for the measurement of fluid velocity inside a tube.

The concentration distribution of the introduced substance is determined by the balance between transport by flow along the tube and by molecular diffusion.

The convective diffusion equation takes the form

$$\frac{\partial c}{\partial t}+v_0(r)\frac{\partial c}{\partial x}=D\left(\frac{\partial^2 c}{\partial r^2}+\frac{1}{r}\frac{\partial c}{\partial r}+\frac{\partial^2 c}{\partial x^2}\right), \tag{21.1}$$

where x is the distance along the tube, r is the distance from the axis of the tube (the radius of the tube is R), and $v_0(r)$ is the flow velocity.

For convenience in what follows, we undertake our computations in a coordinate system that moves with the fluid at a velocity U, where U is the average flow velocity in the tube. In such a system, it is evident that

$$v(r)=v_0(r)-U=2U\left(1-\frac{r^2}{R^2}\right)-U=U\left(1-2\frac{r^2}{R^2}\right). \tag{21.2}$$

The term $\frac{\partial^2 c}{\partial x^2}$ in (21.1) may be omitted if it is assumed that the axial change in concentration is much less than the radial change.

We then obtain

$$\frac{\partial c}{\partial t}+U\left(1-\frac{2r^2}{R^2}\right)\frac{\partial c}{\partial x}=D\left(\frac{\partial^2 c}{\partial r^2}+\frac{1}{r}\frac{\partial c}{\partial r}\right), \tag{21.3}$$

where the derivative with respect to t is taken in the moving coordinate system.

The boundary condition at the tube surface is

$$\left(\frac{\partial c}{\partial r}\right)_{r=R}=0;$$

i.e., there is no flow of the substance through the wall of the tube. Furthermore, at the axis of the tube, the concentration must have a finite value.

Let us first approximate the concentration gradient along the axis of the tube by

$$\frac{\partial \bar{c}}{\partial x}=\text{const.}$$

Then, the solution of equation (21.3) may be written as

$$\left.\begin{aligned} c=c_0+\frac{UR^2}{4D}\frac{\partial \bar{c}}{\partial x}\left(\frac{r^2}{R^2}-\frac{1}{2}\frac{r^4}{R^4}\right),\\ \frac{\partial c}{\partial t}=0,\end{aligned}\right\} \tag{21.4}$$

where $c_0$ is the concentration at the axis of the tube, i.e., where $r = 0$, and

$$\bar{c}=2\int_0^R \frac{cr\,dr}{R^2}=\frac{2\pi\int_0^R cr\,dr}{S}$$

is the average concentration over the cross section of the tube. Hence, we find

$$\left.\begin{aligned} c=\bar{c}+\frac{R^2U}{4D}\frac{\partial \bar{c}}{\partial x}\left(-\frac{1}{3}+\frac{r^2}{R^2}-\frac{1}{2}\frac{r^4}{R^4}\right),\\ \frac{\partial c}{\partial t}=0.\end{aligned}\right\} \tag{21.5}$$

The total flow of substance over the cross section of the tube is equal to

$$Q=2\pi\int_0^R cvr\,dr=-\pi R^2\left(\frac{R^2\cdot U^2}{48D}\right)\frac{\partial \bar{c}}{\partial x}.$$

The flux

$$j=\frac{Q}{\pi R^2}=-\left(\frac{R^2U^2}{48D}\right)\frac{\partial \bar{c}}{\partial x} \tag{21.6}$$

has the same form as the diffusional flux for a concentration gradient $\frac{\partial \bar{c}}{\partial x}$ and has the following effective diffusion coefficient:

$$D_{\text{eff}} = \frac{R^2U^2}{48D}. \quad (21.7)$$

In the next approximation, assuming that $\frac{\partial \bar{c}}{\partial x} \neq$ const, we may write

$$\frac{\partial \bar{c}}{\partial t} = -\frac{\partial}{\partial x} j,$$

or

$$\frac{\partial \bar{c}}{\partial t} = D_{\text{eff}} \frac{\partial^2 \bar{c}}{\partial x^2}. \quad (21.8)$$

Equation (21.8) shows that the average concentration profile along the cross section is governed (in the coordinate system moving at the average velocity) by the diffusion equation for a stationary medium with an effective diffusion coefficient (21.7)

We can now define the conditions under which the above approximations are fulfilled. Our neglect of the axial molecular diffusion as compared to convective diffusion is justified if the following inequality holds:

$$D \ll D_{\text{eff}}$$

or

$$\frac{RU}{D} \gg 7. \quad (21.9)$$

We can regard c as a slowly changing function of x (so that as a first approximation, $\frac{\partial c}{\partial x} \approx$ const) when

$$\frac{\partial c}{\partial x} = \frac{\partial \bar{c}}{\partial x} + \frac{R^2U}{4D} \frac{\partial^2 \bar{c}}{\partial x^2} \left( -\frac{1}{3} + \frac{r^2}{R^2} - \frac{1}{2} \frac{r^4}{R^4} \right) \approx \frac{\partial \bar{c}}{\partial x},$$

or

$$\frac{\partial \bar{c}}{\partial x} \gg \frac{R^2U}{4D} \frac{\partial^2 \bar{c}}{\partial x^2}.$$

If L is the length over which a noticeable change in c can occur then this inequality may be written as

$$\frac{LD}{R^2U} \gg 1$$

or, by using (21.9), as

$$\frac{L}{R} \gg \frac{UR}{D} \gg 7. \quad (21.10)$$

Inequalities (21.9) and (21.10) give the limits over which equation (21.8) is applicable. As an example of its use, we examine the following problem:

A certain amount N of a substance is introduced into the stream at time $t = 0$ and $x = 0$ (at the origin of the coordinate system). The substance occupies a small section of the tube in comparison with the radius, i.e.,

$$c = \frac{N}{\pi R^2} \cdot \delta(x) \quad \text{at} \quad t = 0. \tag{21.11}$$

Find the axial distribution of the substance for $t > 0$.

The solution of equation (21.8) with the initial condition (21.11) and in a stationary medium is

$$c = \frac{1}{2} \frac{N}{\pi R^2} \frac{1}{\sqrt{\pi D_{eff} t}} e^{-\frac{(x-Ut)^2}{4D_{eff} t}}. \tag{21.12}$$

The unknown $D_{eff}$ can be determined from the experimental measurements of the concentration distribution at a certain time interval t and the subsequent calculation of $D_{eff}$ using equation (21.12). Knowing $D_{eff}$, one can find the velocity of the stream (if one knows the value of D) or the diffusion coefficient (if one knows the velocity of the stream U).

Figure 22 illustrates the concentration distribution of $KMnO_4$ 11 minutes after its introduction into a tube 0.0504 cm in diameter, filled with water moving at a velocity of 0.167 cm/sec. The value of D determined from these measurements agrees well with values found by other methods.

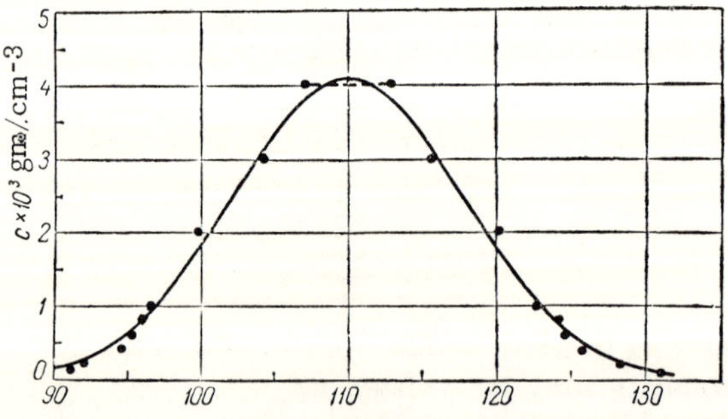


Figure 22. Concentration distribution of $KMnO_4$ in the stream. Time: 11 minutes after injection of $KMnO_4$ (Taylor's experiment).

## 22. CONVECTIVE DIFFUSION IN A TWO-COMPONENT LIQUID SYSTEM NEAR THE CRITICAL SOLUTION POINT

As early as 1899, W. Gibbs [29] established that the driving force of diffusion, i.e., the quantity that determines the value of the mass flux j, is the chemical potential gradient.

Verification of this assertion was extremely important because, in the usually accepted interpretation of diffusion phenomena, the concentration gradient, rather than the chemical potential gradient, was assumed to be the driving force. The concentration gradient undergoes no changes at the critical solution point. Therefore, according to the usual interpretation, no peculiarities in diffusion should be found in the neighborhood of the critical concentration range.

The fact that the diffusion coefficient becomes zero at the critical solution point of a two-component liquid system was first discovered by I. R. Krichevskiy and his collaborators. This phenomenon was investigated in detail in the same laboratory by Yu. V. Tsekhanskaya [30], who studied diffusion in water — triethylamine solutions. Experiments with narrow capillaries fully confirmed the prediction that the diffusion coefficient becomes zero. Cessation of diffusion in the critical region must have an important effect on the rate of heterogeneous chemical reactions in which a diffusing substance participates.

Yu. V. Tsekhanskaya [30] studied experimentally the dissolution of a disk of terephthalic acid, occurring as a result of reaction of the acid with triethylamine to form a water-soluble substance (terephthalic acid itself is insoluble in water).

The critical concentration of triethylamine in water is 246 mg per $cm^3$.

The rate of dissolution was determined by the flow of triethylamine to the surface of a rotating disk. The disk was rotated at 16.6 and 8.4 rps and since the diameter of the terephthalic acid disk was 0.81 cm, the Reynolds numbers were $2 \cdot 10^4$ and $10^4$, respectively. The results of these measurements for $Re = 2 \cdot 10^4$, and at a temperature of 290 °K, are shown in Figure 23. The solid curves 1 represent the experimental data, and the dotted lines 2 represent the flux of triethylamine calculated from equation (11.30). The flux values were equal to the measured rates of dissolution of terephthalic acid. All values on Figure 23 are given as a function of the concentration of triethylamine in the bulk.

We see that in the sub-critical range the actual dissolution rate closely coincides with the calculated rate. On reaching the critical range, however, the proportionality between the flow of triethylamine and its concentration is disrupted. The flux is found to be a very weak function of the triethylamine concentration in the bulk of the solution. It is perfectly clear that the convective diffusion theory set forth above, and formula (11.30) in particular, do not apply in the critical range.

In order to derive a formula * for the mass flux that is valid

---

* The derivation given below is based on the work of reference [34].

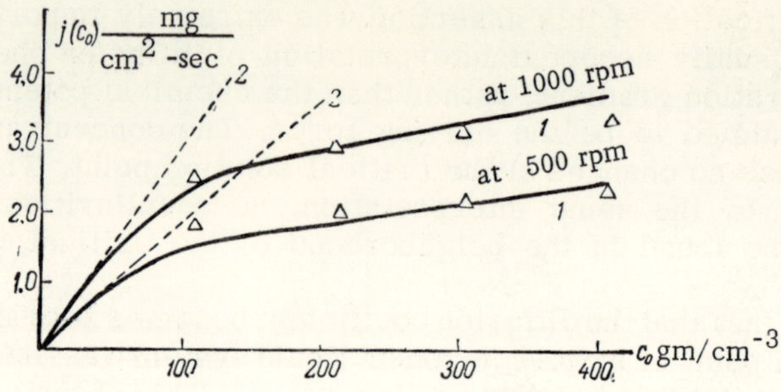


Figure 23. The flux of solute to the surface of the disk as a function of the volumetric concentration of the solution. 1 - experimental curves; 2 - theoretical lines ignoring the critical zone, Δ - theoretically calculated points with consideration of the critical zone.

both in the critical and supercritical concentration regions, the following factors must be taken into consideration:

1. The dependence of the diffusion coefficient on the concentration of solute and, consequently, its dependence on the distance y measured from the surface of the disk.

2. The dependence of the viscosity of the solution on the concentration of the diffusing substance. This dependence may also play an essential role.

Let us examine qualitatively the change in the concentration and in the other quantities as a function of the distance from the rotating disk surface. Let us assume that the concentration of triethylamine in the solution is $c_0 > c_{cr}$, where $c_{cr}$ is the critical concentration, and the concentration at the surface of the disk is $c = 0$ (the reaction, triethylamine + terephthalic acid → soluble salt, and the subsequent removal of the soluble salt, are rapid processes). In the region of concentrations $c > c_{cr}$, the diffusion coefficient is extremely small, and convective transfer predominates over diffusional transfer. However, such predominance cannot exist in all regions, since this would indicate a constant concentration. Consequently, over a certain distance the concentration drops from $c_0$ to $c_{cr}$ (Figure 24). The width of this interval $\delta_1$ is calculated later. Within the interval $\delta_1$, the diffusion coefficient decreases from $D_1$ to almost zero, and the kinematic viscosity drops from $\nu_1$ to $\nu_{cr}$. It is the convective transfer that ensures the passage of the substance through the surface $c = c_{cr}$.

A further change in the concentration from $c_{cr}$ to $c = 0$ occurs within the region $\delta$ as a result of convective diffusion. This change occurs so close to the surface of the disk that the value of the fluid velocity $v_y$ is substantially lower here than far from the disk. The diffusion coefficient in the region $\delta$ rises to a value $D_0$ corresponding

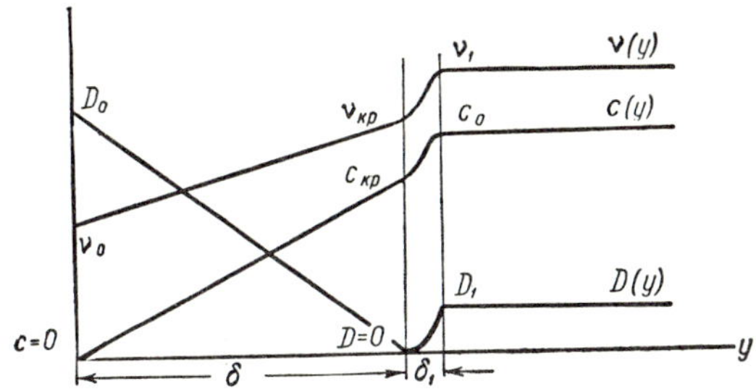


Figure 24. Concentration, kinematic viscosity, and diffusion coefficent as functions of distance from the disk surface.

to a very low triethylamine concentration in water, and the kinematic viscosity drops to a value $\nu_0$. The regions $\delta$ and $\delta_1$, in which the changes in concentration occur, are zones of diffusional resistance. In the following treatment we disregard the resistance in the region $\delta_1$, and assume that all diffusional resistance is concentrated in the region $\delta$. The thickness of the region $\delta_1$ may be estimated from the relation $j = \frac{D(c_0 - c_{cr})}{\delta_1}$, since in this region the diffusion coefficient D does not significantly change with a change in concentration of the solution. It is shown later that $\delta$ is substantially greater than $\delta_1$. Indeed, this means that diffusional resistance is mainly concentrated in the region $\delta$ (the driving forces in both regions are of the same order of magnitude).

An exact solution of the hydrodynamic problem, allowing for the dependence of viscosity on concentration and considering the convective diffusion equation with its variable diffusion coefficient, presents insuperable mathematical difficulties. In view of this, the problem may be simplified in the following manner.

1. It is assumed that in the region $\delta$, the change in concentration from $c = c_{cr}$ to $c = 0$ is approximately linear. Such an approximation is inadequate to permit accurate determination of the flow. It is sufficiently adequate, however, to simplify the functional forms $D(c)$ and $\nu(c)$. Indeed, if c changes linearly, and the functions $D(c)$ and $\nu(c)$ (according to the data of Yu. V. Tsekhanskaya) are also nearly linear, then we may suppose that the following approximations are valid within the limits $0 \le y \le \delta$:

$$D = D_0\left(1 - \frac{y}{\delta}\right), \tag{22.1}$$

$$\nu = \nu_0 + y\frac{\nu_1 - \nu_0}{\delta} = \nu_0 + \alpha y, \tag{22.2}$$

where the value of $\delta$ is to be determined, and $D_0$, $\nu_0$, and $\alpha$ are obtained from the experimental measurements of Yu. V. Tsekhanskaya.

2. It is assumed that the zone $\delta < y < \delta + \delta_1$ is extremely narrow and does not affect the velocity distribution.

3. The entire change in concentration occurs at such distances from the disk that are small compared to the thickness of the hydrodynamic boundary layer.

We first examine the velocity distribution in the neighborhood of the disk. In the zone $y < \delta$, the hydrodynamic equations are

$$F^2 - G^2 + HF' = F'' + \frac{\alpha}{\sqrt{\omega\nu_0}} F' + \frac{\alpha}{\sqrt{\omega\nu_0}} F'' \cdot \xi, \tag{22.3}$$

$$2FG + G'H = G'' + \frac{\alpha}{\sqrt{\omega\nu_0}} G' + \frac{\alpha}{\sqrt{\omega\nu_0}} G'' \cdot \xi, \tag{22.4}$$

$$H' + 2F = 0, \tag{22.5}$$

in which we are using the notation of Section 11.

Since we are interested in the velocity distribution in the neighborhood of the disk, we seek a solution for equations (22.3) to (22.5), that satisfies the boundary conditions on the disk surface

$$F = 0, \quad G = 1, \quad H = 0 \quad \text{at} \quad \xi = 0,$$
$$F = G = 0 \quad \text{as} \quad \xi \to \infty,$$

in the form

$$F = a\xi + b\xi^2 + c\xi^3,$$
$$G = 1 + d\xi + e\xi^2 + f\xi^3,$$
$$H = g\xi^2 + h\xi^3.$$

Substituting these expansions into equations (22.3) to (22.5), we obtain

$$F = a\xi - \frac{1}{2}(1 + \gamma a)\xi^2 - \frac{1}{3}[d - \gamma(1 + \gamma a)]\xi^3 + \ldots, \tag{22.6}$$

$$G = 1 + d\xi - \frac{1}{2}\gamma d\xi^2 + \frac{1}{3}[a + \gamma^2 d]\xi^3 + \ldots, \tag{22.7}$$

$$H = -a\xi^2 + \frac{1}{3}(1 + \gamma a)\xi^3 + \ldots, \tag{22.8}$$

where the following notation has been introduced:

$$\gamma = \frac{\alpha}{\sqrt{\omega\nu_0}} = \frac{\nu_1 - \nu_0}{\delta\sqrt{\omega\nu_0}}. \tag{22.9}$$

The expansions (22.6) to (22.8) should match similar expansions for a liquid having a constant kinematic viscosity $\nu_1$ at a certain (dimensionless) distance from the disk surface

$$\xi_\delta = \delta\sqrt{\frac{\omega}{\nu_0}} \qquad (22.10)$$

Here an error is committed by disregarding the thickness $\delta_1$.

Assuming that the velocity distribution for $\xi > \xi_\delta$ has the same form as in Section 11, we obtain from the conditions implied by item 3 above the relations for determining the three unknown quantities a, d, and $\delta$. These quantities may be expressed in terms of $\nu_0, \nu_{cr}$, $\nu_1$, and $\omega$, which are known from experimental measurements of Yu. V. Tsekhanskaya. Tedious numerical calculations for the triethylamine — water system at 290°K yielded the following values for $\sqrt{\frac{\nu_0}{\nu_1}}$, $\xi_\delta$, and a:

| Concentration $c_0$ | 100 mg/cm³ | 200 mg/cm³ | 300mg/cm³ | 400 mg/cm³ |
|---|---|---|---|---|
| $\sqrt{\frac{\nu_0}{\nu_1}}$ | 0.68 | 0.57 | 0.49 | 0.47 |
| $\xi_\delta$ | 0.17 | 0.28 | 0.34 | 0.35 |
| a | 0.39 | 0.34 | 0.31 | 0.30 |

Passing now to the solution of the convective diffusion equation, we have, in view of (22.1)

$$v_y\frac{dc}{dy} = \frac{d}{dy}\left[D_0\left(1-\frac{y}{\delta}\right)\frac{dc}{dy}\right], \qquad (22.11)$$

where

$$v_y \cong -a\omega^{3/2}\nu_0^{-3/2}y^2 = -Ky^2.$$

Equation (22.11) can be integrated directly. Repeated integration gives

$$c(y) = a_1\int_0^y\left(1-\frac{z}{\delta}\right)^{\left(\frac{K\delta^3}{D_0}-1\right)}\exp\left\{\frac{K\delta^3}{D_0}\left[\frac{z}{\delta}+\frac{1}{2}\left(\frac{z}{\delta}\right)^2\right]\right\}dz + a_2.$$

The boundary conditions

$$c = 0 \quad \text{at} \quad y = 0, \qquad (22.12)$$

$$c = c_{cr} \quad \text{at} \quad y = \delta \qquad (22.13)$$

permit the determination of the constants $a_1$ and $a_2$.

Further integration beyond the limit of the plane $y = \delta$ cannot be made, since the expression used for viscosity loses its meaning in that region.

An approximate calculation shows that convection ensures the maintenance of a constant concentration in the plane $y = \delta$ (in other words, the transport of solute from the plane $y = \delta$ to $y = 0$ is the slow stage of the process; transport from infinity to $y = \delta$ is the rapid stage of the process).

Taking the boundary conditions into account, we obtain, finally

$$c(y) = \frac{c_{cr} \int_0^y \left(1 - \frac{z}{\delta}\right)^{\left(\frac{K\delta^3}{D_0} - 1\right)} e^{\frac{K\delta^3}{D_0}\left[\frac{z}{\delta} + \frac{1}{2}\left(\frac{z}{\delta}\right)^2\right]} dz}{\int_0^\delta \left(1 - \frac{z}{\delta}\right)^{\left(\frac{K\delta^3}{D_0} - 1\right)} e^{\frac{K\delta^3}{D_0}\left[\frac{z}{\delta} + \frac{1}{2}\left(\frac{z}{\delta}\right)^2\right]} dz}. \tag{22.14}$$

The curve c(y) is shown in Figure 25. To a first approximation it may indeed be represented by a straight line.

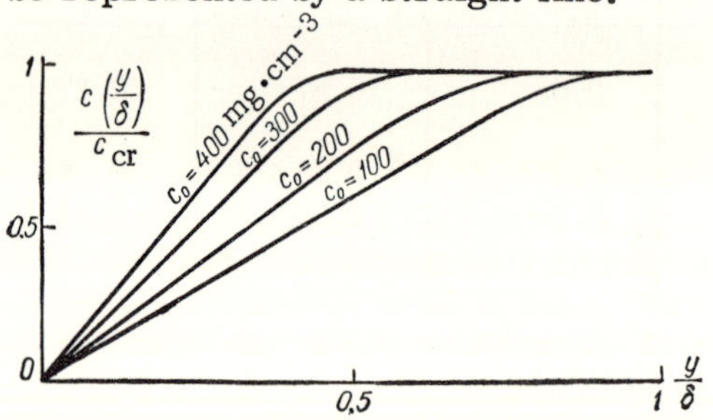


Figure 25. Dependence of the ratio $\frac{c}{c_{cr}}$ on distance from the disk surface at different concentrations $c_0$ of the solution.

The flux of solute to the surface of the disk is given by

$$j = D_0\left(\frac{\partial c}{\partial y}\right)_{y=0} = \frac{D_0 c_{cr}}{\delta I} = D_0 c_{cr} \sqrt{\frac{\omega}{\nu_0}} \frac{1}{\xi_\delta I}, \tag{22.15}$$

where

$$I = \int_0^1 (1 - t)^{\beta - 1} e^{\beta\left(t + \frac{t^2}{2}\right)} dt$$

and for the conditions of Tsekhanovskaya's experiments

$$\beta = \frac{K\delta^3}{D_0} = 0.256 \cdot 10^4 a \xi_\delta^3$$

The integral I was evaluated numerically. At the stated value of $\beta$ it converges very rapidly. The following expression was obtained for the flow of solute to 1 cm$^2$ of the disk's surface:

$$j = AD_0 c_{cr} \sqrt{\frac{\omega}{\nu_0}}, \qquad (22.16)$$

where the numerical coefficient A is a function of the concentration $c_0$ of the solution:

| Concentration $c_0$ in mg/cm³ | 100 | 200 | 300 | 400 |
|---|---|---|---|---|
| Coefficient A | 6.06 | 6.61 | 6.74 | 6.96 |

We see that the flux of solute is a weak function of the concentration $c_0$, and is proportional to $\sqrt{\omega}$.

The dependence of j on $c_{cr}$, rather than on $c_0$, is the result of the independent nature of mass transport ot the plane where $c = c_{cr}$, and from this plane to the disk surface. Mass transport to the plane where $c = c_{cr}$ is basically convective.

Figure 23, illustrates the dependence of the flux j on the concentration $c_0$ of the solution in relative units. The theoretical values in Figure 23 calculated from (22.16) are depicted by triangles.

It is evident that the theoretical values of j in the sub-critical region are proportional to $c_0$. In the supercritical region, they are virtually independent of $c_0$. The quantitative agreement between theory and experiment is valid in both the sub-critical and the supercritical concentration regions. The expression (22.16) for the flux is general for all liquid solutions in the critical concentration range. The numerical coefficient A, however, and its dependence on $c_0$ are related to the numerical values of the constants $\nu_0$, $\nu_{cr}$, $\nu_1$, and $\xi_\delta$, and may have different values for different solutions.

## 23. DIFFUSIONAL FLOW IN NATURAL CONVECTION. THE CASE OF A VERTICAL PLATE.

Natural convection is one of the cases of the mixing of solutions often encountered in actual practice [31], [32]. Motion of the solution in natural convection is spontaneous and arises due to forces originating from heterogeneous reactions. Such forces usually follow the changes in the density of the solution which accompany the reaction*. Changes in the density of the solution may be caused by two phenomena: first, the concentration in the proximity of the reaction surface changes in the course of a heterogeneous reaction and leads to

---

* The problem of heat transfer due to natural convection near a vertical plate has been previously studied by many investigators [32]. The Prandtl number in this problem was taken as equal to unity. Let us also note that the problem examined here is analogous to the problem of thermal convection near a heated plate cooled by a cold flowing fluid.

changes in the density of the solution; second, if the heterogeneous reaction is exothermic and is accompanied by a substantial release of heat, the density of the solution changes from point to point as a result of non-uniform changes in the temperature of the solution.

The examination of the second phenomenon is beyond the scope of this book, which does not cover exothermic reactions. We confine our analysis to the diffusional flow in cases when the convection results from changes in solution density (often very large), which in turn, are associated with changes in the concentration.

Natural convection arises only when the density change occurs in the gravitational field, and only in those cases where the density gradient is perpendicular to the gravitational field or when the density increases from the bottom upward.

The magnitude and distribution of hydrodynamic and diffusional flows depend primarily on the geometry, and, in particular, on the shape and orientation of the reaction surface. In the simplest case, where the reaction surface is a smooth vertical plate placed in the gravitational field, a quantitative computation of the diffusional flow to its surface becomes possible. The case of horizontal orientation of the reaction surface is considerably more complicated. The problem of computing the diffusional flow in that case has not yet been resolved, and we are forced to be satisfied with an analogy between diffusion and heat transfer.

Let us examine a vertical plate in a gravitational field, on which a certain heterogeneous chemical reaction takes place. Let us assume that the chemical reaction rate is very large compared to the rate of mass transport, and that the concentration of the reacting substance on the surface of the plate is zero. As usual, we designate the concentration far from the plate by $c_0$. The density of the solution is a function of its concentration. Therefore, the density is different far from the plate than in proximity to it. Thus, when the density of the solution far from the plate equals $\rho(c_0)$, then it will be equal to $\rho(0)$ at the plate surface. If the density of the solution is a weak function of its concentration, then, approximately,

$$\rho(c_0) \approx \rho(0) + \frac{\partial \rho}{\partial c} c_0.$$

Generally, on the basis of the well-known Bjerknes theorem, we may expect that convection will appear in the fluid: surfaces of equal fluid density are oriented perpendicular to surfaces of equal pressure.

Let us take $y = 0$ as the surface of the plate, directing the y-axis into the solution and the x-axis vertically upward. The lower edge of the plate corresponds to the value $x = 0$. We now consider the equations of fluid motion brought about by gravity. The force acting on a unit of fluid volume, obviously is equal to $\rho g$ and changes form point to point in the solution.

The concentration and the density of the solution are functions both of the distance y from the plate and of the distance x along the plate. It is natural to consider, however, that most of the change in the concentration of the solution occurs in a very thin layer immediately adjacent to the reaction surface in the diffusion layer. This change in concentration is the sole reason for fluid motion. Therefore, we can safely assume that fluid motion also occurs in this layer. Thus, the theories for the hydrodynamic boundary layer can also be applied to fluid motion in natural convection. In this case, the hydrodynamic boundary layer coincides with the diffusion layer.

The equations of fluid motion in a boundary layer located in a field of gravity are

$$v_x \frac{\partial v_x}{\partial x} + v_y \frac{\partial v_x}{\partial y} = \nu \frac{\partial^2 v_x}{\partial y^2} + g \frac{\rho(c_0) - \rho(c)}{\rho(c_0)}, \tag{23.1}$$

$$\frac{\partial v_x}{\partial x} + \frac{\partial v_y}{\partial y} = 0, \tag{23.2}$$

where $\rho(c)$ is the density of the solution at an arbitrary point in the solution, and $\rho(c_0)$ is the density far from the plate (at infinity); $v_x$ and $v_y$ are the tangential and normal components of the fluid velocity.

The concentration distribution in the diffusion boundary layer is determined from the solution of the convective diffusion equation

$$v_x \frac{\partial c}{\partial x} + v_y \frac{\partial c}{\partial y} = D \frac{\partial^2 c}{\partial y^2}. \tag{23.3}$$

In order to solve simultaneously equations (23.1) to (23.3), we assume that the changes in density resulting from changes in concentration are sufficiently small and, therefore, the following approximation for the concentration at any point in the solution can be made.

$$\rho(c) \approx \rho(c_0) + \left(\frac{\partial \rho}{\partial c}\right)_{c=c_0} (c - c_0). \tag{23.4}$$

Substituting this approximated expression for $\rho(c)$ into equation (23.1), we obtain

$$v_x \frac{\partial v_x}{\partial x} + v_y \frac{\partial v_y}{\partial y} = \nu \frac{\partial^2 v_x}{\partial y^2} + \frac{g(c_0 - c)}{\rho(c_0)} \cdot \left(\frac{\partial \rho}{\partial c}\right)_{c=c_0}. \tag{23.5}$$

It is convenient, for the subsequent analysis, to introduce a new dimensionless concentration $\varphi = \frac{c_0 - c}{c_0}$ into equations (23.5) and (23.3). These equations then become

$$v_x \frac{\partial v_x}{\partial x} + v_y \frac{\partial v_x}{\partial y} = \nu \frac{\partial^2 v_x}{\partial y^2} + g\alpha\varphi, \tag{23.6}$$

$$v_x \frac{\partial \varphi}{\partial x} + v_y \frac{\partial \varphi}{\partial y} = D \frac{\partial^2 \varphi}{\partial y^2}, \tag{23.7}$$

$$\alpha = \frac{c_0}{\rho(c_0)} \left(\frac{\partial \rho}{\partial c}\right)_{c=c_0}.$$

The boundary conditions for equations (23.6) and (23.7) are: at the surface of the plate, where y = 0, the velocity of the fluid must reduce to zero. Furthermore, the concentration of the reactant on the reaction surface must also be zero. The definition of the dimensionless concentration $\varphi$ shows, that $\varphi=1$ for y = 0; thus the boundary conditions are

$$\left.\begin{aligned} v_x = v_y &= 0, \\ \varphi &= 1, \end{aligned}\right\} \quad y = 0, \tag{23.8}$$

$$\left.\begin{aligned} v_x &= 0, \\ v_y &= 0, \end{aligned}\right\} \quad y \to \infty. \tag{23.9}$$

The solution for the system of equations (23.6) and (23.7) with boundary conditions (23.8) and (23.9) is found by introducing another dimensionless variable

$$\eta = \left(\frac{g\alpha}{4\nu^2}\right)^{1/4} \frac{y}{x^{1/4}} \tag{23.10}$$

and the stream function

$$\psi = 4\nu \left(\frac{g\alpha}{4\nu^2}\right)^{1/4} x^{3/4} f(\eta). \tag{23.11}$$

In terms of the new variable, the velocity components are

$$v_x = 4\nu \left(\frac{g\alpha x}{4\nu^2}\right)^{1/2} f'(\eta), \tag{23.12}$$

$$v_y = -\frac{\partial\psi}{\partial x} = \nu \left(\frac{g\alpha}{4\nu^2}\right)^{1/4} \frac{(\eta f' - 3f)}{x^{1/4}}, \tag{23.13}$$

where the function f satisfies the equation

$$f''' + 3ff'' - 2(f')^2 + \varphi = 0. \tag{23.14}$$

Also in terms of the new variable, the equation for dimensionless concentration (23.7) takes the form

$$\varphi'' + 3\,\mathrm{Pr}\, f\varphi' = 0. \tag{23.15}$$

Finally, for the boundary conditions of the problem, we obtain the expressions

$$\left.\begin{aligned} f = f' &= 0, \\ \varphi &= 1, \end{aligned}\right\} \quad \eta = 0, \tag{23.16}$$

$$\left.\begin{aligned} f' &= 0, \\ \varphi &= 0, \end{aligned}\right\} \quad \eta \to \infty. \tag{23.17}$$

To solve the equations, we shall take advantage of the fact that the Prandtl number is very large. The solution of equation (23.15) takes the following form:

$$\varphi = 1 - \frac{\int\limits_0^{\eta} \exp\left\{-3\,\mathrm{Pr}\int\limits_0^{\eta} f\,d\eta\right\} d\eta}{\int\limits_0^{\infty} \exp\left\{-3\,\mathrm{Pr}\int\limits_0^{\eta} f\,d\eta\right\} d\eta}. \tag{23.18}$$

In equation (23.18), the function f is unknown. To determine f we now utilize the fact that the integrals in formula (23.18) converge very rapidly due to the extremely large values of the Prandtl number. Therefore, the magnitudes of the integrals are determined mainly by the value of the function f, for small values of the variable $\eta$. On the contrary, the form of the function f at large values of $\eta$ has no noticeable effect on the concentration distribution in the proximity of the plate. Therefore, without introducing any significant error, the boundary condition at infinity may be replaced by a new boundary condition which is fulfilled at a finite distance from the plate. Farther from the plate the fluid is stationary and does not transport any solute. In other words, it is possible to introduce a boundary layer thickness $\eta_0$ beyond which (for $\eta > \eta_0$) the velocity of the fluid will be zero, while at $\eta = \eta_0$ the following condition is fulfilled:

$$\left.\begin{aligned} f' &= 0, \\ \varphi &= 0, \end{aligned}\right\} \quad \eta = \eta_0. \tag{23.19}$$

The thickness of the boundary layer $\eta_0$ has not yet been determined and must be found by further analysis. Owing to the rapid convergence of the integrals in (23.18) we can write a series expansion in powers of $\eta$ for the function f for $\eta < \eta_0$, and retain only the first terms of this expansion. In view of the boundary condition (23.16), the series expansion of f is

$$f = \frac{\beta}{2}\eta^2 + \frac{\gamma}{32}\eta^3 + \dots \tag{23.20}$$

Terms containing the zero and the first powers of $\eta$ must be absent. The coefficients of the series (23.20) still have to be determined.

Substituting the expression for f from (23.20) into the solution (23.18), and using as a first approximation only one term, we find

$$\varphi = 1 - \frac{\int\limits_0^{\eta} \exp\left\{-\beta\,\mathrm{Pr}\,\frac{\eta^3}{2}\right\} d\eta}{\int\limits_0^{\infty} \exp\left\{-\beta\,\mathrm{Pr}\,\frac{\eta^3}{2}\right\} d\eta}.$$

Introducing a new integration variable $t = \left(\frac{\beta \Pr}{2}\right)^{1/3} \eta$, we obtain

$$\varphi = 1 - \frac{\int\limits_0^{\left(\frac{\beta \Pr}{2}\right)^{1/3}\eta} \exp(-t^3)\, dt}{\frac{1}{3}\Gamma\left(\frac{1}{3}\right)}. \tag{23.21}$$

It follows from formula (23.21) that for $\left(\frac{\beta \Pr}{2}\right)^{1/3} \eta > 1$ and owing to the rapid convergence of the integral, its upper limit may be replaced by infinity:

$$\varphi \approx 1 - \frac{\int\limits_0^{\infty} \exp\{-t^3\}\, dt}{\frac{1}{3}\Gamma\left(\frac{1}{3}\right)} = 0.$$

Conversely, for $\left(\frac{\beta \Pr}{2}\right)^{1/3} \eta < 1$, we have the approximation

$$\varphi = 1 - \frac{\int\limits_0^{\left(\frac{\beta \Pr}{2}\right)^{1/3}\eta} (1 - t^3 + \ldots)\, dt}{\frac{1}{3}\Gamma\left(\frac{1}{3}\right)} \approx 1 - \frac{\left(\frac{\beta \Pr}{2}\right)^{1/3}\eta}{\frac{1}{3}\Gamma\left(\frac{1}{3}\right)} = 1 - \frac{\left(\frac{\beta \Pr}{2}\right)^{1/3}\eta}{0{,}89} \tag{23.22}$$

Using the concentration distribution within the boundary layer given by (23.22) and the boundary condition (23.19), we can determine the following:

$$\varphi = 0 \approx 1 - \frac{\left(\frac{\beta \Pr}{2}\right)^{1/3}\eta_0}{0.89},$$

whence, the thickness of the boundary layer on the plate is

$$\eta_0 = \frac{0.89}{\left(\frac{\beta \Pr}{2}\right)^{1/3}}. \tag{23.23}$$

Equation (23.23) contains another unknown quantity $\beta$, whose value may be found in the following manner. Substituting the concentration distribution given by formula (23.22) into equation (23.14), we obtain the following approximate relation:

$$f''' + 3ff'' - 2(f')^2 + \left(1 - \frac{\left(\frac{\beta \mathrm{Pr}}{2}\right)^{1/3}\eta}{0.89}\right) = 0. \tag{23.24}$$

A solution of equation (23.24) may be obtained by successive approximation. Since we are interested only in the first term of the series expansion of f in powers of $\eta$, we may restrict our solution to the lower powers of $\eta$. Thus, an approximate solution of (23.24) may be written in the form

$$f = \frac{\beta\eta^2}{2} - \frac{\eta^3}{6} + \frac{\left(\frac{\beta \mathrm{Pr}}{2}\right)^{1/3}\eta^4}{24 \cdot 0.89}. \tag{23.25}$$

The quantity $\beta$ may be found from the boundary condition (23.19)

$$f' = \beta\eta - \frac{\eta^2}{2} + \frac{\left(\frac{\beta \mathrm{Pr}}{2}\right)^{1/3}\eta^3}{6 \cdot 0.89}.$$

At the point $\eta = \eta_0$, the derivative f′ must become zero. This leads to the relation

$$f' = 0 = \beta\eta_0 - \frac{\eta_0^2}{2} + \frac{\left(\frac{\beta \mathrm{Pr}}{2}\right)^{1/3}\eta_0^3}{6 \cdot 0.89}. \tag{23.26}$$

Finally, $\beta$ may be determined by substituting the value of $\eta_0$ into equation (23.26):

$$\beta = \frac{0.48}{\mathrm{Pr}^{1/4}}. \tag{23.27}$$

Thus, all unknown constants may be evaluated for a typical equation of diffusional flow resulting from natural convection near a vertical plate, by assuming that the fluid motion occurs in a layer of thickness $\eta_0$ adjacent to the wall.

Substituting the value of $\beta$ into equation (23,22) and reverting to the dimensional concentration c, we obtain

$$c = 0.7\,\mathrm{Pr}^{1/4}\eta c_0 = 0.7\,\mathrm{Pr}^{1/4}\left(\frac{g\alpha}{4\nu^2}\right)^{1/4}\frac{yc_0}{x^{1/4}}. \tag{23.28}$$

The diffusional flux in natural convection has the form

$$j = D\left(\frac{\partial c}{\partial y}\right)_{y=0} = 0.7D\,\mathrm{Pr}^{1/4}\left(\frac{g\alpha}{4\nu^2}\right)^{1/4}\frac{c_0}{x^{1/4}} = 0.7\mathrm{Pr}^{1/4}D\left[\frac{gc_0}{4\nu^2\rho}\left(\frac{\partial\rho}{\partial c}\right)_{c=c_0}\right]^{1/4}\frac{c_0}{x^{1/4}} \tag{23.29}$$

The diffusional flux changes rather slowly with distance from the lower edge of the plate. It is also a function of the concentration of the solution $c_0$, the kinematic viscosity $\nu$, the derivative $\left(\frac{\partial \rho}{\partial c}\right)_{c = c}$, and of the Prandtl number.

Integrating expression (23.29) over the surface of the plate, we obtain the total diffusional flow to the plate

$$I = 0.9D\mathrm{Pr}^{1/4}\left[\frac{g\alpha}{4\nu^2}\right]^{1/4} bh^{3/4}c_0 = 0.9\mathrm{Pr}^{1/4}\left[\frac{gc_0}{4\nu^2\rho}\left(\frac{\partial\rho}{\partial c}\right)_{c=c_0}\right]^{1/4} bh^{3/4}c_0 D. \tag{23.30}$$

The total diffusional flow to the surface of the plate is proportional to the fourth root of the Prandtl number, the 5/4 power of the concentration, and inversely proportional to the square root of the fluid viscosity. It is also a function of the derivative $\left(\frac{\partial \rho}{\partial c}\right)_{c = c_0}$ and of the acceleration due to gravity g.

Since the flux changes from point to point along the plate, the total flow to it is not proportional to its area. The value of the derivative $\left(\frac{\partial \rho}{\partial c}\right)_{c = c_0}$ may be easily obtained from available handbooks. If the concentration is expressed in grams per $cm^3$, the value of the ratio $\frac{c_0}{\rho(c_0)}\left(\frac{\partial \rho}{\partial c}\right)_{c = c_0}$ will be approximately unity. In estimating the order of magnitude of diffusional fluxes in natural convection, it can be assumed that

$$j \approx 0.7\, D\, \mathrm{Pr}^{1/4}\left(\frac{gc_0}{4\nu^2}\right)^{1/4}\frac{c_0}{x^{1/4}}. \tag{23.31}$$

Correspondingly, the effective thickness of the diffusion layer is

$$\delta \approx \frac{x^{1/4}}{0.7\,\mathrm{Pr}^{1/4}\left(\frac{gc_0}{4\nu^2}\right)^{1/4}}. \tag{23.32}$$

For example, substitution of the numerical values for a solution whose concentration $c_0 \approx 10^{-3}$ gm per $cm^3$, and for a plate whose height h $\approx$ 1 cm, yields the following thickness for the diffusion layer:

$$\delta \approx 0.03 \text{ cm}.$$

Diffusion boundary layer thicknesses in natural convection are found to be substantially greater than in forced convection, and the values of the diffusional fluxes are found to be correspondingly smaller.

It should be noted that many books and monographs, especially in the field of electrochemistry, often contain assertions that, in natural convection, the quantity $\delta$ always has a value of approximately 0.05 cm [33]. This assertion, as may be seen from the above, may lead to substantial errors in the numerical evaluation of diffusional fluxes.

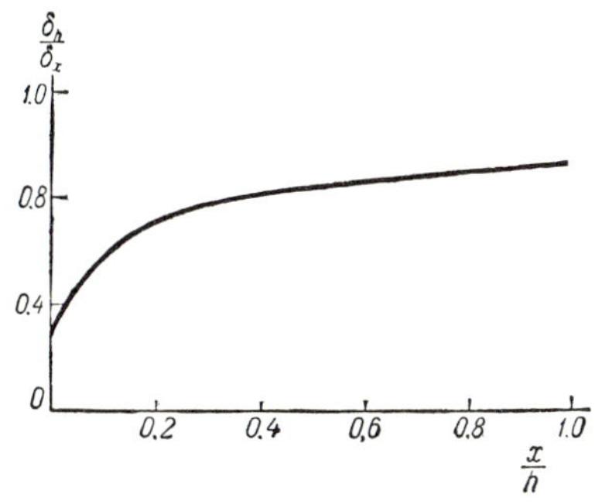


Figure 26. Thickness of the diffusion layer on a vertical plate in the presence of natural convection.

The equation for the diffusional flux can also be written in a dimensionless form. The dimensionless quantity $Gr = \frac{g\alpha h^3}{4\nu^2}$ may also be called, by analogy to heat transfer, the Grashof number for mass transfer. It is the only dimensionless quantity* that enters into the expression for velocity and diffusional flow. By means of this dimensionless group, the expression for the dimensionless diffusional flux can be rewritten in the form

$$Nu = \frac{hI}{DSc_0} = 0.9\, Pr^{1/4} Gr^{1/4}. \qquad (23.33)$$

It is interesting to compare this expression with an analogous one for the diffusional flux induced by changes in fluid density arising from temperature differences. In this case, the thermal Prandtl number is of the order of unity. Thus, the rapid convergence of the integrals cannot be used to simplify the corresponding formulas. Within an order of magnitude, however, the following expression may be written for the dimensionless diffusional flux of heat

* Editor's note: besides the Prandtl number.

$$\mathrm{Nu}_{\mathrm{therm}} \approx \left[\frac{gh^3}{4\nu^2}\left(\frac{\partial\rho}{\partial T}\right)\frac{\Delta T}{T_0}\right]^{1/4} = \mathrm{Gr}^{1/4}_{\mathrm{therm}} \tag{23.34}$$

where $\mathrm{Gr}_{\mathrm{therm}}$ is the thermal Grashof number.

The ratio of the diffusional fluxes, arising from the changes in density caused, respectively by concentration and temperature differences, as given by equations (23.33) and (23.34), is

$$\frac{\mathrm{Nu}}{\mathrm{Nu}_{\mathrm{therm}}} \approx \left[\frac{\mathrm{Pr}}{\mathrm{Pr}_{\mathrm{therm}}}\right]^{1/4}\left[\frac{\partial\rho}{\partial c}\frac{c_0}{\rho}\right]^{1/4} : \left[\frac{\partial\rho}{\partial T}\frac{\Delta T}{T_0}\right]^{1/4}. \tag{23.35}$$

Since the fluid density is less sensitive to temperature changes than to concentration changes, formula (23.35) shows that for small temperature gradients the diffusional flux brought about by changes in concentration is substantially greater than that caused by temperature changes.

It should also be noted that the formulas set forth above are valid only for laminar flow in the boundary layer. It is known, however, that with natural convection, the transition to turbulent flow is relatively easy. A laminar flow regime is replaced by a turbulent regime at values of the Grashof number greater than $10^8$. Thus, the formulas presented above are valid only for $\mathrm{Gr} \leq 10^8$.

## BIBLIOGRAPHY


1. Shchukarev, A. N., Zhur. Russ. Fiz.-Khim. Obshch, 28, 604 (1896); Z. phys. Chem. 8, 76 (1891).
2. Nernst, Z. phys. Chem. 47, 55 (1904).
3. Nernst and Merriam, Z. phys. Chem. 53, 235 (1905).
4. Bruner, Z. anorg. Chem. 28, 314 (1901); Z. phys. Chem. 35, 283 (1900).
5. Samartsev, A. G., Trudy GOI Transactions of the State Institute of Oceanography 9, 16 (1933); Zhur. Fiz. Khim. 5, 1424 (1934).
6. Fage and Townsend, Proc. Roy, Soc. A135, 656 (1932).
7. Moelwyn-Hughes, Kinetika reaktsii v rastvorakh, All-Union State Publishing House of Science and Technology, 1938 see also: The Kinetics of Reaction in Solution, 1st edn, Clarendon, Oxford, 1933, cf. 2nd edn, 1947.
8. Miyamoto, Trans. Farad. Soc. 29, 789 (1933).
9. Langmuir, Phys. Rev. 34, 401 (1912).
10. Walker, Lewis, and McAdams, Tipovaya khimicheskaya apparatura: Typical Chemical Equipment, Khimteoretizdat, Leningrad 1935, p. 29 see also: Principles of Chemical Engineering, McGraw-Hill, 1923; cf. later editions.
11. Kasatkin, A. G., Osnovnye protsessy i apparaty khimicheskoy tekhnologii: Basic Processes and Instruments of Chemical Engineering, Goskhimizdat, 1948, p. 533.

12. Frank-Kamenetskiy, D. A., Diffuziya i teploperedacha v khimicheskoy kinetike, Acad. Sci. USSR Press, 1947; see aslo English translation: Diffusion and Heat Exchange in Chemical Kinetics, Princeton University Press, 1955.
13. Eucken, A., Z. Electrochem. 38 341 (1932).
14. Kirpichev, M. V., Mikheev, M. A., and Eygenson, L.S., Teploperedacha, Heat Transfer, State Power Engineering Publishing House, 1940.
15. Sb. "Sovremennye problemy gidroaerodinamiki" Current Problems in Hydro-Aerodynamics, vol. I, p. 122; IL, 1948; T. Karman, Z. angew, Math. i. Mech. 1, 244 (1921); see also: Goldstein, Modern Developments in Fluid Dynamics, vols. I and II, Clarendon, Oxford.
16. Cochran, Proc. Cambr. Phil. Soc. 30 365 (1934).
17. Levich, V. G., Zhur. Fiz. Khim. 18 335 (1944) see also English version: Acta Physicochimica URSS 17 257 (1942), 19 117 (1944); Zhur. Fiz. Khim. 22 575 (1948).
18. Kochin, N.E., Kibel', I.A., and Roze, N. V., Teoreticheskaya gidromekhanika: Theoretical Hydromechanics, vol. II, Gostekhizdat, p. 338 (1948) see also German translation: Theoretische Hydromechanik, Bd. I and II, Akademie-Verlag, 1954.
19. Aksel'rud, G.A., Zhur. Fiz. Khim. 27 10 (1953).
20. Kruzhilin, Zhur. Tekhn. Fiz. 6 561 (1936).
21. Levich, V. G. and Meiman, N. N., Doklady Akad. Nauk SSSR 79 97 (1951).
22. Sutton, W. Proc. Roy. Soc. A182 48 (1943).
23. Damkohler, Eucken-Jakob Chemie-Ingenieur III, 1, 448 (1937).
24. D'yakonov, G. K., Doklady Akad. Nauk SSSR 39, 143 (1943); 57, 701 (1947); Izvestiya Akad. Nauk SSSR, Otd. tekhn. nauk, p. 473 (1944).
25. Zhavoronkov, N. M., Gidravlicheskie osnovy skrubbernogo protsessa i teploperedachi v skrubberakh: Hydraulic Bases of the Scrubber Process and Heat Transfer in Scrubbers "Sov. nauka" (Soviet Science) Press, 1944.
26. Ramm, V. M., Absorbtsionnye protsessy v khimicheskoy promyshlennosti: Absorption Processes in the Chemical Industry, Goskhimizdat, 1951.
27. Leveque, Ann. des Mines, ser. 12, 13, 201, 305, 381 (1928).
28. Taylor, G. Proc. Phys. Soc. B67, 857 (1954); Proc. Roy. Soc. A219, 186 (1953); A225, 473; A223, 446 (1954).
29. Gibbs, J. W., Termodinamicheskie raboty: Thermodynamic papers, Gostekhizdat, 1950 see also: Scientific Papers, vols. I and II, Dover reprint, New York, 1961.
30. Tsekhanskaya, Yu. V., Zhur. Fiz. Khim. 30, 2315 (1956).
31. Levich, V. G., Zhur. Fiz. Khim. 22, 575 (1948); Acta physicochimica URSS 19, 117 (1944).
32. Glasstone, S., Vvedenie v elektrokhimiyu, IL, 1951, p. 585 see also: An Introduction to Electrochemistry, Van Nostrand,

1942, p. 451; G. Kortum and J. Bockris, Textbook of Electrochemistry, v. II, p. 400, 1951.
34. Ivanov, Yu. B. and Levich, V. G. Zhur. Fiz. Khim. 32, 592 (1958).
35. Grober and Erk, Osnovy ycheniya o teploobmene Principles of Heat Exchange, 1958 see also Die Grundgesetze der Warmeubertragung, Springer, Berlin, 1955.
36. Riddiford and Gregory, J. Chem. Soc. London, 1956, p. 3756.

# CHAPTER III

# Diffusion Rates in Turbulent Flow

## 24. GENERAL LAWS GOVERNING MASS TRANSFER IN TURBULENT FLOW

Up to this point in our examination of the rate of diffusion processes, we have confined our treatment to the case of laminar flow. In actual practice, however, it is often necessary to deal with a fluid in turbulent flow.

In the analysis of both diffusion and heat transfer problems, major importance is attached to the flow regime in the vicinity at solid surfaces. These surfaces are known, respectively, as reaction and heated surfaces.

Turbulent flow in fluid layers near such surfaces occurs:

1) for an interior problem when flow in the conduit is at a sufficiently high Reynolds number.

2) for an exterior problem, when the surface of the body has a non-streamlined shaped (a sphere or a cylinder), or the surface is rough.

3) in the case of flow along a streamlined body (a disk or plate, for example), but at very high Reynolds numbers, of the order of $10^5$.

Turbulent flow in the neighborhood of solid surfaces is generally encountered in actual practice.

In Sections 55 and 56 we show that under certain conditions turbulent flow almost always appears at a wall, unless special measures are taken to prevent it.

Since the kinematic viscosity of the fluid is usually small, high Reynolds numbers occur very easily in flowing fluids.

In order to increase the rate of heterogeneous conversions, transition to a turbulent flow regime is usually required. Such a regime is encountered very often in natural processes. The study of convective diffusion phenomena in a turbulent flow regime, therefore, has practical importance. As will be seen from what follows, this importance is emphasized by the fact that the study of convective diffusion is essential to resolve problems of heat transfer in turbulent flow. Finally, it is found that study of the diffusion process provides important information on the nature of turbulent flow near a solid surface. The experimental and theoretical study of the convective diffusion process in fluids is of major importance to the modern theory of turbulence. In what follows, therefore, we cover in detail the problem of convective diffusion in turbulent flow. The theoretical study is based on the modern theory of turbulence, discussed briefly in Section 4.

We saw that in a turbulent flow regime, an extremely energetic and random agitation of the fluid takes place. Turbulence eddies carry the fluid momentum to the surface of the body around which the stream passes, leading to an average momentum transfer to the surface.

The existence of this average transfer to the solid body means that the surface is subject to surface drag. In Section 4 we saw that the momentum transfer in the fluid may be characterized by the turbulence viscosity $\nu_{turb}$. The shear stress in a turbulent stream, moving past a solid surface is given by formula (4.8)

$$\tau = \rho \nu_{turb} \frac{\partial U}{\partial y}.$$

In the case of motion past a solid surface, it may be assumed that the change in velocity occurs over such a small distance from it that t has a constant value, equal to the shear stress at the wall ifself.

In the same way, it may be assumed that turbulence eddies transport the matter, dissolved in the fluid.

The transfer of solute by the turbulent flow may be characterized by the turbulence diffusion coefficient $D_{turb}$.

Let us begin by assuming that variations in concentration occur in a solution in turbulent flow. Suppose, for example, that the concentration of substance at the reaction surface is maintained at $c = 0$, while in the body of the solution it equals $c_0$. Then a net mean flow of solute to the reaction surface exists which is proportional to the gradient of the average concentration $\partial c/\partial y$. This flow of solute will be transferred by turbulence eddies, despite their wholly random nature.

The transfer of solute by random turbulence eddies is analogous to the transfer of matter by molecular diffusion in a gas: the existence of a concentration gradient leads to a situation in which a net motion in the direction of decreasing concentration is superimposed on the chaotic motion of gas molecules. In the same way, the existence of a concentration gradient in a chaotically agitated fluid leads to a net transfer of matter in the direction of decreasing concentration.

If $j_{turb}$ designates the average flow of substance to 1 cm$^2$ of surface transported by turbulence eddies, then, in view of the foregoing, we may state:

$$j = D_{turb} \frac{\partial c}{\partial y}, \qquad (24.1)$$

where $\partial c/\partial y$ is the average concentration gradient, and $D_{turb}$ is the coefficient of turbulent diffusion. The positive sign in equation (24.1) corresponds to the fact that the decrease in concentration occurs in the direction of negative y. The proportionality coefficient $D_{turb}$ in formula (24.1) plays the same part as the diffusion coefficient in the usual expression for mass transport by molecular diffusion, and has the same dimensions ($cm^2/sec$). Naturally, the coefficient of turbulent diffusion and the coefficient of molecular diffusion have nothing in common. The former characterizes the transfer of matter by chaotic turbulent motion, while the latter characterizes transfer by random molecular motion. Turbulent fluid motion varies with distance from the solid surface. Therefore, the transfer of matter by turbulence eddies also varies with the distance. Thus, the coefficient of turbulent diffusion $D_{turb}$ must be a function of the distance from the reaction surface.

The coefficient of turbulent diffusion may be related to quantities that characterize turbulence. Thus, since the properties of a turbulent fluid are determined by the quantities $\rho$, $\Delta U$ and $l$, and since the only composite quantity among them having the dimensions ($cm^2/sec$) is the product $\Delta U \cdot l$, we must have

$$D_{turb} \approx \Delta U \cdot l. \tag{24.2}$$

The meaning of formula (24.2) is simple: it is the analog of the formula from the kinetic theory of gases

$$D \approx \lambda v. \tag{24.3}$$

where $\lambda$ is the mean free path and v is the average velocity due to thermal agitation of the molecules.

Let us rewrite the expression for the coefficient of turbulent diffusion in a more convenient form.

The velocities $v'$ of the turbulence eddies have an order of magnitude equal to the change in the average velocity over distances equal in order of magnitude to the scale of the turbulence eddies, i.e.,

$$v' \approx \Delta U \approx \frac{\partial U}{\partial l} \cdot l. \tag{24.4}$$

For $D_{turb}$, therefore, we may write

$$D_{turb} \approx v'l \approx l^2 \frac{\partial U}{\partial l}. \tag{24.5}$$

In what follows, we require an expression for diffusional flow in the turbulent boundary layer which exists at the solid surface. It is found that this may be obtained only on the basis of certain hypotheses concerning the nature of the turbulent motion immediately adjacent to the solid surface.

Comparing (4.2) with formula (4.5) for $\nu_{turb}$

$$\nu_{turb} \sim \Delta U l \sim D_{turb}, \tag{24.6}$$

we see that the coefficient of turbulent diffusion is the analog of turbulence viscosity and, generally speaking, has the same order of magnitude. Exactly the same situation exists in the kinetic theory of gases.

Since $D_{turb}$ is proportional to the macroscopic scale of turbulence eddies, it greatly exceeds the coefficient of molecular diffusion D in a fluid with developed turbulence.

The large value of $D_{turb}$ ensures a virtually constant concentration of the solution down to very small distances from the reaction surface. At short distances from the solid surface, however, as we have seen in Section 4, the retarding effect of the wall begins to be felt. Here, the scale of the motion becomes a function of the distance from the solid surface in accordance with the law established in Section 4.

We have seen in Section 4 that a diminution in the scale of turbulence eddies leads to the establishment of a logarithmic velocity profile (4.17).

In an analogous fashion, the change in the scale also affects the distribution of average concentration in the boundary layer.

The transfer of heat to a fluid which is heated nonuniformly occurs in the same manner. Turbulent agitation ensures a transfer from the hotter to the cooler regions of fluid instead of molecular thermal conduction.

If a temperature difference is maintained in the fluid — for example, by immersing a body in a fluid of different temperature — a steady flow of heat takes place from the hotter to the cooler regions.

The heat flux may be characterized by the coefficient of turbulent transfer, in the same way as the momentum transfer or the mass transport.

We will deal with problems of heat transfer in more detail later. Here, we state only that all subsequent discussions of diffusional phenomena may be directly applied to the process of heat transfer.

The change in scale of the eddy motion is a function of distance from the solid surface, according to equation (4.14); thus, there is a decrease in the coefficient of turbulent diffusion according to the relation

$$D_{turb} \sim l^2 \frac{\partial U}{\partial y} \sim y^2 \frac{\partial U}{\partial y}. \tag{24.7}$$

Correspondingly, we find for the flow of solute transported by turbulence eddies

$$j = \beta_0 y^2 \frac{\partial U}{\partial y} \frac{\partial c}{\partial y}. \tag{24.8}$$

where $\beta_0$ is a certain constant.

Let us examine the simplest case of mass flow across an infinite plane $y = 0$ from an infinite volume — the half-space $y > 0$ above the plane. The case of a reaction surface of finite dimensions is treated later. In this case, the flux j of solute obviously can not be a function of the distance from the plane $y = 0$, nor of coordinate x. If this were not so, the law of conservation of matter would be violated.

Thus, in the case under examination, $j = \text{const}$. Substituting the expression for the distribution of average velocity (4.17) into (24.8), we obtain

$$j = \beta v_0 y \frac{dc}{dy}, \qquad \beta = \frac{1}{\sqrt{\alpha}} \beta_0. \tag{24.9}$$

Here we employ the total derivative, since the distribution of average concentration c cannot change along an infinite plane.

Integrating (24.9), we obtain the distribution of average concentration in turbulent flow

$$c_{II} = \frac{j}{\beta v_0} \ln y + a_1, \tag{24.10}$$

where a is a constant of integration. The meaning of the subscript II is indicated later (see Table 2).

Equation (24.10) shows that in turbulent flow along an infinite plane, the distribution of the average concentration is a logarithmic function of distance from the plane. This logarithmic law is analogous to the distribution of average velocity (4.17). However, the constant $\beta$ that enters the relation in the case of concentration has nothing in common with the constant $\alpha$ that enters into (4.17). We stress this point because, in many theoretical studies in heat transfer theory, in which a logarithmic law analogous to (24.10) is established for the temperature, the constant has been set equal to $\sqrt{\alpha}$ without sound basis. The logarithmic equation (24.10) was derived for an idealized condition of a fluid flowing along an infinite plane $y = 0$.

Passing to the case of exterior flow past a semi-infinite plate ($y = 0$, $x \geq 0$), on the basis of results set forth in Section 4, we may assume that equation (24.10) is applicable for finite values of y within the turbulent boundary layer, i.e., in the interval $\delta_0 < y < d$ — where d is the thickness of the turbulent boundary layer determined by equation (4.35), and $\delta_0$ is the thickness of the viscous sublayer given by formula (4.16).

The scale of motion beyond the edge of the boundary layer, i.e., in the mainstream of the turbulent flow, may be regarded as independent of the distance from the wall. Within the mainstream zone of the turbulent flow, both the velocity and the concentration have constant values. If $c_0$ is the value of the concentration in the volume, then, at $y = d$, the following equality must hold:

$$c_{II} = \frac{j}{\beta v_0} \ln d + a_1 = c_0,$$

which permits the value of the integration constant to be determined. The logarithmic velocity profile in a turbulent boundary layer may be written in the following form:

$$c_{II} = \frac{j}{\beta v_0} \ln \frac{y}{d} + c_0. \tag{24.11}$$

The region of the turbulent boundary layer is designated by the index II, leaving I for the mainstream of turbulent motion. The concentration $c_{II}$ is a function not only of the distance from the surface of the plate, but also of the coordinate x along the plate. The dependence on x of $c_{II}$ is determined by the appropriate dependence of the quantities $v_0$ and d, given by equations (4.13) and (4.35).

## 25. DIFFUSIONAL FLOW

As we have seen in Section 4, the logarithmic law of velocity distribution is inapplicable in the viscous sublayer very close to the wall. From what follows, however, we will see that at large values of the Prandtl number, this region offers the principal resistance to diffusion and determines the amount of diffusional flow. The understanding of the damping of turbulence in the viscous sublayer, therefore, acquires a fundamental importance.

Two hypotheses of turbulence damping were examined in Section 4.

1) the Prandtl — Taylor hypotheses [1-3] of a complete cessation of turbulence in the viscous sublayer,

2) the hypothesis of L. D. Landau and the author of a gradual damping of turbulence.

The concept of a gradual damping of turbulence in the vicinity of the wall was developed in recent years by different authors, principally in connection with problems of heat transfer (P. L. Kapitsa, C. Lin).

There is, however, no agreement on the form of the damping law.

Starting with the two hypotheses mentioned above, we obtain two different expressions characterizing the flux of solute in the viscous sublayer. By combining solutions for the regions within and without the viscous sublayer, it is possible to find the limiting diffusional flow passing through the boundary layer.

1. If it is considered that at $y < \delta_0$, turbulence eddies disappear entirely and the flow becomes laminar, and that mass transfer proceeds by molecular diffusion, it may be stated that

$$j = D \frac{\partial c}{\partial y}$$

and, since j = const,

$$c_{\text{III}} = \frac{j}{D} y + a_2,$$

where $a_2$ is the integration constant which may be derived from the condition $c=0$ at $y=0$. It follows that $a_2=0$, so that

$$c_{\text{III}} = \frac{j}{D} y. \tag{25.1}$$

At the edge of the viscous sublayer, at $y \sim \delta_0$, the distribution (25.1) must change to the distribution (24.10) in a continuous manner.

Thus, at $y \sim \delta_0$, we have

$$\frac{j}{\beta v_0} \ln \frac{\delta_0}{d} + c_0 = \frac{j}{D} \delta_0. \tag{25.2}$$

From equation (25.2) we can derive the magnitude of the limiting flow in the form

$$j = \frac{Dc_0}{\delta_0 - \dfrac{D}{\beta v_0} \ln \delta_0/d}. \tag{25.3}$$

Substituting $\dfrac{10\nu}{\delta_0}$ for $v_0$, according to (4.16) we obtain

$$j = \frac{Dc_0}{\delta_0 - \dfrac{D\delta_0}{10\nu\beta} \ln \delta_0/d} = \frac{Dc_0}{\delta_0\left(1 - \dfrac{1}{10\beta \,\mathrm{Pr}} \ln \dfrac{\delta_0}{d}\right)}. \tag{25.4}$$

At a high value of the Prandtl number,

$$j \approx \frac{Dc_0}{\delta_0} = \frac{Dc_0 v_0}{10\nu} \sim \frac{c_0 v_0}{\mathrm{Pr}}. \tag{25.5}$$

In this instance, it is found that the major "diffusion resistance" is in the laminar sublayer, i.e., the transport of solute through the viscous sublayer is limited by molecular diffusion. The diffusional flow is proportional to the characteristic velocity $v_0$ of turbulence eddies, and is inversely proportional to the Prandtl number (i.e., directly proportional to the diffusion coefficient D).

The total stream may be divided into three zones according to the velocity and mass distribution, as shown in Table 2.

Disregarding the dependence of $v_0$ on x, which is of no vital importance to the general conclusion, it may be said that relation (25.5) gives the diffusional flux at any point on the plate.

Equation (25.5) may be rewritten in dimensionless form as follows

$$\mathrm{Nu} \sim \mathrm{Re}. \tag{25.6}$$

Let us stress that (25.5) expresses the limiting case for j, corresponding to high Prandtl numbers $\mathrm{Pr} \gg 1$.

Table 2

| Zone No. | Value of the Coordinate y | Characteristics of the Zone | Mechanism of Transfer | | Law of Concentration Distribution |
|---|---|---|---|---|---|
| | | | of momentum | of mass | |
| I | $y > d$ | Developed Turbulence | Turbulent | Turbulent | $c_I = \text{const}$ |
| II | $\delta_0 < y < d$ | Turbulent Boundary Layer | Turbulent | Turbulent | $c_{II} = \frac{j}{\beta v_0} \ln \frac{y}{d} + c_0$ |
| III | $0 < y < \delta_0$ | Viscous sublayer | Molecular Viscosity | Molecular Diffusion | $c_{III} = \frac{j}{D} y$ |

The limiting relation (25.5) shows that Nu is not a function of Pr, i.e., is not a function of the diffusion coefficient of the solute.

Von Karman criticized the concepts of Prandtl and Taylor on the grounds that a sharply defined border between the turbulent boundary layer and the laminar sublayer was impossible. He then proposed [3] a so-called buffer layer in which the flow of fluid near the wall is subdivided into three zones: 1) a turbulent boundary layer, in which momentum and the substance are transferred by turbulence eddies; 2) a buffer layer, in which momentum and the substance are transferred by the turbulence eddies and the molecular mechanism to an equal degree; 3) a laminar layer, in which both momentum and substance are transferred by the molecular mechanism.

It should be noted that the original studies of Prandtl, Taylor and von Karman examined the heat transfer process rather than the transfer of matter. From what has been said above (see also Section 36), it is clear that this differentiation is not meaningful in principle.

Von Karman's concept was widely accepted in heat transfer theory. The actual expression for the mass flux, with the assumption of a buffer zone is somewhat more complicated than (25.4). In the limiting case of $Pr \gg 1$, however, the Prandtl and von Karman formulas are equal. The meaning of this agreement is simple: at $Pr \gg 1$, i.e., at low values of the diffusion coefficient, transport of matter through the laminar sublayer is the slow stage of the transfer process, as is made especially clear by the change from (25.3) to (25.5). The resistance to diffusion of the turbulent boundary layer, as well as that of the buffer layer, at $Pr \gg 1$, is found to be smaller than the resistance of the laminar sublayer. The value of the Prandtl diffusion number in fluids ($Pr \sim 10^3$) is sufficiently large, in fact, for complete agreement of the Prandtl (25.4) and von Karman formulas.

2. If the above-mentioned hypothesis of a smooth damping of the turbulent flow in the viscous sublayer is adopted, then the structure of the diffusion boundary layer is found to be more complex. Namely, at $d > y > \delta_0$, the logarithmic law of concentration distribution, derived above, is valid. At $y < \delta_0$, we may write for the diffusional flux

$$j = D_{\text{turb}} \frac{\partial c}{\partial y}$$

and assume that the mass transfer by turbulence eddies is significantly greater than the transfer by molecular diffusion, as is the case in the outer portion of the boundary layer.

In order to find $D_{\text{turb}}$ in the viscous layer, we may employ the general expression (24.5). By expressing the quantities entering into it in terms of $v_0$, $\delta_0$, and y, we obtain

$$D_{\text{turb}} \approx \frac{v_0 y^4}{\delta_0^3}. \tag{25.7}$$

As we might have expected, $D_{turb}$ is the same as $\nu_{turb}$, as determined by formula (4.26), to an order of magnitude. But $\nu_{turb}$ in the viscous sublayer is smaller than molecular viscosity $\nu$. Since we are examining the case of $D \ll \nu$, the converse is true for $D_{turb}$ and D.

Substituting the value of $D_{turb}$ from (25.7) in the expression for the diffusional flux, we arrive at an expression for the diffusional flux in the viscous sublayer

$$j = \gamma v_0 \frac{y^4}{\delta_0^3} \frac{\partial c}{\partial y}, \tag{25.8}$$

where $\gamma$ is a certain unknown numerical coefficient. We have already considered in Section 4, along with relation (4.24) which accounts for the damping of turbulence eddies, the assumption that $l \sim y$ inside the viscous sublayer. This assumption leads not to (25.7), but to the expression

$$D_{turb} \sim \frac{v_0 y^3}{\delta_0^2}. \tag{25.9}$$

Integrating (25.8), we find the concentration distribution in the zone $y < \delta_0$ given by

$$c_{III} = -\frac{j\delta_0^3}{3\gamma v_0} \cdot \frac{1}{y^3} \cdot + a_2. \tag{25.10}$$

It is necessary, however, to account for the fact that this distribution can not be valid at the wall itself, where $y = 0$. Indeed, the coefficient of turbulent diffusion $D_{turb}$ in the viscous sublayer is proportional to $y^4$, and therefore decreases rapidly as the wall is approached. At a certain distance from the wall $\delta$, therefore, the coefficient of turbulent diffusion must become so small that the following equality will be valid:

$$D_{turb}(\delta) = \gamma v_0 \frac{\delta^4}{\delta_0^3} = D. \tag{25.11}$$

At $y < \delta$, the coefficient of turbulent diffusion becomes smaller than the molecular diffusion coefficient. Correspondingly, at $y < \delta$ molecular diffusion becomes the mechanism of mass transfer.

We find in this way that a turbulent flow has a four-layered structure as shown in Figure 27.

Far from the surface of the solid body there is a zone of developed turbulence, in which the concentration remains constant. Closer to the body, in the turbulent boundary layer, both the average velocity and the average concentration decrease slowly according to a logarithmic law. In this zone, neither molecular viscosity nor diffusion play a noticeable part. Both momentum and matter are transferred by means of turbulence eddies. Still closer to the wall, in the viscous sublayer, turbulence eddies become so small that the momentum transferred by molecular viscosity exceeds that transferred by turbulence eddies. Since the diffusion coefficient D, however, is one thousand times smaller than kinematic viscosity, the remaining turbulence eddies still transfer substantially more solute than molecular diffusion. Only in the innermost portion of the viscous sublayer, at $y < \delta$, does the molecular diffusion mechanism predominate over the turbulence mechanism.

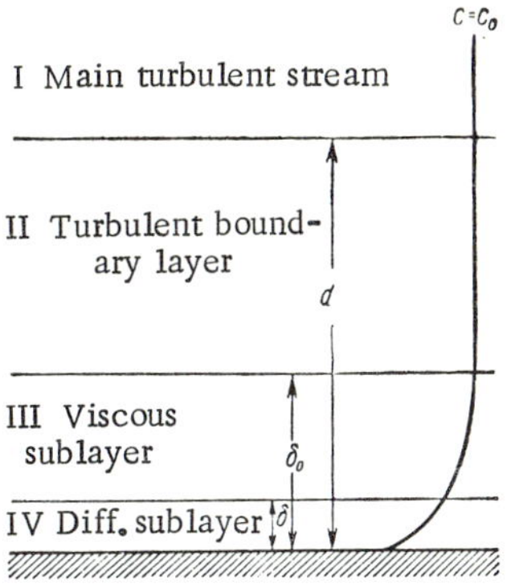


Figure 27. Distribution of the diffusing substance in a turbulent stream.

Here

$$j = D\frac{\partial c}{\partial y},$$

whence

$$c_{IV} = \frac{j}{D}y \qquad (y < \delta). \tag{25.12}$$

The constant $a_2$ in equation (25.10) is derived from the condition that at the point $y = \delta$, formulas (25.10) and (25.12) must give the same value for concentration

$$a_2 = \frac{j\delta_0^3}{3\gamma v_0}\cdot\frac{1}{\delta^3} + \frac{j\delta}{D},$$

whence

$$c_{III} = \frac{j\delta}{D} + \frac{j\delta_0^3}{3\gamma v_0}\left(\frac{1}{\delta^3} - \frac{1}{y^3}\right). \tag{25.13}$$

Since $c_{II}$ and $c_{III}$ are equal at $y = \delta_0$, we find the flux j of the solute

$$\frac{j}{\beta v_0}\ln\frac{\delta_0}{d} + c_0 = \frac{j\delta}{D} + \frac{j\delta_0^3}{3\gamma v_0}\left(\frac{1}{\delta^3} - \frac{1}{\delta_0^3}\right), \tag{25.14}$$

whence

$$j = \frac{Dc_0}{-\frac{D}{\beta v_0}\ln\frac{\delta_0}{d} + \frac{\delta_0^3 D}{3\gamma v_0}\left(\frac{1}{\delta^3} - \frac{1}{\delta_0^3}\right) + \delta}. \tag{25.15}$$

The value of $\delta$ is derived from (25.11), and using (4.16) we obtain

$$\delta = \left(\frac{D\delta_0^3}{\gamma v_0}\right)^{1/4} = \frac{\delta_0}{\mathrm{Pr}^{1/4}\sqrt[4]{10\gamma}} \sim \frac{10^{3/4}\nu}{\mathrm{Pr}^{1/4}\sqrt[4]{\gamma}\, v_0}. \tag{25.16}$$

According to (25.16) the value of at $\mathrm{Pr} \sim 10^3$ is about 1/6 of $\delta_0$. The larger the value of Pr, i.e., the smaller the value of D, the smaller is the ratio of the diffusion sublayer $\delta$ to $\delta_0$. In the Prandtl model, the zone of most of the diffusional resistance was assumed to be independent of D, and always equal in thickness to $\delta_0$.

Let us transform formula (25.15), by using the value of $\delta$ from (25.16):

$$j = \frac{Dc_0}{\delta\left[1 + \frac{1}{30\gamma\,\mathrm{Pr}}\left(\frac{\delta_0^4}{\delta^4} - \frac{\delta_0}{\delta}\right) - \frac{D}{\beta v_0 \delta}\ln\frac{\delta_0}{d}\right]} =$$

$$= \frac{Dc_0}{\delta\left[1 + \frac{1}{30\gamma\,\mathrm{Pr}}\left(10\gamma\,\mathrm{Pr} - \sqrt[4]{10\gamma\,\mathrm{Pr}}\right) - \frac{\delta_0}{10\beta\,\mathrm{Pr}\,\delta}\ln\frac{\delta_0}{d}\right]}.$$

At high values of the Prandtl number $\mathrm{Pr} \gg 1$

$$j \approx \frac{Dc_0}{\delta\left[\frac{4}{3} - \frac{\sqrt[4]{\gamma}}{10^{3/4}\beta\,\mathrm{Pr}^{3/4}}\ln\frac{\delta_0}{d}\right]}. \tag{25.17}$$

The expression for j contains two unknown constants: $\beta$ and $\gamma$. If it is assumed that their values are not greatly different from unity (which will be substantiated later by experimental data), it is found that the second term in the denominator is substantially smaller than the first (see the next section). Then (25.17) may be written in the form

$$j = \frac{Dc_0}{\frac{4}{3}\delta} = \frac{c_0 v_0}{\alpha\,\mathrm{Pr}^{3/4}}, \text{ where } \alpha = \frac{4}{3}\,10^{3/4}\cdot\gamma^{-1/4}. \tag{25.18}$$

The expression for the diffusional flux (25.17) contains the characteristic velocity of the turbulent stream and the thickness d of the turbulent boundary layer. Formula (25.18) differs from (25.5) primarily in the power to which the Prandtl number is raised.

We may express v and d by average quantities, which are characteristic of the flow in the following three cases: when the inner walls of a tube serve as the reaction surface; when the flow is past a smooth plate, and when the reaction surface is a rotating disk. These are considered in the next section. Here, we take up certain additional considerations of a general nature.

In a dimensionless form, we have for (25.18)-

$$\mathrm{Nu} \sim \frac{v_0 L}{D\,\mathrm{Pr}^{3/4}} \approx \sqrt{k_f}\,\mathrm{Re}\,\mathrm{Pr}^{1/4}. \tag{25.19}$$

As in the case of the Prandtl model, at $Pr \gg 1$, the main resistance to diffusion lies in the diffusion layer $\delta$, in which the mass transfer occurs by molecular diffusion. We see, however, that in our picture of the diffusion process, the diffusional flux j is proportional to $D^{3/4}$ and not to D, as in Prandtl's model. The lower power for the diffusion coefficient has a definite meaning (compare Section 14): the smaller the diffusion coefficient D, the closer to the wall is the region in which turbulence eddies predominate over molecular diffusion in mass transfer. Turbulent transfer is more effective than molecular, the increase in the role of turbulence in the transfer of mass compensates partially for the decrease in D. This effect, usually overlooked, is included in our model.

The same point may be expressed differently. In mass transfer, by accounting for the role of turbulence eddies, we conclude that the zone of main diffusional resistance — the diffusion sublayer $\delta$ — occupies only a portion of the viscous sublayer $\delta_0$.

If we adopt the hypothesis concerning damping in the region of the viscous sublayer according to which $l \sim y$, and if we employ equation (25.9) for the coefficient of turbulent diffusion in that region, then instead of formulas (25.16), (25.18) and (25.19), we obtain, respectively;

$$\delta \sim \delta_0 \, Pr^{-1/3}, \tag{25.20}$$

$$j = \frac{c_0 v_0}{\text{const} \cdot Pr^{2/3}} \tag{25.21}$$

and

$$Nu \sim \sqrt{k_f} \cdot Re \cdot Pr^{1/3}. \tag{25.22}$$

These formulas contain the Prandtl number to different powers. The relations (25.20) to (25.22) were obtained by Lin, Moulton and Putnam [4], and by Vielstich [5].

In the first of these works, the authors, without explanation, introduce a definite numerical coefficient into (25.20). In the works of Linton and Sherwood [6], Chilton and Colburn [7] and Lin et al [8], which are widely quoted in both chemical and heat transfer literature, it is pointed out that agreement between theory and experiment may be obtained if (25.20) is taken as some sort of empirical relation.

It must be admitted that the difference between the formulas derived on the one hand from our hypothesis of the damping of turbulence and on the other from the hypothesis that $l \sim y$ in the viscous sublayer, is not great. It reduces to a difference in the exponent of the Prandtl number in the formulas for $D_{turb}$, $\delta$ and j of only 1/12.

The measurement of the diffusional flux and, to a greater extent, of the heat flux, is not sufficiently accurate to permit choice of any one of the hypotheses concerning the damping of turbulence in the viscous sublayer.

Measurements of the concentration distribution as a function of distance from the wall, likewise are not sufficiently accurate. It is clear, however, that for the development of modern concepts concerning the nature of turbulent flow near a solid surface, it is not only important to establish whether or not eddies penetrate into the viscous sublayer [formula (25.5) vs. (25.18)], but also to choose between the two hypotheses of turbulence damping within the viscous sublayer [formula (25.18) vs. (25.21)].

Experimental data of the necessary degree of accuracy have been derived from electrochemical measurements.

The dependence of the flux j on the diffusion coefficient D has been measured (experiments of I. A. Bagotskaya) and so has the temperature dependence of the flux j (experiments by A. I. Fedorova and G. L. Vidovich, see Section 57). The latter is determined from the temperature dependence of the two physical constants in the expression for diffusional flux j, namely the coefficients of viscosity and diffusion. Since these enter into j in dimensionless form as $\Pr = \nu/D$ to different powers, the temperature dependence of j is directly determined by the temperature dependence of Pr to the appropriate power.

These measurements are described in more detail in Chapter VI. Here, we merely point out that the results of the measurements point to the validity of formula (25.18), i.e., to the validity of our hypothesis regarding the nature of the damping of turbulence eddies in the viscous sublayer. In the future, therefore, we shall examine only those relations which are derived from that hypothesis.

It is also interesting to compare, from a qualitative standpoint, the formulas for the distributions of concentration and diffusional flux in laminar and turbulent flow regimes. In the case of a turbulent regime, the thickness of the diffusion sublayer is very small compared to the thickness of the laminar boundary layer. In accordance with this, the relation for distribution of concentration as a function of distance from the solid wall, expressed by formulas (25.12) and (25.13), leads to a very sharp change in concentration near the wall. At distances exceeding the thickness $\delta_0$, concentration is actually a weak function of distance from the wall [in accordance with formula (24.11)]. Thus, on transition from laminar to turbulent flow, the variation of concentration with distance from the wall must become more pronounced near the wall. The value of the concentration gradient must increase rapidly with increase in the Reynolds number of the turbulent flow. This assertion is best illustrated by the photographs in Figures 28 and 29. These are taken from the work of Lin, Moulton and Putnam, cited above, and represent an interferometric measurement of the concentration distribution in turbulent flow near an electrode. The electrode was a portion of the wall of the tube through which the solution was pumped. The first photograph is for a Reynolds number $\mathrm{Re} = 2{,}000$; the second is for $\mathrm{Re} = 7{,}200$.

Passing to a comparison of the formulas for the diffusional flux in laminar and in turbulent regimes, we should first point out that the diffusional flux in a turbulent regime is proportional to a higher power of velocity (or Reynolds number), closer to unity, while in a laminar regime it is proportional to the square root of the velocity. Furthermore, it will be noted from (25.18) that the diffusional flux in a turbulent regime is proportional to the 3/4 power of the diffusion coefficient (and not 2/3, as in a laminar regime). We see that the dependence of the diffusional flux on the molecular diffusion coefficient not only does not disappear, but actually becomes more pronounced.

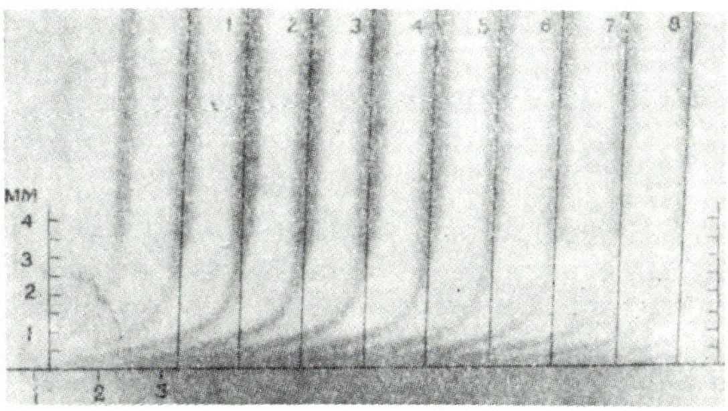


Figure 28. Velocity Distribution at the Wall
Re = 2,000.

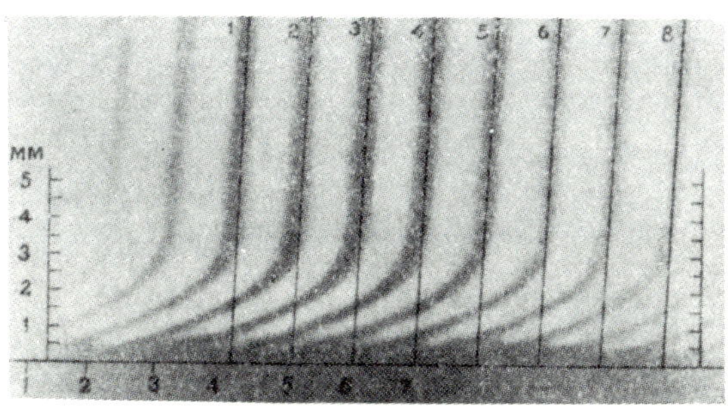


Figure 29. Velocity Distribution at the Wall
Re = 7,200.

As we have noted above, this has a simple meaning: more favorable mixing conditions decrease the role of the flow zones farthest from the wall in the total resistance to diffusion. This resistance, at $Re \gg 1$ and $Pr \gg 1$, is located wholly within the diffusion sublayer. Similar considerations were expressed at the Eighth Mendeleyev Meeting in Moscow (1959) by Sherwood (USA).

We have especially stressed this result because contrary opinions are still voiced in papers published on the subject. In particular, it is often asserted that under conditions of developed turbulence, the diffusional flux ceases to be a function of the coefficient of molecular diffusion of the diffusing substances. Thus, for example, D. A. Frank-Kamenetskii [9], on page 27 of his monograph "Diffusion and Heat Transfer in Chemical Kinetics" writes: "The conclusion, to the effect that in the limiting regime of pure turbulence, the Margules criterion (equivalent to $j/U_0 c_0$ in our notation — V. L.) must tend toward a constant value follows from the fact that under these conditions the course of the process ceases to be a function of molecular constants, and is determined solely by quantities that are characteristic of turbulent flow."

A similar point of view has been expressed in studies by V. V. Kafarov [10] and, in a series of papers by Kolthoff [11], data were cited to support it. These data were derived from experiments on diffusion to a reaction surface of extremely small dimensions (a so-called "microelectrode").

If the diffusional flux j were not a function of molecular constants and, in particular, of the diffusion coefficient D, this would require that the main diffusional resistance lie in the region of developed turbulence.

Such a conclusion contradicts all concepts of turbulent flow.

The nature of the zone of retarded motion with respect to its diffusional resistance becomes especially clear at relatively small values of the diffusion coefficients (i.e., $Pr \gg 1$). It should be noted that the general considerations of D. A. Frank-Kamenetskii and others mentioned here have not been supported by actual calculations.

As to Kolthoff's experimental data — they contradict all other experimental work. It would appear that his measurements suffer from experimental inaccuracies. It should be noted that in recent years there have been no further studies published by Kolthoff, relating to the application of the above-mentioned technique.

## 26. DIFFUSIONAL FLOW TO THE INSIDE SURFACE OF A TUBE AND TO THE SURFACE OF A PLATE

Let us first examine the case of the inside surface of a tube serving as the reaction surface with turbulent flow through the tube. In this case the concentration distribution may be considered logarithmic right up to the center of the tube. In other words, we may say that $d \sim R$ (see Section 4).

The characteristic velocity $v_0$, according to formula (4.38), may be taken as the mean velocity $U_0$ in the tube:

$$v_0 = U_0 \cdot \sqrt{\frac{k_f}{2}},$$

where $k_f$ is the friction factor.

Let us first estimate the relative importance of the two terms in the denominator of formula (25.17).

Evidently, we may write

$$\ln\frac{\delta_0}{R}\approx\ln\frac{10\nu}{v_0R}=\ln\left\{\frac{10}{\sqrt{\frac{k_f}{2}}}\frac{\nu}{U_0R}\right\}=\ln\frac{10}{\mathrm{Re}\sqrt{\frac{k_f}{2}}}.$$

The following empirical formula may be used for $k_f$

$$k_f=\frac{0.32}{\mathrm{Re}^{1/4}},\qquad \mathrm{Re}\leqslant 10^6.$$

At

$$\mathrm{Re}\sim 10^5 \text{ we have } \ln\frac{10}{\mathrm{Re}\sqrt{\frac{k_f}{2}}}\approx\ln 400\sim 6.$$

At $\mathrm{Pr}\sim 10^3$, the first term in the denominator of (25.17) is about 40 times greater than the second (if it is assumed that the constants $\alpha$ and $\frac{1}{\beta}$ are of the order of unity). Therefore, equation (25.17) may be written in the form

$$j=\sqrt{\frac{k_f}{2}}\frac{U_0c_0}{\alpha\,\mathrm{Pr}^{3/4}}. \tag{26.1}$$

At $\mathrm{Re}>10^6$, the empirical correlation for $k_f$ is more complex:

$$k_f=0.0032+\frac{0.221}{\mathrm{Re}^{0.237}}, \tag{26.2}$$

and computations by formula (26.1) are somewhat more tedious. In principle, however, the general nature of the dependence of the diffusional flux on velocity remains unchanged. The flux against the inside surface is not a function of the distance along the tube, except in the inlet section, which extends for a distance of about 100 diameters.

It will be seen from formula (26.1) that the thickness $\delta$ of the diffusion layer in turbulent flow may be written in the following form:

$$\delta=\frac{1.41\alpha\,\mathrm{Pr}^{3/4}.\,D}{\sqrt{k_f}U_0}. \tag{26.3}$$

Comparing expressions (26.3) and (4.16) we see that

$$\delta\approx\frac{\alpha\delta_0}{\mathrm{Pr}^{1/4}}. \tag{26.4}$$

Thus, the thickness of the diffusion layer satisfies the inequality

$$\delta<\delta_0<R.$$

The meaning of this inequality is clear: due to the small value of the diffusion coefficient D, turbulent mixing ensures a constant concentration throughout the entire turbulent boundary layer and in part of the outer zone of the viscous sublayer. As shown below, it may in fact be assumed that $\delta$ constitutes 1/4 or 1/5 of $\delta_0$.

Analogous relations may be written for the case of a plate in a turbulent stream. The only difference in this case is variation of the friction factor $k_f$ and $v_0$ from point to point along the plate. The diffusional flux, therefore, given by formula (26.1), in which $k_f$ now refers to a plate, is found to be a weak function of the coordinate x along the plate.

In an experiment the total diffusional flow reaching the entire reaction surface is measured rather than the local flux,

$$I = \frac{bc_0 U_0}{1.41\alpha \, \mathrm{Pr}^{3/4}} \int_0^l \sqrt{k_f}\, dx. \tag{26.5}$$

The quantity $\sqrt{k_f}$ changes slowly with distance along the plate, as determined by the relation

$$\frac{1}{\sqrt{k_f}} = 4.1 \lg (k_f \cdot \mathrm{Re}) + 1{,}7. \tag{26.6}$$

The integral friction factor per unit length of plate is used together with the drag coefficient to form the law expressing resistance. The latter is determined as the ratio of the total force exerted on one side of the plate, to the pressure and the length of the plate, i.e.,

$$K_f = \frac{F}{\frac{\rho U_0^2}{2} \cdot l} = \frac{\int_0^l k_f\, dx}{l}.$$

It is possible to show [12], to a logarithmic accuracy, at large values of the Reynolds number that

$$K_f = k_f \left(1 + \frac{1}{\ln (\mathrm{Re}\, k_f \cdot 4{,}15)}\right), \tag{25.7}$$

The value of $k_f$ is obtained from equation (26.6), at $x = l$.

It is also possible to show that

$$\sqrt{K_f} = \sqrt{k_f} \left(1 + \frac{1}{\ln 4{,}15 k_f \, \mathrm{Re}}\right). \tag{26.8}$$

Therefore, within logarithmic accuracy

$$I = \frac{blc_0 \sqrt{K_f} U_0}{1.41\alpha \, \mathrm{Pr}^{3/4}} = \frac{\sqrt{K_f} c_0 U_0 S}{1.41\alpha \, \mathrm{Pr}^{3/4}}. \tag{26.9}$$

Thus, the total diffusional flow to the surface of a plate is found to be proportional to the area S, to $U_0$ and to $Pr^{-3/4}$. The value of the drag coefficient is determined from equation (26.7), and, at high values of the Reynolds number, is given approximately by the semi-empirical formula [13]

$$\sqrt{K_f} = \frac{1}{0{,}8 \ln \mathrm{Re}^{1.3}} \approx \frac{0{,}27}{\mathrm{Re}^{0.1}}. \tag{26.10}$$

With a turbulent regime in the boundary layer, the diffusional flow to a plate is dependent on velocity to a power somewhat smaller than unity (about 0.9). Formula (26.9) shows that, in contrast to the case of a laminar boundary layer [compare formula (16.3)], there is no analogy between the transfer of mass and momentum. However, there is an analogy involving the square root of the drag coefficient $K_f$. Under conditions of a turbulent boundary layer, equation (26.9) for the diffusional flow may be applied qualitatively to a body of arbitrary shape.

A direct quantitative verification of formula (26.9) for diffusional flow to a plate, and its application in actual practice, is made difficult by the following circumstance: a turbulent boundary layer does not immediately appear at the point of incidence of the stream on the body, but only at a certain distance downstream, where the laminar boundary layer that forms over the upstream portion of the body has had time to become turbulent.

Consequently, the total mass flow to the plate will be expressed by a certain combination of transport across the laminar layer [formula (15.13)] and transport across the turbulent layer. An exponent n therefore appears on the velocity $U_0$ in the expression for j. The value of n varies between 0.5 and 0.9.

As the Reynolds number and the degree of turbulence increase, the exponent n approaches 0.9.

This difficulty is identical with the well-known complication that arises in hydrodynamics in any attempt to verify formula (4.37).

The expression derived for I may be rewritten in the following dimensionless form:

$$\mathrm{Nu} = \frac{Il}{Dc_0 S} = \frac{\sqrt{K_f}}{1.41\alpha} \mathrm{Re}\, \mathrm{Pr}^{1/4}. \tag{26.11}$$

The numerical value of the constant $\alpha$ has been determined experimentally.

## 27. DIFFUSIONAL FLOW TO A ROTATING DISK

A rotating disk is very convenient for experimental verification of the theory set forth here as to the laws governing diffusional flow.

The turbulent boundary layer on the surface of such a disk has been studied in detail both experimentally and theoretically.

Such a layer appears on well-polished and centered disks at values of the Reynolds number

$$\mathrm{Re}=\frac{a^2\omega}{\nu}\approx 5\cdot 10^4.$$

Usually it appears on a disk at $\mathrm{Re}\sim 10^4$.

For determining the total diffusional flow, we can conveniently utilize equation (25.18), in which eddy velocity $v_0$ is a function of the distance r from the axis of rotation.

If a is the radius of the disk, then the total flow to one of its surfaces is

$$I=\int_0^{2\pi}\int_0^a jr\,dr\,d\varphi=\frac{2\pi c_0}{\alpha \mathrm{Pr}^{3/4}}\int_0^a v_0 r\,dr. \tag{27.1}$$

The eddy velocity $v_0$ may be removed from the integrand in formula (27.1) by means of the relations derived in the theory of the turbulent boundary layer on a rotating disk 14 .

It follows from the theory of a turbulent fluid on a rotating disk that the distance r from the axis of rotation, the eddy velocity $v_0$, the Reynolds number and the moment of drag M may be expressed in terms of a certain parameter $\xi$. At high Reynolds numbers, the corresponding expressions have the form

$$r=\left(\frac{\nu}{\omega}\right)^{1/2} e^{\xi/2}\xi^{1/2}\sqrt{3A^3C}, \tag{27.2}$$

$$v_0=\frac{\omega r}{A\xi}, \tag{27.3}$$

$$M=\frac{8\pi AC}{3}\,\nu\rho a^3\omega\,\frac{e^{\xi}}{\xi}, \tag{27.4}$$

where A and C are constants and the parameter $\xi$ is very large compared to unity. We introduce the dimensionless coefficient of the moment $K_M$, defined as

$$K_M=\frac{M}{\rho\omega^2 a^5},$$

which may be expressed in terms of $\xi$,

$$K_M=\frac{8\pi}{9A^2\xi^2}. \tag{27.5}$$

By means of equations (27.2) to (27.5), it is possible to compute the integral in formula (27.1). Substituting the eddy velocity $v_0$ from (27.3) in this expression, we obtain

$$I = \frac{2\pi c_0 \omega}{\alpha\, \mathrm{Pr}^{3/4} A} \int_0^a \frac{r^2\,dr}{\xi}. \qquad (27.6)$$

The expression in the integrand for the distance r from the axis of rotation may be replaced by its value at high Reynolds numbers, which is given by formula (27.2).

In further calculations we assume that, at Re ≫ 1, $\xi$ is also large.

From equations (27.2), (27.4), and (27.5) it is possible to derive

$$\frac{\omega}{A} \int_0^a \frac{r^2\,dr}{\xi} \approx \frac{\mathrm{Re}^{3/2}\sqrt{K_M}}{\sqrt{18\pi}} \cdot \omega^{-1/2} \nu^{3/2}. \qquad (27.6)$$

In deriving equation (27.7), we use the fact that $\xi \gg 1$ so that

$$\frac{\partial}{\partial \xi}\left(e^{\xi/2}\xi^{1/2}\right) \approx \frac{1}{2}\,\xi^{1/2} e^{\xi/2}$$

and

$$\int e^{3\xi/2}\xi^{1/2} d\xi \approx \frac{2}{3}\,\xi^{1/2} e^{3\xi/2}.$$

Therefore, finally,

$$I \approx \frac{2}{\sqrt{18\pi}} \frac{\sqrt{K_M} c_0 S\,(a\omega)}{\alpha\,\mathrm{Pr}^{3/4}}. \qquad (27.8)$$

A comparison of formula (27.8) with experimental data allows us to compute the numerical value of the constant $\alpha$ (see Section 43).

This formula is identical with (26.9) in its general form, and differs from it only in that $\sqrt{K_M}$ replaces $\sqrt{K_f}$ and a numerical coefficient $\frac{1}{\sqrt{2\pi}}$ has been introduced.

An approximate equation for the moment coefficient [14] is

$$K_M = \frac{0.1}{\left(\frac{a^2\omega}{\nu}\right)^{1/5}}. \qquad (27.9)$$

Therefore, finally,

$$I \approx \frac{0.01 c_0 S}{\alpha\,\mathrm{Pr}^{3/4}} (a\omega) \cdot \left(\frac{\nu}{a^2\omega}\right)^{1/10}, \qquad (27.10)$$

or, in dimensionless form

$$\mathrm{Nu} = \frac{Ia}{SDc_0} \approx \frac{0.01}{\alpha} \mathrm{Pr}^{1/4} \mathrm{Re}^{0.9}. \qquad (27.11)$$

## 28. DIFFUSIONAL FLOW TO THE SURFACE OF A NON-STREAMLINED BODY

We have confined the above treatment to the calculation of diffusional flow to a streamlined body.

Diffusion to non-streamlined bodies has certain features that differ from those discussed above.

Let us take as an example the diffusional flow to the surface of a cylinder whose axis is transverse to the stream. The total surface of the cylinder may be subdivided into two zones, for which the conditions for diffusion are quite different. The upstream portion of the cylinder is in a laminar regime in which a boundary layer is formed. The downstream zone, beyond the line of separation, is in a turbulent regime.

On the basis of the general considerations set forth in Section 9, it is clear that the diffusional flow to the upstream portion, to an order of magnitude, is no different from the diffusional flow to the surface of a plate, which has been determined previously. This is also confirmed by direct computations that may readily be made for the neighborhood of the point of incidence (forward stagnation point). Here, the solution of the boundary layer equations leads to the following expressions for the velocity components:

$$v_t = \frac{4U_0 x}{d}\eta,$$

$$v_n = -\left(\frac{4\nu U_0}{d}\right)^{1/2}\eta^2,$$

where $U_0$ is the velocity of the undisturbed flow far from the cylinder, d is the cylinder diameter, x is the distance from the point of incidence, and

$$\eta = \left(\frac{4U_0}{\nu d}\right)^{1/2}\frac{y}{2}.$$

The convective diffusion equation in the boundary layer, after the introduction of the expressions for $v_t$ and $v_n$ and the new variable $\eta$, takes the form

$$c'' + \Pr \eta^2 c' = 0. \tag{28.1}$$

Its solution, that satisfies the conditions

$$\begin{aligned} c &= c_0 \text{ at } y \to \infty, \\ c &= 0 \text{ at } y = 0, \end{aligned} \tag{28.2}$$

is

$$c = \frac{c_0 \int\limits_0^{\eta} \exp\left\{-\Pr \int\limits_0^{\eta} \eta^2 \, d\eta\right\} d\eta}{\int\limits_0^{\infty} \exp\left\{-\Pr \int\limits_0^{\eta} \eta^2 \, d\eta\right\} d\eta}. \tag{28.3}$$

The diffusional flux to the surface of the cylinder, consequently, is equal to

$$j = D\left(\frac{\partial c}{\partial y}\right)_{y=0} = D\left(\frac{4U_0}{\nu d}\right)^{1/2} \frac{c_0}{\int\limits_0^{\infty} \exp\left\{-\frac{1}{3} \Pr \eta^3\right\} d\eta}. \tag{28.4}$$

Since $\Pr \gg 1$, this equation may be approximated by

$$j \approx D\left(\frac{U_0}{\nu d}\right)^{1/2} \frac{c_0}{\Pr^{2/3}}. \tag{28.5}$$

The coefficient used in this equation may be computed accurately only in the neighborhood of the point of incidence. Within an order of magnitude, however, formula (28.5) is valid for the entire surface across which a fluid in laminar flow streams. The diffusional flow to the downstream surface of the cylinder is determined by formula (26.9). The total diffusional flow to the surface of the cylinder, therefore, may be represented as follows:

$$I = C_1 \frac{196}{360} \cdot \frac{\pi d^2}{4} \cdot D\left(\frac{U_0}{\nu d}\right)^{1/2} \frac{1}{\Pr^{2/3}} + \frac{164}{360} C_2 \frac{\pi d^2}{4} \cdot \frac{\sqrt{K_f}\, U_0 c_0}{\Pr^{3/4}}. \tag{28.6}$$

Here, the first term represents the flow to the upstream side of the cylinder (before the separation point), and the second is the flow to the downstream portion. While the constants $C_1$ and $C_2$ can not be determined precisely, there is no reason to believe that they are appreciably different from unity. Formula (28.6) shows that the diffusional flows to the upstream and downstream portions of the cylinder are comparable. As the velocity of the incident flow $U_0$ increases, however, the first term in (28.6) increases as $U_0^{1/2}$ and the second as $U_0^{0.9}$. It may be expected, therefore, that with increasing velocity, the relative value of the second portion of the diffusional flow will also increase. The diffusional flow to the downstream portion of the cylinder, at sufficiently high flow velocities, becomes greater than to the upstream portion.

Such a conclusion appears paradoxical, because the solution is not impoverished when it reaches the upstream face (or is not enriched, if the case involves dissolution). It should be remembered, however, that highly energetic agitation of the solution occurs in the turbulent zone and turbulence eddies embrace relatively large fluid volumes; and because beyond the separation line the stream of impoverished solution is directed away from the surface of the body toward the mainstream of the fluid, it does not pass over the downstream portion of the surface of the cylinder. Our conclusion regarding the distribution of the diffusional flow over the surface of a cylinder is very general in nature, and pertains to all non-streamlined bodies. Unfortunately, we do not have data at present to permit quantitative verification. We can only point to qualitative confirmation from heat transfer data for the surfaces of a cylinder and a sphere (see Section 56). This leads to the recommendation that non-streamlined bodies be used as reaction surfaces in cases where the greatest possible diffusional flow is desirable. Although the quantitative expressions for diffusional flow derived above have been determined for convective diffusion in liquids, they may also be used qualitatively to describe the diffusion in the gas phase. In particular, the results obtained may be utilized in an analysis of the evaporation of droplets. We note in this connection that, in principle, the result obtained contradicts the work of Academician Leibenzon [15], who computed the diffusional flux at the surface of a sphere immersed in a flowing gas ($Re \gg 1$). The calculation in this work applied only to the region preceding the point of separation. At the same time, the unfounded assumption was made that the diffusional flux downstream of the point of separation is the same as that upstream.

It should also be pointed out that an error crept into the calculations of the diffusional flux upstream of the separation point in the work cited above. The term $v_y \frac{\partial c}{\partial y}$ was omitted from the diffusion equation as being negligible. In actuality, it has the same order of magnitude as the term $v_x \frac{\partial c}{\partial x}$, which was retained. The simplest way to see this is by means of direct computation using formula (20) of Academician Leibenzon's work quoted above.

This circumstance led to inclusion of $U_0$ to an incorrect power in the formula that determines the diffusional flux upstream of the separation point.

## 29. DIFFUSIONAL FLOW TO A ROUGH SURFACE IN A LAMINAR STREAM. DISSOLUTION FROM THE CORNER OF A BODY

So far, in computing diffusional flows, we have assumed the reaction surface to be smooth.

A rough surface changes the nature of the flow past that surface very materially and, consequently, changes the size of the diffusional flow to the surface. By roughness we mean protrusions above the surface of a body of arbitrary shape. These protrusions are so widely separated from each other, that the average distance between them is large compared with their size. We characterize the latter dimension by the height h.

In examining the effect of protrusions on the nature of the flow past a surface, three cases of laminar and turbulent flow near the reaction surface must be distinguished.

Let us first examine the case of laminar flow. We begin by assuming that the size of the protrusion is large compared with the hydrodynamic boundary layer $\delta_0$. These protrusions then rise above the boundary layer and the flow past them will be the main-stream, unrestrained by proximity to the surface.

Let us construct the Reynolds number for the flow past the protrusion. It is, evidently, $Re_{pro} = \frac{U(h)\,h}{\nu}$. If $Re_{pro}$ exceeds 20 to 50, the flow past the protrusion, as in the case of any other non-streamlined body, will be accompanied by separation and the appearance of turbulence.

Figure 30 illustrates schematically the flow past a protrusion. On its downstream side, turbulence appears, while the upstream side is subject to laminar flow. The turbulent flow downstream from the protrusion is gradually damped, but this damping takes place over distances considerably greater than the dimension of the protrusion.

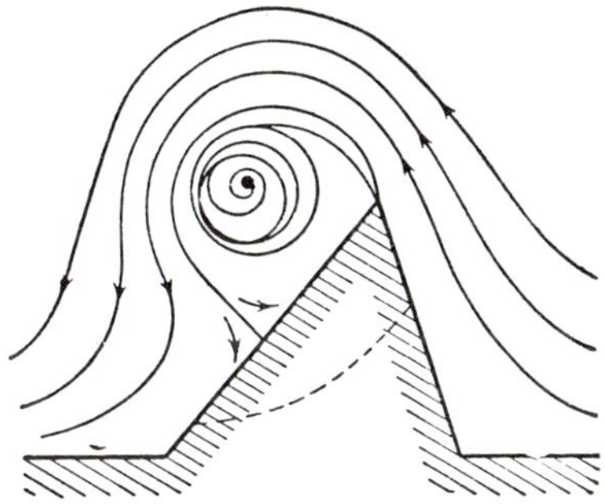

Figure 30. Flow past a protrusion.

The dotted line indicates the shape assumed by a dissolving protrusion.

Another example of unevenness is illustrated in Figure 31. Here, the pressure rises not only on the downstream side of the protrusion but also in front of it — in the region of the corner. The fluid is retarded and the pressure accordingly increases. As a result, the same separation occurs upstream from the protrustion and, as with the downstream side, turbulence appears.* As a result of the appearance of turbulence downstream from the protrusions, there

*Note added in proof. This effect is quantitatively examined in reference [21].

is more intense agitation of the fluid and an increase in the diffusional flow. The order of magnitude of the diffusional flow at an individual protrusion is determined by expression (28.6). Since the number, shapes and sizes of the protrusions may not be assumed to be exactly alike, only qualitative conclusions may be drawn:

1) The existence of protrusions contributes to the establishment of turbulence adjacent to the surface of a solid body, and promotes better fluid mixing conditions;

2) when bodies with rough surfaces dissolve in fluids, the dissolution of protrusions must proceed more rapidly than that of the rest of the body.

The dotted line on Figure 30 shows schematically the way in which a protrusion changes shape as it dissolves.

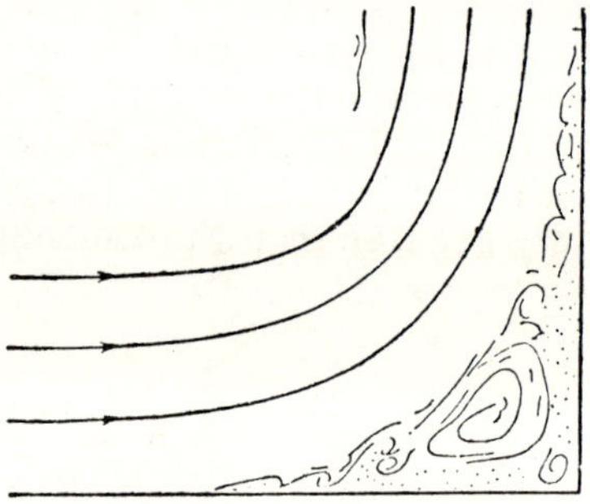

Figure 31. Appearance of turbulence in a corner.

The improved mixing conditions at protrusions causes their being leveled during dissolution. This general conclusion is applicable not only to the large protrusions examined above, but also to very small ones, the dimensions of which are minute in comparison with the thickness $\delta$ of the effective diffusion layer. This is shown in Figure 32, where the dotted lines represent the paths of the particles as they enter the solution both from the protrusion and from the smooth portion of the surface. In the latter case the paths

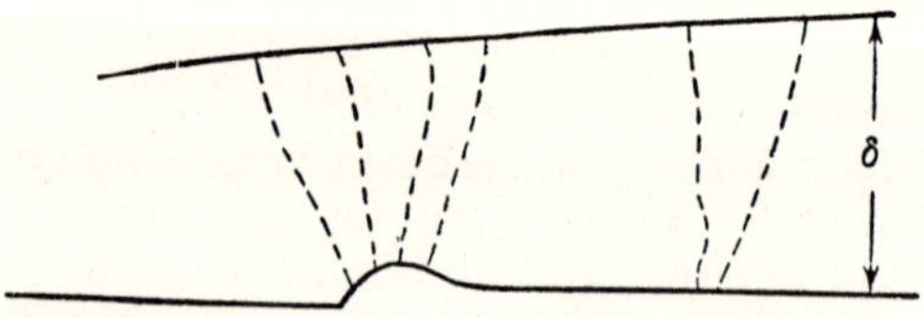


Figure 32. Paths of dissolving particles from a protrusion and from a smooth surface.

are longer, and the dissolution occurs more slowly. The dissolution of protrusions and the smoothing out of dissolving bodies is well known and is used in the so-called anode polishing of metals.

Let us now examine the problem of the dissolution of a convex or a concave corner of a body. In the case of a low rate of flow along one of the planes forming the side of a convex corner (Figure 33), the fluid stream turns smoothly around the corner, and flows past the second plane. At high Reynolds numbers, however, the picture changes considerably. The stream in this case leaves the solid surface and continues in its direction, forming a jet with motion relative to the stationary fluid mass. However, such a flow regime is unstable. The surface separating the moving and stationary fluid disappears and the laminar regime is supplanted by a turbulent regime. Figure 34 pictures the flow that actually appears at the point where the stream separates from the corner. The dotted lines in the boundary zone indicate the turbulent region (a turbulent jet). The arrows parallel to the plane OB indicate the fluid flowing into the turbulent zone.

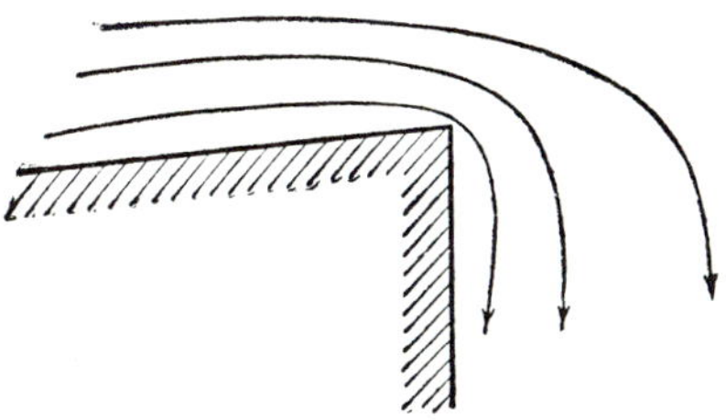

Figure 33. Fluid streamlines around a corner at small Reynolds numbers.

The results given here are useful in considering dissolution in a depression in a body past which a fluid is streaming. Let us assume that there is a depression in the dissolving body, the dimension h of which is considerably greater than the thickness of the boundary layer on the body.

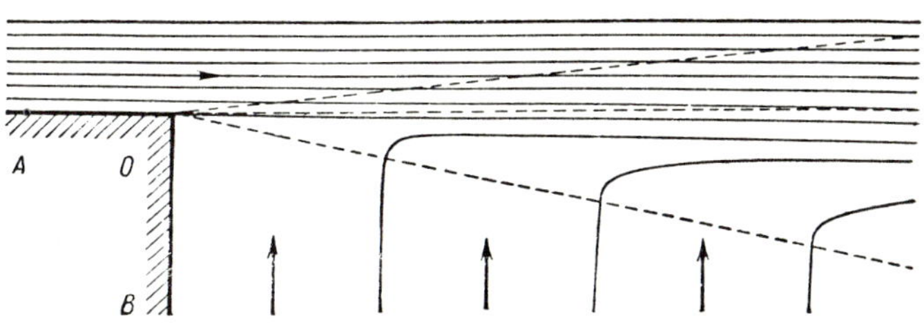


Figure 34. Fluid streamlines past a corner at high Reynolds numbers.

The turbulent zone is indicated by dotted lines.

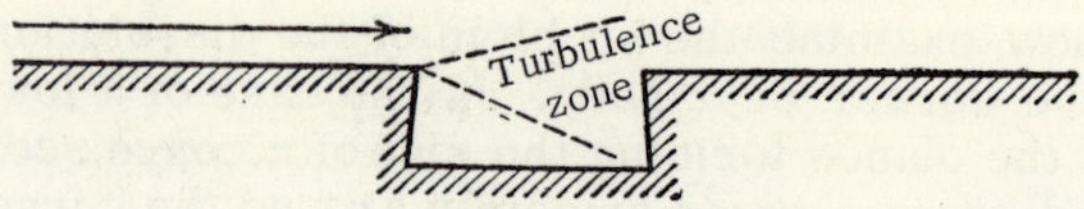


Figure 35. Formation of turbulence inside a depression.

If the condition $\frac{U_0 h}{v} \gg 1$ is satisfied, then an intense agitation of the fluid occurs in the depression along with the appearance of a turbulent zone. This appears to be the phenomenon that causes more rapid corrosion of metals in very rapidly flowing liquids and the appearance of cavitation. Cavitation eats into the surface of the metal creating noticeable depressions in it (of the order of 1 mm). The depressions, once created, are subjected to especially rapid dissolution from the above-mentioned effect. Accelerated dissolution must also occur at sufficiently high velocities past a concave surface (Figure 36). A stagnant zone first appears near the corner. Here, as usual, a reduction in the fluid velocity is accompanied by increased pressure. The latter prevents the flow of fluid in the boundary layer. As a result, the stream separates from the wall at a certain point O, and the flow becomes turbulent. Here, the "stagnant zone" at the top side of the corner becomes a zone of turbulent agitation. Dissolution in this zone, as a consequence, occurs more rapidly than dissolution along the plane AB. More intense dissolution (corrosion) of metals in corners is observed in practice. It should be noted that turbulence in the fluid at the top of the corner may appear not only in the case of forced convection, but in the case of natural convection as well.

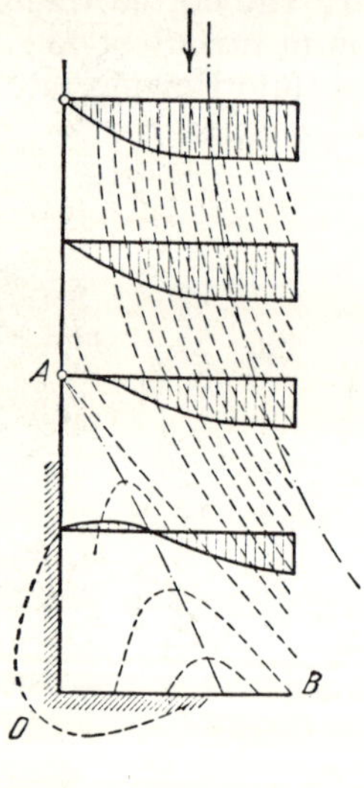


Figure 36. Flow near a corner.

Dotted lines indicate the streamlines. The dashed line curves indicate the velocity profiles. The corner zone is filled with turbulent fluid.

## 30. DIFFUSIONAL FLOW AT A ROUGH SURFACE FROM A TURBULENT STREAM NEAR THE SURFACE

Let us now examine the case of turbulent flow near a solid surface.

If the dimensions of the protrusions are large compared to the thickness d of the boundary layer, the flow past them is the main stream, unrestrained by proximity to the surface, as in the case of individual non-streamlined bodies.

Of great practical interest, however, is the case of protrusions whose height h is small compared to d, but large compared to the thickness $\delta_0$ of the viscous sublayer (i.e., $\delta_0 \ll h \gg d$).

The presence of protrusions has a substantial effect on the nature of the turbulent flow. Namely, the fluid is in turbulent motion around the protrusions, and this motion may be characterized by the quantities $\rho$, $\tau$ and h (but will not be a function of the viscosity $\nu$, since, according to assumption, $Re_{pro} = \frac{v_0 h}{\nu} \gg \frac{v_0 \delta_0}{\nu} \approx 1$). It is clear from dimensional considerations that the velocity of this motion, to an order of magnitude, is $v_0 \approx \sqrt{\frac{\tau}{\rho}}$. Thus, for turbulent flow past a rough surface, as contrasted with that along a smooth surface, the velocity becomes small (of the order of $v_0$), at $y \approx h$, rather than at a distance equal to the thickness of the viscous sublayer $y \approx \delta_0$.

The scale of turbulence eddies, therefore, may be written

$$l \sim y + h.$$

Here, the logarithmic distribution law of the average velocity near the rough surface takes the form

$$u = v_0 \ln \frac{y+h}{h}.$$

By introducing a new variable $y' = y + h$, we may rewrite the expression for u in the form

$$u = v_0 \ln \frac{y'}{h}. \tag{30.1}$$

Here, to avoid new notation, we designate the new coordinate by y'.

With a change in velocity distribution, there is a corresponding change in the drag law. A simple calculation [16] shows that the drag coefficient, in the case of a rough plate, is not a function of the Reynolds number and is determined by the formula

$$\sqrt{\frac{0{,}34}{K_f^{\text{rgh}}}} = \ln\left(\frac{x}{h}\sqrt{K_f^{\text{rgh}}}\right). \tag{30.2}$$

Indeed, eliminating v from formulas (4.13) and (30.1) for the logarithmic profile adjacent to a rough surface, and also from the general expression (4.35) for the thickness of the turbulent boundary layer d, we arrive at formula (30.2).

Since the drag coefficient in a fluid streaming past a rough surface (of a plate) is not a function of Re, the drag is exactly proportional to the square of the velocity $U^2$.

The hydrodynamic formulas (30.1) and (30.2) agree well with actual practice. They are insufficient, however, for resolving the problem that interests us — a relation for the diffusional flow. To derive the diffusional flux when $Pr \gg 1$, the distribution of velocities in the zone $y < h$ must be examined. The velocity distribution in this zone has not previously been of interest in hydrodynamics and has not been studied. Let us first examine the case of large protrusions, for which $Re_{pro} = \frac{hv_0}{v} \gg 1$. For such protrusions it is natural to assume that in the zone $y \approx h$ there is turbulent motion, the scale of which is determined by the dimension of the protrusion h, i.e.,

$$l \approx h. \tag{30.3}$$

The following arguments in favor of such an assumption may be advanced. A separation in the flow at the protrusions occurs, since $Re_{pro} \gg 1$. The motion in the region $y \approx h$, therefore, must be turbulent. The nature of the motion in this zone can not be a function of the viscosity v, but must be a function of the dimension h of the protrusion. Here, the only dimensional quantity (cm) is the dimension h itself; hence, the scale of turbulence eddies must be proportional to h. This may be visualized as follows: the larger the dimensions of the protrusions, the larger is the scale of the eddies that separate from it. The scale of these eddies must be such that they fill the entire zone equal in order of magnitude to the protrusion (i.e., the zone of the turbulent wakes downstream from the protrusion). If we were to assume in the zone $y \lessdot h$ that the scale of motion is $l \approx y$, we would obtain a logarithmic velocity profile and would have an incorrect result, for we would find that the velocity distribution near a rough surface has the same form as for a smooth surface.

All of the above considerations render the hypothesis (30.3) clear and plausible. However, since any law of the form $l \approx h^n y^{1-n}$ cannot be set aside the problem can only be resolved on the basis of experimental data.

Having adopted the hypothesis (30.3), we may, by the use of general formulas (4.5) and (4.8), formulate an expression for an effective coefficient of turbulent viscosity in the zone $y \gtrless h$, and for the corresponding shear stress, as follows:

$$\nu_{turb} \approx h^2 \frac{\partial U}{\partial y}, \tag{30.4}$$

$$\tau \approx \rho \nu_{turb} \frac{\partial U}{\partial y} \approx \rho h^2 \left(\frac{\partial U}{\partial y}\right)^2. \tag{30.5}$$

The distribution of the average velocity of a turbulent flow in the zone $y \leqslant h$ is derived from (30.5) with the condition

$$U \approx v_0 \text{ at } y = h. \tag{30.6}$$

Integrating expression (30.5), expressing $\tau$ in terms of $v_0$, and taking condition (30.6) into account, we obtain

$$U = v_0 \frac{y}{h}. \tag{30.7}$$

The velocity distribution (30.7) is valid, but not for all $y < h$. It is valid only when the corresponding Reynolds number $\frac{U(\delta_0) \cdot \delta_0}{\nu}$ is greater than or equal to 1. Thus, a viscous sublayer always appears in the fluid at the wall itself. The thickness $\delta_0$ of the viscous sublayer is determined by the condition

$$\frac{U(\delta_0) \cdot \delta_0}{\nu} = \frac{v_0 \delta_0^2}{h\nu} \approx 1,$$

whence

$$\delta_0 \approx \left(\frac{\nu h}{v_0}\right)^{1/2} = \frac{h}{\sqrt{\mathrm{Re}_{\mathrm{rgh}}}}. \tag{30.8}$$

The velocity distribution (30.7) in a turbulent flow is valid at distances from the wall that satisfy the inequality $\delta_0 \leq y \leq h$. At smaller values of y in the viscous sublayer, the velocity distribution is determined by the linear law (4.31), but with a coefficient that is a function of the molecular viscosity.

Knowing the distribution of average velocities, we can now turn to an examination of the problem of finding the distribution concentration. This problem is of special interest to us. The distribution must satisfy the same conditions as in Section 25:

$$\begin{aligned} c &= c_0 \quad \text{at } y \approx d, \\ c &= 0 \quad \text{at } y = 0. \end{aligned} \tag{30.9}$$

In the zone $d > y > h$, the concentration distribution is given by the formula (24.11) derived for a smooth surface. In the zone $\delta_0 < y < h$, on the basis of general formulas (24.3) and (24.4), we may derive the following expression for an effective coefficient of turbulent diffusion:

$$D_{\text{turb}} \approx h^2 \frac{\partial U}{\partial y} \approx v_0 h, \tag{30.10}$$

Therefore, for concentration in that zone, we have

$$c = \frac{j}{v_0 h} y + \text{const.} \tag{30.11}$$

In the zones $\delta < y < \delta_0$ and $0 > y < \delta$, the concentration distribution is given by the formulas derived in Section 25. Let us make use of the approximation

$$j \approx \frac{Dc_0}{\delta} \approx \frac{Dc_0 \Pr^{1/4}}{\delta_0}, \tag{30.12}$$

derived earlier, for diffusional flux to a plate. At $\Pr \gg 1$, where the diffusional flux is controlled by mass transfer across the viscous sublayer, a change in the law of distribution of concentration outside the viscous sublayer is not important, to a first approximation. Of far greater importance is a change in the thickness of the viscous sublayer itself, and of the quantity $v_0$. Substituting the value of $\delta_0$ in formula (30.12) from expression (30.8), we obtain

$$j = \frac{Dc_0 v_0^{1/2} \Pr^{1/4}}{\nu^{1/2} h^{1/2}} \approx \frac{c_0 D^{3/4} v_0^{1/2}}{\nu^{1/4} h^{1/2}}. \tag{30.13}$$

By making use of the relation $v_0 = \sqrt{\frac{K_f^{\text{pro}}}{2}}\, U$, let us rewrite (30.13) in the form

$$j \approx \frac{\sqrt[4]{K_f^{\text{rgh}}}\, U^{1/2} D^{3/4} c_0}{\nu^{1/4} h^{1/2}}. \tag{30.14}$$

It is found that the diffusional flux is proportional to the square root of the fluid velocity.

In dimensionless form, expression (30.14) may be written as follows:

$$\mathrm{Nu}_{\text{rgh}} \approx \frac{jl}{Dc_0} \approx \sqrt[4]{K_f^{\text{rgh}}}\, \mathrm{Re}^{1/2} \Pr^{1/4} \left(\frac{l}{h}\right)^{1/2}. \tag{30.15}$$

Formula (30.15) shows that in the case of a rough surface, the dimensionless diffusional flow (per unit of actual surface) is proportional to $\mathrm{Re}^{1/2}$, $\Pr^{1/4}$ and $\left(\frac{l}{h}\right)^{1/2}$.

It should be pointed out here that formula (30.15) errs in giving only the flow of matter at the "valleys" below the "peaks" — the protrusions. Its derivation does not take account of the fact that the protrusions themselves, extending into the boundary layer, are in a somewhat favored position from the standpoint of diffusion. Formula (30.15), therefore, gives a somewhat low value for the diffusional flux. It should also be borne in mind that the total true surface of a rough body, in the case of a large number of protrusions, may greatly exceed the apparent surface. In computing the total diffusional flow, the value determined from formula (30.15) should be multiplied by the true surface. For a small number of protrusions, the difference between the true and apparent surface is insignificant, but, in this case, formula (30.15) loses its meaning. In the case of a very small number of protrusions per unit surface, the velocity distribution is not altered.

Along with the case of large protrusions, just examined, the case of small protrusions $Re_{pro}=\frac{hv_0}{\nu}\sim 1$ should also be examined. In this case it is natural to assume that the spectrum of turbulence eddies stops at a level $y\sim h$ from the wall. The gradual damping that would take place at a smooth wall is distorted by the presence of the protrusions. As a limiting case we may examine the case of total cessation of flow in the depressions between protrusions. This limiting case leads to a mass transfer relation different from that derived above. That is, by assuming the fluid to be static, we may take the distribution (30.11) for the zone $0\leqslant y\leqslant h$ and match it with the logarithmic distribution (24.10). This gives

$$j=\alpha\gamma v_0 c. \tag{30.16}$$

It should be stressed, however, that this relation applies only to the depressions between protrusions. When integrating over the entire surface, the portion directly covered with protrusions must be accounted for. On the whole, it is to be expected that the diffusional flux at a surface covered by small protrusions may be expressed by a formula intermediate between that for a smooth surface (26.9) and equation (30.16). The diffusional flow will be proportional to $U^n$, where n is somewhat closer to 1 than is the case for a smooth surface. The derivation of a quantitative formula, however, is difficult, since it requires an exact profile of the surface.

Available experimental data does not permit convincing comparisons to be made of the derived formulas with experiments. This would undoubtedly be of interest. Let us note in conclusion that, in the derivation of formula (30.15), the assumption that $Pr\gg 1$ is very important. In the case of $Pr\approx 1$, transfer processes outside the viscous sublayer assume greater significance.

## 31. CONCERNING THE ANALOGY BETWEEN CONVECTIVE DIFFUSION AND SURFACE FRICTION IN TURBULENT FLOW

We have repeatedly pointed out the analogy between momentum transfer and mass transfer in a moving fluid.

Mass transfer of substance may be interpreted as transfer of both matter and heat. The latter is discussed later. At the outset, it should be stressed that in cases where a definite, but approximate, solution of the mass transfer problem can be obtained, it is preferable to use it rather than rely on the imperfect analogy between momentum and substance transfer.

The danger in the direct use of this analogy may be clearly illustrated by the example of flow past a non-streamlined body, which was examined in Section 28.

We saw that in the vicinity of the upstream critical point (the point of incidence of the flow), the diffusional flux was found to be constant [cf. (28.5)]. If we compute the drag in this zone, however, we find that it is zero at the point of incidence, but in the vicinity of that point it increases with distance downstream. Thus, no analogy exists between the transfer of momentum and of mass.

We now discuss this analogy in the case of turbulent fluid motion, to which it is considered applicable.

Let us write the fluxes of mass and momentum at a certain infinite plane in the form

$$j = (D + D_{\text{turb}})\frac{\partial c}{\partial y} = (D + \beta\overline{lv})\frac{\partial c}{\partial y}, \tag{31.1}$$

$$\tau = \rho(\nu + \nu_{\text{turb}})\frac{\partial u}{\partial y} = \rho(\nu + \alpha\overline{lv})\frac{\partial u}{\partial y}, \tag{31.2}$$

where $l$ is the scale of the eddies and v is the eddy velocity. The line above the $l$v indicates average values. The quantities $\alpha$ and $\beta$ are constants that, in general, are not the same. Outside the viscous sublayer, relations (31.1) and (31.2) yield a logarithmic profile with constants of differing values.

The concept regarding similarity in the concentration and velocity distributions in turbulent flow was first developed by Reynolds for application to temperature distribution. This is the so-called Reynolds analogy. We have seen that there is no theoretical basis for the assumption that this analogy is valid even when $\Pr = 1$.

Therefore, even when $\Pr = 1$ as is evident from the above formulas, there is no complete identity between the distributions of velocity and concentration.

Only in the unlikely case of $\alpha = \beta$, is there complete similarity of these fields.

At $\Pr \neq 1$, the situation is even more complex.

At $\Pr \approx 1$, the distributions of velocity and concentration may be written in approximate form

$$\left.\begin{aligned} v &= v_0 f\left(\frac{y}{\delta_0}\right), \\ c &= \frac{j}{c_0 v_0}\varphi\left(\frac{y}{\delta_0}\right), \end{aligned}\right\} \tag{31.3}$$

where f and $\varphi$ are different functions of the same argument $\frac{y}{\delta_0}$ ($\delta_0$ is the thickness of the viscous sublayer).

In some papers (e.g., see [16]) this hypothesis has been assumed to be universal. Actually, the relation (31.3) itself assumes that the mechanisms of momentum and mass transfer have a common nature.

We have seen, however, that at $Pr \gg 1$, this common nature is nonexistent.

At $Pr \gg 1$, not only is the similarity between concentration and velocity distributions destroyed, but, even more important, there exist zones in which matter and momentum are subject to different transfer mechanisms. It is no longer possible to assume, therefore, that the distributions of velocity and concentration are functions, albeit different ones, of the same argument $\frac{y}{\delta_0}$.

Thus, it is clear from the relations derived in Sections 4 and 24 that, in the outer region of the boundary layer, $\nu = f\frac{y}{\delta_0}$, and $c = \frac{j}{c_0\nu_0}\varphi\frac{y}{\delta_0}$ are valid. In a region closer to the wall, v, as before, is determined only by the value of $f\frac{y}{\delta_0}$, while c is determined not only by $\frac{y}{\delta_0}$, but also by $\frac{\delta}{\delta_0}$. Some authors, therefore, have determined temperature and concentration functions $\varphi\frac{y}{\delta_0}$ based on experimental velocity distributions $f\frac{y}{\delta_0}$. On the basis of the foregoing, however, these prove to be inapplicable for $Pr \gg 1$.

It is often of interest to compare diffusional (or heat) flux with drag.

In the case of streamlined bodies in laminar flow, as we have seen in Section 16, a proportionality exists between these two quantities.

The situation is somewhat more complex in the case of turbulent flow.

If the distributions of velocity and concentration are similar in nature, it is evident that a proportionality will exist between j and $\tau$.

The assumption of a difference in the values of the coefficients $\alpha$ and $\beta$ destroys this proportionality.

We confine our discussion to the case of $Pr \gg 1$, which is of primary interest to us. In the simplest case of an infinite plate, the diffusional flux may be compared to drag.

Employing (4.38) and (4.13), we have

$$j \sim \sqrt{k_f}\frac{U_0 c_0}{Pr^{3/4}} \sim \sqrt{\frac{\tau}{\rho}}\frac{c_0}{Pr^{3/4}},$$

where $\tau$ is the shear stress that is a function of the distance from the edge of the plate. By means of (26.9), we may write an analogous relation between the average flow I/S and the average shear stress F/S:

$$\frac{I}{S} \sim \sqrt{k_f} \frac{U_0 c_0}{\mathrm{Pr}^{3/4}} \sim \sqrt{\frac{F}{\rho S}} \frac{c_0}{\mathrm{Pr}^{3/4}}.$$

This relation is valid only to a logarithmic degree of accuracy.

We see that I is proportional to $\sqrt{F}$, rather than to F. in addition, the proportionality coefficient is a significant function of Pr.

These observations indicate the impropriety of a quantitative analogy between convective diffusion and fluid friction at $\mathrm{Pr} \gg 1$.

## 32. MOTION OF PARTICLES SUSPENDED IN A TURBULENT FLOW

In a series of problems that are discussed below (see Sections 33 and 41), the motion of particles suspended ("wandering") in a turbulent fluid is of major interest.

We will assume that the number of such particles in a unit of fluid volume is sufficiently small that their effect on the motion of the fluid may be disregarded.

The suspended particles are entrained by turbulence eddies and describe complicated trajectories in the fluid. We assume the turbulent flow to be homogeneous and isotropic. For future purposes we require certain additional characteristics of the velocity and acceleration fields of a homogeneous and isotropic flow.

In Section 4 we examined the change in the velocity over a distance $\lambda$, given by the formula (4.9)

$$v_\lambda \approx \left(\frac{\varepsilon \cdot \lambda}{\rho}\right)^{1/3} \quad \text{at} \quad \lambda > \lambda_0. \tag{32.1}$$

The quantity $v_\lambda$ also represents the velocity of turbulence eddies of scale $\lambda$. The velocity $v_\lambda$ decreases with $\lambda$ according to the "one-third" law.

The characteristic period $T_\lambda$ corresponds to motion of scale $\lambda$ ($\lambda > \lambda_0$) occurring at velocity $v_\lambda$. The period $T_\lambda$ can be a function only of the scale of motion $\lambda$ and the characteristics of the turbulent flow $\left(\frac{\epsilon}{\rho}\right)$. Since $\lambda > \lambda_0$, $T_\lambda$ cannot be a function of the fluid viscoity.

Only one combination may be made up of the quantities $\left(\frac{\epsilon}{\rho}\right)$ and $\lambda$ having the dimension of time:

$$T_\lambda = \left(\frac{\lambda^2 \rho}{\varepsilon}\right)^{1/3} \sim \frac{\lambda}{v_\lambda}. \tag{32.2}$$

This period increases with the scale of motion as $\lambda^{2/3}$, so that large-scale eddies have the greatest periods. Let us find the change of velocity with time of a given volume of fluid that is in motion.

It is evident that the eddy velocities may be written in the form

$$v_\lambda = \frac{d\lambda}{dT} \sim \frac{\lambda}{T_\lambda}. \tag{32.3}$$

Let us now determine the acceleration of eddies of scale $\lambda$. Acceleration may be written in the form

$$w_\lambda = \frac{dv_\lambda}{dT} \sim \frac{v_\lambda}{T_\lambda} \sim \frac{\lambda}{T_\lambda^2} \sim \left(\frac{\varepsilon}{\rho}\right)^{2/3} \cdot \frac{1}{\lambda^{1/3}}. \tag{32.4}$$

We see that as the scale of the eddies decreases, their acceleration increases.

Small-scale eddies have large accelerations, and the largest acceleration occurs in eddies with a scale equal in order of magnitude to the scale of the inner turbulence zone $\lambda_0$. Formula (32.4), of course, is not applicable to an eddy acceleration of scale $\lambda \sim \lambda_0$, since viscosity effects become significant in such eddies. Eddies of scale $\lambda \sim \lambda_0$ have a velocity equal in order of magnitude to

$$v_{\lambda 0} \sim \frac{\nu}{\lambda_0}, \tag{32.5}$$

because $\mathrm{Re}_{\lambda 0} \sim 1$. The period for eddies of scale $\lambda_0$, to an order of magnitude, is equal to

$$T_{\lambda 0} \sim \frac{\lambda_0}{v_{\lambda 0}} \sim \frac{\lambda_0^2}{\nu} \sim \left(\frac{\nu}{\varepsilon_0}\right)^{1/2}, \tag{32.6}$$

where $\epsilon_0 = \frac{\epsilon}{\rho}$.

The acceleration of turbulence eddies of scale $\lambda_0$ equals

$$w_{\lambda 0} \sim \frac{\partial v_{\lambda 0}}{\partial t} \sim \frac{v_{\lambda 0}}{T_{\lambda 0}} \sim \frac{v_{\lambda 0}^2}{\lambda_0} \sim \frac{v_{\lambda 0}^3}{\nu} \sim \frac{\varepsilon_0^{3/4}}{\nu^{1/4}}. \tag{32.7}$$

Formula (32.7) is accurate to an order of magnitude. On the basis of A. I. Kolmogorov's theory of local isotropic turbulence A. M. Yaglom [17] obtained an exact expression for $W_{\lambda 0}$, which differs from (32.7) by the factor $\sqrt{3}$:

$$w_{\lambda 0} = \sqrt{3}\, \frac{\varepsilon_0^{3/4}}{\nu^{1/4}}. \tag{32.8}$$

In the zone $\lambda < \lambda_0$, the motion becomes quasi-viscous in nature. The quadratic terms of the Navier-Stokes equation, at the scale of the inner turbulence zone, are small compared to the linear terms. Therefore, the interaction between separate turbulence eddies ceases, and each eddy is damped independently of the others.

Eddy periods are no longer a function of the scale ($\lambda < \lambda_0$) of motion, and maintain the value (32.6). Turbulence eddy velocities are a function of the scale $\lambda$ according to the relation

$$v_\lambda \sim \frac{v_{\lambda_0}}{\lambda_0} \cdot \lambda \sim \frac{\lambda}{T_{\lambda_0}}. \tag{32.9}$$

Correspondingly, the acceleration in the inner turbulence zone is

$$w_\lambda \sim \frac{v_\lambda}{T_{\lambda_0}} \sim \frac{v_{\lambda_0}^2}{\lambda_0^2} \cdot \lambda. \tag{32.10}$$

Accelerations, as well as velocities, decrease linearly with $\lambda$.

Thus, turbulence eddies of scale $\lambda \sim \lambda_0$ have the highest accelerations. They also have the shortest period $T_{\lambda_0}$.

On the basis of the formulas derived here, it is possible to study the motion of particles suspended in a turbulent fluid.

## 33. DIFFUSION TO PARTICLES SUSPENDED IN A TURBULENT FLUID. THE ELEMENTARY EXTRACTION PROCESS

An important category of heterogeneous physicochemical reactions is that of processes at the surface of small particles suspended in a turbulent fluid. Many technological processes require the dissolution of small particles in a fluid violently agitated by means of rotating mixers. Removal of the soluble constituent from solid or fluid bodies by dissolution in a fluid is termed extraction. In particular, if water is used for extraction from a solid body, the process is known as leaching. During extraction the soluble constituent passes from the suspended solid or liquid particles into the dissolving medium. It is clear that the usual dissolution process is a limiting case of extraction, since no soluble residue remains. The complete removal of soluble matter from the solution is accomplished by means of distillation or crystallization, and constitutes another process.

We confine our examination to extraction from a suspension of fine solid particles. The problem of extraction from liquid particles is broached in Section 72. Extraction from large particles lying motionless at the bottom of an apparatus and washed by a flowing solution is one instance of the process of dissolution of nonstreamlined particles.

Extraction from extremely finely ground particles which are suspended in the dissolving medium is often encountered in hydrometallurgy. As examples, we may cite the extraction of gold from ore by means of cyanide solutions, the formation of aluminum solutions from bauxite treated with sulphuric acid, etc. [18]. The extraction of sugar from sugar beets is the principal process for the production of sugar.

The idea behind the technological development of the extraction process is that the dispersed solid particles are held for more or less lengthy periods of time in a strongly agitated dissolver. The particle surfaces are washed by the dissolver, and the soluble portion migrates from the particle into the solution. The slow stage of the process is the dissolving stage, which depends on the rate at which solute is carried away from the solid surface into the body of the solution. Dissolution is from particles of relatively large dimensions — considerably larger than the scale of the inner turbulence zone.

Another example of a heterogeneous reaction at the surface of particles suspended in a turbulent flow is that which takes place on the surface of finely ground particles of catalyst. Such reactions include the process of hydrogenating fats. S. Yu. Elovich and G. M. Zhabrova [19] examined the enormous amount of experimental data available in the field of fat hydrogenation, and established the fact that the process may proceed both in the kinetic and in the diffusion stages. For a given catalyst concentration and given agitation conditions, the slow stage of the process is the diffusion of hydrogen, which had dissolved in the fluid, to the surface of the catalyst particles.

In hydrogenation and other catalytic reactions in the liquid phase, the average catalyst particle dimensions vary within wide limits ($10^{-2}$ to $10^{-5}$ cm).

Thus, widely varying dimensions of suspended particles are encountered in actual practice.

Usually the liquid is agitated by some type of mixer which produces fully developed turbulence. The maximum velocities of turbulence eddies, in orders of magnitude, are equal to the maximum velocity of the mixer: $U_0 \approx \omega R$, where $\omega$ is the angular velocity and R is the radius of the mixer. The greatest scale of turbulence eddies is $l \approx R$.

Large-scale eddies entrain the particle together with portions of fluid adhering to it, and transfer both as a single unit.

However, if the density $\rho$ of the particle differs from the density $\rho_0$, of the fluid, the entrainment can not be complete. The smaller-scale fluid motions are unable to entrain the particle, and, in relation to them, the particle acts as a motionless solid body. The fluid participating in these small-scale motions flows over the surface of the particle, and an agitation of a corresponding scale appears in the fluid adhering to the surface of the particle. Intermediate-scale eddies do not completely entrain the particle in their motion.

It is clear from what has been said above that the problem of the entrainment of particles by turbulence eddies (when the density of the particle differs from that of the fluid) must be examined separately.

Let us formulate an equation for the motion of the fluid and the particle. Let us designate by F the force exerted by the fluid on the particle. The particle and the fluid form a closed system. Let $\mathbf{v}_1$ be the velocity vector of the fluid where the particle is located.

If the particle were completely entrained by the fluid, so that $\mathbf{v}_1 = \mathbf{v}_0$, then the particle would experience the same force as would fluid enclosed in the same volume, i.e., the force would be $\rho_0 V \frac{d\mathbf{v}_0}{dt}$, where $\rho_0$ is the density of the fluid and V is the volume of the particle.

In a case of partial entrainment, the fluid flows past the particle which, in this case may be considered moving relative to the fluid at the velocity $\mathbf{u} = \mathbf{v}_1 - \mathbf{v}_0$. It thus experiences the drag $\mathbf{F}_{drag}$. Therefore, the total force f, the force exerted on a particle partially entrained by the fluid, is

$$\mathbf{f} = \mathbf{F} + \mathbf{F}_{drag} = \rho_0 V \frac{d\mathbf{v}_0}{dt} + \mathbf{F}_{drag}.$$

The equation of motion of the particle is

$$\rho V \frac{d\mathbf{v}_1}{dt} = \rho_0 V \frac{d\mathbf{v}_0}{dt} + \mathbf{F}_{drag}. \tag{33.1}$$

Introducing relative velocity, we have

$$\rho V \frac{d\mathbf{u}}{dt} = (\rho_0 - \rho) V \frac{d\mathbf{v}_0}{dt} + \mathbf{F}_{drag}. \tag{33.1'}$$

Since the dimension a of the particle is large compared to $\lambda_0$, when its motion in the fluid is at velocities u greater than $v_{\lambda_0}$, the corresponding Reynolds number is $Re = \frac{ua}{\nu} \gg 1$. Therefore, $F_{drag}$ may be written as follows:

$$\mathbf{F}_{drag} = -K_f \rho_0 S u \mathbf{u},$$

where S is the area of the particle cross section.

In the same way, for motion of a particle at $Re \gg 1$, we have

$$\rho V \frac{du}{dt} = (\rho - \rho_0) V \frac{dv_0}{dt} - K_f \rho_0 S u^2 \tag{33.2}$$

(since the vectors $\mathbf{u}$ and $\mathbf{v}_0$ are parallel).

In formulating equation (33.2), we omitted a term characteristic of the supplemental reaction of the body caused by its motion relative to the fluid. It is known [20] that the latter is proportional to the relative acceleration and to the density of the medium. Estimates show that in the case of a medium of low density (as compared to the particle) — and this, in essence, is precisely our case and the one that will continue to interest us — this term is small. The acceleration of the particle $\frac{du}{dt}$, may be estimated as

$$\frac{du}{dt} \sim \frac{u}{T^*}, \tag{33.3}$$

where T* is the period of motion of the particle corresponding to the velocity u and the scale of motion $\lambda$ of the particle under the influence of the entraining turbulence eddy, i.e.,

$$T^* \sim \frac{\lambda}{u}. \tag{33.4}$$

Therefore, the particle acceleration may be written as

$$\frac{du}{dt} \sim \frac{u^2}{\lambda}. \tag{33.5}$$

Acceleration of an eddy of scale $\lambda$ is given by formula (32.4). Introducing (33.5) and (32.4) in the equation of motion of the particle (33.1), we obtain

$$\rho V \frac{u^2}{\lambda} \sim (\rho - \rho_0) V \left(\frac{\varepsilon}{\rho_0}\right)^{2/3} \frac{1}{\lambda^{1/3}} - K_f \rho_0 S u^2. \tag{33.6}$$

Whence we find the velocity of relative motion to be

$$u \sim [(\rho - \rho_0) V]^{1/2} \frac{\varepsilon^{1/3} \lambda^{1/3}}{\rho_0^{1/3} [\rho V + K_f \rho_0 S \lambda]^{1/2}}. \tag{33.7}$$

At a certain value of $\lambda$, the relative velocity has a maximum value $u_{max}$. The corresponding value for $\lambda$ is derived from the condition

$$\frac{\partial u}{\partial \lambda} = 0.$$

A simple calculation gives the scale

$$(\lambda)_{u_{\max}} = \frac{2}{K_f} \frac{\rho}{\rho_0} \frac{V}{S}, \tag{33.8}$$

and the maximum velocity

$$u_{\max} \sim \frac{1}{\sqrt{3}} \left(\frac{\rho - \rho_0}{\rho}\right)^{1/2} \left(\frac{V}{S}\right)^{1/3} \left(\frac{\varepsilon}{\rho_0}\right)^{1/3} \left(\frac{\rho}{\rho_0}\right)^{1/3} \left(\frac{2}{K_f}\right)^{1/3}. \tag{33.9}$$

If the density of the particle is large compared to that of the medium as is the case, for example, in the motion of solid particles or droplets in a gas, then $\rho \gg \rho_0$, and we have

$$u_{\max} \sim \frac{1}{\sqrt{3}}\left(\frac{\rho}{\rho_0}\right)^{1/3}\left(\frac{\varepsilon}{\rho_0}\right)^{1/3}\left(\frac{V}{S}\right)^{1/3}\left(\frac{2}{K_f}\right)^{1/3}. \tag{33.10}$$

In the case of spherical particles of radius a, equation (33.10) becomes

$$u_{\max} \sim \frac{1}{3^{5/6}\sqrt[3]{K_f}}\left(\frac{\rho}{\rho_0}\right)^{1/3}(2a)^{1/3}\left(\frac{\varepsilon}{\rho_0}\right)^{1/3}. \tag{33.11}$$

The numerical coefficient in (33.11), probably has no great significance. It is set forth here only in order to stress the absence in this formula of large numerical coefficients.

$u_{max}$ is that limiting relative velocity at which a fluid will flow around a particle suspended in it.

Let us also compare the term $(\rho - \rho_0) V \cdot \frac{dv_0}{dt}$ to the omitted term associated with the supplemental reaction which, in turn, is related to the relative acceleration. The latter, for a spherical particle, is equal to

$$\frac{\rho_0 V}{2} \frac{du}{dt} \sim \frac{\rho_0 V}{2}\left(\frac{u}{T^*}\right)_{\lambda \sim \lambda_{\max}}.$$

Forming the ratio, we find

$$(\rho - \rho_0) V \frac{dv_0}{dt} : \frac{\rho_0 V}{2} \frac{du}{dt} \sim \frac{\rho - \rho_0}{\rho_0} \sim \frac{\rho}{\rho_0} \gg 1,$$

which justifies our approximation. Therefore at $\rho \gg \rho_0$ the so-called "adhering mass" $\frac{\rho_0 V}{2}$ may be ignored.

In deriving formula (33.11), we disregarded the effect of gravity on particle motion.

It is of interest to compare the expression derived for fluid velocity relative to the particle, with the change in the velocity of turbulence eddies over a distance $\lambda$ equal to the dimension 2a of the particle. A comparison of (33.11) and (4.9) shows, that, with

$$\rho \sim \rho_0 \qquad u_{\max} \sim v_{\lambda = 2a}. \tag{33.12}$$

Thus, owing to the presence of inertial effects and to the change in turbulence velocities, the velocities at which the fluid flows past the particle are found to be of the same order of magnitude. In contrast, at $\rho \gg \rho_0$

$$\frac{u}{v_{\lambda=2a}} \sim \left(\frac{\rho}{\rho_0}\right)^{1/3}. \qquad (33.13)$$

For example, in a gas, where $\rho/\rho_0 \sim 10^3$, u exceeds $v_\lambda = 2a$ by an order of magnitude. The same expression may be written in the form

$$u \sim \left(\frac{\rho}{\rho_0}\right)^{1/3}\left(\frac{\varepsilon}{\rho_0}\right)^{1/3}(2a)^{1/3} \sim v_{\lambda=\lambda_{u_{max}}}.$$

Thus, the eddies flowing past the particle are not of scale 2a, but of a substantially larger scale $\lambda = \lambda_{u_{max}}$.

Let us now compare the periods T* of the particle motion and those T of the turbulence eddies of scale $\lambda \sim 2a$. It follows from (32.2), (4.9) and (33.11) that, with $\rho \gg \rho_0$ the ratio of these periods is

$$\frac{T^*}{T} \sim \left(\frac{\rho_0}{\rho}\right)^{1/3} \sim \frac{v_{\lambda=2a}}{u}.$$

Let us now examine the analogous problem of particle motion in a turbulent flow whose dimensions a are substantially less than the scale of the inner turbulent zone. In this case, we may write the maximum value of the acceleration $\frac{dv_0}{dt}$ of the fluid, using formula (32.8). It is precisely this acceleration that contributes the main force exerted on the particle. Since the particle dimension is less than $\lambda_0$ and its relative velocity, as we will see below, is less than the velocity $v_\lambda$, the Reynolds number $\frac{ua}{\nu} < \frac{v_{\lambda_0}\lambda_0}{\nu} \sim 1$.

Therefore, the drag exerted on the particle due to its relative motion may be represented by the Stokes equation

$$\overline{F} = -\alpha \cdot a\mu u, \qquad (33.14)$$

where $\alpha$ is a numerical coefficient, which equals $6\pi$ for a spherical particle. Then the equation of motion for the particle becomes

$$\rho V\frac{du}{dt} = (\rho - \rho_0)Vw_{\lambda_0} - \alpha \cdot a\mu u. \qquad (33.15)$$

Let us estimate the magnitude of the terms $\rho V\frac{du}{dt}$ and $(\rho - \rho_0)Vw_{\lambda_0}$ in (33.15). Evidently we may write

$$\rho V\frac{du}{dt} \sim \rho V\frac{u}{T^*} \leqslant \rho V\frac{u}{T_\lambda}, \qquad (33.16)$$

$$(\rho - \rho_0)Vw_{\lambda_0} \sim (\rho - \rho_0)V\frac{v_{\lambda_0}}{T_{\lambda_0}} \sim \rho V\frac{v_\lambda}{T_{\lambda_0}}. \qquad (33.17)$$

If we assume that the velocity of the fluid motion relative to a particle having a dimension $a < \lambda_0$ is small in comparison with $v_\lambda$, the first term in (33.15) may be omitted, and it may be written in the form

$$\alpha a \mu u \sim (\rho - \rho_0) V w_{\lambda_0} \tag{33.18}$$

or

$$u \sim \frac{\rho - \rho_0}{\rho_0} \frac{V}{\alpha \nu a} w_{\lambda_0}. \tag{33.19}$$

For spherical particles

$$u \sim \frac{2}{3\sqrt{3}} \frac{(\rho - \rho_0)}{\rho_0} a^2 \frac{\varepsilon_0^{3/4}}{\nu^{5/4}}. \tag{33.20}$$

Formula (33.20) means that the particle motion is quasi-steady. In the field of turbulent accelerations equal to $\omega_0$, the particle moves with a constant velocity in precisely the same manner as if it were moving in some other steady acceleration field, for example, in the field of gravity. Of course, the velocity u, in this case, varies randomly in direction.

The relative velocity is proportional to the square of the particle dimensions and to the difference in densities $\rho - \rho_0$ between the particle and the medium. The dependence of u on the intensity of turbulence is determined by the factor $\epsilon_0{}^{3/4}$. According to formula (4.3), this indicates that u increases with the macroscopic motion $U_0$ according to the relation

$$u \sim U_0^{9/4}. \tag{33.21}$$

The inequality $u \ll v_{\lambda_0}$ indicates the condition under which equation (33.20) may be applied. Substituting the value of u, we find ($\rho_0 \ll \rho$):

$$\frac{\rho}{\rho_0} \frac{a^2 w_{\lambda_0}}{\nu} = \frac{\rho}{\rho_0} \frac{a^2}{\nu} \frac{v_{\lambda_0}^2}{\lambda_0} \ll v_{\lambda_0}, \tag{33.22}$$

or, since $v_{\lambda_0} \cdot \lambda_0 \sim \nu$,

$$\frac{\rho}{\rho_0} \frac{a^2}{\lambda_0^2} \ll 1. \tag{33.23}$$

It follows from this that equation (33.20) is valid at $a < \lambda_0 \sqrt{\frac{\rho_0}{\rho}}$. For particles in a gas, where $\frac{\rho_0}{\rho} \sim 10^{-3}$, this corresponds to $a < \frac{\lambda_0}{30}$.

It should be noted that in a gaseous medium the velocity of relative motion may exceed the change in velocity of turbulence eddies over the distance a for small particles also.

Comparing (33.20) and (32.9), we find

$$\frac{u}{v_\lambda} \sim \frac{\rho}{\rho_0} \frac{a^2 \nu}{\lambda_0^3} : \frac{v_\lambda}{\lambda_0} a \sim \frac{\rho}{\rho_0} \frac{a^2}{\lambda_0^3} \nu : \frac{\nu a}{\lambda_0^2} \sim \frac{\rho}{\rho_0} \frac{a}{\lambda_0} > 1,$$

i.e., at $a > \lambda_0 \frac{\rho_0}{\rho} \sim \frac{\lambda_0}{1000}$. In the interval $\frac{\lambda_0}{30} > a > \frac{\lambda_0}{1000}$ the velocity of relative motion exceeds the change in eddy velocity over distances of the order of a and is expressed by equation (32.20). If the velocity of the eddies flowing past the particle is known it is possible to compute the diffusional flow at its surface. Thus, we can assume, to an order of magnitude, that the diffusional flux to the surface of the particle is equal to the diffusional flux to the surface of a motionless particle around which a fluid moves at a velocity u.

The order of magnitude of diffusional flows is determined by formulas (25.18) and (14.19) for flows to particles whose dimensions are respectively larger and smaller than $\lambda_0$.

Thus, at $a \gg \lambda_0$,

$$I \sim \frac{\sqrt{K_f}\, S \cdot u_{\max} c_0}{\mathrm{Pr}^{1/4}} \sim 4\pi \left(\frac{\rho}{\rho_0}\right)^{1/3} (2a)^{1/3} \left(\frac{\varepsilon}{\rho_0}\right)^{1/3} a^2 c_0 \sim \sim 4\pi \left(\frac{\rho}{\rho_0}\right)^{1/3} \sqrt[3]{2}\, a^{1/3} \frac{u}{L^{1/3}} c_0, \tag{33.24}$$

and at $a \ll \lambda_0$,

$$I \sim 8 c_0 D^{2/3} a^{4/3} u_{\max}^{1/3} \sim 8 c_0 D^{2/3} a^2 \cdot \left(\frac{\rho}{\rho_0}\right)^{1/3} \left(\frac{2}{3\sqrt{3}}\right)^{1/3} \frac{\varepsilon_0^{1/4}}{\nu^{5/12}} \sim \sim 8 \left(\frac{2}{3\sqrt{3}}\right)^{1/3} D^{2/3} \left(\frac{\rho}{\rho_0}\right)^{1/3} \frac{a^2}{\nu^{5/12}} \frac{u^{3/4}}{L^{1/4}} c_0, \tag{33.25}$$

where u and L are the velocity and the scale of motion of the agitated fluid as a whole.

Unfortunately, available experimental data are insufficient to permit a comparison between the theoretical formulas (33.24) and (33.25) and values of diffusional flows determined experimentally. In actual practice, the total diffusional flow at all the particles in the reaction vessel is measured. The particles usually vary widely in size and shape. In addition, turbulence in a reaction vessel may not always be viewed as homogeneous and isotropic. The regions near the walls, in some measurements, play an essential part.

## BIBLIOGRAPHY


1. Taylor, G. I., A.R.C. Reports and Memoranda, No. 272 (1919).
2. Prandtl, L., Phys. Zeit. 11, 1072 (1910); 29, 487 (1928).
3. Von Karman, T., Proc. 4th Internat. Congress for Applied Mechanics, Cambridge, 1934, p. 77.
4. Lin, C. R. Moulton and Putnam, Ind. and Eng. Chem. 45, 636 (1953).
5. Vielstich, W., Zs. f. Electrochem., 57, 646 (1953).
6. Linton, W., and T. Sherwood, Ind. and Eng. Chem. 26, 516 (1934).
7. Chilton, T. and A. Colburn, Ind. and Eng. Chem., 26, 1183 (1934).
8. Lin, C., E. Denton, H. Gaskill, and G. Putnam, Ind. and Eng. Chem., 43, 2136 (1951).

9. Frank-Kamenetskiy, D. A., Diffuziya i teploperedacha v khimicheskoi kinetiki (Diffusion and Heat Transfer in Chemical Kinetics), AS USSR Press, 1947. Frank-Kamenetskiy, Diffusion and Heat Exchange in Chemical Kinetics, Translated by N. Thon, Princeton University Press, 1955.
10. Kafarov, V. V., Zhurn. prikl. khim., 29, 40 (1956); Chem. Industry, 5 (1953).
11. Kolthoff, JACS, 75, 4869 (1953); 76, 5221 (1954).
12. Landau, L. D., and E. M. Lifshits, Mekhanika sploshnykh sred (Mechanics of Continuous Media), Gostekhizdat, 1953, p. 168. Landau and Lifshits, Fluid Mechanics, Pergamon, London, 1959.
13. Sovremennoe sostoyanie gidroaerodinamiki vyazkoi zhidkosti (Current Aspects of the Hydro-Aerodynamics of Viscous Fluids), ed. by Gol'dshtein, IL, 1948. Goldstein, Modern Developments in Fluid Dynamics, Vols. I and II, Clarendon, Oxford, 1938.
14. Goldstein, Proc. Camb. Phil. Soc., 31, 232 (1935); von Karman, Z. ag new. Math. and Mech., 1, 247 (1921); J. Aeronaut., 1, 1 (1934).
15. Leibenzon, L.S., Izvestiya AN SSSR, ser. Geofiz., 285 (1940).
16. Sovremennoe sostoyanie aerodinamiki bol'shikh skorostei. (Current Aspects of High Speed Aerodynamics), ed. by Howarth, Vol. II, IL, 1955. Howarth, Modern Developments in Fluid Dynamics: High Speed Flow, Vols. I and II, Glarendon, Oxford, 1953.
17. Yaglom, A. M., Doklady AN SSSR, 67, No. 5 (1949).
18. Badger and MacCabe, Osnovnye protsessy i apparaty khimicheskikh proizvodstv (Basic Processes and Equipment of Chemical Manufacture), Goskhimizdat, 1933, p. 314. Badger and McCabe, Elements of Chemical Engineering, 1st ed., McGraw-Hill, New York, 1931.
19. Elovich, S. Yu., and G. M. Zhabrova, Teoreticheskie osnovy gidrirovaniya zhirov (Basic Theories of Hydrogenation of Fats), Izvestiya AN SSSR, 1943.
20. Prandtl, L., Gidroaerodinamika (Hydro-Aerodynamics), IL, 1954. Prandtl-Tietjens, Hydro- and Aeromechanics, Dover Publishing Co., New York, 1957. Cf. also: Prandtl, Essentials of Fluid Mechanics, Blackie, London, 1952.
21. Ivanov, Yu. B., article in the collection "Nekotorye voprosy teoreticheskoi fiziki" (Certain Problems in Theoretical Physics), Trudy MIFI, 1958.

CHAPTER IV

# Heat Transfer in Fluids

## 34. HEAT TRANSFER IN FLUIDS

Heat transfer processes in fluids are of great importance in modern engineering. For this reason they have attracted and continue to attract the attention of a large number of investigators. From an experimental standpoint, heat transfer has been the object of very careful study under highly varied and complex conditions. Developments of theoretical concepts and the rationalization of quantitative relations for heat flows in flowing fluids, however, are far from complete.

The process of heat transfer in a moving fluid is similar in many ways to the process of convective diffusion. As with diffusing matter, heat may be conceived as a substance that is transferred by means of convective and molecular mechanisms. As will become clear from what follows, however, along with this similarity there is a substantial difference between the two processes.

Let us first find the differential equation that describes the temperature field in a moving fluid. The complete derivation of this equation is a somewhat unwieldy procedure. Therefore, we refer the interested reader to the book by L. D. Landau and E. M. Lifshitz entitled "Mechanics of Continuous Media", (cf. their "Fluid Mechanics", English translation, 1959) and confine our treatment here to certain simplified discussions, that lead us, nevertheless, to equally valid results. Let us examine a non-uniformly heated fluid, characterized by a variable temperature $T(x, y, z)$. In such a fluid, heat is transferred by molecular thermal conduction. The heat flux $q$, per 1 $cm^2$ per second, obviously will depend on the difference in temperature of the fluid from one point to another. If the temperature gradient is sufficiently small, then, so far as the heat flux is concerned, it is possible to write with a high degree of accuracy

$$q = -\varkappa \cdot \operatorname{grad} T \qquad (34.1)$$

Inasmuch as the direction of the heat flux is from the more to the less strongly heated regions, the thermal conductivity coefficient

$\varkappa$ is necessarily a positive quantity. The thermal conductivity $\varkappa$, in general, depends on both the temperature and the physicochemical properties of the fluid. In a fluid with variable temperature, the density $\rho$, which is temperature dependent, also varies from one point to another. Therefore, in the choice of thermodynamic variables to characterize the fluid, pressure must be included along with the temperature.

Let us focus attention on an arbitrary volume V in the fluid, and let us write a balance for its enthalpy H(p, T) which depends on temperature and pressure locally. The rate of change in the enthalpy of this volume may be written in the form

$$\frac{dH}{dt}=\frac{d}{dt}\int \rho c_p T\,dV=\int \rho c_p \frac{dT}{dt}\,dV$$

where $c_p$ is the heat capacity.

Changes in the enthalpy of volume V are determined by

1) the heat flow leaving the volume; and,

2) the energy dissipation, which is associated with the motion of the fluid in the volume under examination.

The energy $\dot{E}$ dissipated per unit volume, is given by Equation (1.15). Dissipation of energy leads to the liberation of heat.

It is possible to write, therefore,

$$\int \rho c_p \frac{dT}{dt}\,dV=-\oint \mathbf{q}\,ds+\int \dot{E}\,dV=\oint \varkappa\,\mathrm{grad}\,T ds+\int \dot{E}\,dV.$$

or, passing to a volume integral

$$\int \rho c_p \frac{dT}{dt}\,dV=\int \mathrm{div}\,(\varkappa\nabla T)\,dV+\int \dot{E}\,dV$$

whence

$$\rho c_p \frac{dT}{dt}=\mathrm{div}\,(\varkappa\nabla T)+\dot{E}$$

Expanding the total derivative, with due regard to convection, we obtain

$$\frac{dT}{dt}=\frac{\partial T}{\partial t}+\frac{\partial T}{\partial x}\cdot\frac{\partial x}{\partial t}+\frac{\partial T}{\partial y}\cdot\frac{\partial y}{\partial t}+\frac{\partial T}{\partial z}\cdot\frac{\partial z}{\partial t}=$$

$$=\frac{\partial T}{\partial t}+v_x\frac{\partial T}{\partial x}+v_y\frac{\partial T}{\partial y}+v_z\frac{\partial T}{\partial z}=\frac{\partial T}{\partial t}+(\mathbf{v}\,\mathrm{grad})\,T$$

Therefore, finally,

$$\rho c_p\left(\frac{\partial T}{\partial t}+(\mathbf{v}\,\mathrm{grad})\,T\right)=\mathrm{div}\,(\varkappa\,\mathrm{grad}\,T)+\dot{E} \qquad (34.2)$$

This equation expresses the temperature distribution in the fluid. It is highly complex because the physical properties — density, thermal capacity and thermal conductivity — that enter into it are functions of the temperature. In practice, however, this dependence cannot be taken into account, and all three quantities are taken as constants. Disregard of the change of the physical properties is proper if the temperature change in the fluid from one point to another is sufficiently small. In practice, especially in the case of large heat fluxes, this condition does not obtain. Accounting for the dependence of the physical constants on temperature, however, substantially complicates the equations of heat transfer. In the majority of problems, we are forced to accept approximations, in which all physical properties are taken as independent of the temperature. Their magnitude, however, must not be evaluated at the temperature of the boundary surface, but rather at some averaged temperature of the fluid. Disregarding the temperature dependence of the physical properties and, consequently, their dependence on coordinates, we have,

$$\rho c_p\left(\frac{\partial T}{\partial t}+(\mathbf{v}\,\mathrm{grad})\,T\right)=\varkappa\,\mathrm{div}\,(\mathrm{grad}\,T)+\dot{E}$$

or

$$\frac{\partial T}{\partial t}+(\mathbf{v}\,\mathrm{grad})\,T=\chi\Delta T+\frac{\dot{E}}{\rho c_p} \tag{34.3}$$

where $\chi$ denotes the quantity

$$\chi=\frac{\varkappa}{\rho c_p} \tag{34.4}$$

known as the thermal diffusivity.

Comparing (34.3) with the equation of convective diffusion, we see that both equations are quite similar. Despite this similarity, however, the following differences must be noted:

1) equation (34.3) includes an energy dissipation term that is absent from the equation of diffusion (8.6);

2) the thermal diffusivity of the fluid varies over a wider range than the diffusion coefficient D. For this reason the thermal Prandtl number, defined as

$$\mathrm{Pr}_T=\frac{\nu}{\chi} \tag{34.5}$$

may also have a wide spread of values. Table 3 gives the Prandtl numbers for certain typical fluids;

3) finally, it should be pointed out that the approximation according to which the physical properties are considered to be independent of temperature is often far from adequate. On the other hand, the fact that the diffusion coefficient is a function of the concentration of the solution has no real, practical significance.

In most cases the frictional heat in equation (34.3) may be disregarded. This situation obtains when the motion of the fluid is sufficiently slow (since $\dot{E}$ is proportional to the squares of the derivatives of velocity), and the temperature differences are

Table 3

Prandtl numbers $Pr = \frac{\nu}{\chi}$
for certain typical fluids

| Fluid | Pr |
|---|---|
| Air | 0.733 |
| Water | 6.75 |
| Alcohol | 16.6 |
| Glycerine | 7250 |
| Mercury | 0.024 |
| Sodium | 0.007 |

sufficiently large. In this case the temperature rises in the fluid due to viscous dissipation of energy are small compared with temperature differences that exist. Disregarding the dissipation term, we can rewrite equation (34.3) in a much simplified form

$$\frac{\partial T}{\partial t} + (\mathbf{v}\,\mathrm{grad})\,T = \chi \Delta T$$

This equation is the same as the equation for the distribution of concentration (8.6). In the future we will be interested only in steady-state problems.

At steady-state, the temperature distribution is governed by the equation

$$(\mathbf{v}\,\mathrm{grad})\,T = \chi \Delta T \tag{34.6}$$

The boundary conditions encountered in heat transfer problems are usually among the following.

At the surface of the body either:

a) the temperature must have a given constant value controlled by heating or cooling of the body;

b) there is no heat transfer from the surface of the body, in which case the heat flow in a direction normal to the surface of the body must be equal to zero;

c) the value of the heat flux is given.

Far from the body:

a) in the case of an exterior flow, the temperature must have a given value;

b) in the case of flow inside a tube, the temperature of the fluid is given at the inlet.

Thus, the boundary conditions in heat transfer problems are of the same nature as those in diffusion problems. This permits us to apply certain general results obtained previously to heat transfer problems. In the first place, it is possible to assert that the dimensionless heat flux — the Nusselt number — under conditions of forced convection, is a function of two dimensionless numbers: the Reynolds number and the (thermal) Prandtl number. Similarly, in natural convection, the Nusselt number is determined by the Grashof and Prandtl numbers. These functions in the case of heat transfer, however, may differ substantially from those previously obtained for the case of diffusion. As stressed above, the general reason for these differences is linked to the fact that the Prandtl number in heat transfer problems is sometimes much larger than unity; but it may be of the order of unity, or rather smaller than unity, as in the case of liquid metals. To clarify the physical consequences of this situation, let us compare, for example, the heat transfer and mass transfer in a liquid with a Prandtl number $Pr_T \sim 1$ with laminar flow past a plate at $Re \gg 1$.

A hydrodynamic sub-layer and, correspondingly, a diffusion or thermal sub-layer, are formed at the surface of the plate. Because of the high value of the diffusion Prandtl number, we might regard the thickness of the diffusion sub-layer to be small compared to the thickness of the hydrodynamic layer, and utilize the first terms of the expansion to designate velocities. This approach enabled us to obtain expressions for the distribution of concentration in analytical form. Such an approach, generally speaking, is inapplicable in the case of heat transfer for the following reason. For Prandtl numbers of the order of unity, the thermal sublayer has the same thickness as the hydrodynamic layer, while for small Prandtl numbers the thermal layer may be substantially thicker than the boundary layer. Therefore, it is not possible to obtain general expressions for heat fluxes. Even in the simplest of cases — for example the flow of a coolant stream past a heated plate — to find the temperature distribution, equation (34.6) must be integrated numerically. For more complex problems, as, for example, the case of a turbulent stream, the difference in Prandtl numbers also renders inapplicable the results derived from the diffusion theory.

It is not the aim of the present book to provide a systematic presentation of the theory of heat transfer in a moving fluid. This theory has already been covered in sufficient detail in available textbooks. We confine our treatment here to the presentation of solutions to certain typical problems, by means of which it is possible to demonstrate clearly the similarities and differences between the heat transfer and the diffusion processes in fluids.

Furthermore, we will be concerned especially with the theory of heat transfer in liquid metals and will cover this in considerable detail. We will also show how, by a combination of data on diffusion and heat transfer in liquid metals, it is possible to arrive at a general interpolation formula that embraces the process of the transfer for all Prandtl numbers.

## 35. THE SIMPLEST PROBLEMS OF CONVECTIVE HEAT TRANSFER

The first problem in the theory of heat transfer that is of interest for comparison with diffusion theory, is the problem of the heat transfer in a fluid for an interior problem.

a) Let us take the case of a laminar stream inside a tube, the walls of which are maintained at a certain temperature $T_1$. The fluid entering the tube has a temperature $T_0$. We assume that the tube has a circular cross-section, and designate its radius by R.

The temperature distribution equation in cylindrical coordinates (r, θ, x) is

$$v\frac{\partial T}{\partial x}=\chi\left(\frac{\partial^2 T}{\partial r^2}+\frac{1}{r}\frac{\partial T}{\partial r}+\frac{\partial^2 T}{\partial x^2}\right) \tag{35.1}$$

In this case, the independence of temperature on the angle θ has been accounted for.

The boundary conditions are

$$T=T_1 \quad \text{at} \quad r=R, \quad x>0, \tag{35.2}$$

$$T=T_0 \quad \text{at} \quad x=0, \quad r<R. \tag{35.3}$$

The fluid velocity v at the inlet differs from that in the region of established Poiseuille flow. In the first region, the problem is identical with the diffusion problem discussed in Section 15.

Since the coefficient of thermal diffusivity is comparable, in order of magnitude, to the viscosity of the fluid $\nu$, the lengths $h_v$ and $h_T$, in which the Poiseuille velocity profile and the corresponding fully developed temperature distribution become established, are also of the same order of magnitude.

$$h_v \sim h_T \sim \mathrm{Re}\cdot R$$

Let us now consider the region in the tube where the Poiseuille profile of velocities has already become established, i.e., the region $x > h_v$ [1]. Here, (35.1) becomes

$$v_0\left(1-\frac{r^2}{R^2}\right)\frac{\partial T}{\partial x}=\chi\left(\frac{\partial^2 T}{\partial r^2}+\frac{1}{r}\frac{\partial T}{\partial r}+\frac{\partial^2 T}{\partial x^2}\right) \tag{35.4}$$

Assuming that the Peclet number

$$\mathrm{Pe} = \mathrm{Re} \cdot \mathrm{Pr} = \frac{v_0 R}{\chi}$$

is large compared to unity, we may further assume that the convective transfer of heat along the tube exceeds the molecular transfer. In other words the term $\chi \frac{\partial^2 T}{\partial x^2}$ is small compared with the term on the left-hand side and may be dropped from the equation. The value $v_0$ may be expressed in terms of average velocity in the cross section $\bar{v}$ ($v_0 = 2\bar{v}$).

We then have

$$v_0\left(1 - \frac{r^2}{R^2}\right)\frac{\partial T}{\partial x} = \chi\left(\frac{\partial^2 T}{\partial r^2} + \frac{1}{r}\frac{\partial T}{\partial r}\right) \tag{35.5}$$

Equation (35.5) is an equation with separable variables. It is easy to see that its solution is the series

$$T = T_1 + \sum (T_0 - T_1) A_n \exp\left\{-\beta_n^2 \frac{\chi x}{v_0 R^2}\right\} \Psi_n(r) \tag{35.6}$$

where $A_n$ and $\beta_n$ are arbitrary constants, and the function $\Psi_n(r)$ satisfies the equation

$$\frac{d^2\Psi_n}{dr^2} + \frac{1}{r}\frac{d\Psi_n}{dr} + \frac{\beta_n^2}{R^2}\left[1 - \frac{r^2}{R^2}\right]\Psi_n = 0$$

for a boundary condition

$$\Psi_n = 0 \qquad \text{at} \quad r = R$$

The functions $\Psi_n$ may be expressed as an infinite series having no singularity at the origin of coordinates

$$\Psi_n = 1 - \frac{1}{4}\left(\frac{r}{R}\right)^2 \beta_n^2 + \left(\frac{r}{R}\right)^4 \frac{3}{2\,(4!)}\beta_n^4\left(\frac{1}{4} + \frac{1}{\beta_n^2}\right) + \ldots$$

The boundary condition (35.2) gives

$$\Psi_n(R) = 0$$

This equation has an infinite number of roots, the first of which are

$$\beta_1 = 2.705, \quad \beta_2 = 6.66, \quad \beta_3 = 10.6$$

The constants $A_n$ that correspond to these roots may be found by substituting the series (35.6) in the boundary condition (35.3).

They have the following values:

$$A_1 = 1.48, \quad A_2 = -0.81, \quad A_3 = 0.38$$

In terms of these constants, the temperature distribution in the tube takes the form (Figure 37)

$$T = T_1 + 1.48\,(T_0 - T_1)\exp\left\{-3.65\,\frac{\chi x}{v_0 R^2}\right\}\Psi_0\left(\frac{r}{R}\right) - 0.81\,(T_0 - T_1)\exp\left\{-22.18\,\frac{\chi x}{v_0 R^2}\right\}\Psi_1\left(\frac{r}{R}\right) - \ldots$$

From the preceding formulas, it is possible to find the heat flow along the tube

$$q = -\rho c_p \chi \left(\frac{dT}{dr}\right)_{r=R}$$

The change in q, as a function of the coordinate x along the tube, is shown in Figure 38.

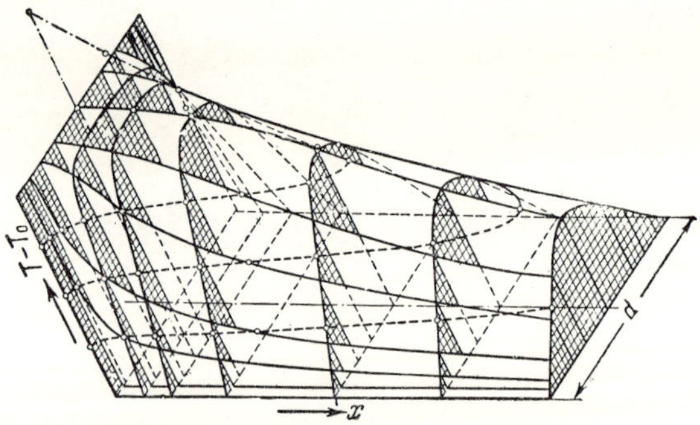


Figure 37. Temperature distribution in the tube. The temperature profiles, shown in solid lines, correspond to the values $\frac{\chi x}{2 v_0 R^2} = 0.005; 0.01; 0.02; 0.04; 0.06; 0.08.$

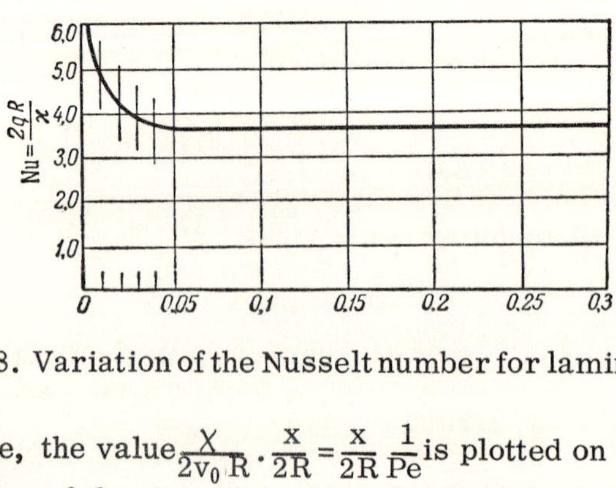


Figure 38. Variation of the Nusselt number for laminar flow.

In this figure, the value $\frac{\chi}{2 v_0 R} \cdot \frac{x}{2R} = \frac{x}{2R}\frac{1}{Pe}$ is plotted on the abscissa, while the value of the Nusselt number for the tube is plotted on the ordinate. We see that, at a distance $x \approx 0.05(2R)\,Pe$ from the tube inlet, the Nusselt number asymptotically approaches

$$\mathrm{Nu} = 3.66 \tag{35.7}$$

b) Convective heat transfer at large Reynolds numbers is also of major interest. Since the Prandtl number in heat transfer problems is usually of the order of unity, and in any event does not exceed 100, the cases which were discussed in Section 14 where $\mathrm{Re} \ll 1$ and $\mathrm{Pe} \gg 1$ are not encountered in heat transfer. Reynolds numbers large compared to unity are encountered more often.

The case of heat transfer in turbulent flow, which is the most important from the standpoint of technological applications, is discussed below. Here we discuss briefly certain instances of heat transfer across a boundary layer to a fluid in laminar flow.

1. Heat transfer at the surface of a rotating disk. The solution to the problem of heat transfer at the surface of a disk was obtained by S. Z. Kibel [2], who accounted for the heat released due to friction. If the latter is neglected, the solution of the heat problem is identical with the solution of the diffusion problem given in Section 11.

The temperature distribution is given by the equation

$$T = \frac{(T_0 - T_1) \int\limits_0^{\sqrt{\frac{\omega}{\nu}} y} \exp\left\{ \int\limits_0^{\xi} \frac{1}{\varkappa} v_y(\xi')\, d\xi' \right\} d\xi}{\int\limits_0^{\infty} \exp\left\{ \int\limits_0^{\xi} \frac{1}{\varkappa} v_y(\xi')\, d\xi' \right\} d\xi} + T_1, \tag{35.8}$$

where $\xi$ is a dimensionless coordinate

$$\xi = \sqrt{\frac{\omega}{\nu}}\, y \quad \text{and} \quad v_y = \sqrt{\nu\omega} H(\xi)$$

so that

$$v_y\, dy = \nu \cdot H(\xi)\, d\xi.$$

In contradistinction to the case of diffusion, however, the value $\chi$, in general, is comparable to the viscosity of the fluid, so that the numerical coefficient in the integrand, $\frac{\nu}{\chi} \sim 1$. For this reason it is not possible to take advantage of the expansion for $H(\xi)$ at small values for $\xi$, as was done in Section 11.

Evaluation of the integral must be done numerically, employing a specific value for the ratio $\mathrm{Pr_T} = \frac{\nu}{\chi}$. The heat flux is given by the formula.

$$q_{\mathrm{heat}} = \frac{\chi (T_0 - T_1)}{\sqrt{\frac{\nu}{\omega}} \int\limits_0^{\infty} \exp\left\{ \int\limits_0^{\xi} \mathrm{Pr} H(\xi')\, d\xi' \right\} d\xi} \tag{35.9}$$

which contains the same integral.

The quantity $\delta_T$, given by the integral

$$\delta_T = \sqrt{\frac{\nu}{\omega}} \int_0^\infty \exp\left\{\Pr \int_0^\xi H(\xi')\, d\xi'\right\} d\xi \tag{35.10}$$

represents the thickness of the thermal boundary layer.

2. Laminar flow along a smooth plate. The temperature distribution is determined by the equation

$$u\frac{\partial T}{\partial x} + v\frac{\partial T}{\partial y} = \chi\frac{\partial^2 T}{\partial y^2} \tag{35.11}$$

for boundary conditions

$$\left.\begin{array}{lll} T = T_1 & \text{at} & y = 0, \quad x > 0, \\ T \to T_0 & \text{as} & y \to \infty. \end{array}\right\} \tag{35.12}$$

The heat transfer problem is thus identical with the problem of distribution of concentration discussed in Section 15. Here too, however, as in the case of a disk, it is not possible to employ the first terms of the series expansions of $v_x$ and $v_y$. Polhausen [3] solved this problem by numerical methods.

On replacement of the variables, as indicated in Section 3, the equation for the velocity distribution in the boundary layer takes the form (3.24). Through this substitution the equation for temperature distribution (35.11) becomes

$$T'' + \Pr f T' = 0, \tag{35.13}$$

where f is determined by expression (3.27).

The solution of equation (35.13) satisfying the boundary conditions (35.12) is

$$T = T_1 + (T_0 - T_1)\frac{\int_0^{\sqrt{\frac{u_0}{4\nu x}}\,y} \exp\left\{-\Pr\int_0^t f\,dt'\right\} dt}{\int_0^\infty \exp\left\{-\Pr\int_0^t f\,dt'\right\} dt} \tag{35.14}$$

Correspondingly, the heat flux is

$$q_{\text{heat}} = \chi\frac{(T_0 - T_1)\cdot\frac{1}{2}\sqrt{\frac{u_0}{\nu x}}}{\int_0^\infty \exp\left\{-\Pr\int_0^t f\,dt'\right\} dt} \tag{35.15}$$

The integrals in formulas (35.14) and (35.15) must be evaluated numerically, since the parameter $\Pr \sim 1$.

The integral

$$\delta_T = 2\left[\int_0^\infty \exp\left\{-\Pr\int_0^t f\,dt'\right\} dt\right]\sqrt{\frac{\nu x}{u_0}} \tag{35.16}$$

gives the thickness of the thermal boundary layer. Its dependence on the coordinate x along the plate and the velocity of the flow $u_0$ is the same as that of the diffusion boundary layer.

Numerical values of the integral

$$\alpha = \int_0^{\infty} \exp\left\{-\Pr \int_0^{t} f(t')\, dt'\right\} dt$$

for various Prandtl numbers are given in Table 4.

Table 4

| Pr | $\alpha$(Pr) | Pr | $\alpha$(Pr) | Pr | $\alpha$(Pr) |
|---|---|---|---|---|---|
| 0.6 | 0.55 | 0.9 | 0.64 | 7 | 1.29 |
| 0.7 | 0.58 | 1 | 0.66 | 10 | 1.46 |
| 0.8 | 0.61 | 1.1 | 0.69 | 15 | 1.67 |

The numerical value of the coefficient in the expression for $\delta_T$ varies rather substantially with changes in the Prandtl number. For $\Pr > 20$, the expressions developed for diffusion under the assumption $\Pr \gg 1$ may be employed to an accuracy of 2 percent. The error which results from neglecting the temperature dependence of the physical constants $\nu$ and $\chi$, apparently exceeds that inaccuracy.

## 36. HEAT TRANSFER IN TURBULENT FLOW

From the outset the principal version of the theory of heat transfer by a turbulent fluid, as proposed by Taylor and Prandtl and discussed by us in Section 25, was not entirely satisfactory. Indeed, the existence of a well defined boundary between the laminar sublayer and the region of turbulent motion was questioned as a matter of principle. Furthermore, it was not possible to secure good agreement between theory and experiment in the area of heat transfer by selecting the single constant in the expression for the heat flux. In this connection a series of improvements to the Taylor and Prandtl theory was proposed.

Von Karman offered the theory of a laminar sublayer with a junction zone (buffer layer). A gradual change in the motion of the fluid from turbulent to strictly laminar was assumed to occur in that zone. In order to obtain the velocity profile observed experimentally, von Karman adopted a logarithmic expression of the type (4.12) for the velocity distribution in the buffer zone. He used, however, different constants from those in the logarithmic velocity distribution for the region of developed turbulence. The introduction of an additional logarithmic expression with two arbitrary constants increases to five the total number of constants required to specify the distribution of velocities.

To determine heat flux, we first evaluate turbulent viscosity from the velocity distribution, using (4.5) $\nu_{\text{turb}} \approx \overline{vl} \approx \text{const} \cdot l^2 \frac{\partial u}{\partial y} = \alpha l^2 \frac{\partial u}{\partial y}$

$$q = \rho c_p (\chi + \chi_{\text{turb}}) \frac{\partial T}{\partial y} \tag{36.1}$$

and

$$\chi_{\text{turb}} \sim \overline{vl} \approx \text{const} \cdot l^2 \frac{\partial u}{\partial y} = \alpha l^2 \frac{\partial u}{\partial y} \tag{36.2}$$

However, values from these do not agree satisfactorily with experimental results.

As stressed by Reichardt [4], there is no reason why the constants in the respective expressions (4.5) and (36.2) for turbulence viscosity and thermal diffusivity must have the same numerical values. On the contrary, as we noted earlier in the example of turbulence diffusion, these constants should have entirely different values. Consequently, formula (36.2) for turbulent thermal diffusivity, according to Reichhardt, should be written as

$$\chi_{\text{turb}} = \gamma \alpha l^2 \frac{du}{\partial y} = \gamma \alpha v_0 y \tag{36.3}$$

where $\gamma$ is a certain constant different from unity. In this case a better agreement between theory and experiment is obtainable, because the temperature distribution is determined not only from the velocity distribution, but also from one additional constant $\gamma$ that may be chosen arbitrarily. *

In evaluating the von Karman — Reichhardt theory, we must note that the introduction of a buffer layer zone and the assumption of a logarithmic distribution for average velocity in that zone cannot be regarded as theoretically sound. In this particular case, agreement between theory and experiment cannot of itself be viewed as corroborating theory. On the one hand, the theoretical expressions contain a large number of arbitrary constants, which make it possible to fit the experimental data to the theoretical equations. On the other hand, experimental data on heat transfer in fluids are relatively inaccurate.

---

* Let us note that all the zones of the boundary layer contribute to the total resistance to heat transfer when the Prandtl numbers are not very large. In contrast, for very large Prandtl numbers, as we have seen above, the main zone of resistance lies within the diffusional (or thermal) sublayer.

The substantial error inherent in these measurements arises because of changes in the physical constants of the fluid with temperature. Thus, with water for example, the Prandtl number at 0° C equals 12.5; at 50° C it drops to 3.5; and at 100° C to 1.7. The fact must also be kept in mind that the quantities constituting the Prandtl number — the viscosity and thermal diffusivity of the fluid — do not vary identically with temperature. This leads to conditions that are different from those postulated for the theoretical formulas. In the well known work of Eagle and Ferguson [5], the experimental data were extrapolated to an infinitely small temperature difference between the heated surface and the liquid coolant. The degree of accuracy for the data thus obtained is satisfactory. However, Eagle and Ferguson's experiments were conducted with water over a very small range of Prandtl numbers ($3 \leqslant \mathrm{Pr} \leqslant 10$). Therefore, their comparison of theory with experiment is quite limited and does not furnish sufficient justification for the belief that the von Karman — Reichhardt theory has been conclusively proved by experiment. However, direct measurements of velocity profiles and temperatures outside the viscous sublayer permitted the constants of the logarithmic profiles to be determined. This yielded a value of 1.25 for $\gamma$ (for $\alpha = 0.17$, as previously noted).

A universal formula, soundly based on theory and correct for any Prandtl number, is difficult, in principle, to imagine: the mechanisms of heat transfer may differ markedly at different Prandtl numbers. For this reason we felt it expedient to approach the problem by a different path. In what follows we will attempt to obtain a formula of heat transfer in fluids for any Prandtl number starting from the following considerations:

a) We derived above an expression for the flux and distribution of matter in a moving fluid in turbulent flow at $\mathrm{Pr} \gg 1$ (in Chapter III the transferred substance was the diffusing solute).

b) Similarly, in the next section, we will find the flux and distribution of "substance" for $\mathrm{Pr} \ll 1$. The latter applies to heat transfer in liquid metals.

c) The existence of expressions for heat flux at $\mathrm{Pr} \gg 1$ and $\mathrm{Pr} \ll 1$ makes it possible to obtain a general expression for heat flux for any Prandtl number by means of interpolation between two limiting formulas.

## 37. THEORY OF CONVECTIVE HEAT TRANSFER IN LIQUID METALS

The problem of heat transfer by means of liquid (usually molten) metals has been studied in detail in recent years, both experimentally and theoretically, in connection with the use of liquid metals as coolants in atomic reactors.

The distinctive feature of liquid metals as coolants is their high thermal conductivity. The mechanism of thermal conduction in liquid as well as solid metals is fundamentally electronic. The viscosity of liquid metals, on the contrary, is no different in order of magnitude, from that of the usual free-flowing liquids. As a result, typical values of the Prandtl number for liquid metals lie within a range of $6 \cdot 10^{-3}$ to $10^{-2}$.

From a formal point of view, the transfer of heat by liquid metals is the opposite limiting case ($Pr \ll 1$) from the transfer of substance by moving ordinary liquids ($Pr \gg 1$).

In this connection, it is of interest to find the limiting laws for the transfer of substance at very high and at very low Prandtl numbers.

Knowledge of these limiting laws leads to the derivation of the universal interpolation formula that applies to the entire range of Prandtl numbers encountered in practice: from $10^{-3}$ to $10^{5}$.

A theoretical discussion of the process of transfer of heat by means of liquid metals is also of general interest.

In further discussions, we do not attempt to account for such effects as the incomplete wetting of the cooled surface by the metal, or the effect on the heat transfer of the oxide films that often form on the surfaces along which the metal streams.

The hydrodynamic properties of liquid metals have been reported in many papers. It was found that the hydrodynamic behavior of liquid metals does not differ from that of ordinary liquids, provided the walls are wetted sufficiently by the flowing metal. Therefore, all hydrodynamic correlations may be directly applied to liquid metals.

The change in the physicochemical properties (viscosity, density, thermal diffusivity, etc.) with temperature for liquid metals is relatively small. When there are variations in temperature the values of these quantities at certain average temperature may be successfully used.

Heat transfer in liquid metals has also been studied in a great number of experiments. At low metal flow velocities, the transfer of heat takes place by means of a conduction mechanism. The convective transfer of heat in liquid metals is important only at sufficiently high Reynolds numbers, in excess of about $5 \cdot 10^{3}$.

In the future, therefore, we will limit ourselves to the study of the turbulent flow of liquid metal.*

Let us assume that the metal flows inside a tube, the radius of which is R. For the sake of simplicity we will disregard the effects of curvature. The velocity distribution in the liquid metal will be

---

* Note added in proof. The theory set forth below is that of the author of this book and is published for the first time. In abbreviated form it was included in the report of S. S. Kutateladze, V. I. Subbotin, V. M. Borishanskii and P. L. Kirillov at the Second International Conference of the UN on the Peaceful Uses of Atomic Energy [10]. The author expresses his thanks to the individuals mentioned.

represented by the usual logarithmic relation (4.12). In view of the slow change of the function ln y, we will apply the logarithmic velocity distribution to the center of the tube, as explained in Section 4.

We assume that the walls of the tube are maintained at a constant temperature; then the boundary condition is

$$T = T_0 \quad \text{at} \quad r = R \tag{37.1}$$

We also assume that the temperature of the metal, at the axis of the tube, is fixed, i.e., that the following condition is valid

$$T = T_R \quad \text{at} \quad r = 0 \tag{37.2}$$

When comparing theory with experimental data, the temperature in the center of the tube is to be taken as the average temperature of the flow. In what follows, we neglect changes in the physical properties — viscosity $\nu$ and thermal diffusivity $\chi$ — with temperature. When comparing the formulas derived with experimental results, these quantities are to be evaluated at the average temperature. We characterize the transfer of heat by turbulence eddies through a certain turbulent thermal diffusivity $\chi_{turb}$, that depends, generally speaking, on the distance from the wall of the tube. The heat transferred by means of turbulence eddies is given by formula (36.1). Once again we should stress that turbulence eddies must not be represented as certain fluid volumes ("lumps" in the parlance of some authors) that transfer heat from point to point isothermally, without interacting with other parts of the fluid. An attempt based on the notion of isolated non-interacting eddies would be contrary to modern concepts of the nature of turbulent motion. All the turbulence eddies that exist in the fluid contribute to the heat flux (36.1). We assume that the Reynolds number is sufficiently high that, at least in the region close to the center of the tube, the heat transfer by turbulence eddies predominates over heat transfer by thermal conduction.

In other words, we assume that the following inequality is valid in the central region of the tube

$$\chi_{turb} \gg \chi \quad \text{at} \quad r \sim 0 \tag{37.3}$$

Inequality (37.3) means that, if a surface in the central portion of the tube coaxial to the walls of the tube be visualized, then the total heat transferred by all turbulence eddies is greater than the heat flow due to thermal conduction.

In the region of the fluid where inequality (37.3) is valid, we can employ formula (36.1) for the heat flow and formula (36.3) for $\chi_{turb}$. Assuming the heat flux to be constant along the radius of the tube (i.e., disregarding for the sake of simplicity the curvature of the tube walls), we can write the logarithmic temperature distribution

$$T = \frac{q}{\gamma \alpha \rho c_p v_0} \ln y + \text{const}, \tag{37.4}$$

where y is the distance from the tube walls.

The constants $\gamma$ and $\alpha$, found by measuring the profiles of the temperatures and the velocities in the tube for fluids with Pr $\sim$ 1, may be employed in (37.4), since that expression relates only to the region of purely turbulent heat transfer. Then

$$\alpha = 0.4, \quad \gamma = 1.2$$

Since ln y is a slowly changing function of logarithmic accuracy, the temperature profile (37.4) is applicable to the center of the tube, where y $\sim$ R. In view of (37.2) the following value for the constant may be obtained:

$$\text{const} = T_R - \frac{q}{\gamma \alpha \rho c_p v_0} \ln R$$

Thus, the logarithmic temperature distribution takes the form

$$T = T_R + \frac{q}{\gamma \alpha \rho c_p v_0} (\ln y - \ln R) \tag{37.5}$$

The temperature distribution (37.5), however, is inapplicable in that region of flow where inequality (37.3) no longer holds.

It is precisely because of the fact that $\chi_{turb}$ decreases toward the wall in accordance with (36.3) that the inequality (37.3) must be replaced by the following approximation at a certain distance $\delta_T'$ from the surface,

$$\chi_{turb} \approx \chi \qquad \text{at} \quad y \approx \delta_T'. \tag{37.6}$$

For $y < \delta_T$ the turbulent thermal diffusivity is lower than the thermal diffusivity of the metal. Utilizing (36.3) for $\chi_{turb}$ and the logarithmic velocity profile (4.12), we find

$$\gamma \alpha v_0 \delta_T' \sim \chi$$

or

$$\delta_T' = b \frac{\chi}{v_0}, \tag{37.7}$$

where b is an arbitrary numerical coefficient. The quantity $\delta_T'$ may be termed the thickness of the thermal sublayer. At low Prandtl numbers $\chi \gg \nu$, and the thickness of the thermal sublayer $\delta_T'$ is large in comparison with the viscous sublayer. For $y < \delta_T'$ the turbulent thermal diffusivity may be neglected as compared with the thermal diffusivity of the metal.

Thus, for a sufficiently high thermal diffusivity, the picture is basically different from the diffusion case. Heat transfer by convection is found to be smaller than the transfer of heat by conduction in a turbulent regime in which the effect of viscosity is negligible (Figure 39).

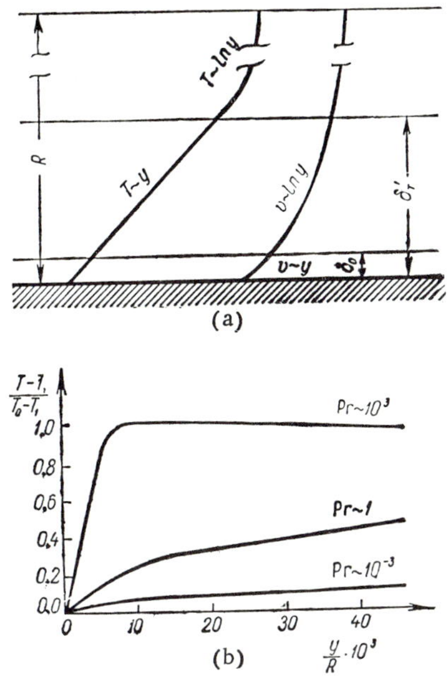


Figure 39. (a) shows the temperature and velocity distributions near the wall. (b) shows temperature profiles for various values of the Pr number.

For $y<\delta'_T$, we may write a simple law for temperature distribution by analogy to formula (25.12)

$$T=\frac{q}{\rho c_p}\frac{y}{\chi}+T_1 \tag{37.8}$$

where the boundary condition (37.1) at the tube walls is fulfilled.

At the point $y\approx\delta'_T$ the distributions (37.5) and (37.8) must match. We then have

$$\begin{aligned} q &= \frac{\rho c_p(T_R-T_1)}{\left[\frac{1}{\alpha\gamma v_0}\ln\frac{R}{\delta'_T}+\frac{\delta'_T}{\chi}\right]} = \frac{\rho c_p(T_R-T_1)}{\left[\frac{1}{\gamma\alpha v_0}\ln\frac{Rv_0}{b\chi}+\frac{b}{v_0}\right]} = \\ &= \frac{\rho c_p(T_R-T_1)\gamma\alpha v_0}{\left[\ln\frac{Rv_0}{\nu}+\ln\frac{\nu}{\chi}-\ln b+\gamma\alpha b\right]} = \frac{0.5\rho c_p(T_R-T_1)v_0}{\left[\ln\frac{Rv_0}{\nu}+\ln \mathrm{Pr}-\ln b+\gamma\alpha b\right]} \end{aligned} \tag{37.9}$$

Let us make note of the fact that equation (37.9) for the heat flux is written so that a negative flux q corresponds to cooling of the surface ($T_1 > T_R$), and a positive flux q corresponds to heating of the surface ($T_R > T_1$). This has been done to preserve the analogy between heat flux q and the mass flux j determined in preceding chapters. The characteristic velocity $v_0$ in (37.9) may be expressed in terms of the average velocity of the stream $\bar{u}$:

$$v_0 = \frac{\sqrt{\lambda_f}}{2\sqrt{2}} \bar{u} \tag{37.10}$$

Introducing the Reynolds number for flow in the tube

$$\mathrm{Re} = \frac{2\bar{u}R}{\nu}$$

and having determined the dimensionless heat flux by the relation

$$\mathrm{Nu} = \frac{2qR}{C_p \chi (T_R - T_1)} \tag{37.11}$$

we may rewrite (37.9) as

$$\mathrm{Nu} = \frac{0.18\,\mathrm{Re}\,\mathrm{Pr}\,\sqrt{\lambda_f}}{\lg \mathrm{Re} + \lg \mathrm{Pr} + \lg 0.35\sqrt{\lambda_f} + A} \tag{37.12}$$

or

$$\mathrm{Nu} = \frac{0.18 \cdot \mathrm{Pe} \cdot \sqrt{\lambda_f}}{\lg \mathrm{Pe} + \lg 0.35\sqrt{\lambda_f} + A} \tag{37.13}$$

where Pe is the Peclet number

$$\mathrm{Pe} = \mathrm{Re} \cdot \mathrm{Pr} = \frac{2R\bar{u}}{\chi}$$

and

$$A = \frac{b - \log b}{2.3} = 0.2b - \frac{\log b}{2.3} \tag{37.14}$$

The constant b may be determined on the basis of the following considerations:*

At Pr = 1, complete similarity must exist in a turbulent flow between the velocity and temperature distributions. Equations (37.7) and (4.16), therefore, must contain the same numerical coefficient. Consequently, we have

$$b \approx 11.7 \tag{37.15}$$

Equation (37.12) applies to the limiting case of heat transfer at very high Reynolds numbers and small Prandtl numbers. Equation (37.13) is applicable only at sufficiently high values of the Peclet number $\mathrm{Pe} > \mathrm{Pe_{cr}}$.

---

* Suggested by S. S. Kutateladze.

Rewriting (37.7) in the form

$$\delta'_T = 11.7 \frac{4\sqrt{2}}{\sqrt{\lambda_f}} \frac{\nu}{2\bar{u}R} \frac{\chi}{\nu} R \approx \frac{66}{\sqrt{\lambda_f}} \frac{R}{\mathrm{Pe}} \tag{37.16}$$

we see that as the Peclet number decreases $\delta'_T$ increases, and for $\mathrm{Pe} = \mathrm{Pe}_{cr}$, the following approximation holds:

$$\delta'_T \approx R$$

It is evident that

$$\mathrm{Pe}_{cr} \approx \frac{66}{\sqrt{\lambda_f}} \tag{37.17}$$

At $\mathrm{Pe} < \mathrm{Pe}_{cr}$, $\delta'_T > R$ and heat transfer by turbulence eddies becomes unimportant.

The values of Re and $\mathrm{Pe}_{cr}$ for which (37.13) is inapplicable are given below:

| Re | $0.5 \cdot 10^4$ | $1 \cdot 10^4$ | $5 \cdot 10^4$ |
|---|---|---|---|
| $\mathrm{Re}_{cr}$ | $3.4 \cdot 10^2$ | $3.7 \cdot 10^2$ | $4.5 \cdot 10^2$ |

In passing to the problem of comparing equation (37.13) with experimental data, it should be noted that the Nusselt number in (37.12) was related by us to the maximum temperature difference $(T_R - T_1)$. In practice the values measured are those related to the average temperature difference $(T_{av} - T_1)$, where

$$T_1 - T_{av} = \frac{2}{R^2} \int_0^R [T_1 - T(r)]\, r\, dr \tag{37.18}$$

In calculating the average, it should be borne in mind that at $r > \delta'_T$, the temperature distribution is given with logarithmic accuracy by (37.5) and at $r < \delta'_T$ by the formula (37.8). The results of the calculations are

$$(T_{av} - T_1) = (T_R - T_1) f\left(\frac{\delta'_T}{R}\right) \tag{37.19}$$

where

$$f\left(\frac{\delta'_T}{R}\right) = 1 + \frac{\left(\frac{\delta'_T}{R}\right)^2 \left(\frac{1}{2} - \frac{\gamma a b}{3}\right) - \frac{1}{2}}{\gamma a b - \ln\left(\frac{\delta'_T}{R}\right)} \tag{37.20}$$

Since $\frac{\delta'_T}{R} < \frac{66}{\sqrt{\lambda_f}} \frac{1}{\mathrm{Pe}}$, for large values of Pe the factor $f\left(\frac{\delta'_T}{R}\right)$ actually is not very different from unity, and it is possible to consider $(T_{av} - T_1) \approx (T_R - T_1)$.

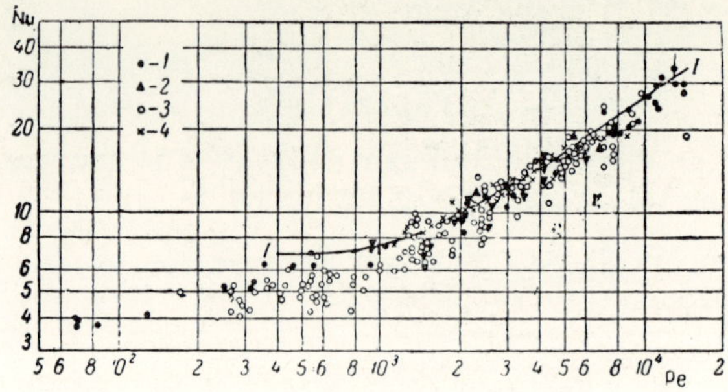


Figure 40. Experimental results of heat transfer to a lead — bismuth alloy and to lead.

1 indicates Pb-Bi; 2 indicates Sn (Mikheev, Baum, Voskresenskii, Fedynskii); 3 indicates Pb-Bi (Borishanskii, Kutateladze, Schneiderman, Ivashchenko); 4 indicates Pb (Ibragimov, Subbotin); I is the computed curve.

Equation (37.12) contains no unknown quantities, and may be compared directly with experimental data. This comparison has been carried out by S. S. Kutateladze. Figure 40 is taken from the report of S. S. Kutateladze et al mentioned above.

The experimental curve includes data obtained by several authors and agrees well with the computed curve.* Bearing in mind that the absence of unknown constants, we must conclude that the the absence of unknown constants, we must conclude that the proposed simple scheme of a division of the heat flow into two regions describes the characteristic peculiarities of heat transfer to liquid metals with sufficient accuracy.

## 38. GENERAL INTERPOLATION FORMULA FOR HEAT FLOW TO FLUIDS AT ANY VALUE OF THE PRANDTL NUMBER.

We have now obtained sufficient information to derive an interpolation formula for heat flow which is valid for any fluid.

Indeed, we may write two relationships for turbulent flow past a tube wall:

$$\mathrm{Nu} = \frac{0.17\,\mathrm{Re}\,\mathrm{Pr}\sqrt{k_f}}{0.17a\,\mathrm{Pr}^{3/4} + \ln\mathrm{Re} + \ln 0.2\sqrt{k_f}}, \tag{38.1}$$

$$\mathrm{Nu} = \frac{0.17\sqrt{k_f}\,\mathrm{Re}\,\mathrm{Pr}}{\ln\mathrm{Re} + \ln\mathrm{Pr} + A + \ln 0.2\sqrt{k_f}}. \tag{38.2}$$

---

* Let us note that measurements of heat transfer to liquid metals were first conducted on mercury by M. A. Styrikovich and I. E. Semenovker [11].

We derived the first for mass transfer at very high Prandtl numbers [Equation (25.17)]. The value of the only unknown quantity in (38.1), a, has been derived from diffusion experiments. According to experiments by I. A. Bagotskaya, given in Section 55, a = 2.6.

The second equation was obtained for very small Prandtl numbers [Equation (3.12)]. The value of the constant A = 2.4, was determined from entirely independent measurements. In deriving both formulas, it has been assumed that the physical properties of the heat carrier are not functions of temperature.

By considering the structure of both equations, it is not difficult to see that the following relation transforms into (38.1) and (38.2) at $Pr \gg 1$ and $Pr \ll 1$, respectively:

$$\mathrm{Nu} = \frac{0.17\sqrt{k_f}\,\mathrm{Re\,Pr}}{\ln \mathrm{Re} + \ln 0.2\sqrt{k_f} + (\ln \mathrm{Pr} + 2.4) + 0.5\,\mathrm{Pr}^{3/4}} \tag{38.3}$$

Since the denominator of formula (38.3) contains ln Pr and $Pr^{3/4}$, the accuracy of the interpolation is relatively high; for example, at $Pr \sim 10^{-2}$, the magnitude of the first is $\sim 5$, and that of the second $\sim 0.03 \ll 5$. At $Pr \sim 10^{2}$, the inequality is reversed.

The interpolation formula (38.3) may be compared with experimental results in the intermediate region, i.e., for Prandtl numbers of the order of unity. Formula (38.3) contains no arbitrary constants. Both constants, A and a, have been evaluated using independent measurements at the two limits: $Pr \ll 1$ and $Pr \gg 1$ respectively.

Figure 41 shows formula (38.3) plotted logarithmically. The crosses denote data obtained from the quoted work by Eagle and Ferguson [5], which are highly accurate and considered to be reliable.

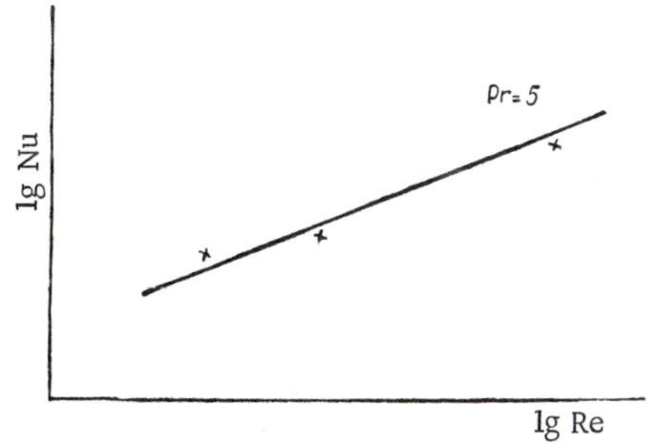


Fig. 41. Comparison of the interpolation formula for Nu with experimental data.

For a comparison with theory, the data for a Prandtl number Pr = 5 were arbitrarily chosen.

Within the limits of accuracy of a logarithmic scale, the agreement between formula (38.3) and the experimental results of Eagle and Ferguson is seen to be quite good. Thus we may assume that the interpolation formula (38.3) is sufficiently accurate and correctly reproduces the dependence of Nu on Re and Pr over the entire range of Prandtl numbers.

The merit of this formula is the feature that both its limiting values are substantiated in theory.

A similar formula may be written for an exterior flow past a plate.

The interpolation formula (38.3) is adaptable to a turbulent regime, provided the Reynolds number satisfies the following inequality:

$$\mathrm{Re} > \left(\frac{140}{\mathrm{Pr}}\right)^{7/8} \tag{38.4}$$

For Pr $\gg 1$, this inequality is immaterial since turbulence appears only at a much higher value of Re. For Pr $\ll 1$, inequality (38.4) is very important.

When using the interpolation formula for heat flow, average values for the physical properties should be employed. Further experimental tests of this formula would be of great interest.

Figure 39b illustrates schematically the temperature distribution for various Pr numbers.

## BIBLIOGRAPHY


1. Grober, G. and S. Erk, Osnovy ucheniya o teploobmene (Principles of Heat Exchange), ONTI, 1936; Grober and Erk, Die Grundgesetze der Warmeubertragung, Springer, Berlin, 1955.
2. Kibel, S. Z., Prikl. Mat. i Mekh., 11, 611 (1947).
3. Pohlhausen, Z.f. angew. Math. u. Mech. 1, 115 - 120 (1921).
4. Reichhardt, H., ZAMM, 297 (1940).
5. Eagle and Ferguson, Proc. Roy. Soc. A127, 540 - 566 (1930).
6. Mikheev, M. A., V. A. Baum, K. D. Voskresenskiy, and O. S. Fedynskiy, Reaktorostroyeniye i teoriya reaktorov (Reactor Construction and Theory of Reactors), Izd. AN SSSR, 1950, p. 139.
7. Novikov, I. I., A. N. Solov'ev, E. M. Khabakhpasheva, V. A. Gruzdev, A. I. Pridantsev, and M. Ya. Vasenina, Atomnaya Energiya (Atomic Energy), 1, 92 (1956).
8. Dwyer, Nucleonics, 12, 30 (1954).
9. Johnson, H., J. Hartnett, and W. Glabauch, Trans ASME 76, 513 (1954).
10. Kutateladze, S. S., V. I. Subbotin, V. M. Borishanskiy, and P. L. Kirillov, Teploobmen pri techenii zhidkogo metalla v trubakh (Heat exchange during the Flow of a Liquid Metal in Tubes), Report at the Second International Conference of the UN on the Peaceful Uses of Atomic Energy.
11. Styrikovich, M. A., and I. E. Semenovker, Zhurnal Teor. Fiz. 10, 1324 (1940).

CHAPTER V

# Certain Problems in the Theory of Coagulation of Dispersions Involving Liquids and Gases

## 39. SMOLUCHOWSKI'S THEORY

The phenomenon of coagulation, in which certain particles suspended in a liquid or gas form aggregates, is of great importance in colloid systems and aerosols.

While we cannot discuss this complicated phenomenon in detail here, we treat those of its aspects which relate closely to the theory of convective diffusion. Thus, we consider in some detail the coagulation of particles in moving dispersions, especially the vital effect of mixing. We begin, however, by reviewing briefly some items from the general theory of coagulation.

The size of particles encountered in dispersed systems varies widely: from approximately $10^{-4}$ cm in coarse dispersions to $10^{-7}$ cm in molecular dispersions. These particles experience Brownian motion, such that the smaller the particle, the more intense its motion. The randomness of this motion frequently brings individual particles in close proximity where Van der Waals forces strongly attract them to each other. In fact, these forces are so great that only in the case of very small particles ($10^{-7}$ cm) can thermal motion disrupt the attraction. Unless special measures are taken to stabilize the dispersion, particles that approach each other within a distance approximating their diameter become bonded to one another, and thus coagulate. When particle bonding occurs at each such meeting, the process is called "rapid coagulation."

The dispersion can be stabilized by:

1) diffuse ion layers (the outer layer of a double layer) that surround lyophobic sol particles [1].

The electrostatic repulsion of ion films under certain conditions is sufficiently strong to prevent the approach of particles to distances within which the Van der Waals attraction would predominate;

2) protective films formed by a layer of the fluid [2].

Protective coatings are formed by certain stabilizers which possess strong viscous properties and possibly a resilience effective against short-lived disturbances;

3) charges, or, possibly, dipoles [3] on the surface of aerosol particles.

The presence of a stabilizer reduces considerably the number of particle encounters which result in union — sometimes to the extent that coagulation becomes negligible. When stabilizing factors are present, the process is known as "slow coagulation."

Our subsequent discussion concentrates on the theory of rapid coagulation and does not further consider the stabilization of dispersions. Our chief concern is the effect of fluid motion on the number of encounters between suspended particles.

To analyze coagulation in a moving medium, we first consider the same phenomenon in a static medium. The theory of the latter phenomenon, developed by M. Smoluchowski [4], can be summarized briefly as follows.

Let us begin by considering a dispersion of $N_0$ spherical particles of equal size. The number of particles per unit volume is assumed to be sufficiently small so that the probability of triple, quadruple or more frequent encounters is negligible, and the dispersing medium (liquid or gas) is static. We are interested in the probability of collision between two dispersed particles as a result of their Brownian motion.

This probability is most easily calculated by taking advantage of the equivalence between Brownian motion and diffusion. Let us consider one of the particles to be stationary. Let us then imagine a sphere of radius R surrounding the particle such that any other particle entering the sphere will be attracted to the first and will become bonded to it. This means that over a spherical surface of radius R the particle concentration remains zero. The diffusional flux across that surface depends on the average rate at which particles cross as a result of Brownian motion. This model is limited, however, to those dispersions in which the concentration of particles is so low that encounters occur only in pairs. In the case of non-interacting particles, the radius of coagulation R will clearly equal the diameter of the particle a, R = a.

In order to find the diffusional flux let us consider the distribution of particles diffusing toward the sphere R. The concentration n of these particles satisfies the diffusion equation

$$\frac{\partial n}{\partial t} = D \frac{1}{r^2} \frac{\partial}{\partial r}\left(r^2 \frac{\partial n}{\partial r}\right), \tag{39.1}$$

where D is the Brownian diffusion coefficient of dispersed particles. (In cases where the Brownian diffusion coefficient might be mistaken for the coefficient of molecular diffusion, we will designate it by $D_{br}$.) The boundary and initial conditions are

$$\begin{aligned} n &= n_0 \quad \text{for} \quad r > R, \ t = 0, \\ n &= 0 \quad \text{at} \quad r = R, \ t > 0, \\ n &= n_0 \quad \text{as} \quad r \to \infty. \end{aligned} \tag{39.2}$$

The solution of equation (39.1) for conditions (39.2) is

$$n = n_0 \left[ 1 - \frac{R}{r} + \frac{2}{\sqrt{\pi}} \frac{R}{r} \int_0^{\frac{r-R}{2\sqrt{Dt}}} \exp\{-z^2\}\, dz \right]. \tag{39.3}$$

The flux of particles across the surface of the sphere is equal to

$$j = D\left(\frac{\partial n}{\partial r}\right)_{r=R} = \frac{Dn_0}{R}\left[1 + \frac{R}{\sqrt{\pi D t}}\right]. \tag{39.4}$$

The total number of particles that cross the surface r = R, i.e., the total number of collisions per unit time, equals

$$M = \int j\, dS = D \cdot 4\pi n_0 R \left[1 + \frac{R}{\sqrt{\pi D t}}\right]. \tag{39.5}$$

In view of what has been stated above, equation (39.5) gives the number of contacts that we seek, i.e., from which we can derive the rate of rapid coagulation.

We now assume that the coagulation process is viewed t seconds after its start, when $t \gg \frac{R^2}{D}$. In the case of usual particle dimensions this means that the elapsed time from the start of the aggregation process exceeds $10^{-3}$ to $10^{-4}$ seconds. The number of encounters per unit time after this initial period equals

$$M = 4\pi D R n_0 . \tag{39.6}$$

In deriving this expression for the number of encounters, we should have accounted for the motion of the particular particles under consideration. It is easy to prove, in the case of like particles, that accounting for this motion leads to doubling of the diffusion coefficient. Thus, for like particles, the total number of contacts per unit volume of dispersion per second is

$$N = 8\pi D R n_0^2 = 8\pi D a n_0^2, \tag{39.7}$$

where a is the diameter of the particle.

For the rate of change of n we obtain

$$\frac{dn}{dt} = -8\pi D R n^2. \tag{39.8}$$

Integration of (39.8) [for constant DR, cf. 39.12 — ed.] gives the reduction of the number of particles in time t due to their coagulation

$$n = \frac{n_0}{1 + \frac{t}{T}}, \tag{39.9}$$

where

$$T = \frac{1}{8\pi D R n_0} \tag{39.10}$$

is known as the coagulation time. The reciprocal value

$$k = \frac{1}{T} = 8\pi D R n_0 \tag{39.11}$$

is known as the coagulation constant. It should be stressed that the above somewhat rough treatment leads to the same results as the rigorous derivation based on the solution of the appropriate stochastic problem by A. N. Kolmogorov and M. A. Leontovich [5].

For comparison with the theory of coagulation in a medium that is in motion, we may consider the above discussion to be adequate. It is not the objective of this book to set forth generalizations for cases of particles of different dimensions [4], or for the case of polydiverse systems [6].

In conclusion let us note that:

1) Brownian diffusion coefficients for colloidal and aerosol particles are extremely small in comparison with molecular diffusion coefficients. If the diffusion coefficient is calculated according to the well-known Einstein formula (which is entirely satisfactory for order of magnitude estimates)

$$D = \frac{kT}{3\pi\mu a}$$

then for colloid particles in water and at room temperature we have

$$D \approx \frac{1.1 \cdot 10^{-13}}{a} \tag{39.12}$$

and for aerosol particles in air under normal conditions

$$D \approx \frac{1 \cdot 10^{-13}}{a} \frac{10^{-2}}{0.15 \cdot 10^{-3}} = \frac{1.3 \cdot 10^{-11}}{2 \cdot a}.$$

Establishing the ratio of the kinematic viscosity of the medium to the coefficient of diffusion, i.e., the analogue of the Prandtl number, we find for water

$$\mathrm{Pr} = \frac{\nu}{D} = \frac{10^{-2}}{1 \cdot 10^{-13}} \cdot a = 2.5 \cdot 10^{10} a$$

and for air

$$\mathrm{Pr} = \frac{0.15a}{2 \cdot 0.13 \cdot 10^{-10}} \approx 10^{10} a.$$

Even for the smallest particles the Prandtl number, in water as well as in air, is found to be much higher than unity. Thus we are justified in applying the methods developed above for diffusion at high Prandtl numbers to diffusional phenomena in dispersions.

2) Because the dispersed particles also possess considerable mass, their behavior exhibits intertial effects not found in purely diffusional processes. A heavy particle, for example, instead of following the curvature of the flow, may move in the straight line dictated by its inertia. This phenomena was first observed by I. Langmuir, who noted its importance in the behavior of aerosols containing particles of high relative density. These intertial effects will be discussed briefly later.

## 40. GRADIENT COAGULATION

Let us examine a dispersed medium in motion — liquid or gas — with a nonzero velocity gradient. The simplest example of such motion might be laminar flow near a solid wall. It is not difficult to see that the existence of a velocity gradient leads to contacts between particles of finite radius. A particle situated far from the wall moves with greater velocity than a particle situated close to the wall. If the distance between the two particles does not exceed the sum of their radii, the particles can collide. For a quantitative computation of the number of encounters, we make the following simplifying assumptions:

1) we assume that the effect of Brownian diffusion of particles toward one another may be disregarded;

2) we do not take into account the hydrodynamic interaction between particles and, in particular, the deformation of their trajectories caused by such interaction.

Let us examine two particles, the radii of which are $R_1$ and $R_2$, in a laminar stream flowing parallel to the wall (along the x axis). Let the distance between their trajectories, taken normal to the wall, be $\Delta y$. Assume that the particles move in straight-line paths that coincide with the streamlines. We may then assume that one is stationary while the other moves relative to it at a velocity of $v_x(\Delta y)$, where $\Delta y$ is the distance between the two particles.

At low values for $\Delta y$ we may write

$$v_x(\Delta y) \approx \frac{\partial v_x}{\partial y} \Delta y = \Gamma \Delta y, \tag{40.1}$$

where $\Gamma$ is the velocity gradient in the fluid.

A collision between particles is possible if the distance between them measured along the y axis does not exceed the quantity

$$\Delta y \leqslant (R_1 + R_2) \sin \theta.$$

The number of encounters occurring in one second is equal to the number of particles that reach the surface of a sphere of radius $(R_1 + R_2)$. This is evidently equal to

$$M = n_0 \int v_x \cdot 2(R_1 + R_2) \cos\theta \, d\,\Delta y =$$

$$= 4n_0\Gamma (R_1 + R_2)^3 \int\limits_0^{\frac{\pi}{2}} \cos^2\theta \sin\theta \, d\theta = \frac{4}{3} n_0 \Gamma (R_1 + R_2)^3.$$

In the case of like particles, i.e., where $R_1 = R_2 = R$, the total number of encounters is expressed by

$$N_{grad} = \frac{32}{3} n_0^2 \Gamma R^3 = \frac{32}{3} n_0^2 \Gamma a^3. \qquad (40.2)$$

This expression was first derived by M. Smoluchowski, and its distinctive feature is the rapid rise in the number of encounters with increasing particle size.

The relative coagulative effect of mixing, as compared with Brownian movement, is given to an order of magnitude determined by the expression

$$\beta = \frac{N_{grad}}{N_{br}} = \frac{4\Gamma a^2}{3\pi D},$$

where $N_{br}$ is the number of contacts resulting from Brownian movement given by (39.7), while $N_{grad}$ is given by (40.2). The ratio increases in proportion to the cube of the dimension of the particles $\left(D = \frac{2.13 \cdot 10^{-13}}{a} \text{ cm}^2/\text{sec}\right)$ and to the velocity gradient of the fluid.

At particle dimensions $R \sim 10^{-4}$ cm in water, we have

$$\beta \sim \frac{10^{-12}}{10^{-12}} \Gamma = \Gamma.$$

At relatively low values of $\Gamma$ ($\Gamma \sim 10$ per sec), $\beta$ is greater than unity, and thus the coagulating effect of mixing predominates over the Brownian effect. For particles on the order of $10^{-6}$ cm, however, gradients 1000 times larger are required to maintain this predominance.

In the case of aerosols, the diffusion coefficient is approximately 100 times greater, thus increasing the scale of magnitude in all relationships to include larger dimensions and velocity gradients.

It must be stressed that (40.2) remains valid not only in the case of laminar fluid flow in a tube, but for every case in which (40.1) holds. This includes any stream in which the velocity gradient is sufficiently small so that, at distances of the order of a particle diameter, the velocity may be regarded as varying linearly according to (40.1). In particular, (40.2) is applicable to the coagulation of particles in a turbulent flow moving along a solid surface.

The average velocity gradient in a turbulent flow, as opposed to a laminar flow, varies with the distance y from the solid surface. Differentiating equation (4.17) with respect to y, we obtain

$$\Gamma \sim \frac{v_0}{y}. \tag{40.3}$$

This shows that the velocity gradient has its greatest value in the viscous sublayer near the solid surface. Here, since velocity is a linear function of the distance from the wall, the velocity gradient has a constant value. This may be determined from (40.2) by letting y equal $\delta_0$. We therefore obtain

$$\Gamma \sim \frac{v_0}{\delta_0} \sim \frac{v_0^2}{\nu}. \tag{40.4}$$

For the number of encounters we derive

$$N_{\text{grad}} \sim \frac{32}{3} n_0 R^3 \frac{v_0^2}{\nu} = \frac{32}{3} n_0 \cdot a^3 \frac{v_0^2}{\nu}. \tag{40.5}$$

The velocity gradient in turbulent streams may attain very high values. Thus, in water at a rate of flow of the order of 15 m/sec, $v_0 \sim 100$ cm/sec and $\Gamma \sim 10^6$.

The total coagulative action, however, is limited to a very thin layer close to the wall and depends on the duration of the contact between the walls and the dispersion. It may be large for passage through long tubes, or for a large area of contact between the solid and the dispersion. The formulas obtained could be generalized to cover cases of polydisperse systems.

In the light of the current state of experimentation in this field, the derivation of more precise formulas would not be of interest.

## 41. THEORY OF COAGULATION OF COLLOIDS IN TURBULENT FLOWS

The mechanism of encounters between particles suspended in a laminar fluid, as discussed in the preceding section, has very limited application in practice.

In the usual mixing of a fluid, the flow is turbulent. The frequency of contacts between particles in such flow increases substantially in comparison with the number of encounters in a motionless medium or in laminar flow. As shown below, the increase in the number of encounters in a turbulent fluid may be caused by two independent and essentially different mechanisms. One of these functions only when the difference between the density of the particles and the density of the medium is great. Thus, for example, it functions in the turbulent mixing of a gas containing aerosol and, in this case, as shown by estimates, even dominates the coagulative process. We reserve discussion of this phenomenon for the following section.

Here we discuss [7] the more general mechanism of dispersions under turbulent agitation, coagulation as seen in the aggregation of colloids in a liquid medium and of aerosols in a gaseous medium (although in the latter instance this mechanism usually plays a secondary role). Because of the latter circumstance, we shall discuss here the aggregation of colloids in a liquid medium.

Let us examine a colloid in a turbulent flow of a liquid and let us assume that the following conditions are fulfilled:

1) The coagulation radius R of the particles is very small in comparison with the microscale of turbulence $\lambda_0$. Since the dimensions of the colloidal particles on the average range from $10^{-5}$ to $10^{-6}$ cm, while the quantity $\lambda_0$, even with very energetic mixing, is about $10^{-2}$ to $10^{-3}$ cm, the condition is in fact always fulfilled;

2) there is complete containment of colloidal particles within the turbulent eddy, and the mechanism of contact between suspended particles occurs on a scale less than the size of the eddy.

This assumption is justified because the density of the colloid particle is not very great, differing from that of the liquid by a factor of two or three. In the case of aerosol particles, where the density is approximately $10^3$ times that of air, full containment of the particles within the eddy does not take place. The second coagulation mechanism, mentioned above, is linked to the presence of inertial effects.

Let us assume that the colloid particles suspended in the turbulent stream have an average concentration n (the number of particles per unit volume). Since the size of the colloidal particles is small compared to the microscale of turbulence, we may assume that they are transferred by isotropic turbulence. Colloidal particles contained in turbulent eddies migrate through the bulk liquid in a chaotic fashion, their movement being similar to that of Brownian motion.

The eddy motion of particles, therefore, may be characterized by a certain turbulent diffusion coefficient $D_{turb}$. The problem of coagulation of colloid particles transported by turbulent eddies may then be reduced to a diffusion problem, in the same way that this is done for Brownian movement in a static medium. We assume that a diffusion of particles occurs toward a sphere of radius R, such that the distribution of the particles may be characterized by the diffusion equation

$$\operatorname{div}(D_* \operatorname{grad} n) = 0. \tag{41.1}$$

In this equation we have accounted for the fact that n represents a temporary average concentration which satisfies the static diffusion equation. $D_*$ designates the effective diffusion coefficient. Particles are diffused over long distances, as will be shown below, by turbulence eddies, in which case $D_*$ becomes the coefficient of turbulent diffusion $D_{turb}$. In regions of short distances, ordinary Brownian diffusion takes place and $D_*$ becomes the coefficient of Brownian diffusion $D_{br}$.

Since the quantity $D_{turb}$ is a function of the eddy scale and varies from point to point, $D_*$ is also in general, a variable quantity.

The boundary conditions are

$$n = 0 \quad \text{at} \quad r = R, \tag{41.2}$$

$$n = n_0 \quad \text{as} \quad r \to \infty. \tag{41.3}$$

Noting the fact that the solution of the boundary value problem may be a function only of the radius-vector r, we can integrate equation (41.1) and write

$$D_* r^2 \frac{\partial n}{\partial r} = \text{const.}$$

The coefficient of turbulent diffusion has already been examined above in connection with the problem of mass transfer across a diffusion boundary layer. Now we must obtain a corresponding expression for $D_{turb}$ in the case of uniform isotropic turbulence.

Having written the relationship (24.5) for $D_{turb}$, and allowing for (32.1) and (32.9), we find

$$D_{turb} \sim v_\lambda \cdot \lambda \sim \begin{cases} \alpha (\varepsilon_0 \lambda)^{1/3} \lambda & \text{for} \quad \lambda > \lambda_0, \\ \beta \sqrt{\frac{\varepsilon_0}{\nu}} \lambda^2 & \text{for} \quad \lambda < \lambda_0. \end{cases} \tag{41.4}$$

The first of these formulas, relating to the region in which eddies exceed (in size) the turbulent microscale, shows that $D_{turb}$ increases with $\lambda$ as $\lambda^{4/3}$. At the same time the value of $D_{turb}$ so greatly exceeds the coefficient of molecular diffusion D that the main zone of diffusion resistance is in the region of small $\lambda$, smaller than $\lambda_0$. In other words, the intense turbulent agitation of the liquid ensures a uniformity of particle concentration in the region $\lambda > \lambda_0$.

We therefore examine the distribution of concentration n only in the region $\lambda < \lambda_0$.

In this region, $D_{turb}$ decreases with decreasing scale of turbulent eddies $\lambda$, as $\lambda^2$. It is quite evident that a sufficiently small-scale $\lambda_1$ may be found at which $D_{turb}$ will become smaller than $D_{br}$. Thus at $\lambda = \lambda_1$,

$$\beta \sqrt{\frac{\varepsilon_0}{\nu}} \lambda_1^2 = D_{br},$$

the quantities $D_{turb}$ and $D_{br}$ are equal, and for $\lambda < \lambda_1$, $D_{turb} < D_{br}$.

Such small scales, however, might never be realized, and this constitutes the difference between the diffusion of dissolved particles of molecular dimensions and the diffusion of macroscopic colloidal particles.

If the coagulation radius R is such that $R > \lambda_1$, then the coefficient of turbulent diffusion $D_{turb}$ exceeds the coefficient of Brownian diffusion $D_{br}$ over the entire region where diffusion takes place.

Let us first examine the case

$$R > \lambda_1 = \sqrt[4]{\frac{D_{br}^2}{\beta^2} \frac{\nu}{\varepsilon_0}}.$$

Then it is possible to state that everywhere

$$D_* = D_{\text{turb}}$$

and equation (41.1) takes the form

$$r^2 D_{\text{turb}} \frac{dn}{dr} = c_1.$$

Substituting the value of $D_{\text{turb}}$ from formula (41.4) we find

$$\beta \sqrt{\frac{\varepsilon_0}{\nu}}\, r^4 \frac{dn}{dr} = c_1 \quad \text{where} \quad r < \lambda_0, \tag{41.5}$$

$$\alpha \sqrt[3]{\varepsilon_0}\, r^{10/3} \frac{dn}{dr} = c_2 \quad \text{where} \quad r > \lambda_0. \tag{41.6}$$

Integrating equations (41.5) and (41.6), using boundary conditions (41.2) and (41.3), and allowing for continuity of n over the surface $r = \lambda_0$, we find

$$n \approx \frac{n_0}{1 + \frac{2}{7}\frac{R^3}{\lambda_0^3}} \left(1 - \frac{R^3}{r^3}\right), \qquad r < \lambda_0.$$

The flux of particles across one $cm^2$ of the surface of the sphere of coagulation r = R is equal to

$$j = D_* \left(\frac{\partial n}{\partial r}\right)_{r=R} = D_{\text{turb}} \left(\frac{\partial n}{\partial r}\right)_{r=R} =$$

$$= \frac{3n_0}{1 + \frac{2}{7}\frac{R^3}{\lambda_0^3}} \frac{R^3}{R^4} \beta \sqrt{\frac{\varepsilon_0}{\nu}} R^2 = \frac{3n_0 R\beta \sqrt{\frac{\varepsilon_0}{\nu}}}{1 + \frac{2}{7}\frac{R^3}{\lambda_0^3}}.$$

With an accuracy of $\left(\frac{R^3}{\lambda_0^3}\right)$, j may be written in the form

$$j = 3n_0 R\beta \sqrt{\frac{\varepsilon_0}{\nu}}. \tag{41.7}$$

This means that the concentration of particles in the region $r > \lambda_0$ may be considered constant, with an accuracy on the order of $\left(\frac{R}{\lambda_0}\right)^3$.

The principal diffusion resistance lies in the region $r < \lambda_0$. The meaning of this last statement is quite clear: relatively large scale eddies $\lambda > \lambda_0 \gg R$ agitate the liquid with such energy that they ensure a uniform distribution of colloid particles in the bulk of the liquid. The total number of encounters per unit time brought about by turbulence agitation, or the rate of coagulation, according to formula (41.7), is equal to

$$N_{\text{turb}} = 4\pi R^2 j n_0 = 12\pi\beta \sqrt{\frac{\varepsilon_0}{\nu}} R^3 n_0^2. \tag{41.8}$$

Expressing $\epsilon$ in terms of the fluid flow velocity v, and in terms of the scale L of large-scale motions, we may rewrite equation (41.8) in the form

$$N_{\text{turb}} \sim \frac{R^3 n_0^2 v^{3/2}}{\sqrt{\nu L}} \sim R^3 \nu n_0^2 \frac{\mathrm{Re}^{3/2}}{L^2}. \tag{41.9}$$

Equations (41.8) and (41.9) show that the rate of coagulation, as in the case of laminar mixing, is proportional to the cube of the radius of coagulation R. Furthermore, $N_{turb}$ increases with the characteristic velocity of the flow as $v^{3/2}$, and is only slightly dependent on the scale of large-scale eddies and the viscosity of the fluid. Equation (41.9) is valid for a monodispersion system in which particles are of such dimensions that

$$R > \sqrt[4]{\frac{D_{\text{br}}^2 \nu}{\beta^2 \varepsilon_0}}.$$

Comparing (41.7) with the analogous expression for the diffusional flux of dissolved particles of molecular dimensions, we see immediately an essential difference between them. While the expression for the flux of dissolved particles always includes the coefficient of molecular diffusion, equation (41.7) does not include the coefficient of Brownian diffusion. The reason for the difference in the two equations is quite evident. In the case of solutions in which the size of the dispersed particles is negligible on a macroscopic scale, a region of molecular diffusion always exists. At the same time, as we saw in Section 25, it is precisely this region that determines the primary diffusion resistance. On the other hand, for sufficiently large colloidal particles the region of Brownian diffusion is totally absent, and the coefficient of Brownian diffusion does not enter the expression for j.

Let us now turn to the examination of small particles for which $R < \sqrt[4]{\frac{D_{\text{br}}^2}{\beta^2} \frac{\nu}{\epsilon_0}}$. In this case, in the region $r < \lambda_1 = \sqrt[4]{\frac{D_{\text{br}}^2}{\beta^2} \frac{\nu}{\epsilon_0}}$, Brownian diffusion predominates over turbulent diffusion. Simple calculations, quite analogous to those just given, give the following expression for the flux to one $cm^2$ of coagulation surface

$$j = \frac{D_{\text{br}} n_0}{R\left[1 + \frac{2}{7}\frac{R^3}{\lambda_1^3}\right]} \approx \frac{D_{\text{br}} n_0}{R}.$$

With an accuracy of the order of $\frac{R^3}{\lambda_1^3}$, therefore, the formula for j coincides with the Smoluchowski formula. This means that the controlling diffusion resistance is located in the region of short distances $r < \lambda_1$. In view of this, the rate of coagulation N is independent of the agitation of the fluid.

Thus, in the case of particles for which $R > \lambda_1$, the rate of coagulation in a turbulent flow is determined by the number of encounters of particles entrained by turbulent eddies. The rate of coagulation of particles for which $R < \lambda_1$, even in a turbulent fluid, is determined by the Brownian diffusion process.

For a quantitative comparison of the rates of coagulation resulting from turbulent agitation and Brownian diffusion, it is of interest to set up the ratio of $N_{turb}$ to $N_{br}$. According to (41.9) and (39.8), we have

$$\frac{N_{turb}}{N_{br}} \sim \sqrt{\frac{\varepsilon_0}{\nu}} \frac{R^2}{D_{br}}$$

In water, at room temperature, this ratio (where R is expressed in centimeters) is

$$\frac{N_{turb}}{N_{br}} \sim 5 \cdot 10^{-13} \sqrt{\varepsilon_0}\, a^3 .$$

Here we have allowed for the fact that $D_{br}$ is a function of a according to equation (39.12).

Under usual conditions of agitation, $\epsilon$ may attain values on the order of $10^3$ or $10^4$ and sometimes greater. At values of $\epsilon_0$ on the order of $10^3$ or $10^4$, the quantity R is greater than $\lambda_1$, and, correspondingly, $N_{turb}$ is greater than $N_{br}$ for particles not exceeding $10^{-5}$ cm in size.

The coagulation of finer particles is brought about primarily by Brownian diffusion.

As the motion of the fluid is intensified, turbulent agitation assumes the major role for the coagulation of ever smaller particles.

The coagulation rate increases with the velocity of large scale motions faster than in the case of a laminar flow.

As in the case of a laminar flow, the pronounced dependence of $N_{turb}$ on the radius of the particles leads to the appearance of a latent period.

The theory set forth here relates mainly to dispersed particles suspended in a liquid phase, i.e., to colloids.

In the case of aerosols, condition $R > \sqrt[4]{\frac{D^2 \nu}{\beta^2 \epsilon_0}}$ (for the same value of $\epsilon_0$) imposes more rigid requirements on particle size, since, at ordinary pressures, the diffusion coefficient D in a gas is much larger than in liquids. By the same token, the assumption as to the complete containment of particles by turbulent eddies in a gas is not valid.

In the next section we examine the coagulation of dispersed particles suspended in a turbulent gas stream.

It must be noted that prior to the publication of our paper, several other papers had been devoted to the problem of the effect of turbulent eddies on the aggregation of colloids and aerosols.

In the work of Wigand and Frankenberg [8], an attempt was made to allow for the effect of turbulence on the coagulation of aerosols by taking the effective velocity gradient in the turbulent fluid to be equal to $\frac{v}{4}$, where v is the average flow velocity. In view of the complete arbitrariness of this assumption, that attempt will not be discussed here.

## 42. A SECOND MECHANISM OF COAGULATION OF AEROSOL PARTICLES IN TURBULENT FLOWS

The diffusion mechanism for encounters between particles suspended in a turbulent fluid, which we examined in Section 41, brings us to the coagulation of aerosol as well as colloid particles. With the diffusion mechanism, however, another mechanism governing the encounters between particles [9] may play an important part in coagulation of aerosols.

Because of the significant differences in density between the medium and the aerosol particles, the latter are not fully entrained by turbulent eddies.

Velocities acquired by the aerosol particles are largely dependent on particle masses. Particles of different sizes move at different velocities and collide with each other as a result.

As shown by computation, the number of encounters brought about by this mechanism in a polydispersed aerosol may substantially exceed the number of encounters resulting from a diffusion mechanism. Therefore, in the polydispersed aerosol examined below, the coagulation mechanism essentially determines the overall rate of the process. As before, we will assume that the size of the aerosol particles is small compared to the microscale of turbulence $\lambda_0$.

As we have seen in Section 33, particles of size $a \lesssim \lambda_0$ suspended in a gas have a relative velocity given by equation (33.19).

For turbulence to produce coagulation, the relative velocity u of the particles must be a function of the particle size a. Since the larger particles in a field of acceleration w acquire greater velocities, they move in relation to the finer particles.

The number of encounters experienced by particles of size a with particles of radius r per unit time is equal to

$$n_0\pi (a+r)^2 (a^2 - r^2)\frac{\rho}{\rho_0}\frac{\varepsilon_0^{3/4}}{\nu^{5/4}}$$

where $\pi (a + r)^2$ in the area, $(a^2 - r^2)\frac{\rho}{\rho_0}\frac{\epsilon_0^{3/4}}{\nu^{5/4}}$ is the height of the cylinder swept out by a particle of radius a in its motion relative to a particle of radius r, and $n_0$ is the average number of particles per unit volume.

The number of encounters experienced by particles of size a with all particles of size $r \leq a$ is equal to

$$\int_0^a \pi (a+r)^2 (a^2 - r^2) n_0 f(r) \frac{\rho}{\rho_0} \frac{\varepsilon_0^{3/4}}{\nu^{5/4}} dr,$$

where f(r) is a function of the size distribution of aerosol particles.

The total number of encounters experienced by aerosol particles per unit time per cm$^3$ (coagulation rate) is equal to

$$N_{accel} = \pi \frac{\rho}{\rho_0} n_0^2 \frac{\varepsilon_0^{3/4}}{\nu^{5/4}} \overline{(a+r)^2 (a^2 - r^2)}, \tag{42.1}$$

where the overbar indicates the average over the entire particle-size distribution of the aerosol. The subscript for N indicates the mechanism that causes the aggregation.

The mechanism examined above results in coagulation only for a polydispersed aerosol. The actual value of

$$\overline{(a+r)^2 (a^2 - r^2)}$$

depends to a great degree on the form of the function of size distribution for the aerosol particles.

Without formulating a specific form for this function, we may set down the following:

$$\overline{(a^2 + r^2)(a^2 - r^2)} = 3\overline{a^4}, \tag{42.2}$$

where $\bar{a}$ is the average size of the particles.

Substituting (42.2) into (42.1), we obtain for the aggregation rate

$$N_{accel} \sim \frac{\rho}{\rho_0} n_0^2 \frac{\varepsilon_0^{3/4}}{\nu^{5/4}} \overline{a^4}. \tag{42.3}$$

Expressing $\epsilon$ in terms of the basic velocity of the turbulent flow given by equation (4.3), we find

$$N \sim v^{9/4};$$

i.e., the aggregation rate increases very rapidly with the velocity v.

The mechanisms of turbulent and Brownian diffusion, of course, are also active in this case, together with the mechanism now under examination.

Let us compare the aggregation rates $N_{accel}$ with $N_{turb}$. Having determined the ratio $N_{accel}/N_{turb}$ given by (42.3) and (41.8), we find

$$\frac{N_{accel}}{N_{turb}} \sim \frac{\rho}{\rho_0} \frac{\bar{a}}{\lambda_0} \sim 10^3 \frac{\bar{a}}{\lambda_0}$$

We see that the predominance of the competing coagulation mechanism is determined in one case by the average size of the particles, and in the other by the intensity of the agitation causing the turbulence.

Generally speaking, the acceleration mechanism predominates in the case of larger particles. For $\lambda_0 \sim 10^{-1} - 10^{-2}$ cm, the acceleration mechanism is dominant for particles greater in size than $\bar{a} > 10^{-4} - 10^{-5}$ cm.

The finer aerosol particles, of sizes less than $10^{-5}$ cm, aggregate mostly as a result of turbulent diffusion.

Brownian diffusion predominates in cases where particles are of the order of $10^{-6}$ cm.

In conclusion, the following important remark must be made: in formula (42.1) we tacitly assumed that each encounter of aerosol particles is effective and leads to coagulation. In other words, we did not allow for Langmuir's collection efficiency. The problem of the effectiveness of collisions is examined separately in Section 43.

## 43. IMPACTION OF AEROSOLS AND COLLOIDS

The precipitation of aerosols and colloids is of great importance from both a theoretical and practical standpoint. By impaction, we mean the process by which a direct contact is established between the dispersed particle and some body with dimensions substantially greater than those of the particle.

We can not, within the framework of this book, examine in detail all the various problems associated with the impaction of aerosols; they are treated, however, in a monograph by N. A. Fuks [10].

We will be interested only in the special case of diffusional impaction of dispersed particles.

However, before discussing the problems of diffusional impaction we will examine briefly the phenomenon of inertial impaction, which plays a very important role in the precipitation of relatively large dispersed particles. Suppose there is a laminar flow past a certain body. Let there be suspended in the laminar flow dispersed particles so large in size that their Brownian diffusion may be disregarded. The suspended particles follow the streamlines of the fluid flowing past the body. Therefore, the dispersed particles could not reach the surface without inertial effect or finite dimensions for the particles.

As shown in Figure 42, for the case of curved streamlines, the inertia of the particles causes them to veer from the streamlines (solid lines) and move in straighter paths (dotted lines).

In this way, dispersed particles that would have by-passed the body by following streamlines that curve around it actually reach the surface of the body. The effectiveness of impaction of particles on the body may be expressed by the quantity E, defined as the ratio of the number of particles that actually reach the body to the number that would have reached it had they continued in a straight line instead of following the streamlines.

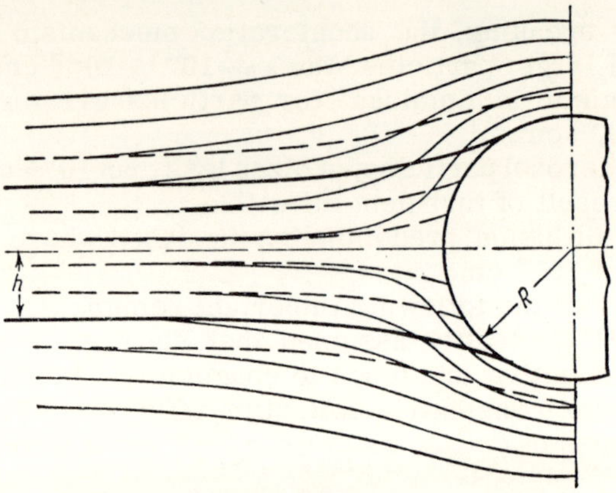


Fig. 42. Inertial impaction of particles.

In order to find the impaction efficiency, it is necessary to find the trajectories of heavy particles in the stream flowing past the body.

If the outermost trajectories of particles that reach the body are known (they are shown by heavy lines in Figure 42), then the precipitation coefficient E will be determined by the ratio of the crosssection of the tubular flow, formed by the outermost paths, to the midsection of the body.

For example, for a sphere

$$E=\frac{\pi h^2}{\pi R^2},$$

where h is the starting distance (Figure 42).

The trajectories of particles in the flowing fluid are determined by equation (33.1)

$$\rho V\frac{d\mathbf{v}_1}{dt}=\rho_0 V\frac{d\mathbf{v}_0}{dt}+\mathbf{F}_{\text{res}}.$$

If the medium is in steady motion and the particle is sufficiently small to allow the use of the Stokes formula for $F_{res}$, then the last equation may be written in the form

$$\rho V\frac{d^2\mathbf{r}}{dt^2}=6\pi\mu a\left(\mathbf{v}_0-\frac{d\mathbf{r}}{dt}\right), \tag{43.1}$$

where r is the radius-vector of the particle.

It is convenient to write equation (43.1) in dimensionless form

$$\text{Stk}\,\frac{d^2\rho}{d\tau^2}+\frac{d\rho}{d\tau}=\mathbf{V}_0(\rho), \tag{43.2}$$

where $\rho = \frac{\mathbf{r}}{R}$ is the dimensionless coordinate of the particle, R is the characteristic dimension of the body around which the flow passes, $\mathbf{V}_0$ is the dimensionless velocity of the fluid $\mathbf{V}_0 = \frac{\mathbf{v}_0}{v_\infty}$, $v_\infty$ is the velocity of the fluid far from the body, $\tau$ is the dimensionless time $\tau = \frac{v_\infty}{R}t$, and Stk denotes the dimensionless quantity

$$\text{Stk}=\frac{4v_\infty\rho a^2}{18\mu R}=\frac{v_\infty m}{6\pi\mu aR}=\frac{2\text{Re}\,\rho a^2}{9\rho_0 R^2}, \tag{43.3}$$

known as the Stokes number (m is the mass of the particle). The Stokes number is the only criterion determining similarity for this problem.

The velocity distribution $V_0(\rho)$ is assumed to be a known function of the coordinates. The particle dimensions are assumed to be small compared to the dimensions of the body. Here, we may disregard the effect of the particles on the flow of fluid past the body.

The integration of equation (43.2) will allow us, in principle, to find the trajectory of the particle $\rho(t)$ and to find, in particular, the outermost trajectories that contact the surface of the body.

From (43.3) and (43.2) where Stk = 0 (particles of small mass), the velocity of the particle is $\mathbf{V}_0 \approx \frac{d\rho}{dt}$; i.e., the particle follows a streamline and never reaches the surface of the body. Similarly, at a sufficiently low value of the dimensionless Stokes number, the particle will also fail to reach the surface of the body.

There exists a certain minimum value of the Stokes numbers, above which the particle's inertia predominates over its entrainment by the fluid and the particle reaches the surface of the body. More precisely, the particle passes at a smaller distance from the surface than the particle's own dimension.

At $\text{Stk} < \text{Stk}_{crit}$ impaction of the particle is completely impossible. At $\text{Stk} > \text{Stk}_{crit}$, impaction is possible, in theory.

The existence of a critical value for the Stokes number, below which inertial impaction is impossible, was first established by Langmuir [11].

The system of equations (43.2) has not yet been integrated in the general case. Only certain simplified cases have been examined:

1) The case of flow around an immersed object for a viscous regime $\text{Re} \ll 1$, with the assumption that the size of the precipitated particle is negligible $\left(\frac{a}{R} \to 0\right)$. The streamlines for any type of fluid flow past a body are deformed at a relatively large distance from the body — a distance of the order of its dimensions. In this

region, the motion of the fluid is retarded and the streamlines curve around the body. Impaction of the particles on the surface of the body, therefore, requires relatively large inertial effects — i.e., a high value for the Stokes number.

Equation (43.2) has, however, been integrated numerically, and a curve has been obtained for the impaction efficiency E as a function of the Stokes number. The critical value of the Stokes number is found to be $Stk \approx 1.2$.

The formulas obtained are relatively limited in application, since the dimensions of the sphere must be large compared to the size of the particle, while the particle itself must be of such magnitude that its diffusion can be neglected.

2) Langmuir and Blodgett [12], by means of numerical integration, computed the value of E for the case of a potential flow past a sphere or circular cylinder. These authors completely disregarded the existence of a boundary layer at the body's surface, assuming the fluid flowing along the surface of the solid body to be ideal. The results of the numerical calculations by these authors can be approximated by the formula

$$E = \begin{cases} \dfrac{(\mathrm{Stk})^2}{(\mathrm{Stk}+0.06)^2}, & \mathrm{Stk} > \mathrm{Stk}_{crit} = 0.08, \\ 0, & \mathrm{Stk} < \mathrm{Stk}_{crit} = 0.08. \end{cases}$$

The computations were also based on the assumption that $\frac{a}{R} \to 0$, and thus the dimensions of the particle could be disregarded.

Figure 43 shows the impaction efficiency E as a function of the Stokes number for a potential flow and a viscous flow past a body.*

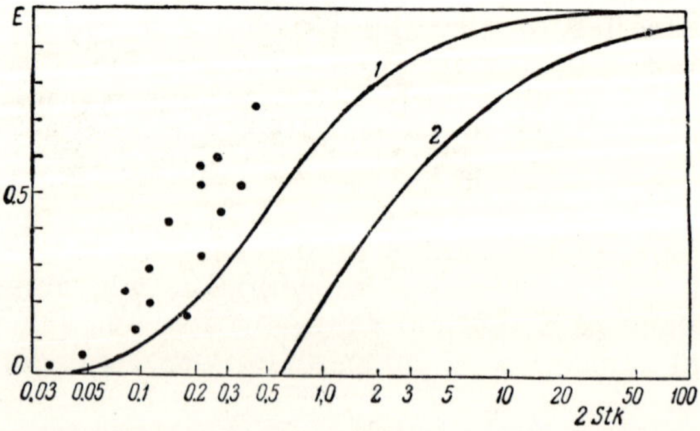


Figure 43. Impaction efficiency for a sphere: 1 — potential flow, 2 — viscous flow; the dots indicate experiments by Ranz and Wong.

*Figure 43 has been taken from the monograph by N. A. Fuks [10]. Cf. Ranz and Wong. Ind. Eng. Chem. 44, 1371 (1952).

The region of displacement in a potential flow, where the streamlines are deformed and curve away from the surface of the body, is substantially smaller than the size of the body. The impaction efficiency is therefore greater at a lower Stokes number for a potential flow than for a viscous flow.

L. M. Levin [13] obtained an analytical solution of equation (43.2) for the case of a potential flow past bodies of various shapes near the stagnation point. Near this point the velocity of the fluid may be written in the form

$$v_{0y} = -ay,$$
$$v_{0x} = bx,$$

where a and b are constants. The coordinate x is plotted along the surface of the body from the stagnation point.

In particular, by combining equations (43.3) and (82.12), given below, we find for the case of a sphere

$$v_{0y} = -3y;$$

i.e., a = 3.

The equation of motion of a particle moving along the stagnation streamline in the direction of a stagnation point takes the form

$$\mathrm{Stk}\frac{d^2y}{d\tau^2} + \frac{dy}{d\tau} + ay = 0,$$

whence

$$y = Ae^{\lambda_{1,2}\tau}, \tag{43.4}$$

where $\lambda_1$ and $\lambda_2$ are the roots of a characteristic equation and are equal to

$$\lambda_{1,2} = -\frac{1}{2\mathrm{Stk}}\left(1 \pm \sqrt{1 - 4a\,\mathrm{Stk}}\right). \tag{43.5}$$

If Stk > $\mathrm{Stk}_{crit} = \frac{1}{4a}$, then the roots $\lambda_{1,2}$ are complex.

The particle will reach the stagnation point y = 0 at the moment in time $\tau_0$ defined by the condition

$$Ay_1(\tau_0) + By_2(\tau_0) = 0 \tag{43.6}$$

This equation has a solution corresponding to a finite value of time $\tau_0$ at any real finite values for A and B. Therefore, the particle for which Stk $> \frac{1}{4a}$ and which is moving along a stagnation streamline may reach the surface of the body within finite time.

But if Stk $< \frac{1}{4a}$, the roots $\lambda_{1,2}$ are found to be real. Here, as shown by analysis, equation (43.6) has no solution at a finite time $\tau_0$. The particle cannot reach the surface of the body even along the central line. Nor will it attain the surface of the body at other points, since the stagnation point is the easiest to reach — along the central streamline the flow has no tangential component of velocity. In particular, for a sphere $Stk_{crit} = \frac{1}{12}$, which is in agreement with the computations of Langmuir and Blodgett.

Therefore, inertial effects cannot guarantee that particles below a certain size will impact on large bodies. Thus, for example, droplets smaller than R = 4 microns will not impact on a drop of radius R = 100 microns freely falling in air. This conclusion led to serious difficulties in the theory of the growth of drops in a rain cloud.

As noted, there are three errors in the Langmuir and Blodgett computations as well as in those by L. M. Levin and other authors: disregard of the effect of the boundary layer, disregard of the dimensions of the precipiating particle, and disregard of the transportation of particles by turbulent fluid beyond the separation point.

The case of particles of finite dimensions was examined by N. A. Fuks [10]. If the inertial effects are disregarded, and if we assume the particle paths follow precisely along the streamlines, then particles will contact the surface of the body when their paths approach to within a distance equal to the particle radius. Thus, the impaction efficiency may also be greater than zero when inertial precipitation does not exist.

N. A. Fuks called this the interception effect. A simple computation [10] shows that as a particle of dimension a passes by a sphere of radius R along a streamline of a potential flow

$$E_{intcp} = \left(1 + \frac{a}{R}\right)^2 - \frac{1}{1 + \frac{a}{R}} \approx 3\frac{a}{R}. \tag{43.7}$$

Of course, such a definition of $E_{intcp}$ is proper only if $\frac{a}{R}$ is so small that the inertial effects may be disregarded. In the case of a limiting inertial effect, when the trajectory of the precipitated particle is a straight line,

$$E_{intcp} = 1 + \left(\frac{a + R}{R}\right)^2 \approx 1 + 2\frac{a}{R}. \tag{43.8}$$

Therefore, by accounting for the interception effect the difficulty noted above is avoided: $E_{intcp}$ is different from zero at any value of $\frac{a}{R}$.

If the dimensions of the particles are small in comparison with the thickness of the boundary layer $\delta$, the approximation of a potential flow is insufficient. Since, however, $\delta$ is small at high Reynolds numbers, particles with dimensions $a \ll \delta$ will possess significant diffusion coefficients.

Precipitation of particles from a turbulent fluid beyond the separation point may be of greater significance. This, too, must be related to diffusional effects.

If the two particles that encounter each other have nearly the same dimensions, the assumption that one of them moves in a stream past the other particle is not justified.

Examination of the entire picture of hydrodynamic interaction between two particles of the same dimensions is extremely difficult. At great distances (great in comparison with the dimensions of the particles), the general nature of hydrodynamic interaction has been established. But for the mutual attraction of bonding of particles, knowledge of motion over short distances is essential. Here the nature of the flow is very complicated and has not been studied in theory.

We are limited to the qualitative assertion that there is little deviation in the impaction efficiency E between the case of deformed paths and the case of nondeformed paths. This follows from the fact that the latter case concerns very small particles, and hence all hydrodynamic interactions are correspondingly small.

Besides the precipitation effects of inertia and interception, there is also a diffusional precipitation of particles. Diffusional precipitation plays a significant part in the case of highly dispersed particles, as well as in the case of particles in a turbulent flow.

Let us first examine diffusional precipitation from a laminar flow. The formulas obtained above for molecular diffusion may be adapted to the process of diffusional precipitation.

For purposes of comparison with inertial and, interception impaction, we introduce the effective impaction efficiency $E^*$. This we define as the ratio of the actual diffusional flux of precipitating particles to the flux which would have taken place for straight-line motion of the particles

$$E^* = \frac{I}{\pi R^2 v_0 n_0}\,. \tag{43.9}$$

Let us, for example, examine the case of precipitation on a sphere. I may be substituted by its value according to equation (14.19). We then obtain

$$E^* \approx 3\left(\frac{D}{\nu}\right)^{2/3} \frac{1}{\mathrm{Re}^{2/3}} \tag{43.10}$$

In contrast to the case of inertial impaction $E^*$ increases with a decrease in the size of the precipitated particles. Diffusional precipitation, therefore, is of interest primarily for highly dispersed particles.

An analogous calculation for the important case of precipitation on the surface of a cylinder from a stream flowing across its generatrix was made by G. L. Natanson [14].

As in the case of a sphere, the diffusion equation at Pe ≫ 1 and Pe ≪ 1 is simplified to the form of (14.10), where the flux function is

$$\psi = 2w\frac{y}{a}\sin\theta \quad \text{and} \quad t = \int_0^\theta (8uaD^2\sin\theta)^{1/2}d\theta,$$

a is the radius of the cylinder and $\omega = \dfrac{u}{2(2 - \ln \mathrm{Re})}$ (e.g., see [15]).

The integration of (14.10) using boundary conditions (14.6) to (14.8) yields

$$c = \frac{3}{\Gamma(1/3)}\left(\frac{4}{9}\right)^{1/3} c_0 \int_0^u e^{-\frac{4}{9}u^3}\,du, \tag{43.11}$$

where

$$u = \frac{\sqrt{\psi}}{t^{1/3}}$$

The diffusional flux over a unit of cylinder length equals

$$J = 2\int_0^\pi D\left(\frac{\partial c}{\partial r}\right)_{r=a} a\,d\theta = \frac{6}{\Gamma\left(\frac{1}{3}\right)}\left(\frac{4}{9}\right)^{1/3} 2.71 c_0 (uaD^2)^{1/2} \tag{43.12}$$

Correspondingly, the precipitation efficiency E* is equal to

$$E^* = \frac{J}{2av_0c_0} = 2.32\left(\frac{u}{v_0}\right)^{1/3}\left(\frac{D}{\nu}\right)^{2/3} \cdot \frac{1}{\mathrm{Re}^{2/3}} \tag{43.13}$$

where $(2av_0c_0)$ is the flow of particles for linear motion per unit of the cylinder length.

Let us compare the phenomenon of diffusional precipitation with impaction caused by interception. Using equations (43.10) and (43.7), we derive for a sphere

$$\frac{E^*}{E_{\mathrm{intcp}}} \sim \left(\frac{D}{\nu}\right)^{2/3} \frac{R}{\mathrm{Re}^{2/3}a} \tag{43.14}$$

Formula (43.11) shows that the efficiency of diffusional precipitation E* is substantially greater than $E_{\mathrm{intcp}}$ at R ≫ a and at low absolute values of a and R. Thus, for the free fall in normal density air of a particle of $R = 10^{-3}$ cm on which particles with $a = 10^{-5}$ cm are precipitating, we have $\mathrm{Re} \sim 10^{-2}$ and $\dfrac{E^*}{E_{\mathrm{intcp}}} \sim 1$.

If the fall takes place at a height of 2000 m, $\frac{E^*}{E_{intcp}} \sim 5$. In the case of particles with a radius $a = 10^{-5}$ cm, the relationship becomes more favorable for diffusional precipitation.

Of even greater significance is diffusional precipitation in a turbulent flow. In this case, to compute the diffusion of dispersed particles to a surface, we use relationships derived in Chapter III. In particular, equation (26.11) defines the diffusional flux of particles to the inside surface of the tube through which the liquid or gas moves.

The conclusions of Section 28 pretain also to diffusional precipitation on the surface of a body of nonstreamlined shape. In particular, the number of particles precipitated per 1 $cm^2$ on the downstream section may exceed the number of particles precipitated on the upstream portion of the body.

In connection with precipitation from turbulent flow, let us direct our attention expecially to the mechanism of precipitation of aerosols.

In Section 42, when determining the number of contacts between larger and finer aerosol particles, we identified actual encounter with aggregation. An allowance for the deformation of the streamlines, however, would seem to require the introduction of the impaction efficiency into the formula for the number of separate occurences of coagulation. This, pointed out by N. A. Fuks, is not, however, true. Aerosol particles in a turbulent flow are entrained by random turbulent eddies; i.e., they always participate in the turbulent diffusion discussed in Section 41.

Therefore a more accurate picture of attraction and bonding of small particles by large ones, as discussed in Section 42, may be stated as follows: finer particles move past a particle of radius R with an average velocity const $(R^2 - r^2)$. When passing the larger particle they are precipitated onto it.

Let us estimate the precipitation coefficient, assuming for the sake of simplicity that $r \ll R$. Obviously we have

$$E^*_{turb} = \frac{4\pi R^2 j_{turb}}{\pi R^2 u(R) n_0}$$

where $j_{turb}$ is given by equation (41.7) and u(R) by equation (33.19).

Substituting the values of these quantities, we have

$$E^*_{turb} \sim 30\beta \frac{\rho_0}{\rho} \frac{1}{R} \sqrt{\frac{\nu^3}{\varepsilon_0}} \qquad (43.15)$$

For $\beta \approx 1$, $\frac{\rho_0}{\rho} \sim 10^{-3}$, $\epsilon_0 \sim 10^3$, $R \sim 10^{-4} - 10^{-3}$ cm, and the quantity $E^*_{turb} \sim 1$.

Therefore, this rough estimate shows that, owing to the very strong effect of turbulent diffusion, the deformation of trajectories has no effect in the precipitation of particles with dimensions $\leq 10^{-3}$ cm.

The physical meaning of this conclusion is very simple: for fine particles in suspension, the deformation of the streamlines by intense turbulence is more significant than deformation by inertial effects of fine suspended particles. In discussing turbulent coagulations, we may therefore assume that each encounter computed on the basis of straight-line trajectories will actually be realized.

## BIBLIOGRAPHY

1. Derjaguin, B. V., Trans. Far. Soc. 36, 730 (1940); B. V. Derjaguin and L. D. Landau, Zhurnal Eksperimental'noy i Teoreticheskoy Fiziki, 15, 663 (1945); B. V. Derjaguin and V. G. Levich, Doklady AN SSSR, 98, 985 (1954).
2. Rebinder, P. A., Predisloviye k knige Kleytona "Emul'sii," (Introduction to Clayton's book "Emulsions"), IL, 1950; V. G. Levich, Doklady AN SSSR, 103, 453 (1955).
3. Tolstoy, N. A., Doklady AN SSSR, 100, 893 (1955).
4. Smoluchowski, M., Coll. Brounovskoye dvizheniye (Brownian Motion), ONTI, 1936.
5. Leontovich, M. A., Statisticheskaya fizika (Statistical Physics), Gostekhizdat, Moscow, 1944.
6. Coll. "Koagulyatsiya colloidov" (Coagulation of Colloids), ONTI, 1936, p. 61.
7. Levich, V. G., Doklady AN SSSR, 99, 809 (1954).
8. Wigand and Frankenberger, Physik. Z., 31, 204 (1930).
9. Levich, V. G., Doklady AN SSSR, 99, 1041 (1954).
10. Fuks, N. A., Mekhanika aerozoley (Aerosol Mechanics), Izd. AN SSSR, 1955.*
11. Langmuir, I., Coll. Fizika obrazovaniya osadkov (Physics of Precipitation Formation), IL, 1951, p. 147.
12. Langmuir, I. and K. Blodgett, Army Air Force, Techn. Rep., No. 5418 (1946).
13. Levin, L. M., Doklady AN SSSR, 91, 1329 (1953).
14. Natanson, G. L., Doklady AN SSSR, 112, 100 (1957).
15. Lamb, H., Gidrodinamika (Hydrodynamics), Gostekhizdat, 1947.

*English translation available as CWL Special Publication 4-12, Technical Information Division, Army Chemical Center, Maryland (1958).

## CHAPTER VI

# Passage of Current Through Electrolytic Solutions

### 44. STATE OF QUASI-EQUILIBRIUM IN AN ELECTROLYTIC CELL

Although a discussion of current flow in electrolytes may be, strictly speaking, beyond the scope of physicochemical hydrodynamics, we feel that a brief treatment of this subject is appropriate for several reasons. As is clear from the following discussion, hydrodynamic factors play a very important part in the passage of current through a solution. Moreover, several electrochemical relations will be needed when we examine the experimental verification of various theories set forth in earlier chapters.

It should also be pointed out that the subject of electrolysis in liquids has been largely ignored by physicists in recent years. The terminology and style of the electrochemical and physicochemical literature have become foreign to physicists. The exposition is usually based on specific chemical circumstances which tend to burden it with details that have no direct bearing on the essentials of the situation. We have attempted below to present the subject in a form that is more understandable to physicists. We should caution that the phenomena occurring during electrolysis are so diverse and complex that we cannot cover all possible events and processes at various electrodes.

We have not attempted a complete presentation, but one that brings out the most important aspects of the subject.

In the following, we examine a system composed of two electrodes, separated by an ionically conducting liquid which may be an aqueous or nonaqueous solution of an electrolyte or a fused electrolyte.

Such a system, consisting of two electrodes joined by a liquid conductor, is generally known as an electrolytic cell, but sometimes as an electrolyzer or a galvanic bath.

If an electric current generated by an external emf flows through an electrolytic cell, an electrolytic process is said to occur within the cell.

Before taking up the mechanism of electrolysis, we must consider two aspects of equilibrium in an electrolytic cell [1]. As a first approximation an electrolytic solution can be considered electrically neutral whether or not a current flows through it. Only at the solution-metal interface (or at the boundary with any other phase, including air) is there a very thin fluid layer where ions with like charges (positive or negative) predominate. Whether this so-called ionic portion of the electric double layer is positively or negatively charged depends upon the sign of the charge on the metal surface.

The ionic portion of the double layer is the same as a Debye ion cloud, except for its geometric form, and has an effective thickness equal to the Debye radius. The double layer is quite important in certain cases of electrolysis (see Section 47) and in some other processes such as electrophoretic and electrocapillary phenomena (see Chapter IX). However, for the thermodynamic problems examined in this section, the properties of the double layer are unimportant.

We characterize the ions in the bulk of the solution by their partial chemical potentials $\mu_i$, where the subscript i indicates the type of ion. A partial ion potential is a function not only of the composition of the solution, but also of its electrical potential. This assertion requires explanation: for a given composition of the solution, the values of all partial ion potentials in the solution are, strictly speaking, exactly determined. However, a negligibly small change in the composition may greatly affect the potential differences in the cell. In this sense we can speak of changes in the potential differences within the cell which are not accompanied by any change in the composition of the solution.

Let us formulate the equilibrium conditions at the metal-solution interface. First we examine a solution of the type $M^+X^-$, where $M^+$ designates the metal ions (cations) and $X^-$ designates the anions in the solution in which the metal M is immersed. We shall consider the metal-solution interface to be impervious to $X^-$ ions but always permeable to $M^+$ ions; i.e., at the metal M-solution interface the $M^+$ ions are discharged and transformed into atoms which then penetrate the crystal lattice of the metal. The metal ions are also transferred from the lattice to the solution. In this way, the interfacial phase boundary serves as a membrane* which is permeable to $M^+$ ions and impervious to $X^-$ ions. Under steady conditions, the phases separated by the membrane attain a partial equilibrium with

---

*In thermodynamics a membrane is a surface between phases which is permeable to some of the components comprising the phases, but not to others.

respect to the component which can penetrate the membrane — the $M^+$ ions in this case.

The condition of thermodynamic equilibrium with respect to $M^+$ ions can be written in the form

$$\mu_M = \mu_{M^+},$$

where $\mu_M$ and $\mu_{M^+}$ are the partial chemical potentials of the atoms in the metal and of the ions in the solution, respectively. Since the particles passing through the metal-solution interface are charged, an electric field exists at this interface, so that the partial potentials can be represented in the form

$$\mu_M = \mu_M^{(0)} + nF\varphi_M; \quad \mu_{M^+} = \mu_{M^+}^{(0)} + nF\varphi_s,$$

where $\mu_M^{(0)}$ and $\mu_{M^+}^{(0)}$ are the partial potentials which would exist for uncharged particles, $\varphi_M$ and $\varphi_s$ are the electric potentials of the respective phases, n is the valence of the metal ions, and F is Faraday's number. By defining $\Delta\varphi = \varphi_M - \varphi_s$ as the potential difference at the metal-solution interface, we can rewrite the equilibrium condition at the phase boundary in the form

$$\mu_M^{(0)} = \mu_{M^+}^{(0)} - nF\,\Delta\varphi. \tag{44.1}$$

For sufficiently low concentrations of the solution, $\mu_{M^+}^{(0)}$ can be replaced by the partial chemical potential of the material in a dilute solution*

$$\mu_{M^+}^{(0)} = RT \ln c_{M^+} + \psi(p, T)$$

and (44.1) can be rewritten in the form

$$\Delta\varphi = \frac{RT}{nF} \ln c_{M^+} + \frac{1}{nF}[(\psi(p, T) - \mu_M^{(0)})] = \frac{RT}{nF} \ln c_{M^+} + \varphi_0^{(0)}, \tag{44.2}$$

where $c_{M^+}$ is the concentration of the metal ions in the solution, $\psi$ is not a function of concentration, and $\varphi_0^{(0)} = \frac{1}{nF}[\psi\ (p, t) - \mu_M^{(0)}]$.

Equation (44.2) relates the concentration and the potential difference at the metal-solution interface. This equilibrium equation has a simple kinetic interpretation: the number of ions crossing the interface in one direction is balanced by the number crossing in the other direction.

It follows from equation (44.2) that the concentration of the solution in the vicinity of the electrode is given by the Nernst relation

---

*In the case of concentrated solutions, concentration should be replaced by activity.

$$c_{M^+} = \text{const}\, e^{\frac{nF\,\Delta\varphi}{RT}}. \qquad (44.3)$$

The sign preceding the interfacial potential jump in equation (44.2) has been chosen so that $\Delta\varphi$ is negative when metal ions discharge on the electrode (the cathode) and positive when metal ions pass into the solution (at the anode).

The words "in the vicinity of the electrode" should be understood in a macroscopic sense; that is, the concentration $c_{M^+}$ indicates the concentration in the neutral solution outside the double layer, and $\Delta\varphi$ indicates the potential difference between the metal and that point in the solution. Both concentration and chemical potential vary from point to point within the double layer.

An amalgam is a commonly used electrode material. This is a solution of a metal M in an inert metal, usually mercury, which does not dissolve in the solution. In this case, the chemical potential $\mu_{M^+}^{(0)}$ of the metal at equilibrium can be written in the form

$$\mu_M^{(0)} = RT \ln \gamma + \psi_{am},$$

where $\gamma$ is the concentration of the metal M in the amalgam and $\psi_{am}$ refers to the amalgam. Then equation (44.2) for the interfacial potential jump can be written in the form

$$\Delta\varphi = \frac{RT}{nF}[\ln c_{M^+} - \ln \gamma] + \varphi_0^{(0)},$$

where the constant $\varphi_0^{(0)} = [\psi\,(p, T) - \psi_{am}] \cdot \frac{1}{nF}$.

It should be stressed that what is actually measured is not the potential difference $\Delta\varphi$ at one interface, but rather the emf between the two electrodes in the cell. Therefore, a relation analogous to (44.2) should be written for a system consisting of two electrodes. Let us examine some of the simplest examples of such systems. For example, let the metal electrode M be immersed in a solution of $M^+X^-$ ions. The second electrode is electrode N, immersed in a solution of $N^+X^-$ ions. Here, we assume that the passage of current is not accompanied by the direct transfer of metal ions across the interface of the solutions. For the potential difference between the electrodes (the emf that corresponds to it), we can write:

$$V = \Delta\varphi_1 - \Delta\varphi_2 = \frac{RT}{n_M F} \ln c_{M^+} - \frac{RT}{n_N F} \ln c_{N^+} + \varphi_1^{(0)} - \varphi_2^{(0)}, \qquad (44.4)$$

where $\varphi_1^{(0)}$ and $\varphi_2^{(0)}$ are unknown constants relating to the two electrodes.

In order to demonstrate the validity of expression (44.4), it is sufficient to examine the circuit illustrated in Figure 44. Here M denotes the first metal and MN denotes the contact between the first and second metals (the membrane that is permeable to electrons $\epsilon$). Under the arrows we have noted the type of charged particle which can pass through the corresponding interfacial membranes. It is clear from the figure that the entire circuit represents a combination of four membranes. By successively formulating the conditions of membrane equilibrium (44.2), and by excluding the intermediate interfaces, we arrive at equation (44.4) directly.

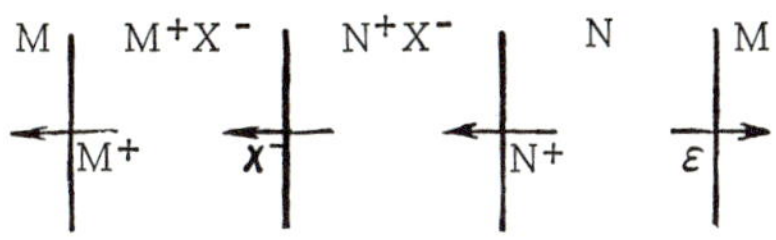


Figure 44. Diagram of a circuit with four membranes.

In actual practice all electrode potential differences are generally referred to a so-called standard, or reference electrode which is usually a calomel or hydrogen electrode. A reference hydrogen electrode is a platinized (hydrogen saturated) platinum sheet immersed in a one normal solution of hydrogen ions. The constant $\varphi^{(0)}$ of this electrode is taken to be zero.

The use of such a scale automatically excludes the interfacial potentials at metal-metal interfaces. Here, each metal is characterized by a constant $\varphi^{(0)}$ which is not a function of the concentration of the solution and which represents the open-circuit potential difference between the electrode and the hydrogen electrode for a concentration of metal ions $c_{M^+} = 1$ (i.e., one mole per liter). It should be noted that the quantities $\varphi^{(0)}$ are purely thermodynamic in nature and have significance at equilibrium conditions within the electrochemical cell. They are a function of the nature of the metal (the partial potential $\mu_M$ of the metal enters into their determination), the nature of the solvent [through $\psi$ (p, T)], and of the temperature and the pressure.

Table 5 shows the measured potential difference in volts between various electrodes and the normal hydrogen electrode (the so-called electromotive series). The minus and plus signs indicate that a given electrode is more negative or more positive than the normal hydrogen electrode.

Connecting two cells to each other* creates an electrolytic cell that is even simpler in nature. Consider, for example, the system

---

*In the circuit

$$M \mid M^+X^- \, (c_1) \mid N \mid M^+X^- \, (c_2) \mid M$$

there are no diffusion potentials.

which includes: metal M in a solution of $M^+X^-$ ions at a concentration $c_1$, and metal M in a solution of $M^+X^-$ ions at a concentration $c_2$. In this case the constants $\varphi_1^{(0)}$ and $\varphi_2^{(0)}$ cancel, and the potential difference across the cell is:

$$V = \frac{RT}{n_M F} \ln \frac{c_1}{c_2}. \tag{44.5}$$

The proviso made concerning the absence of a fluid boundary permeable to metal ions is related to the following: if two $M^+X^-$ solutions of different concentrations are brought directly together, ion diffusion will occur from the more to the less concentrated solution. Unbalanced diffusional processes will appear in the system, and the thermodynamic relations will be invalid. The potential difference at the interface between the solutions across which the charged particles would diffuse is added to the net potential difference determined by the single electrode potential differences. It should be noted, however, that this potential difference is not large and can be reduced by special procedures. Accordingly, we shall ignore it in the subsequent discussion.

Table 5

Electromotive series in aqueous solutions at room temperature

| Electrode | Reaction | Potential, v |
|---|---|---|
| Li, $Li^+$ | $Li \rightleftarrows Li^+ + \epsilon$ | - 3.024 |
| Na, $Na^+$ | $Na \rightleftarrows Na^+ + \epsilon$ | - 2.714 |
| Zn, $Zn^{++}$ | $Zn \rightleftarrows Zn^{++} + 2\epsilon$ | - 0.761 |
| Fe, $Fe^{++}$ | $Fe \rightleftarrows Fe^{++} + 2\epsilon$ | - 0.441 |
| Cd, $Cd^{++}$ | $Cd \rightleftarrows Cd^{++} + 2\epsilon$ | - 0.402 |
| Ni, $Ni^{++}$ | $Ni \rightleftarrows Ni^{++} + 2\epsilon$ | - 0.236 |
| Pb, $Pb^{++}$ | $Pb \rightleftarrows Pb^{++} + 2\epsilon$ | - 0.126 |
| $H_2$, $H^+$ | $\frac{1}{2} H_2 \rightleftarrows H^+ + \epsilon$ | 0.000 |
| Cu, $Cu^{++}$ | $Cu \rightleftarrows Cu^{++} + 2\epsilon$ | + 0.340 |
| Ag, $Ag^+$ | $Ag \rightleftarrows Ag^+ + \epsilon$ | + 0.799 |
| Hg, $Hg_2^{++}$ | $Hg \rightleftarrows \frac{1}{2} Hg_2^{++} + \epsilon$ | + 0.799 |
| $H_2$, $OH^-$ | $\frac{1}{2} H_2 + OH^- \rightleftarrows H_2O + \epsilon$ | - 0.828 |
| $I_2$ (S), $I^-$ | $I^- \rightleftarrows \frac{1}{2} I_2 + \epsilon$ | + 0.536 |

Let us now examine the case of a metallic electrode M immersed in a solution containing $N^+$ ions of another metal.

If the metal-solution interface is impervious to the $N^+$ ions at a given potential difference, then there will be no equivalent exchange of particles between the electrode and the solution. Using Planck's terminology, an electrode whose surface is impervious to ions in the solution is said to be ideally polarizable. On the other hand, an electrode that is in equilibrium with the dissolved ions is said to be unpolarizable.

If a constant current flows across the interface (membrane), between the electrode and a solution of $N^+$ ions, such a membrane is in a steady, but not an equilibrium, state.

Here, we do not examine special cases, such as nonmetallic electrodes or solutions containing ions of different valences.

In practice, very dilute solutions are seldom encountered. Consequently, activity replaces concentration in equation (44.4).

## 45. CURRENT FLOW IN AN ELECTROLYTIC CELL

Let us now pass to a discussion of the mechanism of electrolysis. The passage of current through an electrolytic cell can be viewed as a special case of a heterogeneous chemical reaction. Like any other heterogeneous reaction, it includes three steps:

1) the transfer of ions from the bulk of the solution to the surface of the electrode;

2) the electrochemical reaction itself involving ions or molecules;

3) the formation of the final products of the reaction and their deposition on the surface of the electrode or their removal from that surface.

As an example of the last step we may cite the process of creating $H_2$ molecules from H atoms, which result from the discharge of $H^+$ ions, and the removal of the gaseous hydrogen as bubbles.

It is evident that the over-all rate of the chemical reaction is determined by the slowest step of the process. In different cases, the slow step can be any one of the three mentioned above. In order to maintain a finite rate for the over-all electrochemical process, an emf must be applied.

Very often (see below) the first step is the slowest, i.e., the transfer of ions to the surface of the electrode. The quantitative theory of this stage will be developed later for real conditions. Here we confine our treatment to a somewhat idealized case.

Cases in which the slow step of the process is the second or the third stage are equally important, although they are less frequently encountered in practice. However, cases of precisely this type are of fundamental interest in theoretical electrochemistry. In several processes that are important both in nature and in industry, such as the electrolysis of water and of aqueous solutions in which hydrogen and oxygen are liberated, the slow step is the second stage of the process.

Upon application of an emf, no matter how small, the equilibrium at the metal-solution interface is destroyed. The rate of discharge of ions on one electrode becomes greater than the rate of transfer of ions from the metal to the solution. On the other hand, at the other electrode the rate of dissolution exceeds that of the discharge. As a result, one of the electrodes (the anode) is continuously dissolved, and a discharge of metal ions takes place at the other (the cathode). It may be assumed, in this case, that both processes — discharge and dissolution — occur at any desired rate and do not require an activation energy.

For small currents, the amount of current flowing through the cell is determined by the usual Ohm's law.

For brevity in the subsequent discussion, we examine the phenomena at one of the electrodes, for example the cathode, and assume that the anode of the cell is a reference hydrogen electrode.

When current passes through a cell with a cathode M and an electrolyte $N^+X^-$, two basic cases are possible: firstly, the discharge of "foreign" ions at the electrode, which are removed from the metal surface without hindrance; and, secondly, the discharge of ions to form deposits, actually a new phase, on the electrode surface.

Let us first examine the former case.

As an example of a system involving the removal of the products of the discharge process, we may cite a platinum electrode (or an electrode of some other metal that absorbs hydrogen) in a solution containing hydrogen ions. The latter are discharged on the platinum and combine to form hydrogen molecules. The molecular hydrogen is then dissolved in the liquid.

A second example is a system involving a mercury electrode and ions of a metal which amalgamates with mercury. The discharged metal ions dissolve in the mercury in the form of atoms and are removed from the surface by diffusion. The metal-solution interface in this instance is a membrane permeable to the discharging particles.

In a state of equilibrium, the partial chemical potentials of the particles in both phases are equal, so that relation (44.3) is applicable for this particular interface.

When a potential is applied, current flows through the cell, and in this case, if the departure from equilibrium is not very great, the number of discharged ions (and, therefore, the size of the current) is an exponential function of the potential differences in accordance with formula (44.4).

The current flowing through the cell increases rapidly with the applied emf. At relatively large departures from the equilibrium potentials, new phenomena are manifested which are examined below.

More complex is the case of the formation of metallic ions $N^+$ on the surface of metal M when the discharging ions form a new phase on the surface of the electrode. In this case, in order that ions $N^+$ can be formed on the cathode M, the difference in potential between the anode and the cathode must exceed a certain value $V_0$. As indicated above, the system metal-ionic solution is not in equilibrium if the metal-solution interface is impervious to $N^+$ ions. For small applied emf's across the electrodes, the metal-solution interface remains impervious to $N^+$ ions, and no current passes through the cell.

In order to find the potential difference at which the interface of the metal M-solution of $N^+$ ions becomes permeable to $N^+$ ions which form a new phase on it, let us examine a membrane system represented by the sketch,

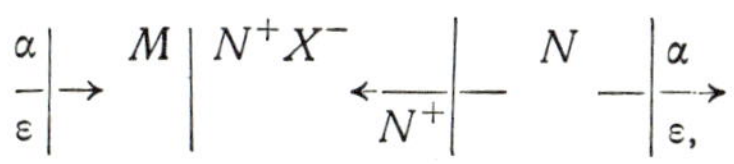


where $\alpha$ is any metal. It is evident that the potential of the metal $\alpha$ is the same on both the right and left sides. If an emf is applied across electrodes M and N,

$$V_0 = \varphi_N^{(0)} - \varphi_M^{(0)}. \tag{45.1}$$

If the applied emf becomes greater than $V_0$ by an infinitesimally small amount (the cathode potential becomes more negative than an equilibrium cathode), the $M \mid N^+$ membrane becomes permeable to the $N^+$ ions, and they start to discharge on the cathode M. An infinitesimally small current will start to flow through the cell. The quantity $V_0$ is known as the discharge (decomposition) potential. The discharge potentials for various ions on a given electrode can be found from formula (45.1) with the help of Table 5.

It is important to stress here that if several types of ions are present in the solution, then, as the applied emf is raised, the ions with a lower discharge potential will begin to discharge first (Figure 45). Thus, for example, with a platinum electrode immersed in a solution containing one normal concentration of $Zn^{++}$ and $Cu^{++}$ ions, as the potential difference across the cell is increased, the $Cu^{++}$ ions discharge first (at $V > V_0 = +0.34$ volts), followed by the $Zn^{++}$ ions (at $V > V_0 = -0.77$ volts).

Thus, for purely thermodynamic reasons, the absence of "native" ions (identical with the electrode material) in a solution leads to a situation in which a current flow starts only upon application of an emf that exceeds the discharge potential $V_0$.

Figures 46 and 47 show the dependence of the current i on the applied emf in both of the cases that have been examined. It should be borne in mind that, although not all but only part of the ions in the

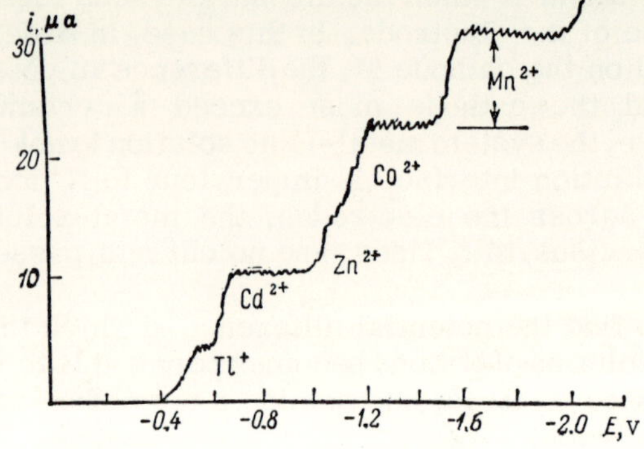


Figure 45. Discharge potentials E (relative to a saturated calomel electrode) for various ions on a mercury electrode.

solution can discharge upon an electrode for a given emf, nevertheless all the ions in the bulk of the solution participate in the transfer of current. Naturally, when the electrolytic process is prolonged so that the $N^+$ ions form a deposit of metal N on the surface of cathode M, the latter will begin to act as a cathode made of metal N.

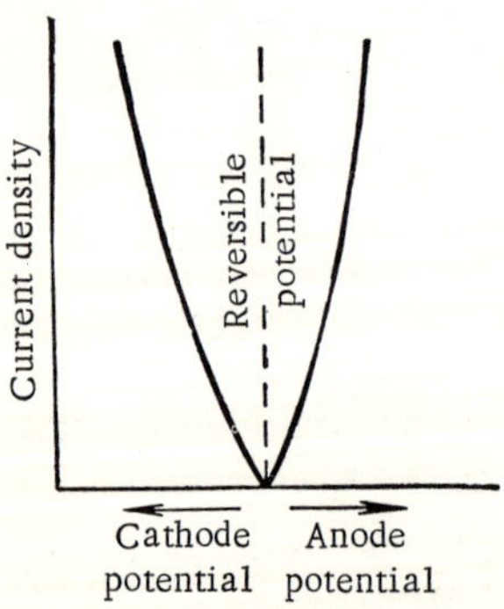


Figure 46. Dependence of the electrode potential on current density.

It should be noted that the difference between ions which form a new phase on an electrode or on an adsorbed layer is, in fact, of no importance. If, for example, the discharged $N^+$ ions diffuse rapidly into the body of the metal, the discharge potential $V_0$ would be absent.

An apparent discharge potential appears in the case in which the liberation of hydrogen occurs with the formation of bubbles that are liberated into the atmosphere. This case, complicated by the incidental bubble formation, is of great practical significance but does not introduce anything essentially new into the substance of the process.

Up to this point we have only been examining steady state electrolytic processes. In some instances we encounter unsteady state processes. Let us assume that in the cell, cathode M-solution $N^+X^-$-reference electrode, a difference in potentials $V < V_0$ is applied, the metal M-solution $N^+X^-$ interface remaining impervious to $N^+$ ions. The change in charge on the surface of electrode M leads to a change in the number of $N^+$ ions that are within the double layer. With an increase in the negative potential of the cathode, the $N^+$ ions migrate to a certain extent from the bulk of the solution into the double layer. The accumulation of $N^+$ ions at the surface of the cathode is equivalent to the passage through the cell of a certain current, known as the charging current. Its density is related to the charge on a unit of cathode surface by the relation

$$i_{\text{charging}} = \frac{\varepsilon}{t}, \tag{45.2}$$

where t is the time of the electrolysis and is considered short.

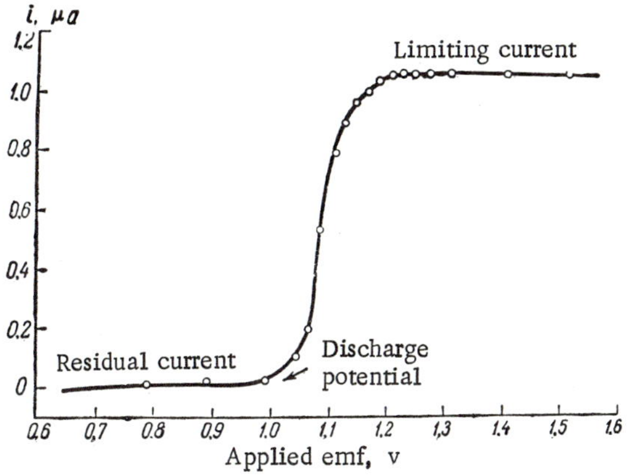


Figure 47. Dependence of the current in an electrolytic cell on the applied emf.

We have limited our above treatment to the case of a very small current flowing through an electrolytic cell, so that the state of the cell can be considered arbitrarily close to equilibrium. Here, the difference in potential applied to the cell must differ from the equilibrium potential (or discharge potential $V_0$) by an arbitrarily small amount.

For the passage of a finite current through an electrolytic cell, i.e., in order to obtain a finite rate for the electrochemical reactions at the electrodes, the applied potential difference must exceed the equilibrium difference by a finite amount.

The difference between the single electrode potential $\Delta\varphi$ (i) for current density i at the electrode-solution interface and at equilibrium $\Delta\varphi_{equ}$ is termed the overpotential $\eta$:

$$\eta = |\Delta\varphi(i) - \Delta\varphi_{equ}|. \tag{45.3}$$

The magnitude of $\eta$, therefore, characterizes the departure from quasi-equilibrium conditions that exists in a cell through which a finite current flows. In electrochemistry, this phenomenon is usually known by the somewhat unfortunate term "polarization."

Evidently the potential difference applied to the cell is related to overpotential by

$$V = \Delta\varphi_{ohm} + \eta(i) + \Delta\varphi_{equ}, \tag{45.4}$$

where $\Delta\varphi_{ohm}$ is the ohmic potential drop in the solution, which is a function of the geometric path of the current, the current density, and the electrical conductivity of the solution.

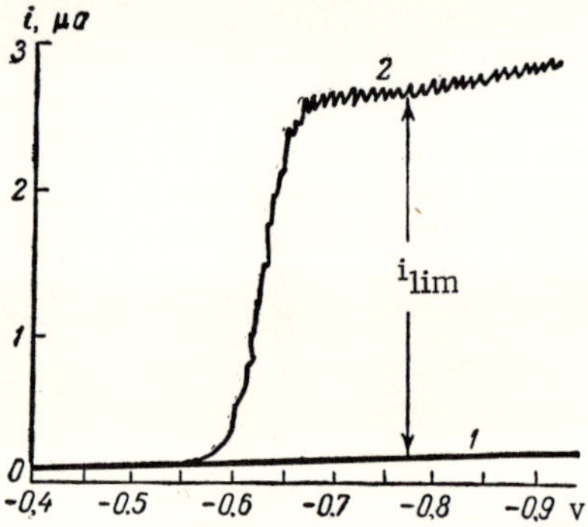


Figure 48. Discharge current. The straight line is an extrapolation.

In expression (45.4) we have omitted the quantity $\eta$ for the second electrode (the anode), and for simplicity have assumed that $\eta = 0$ at that electrode and that $\Delta\varphi = \Delta\varphi_{equ} = 0$. As shown below, this often holds in actual electrolysis.

Equation (45.4) evidently represents the volt-ampere characteristics of the discharge. In order to develop its complete expression, the mechanism of processes which lead to overpotential must be examined, and $\eta$ (i) must be known.

## 46. CONCENTRATION OVERPOTENTIAL

Let us first consider the process by which the reacting particles are transferred to the electrode. Here, it should be borne in mind that it is not always the arrival of the discharging ions themselves which limits the over-all rate of the electrode reaction, for the rate is frequently limited by other particles which take part in the electrode reaction, such as, for example, molecules of dissolved oxygen.

The ions are transferred from the solution to the surface of the electrode by three mechanisms: migration in the electric field applied to the cell; diffusion from regions of higher to lower concentration; and convection or motion of the ions with the solvent. We shall examine each of these processes separately below.

The passage of current through the solution leads to a change in the concentration at one of the electrodes and the appearance of so-called concentration overpotential $\eta_{conc}$ (also known as concentration polarization in electrochemistry).

We have seen above that the potential drop at the metal-solution interface is given, at equilibrium, by equation (44.2). When a potential different from the equilibrium potential is applied, the concentration of ions discharging on the electrode becomes smaller near the cathode than in the bulk of the solution. Let us denote by $c_C$ the concentration of ions at the surface of the cathode. This concentration is related to the cathode potential $\varphi_C$. Since the exchange rate of the reaction on the cathode is high, it may be assumed that when the current is small $c_C$ has a value close to the equilibrium value given by formula (44.2)

$$\varphi_C = \varphi_0 + \frac{RT}{nF} \ln c_C. \tag{46.1}$$

The change in concentration of the solution from $c_0$ to $c_C$ leads to a change in the electrode potential of

$$\eta_{conc} = \left(\varphi_0 + \frac{RT}{nF} \ln c_C\right) - \left(\varphi_0 + \frac{RT}{nF} \ln c_0\right) = -\frac{RT}{nF} \ln \frac{c_0}{c_C}. \tag{46.2}$$

It is not hard to imagine that the change in the potential of the electrode has a negative sign, i.e., it is derived from the applied emf. Indeed, according to the principle of Le Chatelier, the potential change prevents a further change in concentration. This may also be seen in another way: osmotic pressure is always in the direction from higher to lower concentrations, which, in our case, coincides with the direction of the applied field. On the contrary, the difference in potential that appears has a tendency to compensate for the difference in osmotic pressure and creates a field in the opposite direction.

Thus, in the case we are examining, when the transfer of ions is the slow stage of the process, the drop in potential between the electrodes is composed of three parts — the ohmic drop, the concentration overpotential, and $\Delta\varphi_{equ}$ (if the last quantity is not zero). Accordingly, the applied emf may be expressed as

$$V = \Delta\varphi_{ohm} + \frac{RT}{nF} \ln \frac{c_C}{c_0} + \Delta\varphi^{(0)}_{equ}. \tag{46.3}$$

In this expression we have omitted the term for the concentration overpotential at the anode. It adds nothing of significance. In addition, under actual conditions, the concentration overpotential at one of the electrodes is usually much greater than at the other.

To find the volt-ampere characteristic of the discharge $i = f(V)$, known in electrochemistry as the polarization curve, it is necessary to know the quantity $c_c$ and the value of the ohmic potential drop $\Delta\varphi_{ohm}$.

The concentration $c_c$ of ions discharging at the surface is determined by the rate of arrival of these ions from the bulk of the solution.

For a qualitative evaluation of the type of possible volt-ampere functions, let us examine here the volt-ampere characteristic of an electrolytic process in a binary electrolyte (for example, in a solution of $AgNO_3$ between two silver electrodes), and let us assume that the solution is motionless.

As will be shown below, a motionless solution can be achieved in practice only in exceptional cases, for example, where the solution is immobilized by the addition of gelatine or agar-agar.

Let us assume that the electrodes are infinite planes at $x = 0$ (cathode) and $x = d$ (anode). As in previous cases, we shall consider only the concentration overpotential at the cathode.

The ions in the solution will migrate in the direction of the electrodes under the influence of osmosis (the concentration gradient) and the electric field. Approaching an electrode, in our case the cathode, the ions are discharged, thereby producing a current in the circuit. According to Faraday's law, the current flowing through the electrolytic cell is equal to the number of ions which approach and discharge upon the electrode multiplied by the charge on an ion.*

In a motionless medium the following expressions may be written for the currents carried by each type of ion:

$$-i_1 = -D_1 F z_1 \frac{\partial c_1}{\partial x} - \gamma_1 F z_1^2 \frac{\partial \varphi}{\partial x} c_1, \tag{46.4}$$

$$i_2 = 0 = -D_2 F z_2 \frac{\partial c_2}{\partial x} + \gamma_2 F z_2^2 \frac{\partial \varphi}{\partial x} c_2, \tag{46.5}$$

where $c_1$ and $c_2$ are the cation and anion concentrations and $D_1$ and $D_2$, $\gamma_1$ and $\gamma_2$, and $z_1$ and $z_2$ are the diffusion coefficient, the mobility, and the valence of the ions.

The current flowing toward the cathode in the negative direction of the x axis is assumed to be negative (and equal to $-i_1$). Since the current carried by the second type of ions (anions) is zero, we have set $i_2 = 0$ in (46.5).

The first term in equations (46.4) and (46.5) represents the diffusional flux, and the second term, the flux of ions migrating in the electric field $E = -\frac{\partial \varphi}{\partial x}$.

---

*Here, we neglect the ions transported in charging the double layer and the adsorption of ions on the electrode surface.

Let us assume in a first approximation that the electrolytic solution is electrically neutral throughout its volume, except in the region of the double layer. Then we can write

$$c_1 z_1 = c_2 z_2. \tag{46.6}$$

For the integration of equations (46.4) and (46.5), it is convenient to introduce the quantity c defined by the ratios

$$c = \frac{c_1}{z_2} = \frac{c_2}{z_1}. \tag{46.7}$$

We shall call c the molar concentration of the electrolyte. Then from equations (46.4) and (46.5), we have:

$$-i_1 = -D_1 F z_1 z_2 \frac{\partial c}{\partial x} - F\gamma_1 z_1^2 z_2 c \frac{\partial \varphi}{\partial x}, \tag{46.8}$$

$$0 = -D_2 F z_1 z_2 \frac{\partial c}{\partial x} + F\gamma_2 z_2^2 z_1 c \frac{\partial \varphi}{\partial x}. \tag{46.9}$$

Bearing in mind that the ionic mobilities $\gamma_1$ and $\gamma_2$ are related to the diffusion coefficients $D_1$ and $D_2$ by the ratios

$$\gamma_1 = \frac{D_1 F}{RT}, \qquad \gamma_2 = \frac{D_2 F}{RT},$$

we find, from equation (46.9):

$$\frac{\partial \varphi}{\partial x} = \frac{RT}{z_2 F c} \frac{\partial c}{\partial x}. \tag{46.10}$$

Substituting this value of $\frac{\partial \varphi}{\partial x}$ into expression (46.8), we obtain:

$$i_1 = D_1 z_1 z_2 F \frac{\partial c}{\partial x}\left(1 + \frac{z_1}{z_2}\right). \tag{46.11}$$

Integrating (46.11), and noting that $c = c_0$ at $x = d$, we obtain:

$$c = \frac{i_1}{D_1 F z_1 (z_1 + z_2)}(x - d) + c_0. \tag{46.12}$$

In particular, on the surface of the cathode where $x = 0$, the concentration that is of interest to us is

$$c_C = c_0 - \frac{(i_1)\, d}{D_1 F z_1 (z_1 + z_2)}. \tag{46.13}$$

The volt-ampere characteristic can be obtained by integration of equation (46.8) using (46.12).

The expression for $\varphi$ is:

$$\varphi = \frac{RT\,(i_1)}{F^2 z_1^2 z_2 D_1} \int \frac{dx}{c\,(x)} - \frac{RT}{F z_1} \ln c. \tag{46.14}$$

Substituting here the value of c(x) from (46.12), we derive the formula for the potential drop between the cathode and the anode

$$\varphi_{x=0} - \varphi_{x=d} = \frac{RT\,(z_1 + z_2)}{F z_1 z_2} \ln\left(1 - \frac{i_1 d}{D_1 F z_1 (z_1 + z_2)\, c_0}\right) + \frac{RT}{z_1 F} \ln \frac{c_0}{c_C}. \tag{46.15}$$

Attention should be drawn to an essential feature of this equation. The ohmic potential drop between the electrodes is composed of two parts. The first part represents the usual ohmic potential drop in a solution of varying concentration, equal to $\int \frac{i_1 dx}{\varkappa}$, where $\varkappa = \frac{cF^2 D_1 z_1^2 z_2}{RT}$ represents the electrical conductivity of the solution due to the mobility of the first kind of ions.

The second kind of ions are motionless on the whole and do not produce a current. Nevertheless, the presence of motionless anions leads to an additional potential drop. This additional drop, expressed by the second term on the right side of (46.15), serves to compensate for the gradient of osmotic pressure in the electrolyte.

Thus, the varying concentration of anions leads to the additional term $\frac{RT}{z_1 F} \ln \frac{c_0}{c_C}$ in the complete expression for the ohmic potential drop. This portion of the ohmic potential drop is also present when the current is zero. Its numerical value is equal to the concentration overpotential for the discharge of ions with a valence $z_1$. It should be noted that the existence of this additional ohmic drop in the solution has not been taken into consideration by some authors, even in relatively recent electrochemical papers. This leads to erroneous results.

According to (46.3), the applied emf is

$$V = \varphi_{x=0} - \varphi_{x=d} + \eta_{\text{conc}} + \Delta\varphi_{\text{equ}} = \Delta\varphi_{\text{equ}} + \\ + \frac{RT\,(z_1 + z_2)}{F z_1 z_2} \ln\left(1 - \frac{i_1 d}{D_1 F z_1 (z_1 + z_2)\, c_0}\right). \tag{46.16}$$

The volt-ampere characteristic of the discharge, i = f (V), has the form

$$i_1 = \frac{D_1 F z_1 (z_1 + z_2)\, c_0}{d} \left[1 - \exp\left\{\frac{F z_1 z_2}{RT\,(z_1 + z_2)} \left(V - \Delta\varphi_{\text{equ}}\right)\right\}\right]. \tag{46.17}$$

The curve in Figure 49 shows the dependence of current density on the applied emf (V is negative!).

It is clear from Figure 49 and equation (46.17) that for small values of the quantity

$$\frac{Fz_1z_2}{RT(z_1+z_2)}(V-\Delta\varphi_{\text{equ}})$$

the current is a linear function of the applied voltage. For high values of this quantity, the current reaches a limiting value

$$i_{\text{lim}}=\frac{D_1Fz_1(z_1+z_2)\,c_0}{d} \qquad (46.18)$$

and is no longer a function of the applied emf [2].

As noted from (46.13), the attainment of the limiting current corresponds to zero concentration of the discharging ions at the surface of the cathode,

$$c_{\text{C}}=0. \qquad (46.19)$$

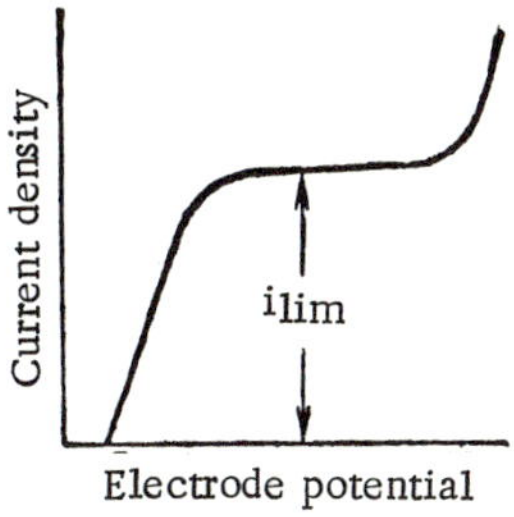


Figure 49. Limiting current density $i_{\text{lim}}$.

Evidently, therefore, $i_{\text{lim}}$ indicates the limiting migration-diffusion current. If a sufficiently strong emf is applied, the ions approaching the surface of the cathode lose their charge rapidly, and the value of the current flowing through the cell is determined by the number of ions conveyed to the cathode by diffusion and migration.

It follows from (46.18) that the ionic diffusion current can be written in the form

$$i_{\text{lim}}=\frac{D_1Fz_1c_1^{(0)}}{d}\left(1+\frac{z_1}{z_2}\right).$$

If neutral particles diffuse instead of ions, the diffusional current is $\left(\frac{D_1Fz_1c_1^{(0)}}{d}\right)$. The migration of ions in an electric field raises it by the factor $\left(1+\frac{z_1}{z_2}\right)$.

Introducing the expression for $i_{\text{lim}}$ into formula (46.17), we can derive the following expression for $i_1$:

$$i_1=i_{\text{lim}}\left(1-\exp\left\{\frac{z_1z_2F(V-\Delta\varphi_{\text{equ}})}{(z_1+z_2)RT}\right\}\right). \qquad (46.20)$$

From (46.13), the following expression applies to the same $i_1$:

$$i_1=\frac{D_1(c_0-c_{\text{C}})Fz_1(z_1+z_2)}{d}; \qquad (46.21)$$

i.e., the current in the cell has the form of a diffusional current between the anode and the cathode, carried by the discharging ions

$$\left[i_{\text{diff}}=\frac{D_1(c_0-c_{\text{C}})Fz_1}{d}\ \text{for a concentration difference of}\ c_0-c_{\text{C}}\right],$$

but multiplied by the factor $(z_1 + z_2)$, which characterizes the migration of ions in an electric field.

Equations (46.17) and (46.18) show that the current density in the cell is inversely proportional to the distance between the electrodes. When the distance between electrodes is large, the current in the cell is very small. However, since the value of the current is determined by the rate at which the ions are conveyed to the surface of the cathode, it is clear that it can be increased considerably by agitation of the solution. In this case, the ions are carried by convection with the flowing solution, in addition to transfer by diffusion and migration.

It is shown below (Section 50), from studies of the flow of current through a cell in which the solution is agitated, that the hydrodynamic regime in most cases determines the current density in an electrochemical cell.

The theory presented above regarding concentration overpotential was first developed in 1907. Until recently, however, errors were permitted to creep into certain specialized works on electrochemistry when this problem was examined. Thus, for example, Agar and Bowden [3] allowed for the ohmic potential drop in a solution of variable concentration but neglected the concentration overpotential. As a result, they derived an erroneous expression for i (V).

We have limited our preceding discussions to the examination of a steady-state regime of current flow through an electrochemical cell.

The choice of the cathode as the electrode at which we examined the concentration polarization was not arbitrary. In principle, concentration polarization at an anode is no different from that occurring at the cathode. Considerations analogous to those employed for the cathode yield the following expression for the volt-ampere characteristic of the cell, with polarization only at the anode (V is positive):

$$i = i_{\text{lim}}\left[\exp\left\{\frac{Fz_1z_2(V - \Delta\varphi_{\text{equ}})}{(z_1 + z_2)RT}\right\} - 1\right]. \tag{46.22}$$

Equation (46.22) indicates that, as the applied emf increases, the current at the anode increases indefinitely and does not tend toward a limiting value, as is the case at the cathode. For this reason, the concentration overpotential on the anode can be neglected in many cases. Let us note, however, that the anode current, in fact, cannot increase indefinitely with an increase in V, due to the formation of a saturated salt solution at the surface and of deposits on the anode or because of the attainment of a so-called passive state of the electrode, when the electrode process is the slow stage of the over-all process.

In conclusion, let us examine a problem that is of theoretical interest and which has not previously been examined in electrochemical literature [62]. We refer here to the validity of the assumption as to electrical neutrality of the solution outside the electric double layer.

The electrical neutrality of electrolytic solutions and the maintenance of this condition during the flow of a current was established by Faraday.

The thickness of the electric double layer, in whose outer sheet there is a concentration of ions of the same sign, is of the same order of magnitude as the Debye length $\frac{1}{\lambda}$ and is $10^{-5}$ cm even for very dilute solutions.

For the rest of the volume in an electrolytic cell, condition (46.6) is accepted.

Knowing the potential distribution, one can test the validity of this condition. Thus, the charge density in the cell is evidently

$$\rho = -\frac{\varepsilon}{4\pi}\frac{d^2\varphi}{dx^2} = -\frac{\varepsilon RT}{4\pi F_2}\left\{\frac{1}{c}\frac{d^2c}{dx^2} - \frac{1}{c^2}\left(\frac{dc}{dx}\right)^2\right\}. \tag{46.23}$$

Here, we have used equation (46.10). Substituting the value of c from (46.12) into (46.23), and using (46.18), we obtain

$$\rho = \frac{c_0^{(0)}Fz_1(z_1+z_2)\left(\frac{1}{\lambda}\right)^2}{d^2}\,\frac{1}{\left[\left(\frac{i_{\text{lim}}}{i}-1\right)+\frac{x}{d}\right]^2}. \tag{46.24}$$

Equation (46.24) shows that, for $i \neq i_{\text{lim}}$, i.e., when the current passing through the solution is less than the limiting value, the charge density in the bulk of the solution (outside the layer $x \leq \frac{1}{\lambda}$) is very small:

$$\rho < \rho_{x=0} = c_1^{(0)}Fz_1(z_1+z_2)\frac{\left(\frac{1}{\lambda}\right)^2}{d^2}\,\frac{1}{\left(\frac{i_{\text{lim}}}{i}-1\right)^2}. \tag{46.25}$$

The difference $(c_1z_1 - c_2z_2)F$ is very small compared to each of the terms separately, the ratio being $(1/\lambda)^2/d^2$.

The situation changes substantially, however, at $i = i_{\text{lim}}$, i.e., when the current passing through the solution has the limiting value. In that case, the charge density in the bulk of the solution may be expressed by:

$$\rho = c_1^{(0)}Fz_1(z_1+z_2)\frac{\left(\frac{1}{\lambda}\right)^2}{x^2}. \tag{46.26}$$

It follows from (46.26) that, when a current of limiting value flows through the cell, a charge appears in the vicinity of the cathode, whose density is relatively very high and decreases slowly (in accordance with $1/x^2$) with distance from the electrode. For $x \sim 10 \cdot 1/\lambda$, the charge in the solution is still great. It is precisely because of the existence of a space charge that currents greater than the limiting value $i_{lim}$ cannot flow in a steady-state manner through the cell.

From a quantitative viewpoint, expression (46.26) for the charge density, which has been derived on the assumption of electrical neutrality in the solution, can be employed only when this density is low, i.e., so long as

$$\rho \ll \begin{cases} Fc_1^{(0)}, \\ Fc_2^{(0)}, \end{cases}$$

and the corresponding potential represents but a small correction to that given by (46.14). In the vicinity of the cathode, for $x \sim 1/\lambda$, (46.26) is inapplicable. At such distances from the electrode, the region of the double layer begins, in which the solution cannot be considered electrically neutral even in the first approximation, as was assumed for the derivation of formula (46.26).

There are many indications in the electrochemical literature of the existence of space charges [61]. In particular, potential distribution was measured in the region between the electrodes. The deviation of this from a linear distribution was interpreted by the authors of these works as proof of the existence of a space charge in the solution.

The values of the charge densities computed from the measured deviations coincide, in terms of order of magnitude, with the values derived by us. The spatial distribution of the charge density in the bulk of the solution is, from a qualitative standpoint, similar to the $1/x^2$ relation. A quantitative comparison of these measurements with the theoretical formulas, however, would hardly be meaningful. From the standpoint of technique, all the experimental works deserve serious criticism since they made no allowance for natural convection. In addition, in computing the deviations of the potential from the linear path, the authors took no cognizance of border effects at the edges of the electrode or the walls of the cell, which led to an apparent dependence of the potential on the coordinates y and z as well as on the distance x between the electrodes. Finally, these studies made no allowance for the effect of the probe electrode on the potential distribution. An exact measurement of the potential distribution and of the space charges in an electrolytic cell would be of great interest.

## 47. CHEMICAL OVERPOTENTIAL. HYDROGEN OVERPOTENTIAL

By chemical overpotential we mean the excess of the potential difference above the equilibrium value that must be applied to the cell in order to maintain a finite rate for the processes occurring directly on the electrode.

Chemical overpotential occurs when the slow stage of the process is:

1) the electrochemical reaction on the electrode, including the discharge of ions;

2) the formation of final products from the discharged ions (for example, the formation of $H_2$ molecules from H atoms);

3) the formation of new phase nuclei on the surface of the electrode (for example, nuclei of crystalline deposits in the deposition of metal ions);

4) growth of a new phase, etc.

The variety of causes of chemical overpotential prevents our devoting space to their review, and we simply refer to pertinent literature [1, 4]. Available quantitative results of measurements of chemical overpotential relate mostly to the case where the discharge of ions is the slow step of the process. The theory of the ion discharge process was developed mainly from the study of this phenomenon for hydrogen ions.

Generally, it can be stated that the rate of chemical reaction (the discharge of ions and the formation of neutral final products), and, consequently, the current density at the electrode is determined by the potential difference at the metal-solution interface. This consists of the chemical overpotential and the equilibrium potential difference. The chemical overpotential $\eta_{chem}$, therefore, is a function of the current density in the cell,

$$\eta_{chem} = \eta_{chem}(i).$$

For small i, the function $\eta_{chem}$ can be expanded in a series, and can be estimated by the first term of the expansion. Thus, for small i we have:

$$\eta_{chem} = ki. \qquad (47.1)$$

At high current densities, the function $\eta_{chem}$ can be determined on the basis of general kinetic considerations. If it is assumed that the rate of the chemical reaction is determined by the over-all rate of the process of ion discharge and of the reverse process — ionization of atoms — then we can write:

$$i = F_1 - F_2, \qquad (47.2)$$

where the rates of discharge and ionization $F_1$ and $F_2$ are unknown functions of the potential difference at the electrode-solution interface.

If it is assumed that the discharge and ionization processes are related to passage through a certain potential barrier (for example, one related to the transition of an ion from a hydrate envelope to a metal atom adsorbed on the surface and back again), then $F_1$ and $F_2$ must be exponential functions of the height U of the potential barrier for the process, i.e.,

$$F_1 \approx ce^{-\frac{U}{RT}}, \qquad F_2 \approx e^{\frac{U}{RT}}, \tag{47.3}$$

where c is the concentration of the ions that lose their charge in the vicinity of the barrier.

The passage of particles through the potential barrier occurs in the body of the double layer. The electric field near the surface of the metal in the inner portion of the double layer is very strong. Its electric field intensity attains values of $10^6$ to $10^7$ v/cm.

If it is assumed that the electrical field has an important effect on the passage of ions through the barrier, by helping their migration in the direction toward the metal from the solution (discharge and hindering their movement in the opposite direction, then in the electrical field $F_2 \ll F_1$ and $U = f(\Delta\varphi)$ where $\Delta\varphi$ is the potential jump between the point in the solution from which the ion starts its migration and the metal surface. In order that theoretical results may coincide with experimental data, it is necessary to state that:

$$U = U_0 + \alpha F \Delta\varphi,$$

where $U_0$ is the height of the barrier in the absence of an electric field and $\alpha$ is a constant.

It should be stressed that the latter expression for U is not a series expansion in powers of $\Delta\varphi$. It is valid when the second term has the same order of magnitude as the first.

Using (47.2) and (47.3), we obtain, for $F_2 \ll F_1$:

$$i \approx ce^{-\frac{\alpha F \Delta\varphi}{RT}},$$

or

$$\Delta\varphi = -\frac{RT}{\alpha F} \ln i + \text{const.}$$

The chemical overpotential can be written in the form

$$\eta_{\text{chem}} = -(\Delta\varphi - \Delta\varphi_{\text{equ}}) = \frac{RT}{\alpha F} \ln i + \text{const.} \tag{47.4}$$

The value of $\alpha$ must be determined from experimental data. For many reactions it is found to be extremely close to 1/2.

The kinetic considerations expressed above show that, at low current densities, the chemical overpotential is a linear function of the current density, and at high densities, it is a logarithmic function. The numerical values of the constants in (47.1) and (47.4) are functions of the nature of the process and of the properties of the metal and the solution, and must be experimentally determined.

A study of different electrochemical processes that take place with noticeable chemical overpotentials showed that both limiting laws — the linear and the logarithmic — are encountered in actual practice.

As examples of processes accompanied by noticeable chemical overpotential, we may point to the liberation of hydrogen and oxygen in the electrolysis of a solution of $H_2SO_4$, to the deposition of certain metal ions as crystals, etc. [4, 1]. The process of hydrogen liberation has attracted especial attention due to its great general and practical significance.

The overpotential of hydrogen liberation on an electrode is often very large. In most cases it exceeds the magnitude of the concentration overpotential, as well as the discharge potentials of ions on various electrodes. The latter fact is of tremendous significance in many electrochemical processes. In the electrolysis of aqueous electrolytes, hydrogen ions are present in the solution to a greater or lesser extent along with ions of the electrolyte. If there were no hydrogen overpotential, hydrogen ions would discharge on the electrode when an emf is applied. Actually, since this overpotential is large, the process starts with the discharge of other ions. A noticeable liberation of hydrogen occurs only when the emf substantially exceeds the discharge potential of other ions (Figure 50).

The meaning of what has been said is obvious. Let us merely repeat that it applies to the operation of storage batteries (liberation of Pb from a solution containing few $Pb^{++}$ ions and many $H^+$ ions), the discharge of ions of alkali metals on a mercury electrode (from a solution containing alkali metal ions and $H^+$ ions), and others. The large emf necessary for the electrolytic decomposition of water is also related to this effect.

It should be noted here that the overpotential required for the discharge of hydrogen on an electrode is not constant, but is a function of the current density.* It has been

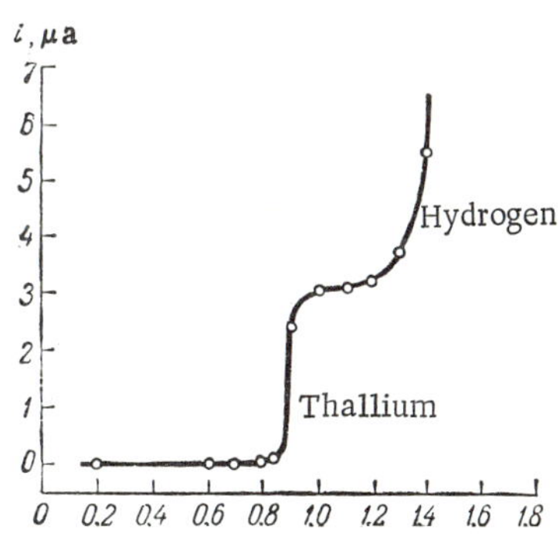


Figure 50. Curve of discharge of thallium and hydrogen ions (on a mercury electrode).

*Strictly speaking, therefore, hydrogen ions always discharge on an electrode, but below an applied emf of $V > \eta_{chem}$ their number is insignificant.

found experimentally that, at extremely low current densities, the hydrogen overpotential is a function of the current density in accordance with the relation

$$\eta_H = \text{const } i. \tag{47.5}$$

At high current densities, the hydrogen overpotential is expressed by the Tafel formula

$$\eta_H = a + \frac{RT}{\alpha F} \ln i. \tag{47.6}$$

Numerous and very accurate studies, conducted by A. N. Frumkin and co-workers, showed that the Tafel law is valid to a high degree of accuracy over an enormous range of current densities: from $i \approx 10^{-1}$ to $i \approx 10^{-7}$ a/cm$^2$ (Figure 51).

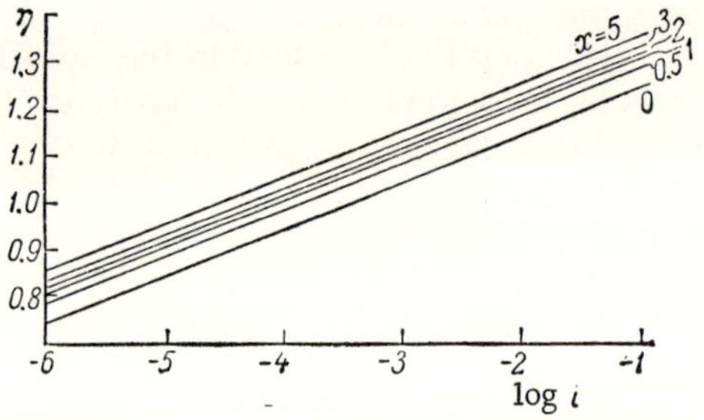


Figure 51. Hydrogen overpotential as a function of current density.

Measurements were conducted in a solution of 0.1 g $H_2SO_4$ + x g $Na_2SO_4$ in 1000 g $H_2O$.

The value of the coefficient $\frac{1}{\alpha}$ before the logarithm is not a function of the nature of the metal. Usually $\alpha$ is close to $\frac{1}{2}$. In certain cases, however, $\alpha$ is significantly different from $\frac{1}{2}$ and, generally speaking, assumes all values in the range from 0 to 1.

The quantity a changes substantially from one metal to another, and, in certain instances (see below), is a function of the concentration of the solution.

Comparing Tafel's empirical formula with (47.4), their similarity is apparent. In the case of hydrogen ions, as shown by A. N. Frumkin, the rate of the process of conversion of a hydrated proton into an adsorbed H atom can be written as:

$$F_1 \approx c_{\mathrm{H}^+}^{(S)} e^{-\frac{U_0+\alpha F\Delta\varphi}{RT}}, \tag{47.7}$$

where $c_{\mathrm{H}^+}^{(\mathrm{S})}$ is the number of protons in the region of the solution directly surrounding the electrode (related to 1 cm$^3$).

Evidently

$$c_{\mathrm{H}^+}^{(S)} = c_{\mathrm{H}^+} e^{-\frac{F\psi_1}{RT}}, \tag{47.8}$$

where $\psi_1$ is the drop in potential in the double layer in the zone extending from the neutral electrolyte to the plane at which the transfer occurs (Figure 52), and $c_{\mathrm{H}^+}$ is the concentration of hydrogen ions at the double layer-neutral solution interface.

Therefore,

$$F_1 \approx c_{\mathrm{H}^+} e^{-\frac{F\psi_1}{RT}} e^{-\frac{U_0+\alpha F\Delta\varphi}{RT}}. \tag{47.9}$$

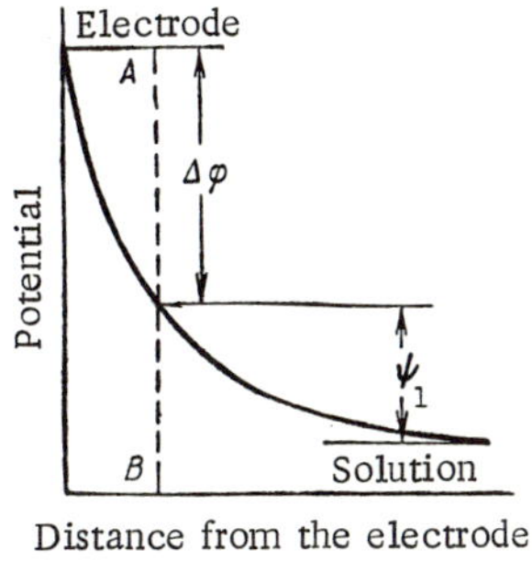


Figure 52. Variation in potential in an electric double layer.

On the basis of general kinetic considerations, the possibility of the reverse transition of atoms into an ionic state is determined by the expression

$$F_2 \approx e^{-\frac{U_0-(1-\alpha)F\Delta\varphi}{RT}} \tag{47.10}$$

Evidently, at equilibrium,

$$F_1 = F_2.$$

In a nonequilibrium state the over-all rate of the process of hydrogen ion discharge on 1 cm$^2$ of the cathode surface is

$$i = F_1 - F_2 = \left(k_1 c_{\mathrm{H}^+} e^{-\frac{F\psi_1}{RT}} e^{-\frac{\alpha F\Delta\varphi}{RT}} - k_2 e^{\frac{(1-\alpha)F\Delta\varphi}{RT}}\right) e^{-\frac{U_0}{RT}}. \tag{47.11}$$

If the potential of the solution is taken as zero, the metal-solution potential difference equals the cathode potential $\Delta\varphi_C$. The cathode potential can be subdivided into the potential drop $\psi_1$ in the outer, or diffuse, portion of the double layer and the potential drop $\Delta\varphi$ in the inner, dense portion of the layer: $\Delta\varphi_C = \Delta\varphi_C + \psi_1$ (Figure 52). Therefore,

$$i = \left(k_1 c_{\mathrm{H}^+} e^{-\frac{\alpha F\Delta\varphi_C}{RT}} e^{-\frac{F(1-\alpha)\psi_1}{RT}} - k_2 e^{\frac{(1-\alpha)F(\Delta\varphi_C-\psi_1)}{RT}}\right) e^{-\frac{U_0}{RT}}. \tag{47.12}$$

When the cathode potential is close to its equilibrium value, the exponents can be expanded in a series and the latter expression becomes equation (47.5). At the other extreme, for a large negative $\Delta\varphi_C$, the second term in the right side of equation (47.12) can be omitted. Then

$$i \approx c_{H^+} e^{-\frac{\alpha F \Delta\varphi_C}{RT}} e^{-\frac{(1-\alpha) F\psi_1}{RT}}. \qquad (47.13)$$

The quantity $\psi_1$ is a function of the concentration of hydrogen ions in the solution. The dependence of $\psi_1$ on $c_{H^+}$ can be found in two cases: when the solution contains a significant concentration of an inert electrolyte, and in a binary solution. In the presence of an excess of neutral electrolyte, the most mass of ions in the double layer are ions of the inert electrolyte. It is therefore possible to neglect the fact that $\psi_1$ is a function of the concentration of hydrogen ions, and to write:

$$\Delta\varphi_C = \frac{RT}{\alpha F} \ln c_{H^+} - \frac{RT}{\alpha F} \ln i + \text{const}. \qquad (47.14)$$

In equation (47.14) the cathode potential is measured with respect to the potential at the solution-double layer interface, which may be taken as zero.

The hydrogen overpotential, computed from formula (47.14), is determined by the expression

$$\begin{aligned} -\eta_H &= \Delta\varphi_K - (\Delta\varphi_C)_{equ} = \Delta\varphi_C - \frac{RT}{F} \ln c_{H^+} = \\ &= -\frac{RT}{\alpha F} \ln i + \left(\frac{1}{\alpha} - 1\right) \frac{RT}{F} \ln c_{H^+}. \end{aligned} \qquad (47.15)$$

The latter coincides with Tafel's empirical formula if $\alpha = \frac{1}{2}$.

It has not heretofore been possible to determine this value of $\alpha$ by convincing theoretical methods. The Frumkin-Volmer theory, therefore, is semiempirical.

On the assumption that $\alpha = \frac{1}{2}$, we can determine the relation existing between the concentration on the cathode, the current density, and the cathode potential from (47.15):

$$i = A c_{H^+} e^{-\frac{F\Delta\varphi_C}{2RT}}. \qquad (47.16)$$

Upon addition of an indifferent electrolyte, $\psi_1$, as is shown by the theory of the double layer [1, 4], is not a function of the hydrogen ion concentration, and

$$A = \text{const} \cdot e^{-\frac{F\psi_1}{2RT}} = \text{const}.$$

In a binary electrolyte, computations based on the theory of the double layer [1, 4] show that

$$\psi_1 = \frac{RT}{F} \ln c_{H^+} + \text{const}$$

and, correspondingly,

$$i = B(c_{H^+})^{1/2} e^{-\frac{F\Delta\varphi_C}{2RT}}. \qquad (47.17)$$

Experimental verification has confirmed the dependence of i on $c_{H^+}$ as predicted by the Frumkin theory. In general, it is therefore possible to state that this theory correctly reflects the mechanism of the process.

Henceforth, in examining electrochemical reactions that follow a course of mixed kinetics, we shall employ only equation (47.16) as the boundary condition for the surface of the cathode; i.e., we shall limit ourselves to an examination of reactions taking place in the presence of additions of an indifferent electrolyte. This equation shows a direct proportionality between the current density and the concentration of hydrogen ions (at a given negative cathode potential). The value of $c_{H^+}$ is determined by the rate of arrival of hydrogen ions at the cathode. As $\Delta\varphi_C$ decreases (with an increase in the applied emf), i increases exponentially to the power $\frac{F\Delta\varphi_C}{2RT}$. In the limit, when $\left|\frac{F\Delta\varphi_C}{2RT}\right| \gg 1$, as may be seen from expression (47.8), $c_{H^+}^{(S)} \to 0$. This means that the rate of discharge of ions is sufficiently great that all of the ions approaching the cathode surface will discharge. In other words, the over-all rate of the process is determined by the rate of arrival of hydrogen ions, so that the reaction becomes one of diffusion.

Other opinions have been advanced regarding the mechanism of the slow process in hydrogen discharge. Thus, for example, it has been assumed that the slow process is the combination of atomic hydrogen into $H_2$ molecules, the formation of hydrogen bubbles from separate molecules, and so on. We refer the reader to the specialized literature [1, 4] for details.

## 48. COMPARISON OF VARIOUS FACTORS THAT DETERMINE THE VALUE OF THE CURRENT IN A CELL

From an electrodynamic viewpoint, the drop in potential across an electrolytic cell may be viewed as comprising the ohmic potential drop in the solution, the concentration potential, and the chemical overpotential. The first item characterizes the potential drop in the bulk of the solution. The concentration potential drop occurs due to a change in ion concentration at the electrode, as compared to the ion concentration in the bulk of the solution. The third represents the potential drop at the very surface of the electrode, or, more exactly, within the double layer on that surface.

The emf applied between the electrodes is determined by the expression

$$V = \Delta\varphi_{equ} + \Delta\varphi_{ohm} + \eta_{conc} + \eta_{chem} \tag{48.1}$$

where $\Delta\varphi_{ohm}$ is the ohmic potential drop, and $\eta_{conc}$ and $\eta_{chem}$ are the concentration and chemical overpotentials.

For a comparison of the part played by each of them, let us examine the effective resistance of each of the processes.

By definition, the effective resistance w of the elements of a linear circuit is equal to $\frac{\partial V}{\partial i}$.

By differentiating (48.1) we obtain:

$$w = \frac{\partial \Delta\varphi_{ohm}}{\partial i} + \frac{\partial \eta_{conc}}{\partial i} + \frac{\partial \eta_{chem}}{\partial i}. \tag{48.2}$$

The ohmic potential drop in a solution of variable concentration was calculated above [see formula (46.15)]. Henceforth, in examining the concentration distribution in an agitated solution, it will be demonstrated that all of the change in the concentration of the solution occurs near the surface of the electrode in a thin fluid layer of a thickness $\delta$. In the balance of the volume of the solution, the concentration is constant and corresponds to an ohmic resistance $w_0$. In the general case, therefore, we can write

$$\frac{\partial \Delta\varphi_{ohm}}{\partial i} = w_0 + \frac{\partial \Delta\varphi_{\delta}}{\partial i}. \tag{48.3}$$

The quantity $\frac{\partial \Delta\varphi_{\delta}}{\partial i}$, which determines the effective resistance of the layer of variable concentration, can be computed from (46.15), taking into account (46.20). Evidently,

$$\frac{\partial \Delta\varphi_{\delta}}{\partial i} = + \frac{RT(z_1 + z_2)}{Fz_1 z_2} \frac{1}{i_{lim}\left(1 - \frac{i}{i_{lim}}\right)} + \frac{RT}{Fz_1} \frac{\partial}{\partial i} \ln \frac{c_0}{c_c}. \tag{48.4}$$

Then, for $\frac{\partial \eta_{conc}}{\partial i}$ we have, according to (46.2),

$$\frac{\partial \eta_{conc}}{\partial i} = - \frac{RT}{Fz_1} \frac{\partial}{\partial i} \ln \frac{c_0}{c_c}. \tag{48.5}$$

The full effective resistance w of the circuit, therefore (ignoring phenomena at the anode), is determined by the equation

$$w = w_0 + \frac{RT\,(z_1 + z_2)}{Fz_1z_2}\frac{1}{\left(1 - \frac{i}{i_{\text{lim}}}\right) i_{\text{lim}}} + \frac{\partial \eta_{\text{chem}}}{\partial i} = w_0 + w_1 + w_2. \quad (48.6)$$

The entire circuit can be regarded as a series of a linear and two nonlinear elements. The linear element is the solution with constant concentration, while the nonlinear elements are respectively the solution with variable concentration and the "zone of chemical reaction."

The potential distribution in the solution is controlled by the largest of the three terms $w_0$, $w_1$, and $w_2$ in equation (48.6). Usually they do not all have the same order of magnitude.

The ohmic resistance $w_0$ is relatively large in electrochemical cells of large dimensions (in galvanic baths, for example, and so on), and also in very dilute solutions.

It is determined exculsively by the dimensions and the geometric shape of the electrolyzer and by the electrical conductivity of a solution with constant concentration. The latter is equal to

$$\varkappa = \frac{F^2}{RT}\sum z_i^2 D_i c_i,$$

where the summation is extended over the types of ions and where $c_i$ is the concentration in the solution.

In the special case of a steady state electrolytic process in a static solution when the electricity carrier is a single type of ion, $\varkappa$ is given by the expression presented in Section 46.

As to the quantities $w_1$ and $w_2$, according to expressions (48.6), (47.5) and (47.6), they are as follows:

$$w_1 = \frac{RT}{F}\left(\frac{1}{z_1} + \frac{1}{z_2}\right)\cdot\frac{1}{i_{\text{lim}} - i},$$

$$w_2 = \begin{cases} \text{const}, \\ \frac{1}{\alpha}\frac{RT}{F\cdot i}. \end{cases}$$

If $i \ll i_{\text{lim}}$, then $w_1 \sim \frac{RT}{F \cdot i_{\text{lim}}}$.

The quantity $i_{\text{lim}}$ of course can vary within very wide limits. Nevertheless, it may be asserted that for $i \ll i_{\text{lim}}$, the resistance $w_1$ is very small. If, in addition, chemical overpotential is present, then $w_2 \gg w_1$ always holds. Indeed, examining the ratio $\frac{w_1}{w_2}$, we have, for $i \ll i_{\text{lim}}$,

$$\frac{w_1}{w_2} = \left(\frac{1}{z_1} + \frac{1}{z_2}\right)\frac{i}{2i_{\text{lim}}}\frac{1}{1 - \frac{i}{i_{\text{lim}}}} \approx \frac{i}{i_{\text{lim}}} \ll 1.$$

Thus, for $i \ll i_{lim}$, resistance $w_1$ can be neglected as compared to $w_2$. The situation changes radically when the current i approaches the limiting value $i_{lim}$. In this case, resistance $w_1$ increases rapidly, and its value may exceed $w_2$.

Thus, the picture can be characterized qualitatively in the following manner: if the current i flowing through the cell is small as compared to the limiting diffusion current $i_{lim}$, the change in the concentration of current-carrying ions can be neglected. Likewise, it is possible to disregard the concentration overpotential and to assume that the emf applied to the cell is equal to the sum of the ohmic potential drop and the chemical overpotential. In that case, the diffusional phenomena described in Section 45 do not play a significant part in the current flow through the cell. The entire solution may be characterized by an electrical conductivity constant $\varkappa$, and distribution of the current in the solution can be determined by solving a purely electrical problem.

If, on the other hand, the current in the cell is comparable to the limiting diffusion current, then the concentration overpotential rises sharply and acquires a leading role in the electrolytic process. In that case, the effect of the electrical and chemical factors on the over-all rate of the process usually is found to be secondary, and the value of the current flowing through the cell will be determined by the rate at which the ions arrive at the electrode surface. The latter rate is usually determined in practice by the nature of the hydrodynamic regime for an agitated solution. It is precisely this case that is the main subject of study in this chapter. Later (Sections 50 to 53) the value of $i_{lim}$ is computed for different cases of mixing.

In conclusion, let us point out that the determination of the value of $i_{lim}$ is significant not only in cases when the processes take place at current densities close to $i_{lim}$, but also when $i \ll i_{lim}$. In the latter instance it is important to know the value of $i_{lim}$ for comparison with the current actually flowing through the cell, so that the validity of the assumption that the value of the latter is small can be verified.

## 49. CURRENT DISTRIBUTION IN AN ELECTROLYTIC CELL IN THE ABSENCE OF CONCENTRATION OVERPOTENTIAL

Let us examine the distribution of current in an electrolytic cell in the absence of noticeable changes in the concentration of the solution at the electrode, i.e., at currents so low that the inequality $i \ll i_{lim}$ can be considered valid.

According to what was said in the previous section, the concentration overpotential can be neglected in this case. The current flowing through the cell is determined by a combination of purely electrical factors (electrical conductivity of the solution, geometric shape of the electrodes and their relative disposition in the cell) and chemical factors (dependence of the chemical overpotential on current density). Since the current distribution in the solution is an example of direct current flow through a distributed-parameter system the potential distribution in the solution must be found in order to determine the current. The potential in the solution satisfies the Laplace equation

$$\Delta\varphi = 0. \qquad (49.1)$$

Let us write the boundary conditions that must be satisfied by the potential at the electrode surface. By analogy with what has been done previously, let us limit our examination to only one of the electrodes, the cathode, for example.

If the electrochemical reaction on the surface of the cathode occurs with chemical overpotential $\eta_{chem}(i)$, then the potential difference at this surface will be:

$$\Delta\varphi_c = \varphi_{equ} + \eta_{chem}(i). \qquad (49.2)$$

The $\Delta\varphi_c$ in (49.2) may be more conveniently presented as the difference in potential between the electrode metal and the solution at the surface of the cathode

$$\Delta\varphi_c = \varphi_{met} - \varphi_c,$$

where $\varphi_{met}$ is the constant potential of the metal and $\varphi_c$ is the potential of the solution at the surface of the cathode.

Considering likewise that $i = \varkappa\left(\frac{\partial\varphi_c}{\partial n}\right)$, where n is the direction of the outward normal to the surface of the cathode and $\frac{\partial\varphi_c}{\partial n}$ is evaluated at the cathode surface, the potential difference can be expressed as:

$$\varphi_{met} - \varphi_c = \varphi_{equ} + \eta_{chem}\left(\varkappa \frac{\partial\varphi_c}{\partial n}\right). \qquad (49.3)$$

The potential of the second electrode can usually be considered constant and taken as the zero potential.

Equation (49.3) represents the boundary condition of a boundary value problem.

Since we are not interested in the change in potential across the double layer, we shall call the potential $\varphi_c$ the potential at the surface of the electrode.

As distinct from the metal potential $\varphi_{met}$, the potential $\varphi_c$ may vary from point to point.

We see that, in the general case, the boundary condition (49.3) has a very complex form.

Indeed, in the general case $\eta_{\text{chem}}\ \varkappa\left(\frac{\partial \varphi_c}{\partial n}\right)$ is a nonlinear function and the boundary condition is also nonlinear.

It is known, however, that there is no general analytic method for the solution of partial differential equations with nonlinear boundary conditions.

Therefore, to solve the very complicated problem which appears here, it is necessary either to obtain a simplification of the boundary conditions, or to develop methods for an approximate solution.

The problem of distribution of potential and current in an electrolytic cell constitutes one of the basic problems of applied electrochemistry.

The current distribution in the electrolyzer plays an important part in electroplating, in the theory of corrosion, in electrochemical protection, in industrial electrochemistry, and so forth. In many cases, the passage of current is not limited by concentration polarization.

The discussion of all problems that arise in connection with the passage of current through an electrolytic cell, when concentration polarization can be disregarded, cannot be an objective of this book. Such problems are not directly related to physicochemical hydrodynamics.

It can be shown, however, that important problems not directly connected with concentration polarization sometimes depend significantly on the convective diffusion of certain products (neutral molecules, for example).

In addition, up to the present, nothing has been published representing a general approach to the solution of the problem of current distribution in an electrolytic cell.

The bulk of the literature on the subject, such as Riemann's interesting work, contains a large number of erroneous studies. Therefore, we shall examine briefly certain problems of a general nature concerning the theory of current distribution in an electrolytic cell in the absence of concentration polarization and shall examine a series of important or typical problems.

First of all, let us mention the problem of simplifying the boundary condition on the surface of the electrode [6].

In principle, such a simplification may be found in three basic cases:

1) If the chemical overpotential on the electrode is absent (in electrochemistry, such electrodes are known as nonpolarizable), and the processes occurring on the electrode are reversible.

For such cases, omitting chemical overpotential in (49.1), let us write the boundary condition in the form

$$\varphi_c = \varphi_{\text{met}} - \varphi_{\text{equ}} = \text{const} \quad \text{(on the surface of the electrode).} \qquad (49.4)$$

2) If current densities are sufficiently low, so that overpotential is expressed by a linear law (mildly polarizable electrode). Here,

the boundary condition can be written in the form of a linear relation between the potential and the current density

$$k\varkappa \frac{\partial \varphi}{\partial n} = \varphi_{\text{met}} - \varphi_{\text{C}} - \varphi_{\text{equ}}, \tag{49.5}$$

where k is a certain constant which is relatively small at low overpotentials. A linear boundary condition is very often used in the design of electrolytic cells.

Actually, however, its area of application is very limited since the linear overpotential law (47.1) can be utilized only at very low current densities.

3) The third instance of possible simplification of the boundary condition is, therefore, of great importance. Such a simplification can be made at high values of overpotential for an electrode of limited dimensions.

For $\eta_{\text{chem}}$, let us utilize the general expression (47.4) and let us rewrite (49.3) in the form

$$\varkappa \frac{\partial \varphi}{\partial n} = \exp\left\{\frac{F}{2RT}(\varphi_{\text{met}} - \varphi_{\text{C}} - \varphi_{\text{equ}} - a\right\}. \tag{49.3'}$$

For cathodes of finite dimensions, the above boundary condition admits of substantial simplification.

Let L be the characteristic dimension of the electrode. Assume that the change of the potential of the solution over this area is very small compared to the quantity $\frac{2RT}{F} = 50 \cdot 10^{-3}$ v. We shall explain below the conditions under which this assumption is valid. Then

$$\varphi_{\text{C}} = \overline{\varphi_{\text{C}}^{(0)}} + \varphi,$$

where $\overline{\varphi_{\text{C}}^{(0)}}$ is the potential at a central point on the cathode and $\varphi$ is the deviation of the potential from that central value.

Substituting $\varphi_{\text{C}}$ into (49.3 ), we obtain

$$\varkappa \frac{\partial \varphi}{\partial n} = \exp\left\{\frac{F}{2RT}(\varphi_{\text{met}} - \overline{\varphi_{\text{C}}^{(0)}} - \varphi_{\text{equ}} - a)\right\}\left(1 - \frac{F\varphi}{2RT}\right).$$

In a zero approximation of the potential deviation,

$$\varkappa \frac{\partial \varphi}{\partial n} = \exp\left\{\frac{F}{2RT}(\varphi_{\text{met}} - \overline{\varphi_{\text{C}}^{(0)}} - \varphi_{\text{equ}} - a)\right\} = i_0. \tag{49.6}$$

Thus, the current density in the zero approximation can be considered constant over the entire surface of the electrode and equal to the current $i_0$ at the central point of the cathode [6].

In a further approximation, on the surface of the cathode, we have

$$\varkappa \frac{\partial \varphi}{\partial n} = i_0 + \left(\frac{\partial i_0}{\partial \varphi_C}\right) \varphi. \tag{49.7}$$

Conditions (49.6) and (49.7) are linear boundary conditions which must be satisfied by the potential and its derivatives.

To clarify the conditions under which the expansion (49.7) is admissible, let us rewrite this condition in dimensionless form, replacing $\frac{\partial \varphi}{\partial n}$ by $\frac{1}{L}\frac{\partial \varphi}{\partial N}$, where N is a dimensionless coordinate plotted along the normal.

Then

$$\frac{\partial \varphi}{\partial N} = \frac{i_0 L}{\varkappa} + \left(\frac{\partial i_0}{\partial \varphi_C}\right) \cdot \frac{L}{\varkappa} \varphi. \tag{49.8}$$

Designating the dimensionless quantity in (49.8) by H, we have

$$H = \frac{L}{\varkappa} \frac{\partial i_0}{\partial \varphi_C},$$

yielding:

$$\frac{\partial \varphi}{\partial N} = \frac{i_0 L}{\varkappa} + H\varphi. \tag{49.9}$$

Evidently the dimensionless quantity H determines the nature of the change in potential across the electrode.

If $H \ll 1$, conditions (49.6) and (49.7) are valid. The current density on the surface of the cathode is virtually constant.

On the other hand, for $H \gg 1$, we may assume, on the surface of the cathode,

$$\varphi \approx 0, \tag{49.10}$$

or

$$\varphi_C = \overline{\varphi_C^{(0)}}.$$

This should be clear from the following considerations. Let us set up the ratio of the ohmic and "chemical" resistances, the latter being equal to $\frac{\partial \varphi_C}{\partial i_0}$:

$$\frac{R_{\text{ohm}}}{R_{\text{chem}}} = \frac{L}{\varkappa} : \frac{\partial \varphi_C}{\partial i_0} = H$$

Thus, the quantity H, formally introduced earlier, has the actual meaning $R_{\text{ohm}}/R_{\text{chem}}$.

The case of $H \gg 1$ relates to a small chemical resistance compared to the ohmic resistance. Therefore, the voltage drop due to the chemical resistance can be neglected; i.e., $\varphi = 0$.

Thus, for $H \gg 1$, the potential on the surface of the electrode has a constant value.

The meaning of these approximations is not difficult to see: in the first instance, when the cathode dimensions are sufficiently small, the distribution of the current density on the surface of the cathode is determined by the course of the electrochemical reaction. Since all points on the cathode are the same with respect to this reaction, the current density is found to be constant on the surface of the cathode.

In the opposite limiting case, which occurs at very large linear dimensions of the electrode, the current distribution on the cathode is not a function of the electrochemical stage of the reaction. In such cases we talk about the absence of chemical polarization.

Several important problems examined below utilize the approximate boundary conditions derived by us.

Now, we shall indicate a possibly general method of approximate calculation of fields in electrolytic cells, valid for the most generalized form of boundary conditions.

It is well known that the distribution of a field in a conducting medium corresponds to a minimum dissipation of energy.

In other words, a current distribution is established for which the generation of heat is least.

Let us set forth here a demonstration of this theorem.

The amount of heat emitted during the flow of a current through a certain volume of a medium with conductivity $\varkappa$ is evidently expressed by the relation

$$W = \int \mathbf{j}\mathbf{E}\, dv = \int \varkappa E^2\, dv = \int \varkappa (\operatorname{grad} \varphi)^2\, dv.$$

For the sake of brevity, let us assume that the region is a plane, so that $\varphi = \varphi(x, y)$ and

$$W = \int \varkappa \left[ \left( \frac{\partial \varphi}{\partial x} \right)^2 + \left( \frac{\partial \varphi}{\partial y} \right)^2 \right] ds. \tag{49.11}$$

Let us assume, to begin with, that the potential on the boundaries of the integration field (the contour L) is maintained at a constant value (metallic walls, without chemical overpotential). Then, it is easy to write the condition of minimum energy dissipation (heat emission). It is known that a necessary condition for a minimum of the integral (49.11) is that its variation by equal to zero*:

*In this case, it can also be easily shown that this condition is sufficient.

$$\delta W = \delta \left\{ \varkappa \int \left[ \left( \frac{\partial \varphi}{\partial x} \right)^2 + \left( \frac{\partial \varphi}{\partial y} \right)^2 \right] ds \right\} = 0.$$

Computing the variation and integrating by parts we have:

$$\delta W = 2\varkappa \int\limits_s \int \left[ \frac{\partial \varphi}{\partial x} \left( \frac{\partial \delta \varphi}{\partial x} \right) + \frac{\partial \varphi}{\partial y} \left( \frac{\partial \delta \varphi}{\partial y} \right) \right] ds =$$

$$= 2\varkappa \left\{ \int\limits_s \int \left[ \frac{\partial}{\partial x} \left( \frac{\partial \varphi}{\partial x} \delta \varphi \right) + \frac{\partial}{\partial y} \left( \frac{\partial \varphi}{\partial y} \delta \varphi \right) \right] dx\, dy - \right.$$

$$\left. - \int\limits_s \int \left( \frac{\partial^2 \varphi}{\partial x^2} + \frac{\partial^2 \varphi}{\partial y^2} \right) \delta \varphi \, dx\, dy \right\} = 0.$$

The first integral can be transformed by Green's theorem into an integral along the contour bounding the surface S.

We then have:

$$\delta W = 2\varkappa \int\limits_L \left( \frac{\partial \varphi}{\partial x} \frac{dy}{dl} - \frac{\partial \varphi}{\partial y} \frac{dx}{dl} \right) (\delta \varphi)_L \, dl - 2\varkappa \int\limits_s \Delta \varphi \, \delta \varphi \, dx\, dy =$$

$$= 2\varkappa \int\limits_L \frac{\partial \varphi}{\partial n} (\delta \varphi)_L \, dl - 2\varkappa \int\limits_s \Delta \varphi \, \delta \varphi \, ds = 0.$$

In the latter expression, $(\delta\varphi)L$ in the first integral indicates variation of the potential along the contour L. Based on the condition of constancy of potential along this contour, $(\delta\varphi)_L = 0$.

Due to the arbitary nature of $\delta\varphi$, it follows from $\delta W = 0$ that

$$\Delta \varphi = 0.$$

This computation can be generalized easily to a three-dimensional case.

Thus, the equation for the distribution of the potential in the conducting medium follows directly from the condition of minimum dissipation of energy during the passage of current.

This is often formulated in another way: current in a spatial system follows lines along which the ohmic resistance is a minimum.

It should be stressed that this principle is not valid for an electrolytic cell.

Indeed, it is valid only to the extent that the potential along the contour of integration may be considered constant. The potential on the electrode surface in an electrolytic cell is constant only in the absence of chemical overpotential. Let us find a new minimal principle that would be valid for an electrolytic cell. It must lead to the boundary value problem formulated above, with boundary conditions on the electrodes of the type of (49.3). We shall assume that the disposition of the electrodes can be characterized by a certain function $\rho(l)$ on the contour bounding the cell. For the sake of brevity, we shall consider the latter to be a plane. The function $\rho(l)$ is usually discontinuous. This means that the electrodes occupy only a portion of the contour.

Let us examine a quantity W* equal to

$$W^* = \varkappa \iint_s \left[\left(\frac{\partial \varphi}{\partial x}\right)^2 + \left(\frac{\partial \varphi}{\partial y}\right)^2\right] dx\, dy + \int_L \rho(l) F(\varphi)\, dl. \qquad (49.12)$$

The potential function F $[\varphi(l)]$ is, for the time being, arbitrary.

Computing the variation of this quantity with respect to the potential, and equating it to zero, we derive the necessary minimal condition

$$\begin{aligned} \delta W^* &= 2\varkappa \int_L \frac{\partial \varphi}{\partial n} (\delta\varphi)_L\, dl - 2\varkappa \iint_s \Delta\varphi\, \delta\varphi\, ds + \int_L \rho(l) \left(\frac{\partial F}{\partial \varphi}\right)_L (\delta\varphi)_L\, dl = \\ &= \int_L \left[2\varkappa \frac{\partial \varphi}{\partial n} + \rho(l) \frac{\partial F}{\partial \varphi}\right]_L (\delta\varphi)_L\, dl - 2\varkappa \iint_s \Delta\varphi\, \delta\varphi\, ds = 0. \end{aligned} \qquad (49.13)$$

Let us now choose the function $F(\varphi)$ in such a way that boundary condition (49.3) is fulfilled along the contour L for an arbitrary $(\delta\varphi)_L$. Namely, let us state:

$$F(\varphi) = -2 \int_0^{\varphi(l)} j(\varphi)\, d\varphi + f(l), \qquad (49.14)$$

where $\varphi(l)$ is the value of the potential at the point $l$, $f(l)$ is any desired function of $l$ independent of the potential, and $j(\varphi)$ expresses the dependence of current density on the surface potential of the electrode. The latter is given by the relation for the chemical overpotential.

We then have

$$\varkappa \frac{\partial \varphi}{\partial n} = \rho(l) j(\varphi) \text{ along the contour L.} \qquad (49.15)$$

This is equivalent to boundary condition (49.3).

On the strength of condition (49.15), for such a choice for $F(\varphi)$, the first integral in (49.13) becomes zero on the contour L. Here, the necessary condition W* will be written as

$$\delta W^* = -2\varkappa \iint_s \Delta\varphi\, \delta\varphi\, ds = 0.$$

In view of the abritrary nature of $\delta\varphi$, this equation again reduces to the Laplace equation (49.1) for $\varphi$.

Thus, the boundary value problem, with boundary conditions of the type of (49.3), or, what amounts to the same thing (49.15), is equivalent to the variational problem of finding the minimum value of W*.

To clarify the meaning of this assertion, let us rewrite W* noting (49.14). We then obtain:

$$W^* = \iint \frac{j^2}{\varkappa} ds - 2\int_L \int_0^{\varphi} j(\varphi)\rho(l)\, d\varphi\, dl. \qquad (49.16)$$

The latter integral represents the amount of heat emitted due to the chemical resistance (on the electrodes).

Thus, W*, as given by (49.16), represents the complete energy dissipation emitted in the form of heat in a spatial cell with a specific resistance $1/\varkappa$ and a chemical resistance along the contour.

It follows from what has been said that the usual considerations of the shortest line of a current is inapplicable, generally speaking, to electrolytic cells. A steady flow of current may occur in them, for which the lines followed by the current are very long and pass to regions of lower chemical overpotential. The picture of the field distribution is complicated further by the fact that the latter is a function of the current density.

The reader is reminded that the results presented here, and, especially, considerations regarding the variational principle, are inapplicable in cases where the concentration overpotential is not negligible.

The importance of the theorem just proved consists in the following: in the calculus of variations, very convenient methods of approximation have been developed, known as direct methods. In accordance with these, instead of deriving the function $\varphi(x, y)$ that is a solution of the boundary value problem and gives the minimum for integral (49.16) accurately, an approximate expression is chosen for $\varphi(x, y)$ for which the integral (49.16) has an approximately minimum value.

The methods of choosing such a function differ. All of them, however, lead in a relatively simple manner to the derivation of $\varphi(x, y)$ which approximates the minimum of the integral (49.16) and which describes the distribution of the field, also approximately.

Thus, the minimal principle* formulated above permits a very effective design of any desired electrolytic cell by taking into account the chemical overpotential on the electrodes.

Passing now to the examination of certain typical problems — problems that admit of an analytic solution — let us first examine a system in which the cathode includes a small amount of metal M in the major metal N.

The major metal represents the plane $y = 0$, and the inclusion represents a disk of radius R. The half-space $y > 0$ is occupied by the electrolyte. Such inclusions play an essential part in the process of anodic dissolution of metals [6].

---

*Its formulation is that of the author and of Yu. A. Chizmadzhev.

The features of the inclusion are the following: two independent processes take place on a homogeneous surface adjacent to the electrode — the first, the anodic process, consists in the transition of the metal into an ionic state in the solution; the second, the cathodic process, consists in, for example, the process of discharge of hydrogen ions.

In a steady state, the rates of the anodic and the cathodic processes are equal. This relates to a definite value of the metal potential known as stationary.*

It should be borne in mind that both the anodic and the cathodic processes include both forward and reverse reactions. The course of the entire process may be written in the form

$$\left.\begin{array}{l} H^+ + e \rightarrow \frac{1}{2} H_2 \\ \frac{1}{2} H_2 \rightarrow H^+ + e \end{array}\right\} \qquad \begin{array}{ll} \text{Rate} & i_1 \\ \text{"} & i_2 \end{array}$$

$$\left.\begin{array}{l} Me^{n+} + ne \rightarrow Me \\ Me \rightarrow Me^{n+} + ne \end{array}\right\} \qquad \begin{array}{ll} \text{"} & i_3 \\ \text{"} & i_4 \end{array}$$

In the stationary state

$$i_1 - i_2 = i_4 - i_3.$$

The presence of an inclusion, on which the hydrogen ions discharge more rapidly than on the major metal under stationary conditions, leads to a change in the potential of the metal. Namely, it changes in such a manner that the rate of hydrogen discharge is lowered, while that of dissolution of the anode rises.

The total amount of hydrogen discharged on the entire surface (major metal + inclusion) remains equal to the amount of metal dissolved.

To compute the rate of dissolution of a nonhomogeneous metal, the distribution of current along its surface must be determined.

For sufficiently small inclusions, on which the rate of hydrogen discharge is relatively rapid, and with a highly conductive solution, the condition $H \ll 1$ can be considered fulfilled. The boundary condition for the inclusion (on the cathode), therefore, has the form

---

*The difference between a stationary and an equilibrium potential consists in the following: an equilibrium potential relates to the thermodynamic equilibrium for a certain process, where the rates of the forward and reverse processes are equal. If, as is the case in the dissolution of metals, several independent processes are possible and the total rate of all direct processes is equal to that of the reverse processes, the state of the entire system remains unchanged in time. It is not in over-all thermodynamic equilibrium, however, since thermodynamic equilibrium requires detailed balancing of each process. Such a system is known as stationary.

$$-\varkappa\left(\frac{\partial\varphi}{\partial n}\right)=i_0=\text{const.}$$

If the chemical overpotential on the major metal is small, its potential can be considered constant

$$\varphi_a=0.$$

Far from the surface of the metal, as $y \to \infty$, the field's potential must approach zero, $\varphi \to 0$ as $y \to \infty$.

For a solution of the boundary value problem, it is convenient to introduce elliptic coordinates $\xi$, $\eta$, and $\zeta$ related to the coordinates x, y, and z by the relations

$$x=r\cos\zeta,$$
$$z=r\sin\zeta,$$
$$y=R\xi\eta,$$
$$r=R\sqrt{(1+\xi^2)(1-\eta^2)},$$

where R is the cathode radius.

The surfaces $\xi$ = const represent the rotation surfaces of confocal ellipsoids, having a focal distance r, around the y axis. For $\xi = 0$, the rotation surface degenerates into that part of the surface $y = 0$ situated inside a circle of radius $r = R$.

The surfaces $\eta$ = const represent the surfaces of confocal hyperboloids. For $\eta = 0$, the surface of the hyperboloid degenerates into that part of the surface $y = 0$ situated beyond the circle of radius $r = R$. For $\eta = 1$, the surface of rotation degenerates into the y axis.

The half-space $y > 0$ is the area of variation of the variables $\xi$, $\eta$, and $\zeta$,

$$0\leqslant\xi<\infty,\quad 0\leqslant\eta\leqslant 1,\quad 0\leqslant\zeta\leqslant 2\pi.$$

It is obvious that in elliptic coordinates the cathode of our cell represents the surface $\xi = 0$, and the anode, the surface $\eta = 0$.

In elliptic coordinates, the Laplace equation takes the form

$$\frac{1}{R^2(\xi^2+\eta^2)}\left\{\frac{\partial}{\partial\xi}\left[(1+\xi^2)\frac{\partial\varphi}{\partial\xi}\right]+\frac{\partial}{\partial\eta}\left[(1-\eta^2)\frac{\partial\varphi}{\partial\eta}\right]+\right.$$
$$\left.+\frac{(\xi^2+\eta^2)}{(1+\xi^2)(1-\eta^2)}\frac{\partial^2\varphi}{\partial\zeta^2}\right\}=0. \qquad (49.17)$$

The boundary conditions are

$$-\varkappa\left(\frac{\partial\varphi}{\partial n}\right)_{\xi=0}=i_0 \quad \text{(on the cathode)}, \qquad (49.18)$$

$$\varphi_{\eta=0}=0 \quad \text{(on the anode)}, \qquad (49.19)$$

$$\varphi_\infty=0 \quad \text{(at infinity)}. \qquad (49.20)$$

where the normal direction is into the surface. Because of the axial symmetry of the problem, the potential can not be a function of $\xi$. To solve the problem, let us suppose that

$$\varphi = P(\eta)\,Q(\xi).$$

Then, for P and Q we find the equations

$$\frac{d}{d\eta}\left[(1-\eta^2)\frac{dP(\eta)}{d\eta}\right] + nP(\eta) = 0, \tag{49.21}$$

$$\frac{d}{d\xi}\left[(1+\xi^2)\frac{dQ(\xi)}{d\xi}\right] - nQ(\xi) = 0, \tag{49.22}$$

where n is the separation parameter.

It is known that equations (49.21) and (49.22) have finite solutions only for $n = l\,(l+1)$, where $l$ is an integer (including zero). Here, (49.22) is the equation for the Legendre function of the variable $(i\xi)$, which must be regular for all values of $\xi$. Therefore, the relation (49.22) is an equation for the Legendre function of the second kind [a Legendre function of the first kind of the variable $(i\xi)$ increases without bound as $\xi \to \infty$, and does not satisfy condition (49.20)].

Expression (49.21) is an equation for the Legendre function of the first kind.

The general solution of equation (49.17) can be written in the form

$$\varphi = \sum A_l P_l(\eta)\,Q_l(i\xi). \tag{49.23}$$

The series includes only terms with odd indices $l$, in order to satisfy conditions (49.19) and (49.20).

In order to satisfy boundary condition (49.18), only the term $n = 1$ can be retained in the solution.

Thus,

$$\varphi = \frac{2i_0 R}{\pi\varkappa} P_1(\eta)\,Q_1(i\xi), \tag{49.24}$$

where $$P_1(\eta) = \eta, \quad Q_1(i\xi) = \xi \operatorname{arccot} \xi - 1.$$

In particular, on the cathode

$$\varphi_c = -\frac{2i_0 R}{\pi\varkappa}\eta = -\frac{2i_0 R}{\pi\varkappa}\sqrt{1-\left(\frac{r}{R}\right)^2}. \tag{49.25}$$

The specific ohmic resistance, in the flow of current from the anode to a point of coordinate r on the cathode, is expresed by the formula

$$w = \frac{\varphi_a - \varphi_c}{i_0} = \frac{2R}{\pi\varkappa}\sqrt{1-\left(\frac{r}{R}\right)^2}. \tag{49.26}$$

The maximum potential difference between the center of the inclusion and the major metal is

$$\Delta\varphi = \varphi_c(0) - \varphi_a = -\frac{2Ri_0}{\pi\varkappa}.$$

The latter relation leads to an important conclusion: for small dimensions of the inclusions ($R \sim 10^{-3}$ to $10^{-4}$ cm), even at relatively high current densities, the potential difference $\Delta\varphi$ is extremely small. This general conclusion was first arrived at by A. N. Frumkin and the author [6] and independently by Wagner [7]. It permits us to regard a nonhomogeneous metal surface as homogeneous, but with different values of characteristic properties, for example, hydrogen overpotential.

The second important problem that requires a derivation of the potential distribution is that of providing metals with anticorrosion protection by means of a protective cover or by electrochemical means.

A device known as a shield is very often used to retard the rate of metal dissolution. This shield consists of an artificially applied coating or application of another metal with special properties to the surface of the metal to be protected. The shield is so chosen that on contact with the electrolyte, a change in potential occurs between it and the protected metal in the direction of the cathode. This means that the anodic process (i.e., dissolution of the substance of the shield) predominates on the shield, while the cathodic process (discharge of hydrogen ions) predominates on the metal that is shielded. Obviously, in this case the rate of dissolution of the protected metal will be slower than would be the case in the absence of a shield.

Thus, the effect of the shield is the reverse of that of inclusions in the electrode metal itself. While inclusions lead to a speeding up of the dissolution (corrosion) of the metal, a shield protects the metal from corrosion.

Naturally, from the standpoint of electrochemical behavior, the pairs metal-inclusion and metal-shield must be antipodes.

The linear dimensions of the shield depend on the size of the area to be protected, but, of course, are considerably greater than the dimensions of accidental local inclusions and may extend dozens of centimeters and even meters.

In an electrolytic cell consisting of a shield and basic metal the former plays the part of the anode and the latter than of the cathode.

In this case it is found that there are two basic instances of electrochemical protection:

1) Processes in the cell as a whole are determined by the electrochemical rate laws on the electrodes and by the electrical conductivity of the medium;

2) The processes depend materially on reactions involving dissolved oxygen on the surface of the protected metal. Here, the over-all rate of the process is most often determined by the rate of convective diffusion of oxygen.

Such a system is known as electrochemical shielding with oxygen depolarization. It is not our objective to devise any specific systems of electrochemical protection. We merely wish to examine its mechanism from a quantitative standpoint.*

Let us initially examine the first case.

Let us assume that the surface of the protected metal is a plane $y = 0$, while the shield is an infinite band $-L \leq x \leq L$ within the plane $y = 0$. The electrolyte occupies the half-space $y > 0$.

Let us assume that the condition $H \ll 1$ is fulfilled on the shield.

Shields satisfying this condition of being strongly polarizable are often used in actual practice. The condition $H \ll 1$ imposes certain limitations on the dimensions and the properties (the quantity $\frac{\partial i_0}{\partial \varphi_c}$) of the shield. If they are fulfilled, the boundary condition on the surface $S_1$ of the shield has the form

$$-\varkappa\left(\frac{\partial \varphi_a}{\partial y}\right)_{\substack{y=0,\\ |x| \leqslant L}} = \text{const} = i_0 \quad (\text{on } S_1). \tag{49.27}$$

The deviation of the potential on the surface of the shielded metal from the stationary value can be considered small if the rate of the processes on its surface is limited by electrochemical reactions. Therefore, the linear relation (49.5) can be used for chemical overpotential, and, with an appropriate choice of a reference for the measurement of potential, the boundary condition on the surface $S_2$ of the shielded metal can be written in the form

$$\varkappa\left(\frac{\partial \varphi_c}{\partial n}\right)_{\substack{y=0,\\ |x| > L}} = H\varphi_c \qquad (\text{on } S_2). \tag{49.28}$$

Here, the constant H can be assumed to be large (H approaching infinity would indicate the absence of a potential deviation from the stationary value).

The potential must approach zero at infinity.

The boundary value problem with boundary conditions (49.27) and (49.28) has been solved by the method of successive approximations. The solution can be presented in the form of a series in powers of the small quantity $1/H$:

$$\varphi = \varphi_0 + \frac{1}{H}\varphi_1 + \frac{1}{H^2}\varphi_2 + \ldots$$

*The theory set forth below is based on the work of R. R. Dogonadze, Yu. A. Chizmadzhev, and the author [60].

In the zero approximation $H \to \infty$, the boundary condition (49.28) reduces to

$$\varphi_C = 0 \qquad (\text{on } S_2). \tag{49.29}$$

In this case the problem essentially coincides with the local inclusion problem examined above, but differs from it in terms of geometry.

A solution of the boundary value problem for boundary conditions (49.27) and (49.29) can be derived by the usual methods of the theory of functions of a complex variable.

Without going into details of the solution, which would require too much space, let us simply give the results. The potential field in the zero approximation has the form

$$\varphi_0(x, y) = \frac{\pi i_0}{\varkappa} Re\left[\sqrt{L^2 - (x + iy)^2} - y\right], \tag{49.30}$$

where Re denotes the real part. Formula (49.30) was first derived by Wagner [8].

In the next approximation, effects of the order of $\frac{1}{H}$ are taken into account.

Expanding the boundary conditions, as usual, in a series in powers of $\frac{1}{H}$, the expression for the potential correction $\varphi_1$ (x, y) can be derived in the first approximation.

From a practical standpoint, of basic interest is the potential distribution on the surface of the shielded metal and, especially, the asymptotic behavior of the potential on the surface $S_2$ at a large distance from the shield. The value of the potential determines its rate of dissolution.

Utilizing the usual techniques of the theory of functions, it can be shown that the surface potential, in a first approximation, is expressed by the relation

$$\varphi_1(x) = \frac{L}{\pi} \int_{-\infty}^{\infty} \frac{\varphi_0(\xi)\, d\xi}{(\xi - x)^2}, \qquad |x| > L.$$

Substituting the value of $\varphi_0$ ($\xi$, y = 0), after certain transformations, we find for $\varphi_1$ (x):

$$\varphi_1(x) = \frac{\pi i_0 L}{H\varkappa}\left[\frac{|x|}{\sqrt{x^2 - L^2}} - 1\right]. \tag{49.31}$$

The asymptotic behavior of the potential for $x \gg L$ is expressed by the formula

$$\varphi_1(x) \sim \frac{\pi i_0 L}{2H\varkappa}\left(\frac{L}{x}\right)^2. \tag{49.32}$$

The law of the change in the potential $\varphi_1(x)$ with distance is found to be invalid for $x \approx L$, where it increases indefinitely. This circumstance has a simple meaning: at $x \approx L$, the chemical resistance becomes greater than the ohmic resistance, and the chemical overpotential can no longer be considered small.

In this region, the dimensionless parameter is $H(x) = \frac{x}{\varkappa} \frac{\partial i_0}{\partial \varphi} \ll 1$, and the potential can no longer be considered close to a constant potential. On the contary, in that region, as well as on the shield, the density of the current is close to a constant value.

Calculations show, however, that this has no no effect on the asymptotic expression for the potential.

Let us now pass to a discussion of the second of the instances mentioned above, when diffusion of oxygen plays a significant part in the mechanism of electrochemical protection.

Without going into the details of an electrochemical mechanism that is discussed in specialized literature, let us merely point out that, if the electrolyte contains a matcrial, oxygen in particular, capable of being reduced electrochemically on the surface of the metal, the transition of metal atoms into the solution may be compensated by the reduction of these substances. For example, instead of the discharge of hydrogen ions during the dissolution process, there is a reduction of oxygen.

In this instance, the mechanism of the protective electrochemical action changes. Dissolution of the metal continues as before on the shield (the anode). The potential on the protected metal (the cathode) shifts in such a way that the oxidation of the particles of the protected metal is reduced. The oxygen molecules that approach the cathode surface are reduced to $OH^-$ ions, which carry the current in the solution in the vicinity of the electrode.

Thus, the steady-state flow of current in the cell comprising the shield and the protected metal requires the diffusion of neutral oxygen molecules to the surface of the latter if it is to be maintained.

The potential distribution in the solution is still determined by the Laplace equation (49.1), but the system of boundary conditions changes.

The fact that the current density is constant can still be employed as the boundary condition on the surface of the shield (the anode).

The number of ions appearing on the surface of the protected metal over an area of 1 $cm^2$ in 1 sec in obviously equal to

$$i_1 = \varkappa \frac{\partial \varphi}{\partial n} = F\alpha Q,$$

where n is the normal direction outward from the electrode, Q is the number of molecules reduced, and $\alpha$ is a numerical coefficient which represents the number of ions produced in the course of the reduction of one molecule of the substance being reduced.

For the reaction

$$O_2 + 4e + 2H_2O \rightarrow 4OH^-$$

the coefficient $\alpha = 4$.

The rate of oxygen reduction is given by the formula (see the cited monograph of A. N. Frumkin et al.)

$$Q = \text{const}\; c_1 e^{-\frac{F\varphi}{2RT}}$$

and is first order in concentration $c_1$ of the oxygen near the surface of the metal.

Here, a balanced condition exists between the approaching and reacting particles

$$j_1 = D\left(\frac{\partial c}{\partial n}\right)_{y=0} = Q.$$

The solubility of oxygen in electrolytic solutions is relatively small. Consequently, the over-all rate of reduction of oxygen, which determines the potential distribution in the cell, is usually controlled by the rate of convective diffusion to the surface of the metal.

However, we have seen that the diffusional flux is constant over the entire surface only on the uniformly accessible surface of a disk, where the diffusion problem is solved in a relatively simple manner. In a study of the mechanisms of the complex processes occurring in electrochemical shielding installations, it is possible to hope for results that are quantitatively comparable with calculated results only in cases of relatively simple systems.

Up to the present time, all experiments in which electrochemical protective action has been studied in the presence of an oxidizer were conducted under undetermined conditions of mixing, thus precluding a calculation of the diffusional flux and the derivation of a quantitative theory for the process. It would appear useful to conduct the study on a rotating disk. The theory of electrochemical protection in such a case is briefly set forth below. Let us assume that a shield of sufficiently small dimensions ($H \ll 1$) is placed in the center of a rotating disk.

The following boundary condition is valid, for $H \ll 1$, on the surface of the shield

$$-\varkappa\left(\frac{\partial \varphi}{\partial y}\right) = i = \text{const} \qquad (\text{at}\quad y = 0,\ r \leqslant R_0).$$

The protected metal constitutes an outer circular surface, In this case, in order to express the density of oxygen flow to the surface, we can utilize formula (12.8)

$$j_1 = D\left(\frac{\partial c}{\partial y}\right)_{y=0} = \frac{Dc_0}{\delta + \frac{D}{k}}.$$

If we designate the value of the limiting diffusional flux by $j_{diff}$

$$j_{diff} = \frac{Dc_0}{\delta},$$

then

$$j_1 = \frac{j_{diff}}{\left(1 + \frac{j_{diff}}{c_0 k(\varphi)}\right)}.$$

On the surface of the protected metal, therefore, the boundary condition for the potential is

$$\varkappa\left(\frac{\partial \varphi}{\partial y}\right)_{y=0} = \alpha F j_1 = \frac{\alpha F D c_0}{\delta + D/k(\varphi)}. \tag{49.33}$$

This boundary condition is quite complex. It is not difficult to see, however, that it can be simplified considerably.

Namely, in the region that is close to the shield, the reaction rate is large compared to the rate of diffusion of oxygen to the reaction site, i.e.,

$$k \gg \frac{D}{\delta}.$$

Therefore, in this case, (49.33) can be rewritten in the form

$$\varkappa\left(\frac{\partial \varphi}{\partial y}\right) \simeq \frac{\alpha F D c_0}{\delta} = i_1 = \text{const} \qquad (\text{at} \quad y = 0, \ R_0 < r \leqslant R_1).$$

Far from the shield, for $r > R_1$, the rate of the reaction is retarded, and the potential of the procted metal becomes small compared with its stationary value.

Here we can, therefore, write $j_{diff} \gg c_0 k(\varphi)$ and $c_1 \simeq c_0$. The boundary condition, therefore, takes the form

$$\varkappa\left(\frac{\partial \varphi}{\partial n}\right)_{y=0} = \alpha F c_1 k(\varphi). \tag{49.34}$$

Since the chemical overpotential is small, the rate of the electrolytic process can be written in the form

$$Q = c_1 \gamma \varphi, \tag{49.35}$$

and confining this to the case of large $\gamma$, (49.34) can be written in final form as

$$\varphi = 0 \qquad (\text{at} \quad y = 0, \ r > R_1). \tag{49.36}$$

The meaning of the latter is that, for $r > R_1$, the potential of the protected metal can be considered to have the stationary value, which can be conveniently chosen as the zero of potential. The magnitude of $R_1$ will be determined from the condition $\gamma\varphi(R_1) \approx \frac{D}{\delta}$. At infinity, the field obviously cannot exist. The following technique can be used for a solution of the boundary value problem formulated above: let us select elliptic coordinates. Here, the inside of the circle $R_1$ includes both the cathode and part of the anode. Inside the circle $r < R_1$, the quantity $\frac{\partial\varphi}{\partial n}$ is expressed by the step function

$$\left(\frac{\partial\varphi}{\partial n}\right)_{y=0} = \Phi(r) = \begin{cases} \frac{i_0}{\varkappa}, & r < R_0, \\ \frac{i_1}{\varkappa}, & R_0 \leqslant r \leqslant R_1. \end{cases} \tag{49.37}$$

The solution of the boundary value problem, vanishing at infinity and satisfying boundary condition (49.36) in the outer region ($r > R_1$), has the form of the series (49.23), with odd value of the index $l$.

The derivative along the normal at the surface ($y = 0$, $r \leq R_1$) has the form

$$\left(\frac{\partial\varphi}{\partial n}\right)_{\xi=0} = \frac{1}{R_1\eta} \sum A_{2k+1} P_{2k+1}(\eta)\, i Q_{2k}(0). \tag{49.38}$$

The coefficients for the last series must be so selected as to insure that condition (49.37) is fulfilled.

The coefficients of the series can be derived due to the orthogonality of the Legendre polynomials $P_{2k+1}(\eta)$.

The appropriate relations are given in more detail in [60].

## 50. FLOW OF CURRENT THROUGH AN AGITATED ELECTROLYTE

Let us now turn to the examination of a case often encountered in actual practice, that is, where the current through an electrolytic cell is close to the limiting current. The value of the current through the cell, in this case, is determined by the rate at which the ions are brought to the surface of the electrode [9].

In Section 45 we set forth detailed computations of the volt-ampere characteristics of the discharge in a static solution. The value of the limiting current through the cell is found to be inversely proportional to the distance between the electrodes and is, in fact, very small. To raise the limiting current which can flow through the cell, it is necessary to improve conditions for the transfer of ions to the surface of the electrodes. In actual practice, this is usually accomplished by means of a more or less energetic agitation of the electrolyte.

A feature of the transfer of ions in a moving solution that differentiates it from the transfer of dissolved neutral particles is the fact that the motion of ions is affected by the electric field in the solution. The ion transfer in the solution is produced by convection, by the diffusion of ions to the electrode, and by the migration of ions in the electric field.

Let us write the equation of ion transfer in the solution. Let it contain an ion of a charge $z_i$ with a diffusion coefficient $D_i$, and let the concentration be $c_i$. If E is the electric field, the force it exerts on the ion is $z_i eE$. The ion's motion in the medium is caused by this force. The velocity of its motion is related to the force exerted on the ion by the field by the usual relation for particle motion in a viscous medium

$$\mathbf{u} = \gamma_i z_i \mathbf{E},$$

where $\gamma_i$ is the mobility of the ion. The latter may be expressed in terms of the diffusion coefficient by means of the well-known Einstein relation

$$\gamma_i = \frac{D_i F}{RT}.$$

The motion of ions under the influence of an electric field is known as migration. The migration of i-type ions can be written in the form

$$(\mathbf{j}_i)_{\text{migr}} = \frac{D_i z_i e \mathbf{E} c_i}{kT} = \frac{D_i z_i F \mathbf{E} c_i}{RT}. \qquad (50.1)$$

The total flux of particles of the i-th species in the moving medium, in the presence of diffusion, is equal to

$$\mathbf{j}_i = c_i \mathbf{v} - D_i \operatorname{grad} c_i + \frac{D_i z_i \mathbf{E} F}{RT} c_i. \qquad (50.2)$$

By introducing the vector $\mathbf{j}$ into equation (8.3), which expresses the law of conservation of matter, and by use of the mobility relation, we obtain:

$$\frac{\partial c_i}{\partial t} + (\mathbf{v} \operatorname{grad}) c_i = D_i \Delta c_i - \frac{D_i z_i F}{RT} \operatorname{div}(c_i \mathbf{E}). \qquad (50.3)$$

The values of the index i in (50.3) cover all ion types.

The value $\mathbf{v}$ of the liquid velocity is considered to be given. In deriving equation (50.3) we assumed that the diffusion coefficients were not a function of the concentration and, therefore, did not vary from point to point. We did not assume that the solution is isothermal, but we ignored the possibility of ion transfer by thermal diffusion.

As we have seen in Section 46, the field intensity E in the solution depends on the concentration distributions in the solution. The system of equations (50.3), therefore, is incomplete. One more equation must be added which establishes the relation between the concentrations of dissolved substances.

Since the total charge in the bulk of the solution does not exceed that of the double layer (it may actually be smaller if the double layers were formed on differently charged surfaces), the charge density in the bulk of the solution is very low. It may be assumed, in this connection, that the bulk of the solution is electrically neutral, so that

$$\sum z_i c_i = 0. \tag{50.4}$$

The resulting error was evaluated above (Section 46).

The solution of the system of equations (50.3) and (50.4) in general form presents serious difficulties. In what follows, we cover only two basic cases:

1) a binary electrolyte solution containing only two types of ions;
2) a solution of ions which discharge or are formed on the electrodes, and in addition a significant amount of a so-called foreign or indifferent electrolyte whose ions do not participate in the electrochemical reactions on the electrodes.

## 51. CURRENT IN A BINARY ELECTROLYTE

Let us examine the case of a current flowing through a binary electrolyte. The system of equations (50.3) may be written in the form

$$\frac{\partial c_1}{\partial t} + \mathbf{v} \operatorname{grad} c_1 = D_1 \Delta c_1 - \frac{z_1 F D_1}{RT} \operatorname{div} (c_1 \mathbf{E}), \tag{51.1}$$

$$\frac{\partial c_2}{\partial t} + \mathbf{v} \operatorname{grad} c_2 = D_2 \Delta c_2 + \frac{z_2 F D_2}{RT} \operatorname{div} (c_2 \mathbf{E}), \tag{51.2}$$

$$z_1 c_1 - z_2 c_2 = 0. \tag{51.3}$$

The signs in (51.2) and (51.3) have been so chosen that F represents the absolute value of the charge.

In the place of concentrations $c_1$ and $c_2$, let us introduce the molecular concentration, related to $c_1$ and $c_2$ by the ratios

$$c = \frac{c_1}{z_2} = \frac{c_2}{z_1}. \tag{51.4}$$

Expressing $c_1$ and $c_2$ in (51.1) and (51.2) by the molecular concentration, we obtain

$$\frac{\partial c}{\partial t} + \mathbf{v} \operatorname{grad} c = D_1 \Delta c - \frac{z_1 F D_1}{RT} \operatorname{div} (c \mathbf{E}), \tag{51.5}$$

$$\frac{\partial c}{\partial t} + \mathbf{v} \operatorname{grad} c = D_2 \Delta c + \frac{z_2 F D_2}{RT} \operatorname{div} (c \mathbf{E}). \tag{51.6}$$

Subtacting the second equation from the first, we have:

$$F(z_1D_1+z_2D_2)\,\mathrm{div}\,(c\mathbf{E})=RT(D_1-D_2)\,\Delta c. \tag{51.7}$$

Substituting the value of div ($c$E) into (51.5) and (51.6), we derive the following expression for the function c:

$$\frac{\partial c}{\partial t}+\mathbf{v}\,\mathrm{grad}\,c=D_{\text{eff}}\,\Delta c, \tag{51.8}$$

or, in the case of a steady state problem,

$$\mathbf{v}\,\mathrm{grad}\,c=D_{\text{eff}}\,\Delta c, \tag{51.9}$$

where $D_{\text{eff}}$ designates the quantity

$$D_{\text{eff}}=\frac{D_1D_2(z_1+z_2)}{z_1D_1+z_2D_2}, \tag{51.10}$$

known as the salt diffusion coefficient, or the effective diffusion coefficient of a binary electrolyte.

Equation (51.8) determines the dependence of the molar concentration on the coordinates and the time.

The distribution of molar concentration is not an explicit function of the electric field and is determined by a purely diffusional equation which, however, does not contain the true diffusion coefficient, but the effective diffusion coefficient $D_{\text{eff}}$. It should be borne in mind, however, that, although the effect of the electric field is apparent and is not taken into account in equation (51.8) for the concentration distribution, it may enter into the boundary conditions which, along with the indicated equation, determine the actual concentration distribution. These results constitute a generalization of the exactly analogous conclusion by Eucken [2] for the case of ions migrating in a static medium.

Knowing, from the solution of equation (51.8), the molar concentration as a function of the coordinates and of time, one can determine the potential distribution in the electrolyte by employing equation (51.7).

Integrating (51.7), we obtain:

$$\frac{F(z_1D_1+z_2D_2)}{RT}\,c\mathbf{E}-(D_1-D_2)\,\mathrm{grad}\,c=\mathbf{P}, \tag{51.11}$$

where **P** is some vector satisfying the condition

$$\mathrm{div}\,\mathbf{P}=0.$$

To clarify the physical meaning of the vector **P**, let us write the expression for the vector of the density of the current, carried by diffusion, convection, and migration of ions in the moving medium.

It is obviously equal to the sum of the vectors of current density transferred by each ion type,

$$\mathbf{i} = \mathbf{i}_1 + \mathbf{i}_2, \tag{51.12}$$

where

$$\mathbf{i}_1 = -D_1 F z_1 \operatorname{grad} c_1 + \frac{D_1 F^2 z_1^2}{RT} c_1 \mathbf{E} + F z_1 c_1 \mathbf{v}, \tag{51.13}$$

$$\mathbf{i}_2 = D_2 F z_2 \operatorname{grad} c_2 + \frac{D_2 F^2 z_2^2}{RT} c_2 \mathbf{E} - F z_2 c_2 \mathbf{v}. \tag{51.14}$$

Expressing the concentration in terms of molar concentration, we obtain:

$$\mathbf{i} = -D_1 F z_1 z_2 \left(\operatorname{grad} c - \frac{F z_1}{RT} c \mathbf{E}\right) + D_2 F z_1 z_2 \left(\operatorname{grad} c + \frac{F z_2}{RT} c \mathbf{E}\right) =$$

$$= -(D_1 - D_2) F z_1 z_2 \operatorname{grad} c + \frac{F^2 z_1 z_2}{RT} (z_1 D_1 + z_2 D_2) c \mathbf{E}.$$

Here, the terms $Fz_1c_1\mathbf{v}$ and $Fz_2c_2\mathbf{v}$ in equations (51.13) and (51.14), which represent the convective portion of the total current carried by ions of all types, cancel one another. This is a direct result of the electrical neutrality of the solution, because the liquid in motion carries equal amounts of ions of opposite sign. Therefore, the convective component in the total current is absent. The vector of the total current in a moving medium has the same form as in a static medium. The role of convection consists only in that it determines the distribution of concentration.

It is easy to see that $\mathbf{P}$ and $\mathbf{i}$ are linked by the following simple relation

$$\mathbf{P} = \frac{\mathbf{i}}{F z_1 z_2}. \tag{51.15}$$

Thus, expression (51.11) represents the vector of current density with an accuracy up to a proportionality factor. Equation (51.11) can be presented in the form

$$\frac{\mathbf{i}}{F z_1 z_2} = -(D_1 - D_2) \operatorname{grad} c + \frac{F(z_1 D_1 + z_2 D_2)}{RT} c \mathbf{E}. \tag{51.16}$$

From the latter, the distribution of the field in the solution can be derived

$$\mathbf{E} = \frac{RT}{F^2 (z_1 D_1 + z_2 D_2) z_1 z_2} \frac{\mathbf{i}}{c} + \frac{(D_1 - D_2) RT}{F (z_1 D_1 + z_2 D_2)} \frac{\operatorname{grad} c}{c}. \tag{51.17}$$

Integrating (51.17), we obtain the potential distribution in the solution

$$\varphi = -\frac{RT(D_1 - D_2)}{F(z_1D_1 + z_2D_2)} \int \frac{\operatorname{grad} c}{c}\, d\mathbf{l} - \\ - \frac{RT}{F^2(z_1D_1 + z_2D_2)\, z_1z_2} \int \frac{\mathbf{i}\, d\mathbf{l}}{c} + \varphi_0, \tag{51.18}$$

where $\varphi_0$ is a constant.

In the integration of the equations for the concentration and potential distributions in the solution, we neglected the fact that the diffusion coefficients are functions of concentration. For the problems usually encountered in actual practice, this error is not important.

Equations (51.8) and (51.18) show that the solution of the problem of the flow of current through the electrolyte should be accomplished in two stages. From the generalized solution of the convective diffusion equation, we first derive the concentration distribution in the electrolyte. The subsequent integration of (51.18), with the known function c, leads to a definite distribution of the potential.

For a complete solution of the problem posed, besides equations of concentration distribution, a system of boundary and initial conditions must be given. We shall now pass to a formulation of these conditions.

If the electrolysis is not a steady state process, then it is common for the distribution of concentration c in the electrolytic cell to be given at the initial moment in time:

$$c = c^0 \quad \text{at} \quad t = 0,$$
$$\varphi = \varphi^{(0)} \quad \text{at} \quad t = 0.$$

For a steady state electrolytic process, only the boundary conditions need be given.

In the case of a binary electrolyte, the following may serve as boundary conditions.

1) Far from the electrode, where the reaction occurs, the concentration of the solution is given as

$$c = c_0. \tag{51.19}$$

2) Only one of the ions can discharge on a given electrode — the cation on the cathode, for example. The current carried by the ion that does not discharge on the electrode and does not participate in the electrochemical reaction must become zero on the surface of that electrode.

For example, suppose that ions of the second type do not lose their charge on the surface of the electrode. Then the flow of these particles through this surface beomes zero. In the general case of nonstationary electrolysis, in order to derive the boundary condition, let us examine the net number of ions in an infinitely small parallelepiped of volume $\delta V$, of height $\delta h$, and of area 1 $cm^2$ set on a surface S. Conservation of the total number of particles indicates that

$$(j_{2n})_S - (j_{2n})_{S+\delta h} = -\int \frac{\partial c_2}{\partial t} dV \approx -\frac{\partial c_2}{\partial t} \delta V = -\frac{\partial c_2}{\partial t} \delta h,$$

where $(j_{2n})_S$ is the normal component of the flow of second type of particles on the surface of the electrode itself. By assumption, this flow is zero, and $(j_{2n})_{S+\delta h}$ is the component of the flow of these particles through a surface at a distance $\delta h$ from the electrode. As we let $\delta h$ approach zero, we find, with an accuracy to second order infinitesimals

$$i_{2n} = zFj_{2n} = 0 \text{ on the surface S.}$$

Since the normal component $v_n$ of the liquid velocity becomes zero on the surface of the solid electrode, the final form of this boundary condition for $i_{2n}$ is written as follows:

$$i_{2n} = D_2 z_2 F \frac{\partial c_2}{\partial n} + \frac{D_2 F^2 z_2^2 c_2 E_n}{RT} = 0. \tag{51.20}$$

3) The final boundary condition, still missing for a complete formulation of the problem, depends vitally on the manner in which the electrolytic process is conducted.

As a rule, the potential difference V across the cell electrodes is given. In this case, the boundary condition on the surface of the electrode is the potential given for that surface. Let us express it in terms of the applied voltage V. If $\Delta\varphi$ is the change in the potential of the solution given by formula (51.18), then

$$V = \Delta\varphi + \eta_{\text{conc}}$$

where $\eta_{\text{conc}}$ is the concentration overpotential. If, as a result of electrolysis, there is established near the electrode a certain concentration $c_1^{(1)}$ of current carrying ions, discharging on the electrode, which is lower then their concentration far from the electrode, then a concentration emf appears in the solution, directed from areas of lower to areas of higher concentration and equal to

$$\eta_{\text{conc}} = \frac{RT}{Fz_1} \ln \frac{c_1^{(1)}}{c_0^{(1)}} = \frac{RT}{Fz_1} \ln \frac{c_1}{c_0}, \tag{51.21}$$

where $c_1$ is the value of molar concentration on the surface of the electrode. If we designate by $\varphi_0$ the potential of the electrode on which there is no concentration polarization, then the potential of the polarized electrode may be considered to be equal to V, and thus related to $\varphi$, and capable of being calculated from the concentration $c_1$ at that electrode according to the following relation:

$$V=\varphi-\varphi_0+\eta_{\text{conc}}=\frac{-RT}{F^2(z_1D_1+z_2D_2)z_1z_2}\int\frac{i\,dl}{c}-\\-\frac{RT(D_1-D_2)}{F(z_1D_1+z_2D_2)}\int\frac{\operatorname{grad} c}{c}\,dl+\frac{RT}{Fz_1}\ln\frac{c_1}{c_0}. \qquad (51.22)$$

The first two terms in (51.22) represent the ohmic potential drop of the solution, both over the region where the concentration may be considered constant and over the region of substantial variation in concentration. The last term in (51.22) expresses the concentration overpotential at the electrode.

In certain instances a given current is passed through the electrolytic cell. This is accomplished by employing a very high ohmic resistance in the circuit or an appropriate electronic circuit. Then, the boundary condition at the surface of the electrode under examination has the form

$$I=i_1S=z_1z_2FS\left[-D_1\frac{\partial c}{\partial n}+\frac{Fz_1D_1cE_n}{RT}\right]_S.$$

Eliminating the value of the field $E_n$ on the surface of the electrode from the equation above and (51.20), we obtain

$$\left[\frac{\partial c}{\partial n}\right]_S=\frac{I}{a}, \qquad (51.23)$$

where

$$a=-z_1z_2FS\left(1+\frac{z_1D_1}{z_2D_2}\right)D_{\text{eff}}.$$

It should be pointed out that the strict integration of the equation of unsteady state convective diffusion with the boundary conditions given above presents great difficulties. In the same way, substantial difficulties are encountered in the simpler case of a steady state process if the electrode surface is not uniformly accessible from a diffusional standpoint. In what follows, therefore, we confine our treatment to the case of a steady state electrolysis on an electrode with a uniformly accessible surface.

In conclusion, let us note that, if the total current in the system is zero, then a nonuniform distribution of concentration in a binary electrolyte leads to the appearance of a so-called diffusion field, the intensity of which is given by formula (51.17) with i = 0:

$$\mathbf{E}=\frac{-(D_2-D_1)RT}{F(z_1D_1+z_2D_2)}\frac{\operatorname{grad} c}{c}. \qquad (51.24)$$

The integration of the latter equation leads to the well known expression for the diffusion potential (the Nernst equation)

$$\varphi = \varphi_0 + \frac{(D_2 - D_1) RT}{(z_1 D_1 + z_2 D_2) F} \ln c. \tag{51.25}$$

However, it should be stressed that the field-intensity equation is general in nature and does not depend on the manner in which the nonuniform distribution of concentration is brought about. In particular, when it is brought about by the convective transfer of ions, a field appears in the electrolyte which may be termed a convective-diffusion field.

The intensity of a convective-diffusion field, to an order of magnitude, may be written in the form

$$E \sim \frac{(D_2 - D_1) RT}{F (z_1 D_1 + z_2 D_2)} \left| \frac{\Delta c}{\delta} \right| \cdot \frac{1}{\bar{c}}, \tag{51.26}$$

where $\Delta c$ is the concentration change over a distance of the order of the thickness of the diffusion boundary layer $\delta$, and $\bar{c}$ is the average concentration.*

## 52. THEORY OF A DISK ELECTRODE IN A BINARY ELECTROLYTE

As we have seen in Section 11, the surface of a disk, rotating around an axis through its center, is a uniformly accessible surface from a diffusional standpoint. This permits a substantial simplification in the formulas of the preceding section and allows us to derive final relations in closed form.

Let us assume that one of the electrodes (the cathode, for example) is a rotating disk. We may omit examining the phenomena occurring at the anode, and we shall consider it a plane at $y = l$.

In accord with the conclusions of the preceding section, the concentration distributions $c_1$ and $c_2$ of ions and the molar concentration $c$ of the solution are determined by the usual diffusion equation.

The solution of the diffusion equation for a steady state distribution of concentration was derived in Section 11. It was shown there that concentration is a function only of the distance $y$ from the surface of the disk and not of the coordinates $r$ and $\varphi$. In view of this, we assume that the distribution of the potential likewise is a function

---

*Proofreading note. The effects related to a convective-diffusion field for the motion of drops in air and bubbles in a solution have been examined independently of us by B. V. Deryagin and S. S. Dukhin [69].

only of distance from the disk. Thus, we must derive a solution for equations (51.18) and (51.19) in which both concentration and potential are functions of y only,

$$c = c(y) \text{ and } \varphi = \varphi(y). \tag{52.1}$$

The system of boundary conditions for the problem will have the following form:

1) At the anode, for y = $l$, the molar concentration is

$$c = c_0 \tag{52.2}$$

and the potential is

$$\varphi = \varphi_a = 0. \tag{52.3}$$

2) Ions of the first type discharge on the surface of the disk electrode (the cathode). Ions of the second type are not discharged, and the current carried by them becomes zero.

This gives, at y = 0,

$$i_2 = 0 = -D_2 F z_2 \frac{\partial c_2}{\partial y} + \frac{D_2 F^2 z_2^2}{RT} c_2 \frac{\partial \varphi}{\partial y}. \tag{52.4}$$

The cathode (disk) potential is equal to the applied potential difference V.

In accordance with the general approach noted in the preceding section, it is first necessary to find the distribution of the molar concentration c given by the convective diffusion equation (51.9). Its solution for the case of a rotating disk was derived in Section 11.

The general solution of equation (51.9) has the form of expression (11.26). Changing the value of the constant, it may be written more conveniently in the form

$$c = a_1 \left[ \int_l^y \exp\left\{ \frac{1}{D} \int_0^{t'} v_y(t)\, dt \right\} dt' \right] + a_2. \tag{52.5}$$

Boundary condition (52.2) gives:

$$a_2 = c_0. \tag{52.6}$$

The potential $\varphi$ at an arbitrary point in the solution has the form

$$\varphi = -\frac{RT}{F^2(z_1 D_1 + z_2 D_2)} \frac{i}{z_1 z_2} \cdot \int_l^y \frac{dy}{a_1 \int_l^y \exp\left\{ \int_0^{t'} \frac{v_y}{D}\, dt \right\} dt' + a_2} - \frac{(D_1 - D_2) RT}{F(z_1 D_1 + z_2 D_2)} \int \frac{\operatorname{grad} c}{c}\, dl. \tag{52.7}$$

Substitution of the values (52.5) and (52.7) into boundary condition (52.4) permits a determination of $a_1$. Namely, after very simple operations, we obtain

$$a_1 = \frac{i}{z_1 F D_1 (z_1 + z_2)}. \tag{52.8}$$

Thus, we obtain:

$$c = \frac{i}{(z_1 + z_2) z_1 F D_1} \int_l^y \exp\left\{\frac{1}{D} \int_0^{t'} v_y \, dt\right\} dt' + c_0, \tag{52.9}$$

$$\varphi = \frac{-RT(D_1 - D_2)}{F(z_1 D_1 + z_2 D_2)} \ln \frac{c}{c_0} -$$

$$- \frac{RTi}{F^2 z_1 z_2 (z_1 D_1 + z_2 D_2)} \int_l^y \frac{dt'}{\frac{i}{(z_1 + z_2) z_1 F D_1} \int_l^{t'} \exp\left\{\int_0^{t'} \frac{v_y}{D} \, dt\right\} dt' + c_0}. \tag{52.10}$$

In accordance with (51.22), the applied voltage is determined from the expression

$$V = \frac{RTi}{F^2 z_1 z_2 (z_1 D_1 + z_2 D_2)} \int_l^0 \frac{dt'}{\frac{i}{(z_1 + z_2) z_1 F D_1} \int_l^{t'} \exp\left\{\frac{1}{D} \int_0^{t'} v_y dt\right\} dt' + c_0}$$

$$- \frac{RT(D_1 - D_2)}{F(z_1 D_1 + z_2 D_2)} \ln \frac{c_1}{c_0} + \frac{RT}{F z_1} \ln \frac{c_1}{c_0}, \tag{52.11}$$

where $c_1$ is the value of the molar concentration on the surface of the cathode.

Let us compute the integral in expression (52.11),

$$I = \int_l^0 \frac{dy}{a_1 \int_l^y \exp\left\{\int_0^{t'} \frac{v_y}{D} \, dt\right\} dt' + c_0}. \tag{52.12}$$

The value of $v_y$ is given by (11.17) and (11.18); $v_y$ increases rapidly with distance from the surface of the disk. The inner integral, therefore, converges rapidly. Hence,

$$I = \int_l^0 \frac{dy}{a_1 \int_l^y \exp\left\{\int_0^{t'} \frac{v_y}{D} \, dt\right\} dt' + c_0} +$$

$$+ \int_c^0 \frac{dy}{a_1 \int_l^y \exp\left\{\int_0^{t'} \frac{v_y}{D} \, dt\right\} dt' + c_0} = I_1 + I_2,$$

where $\delta$ is the thickness of the diffusion boundary layer determined by (11.31). Integration of the first integral is in the region $y > \delta$, in which the concentration is constant and is equal to $c_2$. Here, the integral in the denominator is virually equal to zero, so that

$$I_1 = \int_l^\delta \frac{dy}{a_1 \int_l^y \exp\left\{\int_0^{t'} \frac{v_y}{D}\,dt\right\} dt' + c_0} \approx -\frac{l-\delta}{c_0} \approx -\frac{l}{c_0}.$$

Since the value of the upper limit of the integral in the denominator in $I_2$ is certainly less than $\delta$, $I_2$ may be written approximately as

$$I_2 = \int_\delta^0 \frac{dy}{a_1 \int_l^y \exp\left\{\int_0^{t'} \frac{v_y}{D}\,dt\right\} dt' + c_0} \approx$$

$$\approx \int_\delta^0 \frac{dy}{a_1 \int_\infty^y \exp\left\{\int_0^{t'} \frac{v_y}{D}\,dt\right\} dt' + c_0}.$$

The integral in the denominator of this expression was computed in Section 11. It is equal* to $\delta$, so that

---

*Editor's note: This integral is not equal to $\delta$. Here Levich is using the concentration $c = c_1$ over the entire distance $\delta$, which would predict that V approaches infinity too fast as the limiting current is approached. A better approximation would be

$$c = c_1 + \frac{c_0 - c_1}{\delta} y = c_1 + a_1 y \text{ for } y < \delta.$$

Then we have

$$I_2 = \int_\delta^0 \frac{dy}{c_1 + a_1 y} = \frac{1}{a_1} \ln \frac{c_1}{c_1 + a_1\delta} = \frac{1}{a_1} \ln \frac{c_1}{c_0}.$$

Equation (52.11) becomes

$$V = \frac{RT\,(z_1 + z_2)}{F z_1 z_2} \ln \frac{c_1}{c_0} - \frac{RTil}{z_1 z_2 F^2 (z_1 D_1 + z_2 D_2)\, c_0}.$$

$$I_2 \approx \int_{\delta}^{0} \frac{dy}{c_0 - a_1\delta} = -\frac{\delta}{c_0 - a_1\delta}.$$

Therefore, finally,

$$I = -\frac{l}{c_0} - \frac{\delta}{c_0\left(1 - \frac{a_1\delta}{c_0}\right)}. \tag{52.13}$$

Substituting this value of I into (52.11), we find:

$$V = -\frac{RT\,(D_1 - D_2)}{F\,(z_1D_1 + z_2D_2)} \ln\frac{c_1}{c_0} + \frac{RT}{Fz_1}\ln\frac{c_1}{c_0} -$$
$$-\frac{RTi}{z_1z_2F^2\,(z_1D_1 + z_2D_2)}\left[\frac{l}{c_0} + \frac{\delta}{c_0\left(1 - \frac{a_1\delta}{c_0}\right)}\right] = \frac{D_2\,(z_1 + z_2)\,RT}{Fz_1\,(z_1D_1 + z_2D_2)}\ln\frac{c_1}{c_0} - \tag{52.14}$$
$$-\frac{RTl}{z_1z_2F^2\,(z_1D_1 + z_2D_2)\,c_0}\,i - \frac{RT\delta i}{F^2z_1z_2\,(z_1D_1 + z_2D_2)\,c_0}\cdot\frac{1}{1 - \frac{a_1\delta}{c_0}}.$$

The ohmic resistance of the solution over a length $l$, having a constant concentration $c_0$, is designated by w. Its magnitude is determined by the expression

$$w = \frac{RTl}{F^2\,(z_1D_1 + z_2D_2)\,z_1z_2c_0}.$$

The second term on the right side of (52.14), which represents the ohmic potential drop $\Delta\varphi_{\text{ohm}}$, may be written as wi. The first and last terms on the right side of (52.14) represent the sum of concentration polarization and ohmic potential drop in the solution of variable concentration.

---

Designating the last term by $\Delta\varphi_{\text{ohm}}$ and noting that

$$\frac{c_1}{c_0} = 1 - \frac{i}{i_{\text{lim}}},$$

we can finally write

$$V = \Delta\varphi_{\text{ohm}} + \frac{RT\,(z_1 + z_2)}{Fz_1z_2}\ln\left(1 - \frac{1}{i_{\text{lim}}}\right).$$

This is a much more direct way to obtain the "interpolation formula" (52.22), which is actually more accurate than equation (52.16).

Let us express the last of these in more convenient form. In order to do this, let us note that concentration $c_1$ at the surface of the electrode, at $y = 0$, according to (52.5) and calculations in Section 11, is equal to

$$c_1 \approx a_1 \int_{\infty}^{0} \exp\left\{\int_{0}^{t'} \frac{v_y(t)}{D}\, dt\right\} dt' + c_0 = c_0 - a_1\delta. \tag{52.15}$$

The last component in the first part of (52.14), therefore, may be rewritten as follows:

$$\frac{RTi\delta}{F^2(z_1D_1 + z_2D_2)\, z_1z_2} \cdot \frac{1}{c_1}.$$

Thus, the final equations from which the applied voltage V may be determined will have the form

$$V = \Delta\varphi_{\text{ohm}} + \frac{D_2(z_1 + z_2)\, RT}{Fz_1(z_1D_1 + z_2D_2)} \ln \frac{c_1}{c_0} - \frac{RT\delta i}{F^2 z_1 z_2 (z_1D_1 + z_2D_2)\, c_1}. \tag{52.16}$$

This formula gives the general solution for the problem. It determines the nature of the dependence of the applied voltage V on the current density i flowing through the solution. However, (52.16) is very unwieldy and inconvenient, since the dependence just mentioned is not clearly presented [according to definition (52.15), the quantity $c_1$ is a function of i]. It should be useful, therefore, to replace (52.16) by a more convenient formula from a practical standpoint — by an interpolation formula. In order to do this, let us examine two limiting cases:

$$1)\ c_1 \approx c_0 \ \text{ and }\ 2)\ c_1 \to 0.$$

In the first case, we have, approximately:

$$\ln \frac{c_1}{c_0} = \ln\left(1 - \frac{a_1\delta}{c_0}\right) \approx -\frac{a_1\delta}{c_0}.$$

Then, since $c_1 \approx c_0$, we have:

$$\frac{RT\delta i}{F^2 z_1 z_2 (z_1D_1 + z_2D_2)\, c_1} \approx \frac{RT\delta i}{F^2 z_1 z_2 (z_1D_1 + z_2D_2)\, c_0}.$$

In view of the foregoing approximation, we find

$$\frac{RT\delta i}{F^2 z_1 z_2 (z_1D_1 + z_2D_2)\, c_0} \approx -\frac{RTi}{F^2 z_2 z_1 (z_1D_1 + z_2D_2)\, a_1} \ln \frac{c_1}{c_0}.$$

Substituting the value of $a_1$ from (52.8), we have, finally:

$$\frac{RT\delta i}{F^2 z_1 z_2 (z_1D_1 + z_2D_2)\, c_1} \approx -\frac{RTD_1(z_1 + z_2)}{Fz_2(z_1D_1 + z_2D_2)} \ln \frac{c_1}{c_0}. \tag{52.17}$$

Substituting the value (52.17) into (52.16), we find that the applied voltage is

$$V=\Delta\varphi_{\text{ohm}}+\frac{RT}{F}\frac{z_1+z_2}{z_1z_2}\ln\frac{c_1}{c_0}=$$
$$=\Delta\varphi_{\text{ohm}}+\frac{RT}{F}\frac{z_1+z_2}{z_1z_2}\ln\left(1-\frac{i\delta}{c_0z_1FD_1(z_1+z_2)}\right)= \tag{52.18}$$
$$=\Delta\varphi_{\text{ohm}}+\frac{RT}{F}\frac{z_1+z_2}{z_1z_2}\ln\left(1-\frac{i}{i_{\text{lim}}}\right),$$

where $i_{\text{lim}}$ designates the quantity

$$i_{\text{lim}}=\frac{D_1z_1(z_1+z_2)Fc_0}{\delta}. \tag{52.19}$$

In the second case, when $c_1 \to 0$, (52.14) is found to be inapplicable, since its last term tends to infinity. The value of the current flowing through the solution is determined by the condition

$$c_1=0; \quad V\to\infty. \tag{52.20}$$

This latter condition, which characterizes the concentration $c_1$ at the surface of the electrode, at $y=0$, may be written in the form, noting (52.15),

$$i=\frac{D_1z_1(z_1+z_2)Fc_0}{\delta}=i_{\text{lim}} \tag{52.21}$$

Therefore, the equation

$$V=\Delta\varphi_{\text{ohm}}+\frac{RT}{F}\cdot\frac{z_1+z_2}{z_1z_2}\ln\left(1-\frac{i}{i_{\text{lim}}}\right) \tag{52.22}$$

is the interpolation expression which covers both limiting cases.

For practical purposes, it is useful to rewrite (52.22) in the form

$$i=-i_{\text{lim}}\left(1-e^{\frac{F}{RT}\cdot\frac{z_1z_2}{z_1+z_2}(V-\Delta\varphi_{\text{ohm}})}\right). \tag{52.23}$$

Since V, by convention, is negative, equation (52.23) corresponds to the curve shown in Figure 48.

At low voltages V, when the exponent in (52.23) may be expanded into a series, the current flowing is proportional to the applied voltage

$$i\approx -i_{\text{lim}}\frac{Fz_1z_2}{RT(z_1+z_2)}(V-\Delta\varphi_{\text{ohm}}). \tag{52.24}$$

On the other hand, if the applied voltage is substantially in excess of the quantity

$$\frac{z_1+z_2}{z_1z_2}\cdot\frac{RT}{F}=\frac{z_1+z_2}{z_1z_2}25\cdot 10^{-3}\text{v},$$

the exponent has a very small value compared to unity. The current flowing to the electrode is the saturation value, and is not a function of the voltage V. The value of the saturation current is equal to the diffusional flow that corresponds to the maximum possible concentration variation.

If the current passing through the solution should exceed $i_{lim}$, the electrical neutrality of the solution would be destroyed, and this would be accompanied by the appearance of a very high resistance, as a result of which the density of the current would have to decrease to $i_{lim}$. The saturation current $i_{lim}$ is usually known as the limiting current in the solution.

In summation, it may be said that, in order to find the current-voltage characteristic, or the polarization curve, which gives the dependence of the current density on the applied voltage, it is necessary to compute the limiting current $i_{lim}$ at given conditions of agitation, in accordance with the general rules governing the computation of diffusional flows. Then, equation (52.23) may be used for the calculation of the polarization curve. The quantity $\Delta\varphi_{ohm}$ is found on the basis of the length of the current path in the solution.

In conclusion, let us clarify equation (52.19) by substituting the value of $\delta$ from expression (11.31) and the concentration $c_1^{(0)}$ of current-carrying ions according to equation (51.4). Evidently, $i_{lim}$ may be expressed as

$$i_{lim} = Fz_1D_1\left(1 + \frac{z_1}{z_2}\right) \cdot \frac{c_1^{(0)}\sqrt{\omega}}{1.61D_{eff}^{1/3}\nu^{1/6}}. \tag{52.25}$$

## 53. CURRENT IN THE PRESENCE OF A SUPPORTING ELECTROLYTE

In actual electrochemical practice, solutions containing a significant number of ions that take no part in the electrochemical reactions on the electrodes are often encountered. The totality of these ions is known as a foreign electrolyte. Its role consists in that it raises significantly the electrical conductivity of the solution and thereby decreases the ohmic potential drop. The latter is accompanied by a substantial weakening of the electric field in the solution. From this, it is clear that the addition of a foreign electrolyte reduces the role of migration of ions. It is not difficult to show that, when a foreign electrolyte is added, the convective diffusion of ions occurs in the same way as the diffusion of neutral atoms, if effects of the order of $c_1/c_3$ are ignored, where $c_1$ is the concentration of ions reacting on the electrode and $c_3$ is the concentration of the foreign electrolyte.

The evaluation of the role of migration in a static medium is very simple, but the corresponding computations for a moving medium are much more difficult. Since we are limiting our discussion to an evaluation of the role of migration, it is useful to introduce into the discussion the effective thickness of the diffusion layer, in which the distribution of concentration may be considered to be independent of the agitation of the solution.

Let three types of ions be distributed in a static solution with concentrations $c_1$, $c_2$, and $c_3$, and let $c_3 \approx c_2 \gg c_1$. For the sake of simplicity, let us consider all ions to be monovalent. Ions of the first type are discharged upon the electrode and carry an electric current, the density of which is i. Then the equation of transfer takes the form

$$i = D_1 F \frac{\partial c_1}{\partial y} + \frac{D_1 F^2}{RT} c_1 \frac{\partial \varphi}{\partial y}, \tag{53.1}$$

$$0 = D_2 F \frac{\partial c_2}{\partial y} + \frac{D_2 F^2}{RT} c_2 \frac{\partial \varphi}{\partial y}, \tag{53.2}$$

$$0 = D_3 F \frac{\partial c_3}{\partial y} - \frac{D_3 F^2}{RT} c_3 \frac{\partial \varphi}{\partial y} \tag{53.3}$$

and the condition of electrical neutrality is:

$$c_1 + c_2 - c_3 = 0. \tag{53.4}$$

Here, $\varphi$ designates the potential in the solution.

The following expressions are solutions for equations (53.2) and (53.3)

$$c_2 = c_2^{(0)} e^{-\frac{F\varphi}{RT}} = c_2^{(0)} e^{-\psi}, \tag{53.5}$$

$$c_3 = c_3^{(0)} e^{\frac{F\varphi}{RT}} = c_3^{(0)} e^{\psi}, \tag{53.6}$$

where $\psi = \frac{F\varphi}{RT}$. Presenting the transfer equation (53.1) in the form

$$\frac{i}{D_1 F} = \left(\frac{\partial c_1}{\partial \psi} + c_1\right) \frac{\partial \psi}{\partial y},$$

and integrating it, we obtain as a result:

$$\frac{iy}{D_1 F} = \int \left(\frac{\partial c_1}{\partial \psi} + c_1\right) d\psi.$$

From the condition of electrical neutrality and formulas (53.5) and (53.6), we find:

$$c_1 + \frac{\partial c_1}{\partial \psi} = c_3 - c_2 + \frac{\partial c_3}{\partial \psi} - \frac{\partial c_2}{\partial \psi} = c_3^{(0)} 2e^{\psi}.$$

Therefore,

$$\frac{i}{D_1 F} y = 2c_3^{(0)} e^{\psi} + \text{const.}$$

Selecting the potential on the electrode (the plane y = 0) to be zero, we obtain

$$\frac{iy}{D_1F} = 2c_3^{(0)}\left[e^{\psi} - 1\right],$$

whence we have:

$$\psi = \ln\left(1 + \frac{iy}{2D_1c_3^{(0)}F}\right). \tag{53.7}$$

Thus, the distribution of the concentration of ions of the first type may be presented in the form:

$$c_1 = c_3 - c_2 = c_3^{(0)}\left(1 + \frac{i}{2D_1Fc_3^{(0)}}\,y\right) - \frac{c_2^{(0)}}{1 + \dfrac{iy}{2D_1Fc_3^{(0)}}}. \tag{53.8}$$

Hence, the diffusion current may be expressed as

$$i_{\text{diff}} = FD_1\frac{\partial c_1}{\partial y} = \frac{i}{2} + \frac{c_2^{(0)}}{\left(1 + \dfrac{iy}{2D_1Fc_3^{(0)}}\right)^2}\cdot\frac{i}{2}\cdot\frac{1}{c_3^{(0)}} =$$

$$= \frac{i}{2}\left(1 + \frac{c_2^{(0)}}{\left(1 + \dfrac{iy}{2D_1Fc_3^{(0)}}\right)^2}\cdot\frac{1}{c_3^{(0)}}\right). \tag{53.9}$$

It is known that the density of the current $i \leq \dfrac{D_1Fc_1^{(0)}}{\delta}$. Therefore, the relation

$$\frac{iy}{D_1Fc_3^{(0)}} \approx \frac{c_1^{(0)}y}{c_3^{(0)}\delta} \ll 1.$$

Therefore, the density of the diffusion current

$$i_{\text{diff}} \approx i.$$

The migration current $i_{\text{migr}}$ is determined by the expression

$$i_{\text{migr}} = \frac{D_1F^2c_1}{RT}\cdot\frac{\partial\varphi}{\partial y} \approx \frac{i}{2}\,\frac{c_1^{(0)}}{c_3^{(0)}} \approx 0 \tag{53.10}$$

with an accuracy to the ratio $\dfrac{c_1^{(0)}}{c_3^{0}}$.

With an accuracy to the above ratio, the migration current can be neglected as compared to the diffusion current, and we can assume that the flow of current-carrying ions in the solution, in the presence of a foreign electrolyte, is equal to the diffusional flux of uncharged particles.

In this case the density of the current is

$$i = z_1 F j_{\text{lim}} = nF j_{\text{lim}}.$$

For example, with a disk electrode, we have from equation (11.30):

$$i = \frac{FnD_1 \sqrt{\omega}\, c_1^{(0)}}{1.61 \nu^{1/6} D_1^{1/3}} = \frac{D_1^{2/3} nF \sqrt{\omega}\, c_1^{(0)}}{1.61 \nu^{1/6}}. \tag{53.11}$$

In the case of reversible reactions, whose rate is not great, the general conclusion as to the existence of a limiting current remains valid for electrodes of an arbitrary shape. The value of the limiting current flowing to a unit of electrode surface is determined by the equations derived in Chapters II and III.

The form of the current-voltage characteristic in the presence of an inert electrolyte is no different from that found in Section 52.

In the case of irreversible reactions on the surface of an electrode, when the chemical overvoltage is large, the polarization curves on a uniformly accessible (disk) surface and on a nonuniformly accessible surface are significantly different.

Comparing expressions (17.1) and (47.16), we see that the results obtained in Section 17 may be directly and quantitatively transferred to the case of a current in the presence of chemical overpotential related to hydrogen liberation. Qualitatively they are of the same general nature.

Namely, as we have seen in Section 17, if the surface of the reaction is not uniformly accessible from a diffusional standpoint, the conversion rate on its various sections is governed by different laws.

Near its edge, the surface of the electrode is immersed in a solution that is not depleted, and the value of the current density is controlled by chemical overvoltage.

Far from the edge, the current density is controlled by the diffusion process, and the current attains the limiting value. The volt-ampere characteristic is quite different on different sections of the electrode.

It is evident that the experimental characteristic of the discharge total current on the electrode as a function of the potential difference will have a complex form. In particular, the horizontal plateau which expresses the limiting current, in the case of irreversible reactions, will not be sharply defined.

Up to the present time this fact has not been taken into consideration in electrochemical practice.

The volt-ampere characteristics of irreversible reactions were measured in an agitated solution for electrodes of various shapes.

It may be considered, on the basis of what has been said, that a rotating disk electrode should be used in the study of the kinetics of irreversible electrochemical processes.

## 54. DIFFUSION CURRENT AT THE SURFACES OF DISK ELECTRODES AND PLATES IN THE PRESENCE OF A SUPPORTING ELECTROLYTE

To begin with, let us set forth the expression for the current density on the surface of a disk electrode for an irreversible reaction on its surface.

The current density in a solution that contains an added foreign electrolyte is given directly by formula (53.11) provided that the potential on the surface of the electrode is maintained at a constant value.

Indeed, corresponding to the constant potential in a reversible reaction, a constant concentration exists at the surface of the electrode, according to (44.3). The current density in this case is, therefore

$$i = i_{\text{lim}} nF \left(1 - e^{\frac{ne(V - \Delta\varphi_{\text{ohm}})}{kT}}\right), \tag{54.1}$$

where the limiting flux is

$$i_{\text{lim}} = \frac{D_1^{2/3} \sqrt{\omega}}{1.61 \nu^{1/6}} c_1^{(0)}. \tag{54.2}$$

Here n indicates the charge of the ions discharging upon the electrode.

In the case of an irreversible reaction, the current density is a function of the reaction kinetics and can be derived with the help of relations given in Section 17.

Let us examine the important case of the discharge of hydrogen ions on an amalgamated electrode.

According to the theory of A. N. Frumkin [1], the discharge of hydrogen ions in a solution which contains an indifferent electrolyte is a retarded reaction of the first order in the concentration of the hydrogen ions.

The rate of discharge of hydrogen ions on an amalgamated electrode, according to measurements conducted by V. S. Bagotskiy [1], is expressed by the relation

$$\varphi = -1.460 - \psi_1 + \frac{2RT}{F} \ln c - \frac{2RT}{F} \ln i, \tag{54.3}$$

where i is the current carried by hydrogen ions discharging upon 1 cm$^2$ of cathode surface, c is the hydrogen ion concentration, $\varphi$ is the electrode potential, and $\psi_1$ is the potential of the compact (Helmholtz) section of the double layer. For a given value of the potential

$$i = nFD\left(\frac{\partial c}{\partial y}\right)_{y=0} = kc_1, \tag{54.4}$$

where

$$k = e^{\frac{e(\varphi+\psi_1+1.46)}{kT}} \tag{54.5}$$

and $\varphi < 0$.

Equation (54.5) is identical with (12.11). Utilizing equation (12.8), we obtain:

$$i = \frac{DnFc_0}{\left\{1.61\left(\frac{D}{\nu}\right)^{1/3}\sqrt{\frac{\nu}{\omega}} + D\,nF\exp\left[-\frac{e(\varphi+\psi_1+1.46)}{kT}\right]\right\}}. \tag{54.6}$$

Let us finally examine the case of a constant current at the surface of a disk electrode for an inreversible reaction. With a constant current, we have:

$$\left(\frac{\partial c}{\partial y}\right)_{y=0} = \frac{i}{nFD}. \tag{54.7}$$

A general solution of the convective diffusion equation (11.26), with boundary conditions (11.21) and (54.7), gives:

$$c = \frac{i}{nFD}\int_{\infty}^{y}\exp\left\{\frac{1}{D}\int_{0}^{y_1} v_{y_1}\,dt\right\}dy_1 + c_0.$$

In particular, at the surface of the electrode both concentration and potential are constant. The surface concentration is

$$c_1 = \frac{i}{nFD}\int_{\infty}^{0}\exp\left\{\frac{1}{D}\int_{0}^{y_1} v_y\,dt\right\}dy_1 + c_0 = c_0 - \frac{i\delta}{nFD}. \tag{54.8}$$

It is evident that a constant current will be maintained under the given regime of mixing only at values of i that do not exceed the limiting value of the current $i_{lim}$. At greater values of i ($i > i_{lim}$) the maintenance of a steady-state electrolysis regime is impossible.

Let us now examine an electrode in the form of a plate immersed in a laminar flow. The value of the limiting current, obviously, is determined by equation (15.11). In the case of an irreversible reaction at the surface of the plate, the problem of finding the distribution of the current becomes extremely complex. If, however, the irreversible reaction is first order in concentration, as in the case examined above for the discharge of hydrogen ions at an amalgam electrode, the boundary condition at the surface of the plate acquires the form of the linear expression (17.1). Here, the problem of current distribution on the surface of the plate in the presence of a foreign electrolyte is identical with the problem discussed in Section 17, concerning convective diffusion in mixed kinetics. This distribution is given by (17.13) and (17.15).

Let us now examine the case of a constant current on an electrode, i.e., condition (54.7) prevails on the surface of the plate.

As in Section 17, let us seek a solution in the form (17.8)

$$c(\xi, z) = \int_0^\infty f_1(\lambda) z^{1/3} \lambda^{2/3} \exp\left\{-\frac{z^2+\lambda^2}{4\xi}\right\} I_{1/3}\left(\frac{\lambda z}{2\xi}\right)\frac{d\lambda}{2\xi} + \int_0^\infty f_2(\lambda) z^{1/3}\lambda^{2/3} \exp\left\{-\frac{z^2+\lambda^2}{4\xi}\right\} I_{-1/3}\left(\frac{\lambda z}{2\xi}\right)\frac{d\lambda}{2\xi} = F_1(\xi, z) + F_2(\xi, z). \tag{54.9}$$

The boundary condition (54.7) in terms of the variable z, according to Section 17, takes the form

$$D\left(\frac{\partial c}{\partial y}\right)_{y=0} = 2^{-4/3}\beta^{4/3}\xi^{-1/3}\left\{z^{1/3}\frac{\partial c}{\partial z}\right\}_{z\to 0}. \tag{54.10}$$

Utilizing property 5 of function $F_1$ and property 7 of function $F_2$, we find that boundary condition (54.10) assumes the form

$$\frac{\beta^{4/3}}{2\Gamma\left(\frac{1}{3}\right)}\xi^{-2/3}\int_0^\infty f_1(2\sqrt{p\xi})\, e^{-p}\, dp = \frac{i}{nF}.$$

Introducing, as in Section 17, the notation

$$p\xi = t, \quad \xi = \frac{1}{u}, \quad f_1(2\sqrt{t}) = \varphi(t),$$

we have:

$$\int_0^\infty \varphi(t) e^{-ut}\, dt = \frac{i}{nF}\cdot\frac{2\Gamma\left(\frac{1}{3}\right)u^{-5/3}}{\beta^{4/3}} = \gamma u^{-5/3} \tag{54.11}$$

where $\gamma$ is a constant.

Expression (54.11) is the integral equation for determining the function u.

For its solution, let us utilize the inverse Laplace transformation formula

$$\varphi(t) = \frac{\gamma}{2\pi i}\int_{a-i\infty}^{a+i\infty} u^{-5/3} e^{ut}\, du,$$

where $a > 0$ and the integration is along the straight line parallel to the complex axis. Designating

$$ut = v,$$

we have

$$\frac{1}{2\pi i}\int_{a-i\infty}^{a+i\infty} u^{-5/3}e^{ut}\,du = \frac{t^{2/3}}{2\pi i}\int_{at-i\infty}^{at+i\infty} v^{-5/3}e^{v}\,dv.$$

Utilizing the gamma-function in the form of a complex integral (see, e.g., Smirnov, Kurs vysshey matematiki [Course in Higher Mathematics], vol. III, page 275), we obtain

$$\frac{1}{2\pi i}\int_{at-i\infty}^{at+i\infty} v^{-5/3}e^{v}\,dv = \frac{1}{\Gamma\left(\frac{5}{3}\right)}.$$

Therefore, for $\varphi(t)$, we find

$$\varphi(t) = \gamma\frac{t^{2/3}}{\Gamma\left(\frac{5}{3}\right)} = \frac{\gamma t^{2/3}}{\frac{2}{3}\Gamma\left(\frac{2}{3}\right)} = \gamma\frac{3}{2}\frac{\sin\frac{\pi}{3}}{\pi}\Gamma\left(\frac{1}{3}\right)t^{2/3} = \frac{3\sqrt{3}}{4\pi}t^{2/3}\Gamma\left(\frac{1}{3}\right)\gamma.$$

The function $f_1(\lambda)$, therefore, is

$$f_1(\lambda) = \varphi\left(\frac{\lambda^2}{4}\right) = \frac{3\sqrt{3}}{4\pi}\Gamma\left(\frac{1}{3}\right)\gamma\left(\frac{\lambda^2}{4}\right)^{2/3}. \qquad (54.12)$$

Function $f_2(\lambda)$ has the form

$$f_2(\lambda) = c_0 - f_1(\lambda) = c_0 - \frac{3\sqrt{3}}{4\pi}\Gamma\left(\frac{1}{3}\right)\gamma\left(\frac{\lambda^2}{4}\right)^{2/3}. \qquad (54.13)$$

Substituting the values of the functions $f_1(\lambda)$ and $f_2(\lambda)$ into (54.9), we obtain the spatial distribution of concentration.

We are interested, in particular, in the distribution of the concentration at the very surface of the electrode $y = 0$. In view of property 4 of function $F_1$ and property 6 of function $F_2$, we have:

$$c(x, 0) = \lim_{z\to 0} F_1(\xi, z) + \lim_{z\to 0} F_2(\xi, z) = \lim_{z\to 0} F_2(\xi, z) =$$

$$= \frac{1}{\Gamma\left(\frac{2}{3}\right)}\int_0^\infty f_2(2\sqrt{\xi\rho})\,e^{-\rho}\rho^{-1/3}\,d\rho =$$

$$= \frac{1}{\Gamma\left(\frac{2}{3}\right)}\int_0^\infty \left[c_0 - \gamma\frac{3\sqrt{3}}{4\pi}\Gamma\left(\frac{1}{3}\right)\rho^{2/3}\xi^{2/3}\right]e^{-\rho}\rho^{-1/3}\,d\rho =$$

$$= c_0 - \xi^{2/3}\frac{\gamma\frac{3\sqrt{3}}{4\pi}\Gamma\left(\frac{1}{3}\right)}{\Gamma\left(\frac{2}{3}\right)}\int_0^\infty e^{-\rho}\rho^{1/3}\,d\rho =$$

$$= c_0 - \frac{\gamma\frac{3\sqrt{3}}{4\pi}\Gamma\left(\frac{1}{3}\right)\Gamma\left(\frac{4}{3}\right)}{\Gamma\left(\frac{2}{3}\right)}\xi^{2/3} = c_0 - \frac{3\gamma}{8\pi^2}\left[\Gamma\left(\frac{1}{3}\right)\right]^3\xi^{2/3}.$$

Since $\xi = \frac{1}{3}\beta x^{3/4}$, we have, finally,

$$c(x, 0) = c_0 - \left(\frac{3}{2}\right)^{1/3} \beta^{1/3} \Gamma\left(\frac{1}{3}\right)\sqrt{x}\left[\Gamma\left(\frac{1}{3}\right)\right]^3 = \\ = c_0 - \frac{i}{nF}\frac{3^{1/3}\beta^{-2/3}\left[\Gamma\left(\frac{1}{3}\right)\right]^4}{4\pi^2}\sqrt{x} = c_0\left(1 - \frac{i}{i_{\lim}(x)}\right), \tag{54.14}$$

where $i_{\lim}(x)$ is the limiting diffusional flow on the surface of the plate at a distance x from its edge.

Equation (54.14) shows that the concentration c(x, 0) (and, consequently, the potential as well) on the surface of the electrode changes with the distance from its edge. This is perfectly natural, since the surface of the plate is not uniformly accessible from a diffusional standpoint. The solution is increasingly impoverished with an increase in the distance from the edge of the plate. At a sufficiently great distance from it, the limiting current density attains the value i. At still greater distances from the edge of the plate, the stationary maintenance of a constant current is found to be impossible.

## 55. COMPARISON OF THEORY AND EXPERIMENT

Let us summarize the results obtained in Chapters II, III, and VI, which will be compared later with experimental data.

1. In heterogeneous (including electrochemical) reactions at a solid body — solution interface in a moving fluid, a thin fluid layer exists near the surface of the solid body within which the principal variation in the concentration of the reacting substance takes place. This layer is known as the diffusion boundary layer.

2. The quantitative computation of the value of the density of diffusional flux in a laminar fluid showed that the diffusional flow is a function of the characteristic fluid velocity U, in accordance with the law $U^{-n}$, for a great variety of geometrical shapes of streamlined bodies and for different types of surface — namely, when the flow is across the surface: a) of a spherical solid particle falling slowly in a fluid (14.19); b) of a rotating disk (11.32); c) of a plate in a fluid under forced convection (15.11); d) of a vertical plate in natural convection (23.29); e) of a tube in which a solution is in motion (20.9).

The exponent n is equal to 1/3 for problems a) and e) and 1/2 for problems b) and c).

---

*Editor's note: Equation (54.14) is in error and should read: $c(x,\ 0) = c_0\left(1 - 0.73\frac{i}{i_{\lim}(x)}\right)$. Certain of the statements following equation (54.14) should therefore be modified.

3. Along with velocity, the diffusional flow is a function of the geometric conditions of the flow, the fluid viscosity, and the diffusion coefficients of the reacting substances. The latter shows that the thickness of the diffusion boundary layer depends on the nature of the diffusing particles.

4. The same indications are given by computation of diffusional flows under conditions of mixed kinetics: the magnitude of the diffusional flux is found to be a function not only of the factors enumerated above, but of the rate of the heterogeneous reaction (12.6) and (17.11).

5. A study of the relaxation of the diffusion layer shows that, contrary to the conclusions of the Nernst theory, the effect of a coated part is felt over great distances, considerably exceeding the thickness of the diffusion layer. This indicates that, if a portion of the reaction surface is coated, the density of the diffusional flow over the remainder of the surface, the uncoated portion immersed in an undepleted flow, increases substantially.

6. In a turbulent fluid flow, the improved mixing conditions lead to an increase in the limiting diffusional flow.

Computations of the limiting diffusional flow to a plate and to a disk show that such flows may be considered approximately proportional to the 0.8 to 0.9 power of the fluid velocity. The dependence of the diffusional flow on the Prandtl number varies with the different models of the viscous sublayer — (25.18), (25.5) and (25.21).

7. The diffusional flow to the surface of a nonstreamlined body is quite different over the upstream and downstream portions. At high Reynolds numbers, the diffusional flow to the downstream portion, beyond the separation line, exceeds the flow to the upstream portion, despite the fact that the fluid on the upstream portion of the body is not depleted.

8. The diffusional flow to a particle suspended in a turbulent flow is determined by formulas (33.24) and (33.25).

9. Limiting currents in electrolytic processes in a binary solution can be computed by means of the same formulas as diffusional flows of uncharged particles to uniformly accessible surfaces. In the case of a rotating disk, in particular, the current is given by formula (52.25).

10. In the presence of an indifferent electrolyte, the migration of ions in the electric field does not play a material role, and the diffusion currents of ions coincide with flows of uncharged particles.

11. Under unsteady-state conditions, the effect of mixing is felt for a relatively considerable duration of the process. The relaxation period of the diffusion process is determined by formula (62.22).

12. There is a qualitative similarity between diffusional and heat flows. However, quantitative coincidence between these quantities does not exist.

The above-noted conclusions, which result from a theoretical examination of the phenomena occurring in the diffusion stage of heterogeneous reactions at a fluid-solid body interface, will be compared later with experimental data. Despite the large amount of literature available on the subject of dissolution of metals and nonmetals in liquids, on problems of heterogeneous reactions, etc., as well as the large number of electrochemical studies in which the magnitude of the limiting diffusion current was obtained, this comparison is accompanied by many difficulties.

Almost all of the measurements reported were obtained without sufficient consideration of the hydrodynamic picture of the flow under complex geometrical conditions, so that a quantitative comparison with the results of theoretical computations is precluded. Experimental errors were large in most instances, since the authors usually attempted to obtain only the qualitative characteristics of the process. A direct proportionality between the amount of diffusing substance, the size of the reaction surface and the difference in concentrations was established in the work of A. N. Shchukarev, cited above, and confirmed by measurements made by Nernst and Merriam [10] and in more recent studies [11, 12]. The most accurate measurements made by these authors relate to the measurement of limiting currents. The existence of a direct proportionality between the rate of reactions in the diffusion region and variations in the concentration and size of the reaction surface may be considered as established, on the basis of these measurements. Of greatest interest is the dependence of the diffusion current on the velocity of the fluid motion. As we have seen above in the case of an exterior flow, the theory leads to a proportionality of the limiting current to the square root of the fluid velocity in laminar flow, and to a higher power of velocity in turbulent flow. The power exponent (n) of the fluid velocity increases in the latter case from a value of 0.8 to almost unity with an increase in the Reynolds number.

It should be pointed out here that published material frequently refers to the existence of noticeable differences between the results presented by different workers in this field. Thus, the data obtained by Eucken [13] agree well with an exponent of 1/2. Nernst and Merriam [10] arrive at an empirical formula for the limiting current with velocity to the 0.66 power.

$$i_{\text{lim}} = \frac{DFc_0}{0.003} Z^{2/3}, \tag{55.1}$$

where Z is the number of revolutions per second of a rotating electrode.

Brunner [14] found a somewhat higher exponent. Eucken [2] states the proportionality of the limiting current to the rotating velocity of the electrode is to the 0.5 to 0.6 power. In Van Name's experiments [15], the diffusion current was found to be proportional to velocity to the 0.7 to 0.9 power. King [16] studied the dissolution of zinc. The peripheral rotating velocity of the dissolving cylinder was brought to 27 m/sec, which corresponded to Reynolds numbers of approximately $2.7 \cdot 10^5$. In this case, the results reflected a proportionality of the diffusional flow to the 4/5 power of the rotating velocity. Wildermann [17] indicated a proportionality of the flow to velocity to the first power.

We advanced the proposal that the difference between the experimental results was related to different degrees of turbulence of the fluid at the surface of the reaction.

In Eucken's [13] more recent experiments, special measures were taken to guarantee a laminar flow regime. The lowest value for the power exponent of velocity ($n = 1/2$) was obtained, which is in good agreement with theory. In the other experiments, no measures were taken to assure a laminar regime of the electrolyte in the vicinity of the electrode.

In the experiments of Nernst and Merriam, the rotating electrodes were thin wires. The rate of revolution was relatively small (up to 600 rev/min), so that the Reynolds numbers must have been relatively small, and much turbulence in the flow would not be expected.

In the experiments of Brunner and Van Name, agitation was by means of a mixer located near a flat electrode.

In Brunner's experiments the Reynolds number was of the order of $Re \approx 2{,}000$, so that turbulence should have been minor. However, judging by the results ($n = 0.66$) it was noticeable due to poor geometrical conditions.

In the case of van Name's experiments, strong turbulence apparently existed. The Reynolds number in these experiments attained a value of $Re \approx 10^4$ to $10^5$, and the geometric conditions also contributed to turbulence in the fluid.

King's data relate wholly to turbulence on the electrode. The Reynolds numbers are in a range of $3 \cdot 10^4$ to $2 \cdot 10^5$, and the dissolving sample in these experiments was cylindrical in shape. It is known, however, that for a rotating cylinder at relatively low Reynolds numbers the fluid motion is unstable and becomes turbulent. Thus, although systematic measurements of the limiting current under simple geometric conditions (a flat electrode, for example) were not made over a sufficiently wide range of Reynolds numbers, embracing laminar as well as turbulent regimes, the aggregate of all the data obtained by various authors clearly indicates that the dependence of the limiting current on fluid velocity is in good agreement with theoretical conclusions.

The dependence of the limiting current on diffusion coefficients is somewhat complex. As shown in Sections 11 and 10, the diffusional flow of the substance is proportional to $D^{2/3}$. In the case of a binary electrolyte, the limiting current, according to formula (52.25), is proportional to the effective diffusion coefficient to the minus 1/3 power and to the diffusion coefficient $D_1$ of the current-carrying ion to the first power. If the solution contains several types of ions this dependence is still more complex.

However, since the diffusion coefficients of different ions, with the exception of $H_3O^+$ and $OH^-$ ions, do not differ much from each other, it may be assumed that the limiting current $i_{lim}$ is approximately proportional to D to the 0.66 power,

$$i_{\lim} \approx D^{0.66}. \tag{55.2}$$

This form of dependence is in good agreement with Eucken's measurements, given in [13], as well as with King's data [16] regarding dissolution, in which it was found that the rate of dissolution was proportional to the diffusion coefficient of different ions to the 0.7 to 0.83 power.

Although the 0.83 power is too great, we must bear in mind the fact that diffusion coefficients are functions of the concentration of the solution, and, in certain instances — such as HCl, for example — change drastically in the presence of added salts, so that such experiments do not provide highly accurate data.

The dependence of the limiting current on viscosity in a laminar flow, noting the semiempirical formula (8.15), may be presented in the form

$$i_{\lim} \approx \frac{1}{\nu^{5/6}}. \tag{55.3}$$

The above formula is in good agreement with the data obtained by King for the dissolution of zinc in acid. The viscosity of the solution was varied by sugar additions. In this case it was found that the rate of dissolution changes approximately inversely with viscosity. Bearing in mind the relatively crude experimental conditions, it is doubtful whether a power of 5/6 could have been distinguished from unity.

Measurements of the dependence of $i_{lim}$ on temperature permit only qualitative conclusions as to the nature of the factor which determines the rate for a given heterogeneous reaction, and do not allow any quantitative conclusions regarding the dependence of $i_{lim}$ on the diffusion coefficient or the viscosity of the solution.

The relatively weak dependence of the rate of a heterogeneous reaction on temperature indicates that this rate is limited by the supply of reactant at the reaction site, and not by its kinetics.

No conclusions may be drawn from the proportionality of $i_{lim}$ to any given power of the diffusion coefficient or the viscosity in view of the strong temperature dependence of both these quantities, which incidentally enter the expression for $i_{lim}$ in a complex combination.

It can be concluded from what has been said that qualitatively the theory of a diffusion boundary layer, developed above, is in very good agreement with experiment. A quantitative comparison, however, can be made only with the experiments conducted by Eucken and by Nernst and Merriam. In the case of Eucken's experiments, the current passed between two platinum electrodes in a solution of KI and KCl in the presence of $I^-$, $Br^-$, and $H_3O^+$ as depolarizers. One of the electrodes was a fixed plate of large size, so that the current flowing through the solution was very small compared to the limiting current on this plate. The polarization phenomena on this electrode, therefore, could be neglected. The second electrode was a plate 0.28 cm high and 0.089 cm wide. Mixing was accomplished by means of a cylinder, the outer wall of which rotated past the second electrode. This cylinder contained both the solution and the electrodes. The distance between the electrode and the moving wall was 0.05 to 0.4 cm and was, therefore, always greater than the corresponding thickness of the boundary layer. The Reynolds numbers attained a value of $10^4$. Here, as stressed by Eucken, the flow around the plate remained strictly laminar.

Table 6 gives the values of the current according to Eucken's data and as computed by formula (15.13).

The figures cited show that satisfactory agreement exists between the theoretical and experimental values of the limiting current.

In the case of the experiments of Nernst and Merriam, introduction of the figures into formula (11.31) for a rotating disk leads to the expression

$$\delta = \frac{0{,}005}{\sqrt{Z}}. \tag{55.4}$$

Bearing in mind the different geometric conditions (in the experiments of Nernst and Merriam, there was a rotating wire rather than a disk), the agreement between expression (55.4) and the empirical formula (55.1) must be regarded as good.

Table 6

Values of Limiting Currents

| Absolute Values | | |
|---|---|---|
| $\omega_1 = 88.3$ | $i_{calc} = 1.33 \cdot 10^{-4}$ | $i_{obs} = 1.15 \cdot 10^{-4}$ |
| $\omega_2 = 19.3$ | $i_{calc} = 0.67 \cdot 10^{-4}$ | $i_{obs} = 0.68 \cdot 10^{-4}$ |
| **Relative Values** | | |
| $\frac{\omega_1}{\omega_2}$ | $\frac{(i_{calc})_1}{(i_{calc})_2}$ | $\frac{(i_{obs})_1}{(i_{obs})_2}$ |
| 1.27 | 1.14 | 1.09 |
| 1.5 | 1.22 | 1.15 |
| 2.39 | 1.55 | 1.33 |

Although all the data given indicate a good agreement between theory and experiment, the establishment of accurate measurement techniques for diffusional flows would have been of great interest for two reasons.

1) A coincidence of theoretical formulas with experimental data would have allowed the use of theoretical expressions for diffusional flows for purposes of analysis of heterogeneous process and thus the determination of the rates of the heterogeneous processes themselves.

2) It would have permitted a verification of the formulas for the case of turbulent flow, whose derivation was based on definite hypotheses regarding the nature of turbulent motion near a wall. Clarification of the nature of turbulent motion in the layer immediately adjacent to the wall would have been of material interest to the theory of turbulent fluid motion.

Finally, in addition to these considerations, an experimental confirmation of theory would have practical significance. It is known (see Chapter X) that measurements of diffusion currents on a dropping mercury electrode are used for accurate quantitative and qualitative analyses of solutions (polarographic analysis). Although this analysis has received very wide recognition and a dropping mercury electrode has many advantages (due to a renewal of the surface of the drop), the substitution of a solid electrode would have been of substantial practical interest.

A solid electrode has advantages over a dropping electrode, including simplicity of construction and steady state operating conditions. The introduction of solid electrodes into the technique of polarographic analysis requires knowledge of the exact expression for the limiting diffusion current.

Experimental confirmation of the theoretical results obtained would have permitted recommendation of the use of solid rather than dropping electrodes in many instances.

## 56. QUANTITATIVE VERIFICATION OF THE THEORY. LAMINAR FLOW REGIME

The theory of convective diffusion set forth above and the theory of concentration polarization have been subjected to very careful experimental verification in many studies by Soviet and foregin investigators. Quantitative verification of the theory is of material practical significance, since it can assure the utilization of theoretical relationships for computation of heterogeneous reaction rates. The disk electrode has been especially carefully studied, because its surface provides an example of a uniformly accessible diffusion, as was stressed in Section 12. This has permitted a degree of accuracy in the measurement of currents on the surface of a disk, so that it has been possible to use a disk electrode for a quantitative chemical analysis of solutions and as an instrument for measuring ion diffusion coefficients (see below). We cannot set forth in this book all the details of the experimental work done but merely provide a general survey of it.

The first quantitative verification of the theory of convective diffusion at the surface of a rotating disk in a laminar fluid was conducted in two studies by B. N. Kabanov and Yu. G. Siver [18]. They measured the diffusional flow of dissolved oxygen to a rotating disk electrode on whose surface cathodic regeneration of oxygen in weak sulfuric acid solutions took place. The disk electrodes were made of silver and amalgamated copper. The surface and axis of the silver disk were coated with a glyptal film. The copper cathode was in the form of a cone whose base (the working surface) had a diameter of 2.5 cm. The sides of the cone and the axis were covered by a tightly fitting glass funnel. The anode was a platinum wire ring. The number of revolutions of the electrode varied from 0.5 to 50 rev/sec. The corresponding Reynolds numbers were within the limits $5 \cdot 10^2$ to $5 \cdot 10^4$.

Figure 53 shows the results of the measurements of limiting diffusion currents as a function of the speed of rotation $\omega$. The straight lines represent computed values of the current, according to (53.11), and the circles show the experimental values of $i_{lim}$. Agreement between theory and experiment is found to be excellent. The divergence between computed and measured values does not exceed 3%, which is no greater than the total error in the quantities used in the computation (primarily in the values of the oxygen diffusion coefficient in 0.05 N sulphuric acid).

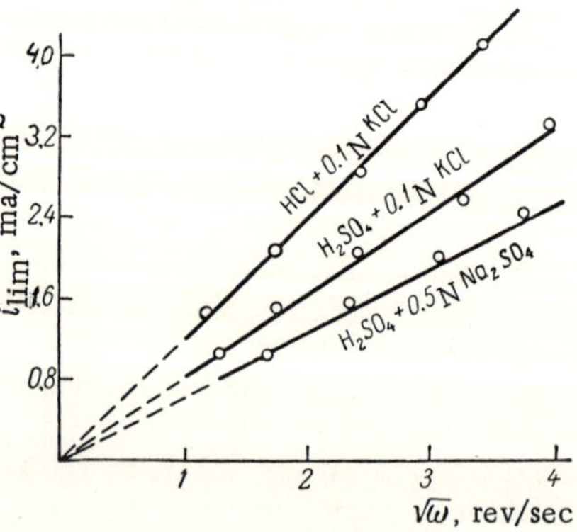


Figure 53. Dependence of the limiting current of hydrogen ion discharge on the square root of the rotation speed of the electrode.

It is interesting to note that in verifying the theoretical results, a silver electrode was used to clarify the nature of the heterogeneous process on its surface. It was clear in this case that agreement between theory and experiment may be obtained only if it could be proved that, in the course of oxygen regeneration from the oxygen diffusing to the silver electrode, hydrogen peroxide rather than water forms on its surface. A quantitative analysis of the solution for hydrogen peroxide confirmed this assumption.

B. N. Kabanov and Yu. G. Siver also conducted an experimental verification of theoretical formulas concerning the dependence of the diffusion current at the rotating disk on the ion diffusion coefficient. According to formula (53.11), the diffusional flow on the surface of a rotating disk under conditions in which there is an excess of indifferent ions in the solution must be proportional to the 2/3 power of the ion diffusion coefficient for the ions that discharge on the electrode. This is not in agreement with the conclusions of the Nernst theory, which holds that the current is proportional to the first power of the diffusion coefficient (since that theory assumes the thickness of the diffusion layer to be independent of the diffusing ions).

In the studies conducted by B. N. Kabanov and Yu. G. Siver, the limiting diffusion current on a rotating amalgamated copper electrode was measured from the hydrogen and oxygen liberated. The electrolyte was a $2 \cdot 10^{-3}$ N salt solution in the presence of 0.1 N potassium chloride. The speed of the disk was varied from 1.5 to 12 rev/sec.

The choice of hydrogen and oxygen as subjects of study was based on the fact that hydrogen ions and oxygen molecules have very different diffusion coefficients. This assured measurement accuracy. The values of the diffusion coefficients of hydrogen ions and oxygen atoms were chosen from published data and were not very accurate.

The results of the measurements made by B. N. Kabanov and Yu. G. Siver are given in Table 7. The table includes the values of the ratios of the diffusion currents carried by hydrogen ions and oxygen molecules at varying electrode rotation rates. According to Nernst's theory, this ratio must equal that of the diffusion coefficients. But according to formula (53.11), it is equal to the ratio of the diffusion coefficients to the 2/3 power. K in Table 7 indicates the ratio

$$\frac{c_{H^+} z_{H^+}}{c_{O_2} z_{O_2}} = 3.85.$$

As noted from the figures in column six, the measured ratio of diffusion currents is not a function of the rotating speed of the disk, and has a value close to that computed on the basis of (53.11). The observed discrepancies between the measured values of the current ratios and those computed on the basis of the Nernst theory lie far outside the limits of accuracy of the experiment. Taking into account the fact that the value of the ion diffusion coefficients of hydrogen and oxygen, taken from published data, were determined rather roughly, the agreement between experiment and theoretical formula (53.11) is regarded as quite good.

Table 7

Dependence of $i_{lim}$ on the diffusion coefficient

| $\omega$, rev/sec | $i_{lim}^{H^-}$, a/cm² exptl. × 10⁵ | $i_{lim}^{O_2}$, a/cm² exptl. × 10⁵ | $K \cdot i_{lim}^{O_2}$ | $\frac{D^{H^+}}{D^{O_2}}$ | $\frac{i_{lim}^{H^-}}{K \cdot i_{lim}^{O_2}}$ | $\left(\frac{D^{H^-}}{D^{O_2}}\right)^{2/3}$ |
|---|---|---|---|---|---|---|
| 1.5 | 140.7 | 14.46 | 55.6 | | 2.52 | |
| 3.17 | 201.0 | 20.98 | 81.2 | | 2.48 | |
| 6.0 | 273.4 | 29.14 | 112.0 | 3.76 | 2.44 | 2.42 |
| 9.0 | 340.7 | 35.76 | 137.7 | | 2.47 | |
| 12.0 | 386.9 | 41.91 | 161.3 | | 2.39 | |

As an additional verification of theoretical computations, the authors mentioned above measured the dependence on disk speed of the limiting current for hydrogen deposition. The results of their observations in this case also confirmed the correctness of formula (53.11), according to which the diffusion current upon a rotating disk is proportional to the square root of the angular velocity. The same study describes the results of a comparison made by the authors between computed and an experimentally determined polarization curve (see below).

Among the other studies conducted especially for the purpose of verifying the formula for the diffusion current on a disk in the laminar flow regime, let us first note the study by Hogge and Kraichman [20].

The cathode in that study was a rotating platinum disk about 2 cm in diameter. It was immersed in a container 0.6 liter in volume and about 9 cm in diameter. The anode area was about 20 times that of the cathode and was situated 5 cm apart from the latter. The speed of rotation varied between 1.5 and 44.3 rev/sec. At the maximum velocity the Reynolds number was about 5000. The dissociation of an iodine ion was studied

$$I_3^- + 2e^- \to 3I^-$$

in the system $KI - KI_3$, in which KI was the indifferent electrolyte. The container and the solution were carefully freed of oxygen. The dependence of the limiting current density on the rotating speed of the disk was studied at various concentrations of $I_3^-$ in the solution. The accuracy of the measurements, as given by the authors, was 5%.

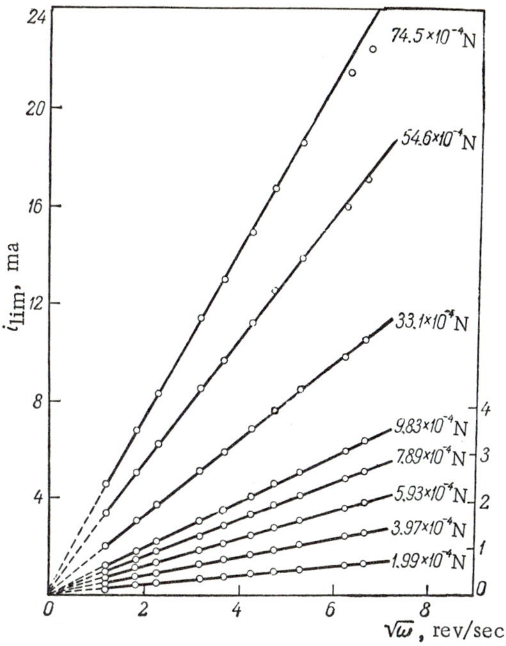


Figure 54. Comparison of the current density on a disk, as determined by (53.11), with Hogge and Kraichman data. The different lines correspond to different solution concentrations.

The results of the measurements are given in Figure 54, which shows the theoretical curves computed by formula (53.11).

We see that, as in the case of the experiments of B. N. Kabanov and Yu. G. Siver, agreement between theory and experiment is very good.

In the study by E. A. Aykazyan and A. I. Fedorova [21], formula (53.11) was subjected to verification using anodic ionization of hydrogen atoms in different electrolytes as the example. The anode was a platinum disk 5 mm in diameter welded to a brass cone fixed on a brass rod. The brass parts of the instrument were covered with Plexiglas. The Reynolds numbers in these measurements were within the range 80 to 1700. The results of the measurements are shown in Figure 55.

Agreement with theory is found to be very good. The points corresponding to 1.5 and 2.0 rev/sec, however, are somewhat above the theoretical line. Apparently, at these points, the flow of particles carried by natural convection adds to the forced convection flow.

The last point, obtained at 43.7 rev/sec, is below the theoretical line. At high convection velocities, kinetic limitations begin to be effective (finite rate of the ionization of hydrogen atoms).

Several other studies conducted on a rotating disk are examined below.

Data are also available for the experimental verification of other instances of convective diffusion in a laminar flow regime, for which the expressions for diffusional flows were derived: the case of diffusion to the surface of a sphere at $Re \ll 1$ and $Pe \gg 1$, and the case of diffusion on the surface of a plate. In the study of G. A. Aksel'rud [22], the rate of dissolution of a ball of compressed benzoic acid in a vegetable oil was studied. The ball was attached to a shaft connected to a rotating upright rod so that it described circular paths. In order to avoid having the ball pass through its own wake, enriched by benzoic acid, a rotating mixer moved along the path behind the ball. This mixer, by energetically agitating the solution in back of the ball, assured the flow of fresh liquid to the latter. Since the liquid viscosity was high ($\nu \sim 0.7$) the disturbances caused by the mixer were rapidly damped, so that when the ball subsequently approached a given point, the liquid was at rest. The diameters of the dissolving balls were in a range 0.4 to 0.7 cm. The velocity of the motion ranged from 0.2 to 2.5 cm/sec. The Reynolds numbers were from 0.1 to 2.5, while the Peclet numbers ranged approximately from 4900 to 29,000. The results of the measurements are shown in Figure 56. Here $Nu' = \frac{2Ia}{D}\frac{1}{S}$, $Pe' = \frac{2Ua}{D}$, $D = 0.32 \cdot 10^{-6}$ cm²/sec. The theoretical expression for $Nu' = 1.1\sqrt[3]{Pe}$ is shown in Figure 56 as a straight line. There is complete agreement between theory and experimental data.*

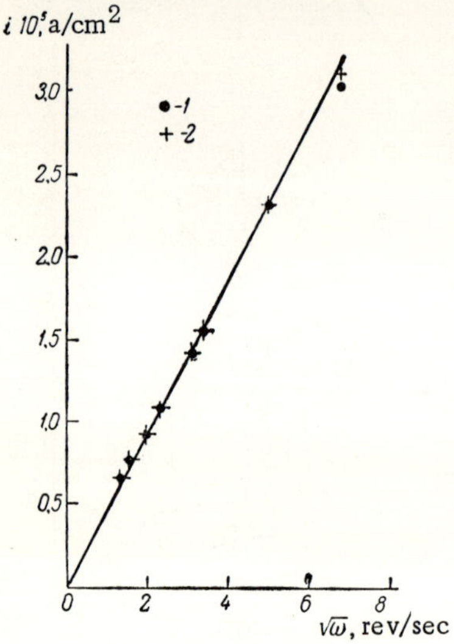


Figure 55. Dependence of the density of the hydrogen diffusion current at the surface of a platinum disk electrode on angular velocity. 1 — smooth platinum electrode; 2 — platinized electrode.

A study by Trumpler and Zeller [23] was conducted for the purpose of verifying the formula for forced liquid flow past a plate.

---

*Note in proofreading. A formula very similar to (14.19) was derived by means of numerical integration by Friedlander [64]. The data given in studies by Ranz [65] and Kramers [66] agree well with (14.19).

In this study the electrolyte was forced to flow past the plate. The current was measured by means of special probes. The velocities of the electrolyte varied from 0.7 to 27.6 cm/sec, which corresponds to Reynolds numbers in that range 104 to 6100. The results of the measurements are described by the empirical formula

$$I \sim U^{0.43}.$$

Considering the disturbances in the liquid stream caused by the presence of the probes, agreement with theory, as pointed out by the authors, is fully satisfactory.

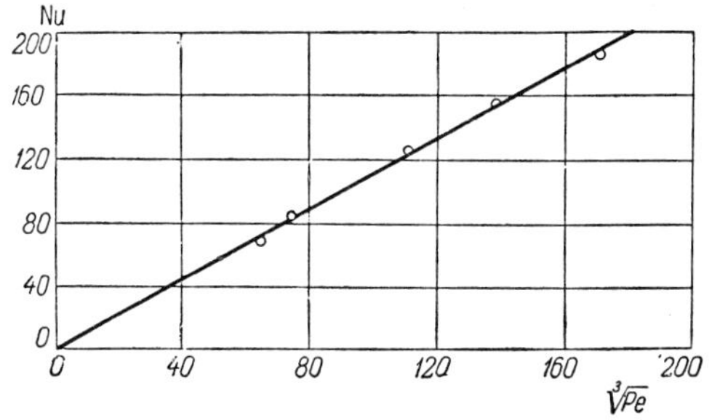


Figure 56. Kinetics of the dissolution of a ball in the laminar flow regime.

An important conclusion of the general theory of the diffusion kinetics of heterogeneous reactions (Section 18) relating to the relaxation length of the diffusion layer was verified experimentally by M. M. Nikiforova under the supervision of B. N. Kabanov. The diffusion current on the surface of a disk electrode was measured. The central portion of the disk was coated with glyptal. Thus, the working surface of the electrode became a ring surface with a width of $R - r_i$ which varied within the range 0.1 to 1.25 cm (the disk radius was $R = 1.25$ cm).

The current for the reduction of dissolved oxygen was measured. In addition to dissolved oxygen ($2.55 \cdot 10^{-4}$ mole/liter), the solution contained sulfuric acid ($5 \cdot 10^{-3}$ m/$l$), potassium sulfate ($10^{-1}$ m/$l$), and hydrogen peroxide ($3 \cdot 10^{-3}$ m/$l$). The speed of rotation of the electrode in all experiments was maintained constant at 422 ± 2 rev/min. The results of the measurements of the current for oxygen are shown in Figure 57.

For a quantitative comparison of theory and experiment, a curve has been drawn, which is shown in Figure 58. The values of the ratio $r_i/R$ are plotted along the abscissa, while the values of the function $f(r_i)$ are plotted on the ordinate. This function is defined by the equality

$$f(r_i) = \frac{(i_i)_{av}}{(i_0)_{av}}, \tag{56.1}$$

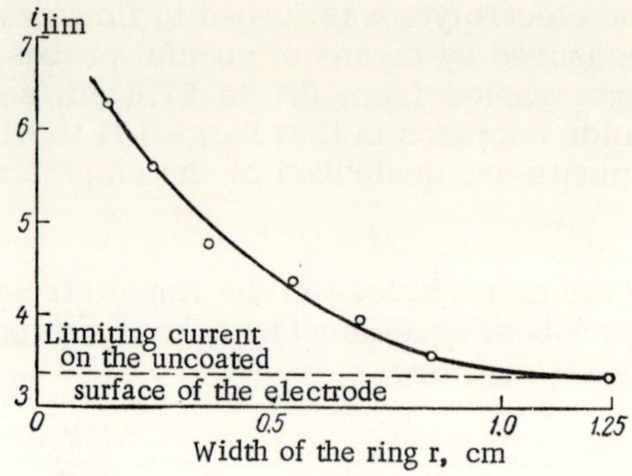


Figure 57. Limiting current $i_{lim}$ upon the surface of a coated electrode

where the average value of the current density is defined as

$$(i_i)_{av} = \frac{2\pi \int_{r_i}^{R} i_i r\, dr}{S_i}. \qquad (56.2)$$

Since $i_i$ is given by formula (18.24), and the uncoated area is equal to $S_i = \pi(R^2 - r_i^2)$, we may write for $(i_i)_{av}$:

$$(i_i)_{av} = \frac{A\left(R^3 - r_i^2\right)^{2/3}}{R^2 - r_i^2} \qquad (56.3)$$

and $(i_0)_{av} = A$, where $A = \dfrac{Dc_0}{1.61\left(\frac{D}{\nu}\right)^{1/3}\sqrt{\frac{\nu}{\omega}}}$. Therefore, we finally obtain the theoretical expression for $f(r_i)$:

$$f(r_i) = \frac{\left(R^3 - r_i^3\right)^{2/3}}{R^2 - r_i^2}. \qquad (56.4)$$

The experimental values of $f(r_i)$ were determined from the measured values of limiting current densities at six different values of $r_i$. The experimental results, accounting for errors, are shown in Figure 58.

Agreement between theory and experiment is found to be entirely satisfactory.

It follows from Figure 58 that the current density increases significantly with decrease in the width of the ring. This is in complete accord with the theoretical prediction. The smaller the width of the ring, the larger is the working portion of the electrode surface that is advantageously located to receive undepleted solution. The current density upon the electrode must increase accordingly. If there were no tangential transfer of substance on the electrode or if the latter did not play a material part in the total amount of substance arriving at the electrode — which is the case under conditions of diffusion in a motionless medium — the current density at the ring surface would remain practically constant upon reduction in the width of the ring.

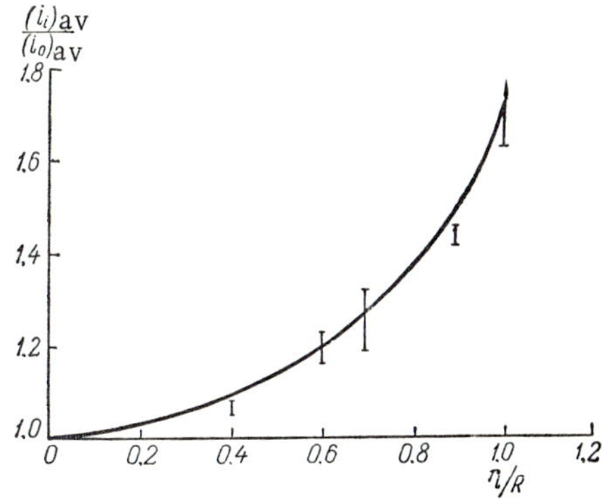


Figure 58. Comparison of theory and experiment for a coated electrode.

Measurements of the limiting current upon the surface of a ring electrode were conducted in the laboratory of Prof. Ya. V. Durdin. A thin circular band served as the working surface on the surface of the disk. The main portion of the disk was coated. Measurements were carried out by V. A. Avdeyeva and were of a preliminary nature. In one series of measurements, the ring radius was 0.825 cm and the width of the ring was 0.028 cm. In another series, the ring radius was 1.1 cm and the width was 0.15 cm.

In both instances the theoretical requirement for increasing limiting current compared to the limiting current on the uncoated portion of the disk (see Section 18) was fulfilled.

Complete quantitative agreement of the limiting current with computed values, however, was not observed in these experiments.

In the first instance of smaller ring width, the measured value of the ratio of the average limiting current on the surface of the ring to the limiting current on the uncoated surface of the disk was 2.2. According to formula (18.25), the theoretical value of this ratio should be 3. In the second instance, the measured and computed values of the ratio were respectively 1.6 and 1.9.

In both cases the computed value of the ratio was 20 to 30% greater than the measured value. It is difficult to find the possible reason for this discrepancy.

The data derived by Brenner [24] may be utilized for the purpose of verifying the theory of the diffusional flow in natural convection. In his studies, the solution in a state of natural convection was suddenly frozen. Thin layers of the frozen solution were then sliced from the region adjacent to the electrode surface and quantitatively analyzed to determine their composition.

In the study of Keilegan, mentioned in Section 23, an analysis of Brenner's data was made, and they were found to agree well with theory. This is illustrated in Table 8, in which the diffusion layer thicknesses are shown for the deposition of different metals at various current densities.

Table 8

Thickness of the diffusion layer (cm)

| Measured | Computed |
|---|---|
| 0.032 (deposition of copper) | 0.030 |
| 0.037 } (deposition of nickel) | 0.035 |
| 0.034 } (deposition of nickel) | 0.032 |

The studies by N. Ibl [25] constitute a much more thorough verification of the theory. Ibl established the fact that the formation of dense (shiny) metallic deposits on the electrode occurs during transition of the electrolysis regime toward a limiting diffusion current.

Measurement of the thickness of the deposits, therefore, permits the value of the limiting current attainable in natural convection to be determined.

To avoid scattering effects and various types of interference, the vertical electrode was surrounded on all sides by a special shield. Ibl's measurements established the dependence of the diffusion current density on the concentration of a solution of $CuSO_4$, over a wide concentration range (0.01 to 1 mole/liter). The results of these measurements are shown in Figure 59 and are described by the empirical formula

$$i_{lim} \sim c_0^{1,20}.$$

This agrees with the theoretical relation (23.29) to a considerable degree of accuracy of measurements attained in Ibl's studies permitted his verifying formula (23.29) quantitatively.

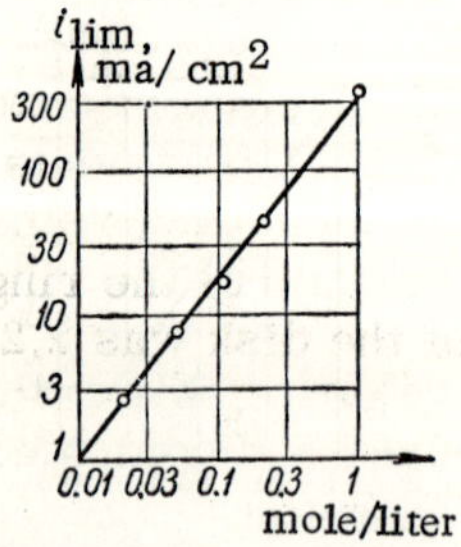


Figure 59. Dependence of the limiting current on concentration, from experiments by Ibl.

The average limiting current on a plate of height h, according to formula (23.30), is

$$\bar{i}_{lim} = \frac{4}{3} i_h.$$

For a 0.1 m/$l$ $CuSO_4$ solution at a temperature of 22.5°C, computation by the formula $\bar{i}_{lim} = \frac{4}{3} i$ leads to the value $i_{lim} = 21.6$ ma/cm². The experimental value of current density at which the formation of a dense deposit begins, and which corresponds to a limiting current regime, amounts to 18.0 ma/cm² in the presence of a shield and 20.3 ma/cm² without it. Thus, the quantitative agreement of theory and experiment is found to be quite satisfactory. The same author, using an electrode composed of bands up to 3 cm long, verified the law of current distribution on the surface of an electrode experimentally. The theoretical formula $\frac{i_h}{i_{lim}} = \frac{3}{4}$ was satisfactorily confirmed by his experimental data.

## 57. QUANTITATIVE VERIFICATION OF THE THEORY. TURBULENT FLOW REGIME

In practice, besides laminar flow, turbulent motion is often encountered near the reaction surface.

A turbulent regime is observed in the boundary layer under conditions where extremely energetic agitation of the solution is employed for the purpose of attaining especially high diffusional flow values. This is the reason why experimental verification of the theory for very high Reynolds numbers is of major interest. We have pointed out previously that, in establishing the correctness of the theoretical characteristics of turbulence near the surface of a solid body, the verification of computations by experimental measurements was the decisive factor.

The study by I. A. Bagotskaya [26] of diffusional flows in the region of the turbulent fluid in the boundary layer permitted a single-valued solution of the problem of structure of the turbulent fluid near the surface of a solid body. Her experiments measured the diffusional flow on the surface of a rotating disk. Measurements were made for the electrochemical reduction of oxygen and liberation of hydrogen. The limiting diffusion current on the surface of a rotating disk electrode was the basis for the determinations. The electrode was an amalgamated copper disk 6 cm in diameter. As in the case of the study by B. N. Kabanov, the upper side of the disk and its axis of rotation were isolated by means of a Plexiglas cover, so that the reaction occurred only on the lower surface of the disk. The disk was placed on the end of the shaft of an electric motor, and rotational speeds were carefully controlled. The disk speeds varied over the range 500 to 1750 rev/min. The corresponding Reynolds numbers were within the range $4.5 \cdot 10^4$ to $1.6 \cdot 10^5$. The anode was a platinum grid 6 × 6 cm and was situated on the bottom of the vessel containing the solution. The volume of the solution was 5.3 $l$. The electrolytes were 0.1 N KOH and 0.2 N KCl + $7 \cdot 10^{-4}$ N HCl.

The upper limit of the rotational velocity was established by the fact that at high rotating speeds, the funnel that formed at the axis of the rotating disk caused the introduction of air bubbles under the disk, which disturbed the mixing regime. Measurements of the limiting current on a disk electrode were based on oxygen molecules and hydrogen ions. The diffusion current carried by oxygen molecules was measured by the limiting current for the reaction

$$O_2 + 4H^+ + 4e = 2H_2O$$

in a solution of 0.1 N KOH. Oxygen concentration in the solution was measured by the comparison method with the limiting current in the same solution being determined polarographically. Maintenance of constant concentration of oxygen in the solution was assured by dissolution of atmospheric oxygen.

The diffusion current carried by hydrogen ions was measured in a solution of 0.1 N KCl $+ 7 \cdot 10^{-4}$ N HCl. A constant concentration of hydrogen was maintained since the anode was hydrogen-saturated, palladium-coated platinum.

The results of the measurements established the dependence of limiting diffusion currents on the rotational velocity of the disk electrode (Figure 60) and the dependence of these limiting currents on the diffusion coefficient.

As will be noted from Figure 60, the diffusion current is found to be proportional to the angular velocity of the electrode rotation to a power equal to about unity. The errors in the experiments were, of course, too great to allow a distinction to be made between the first power of the velocity and the theoretically determined value of 0.8 to 0.95. Nevertheless, the theoretical prediction of the increase in the exponent of velocity derived in the expression for the limiting current, from 0.5 in a laminar boundary layer regime to 0.9 to 0.95, is quite clearly shown. Let us note, in this connection, that the transition to a turbulent regime in the case of a very accurately centered disk and in the absence of disturbances is observed only at Reynolds numbers of the order of $3 \cdot 10^4$. This value decreases noticeably, however, in the presence of such disturbances as were observed in the experiments conducted by Bagotskaya. Such disturbances could be vibrations of the shaft of the disk, effects due to the Plexiglas cover, proximity to the free surface of the liquid, and the like. In view of this, the transition to a turbulent regime began at Reynolds numbers of the order of $2 \cdot 10^4$.

The diffusion current values for different diffusing particles are most informative. In accordance with what was said in Section 25, the different models of a turbulent boundary layer determine the varying dependence of the diffusional flow on the diffusion coefficient of the diffusing particles. The assumption of a laminar nature of fluid motion in the viscous sublayer leads to formula (25.5) for the diffusional flux, in which the latter is proportional to the first power of the diffusion coefficient. Therefore, in a comparison of currents carried by oxygen and by hydrogen ions, according to this theory, it would be expected that diffusion currents by oxygen and hydrogen ions differ by about a factor of $D_{H^+}/D_{O_2}$.

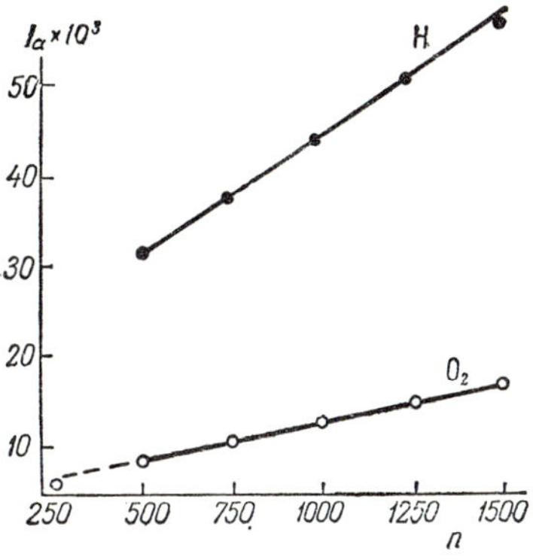


Figure 60. Limiting currents for molecular oxygen and hydrogen ions on a disk electrode as functions of disk rpm.

According to the theory that we have developed which accounts for the diminution of turbulence in the boundary layer and leads to expression (25.15) for the diffusional flux, the latter is found to be proportional to the 3/4 power of the diffusion coefficient, so that the ratio of the diffusion currents carried by oxygen and hydrogen ions must be proportional to $(D_{H^+}/D_{O_2})^{3/4}$. The measured ratio of the two diffusion currents was found to vary from 0.67 to 0.72. Thus, the experimental values of this ratio are very close to those we computed theoretically and differ radically from those resulting from formula (25.18) — which is based on the theory of a laminar sublayer.

I. A. Bagotskaya's work leads to the conclusion that the model of a turbulent boundary layer proposed by us and the quantitative agreement between formula (25.15) and experiment are correct.

I. A. Bagotskaya's measurements permit a determination of the numerical value of the constant appearing in the expression for the diffusional flow upon the surface of a rotating disk.

This constant was found to be equal to

$$a = 2.6.$$

Additional progress of note in the experimental study of turbulence near a solid surface was made in the study by A. I. Fedorova and G. L. Vidovich [27].

This study measured the temperature dependence of the limiting diffusion current during turbulent motion of the solution.

The electrode was a plate that moved through the fluid. It consisted of a piece of platinum wire sealed into glass. The moving part of the instrument had the shape of a Γ-shaped rod (Figure 61), rotating around the long axis.

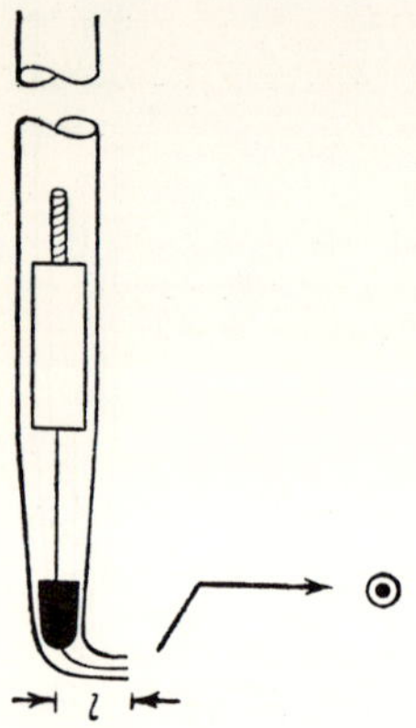

Figure 61. Non-streamlined electrode. The cross section of the electrode is shown on the right.

The Reynolds number reached values of the order of $7.5 \cdot 10^4$. The temperature dependence of the limiting diffusion current was determined by measuring polarization curves at temperatures of 0°, 20°, and 40°C.

The reaction studied was that of cathodic iodine reduction $I_3' + 2e^- = 3I'$ in a solution of 0.1 $MI_2$ + 1 MKI, without formation of a gaseous phase.

In order to determine the diffusion coefficients, auxiliary measurements of limiting currents on a rotating disk in a laminar regime were made.

The diffusion coefficients were computed by formula (11.33) at the three temperatures stated, account being taken of the change in the kinematic viscosity of the solution with temperature. Kinematic viscosity was measured in a special way. Auxiliary measurements permitted the establishment of the values of the energy of activation in the formulas

$$D = D_0 e^{-\frac{E_D}{RT}},$$

$$\nu = \nu_0 e^{+\frac{E_\nu}{RT}}.$$

It was found that to these energies could be assigned the values stated in Table 9.

Table 9

| Temperature in °C | Energy of activation in cal/mole | | |
|---|---|---|---|
| | $E_D$ | $E_\nu$ | $E_{lam}$ |
| 20<br>40 | 2870<br>3200 | 3500<br>3750 | 2520<br>2830 |

The quantity $E_{lam}$ represents the temperature coefficient in the temperature dependence relation of the limiting current in laminar flow (11.35)

$$j \sim \frac{D^{2/3}}{\nu^{1/6}} \sim j_0 e^{-\frac{4E_D + E_\nu}{6RT}}.$$

The basic measurements determined the dependence of the limiting diffusion current density on the electrode rpm at different temperatures. The results of the measurements are given in Figure 62.

The proportionality theoretically required between the current density and the rpm is very clearly shown.

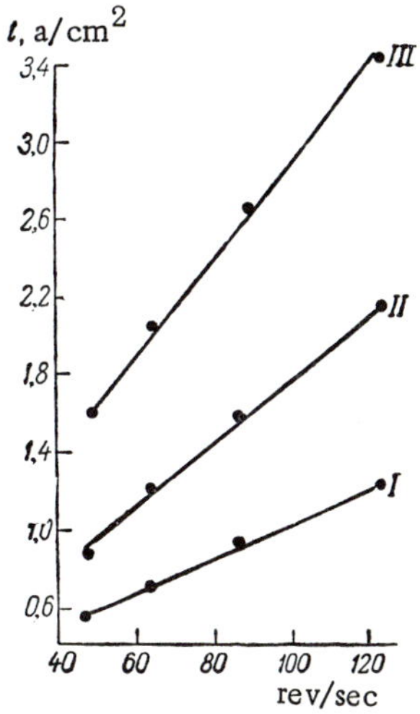


Figure 62. Dependence of the limiting diffusion current on electrode rpm in a solution of 0.1 M $I_2$ + 1 M KI in a turbulent agitation regime at various temperatures: I—0°, II—20°, III—40°C.

By formulas (27.8) and (27.9), the dependence of the limiting current density, in a turbulent regime, on the energies of activation $E_D$ and $E_\nu$ has the form

$$j \sim \frac{\sqrt{k_f}}{\mathrm{Pr}^{3/4}} \sim \frac{D^{0.75}\,\nu^{0.10}}{\nu^{0.75}} \sim e^{-\frac{E^{L}_{turb}}{RT}},$$

where

$$E^{L}_{turb} = 0.75\,E_D - 0.65E_\nu.$$

If, instead of (27.8) we had utilized formula (25.2), we would have obtained:

$$j \sim \frac{\sqrt{k_f}}{\mathrm{Pr}^{2/3}} \sim \frac{\nu^{1/40} D^{2/3}}{\nu^{2/3}} \sim e^{-\frac{E^{V}_{turb}}{RT}},$$

where

$$E^{V}_{turb} = 0.66E_D - 0.56E_\nu.$$

Computations according to formula (25.5), derived on the assumption of the absence of turbulence eddies in the viscous sublayer, gives:

$$j \sim \frac{\sqrt{k_f}}{\mathrm{Pr}} \sim \frac{D}{\nu^{9/10}} \sim e^{-\frac{E^{P}_{turb}}{RT}},$$

where

$$E^{P}_{turb} = E_D - 0.9E_\nu.$$

The results of measured and theoretical values of the activation energy $E_{turb}$ obtained on the basis of our theory $E^{L}_{turb}$, the Vielstich theory $E^{V}_{turb}$, and others are given in Table 10.

Table 10

Activation Energies, cal/mole

| Temperature in °C | $E_{turb}$ (meas) | $E^L_{turb}$ | $E^V_{turb}$ | $E^P_{turb}$ |
|---|---|---|---|---|
| 20 | 4270 | 4440 | 3850 | 6020 |
| 40 | 4870 | 4840 | 4200 | 6575 |

The accuracy of the measurements of $E_{turb}$ is within the range ± 200 cal/mole.

We see from this table that the experimental data show our hypothesis regarding the nature of the damping of turbulence near a solid surface to be preferable. The agreement between our theoretical formula (27.8) and experiment is quite satisfactory.

The values of the energy of activation $E^V_{turb}$, derived on the basis of the assumption of an identical law of turbulence damping in the turbulent layer and in the viscous sublayer $l \sim y$, are lower than measured values. The discrepancy is greater than possible measurement errors, especially within the temperature range 20 - 40°C.

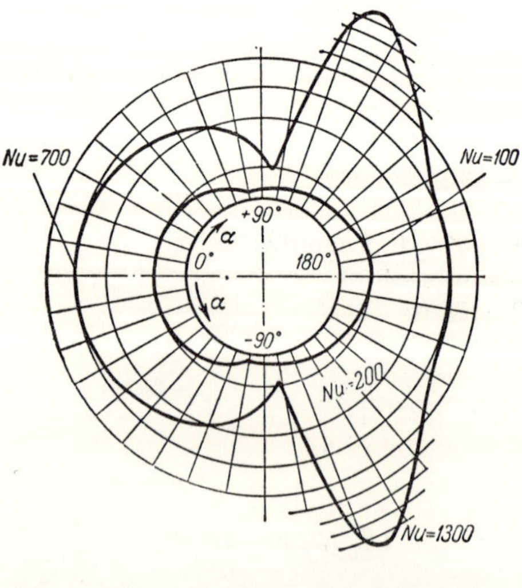


Figure 63. Heat flux from the surface of a cylinder. Inside curve corresponds to values of Nu at Re = 39,800; the outer curve corresponds to values of Nu at Re = 426,000.

The hypothesis regarding the absence of turbulence eddies in the viscous sublayer leads to values for $E^{P}_{turb}$ that exceed measured values by almost 50%. This exceeds all the experimental errors.

Up to the present time, the theoretical conclusions regarding diffusion to a nonstreamlined body have not been experimentally verified. This problem, however, has been studied experimentally in connection with heat transfer for flow past spheres and cylinders [28].

From a qualitative standpoint, the heat transfer data may be carried over into the field of diffusion. Figure 63 shows the results of the measurements of a heat flow per unit of cylindrical surface, for a fluid stream perpendicular to its axis.

The increase in heat flux at the downstream end of the cylinder is clearly shown at high Reynolds numbers. In the case of diffusion, the increase in diffusion current is expressed more sharply and should be manifested at considerably smaller Reynolds numbers.

## 58. APPLICATION OF THE THEORY OF CONVECTIVE DIFFUSION TO THE SOLUTION OF ELECTROCHEMICAL PROBLEMS

The theory developed for concentration polarization and its confirmation by experimental data permit its use in the solution of various practical electrochemical problems.

1. Finding the shape of the polarization curve (the volt-ampere characteristic of the reaction). Prior to the development of a quantitative theory of concentration polarization, no method existed for the computation of polarization curves. Such a curve could only be obtained experimentally by measuring the entire curve $\varphi(i)$.

The shape of these curves can now be determined on the basis of the theory set forth above, for reversible or for rapid, irreversible electrochemical reactions.

For irreversible reactions whose rate is comparable to the rate of transfer of reagents to the surface, the shape of the polarization curve depends on the reaction kinetics. If the latter are known, the volt-ampere characteristic can be established theoretically. If the kinetics are not known, then, proceeding from the measured characteristics and knowing its shape without correction for the kinetics, one can determine the

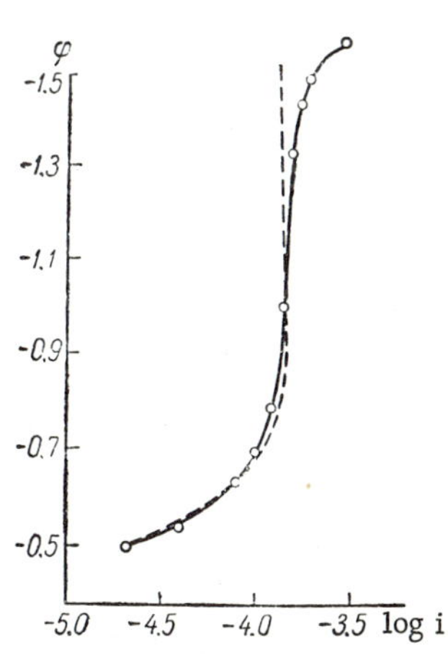


Figure 64. Theoretical (dotted line) and experimental polarization curves for an electrode rotation of 1.5 rev/sec.

nature of the electrochemical kinetics. As an example, we may cite the computation of the polarization curve for the cathodic reduction of dissolved oxygen with the formation of hydrogen peroxide. The computed curve is shown by a broken line in Figure 64. The solid line on the same figure shows the measured polarization curve. The circles indicate experimental points. With due regard for the fact that the theoretical curve was constructed without a single experimental constant, the agreement between theory and experiment should be considered beyond question.

2. Use of solid electrodes for the chemical analysis of solutions. By determining the volt-ampere characteristic of the process in a solution of given composition, a quantitative analysis of the solution may be made. Indeed, since the current on the electrode for a given type of ions begins at a certain electrode potential (with respect to the reference electrode), the volt-ampere discharge characteristic in a solution of complex composition has the form of a step curve (see Figure 47). At a potential that corresponds to the initiation of a new electrochemical process on the electrode, the current in the cell increases sharply. A further increase in the potential difference applied across the cell leads to the establishment of a limiting current that is proportional to the concentration of the given type of ions in the solution. Therefore, from a qualitative standpoint, the composition of the solution may be determined by the position on the volt-ampere curve of the points at which the corresponding electrode reactions begin.

The quantitative composition of the solution can be determined by the magnitudes of the corresponding limiting currents.

The use of electrolysis for purposes of quantitative and qualitative analysis has found the widest acceptance in actual practice and is known as polarographic analysis.

For many reasons, such an analysis of solutions is made mostly be means of a dropping mercury electrode.

We shall indicate the advantages and disadvantages of this method in Chapter X, which is devoted to the theory of polarographic analysis. It should be noted, however, that many recent studies have indicated the usefulness of solid electrodes in polarographic analysis of solutions. Solid electrodes are simpler in many ways and more convenient from a practical standpoint than mercury electrodes. The greatest disadvantage of solid electrodes is that, during the electrolytic process, a deposit forms on their surface and changes the properties of the electrode. In addition, quantitative analysis of solutions has been difficult heretofore owing to the absence of a theoretical expression for limiting current. The good agreement between the theory of convection currents in solutions and experiment does away with this disadvantage of solid electrodes.

Certain authors have now published a large number of analytical studies conducted by means of solid electrodes [29].

3. Measurement of diffusion coefficients. The accurate measurement of diffusion coefficients is associated with the need to overcome experimental difficulties. In the studies of B. N. Kabanov, cited previously, the electrochemical method of measurement on the basis of the magnitude of the diffusion current was successfully used for measuring diffusion coefficients of certain ions and atoms.

In the study by Hogge and Kraichman, mentioned in Section 53, the diffusion coefficient of $I_3^-$ was measured. The study by E. A. Aykazyan and A. I. Fedorova measured the diffusion coefficient of hydrogen molecules. V. D. Yukhtanova et al., in their study [30], obtained the values of the diffusion coefficients of bivalent cations, which are difficult to measure. Yu. V. Pleskov and E. A. Aykazyan, in their study [31], measured the diffusion coefficients of quinone and hydroquinone.

Tables 11 and 12 show the diffusion coefficient values obtained in the studies noted. These data give a picture of the accuracy of measurement that can be achieved by means of a disk electrode.

Table 11 gives the data of B. N. Kabanov, Yu. G. Siver, A. I. Fedorova, E. A. Aykazyan, and Yu. V. Pleskov. Table 12 presents the data obtained by V. D. Yukhtanova. They are compared with the most accurate measurements of diffusion in capillaries obtained by Stackelberg [32].

4. Determination of the order and mechanism of a reaction. The study of the kinetics of electrochemical reactions is an important field of application for the results obtained.

Table 11

| Material | $O_2$[1] | $H_2$[2] | Quinone[3] | Hydroquinone[4] |
|---|---|---|---|---|
| D, cm²/sec, exptl. with disk electrode | $1.93 \cdot 10^{-5}$ | $3.83 \cdot 10^{-5}$ | $1.27 \cdot 10^{-5}$ | $0.85 \cdot 10^{-5}$ |
| D, cm²/sec by diffusion in capillaries and computed on basis of mobilities | $1.98 \cdot 10^{-5}$ | $3.90 \cdot 10^{-5}$ | $1.30 \cdot 10^{-5}$ | $0.87 \cdot 10^{-5}$ |

[1]Electrochemical and published data refer to 0.05 N $H_2SO_4$ at 20°.

[2]Electrochemical data for 1 N $H_2SO_4$ at 23°; published data for 2% agar-agar at 23°.

[3]Electrochemical data for 2 N KCl at 26°; published data for a phosphate buffer at pH = 6.8 at 27°.

[4]Electrochemical data for 0.1 N KCl $+10^{-3}$N $H_2SO_4$ at 16°; published data for water at 16°.

Table 12

| Source | $Cd^{++}$ | | $Tl^{++}$ | | $Pb^{++}$ | | $Zn^{++}$ | |
|---|---|---|---|---|---|---|---|---|
| | On rotating electrode | Stackelberg | On rotating electrode | Stackelberg | On rotating electrode | Stackelberg | On rotating electrode | Stackelberg |
| 1 N KCl | 7.73 | 7.89 | 15.3 | 15.67 | 8.95 | 9.23 | 7.30 | 7.23 |
| 0.1 N KCl | 7.10 | 7.16 | 17.5 | 17.44 | 8.69 | 8.65 | 6.62 | 6.73 |

As an example, let us examine the study by A. N. Frumkin and E. A. Aykazyan [58] of the kinetics of ionization of molecular hydrogen on a platinum anode. The anode was a disk similar to that used in earlier experiments by E. A. Aykazyan and A. I. Fedorova (see Section 56).

At not too great rotational speeds, the density of the ionization current of hydrogen was determined by the rate of diffusion of the latter in the solution.

Increasing the rotational velocity permitted diffusional limitations to be removed and led to the attainment of a limit for which a further increase in the rotational speed caused no further increase in the current density. Such a case was realized in the experiments by A. N. Frumkin and E. A. Aykazyan at a sufficiently high anodic overpotential.

The measured values of current density as functions of the square root of the disk rpm $\sqrt{\omega}$, for relatively high anodic overpotential, were given in Figure 15. Curve 1 of that figure gives the dependence of the current on $\sqrt{\omega}$ as given by formula (12.8) for a first order reaction to the dissolved hydrogen concentration. The experimental curve, as distinct from the smooth curve 1, shows a sharp break. In accordance with what has been said in Section 12, this corresponds to a very low, almost zero, order of the reaction. The curve which corresponds to a zero order of reaction is shown in the same figure (curve 2). Thus, measurements of the current as functions of the rotation velocity enabled the order of the hydrogen reaction to be established. This made it possible for the authors to come to definite conclusions regarding the mechanism of the reaction.

Let us note that if analogous reactions had not been conducted on a disk electrode (i.e., on a uniformly accessible surface) but on an electrode of a different shape, the break in curve 2 would have been less pronounced, and conclusions regarding the order of the reaction would have been difficult.

5. Study of intermediate products in electrochemical reactions by means of a nonhomogeneous disk electrode. A recent and very important application of the rotating disk electrode is the study of stage and step-type electrode reactions that may be conducted with it.

It is well known that many electrode reactions pass through an intermediate stage, i.e., as

$$\begin{aligned} A &\rightarrow B^*, \\ B^* &\rightarrow C, \end{aligned} \tag{58.1}$$

where A is the initial, C the final, and B* the intermediate products of the reaction. This relates primarily to electrode reactions that involve the addition or loss of several electrons, In many instances the intermediate products are thermodynamically stable and may accumulate to considerable concentrations.

They can then be easily detected. However, there are as many cases where the intermediate product B* is thermodynamically unstable. Specific examples of such reactions are given in the monograph by A. N. Frumkin et al. The intermediate product particles in this case do not accumulate at the electrode in noticeable quantities. Here, the first and second stages of the reaction, under normal electrolysis conditions, appear to take place on the electrode as a single reaction.

It is very important to find a general experimental method for separating the stages of the reaction and determining the rate of each stage separately.

This is all the more important since the rate of the second stage of the electrode reaction in many instances is very rapid, and the nature of the stage for many reaction types has not been proven to date.

Academician A. N. Frumkin [67] expressed the view that unstable intermediate product particles can be detected if the electrolysis takes place under conditions of a sufficiently vigorous stirring of the electrolyte. With proper selection of geometric conditions, the liquid flow will carry away the intermediate product particles, which may then be collected at another electrode. In this case, two possibilities exist.

1) When the discharge potential of the intermediate product particles is lower than that of particles A, a lower potential may be applied to the second electrode than to the first, so that only the reaction B* → C takes place on it.

2) In the more general case of an arbitrary discharge potential of the intermediate product, a potential may be applied to the second electrode that is lower than that of the discharge of the initial product A. Then only the reverse reaction B* → A will occur on it. The presence of a current on the second electrode may serve as convincing proof of the existence of intermediate product particles. Moreover, in measuring the electrode current resulting from the

electrode reaction, in which only particles of the unstable intermediate product participate, and having at our disposal the theory connecting this current with the parameters of the process and the decomposition constant k, the value of the latter may be determined.

Since the lifetime of unstable intermediate product particles is short, and the electrode and the spacing between electrodes must have a finite thickness (to ensure the electrical isolation of the electrodes), the velocity of the liquid that carries away the intermediate product particles must be fairly large.

On the other hand, the motion of the liquid must remain laminar. If the particles remain too long in a turbulent fluid, they will be returned to the point where they were formed and will decompose without reaching the second electrode.

Tests have shown that, from an experimental standpoint, the optimum system is one in which both electrodes are located on the surface of a rotating disk.*

The first electrode (designated below as zone I) is the inside portion of the disk. In zone I the electrode reaction $A \rightarrow B^*$ takes place. Particles of the intermediate product $B^*$ here partially enter the second stage of the reaction, being converted into product C, and are partially carried away by the liquid flow. The latter particles arrive at the second electrode, which is an annular zone (zone III) at the outer portion of the disk.

Between the electrodes there is an annular insulating ring (zone II).

The potential applied to the surface of the second electrode is such that the reaction $B^* \rightarrow C$ or $B^* \rightarrow A$ is very rapid, so that the concentration of the intermediate product on the surface of the disk in zone III is zero. The change in concentration of the intermediate product between zones I and III, and the fact that particles are carried away by the liquid, assures their transfer from zone I to zone III by convective diffusion.

The virtue of a disk lies in the fact that, first of all, its surface is uniformly accessible from a diffusion standpoint. This assures a constant flux of particles of the intermediate product $B^*$ that form on the inner portion of the disk (zone I). In the second place, turbulence of the liquid flow entrained by the rotating disk occurs at relatively high Reynolds numbers ($Re > 4 \cdot 10^5$). Distribution of the intermediate product concentration on the surface of the disk is determined by the convective diffusion equation (11.20), in which the concentration in this case is a function of the variables (y, z). Here, it is assumed that in cases where the particles $B^*$ are ions, the solution contains an excess of indifferent electrolyte and there is no migration.

The system of boundary conditions has the following form.

---

*The discussion that follows is based on [68].

Reaction A → B* takes place in zone I. B* particles form evenly over the surface of the disk in zone I at a constant rate. The number of B* particles formed on 1 cm$^2$ of surface in 1 sec is equal to the diffusional flux of A particles, given by formula (11.30).

Disappearance of the intermediate product particles as a result of the electrochemical reaction may be characterized by the constant k, so that the number Q of decomposing particles equals $Q = kc\big|_{y=0}$, where $c\big|_{y=0}$ is the surface concentration of B*.

Intermediate product particles that have no time to become converted on the surface of the disk in zone I diffuse into the flowing stream.

The material balance for the intermediate product particles is:

$$j_0 = kc\Big|_{y=0} - D\frac{\partial c}{\partial y}\Big|_{y=0}, \qquad r < r_1. \tag{58.2}$$

The minus sign indicates that the particles diffuse outward from the surface of the disk.

No intermediate product particles are formed or disappear in zone II, so that the current here is zero.

$$\frac{\partial c}{\partial y}\Big|_{y=0} = 0, \qquad r_1 < r < r_2. \tag{58.3}$$

In zone III, the rate of the electrode reaction involving intermediate product particles is very high compared to the rate of their transfer to the surface of the electrode. Therefore, their concentration on the disk surface is zero,

$$c\,\big|_{y=0} = 0, \qquad r_2 < r < r_3. \tag{58.4}$$

Far from the surface of the disk, as y → ∞, the concentration of the intermediate product must become zero everywhere

$$\lim_{y \to \infty} c = 0, \qquad 0 < r < r_3. \tag{58.5}$$

Solution of the convective diffusion equation for diffusion to the surface of a "multilayer" disk, with boundary conditions (58.2) to (58.5), is found to be possible by means of Sutton's method, described in Section 17.

By introducing new variables, as was done in the case of the plate examined above,

$$z = \frac{2}{3}\Psi^{3/4} = \sqrt{\frac{\gamma^3}{18}}\, r^{3/2} y^{3/2}, \tag{58.6}$$

$$\xi = \frac{D\gamma}{12} r^3, \tag{58.7}$$

where

$$\gamma = \sqrt{2 \cdot 0.51\omega}\sqrt[4]{\frac{\omega}{\nu}}, \tag{58.8}$$

the convective diffusion equation can be transformed into the form (17.5). Let us now pass to the formulation of boundary conditions in the new variables.

The boundary conditions in zone I will be written in the form

$$\lim_{z \to 0}\left\{-z^{1/3}\frac{\partial c(z, \xi)}{\partial z} + h(\xi)\, c(z, \xi)\right\} = g(\xi), \qquad 0 \leqslant \xi < \xi_1, \tag{58.9}$$

where

$$h(\xi) = \frac{1}{\xi^{1/3}}\frac{k}{D\sqrt[3]{\frac{\gamma^2}{D}}}\sqrt[3]{\frac{4}{9}}, \tag{58.10}$$

$$g(\xi) = \frac{1}{\xi^{1/3}}\frac{j_0}{D\sqrt[3]{\frac{\gamma^2}{D}}}\sqrt[3]{\frac{4}{9}}. \tag{58.11}$$

In addition to boundary condition (58.9), in order to completely specify one problem, the boundary condition for infinity must be given. Obviously, we have

$$\lim_{z \to \infty} c(z, \xi) = 0. \tag{58.12}$$

On the plane $(z, \xi)$, the point $z = 0$, $\xi = 0$ is a singular point of equation (17.5). Noting that $z \sim y^{3/2}\xi^{1/2}$, we see that as $\xi \to 0$, with $z \neq 0$, $y$ increases without limit.

Therefore,

$$\lim_{\substack{\xi \to 0 \\ z \neq 0}} c(z, \xi) = 0. \tag{58.13}$$

The boundary conditions in zone II take the form

$$\lim_{z \to 0}\left(z^{1/3}\frac{\partial c}{\partial z}\right) = 0, \tag{58.14}$$

$$\lim_{z \to \infty} c(z, \xi) = 0, \qquad \xi_1 < \xi < \xi_2. \tag{58.15}$$

$$\lim_{\xi \to \xi_1} c(z, \xi) = c(z, \xi_1), \tag{58.16}$$

The last condition expresses the uninterrupted nature of the concentration of the intermediate product in its transit from zone I to zone II but not on the actual surface of the disk.

Finally, in zone III, we have:

$$\lim_{z \to 0} c(z, \xi) = 0, \qquad (58.17)$$

$$\lim_{z \to \infty} c(z, \xi) = 0, \qquad \xi_2 < \xi < \xi_3. \qquad (58.18)$$

$$\lim_{\xi \to \xi_2} c(z, \xi) = c(z, \xi_2), \qquad (58.19)$$

We see that the solution for each inner zone enters the solution for the next outer zone in the form of a boundary condition. Namely, the value of $c(z, \xi)$ for zone I (the inner electrode), as $\xi \to \xi_1$, is the boundary condition for zone II (the isolation ring), while a solution $c(z, \xi)$, as $\xi \to \xi_2$, in the insulating zone II, is the boundary condition for zone III (outer electrode).

The general solution of the boundary value problem can be expressed by the contour integral (17.8).

To satisfy the boundary conditions in zone I, let us select the semi-infinite straight line $\lambda = 0$ as the contour of integration. As will be made clear later, in this case an expression is derived for $c(z, \xi)$, which, together with its derivative, differs from zero, at $z = 0$.

Assuming that in (17.8) $A(\lambda, \rho) = 0$, and designating $B(0, \rho)Q(0, \rho) = f(\rho)$, we have

$$c_{\mathrm{I}}(z, \xi) = \frac{1}{2^{1/3}\Gamma\left(\frac{2}{3}\right)} \int_0^{\xi} f(\rho)\, \chi_2(z, \xi, \rho, \lambda = 0)\, d\rho =$$

$$= \frac{1}{2^{1/3}\Gamma\left(\frac{2}{3}\right)} \int_0^{\xi} f(\rho) \exp\left\{-\frac{z^2}{4(\xi - \rho)}\right\} \frac{d\rho}{(\xi - \rho)^{2/3}}. \qquad (58.20)$$

The integrand can easily be derived by an expansion of the Bessel function $I_{-1/3}$, which enters into the expression $\chi_2(z, \xi, \rho, \lambda)$ as $\lambda \to 0$.

The function $c_{\mathrm{I}}(z, \xi)$ satisfies the convective diffusion equation (11.20) and the limit properties

I $$\lim_{\xi \to 0} c_{\mathrm{I}}(z, \xi) = 0,$$

II $$\lim_{z \to \infty} c_{\mathrm{I}}(z, \xi) = 0,$$

III $$\lim_{z \to 0} c_{\mathrm{I}}(z, \xi) = \frac{1}{2^{1/3}\Gamma\left(\frac{2}{3}\right)} \int_0^{\xi} f(\rho) \frac{d\rho}{(\xi - \rho)^{2/3}},$$

IV $$\lim_{z \to 0} \left\{-z^{1/3} \frac{\partial c_{\mathrm{I}}(r, z)}{\partial z}\right\} = f(\xi).$$

The first three are obvious. The last was demonstrated in the work of Sutton cited previously.

Comparing properties I and II with the boundary conditions (58.12) and (58.13), we see that $c_I(z, \xi)$ satisfies them automatically.

It remains for us to select the function $f(\rho)$ so that boundary condition (58.9) can be satisfied.

Introducing the function $c_I(z, \xi)$ into it, and taking the above limit properties into account, we find that the following integral expression corresponds to that condition

$$f(\xi)+\frac{2^{-1/3}}{\Gamma\left(\frac{2}{3}\right)}h(\xi)\int_0^\xi f(\rho)(\xi-\rho)^{-2/3}d\rho=g(\xi). \tag{58.21}$$

Substituting the values of $h(\xi)$ and $g(\xi)$ from (58.10) and (58.11), we have:

$$f(\xi)+a\int_0^\xi f(\rho)\frac{(\xi-\rho)^{-2/3}}{\xi^{1/3}}d\rho=\frac{b}{\xi^{1/3}}, \tag{58.22}$$

where

$$a=\frac{2^{-1/3}k}{\Gamma\left(\frac{2}{3}\right)D\sqrt[3]{\frac{9\gamma^2}{4D}}}, \qquad b=\frac{j_0}{D\sqrt[3]{\frac{9\gamma^2}{4D}}}. \tag{58.23}$$

The integral equation (58.22) admits of an exact solution that has the form

$$f(\xi)=\frac{b}{1+a\Gamma\left(\frac{1}{3}\right)\Gamma\left(\frac{2}{3}\right)}\frac{1}{\xi^{1/3}}=\frac{\alpha}{\xi^{1/3}}, \tag{58.24}$$

where

$$\alpha=\frac{b}{1+a\Gamma\left(\frac{1}{3}\right)\Gamma\left(\frac{2}{3}\right)}.$$

The validity of the solution (58.24) is confirmed by its direct substitution into (58.22).

Taking (58.24) into account, we can write the solution of equation (11.20) which satisfies boundary conditions (58.9), (58.12), and (58.13) in zone I in the form

$$c_I(z, \xi)=\Phi(z, \xi)=\frac{\alpha}{2^{1/3}\Gamma\left(\frac{2}{3}\right)}\int_0^\xi\frac{(\xi-\rho)^{-2/3}}{\rho^{1/3}}\exp\left[-\frac{z^2}{4(\xi-\rho)}\right]d\rho. \tag{58.25}$$

The integral entering into the determination of the function $c_I$ can be transformed, for the sake of convenience, by the introduction of the variable $v=\frac{\xi}{\xi-\rho}$ and the parameter $t=\frac{z^2}{4\xi}$. By designating

$$I=\int_0^{\xi}\frac{(\xi-\rho)^{-2/3}}{\rho^{1/3}}\exp\left[-\frac{z^2}{4(\xi-\rho)}\right]d\rho=\int_1^{\infty}\frac{e^{-tv}}{v(v-1)^{1/3}}\,dv,$$

and, on differentiating with respect to the parameter t, we obtain:

$$\frac{dI}{dt}=-e^{-t}\int_1^{\infty}\frac{e^{-t(v-1)}}{(v-1)^{1/3}}\,dv=-\frac{e^{-t}}{t^{2/3}}\Gamma\left(\frac{2}{3}\right).$$

Integrating and taking account of the fact that $\lim\limits_{t\to\infty} I(t) = 0$, we obtain

$$I(t)=\Gamma\left(\frac{2}{3}\right)\int_t^{\infty}\frac{e^{-x}}{x^{2/3}}\,dx=\Gamma\left(\frac{2}{3}\right)\frac{e^{-t/2}}{t^{1/3}}W_{-1/3,\,1/6}(t),$$

where $W_{-1/3,\,1/6}(t)$ is the Whittaker function.

Thus, the concentration distribution in zone I has the form

$$c_{\rm I}(z,\,\xi)=c_{\rm I}(t)=\frac{\alpha}{2^{1/3}}\,t^{-1/3}e^{-t/2}W_{-1/3,\,1/6}(t). \tag{58.26}$$

In the coordinates (r, y), we have $t=\frac{\gamma^2}{6D}y^3$, so that for c(r, y), we find

$$c_{\rm I}(r,\,y)=\frac{\alpha}{2^{1/3}}\left[\frac{\gamma^2}{6D}y^3\right]^{-1/3}\exp\left[-\frac{\gamma^2}{2\cdot 6\cdot D}y^3\right]W_{-1/3,\,1/6}\left[\frac{\gamma^2}{6D}y^3\right]. \tag{58.27}$$

The fact that the concentration of the intermediate product is not a function of the coordinate in zone I is the natural consequence of the "uniform accessibility" of the disk surface from a diffusional standpoint.

Let us pass to a derivation of the function $c(z,\,\xi)$ in zone II. In order to do this, a path of integration must be found for (17.8) so as to obtain an expression that satisfies boundary conditions (58.14) to (58.16), in particular, with a derivative that becomes zero. For this purpose let us select the straight line $\rho = \xi_1 = \text{const}$ as the integration path. Assuming in (17.8) that:

$$A(\lambda,\,\xi_1)=0,\quad B(\lambda,\,\xi_1)P(\lambda,\,\xi_1)=\varphi(\lambda,\,\xi_1),$$

let us write the concentration in zone II in the form

$$c_{\rm II}(z,\,\xi)=\int_0^{\infty}\varphi(\lambda,\,\xi_1)\chi_2(z,\,\xi,\,\lambda,\,\xi_1)\,d\lambda=$$

$$=\frac{1}{2(\xi-\xi_1)}\int_0^{\infty}z^{1/3}\lambda^{2/3}\varphi(\lambda,\,\xi_1)\exp\left[-\frac{z^2+\lambda^2}{4(\xi-\xi_1)}\right]I_{-1/3}\left(\frac{\lambda z}{2(\xi-\xi_1)}\right)d\lambda \tag{58.28}$$

$$(\xi_2>\xi>\xi_1).$$

The function $c_{II}(z, \xi)$ satisfies the limit properties

I $$\lim_{z \to \infty} c_{II}(z, \xi) = 0,$$

II $$\lim_{z \to 0} \left\{ z^{1/3} \frac{\partial c_{II}(z, \xi)}{\partial z} \right\} = 0,$$

III $$\lim_{\xi \to \xi_1} c_{II}(z, \xi) = \varphi(z, \xi_1).$$

In this way $c_{II}(z, \xi)$ automatically satisfies conditions (58.14) and (58.15). To satisfy condition (58.16), it is sufficient to state

$$\varphi(\lambda, \xi_1) = c_I(\lambda, \xi_1).$$

Substituting this value into (58.28), and using (58.26), we find $c_{II}(z, \xi)$. The function $c_{II}(z, \xi)$ can be computed to the end, if $c_I$ is presented in the form of an exponential

$$c_I(t) = \Gamma\left(\frac{1}{3}\right) \frac{\alpha}{2^{1/3}} e^{-\beta t/2},$$

by selecting the parameter $\beta \approx 7.9$ in such a way as best to approximate $c_I(t)$, given in (58.26). Here

$$c_{II}(z, \xi) \approx \alpha \cdot 2^{-5/3} \Gamma\left(\frac{1}{3}\right) \left( \frac{\xi_1}{\xi - \frac{3}{4}\xi_1} \right)^{2/3} \exp\left\{ -\frac{z^2}{4\xi\left(1 - \frac{3}{4}\frac{\xi_1}{\xi}\right)} \right\}. \qquad (58.29)$$

In the coordinates (r, y), we have in this case:

$$c_{II}(r, y) \approx \alpha \cdot 2^{-5/3} \Gamma\left(\frac{1}{3}\right) \frac{1}{\left[1 - \frac{3}{4}\frac{r^3}{r_1^3}\right]^{2/3}} \exp\left\{ -\frac{\gamma^2}{6D} \frac{y^3}{1 - \frac{3}{4}\frac{r_1^3}{r^3}} \right\}. \qquad (58.30)$$

In a completely analogous manner, we can derive the function $c_{III}(z, \xi)$ in zone III. By stating in (17.8) that $B(\lambda, \rho) = 0$, and selecting the straight line $\xi = \xi_2$ as the path of integration, and proceeding as before, we have:

$$c_{III}(z, \xi) = \frac{1}{2(\xi - \xi_2)} \int_0^\infty \lambda^{2/3} z^{1/3} c_{II}(\lambda, \xi_2) \exp\left\{ -\frac{z^2 + \lambda^2}{4(\xi - \xi_2)} \right\} I_{1/3}\left( \frac{\lambda z}{2(\xi - \xi_2)} \right) d\lambda. \qquad (58.31)$$

The function $c_{III}$ satisfies the convective diffusion equation and the boundary conditions (58.17) to (58.19).

In actual practice, however, it is more convenient to measure the total current created by the intermediate product particles over the entire surface of zone III. The flux of particles of the intermediate product over 1 cm$^2$ of the disk's surface in zone III is

$$j = D\left(\frac{\partial c_{III}}{\partial z}\right)_{z=0} = D \sqrt[3]{\frac{9\gamma^2}{4D}} \xi^{1/3} \left\{ z^{1/3} \frac{\partial C_{III}(z, \xi)}{\partial z} \right\}_{z=0}. \qquad (58.32)$$

To compute $j(\xi)$, it is more convenient not to differentiate $c_{III}(z, \xi)$, but to utilize the following relation, the derivation for which is given in the previously cited work by Sutton:

$$\lim_{z\to 0}\left\{z^{1/3}\frac{\partial c_{III}(z,\xi)}{\partial z}\right\}=\frac{2^{1/3}}{\Gamma\left(\frac{1}{3}\right)(\xi-\xi_2)^{1/3}}\int_0^\infty c_{II}\left[2\sqrt{s(\xi-\xi_2)},\ \xi_2\right]e^{-s}\,ds. \quad (58.33)$$

Substituting the expression for $c_{II}$ from (58.29) into (58.33), and noting that

$$c_{II}\left[2\sqrt{s(\xi-\xi_2)},\ \xi_2\right]=2^{-5/3}\Gamma\left(\frac{1}{3}\right)\alpha\left(\frac{\xi_1}{-\frac{3}{4}\xi_1+\xi_2}\right)^{2/3}\exp\left\{-\frac{s(\xi-\xi_2)}{-\frac{3}{4}\xi_1+\xi_2}\right\},$$

we obtain

$$\lim\left\{z^{2/3}\frac{\partial c_{III}(z,\xi)}{\partial z}\right\}=2^{-4/3}\alpha\left(\frac{\xi_1}{-\frac{3}{4}\xi_1+\xi_2}\right)^{2/3}\frac{-\frac{3}{4}\xi_1+\xi_2}{(\xi-\xi_2)^{1/3}\left(\xi-\frac{3}{4}\xi_1\right)}. \quad (58.34)$$

Utilizing (58.7), (58.23), (58.32), and (58.34), it is easy to find the flow density of intermediate product particles per unit of surface in zone III

$$j(r)=\frac{0.4j_0}{1+\frac{k\delta'}{D}}\frac{r_1^2r_2}{r^3}\frac{\left(1-\frac{3}{4}\frac{r_1^3}{r_2^3}\right)^{1/3}}{\left(1-\frac{r_2^3}{r^3}\right)^{1/3}\left(1-\frac{3}{4}\frac{r_1^3}{r^3}\right)}. \quad (58.35)$$

Under actual conditions, the width of the isolation ring is usually small, i.e., $r_2 - r_1 \ll r_1$. Let us also assume that $r_3 - r_2 \ll r_2$. Expression (58.35) becomes simplified:

$$j(r)\approx\frac{j_0}{1+\frac{k\delta'}{D}}\frac{r_1^2r_2}{r^3\left[1-\frac{r_2^3}{r^3}\right]^{1/3}}. \quad (58.36)$$

Let us find the value of the total intermediate product flow on the entire circular surface of zone III

$$I=2\pi\int_{r_2}^{r_3}j(r)\,r\,dr\approx\frac{0{,}8\left(1-\frac{3}{4}\frac{r_1^3}{r_2^3}\right)^{1/3}I_0}{1+\frac{k\delta'}{D}}\int_1^{r_3/r_2}\frac{y^2\,dy}{(y^3-1)^{1/3}\left(y^3-\frac{3}{4}\frac{r_1^3}{r_2^3}\right)}. \quad (58.37)$$

As should have been expected, this flow is proportional to the total amount of the intermediate product discharged in zone I (i.e., $\sim I_0$). The dependence of the flow on the quantities $\nu$ and $\omega$, which determine the convective motion of the liquid, is conditioned by the thickness $\delta'$ of the diffusion boundary layer.

The greater is $\delta'$, all other conditions being equal, the weaker is the transfer of substance and the smaller is the flow to the surface of zone III.

An increase in the decomposition constant k also is followed by a decrease in total flow onto zone III.

The constant k can be determined from formula (58.37) by means of direct measurement of the quantities.*

## 59. DISSOLUTION OF HOMOGENEOUS METALS IN ACIDS. DISSOLUTION OF INCLUSIONS

The problem of dissolution of metals in acids and problems related to corrosion processes are extremely complex. Those who are interested in a more profound study of the modern theory of dissolution as well as of the theory of corrosion are referred to specialized publications [33]. Here we shall merely examine certain problems of the dissolution of metals that, by their nature, are directly related to physicochemical hydrodynamics.

If a metal M is immersed in an electrolytic solution containing ions of another metal N (including hydrogen ions), then the transfer of M ions into the solution is possible in principle. This is accompanied by the displacement of N ions from the solution. For this to occur, it is necessary for the equilibrium potential of metal M to be in a range of voltages higher (it should have a more negative potential) than metal N in the same solution.

Of greatest interest is the case of dissolution accompanied by the liberation of hydrogen (dissolution of metals in acids). If the dissolution occurs in an electrolytic cell in which the dissolving metal serves as the anode, then the metal is said to be dissolved in a current. If the dissolution is not accompanied by electrolysis, as, for example, in the case of a metal immersed in an acid, then the metal is said to be self-dissolving. The latter case is the one to which we limit our study, since electrolysis constitutes an additional complicating circumstance that does not alter the principles of the phenomenon.

In discussing the problem of self-dissolving metals we shall avoid complications which arise in the dissolution of metals whose surface is not homogeneous, but contains various admixtures and

---

*Note in proofreading. Formula (58.37) is in agreement with experimental data obtained in reference [67].

inclusions. In actuality, metals with a homogeneous surface do not exist. However, as shown by N. D. Tomashov [34], when the dimensions of the inclusions and nonhomogeneities on metal surfaces are very small compared to the thickness of the diffusion layer, the metal will behave from a diffusion standpoint as a completely homogeneous metal. The dissolution rate of a homogeneous (in that sense of the word) metal in a solution without an oxidizer is determined by the slow step of the process, which is composed of a kinetic stage — phenomena which occur on the very surface of the metal — and a diffusion stage, which is related to the removal of the reaction products from the surface.

The first systematic study of the dissolution of metals in acids was conducted by N. N. Kayander [35]. He determined that the slow step, which controls the dissolution rate for a series of metals in different acids, is the kinetic stage in the process of transfer of metal particles into the solution. This conclusion was verified later many times experimentally. At the present time, we may consider it as an established fact that the dissolution rate of most homogeneous metals in acids, in the absence of an oxidizer, relates to the kinetic stage of the process.

The theory of dissolution of metals having a homogeneous surface was first propounded by A. N. Frumkin [36]. Further theoretical and experimental studies of that theory by the author and others [36-40] have, in the main, served to confirm the correctness of the indicated theory.

The study of the diffusion kinetics of dissolving metals was first undertaken by A. P. Shchukarev, in a previously cited study (see Chapter II). The systematic studies by Van Name [41] established for certain conditions that the rate of dissolution in acids of such metals as cadmium, zinc, copper, silver, nickel, and others is limited by the diffusion stage of the process.

Recently, in the studies by Ya. V. Durdin and Z. U. Dukhnyakova [42] and L. Bircumshaw and A. Riddiford [43], a systematic and very complete study of the dissolution of magnesium, manganese, and zinc was conducted.

The study by Ya. V. Durdin and Z. U. Dukhnyakov [42] involved the dissolution of manganese and magnesium in hydrochloric acid and in organic acids, namely acetic and formic.

The liberation of gaseous hydrogen and the transfer of metal atoms into the solution are the rapid stages of the process, so that the over-all rate of the process is determined by the rate at which the hydrogen ions in the case of hydrochloric acid, and of acid in the case of an organic acid, are conveyed to the surface of the dissolving sample. The rate of dissolution of the sample is determined by the formula

$$I = \frac{D_a c_a}{\delta_a},$$

where $c_\alpha$ and $D_\alpha$ are the concentration and the diffusion coefficient of hydrogen ions in the first case and of the organic acid in the second. The value of $\delta_\alpha$, in the case where the dissolving sample is a rotating disk, is given by formula (11.31), in which the diffusion coefficient has a corresponding meaning. The dissolving magnesium and manganese disks were rotated at velocities of 5 to 25 rev/sec. An exact centering of the disk assured good reproducibility in the experiments, much better than in the case of experiments involving the dissolution of motionless samples. The average error of the measurements was about 2 percent.

The authors observed corrosion contours on the disks that accurately reproduced the fluid streamlines. Figure 11b (see p. 66) illustrates the corrosion contours on the surface of a disk of manganese dissolving in 0.1 N HCl.

What was studied was the dependence of the rate of dissolution on the concentration of the acid, on the rotating velocity of the disk, and on the magnitude of the diffusion coefficient. A quantitative comparison of the absolute magnitudes of theoretical and experimental data was also made. A close proportionality was observed between the rate of dissolution and the concentration of hydrochloric acid for the dissolution of magnesium and of manganese in HCl, and in organic acids analogous results were obtained.

The dependence of the rate of dissolution on the rate of rotation of the disk is illustrated in Table 13, where v is the rate of dissolution based on the amount of hydrogen liberated per 1 cm$^2$ of the disk surface, and n is the exponent for the angular velocity in formula (11.30).

Table 13

| Solution | Metal | v · 100 average | | | | | | n average |
|---|---|---|---|---|---|---|---|---|
| | | $\omega$ = 300 rpm | 500 rpm | 750 rpm | 900 rpm | 1200 rpm | 1500 rpm | |
| 0.1 N HCl + 0.5 N KCl | Mg | 107 | 132 | 164 | 172 | 207 | 225 | 0.52 |
| | Mn | 95 | 121 | 151 | 168 | 211 | 221 | |
| 0.1 N HCl | Mg | 88 | 109 | 133 | 146 | 170 | 189 | 0.49 |
| 0.15 N HCOOH | Mg | | 73.5 | 94.5 | 108 | 122 | | 0.55 |
| 0.15 N $CH_3COOH$ | Mg | 43 | 57 | 69 | 76 | 80 | 94 | 0.51 |

It will be seen from Table 13 that the value of the exponent n is not a function of the nature of the metal or the composition of the acid, and is in quite good agreement with the theoretical value n = 1/2. The dependence of the dissolution rate of magnesium in various acids on the diffusion coefficient is illustrated in Figure 65.

According to (8.15) and (11.32) the dissolution rate should be proportional to $D^{2/3}$ – the straight line in Figure 65. All experimental points fit well on the theoretical line.

The study by Ya. V. Durdin and Z. U. Dukhnyakova was the first absolutely quantitative comparison made for theoretically computed and measured values of dissolution rates of magnesium in various acids. The results of this comparison are given in Table 14. The main difficulty in such a comparison is the absence of sufficiently trustworthy values for the diffusion coefficients.

It will be seen from Table 14 that the computed and observed absolute values for dissolution rates are close. The measured values, however, are higher than the computed values by 15%. The authors ascribe the systematic difference between theoretical and measured values to an increase in the true surface area of the dissolving sample because of the appearance of corrosion contours, and to insufficient accuracy in the diffusion coefficient values utilized in the derivation of formula (8.15).

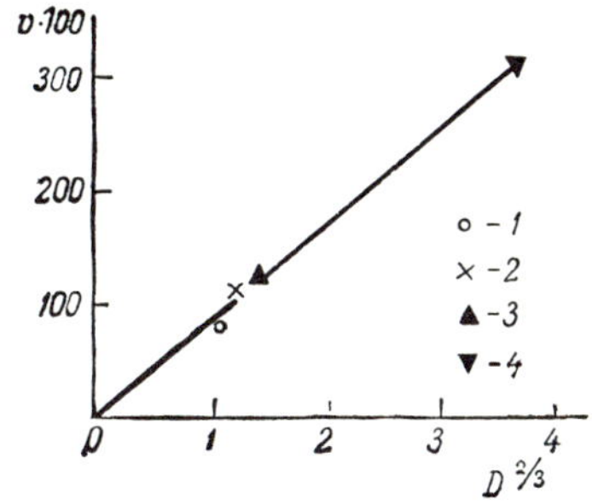

Figure 65. Dependence of the dissolution rate of Mg in acids on the diffusion coefficients of the acids: 1 - $CH_3COOH$; 2 - HCOOH; 3 - $(COOH)_2$; 4 - HCl.

The dissolution of zinc in an aqueous solution of iodine was studied by L. Bircumshaw and A. Riddiford. The dissolving sample was a disk of rather considerable thickness rotating on a shaft.

Table 14

| Acid | Acid normality | $v \cdot 10^7$ theoretical | $v \cdot 10^7$ experimental | Δ,% |
|---|---|---|---|---|
| HCl . . . . . . . . | 0.1 | 26.3 | 31.2 | 16 |
| $CH_3COOH$. . . . | 0.15 | 11.0 | 11.8 | 6 |
| $(COOH)_2$. . . . . | 0.15 | 13.7 | 18.3 | 22 |
| HCOOH . . . . . | 0.15 | 13.1 | 16.6 | 18 |

Changing the composition of the solution and the velocity of the rotation permitted variations in the Reynolds and Prandtl numbers over very wide limits. The Reynolds number varied from 142 to 5420. The Prandtl numbers varied from 367 to 87,300. The dependence of the dissolution rate on the Reynolds and Prandtl numbers was measured, as well as its dependence on the angular velocity $\omega$ and on the viscosity of the solution – quantities which enter into the dimensionless Reynolds and Prandtl numbers. The average measurement error was 3.4%. The results obtained by the authors are given in Table 15 and Figure 66.

Table 15

| $\omega$, rad·sec$^{-1}$ ...... | 5.23 | 7.85 | 10.5 | 15.7 | 20.9 |
|---|---|---|---|---|---|
| $10^3\ \delta$, cm (exp.) .... | 7.98 | 6.40 | 5.61 | 4.37 | 3.72 |
| $10^3\ \delta$, cm (theor.) ... | 7.36 | 6.01 | 5.20 | 4.25 | 3.68 |

The table gives the values of the diffusion layer thicknesses computed by formula (11.31) by using experimentally determined dissolution rates.

A comparison of the results shows that good agreement is often obtained between theoretical and experimental quantities. At high rotational velocities, the discrepancy is within the limits of the margin of error of the measurements.

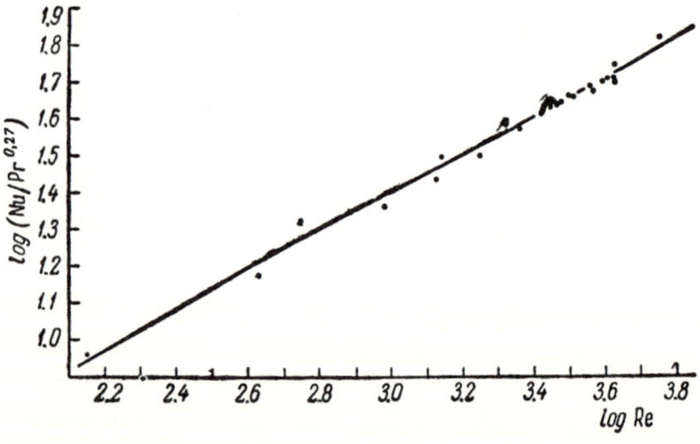


Figure 66. Bircumshaw and Riddiford experiments for the dissolution of zinc.

Changing the composition of the solution (from an iodine concentration of 0.002 to 0.16 gram-equivalents per liter) permitted varying its viscosity over a wide range. In this case the observed change in dissolution rate shows it to be proportional to $\nu^{-0.76}$. The theoretical value. according to (11.32) is $\sim\nu^{-0.83}$. The results of the measurements have been reduced by the authors to the empirical correlation

$$\mathrm{Nu} = 0.56\,\mathrm{Re}^{0.56}\,\mathrm{Pr}^{0.27}.$$

Comparing this expression with (11.32), we see that the qualitative agreement is fully adequate. The power of Re, however, is somewhat higher and that of Pr is somewhat lower than the theoretical values.

As noted by the authors, these differences are related to a certain (although slight) amount of turbulence in the fluid that was observed under the conditions of the experiment. This turbulence was created, in particular, by the holder to which the dissolving sample was attached.

In the studies cited above, the rate of dissolution of the metals was controlled by purely diffusional factors. Very often, however, changing the conditions of the experiment leads to a transition of the process from the diffusion zone into the kinetic zone. For example, in the study by Halpern [44], in which the dissolution of copper in an aqueous solution of ammonia was studied [$Cu + 4NH_3 + H_2O + \frac{1}{2}O_2 \rightarrow 2OH^- + Cu(NH_3)_4^{++}$], the dissolving sample was a disk of $\approx$ 0.8 cm radius, rotating at angular velocities from ~ 8 to 13.7 rev/sec. Depending on the pressure of oxygen in the vessel above the surface of the solution, the nature of the kinetics of the dissolution varied. Table 16 shows the relative dissolution rates measured for the two pressures as computed by formula (11.32).

Table 16

| Disk rotational velocity, rev/sec | Relative dissolution rate (measured) | | Computed |
|---|---|---|---|
| | low pressures $P_{O_2}$ = 1.4 atm | high pressures $P_{O_2}$ = 7.8 atm | |
| 8 | 1 | 2 | 1 |
| 9.1 | 1.2 | — | — |
| 11 | 1.3 | 2 | 1.2 |
| 13.7 | 1.4 | 2 | 1.3 |

It will be seen from the table that, within the limits of accuracy of the experiment, the dissolution rate at low pressures is proportional to the square root of the angular velocity and is not a function of the velocity at high pressures (the process is transferred into the kinetic zone).

If the solution contains a substance capable of reacting with hydrogen ions on the electrode (oxidizer), the picture of the dissolution changes radically. In this case, the slow discharge of hydrogen ions is replaced by the ionization of oxygen.

For the same potentials, the rate of oxygen ionization is much greater than that of the discharge of hydrogen ions. Therefore, in the presence of an oxidizer, the dissolution rate for many metals is determined by the rate at which the oxidizer is conveyed to the surface of the metal. As an example, the dissolution of zinc in acid may be cited.

In these cases, the dissolution rate is equal to the diffusional flow of the oxidizer to the surface of the dissolving sample, taking into account the number of oxidizer atoms required to oxidize an ion of hydrogen which must be formed on the surface for an atom of the metal to pass into the solution. This was experimentally confirmed by King and his collaborators [16] and by Z. U. Dukhnyakova, who showed that the rate of dissolution of zinc in the presence of an oxidizer, under the same conditions of stirring, agrees with that of the dissolution of magnesium and manganese and is determined by formula (11.32).

Another instance of dissolution of homogeneous metals in which diffusional processes play an important role is the process of dissolution for an initial zero concentration of metal ions in the solution, which has been studied by Ya. M. Kolotyrkin [45]. It was established that, under stationary conditions, the dissolution in this instance occurs in the mixed zone, where the rate of exchange of particles between the solid (metal) and fluid phases is equal to the rate of removal of the dissolved metal particles

$$I = j_{\text{diff}} = \frac{Dc_1}{\delta},$$

where I is the over-all rate at which the metal particles pass into the solution, $j_{\text{diff}}$ is the diffusional flux of dissolved particles, and $c_1$ is the concentration of a saturated solution at the surface of the metal.

The concentration of metal ions in the bulk of the solution is set equal to zero by convention. The dependence of the concentration $c_1$ on the potential difference $\Delta\varphi$ at the metal-solution interface is determined by the equilibrium formula (44.3).

At the same time, the dissolution rate I is equal to the rate of liberation of hydrogen, determined by the Tafel formula. Therefore, we may write

$$\frac{Dc_1}{\delta} = \text{const } e^{-\frac{F\,\Delta\varphi}{2RT}}. \tag{59.1}$$

Introducing the value of $c_1$ expressed through $\Delta\varphi$ by formula (44.3), assuming that the dissolving sample is in the shape of a disk rotating at an angular velocity $\omega$, and taking into account the fact that $\delta$, according to expression (11.31), is a function of $\omega$, we obtain:

$$\sqrt{\omega}\, e^{\frac{n_{\text{M}} F\,\Delta\varphi}{RT}} = \text{const } e^{-\frac{F\,\Delta\varphi}{2RT}}.$$

From this we can derive the steady-state value of the potential difference of the dissolving metal with a valence of $n_{\text{M}}$

$$\Delta\varphi = -\frac{RT}{F}\frac{1}{n_{\text{M}} + \frac{1}{2}} \ln \sqrt{\omega} + \text{const.} \tag{59.2}$$

The quantity $\Delta\varphi$ is found to be proportional to $l\text{n}\sqrt{\omega}$.

Thus, the theoretical formulas derived by us may be successfully applied to the computation of dissolution rates, in cases where these rates are controlled by the diffusion stage of the process.

In actual practice one seldom meets the dissolution of absolutely pure metals containing no admixtures or foreign inclusions.

The presence of the latter in the body of the metal materially affects its solubility.

The significance of inclusions consists in that they form a galvanic cell with the major metal. If the overpotential for the discharge of hydrogen on the inclusion is less than on the major metal, the inclusion will act as the cathode of that cell. The discharge of hydrogen takes place on the cathode. The major metal, acting as the anode, passes into the solution.

The action of such galvanic cells, known as local cells, constitutes the chief mechanism for the dissolution of nonhomogeneous and industrial metals. The concept of local cells was first put forward by N. N. Kayander. The rate of dissolution is determined by the value of the total current flowing in the system of local cells acting on metal surfaces.

We cannot here set forth all of the problems concerning the mechanism of local cells. There is much published literature on the subject [46]. Let us merely note that the dimensions of the inhomogeneities encountered in actual practice vary over very wide limits — from molecular to macroscopic (of the order of several centimeters). The dissolution of inhomogeneous bodies and compositions of several metals are also frequently encountered. In this connection, the value of the current flowing through the local element varies over wide limits. In the case of large inclusions, the size of the ohmic resistance of the current path is very large, and the current flowing through the local cell is small compared to the limiting current. The value of the current, in this case, is determined by purely electrical factors. On the other hand, when the dimensions of the inclusions are not large, but very small compared to the dimensions of the dissolving sample, the ohmic resistance of the path is relatively low. The mechanism of dissolution of industrial metals is determined by a multitude of factors. In the case of large inhomogeneities, the rate of diffusion of hydrogen ions or of the oxidizer atoms to the inclusion on the surface of the metal may become the controlling factor.

Let us assume that the inclusion is a band of width h in the body of the metal. We shall assume the latter to be a semi-infinite plate in a laminar flow regime. Let us designate by $l$ the distance from the leading edge of the plate to the edge of the inclusion.

Under these conditions, the problem of finding the diffusional flow on the inclusion is equivalent to the problem of the flow of substance on a coated plate, examined in Section 18.

Formula (18.18) gives:

$$j=\frac{j_{\lim}}{\left[1-\left(\frac{l}{x}\right)^{3/4}\right]^{1/3}},$$

where $j_{\lim}$ is determined by formula (15.11):

Therefore

$$j=0.3\frac{Dc_0\sqrt{U}}{\sqrt{\nu}\left(\frac{D}{\nu}\right)^{1/8}\sqrt{x}}\frac{1}{\left[1-\left(\frac{l}{x}\right)^{3/4}\right]^{1/3}}. \tag{59.3}$$

The total flow onto a unit of inclusion surface is

$$I=\frac{0.3\,Dc_0}{\left(\frac{D}{\nu}\right)^{1/3}\sqrt{\nu}}\int_l^{l+h}\frac{\sqrt{U}}{\sqrt{x}}\cdot\frac{1}{\left[1-\left(\frac{l}{x}\right)^{3/4}\right]^{1/3}}\,dx. \tag{59.4}$$

Introducing the variable $x = l + y$, we obtain:

$$I=\frac{0.3\,Dc_0\sqrt{U}}{\left(\frac{D}{\nu}\right)^{1/3}\sqrt{\nu}}\cdot\frac{1}{\sqrt{l}}\int_0^h\frac{dy}{\sqrt{1+\frac{y}{l}}\left[1-\frac{1}{\left(1+\frac{y}{l}\right)^{3/4}}\right]^{1/3}}. \tag{59.5}$$

Assuming that $h \ll l$, the integrand can be expanded into a series, and we can write:

$$I\approx\frac{0.3\,Dc_0\sqrt{U}}{\sqrt{\nu}\left(\frac{D}{\nu}\right)^{1/3}}\cdot\frac{h^{2/3}}{\sqrt{l}}\cdot\frac{3l^{1/3}}{2\left(\frac{3}{4}\right)^{1/4}}\approx\frac{Dc_0\sqrt{U}}{\sqrt{\nu}}\cdot\frac{h^{2/3}}{l^{1/6}}\cdot\frac{1}{2}\left(\frac{\nu}{D}\right)^{1/3}. \tag{59.6}$$

In actual practice a local cell often functions under conditions of natural convection. In that case, substituting, instead of $j_{\lim}$, its value according to expression (23,29), we obtain, similarly to (59.6), the diffusional flow on the inclusion in a metal plate oriented vertically,

$$I\approx 0.7\,\mathrm{Pr}^{3/4}D\left[\frac{gc_0}{4\nu^2\rho}\left(\frac{\partial c}{\partial\rho}\right)\right]^{1/4}c_0\cdot\frac{3}{2}\cdot\frac{l^{\frac{1}{12}}h^{2/3}}{\left(\frac{3}{4}\right)^{1/3}}.$$

The diffusional flow on an inclusion in a laminar flow regime depends on the position of the inclusion (its distance from the edge of the plate), and is proportional to $h^{2/3}$ under both forced and natural convection.

The situation changes under conditions of a turbulent stirring of the fluid. Relaxation phenomena here are blurred by random turbulence eddies, so that the diffusional flow on the inclusion in this case is proportional to its area. It remains dependent, however, on the position of the inclusion on the metal surface. In cases where the inclusions are contained within a dissolving plate, this dependence may be determined by formula (26.1). The dependence of the rate of dissolution on the position of the inclusion is less pronounced than in a laminar flow. The characteristic expressions derived above for the flow to an inclusion determine the rate of dissolution of an inhomogeneous metal, in cases where the rate is controlled only by diffusion of the particles to the inclusions. Certain other cases of dissolution — dissolution of protuberances, corners, and cavities — have been examined in Section 29.

## 60. STUDY OF KINETIC AND CATALYTIC ELECTROCHEMICAL PROCESSES BY MEANS OF A ROTATING DISK ELECTRODE*

Development of the kinetics of electrochemical reactions in recent years has been marked by the start of a systematic study of electrode reactions, the course of which is closely related to accompanying chemical reactions in the bulk of the solution. The latter are usually divided into two classes.

The first are the so-called kinetic reactions. These are usually known as processes in which the electrode reaction is partially supported by products of the chemical reaction in the bulk of the solution. One of these products diffuses to the electrode and there enters into an electrode reaction.

The principle of such a process may be written in the form

| bulk of the solution | electrode |
|---|---|
| $n\mathrm{A} \underset{k_2}{\overset{k_1}{\rightleftarrows}} m\mathrm{B}$ | $\mathrm{B} \longrightarrow \mathrm{C},$ |

where A, B, and C represent substances entering into the reaction in the appropriate stoichiometric ratios determined by the coefficients n and m. The B substance is reduced on the electrode. The A substance, at the given potential, is not reduced on the electrode. Finally, $k_1$ and $k_2$ are reaction rate constants.

Catalytic reactions are usually known as reactions in which the substance participating in the electrode reaction is partially

---

*The contents of this section are based on the joint study by J. Koutecký and the author [70].

regenerated as a result of reaction in the bulk of the solution, in which the products of the electrode reaction participate.

The principle of such reactions may be written in the form

| bulk of the solution | electrode |
|---|---|
| $nA + mB \rightarrow C$ | $C \rightarrow lA.$ |

Substance C is transformed into substance A as a result of the electrode reaction. Substance A passes into the bulk of the solution and reacts there with substance B, regenerating C.

The net rate of the process of reduction of substance C is determined by the diffusion of the corresponding reactants and the rate at which A is oxidized by B.

The systematic study of kinetic and catalytic processes, started by Czech researchers, has so far been conducted by means of a dropping mercury electrode.

The general theory of the processes that take place in this device has been developed in the studies of J. Koutecký. These are briefly described in Section 104.

There are reasons to believe, however, that a rotating disk electrode, as a device for the study of kinetic and catalytic currents, has certain advantages as compared to a dropping electrode: the processes on a rotating disk take place under conditions of a stationary regime.

This latter circumstance permits computed formulas to be derived with ease for relatively complex reactions. The existence of such formulas is a necessary prerequisite in the study of the rate of chemical reactions. At the same time, the study of stationary processes is simpler from an experimental standpoint.

Our objective is not to cover all the various types of kinetic and catalytic processes. However, we do cover typical techniques for the solution of problems that arise in the study of kinetic and catalytic processes, in the light of a few examples.

As the first example, let us examine the kinetic process that follows the following course

| bulk of the solution | electrode |
|---|---|
| $A \underset{k_2}{\overset{k_1}{\rightleftarrows}} B$ | $A \longrightarrow C.$ |

The reaction is of pseudo-monomolecular nature in the sense that the rate constants of the reaction $A \rightleftarrows B$, in this case, may be functions of the concentration of a certain substance D present in the solution:

$$k_2 = k_2 [D].$$

Later we introduce the designations

$$k_2 = \rho, \quad k_1 = \rho\sigma, \text{ where } \sigma = \frac{k_1}{k_2\,[D]}.$$

If $c_1$ and $c_2$ are the concentrations of substances A and B, respectively, in the bulk of the solution far from the electrode, then the equilibrium condition can be written in the form

$$c_2^{(0)} - \sigma c_1^{(0)} = 0. \tag{60.1}$$

Here, the over-all concentration of the solution has a given value

$$c_1^{(0)} + c_2^{(0)} = c_0. \tag{60.2}$$

Let us now examine the equations for the distribution of concentration near the reaction surface, that is, at the rotating disk $y = 0$.

A convective transfer of particles occurs here simultaneously with the reaction in the bulk of the solution, during the course of which molecules A and B appear and disappear.

The equilibrium near the reaction surface will be disturbed since substance A disappears at the surface itself as a result of the electrode reaction.

The strength of the source terms of A and B, respectively, in the solution proper can be written

$$Q_1 = \rho c_2 - \sigma\rho c_1, \qquad Q_2 = -\rho c_2 + \sigma\rho c_1,$$

where $\rho c_2$ is the number of B molecules that disappear (equal to the number of A molecules appearing) in a given place in the solution, per 1 $cm^3$ in 1 sec, and $\rho\sigma c_1$ has a similar meaning.

The convective diffusion equations, in the presence of sources in the bulk of the solution, according to (8.10), can be written in the form

$$v\frac{dc_1}{dy} = D\frac{d^2c_1}{dy^2} + \rho(c_2 - \sigma c_1), \tag{60.3}$$

$$v\frac{dc_2}{dy} = D\frac{d^2c_2}{dy^2} - \rho(c_2 - \sigma c_1). \tag{60.4}$$

In writing (60.3) and (60.4), we took advantage of the special property of a rotating disk, the surface of which is uniformly accessible from a diffusional standpoint provided edge effects are neglected. The concentrations $c_1$ and $c_2$ can be considered independent of the coordinates r and $\varphi$ in the cylindrical coordinate system (y, r, $\varphi$).

Furthermore, to simplify computations, we have stated that the diffusion coefficients of particles A and B are equal. In actual practice, they often have values that are close to each other. The case of $D_1 \neq D_2$, in principle, does not differ from the case examined below, but it requires more lengthy computations.*

The relations (60.1) and (60.2) serve as boundary conditions at infinity. Let us now formulate the boundary conditions on the surface of the disk. We shall suppose that the electrode reaction is very rapid, so that concentration $c_1$ of A at the surface of the disk satisfies the condition

$$c_1 = 0 \quad \text{at} \quad y = 0; \tag{60.5}$$

i.e., we confine the case to a limiting diffusional flow regime.

Substance B does not appear nor does it disappear on the electrode. Its concentration $c_2$, therefore, satisfies the condition

$$\frac{dc_2}{dy} = 0 \quad \text{at} \quad y = 0. \tag{60.6}$$

The solution of the system of equations (60.3) and (60.4) can be easily obtained for the indicated boundary conditions if the chemical reaction in the bulk of the solution is sufficiently rapid. The meaning of this condition will be presented with greater rigor below.

It is convenient, for a solution of the system, to introduce the dependent variables:

$$\varphi = \rho(c_2 - \sigma c_1), \tag{60.7}$$

$$\psi = c_1 + c_2. \tag{60.8}$$

Adding (60.3) and (60.4), we find that the function $\psi$ satisfies the equation

$$v\frac{d\psi}{dy} = D\frac{d^2\psi}{dy^2}. \tag{60.9}$$

Multiplying (60.3) by $\sigma$ and subtracting it from (60.4), we find

$$v\frac{d\varphi}{dy} = D\frac{d^2\varphi}{dy^2} - \alpha\varphi, \tag{60.10}$$

where

$$\alpha = \rho(1 + \sigma) = k_2\left(1 + \frac{k_1}{k_2}\right). \tag{60.11}$$

Let us now write the boundary conditions for the new unknown functions $\varphi$ and $\psi$. It follows from (60.2) that:

$$\psi \to c_0 \quad \text{as} \quad y \to \infty, \tag{60.12}$$

while (60.1) gives:

$$\varphi \to 0 \quad \text{as} \quad y \to \infty. \tag{60.13}$$

---

*The case of $D_1 \neq D_2$ has been examined by R. R. Dogonadze [63].

On the reaction surface, in view of (60.5), we find:

$$\varphi=\rho\psi \quad \text{at} \quad y=0. \tag{60.14}$$

From (60.6) we derive:

$$\left(\frac{d\psi}{dy}\right)_{y=0}=\left(\frac{dc_1}{dy}\right)_{y=0},$$
$$\left(\frac{d\varphi}{dy}\right)_{y=0}=-\rho\sigma\left(\frac{dc_1}{dy}\right)_{y=0}.$$

Therefore

$$\left(\frac{d\varphi}{dy}\right)=-\rho\sigma\left(\frac{d\psi}{dy}\right) \quad \text{at} \quad y=0. \tag{60.15}$$

On the strength of formula (11.26), the general solution for (60.9) can be written in the form

$$\psi=a_1\int_0^y \exp\left\{\int_0^t \frac{v(t')\,dt'}{D}\right\}dt+a_2,$$

where $a_1$ and $a_2$ are arbitrary constants.

Taking into account the boundary condition (60.12) and introducing the notation of Section 11, we have:

$$a_2=c_0-a_1\delta,$$

so that

$$\psi=a_1\int_0^y \exp\left\{\int_0^t \frac{v(t')\,dt'}{D}\right\}dt+(c_0-a_1\delta). \tag{60.16}$$

In order to solve (60.10), let us assume that the rate of the chemical reaction in the bulk of the solution is so great that the following inequality is always valid

$$v\frac{d\varphi}{dy}\ll\alpha\varphi. \tag{60.17}$$

Instead of (60.10), we may then write $\varphi\approx\varphi_0$, where $\varphi_0$ satisfies the equation without the convection term

$$\frac{d^2\varphi_0}{dy^2}-\frac{\alpha}{D}\varphi_0=0. \tag{60.18}$$

The solution of (60.18), taking account of (60.13), has the form

$$\varphi_0=a_3e^{-\sqrt{\frac{\alpha}{D}}\,y}. \tag{60.19}$$

Boundary conditions (60.14) and (60.15) give:

$$a_1 = \frac{c_0}{\left(\frac{\sigma}{\delta}\sqrt{\frac{D}{\alpha}}+1\right)\delta}, \quad a_3 = \rho(c_0 - a_1\delta)$$

and

$$\psi = \frac{c_0\left(\frac{1}{\sigma}\sqrt{\frac{\alpha}{D}}\right)}{\left(1+\frac{\delta}{\sigma}\sqrt{\frac{\alpha}{D}}\right)} \int_0^y \exp\left\{\frac{1}{D}\int_0^t v(t')\,dt'\right\} dt + (c_0 - a_1\delta), \qquad (60.20)$$

$$\varphi_0 = \rho(c_0 - a_1\delta)\exp\left\{-\sqrt{\frac{\alpha}{D}}\,y\right\}. \qquad (60.21)$$

The distribution of concentrations $c_1$ and $c_2$ can be easily derived from formulas (60.20) and (60.21):

$$c_1 = \frac{\rho\psi - \varphi_0}{\alpha},$$

$$c_2 = \frac{\rho\sigma\psi + \varphi_0}{\alpha}.$$

Of fundamental interest is the over-all rate of the reaction, expressed by the formula

$$j = D\left(\frac{dc_1}{dy}\right)_{y=0} = D\left(\frac{d\psi}{dy}\right)_{y=0} = Da_1 = \frac{Dc_0}{\left(\frac{\sigma}{\delta}\sqrt{\frac{D}{\alpha}}+1\right)\delta} = \frac{j_0}{1+\frac{\sigma}{\delta}\sqrt{\frac{D}{\alpha}}},$$

where $j_0$ is the diffusional flux determined by formula (11.30).

By analogy with $\delta$, the quantity $\sqrt{\frac{D}{a}} = \delta_k$ can be termed the thickness of the kinetic layer. The quantity $\varphi$ becomes practically zero at a distance $\delta_k$ from the reaction surface $y = 0$. This indicates that outside a layer of thickness $\delta_k$ the chemical reaction in the bulk of the solution proceeds in an equilibrium manner. It is only within a layer of the solution whose thickness is $\delta_k$ on the surface of the electrode that a kinetic (nonequilibrium) chemical process takes place.

In the case of a disk, both $\delta_k$ and $\delta$ have constant values over its entire surface. By means of $\delta_k$ and $\delta$, j can be written in the form

$$j = \frac{j_0}{1+\sigma\frac{\delta_k}{\delta}}.$$

Substituting the values of $j_0$ and $\delta$, we obtain:

$$j = \frac{Dc_0}{1.61\left(\frac{D}{\nu}\right)^{1/3}\sqrt{\frac{\nu}{\omega}}\left(1+\frac{\sigma}{1.61\left(\frac{D}{\nu}\right)^{1/3}\sqrt{\frac{\nu}{\omega}}}\sqrt{\frac{D}{\alpha}}\right)} = \tag{60.22}$$

$$= \frac{Dc_0}{1.61\left(\frac{D}{\nu}\right)^{1/3}\sqrt{\frac{\nu}{\omega}}\left[1+\frac{\sigma}{1.61}\sqrt{\frac{\omega}{\alpha}}\left(\frac{D}{\nu}\right)^{1/3}\right]}.$$

To clarify the actual meaning of condition (60.17), a correction for $\varphi$ must be computed in the next approximation. Let us state

$$\varphi = \varphi_0 + \varphi_1,$$

where $\varphi_1 \ll \varphi_0$ and satisfies the equation

$$D\frac{d^2\varphi_1}{dy^2} - \alpha\varphi_1 = v\frac{d\varphi_0}{dy}. \tag{60.23}$$

Substituting for v(y) its value from formula (11.18), we find for $\varphi_1$ the expression

$$\varphi_1 = -\frac{1}{4}\frac{\omega^{3/2}}{\nu^{1/2}\alpha}a_3 y e^{-\sqrt{\frac{\alpha}{D}}y}\left[\frac{1}{2}+\frac{1}{2}\sqrt{\frac{\alpha}{D}}y+\frac{1}{3}\left(\frac{\alpha}{D}\right)y^2\right].$$

With the help of $\varphi_1$ a more exact expression for the boundary condition on the surface of the disk may be obtained.

Elementary calculations lead to the condition

$$-\rho\sigma\left(\frac{d\psi}{dy}\right)_{y=0} = \left(\frac{d\varphi_0}{dy}\right)_{y=0}\left[1+\frac{1}{8}\sqrt{\frac{\omega^3}{\alpha^3}\left(\frac{D}{\nu}\right)}\right].$$

We see from the latter equation that the solution derived above, in which the convective term was omitted, is valid, provided the following inequality is fulfilled

$$\frac{1}{8}\sqrt{\frac{\omega^3}{\alpha^3}\left(\frac{D}{\nu}\right)} \ll 1. \tag{60.24}$$

Its meaning is simple: if the fact is taken into account that the major variation of the function $\varphi$ occurs in the thickness of the kinetic sublayer $\delta_k$, the inequality (60.17), taking account of (11.18), can be presented in the form

$$\sqrt{\nu\omega}\left(\sqrt{\frac{\omega}{\nu}}\delta_k\right)^2\frac{\varphi_0}{\delta_k} \ll \alpha\varphi_0,$$

or, substituting for $\delta_k$ its value, we find

$$\left(\frac{\omega}{\alpha}\right)^{3/2}\left(\frac{D}{\nu}\right)^{1/2} \ll 1, \tag{60.25}$$

which, in terms of an order of magnitude, coincides with (60.24).

It follows from inequality (60.24) that in the case of a rapid chemical reaction, the principal change in $\varphi$ occurs at a distance $\delta_k$ from the surface, which is small compared to the thickness $\delta$ of the diffusion layer. Since, at a distance $y \ll \delta$, the convective transfer of substance is small, we can neglect the convective term on the left side of equation (60.10).

It should be noted that the rate of chemical reactions in the bulk of the solution is usually high, so that inequality (60.24) is fulfilled at all reasonable values of disk velocity.

Let us now examine another typical example of a kinetic process. Let us assume that the reaction in the bulk of the solution is determined by

$$2A \underset{k_2}{\overset{k_1}{\rightleftarrows}} B$$

and is, therefore, of a bimolecular nature. In this case, the source strengths in the bulk have the form

$$Q_1 = k_2 c_2 - k_1 c_1^2,$$
$$Q_2 = \frac{1}{2} k_1 c_1^2 - \frac{1}{2} k_2 c_2.$$

The coefficient of 1/2 indicates that 1/2 of a B molecule emerges upon the disappearance of one A molecule.

The distribution of the concentrations of substances A and B, in this case, is determined by the equations

$$v \frac{dc_1}{dy} = D \frac{d^2 c_1}{dy^2} + k_2 c_2 - k_1 c_1^2, \qquad (60.26)$$

$$v \frac{dc_2}{dy} = D \frac{d^2 c_2}{dy^2} - \frac{1}{2} k_2 c_2 + \frac{1}{2} k_1 c_1^2. \qquad (60.27)$$

The boundary conditions at infinity can be written in the form

$$k_1 c_1^2 - k_2 c_2 = 0 \quad \text{as} \quad y \to \infty. \qquad (60.28)$$

The latter equality evidently expresses the equilibrium condition. We shall assume that the rate constant of the reaction B → 2A is high compared to the rate constant of the reverse reaction, so that the following inequality is valid in the bulk of the solution

$$c_2^{(0)} \gg c_1^{(0)}.$$

Then, boundary condition (60.2) can be written in the form

$$c_2^{(0)} \approx c_0 \quad \text{as} \quad y \to \infty. \qquad (60.29)$$

Limiting ourselves, as before, to the case of a very rapid electrode reaction, we may consider that the boundary conditions on the surface of the disk have the form (60.5) and (60.6). To solve the system of equations (60.26) and (60.27), let us introduce a new unknown function:

$$\psi = c_1 + 2c_2.$$

Multiplying (60.27) by 2 and subtracting from (60.26), we find that $\psi$ satisfies equation (60.9), the solution of which is already known. Expressing $c_2$ through $\psi$ and $c_1$, we can write (60.26) in the form

$$v\frac{dc_1}{dy} = D\frac{d^2c_1}{dy^2} + k_2\frac{\psi - c_1}{2} - k_1c_1^2. \qquad (60.30)$$

Assuming, as we have done in the case of the above example of a monomolecular reaction, that the reaction rate in the bulk of the solution is sufficiently high, the convection term in (60.30) can be neglected, and it can be rewritten in the form

$$D\frac{d^2c_1}{dy^2} + k_2\frac{\psi - c_1}{2} - k_1c_1^2 = 0. \qquad (60.31)$$

Assuming that the function $\psi$ varies in the thickness $\delta$ of the diffusion layer, and that $c_1$ varies within the kinetic layer thickness $\delta_k \ll \delta$, the first integral of (60.31) can be obtained. Our assumption permits us to consider the quantity $\psi$ to be constant within the region of variation of $c_1$. Then, multiplying (60.31) by $\frac{dc_1}{dy}$ and integrating, we find:

$$D\left(\frac{dc_1}{dy}\right)^2 + k_2\psi c_1 - \frac{k_2c_1^2}{2} - 2k_1\frac{c_1^3}{3} = \Gamma, \qquad (60.32)$$

where $\Gamma$ is the integration constant. Let us utilize equation (60.32) for the purpose of formulating the boundary condition for the function $\psi$.

In view of (60.5), we have, at the surface $y = 0$:

$$\left(\frac{dc_1}{dy}\right)_{y=0} = \sqrt{\frac{\Gamma}{D}}.$$

In view of (60.6), on the other hand,

$$\left(\frac{d\psi}{dy}\right)_{y=0} = \left(\frac{dc_1}{dy}\right)_{y=0} = \sqrt{\frac{\Gamma}{D}}. \qquad (60.33)$$

In the approximation, in which $\psi$ varies little in the layer $\delta_k$, $\psi$ may be related to the constant $\Gamma$, considering equation (60.32) in the region of values for y that are in the range $\delta > y > \delta_k$.

For $y > \delta_k$, we may evidently consider

$$\psi \approx 2c_2 + c_1^{(0)} \approx 2c_2 \gg c_1^{(0)},$$

$$c_1 \approx c_1^{(0)} \sim \sqrt{\frac{k_2 c_2}{k_1}} \sim \sqrt{\frac{k_2}{2k_1}}\sqrt{\psi} \sim \sqrt{\frac{k_2}{k_1}\, c_0},$$

$$\frac{dc_1}{dy} \approx 0.$$

Furthermore, the following inequality is fulfilled here

$$\psi c_1^{(0)} \sim c_2 c_1^{(0)} \gg \left(c_1^{(0)}\right)^2.$$

We may, therefore, omit the first and third terms in (60.32), and rewrite it in the form

$$k_2 \psi c_1^{(0)} - \frac{2}{3} k_1 \left(c_1^{(0)}\right)^3 = \Gamma.$$

Substituting the value of $c_1^{(0)}$, we obtain:

$$\frac{\sqrt{2}}{3}\,\frac{k_2^{3/2}}{k_1^{1/2}}\,\psi^{3/2} = \Gamma.$$

The latter formula determines the value of $\psi$ at $y \sim \delta_k$. Considering that

$$\psi_{y \sim \delta_k} \approx \psi_{y=0},$$

and substituting the value of $\Gamma$ into (60.33), we obtain:

$$\left(\frac{d\psi}{dy}\right)_{y=0} = k\psi_{y=0}^{3/4}, \tag{60.34}$$

where

$$k = \left(\frac{k_2^3}{k_1 D^2}\right)^{1/4} \sqrt[4]{\frac{2}{9}}.$$

On the strength of (60.29), the boundary value for $\psi$ far from the disk has the simple form

$$\psi \to 2c_0 \quad \text{as} \quad y \to \infty. \tag{60.35}$$

The solution of equation (60.9), for boundary conditions (60.34) and (60.35), is identical with the solution of the problem of mixed kinetics given in Section 12.

We saw in that section that the solution of the problem can be written in the form

$$\psi = \frac{2c_0 - \psi_{y=0}}{\delta} \int_0^y \exp\left\{\frac{1}{D}\int_0^t v(t')\,dt'\right\} dt + \psi_{y=0}, \tag{60.36}$$

where $\psi_{y=0}$ satisfies the algebraic equation

$$(\psi_{y=0})^{3/4} + \frac{1}{k\delta}(\psi_{y=0}) - \frac{2c_0}{k\delta} = 0. \tag{60.37}$$

The solution for the latter can be easily obtained in graphic form, as explained in Section 12. To do this it is necessary to set a numerical value for the dimensionless parameter

$$\mu = \frac{c_0^{1/4}}{k\delta}.$$

The magnitude of the flux to the surface, which controls the rate of the over-all reaction, is

$$j = D\frac{2c_0 - \psi_{y=0}}{\delta}. \tag{60.38}$$

Let us now clarify the validity of the approximations made. For this, it is necessary to estimate the value of the kinetic layer thickness $\delta_k$. An exact evaluation can be derived from a solution of equation (60.30). The latter, however, has no analytic solution. Therefore, we are limited to a rough evaluation, as follows:

$$D\frac{d^2c_1}{dy^2} \sim \frac{Dc_1^{(0)}}{\delta_k^2}$$

and

$$D\frac{d^2c_1}{dy^2} \sim \frac{Dc_1^{(0)}}{\delta_k^2} \sim k_2\psi \sim k_2c_2^{(0)} \sim k_2c_0.$$

We obtain:

$$\delta_k \sim \sqrt{\frac{D}{k_2^{1/2}k_1^{1/2}c_0^{1/2}}} = \sqrt[4]{\frac{D^2}{k_1k_2c_0}}. \tag{60.39}$$

The condition of omission of the convection term in equation (60.30) can be presented in the form

$$v(\delta_k)\frac{c_1^{(0)}}{\delta_k} \ll k_2\psi \sim k_2c_0$$

or

$$\left(\frac{\omega}{\sqrt{k_1k_2c_0}}\right)^{3/2}\left(\frac{D}{\nu}\right)^{1/2} \ll 1. \tag{60.40}$$

The condition of constancy of $\psi$ over the distance $\delta_k$ reduces to the inequality

$$\delta_k \ll \delta_0.$$

The substitution of $\delta_k$ from (60.39) and $\delta$ from (11.31) shows that the latter inequality is identical to (60.40). In both of the samples discussed, we have limited our treatment to the case most often encountered in practice, of a rapid chemical reaction. The reverse case of a slow chemical reaction must be solved by means of series that have an unwieldy form.

In conclusion, let us examine the case of a catalytic reaction, proceeding along pseudo-monomolecular lines, characterized by the course

| bulk of the solution | electrode |
|---|---|
| $A \underset{k_2}{\overset{k_1}{\rightleftarrows}} B$ | $A \rightarrow B.$ |

Here

$$k_2 = k_2[D].$$

The concentration distribution equations agree with (60.3) and (60.4). The boundary conditions far from the disk coincide with (60.1) and (60.2). The boundary conditions at the surface of the disk, however, must be formulated anew. If we examine the limiting regime of a rapid electrode reaction on the surface of the disk, all approaching A particles will enter the reaction, and the concentration $c_1$ on the disk will satisfy equation (60.5).

The second boundary condition is derived on the basis of the following argument: the surface y = 0 serves as a sink of A particles and a source of B particles. In the stationary state, the number of A particles entering the electrode reaction and disappearing must be equal to the number of B particles being formed, i.e.,

$$D\left(\frac{dc_1}{dy}\right)_{y=0} = -D\left(\frac{dc_2}{dy}\right)_{y=0}. \qquad (60.41)$$

The minus sign indicates that the A and B particles flow in opposite directions.

In order to solve the problem posed, let us introduce, as we did in the first example, new unknown functions $\psi$ and $\varphi$. Assuming the chemical reaction to be rapid, for $\varphi$ we may write formula (60.19). The boundary conditions for $\psi$ have the form

$$\left.\begin{aligned} p\psi = \varphi = \text{const} \quad &\text{at} \quad y = 0, \\ \left(\frac{d\psi}{dy}\right)_{y=0} = \left(\frac{dc_1}{dy}\right)_{y=0} + \left(\frac{dc_2}{dy}\right)_{y=0} &= 0, \\ \psi \to c_0 = \text{const} \quad &\text{as} \quad y \to \infty. \end{aligned}\right\} \qquad (60.42)$$

Conditions (60.42) show that $\psi$, together with its derivative, has a constant value at both integration limits. This indicates that everywhere

$$\psi = c_1 + c_2 = \text{const} = c_0.$$

Accordingly

$$\varphi = pc_0 e^{-\sqrt{\frac{\alpha}{D}}y}.$$

The flux on the electrode is equal to

$$j = D\left(\frac{dc_1}{dy}\right)_{y=0} = \frac{D}{\rho}\frac{d}{dy}\left(\frac{\rho\psi - \varphi}{1+\sigma}\right) = -\frac{D}{\alpha}\left(\frac{d\varphi}{dy}\right)_{y=0} = \\ = pc_0\sqrt{\frac{D}{\alpha}} = c_0 k_2 \sqrt{\frac{D}{k_1 + k_2}}. \tag{60.43}$$

We see from (60.43) that, on the basis of the assumptions made, the flow to the electrode is not a function of the stirring and is determined exclusively by the kinetics of the reaction in the bulk of the solution and the diffusion of matter. For this, the following inequality must be fulfilled:

$$\delta_k \approx \sqrt{\frac{D}{\alpha}} \ll \delta \approx \left(\frac{D}{\nu}\right)^{1/3}\sqrt{\frac{\nu}{\omega}}$$

or

$$\left(\frac{\omega}{\alpha}\right)^{3/2}\left(\frac{D}{\nu}\right)^{1/2} \ll 1.$$

The number of examples of such kinetic and catalytic reactions could be multiplied considerably since every chemical reaction mechanism has its own corresponding system of diffusion equations and boundary conditions.

A comparison of the computations for a dropping electrode with a rotating disk shows the latter to be much simpler. This indicates that in the case of a disk electrode more complex process systems can be examined, in particular those involving irreversible electrochemical reactions.

As a reaction surface, a disk has the important advantage of permitting the study of processes under stationary conditions, which can be varied widely by changing the rotational velocity of the disk.

## 61. NONSTATIONARY CONVECTIVE DIFFUSION. TIME REQUIRED TO ESTABLISH A STATIONARY REGIME

Various cases of a stationary regime of convective diffusion and of concentration polarization were examined above.

Nonstationary diffusion processes, however, are encountered in many instances. Thus, for example, when an emf is applied to an electrochemical cell, a current flows through the circuit which exceeds the limiting diffusion current during the time period immediately following the application (at a fixed value of the electrode potential). As the solution becomes depleted in current-carrying ions in the region immediately adjacent to the electrode, the value of the current gradually decreases. After a certain time T, which we term the time during which a transient period is established, the value of the current decreases to a level equal to the limiting current value for a given stirring regime.

In some instances in electrochemical practice, another case of establishment of a stationary regime is encountered. Such is the case of a given current density i through the cell which exceeds the value of the limiting diffusion current (inducing the passage of such a current through the cell is accomplished by introducing a large ohmic resistance in the circuit which controls the total current).

As the current flows through the solution, the latter becomes impoverished near the electrode, and a corresponding increase in the electrode potential occurs. When the concentration of current-carrying ions at the surface of the electrode reaches zero, the electrode potential rises indefinitely and the regime of constant density current flow ceases. In this case, the time T is that time period for the concentration of current-carrying ions on the surface of the electrode to reach zero.

In studying heterogeneous reactions, it is useful, in some instances, to conduct the process under nonstationary conditions. In the initial moments in time, the rate of the reaction is controlled by the reaction step, and the reaction passes to the diffusion stage only after an elapsed time T, the time period for establishment of a stationary regime.

The problem of nonstationary phenomena in heterogeneous processes was first studied in detail by A. N. Sokolov [54]. He studied the nonstationary regime of concentration polarization in a static liquid. It was assumed that the current through the electrochemical cell was limited by the rate of removal (diffusion rate) of the products of the reaction from the surface of the electrode.

It should be pointed out that A. N. Sokolov's study was the first in which a theoretically computed analysis of the phenomena occurring during the passage of current through an electrolytic cell was given.

We examine below certain cases of the establishment of a stationary regime of convective diffusion. In so doing, we confine the discussion to a stationary regime of mixing, i.e., we shall assume that the velocity of motion of the fluid past the reaction surface is constant in time and is a known function of the coordinates.

## 62. THE CASE OF A GIVEN SURFACE CONCENTRATION

This case is an example of electrolysis at a constant potential, where the current density flowing to the electrode varies in time.

The density of the diffusional current at the initial moment may be as great as desired, but as the solution becomes impoverished near the surface of the electrode, that density decreases. At the end of the time for establishment it attains the limiting density value which is a function of the mixing velocity of the fluid.

In order to compute the period of time for the establishment of a stationary regime, let us examine the diffusional flow on the surface of a rotating disk electrode, to which, at a moment in time $t = 0$, a sufficiently large emf is applied, so that the concentration of current-carrying ions at the surface of the electrode is $c_{y=0} = 0$, at any $t > 0$.

For the sake of simplicity, we shall assume that a substantial amount of an indifferent electrolyte has been added to the solution.

The convective diffusion equation will have the form

$$\frac{\partial c}{\partial t} + v_y \frac{\partial c}{\partial y} = D \frac{\partial^2 c}{\partial y^2}. \tag{62.1}$$

The following serve as boundary conditions:

$$c = c_0 \quad \text{at} \quad t = 0, \quad y \geqslant 0, \tag{62.2}$$

$$c = 0 \quad \text{at} \quad y = 0, \quad t > 0, \tag{62.3}$$

$$c = c_0 \quad \text{as} \quad y \to \infty, \quad t > 0. \tag{62.4}$$

Let us evaluate the terms of equation (62.1) to compare orders of magnitude. It is evident that, for $t < T$, the convection term is small compared to the diffusion term. Indeed, a very high concentration gradient exists at the initial moment at the surface of the electrode, and the flow of substance transferred by diffusion is large compared to the convective flow.

If $\delta_{diff}$ designates the thickness of the diffusion layer, for diffusion in a static medium

$$\delta_{diff} = \sqrt{Dt}, \tag{62.5}$$

and, for small values of y, we can write

$$D \frac{\partial^2 c}{\partial y^2} \approx \frac{Dc}{\delta_{diff}^2}.$$

Bearing in mind that for $y < \delta_0$ (where $\delta_0$ is the thickness of the hydrodynamic boundary layer), according to formulas (11.18) and (11.31), we may state $v_y = -v_0 \frac{y^2}{\delta_0^2} = -ay^2$ (where $v_0 = 6.5\sqrt{\nu\omega}$ and $\delta_0 = 3.6\sqrt{\frac{\nu}{\omega}}$. We then obtain:

$$v_y \frac{\partial c}{\partial y} \approx v_0 \frac{\delta_{diff}}{\delta_0^2} c.$$

The condition, on the strength of which the diffusional flow is large as compared to the convective flow, may be written in the form

$$D\frac{\partial^2 c}{\partial y^2} \approx D\frac{c}{\delta^2_{\text{diff}}} \gg v_0 \frac{\delta_{\text{diff}}\, c}{\delta_0^2}.$$

This inequality will be fulfilled at values

$$\delta_{\text{diff}} \ll \left(\frac{\delta_0^2 D}{v_0}\right)^{1/3} \tag{62.6}$$

or at times

$$t \ll \frac{1}{\omega}\left(\frac{\nu}{D}\right)^{1/3}. \tag{62.7}$$

The diffusional and convective flows become comparable (in terms of orders of magnitude), and a stationary regime of convective diffusion becomes established, at

$$t \approx T \approx \frac{1}{\omega}\left(\frac{\nu}{D}\right)^{1/3}. \tag{62.8}$$

Let us now determine the process of establishment of a stationary diffusion regime from a quantitative standpoint, and let us examine the dependence of the diffusional flow on time.

At the start of the process, for $t \ll T$, the convective term in expression (62.1) can be omitted and the latter can be reduced to the usual convective diffusion equation for a static medium

$$\frac{\partial c_1}{\partial t} = D\frac{\partial^2 c_1}{\partial y^2}. \tag{62.9}$$

The solution of equation (62.9) satisfying boundary conditions (62.2) to (62.4) is

$$c_1 = \frac{2c_0}{\sqrt{\pi}} \int_0^{\frac{y}{2\sqrt{Dt}}} e^{-z^2} dz. \tag{62.10}$$

Let us now take account of the existence of convection by the method of successive approximations, considering the convective flow to be a small contribution to the molecular flow. Here, instead of c, the concentration determined by formula (62.10) should be inserted in the derivative $\frac{\partial c}{\partial y}$ of equation (62.1).

We shall seek a solution of equation (62.1) by the successive approximation method, by writing:

$$c = c_1 + c_2, \tag{62.11}$$

where $c_1$ is the solution of the diffusion equation in a motionless medium and $c_2$ is a certain small correction.

This correction will be determined by the equation

$$\frac{\partial c_2}{\partial t} = D\frac{\partial^2 c_2}{\partial y^2} + \frac{v_0 c_0 y^2}{\delta_0^2 \sqrt{\pi D t}} \cdot e^{-\frac{y^2}{4Dt}}. \tag{62.12}$$

The boundary conditions for $c_2$ are:

$$\begin{aligned} c_2 &= 0 \quad \text{at} \quad y = 0, \\ c_2 &= 0 \quad \text{as} \quad y \to \infty. \end{aligned} \tag{62.13}$$

Expression (62.12) is an equation of the type of a thermal conduction equation with heat sources. Since we are henceforth only interested in the flow of substance on the reaction surface, we may confine equation (62.12) to the region of small y (compared to $\sqrt{Dt}$), and assume that the exponent $e^{-(y^2/4Dt)}$ is equal to unity. This is all the more necessary, since the expression for velocity used in the convective diffusion equation (62.12) is valid only at small values of y.

With such a substitution, expression (62.12) becomes considerably simplified and takes the form

$$\frac{\partial c_2}{\partial t} = D\frac{\partial^2 c_2}{\partial y^2} + \frac{D y^2 c_0}{\sqrt{\pi t}}\beta, \tag{62.14}$$

where $\beta = \frac{v_0}{\delta_0^2 D^{3/2}}$.

It is convenient to seek a solution of equation (62.14) for boundary conditions (62.13) by means of the Laplace transformation method.

Multiplying equation (62.14) by $e^{-st}$ and integrating it along the variable t within the limits zero to infinity, we find the following in functional notation:

$$sF = DF'' + D\beta y^2 c_0 \cdot \frac{1}{\sqrt{s}}. \tag{62.15}$$

Expression (62.15) is an ordinary differential equation and can be solved without difficulty. Evidently

$$F = Ae^{-\sqrt{\frac{s}{D}}y} + Be^{\sqrt{\frac{s}{D}}y} + \frac{D\beta y^2 c_0}{s^{3/2}} + \frac{2\beta D^2 c_0}{s^{5/2}}. \tag{62.16}$$

The first two terms of this expression represent the general solution of the homogeneous equation (62.15). The second two terms are a particular solution of the inhomogeneous equation (62.15).

Starting from the boundary conditions (62.13), the following conditions can be written for F:

$$F(0) = 0; \qquad F(\infty) = 0. \tag{62.17}$$

Taking the latter into account, we find that

$$A = -\frac{2\beta D^2 c_0}{s^{5/2}}, \quad B = 0.$$

Finally, we obtain for F:

$$F = \left[ -\frac{2\beta D^2}{s^{5/2}} e^{-\sqrt{\frac{s}{D}}\,y} + \frac{D\beta y^2}{s^{3/2}} + \frac{2\beta D^2}{s^{5/2}} \right] c_0. \tag{62.18}$$

To pass from F to $c_2$, it is not necessary to perform the inverse transformation of the Laplace expression. We can utilize tables which are available in many manuals. Utilizing, for example,* the table in the monograph by A. V. Lykov entitled "Teploprovodnost' nestatsionarnykh protsessov" [Thermal Conductivity of Nonstationary Processes], we obtain an expression for the original function of $c_2$ that we are seeking:

$$c_2(y, t) = \frac{c_0 8}{3\sqrt{\pi}} \beta D^2 t^{3/2} + \frac{2c_0\beta D y^2}{\sqrt{\pi}} \sqrt{t} - 2c_0\beta D^2 \int_0^t e^{-\frac{y^2}{4Dz}} \frac{t-z}{\sqrt{\pi z}}\, dz. \tag{62.19}$$

This solution for $c_2$ evidently satisfies the boundary condition on the surface of the disk, y = 0, but does not satisfy the condition for infinity. This is due to the fact that, in going from equation (62.12) to (62.14), we have replaced the rapidly decreasing expression $y^2 e^{-(y^2/4Dt)}$ by the function $y^2$ which increases with increasing y. This limits the area of application of solution (62.19). Expression (62.19), therefore, may be used to find the flow that interests us on the reaction surface at the plane y = 0. The flow of substance per 1 cm$^2$ of disk surface is

$$j = D\left(\frac{\partial c}{\partial y}\right)_{y=0} = D\left(\frac{\partial c_1}{\partial y}\right)_{y=0} + D\left(\frac{\partial c_2}{\partial y}\right)_{y=0}.$$

Substituting for $c_1$ and $c_2$ their values from (62.10) and (62.19), we obtain

$$j = \frac{Dc_0}{\sqrt{\pi D t}} e^{-\frac{y^2}{4Dt}} + \frac{\beta c_0 D^2 t}{\sqrt{\pi}} \int_0^t \frac{y}{z^{3/2}} e^{-\frac{y^2}{4Dz}}\, dz - \\ - \frac{D^2\beta c_0}{\sqrt{\pi}} \int_0^t e^{-\frac{y^2}{4Dz}} \frac{y\, dz}{\sqrt{z}} + \frac{4}{\sqrt{\pi}} c_0 D^2 \beta y \sqrt{t} \tag{62.20}$$

*Editor's note: or, for example, R. V. Churchill, Operational Mathematics, McGraw-Hill, 1958, pp. 323-328.

Before passing on to the limit y = 0, the integrals in expression (62.20) should be transformed. For this, let us introduce a new variable

$$\frac{y}{2\sqrt{Dz}} = u.$$

After simple transformations, we have, finally:

$$\lim_{y \to 0} \int_0^t \frac{y}{z^{3/2}} e^{-\frac{y^2}{4Dz}} dz = 2\sqrt{\pi D},$$

$$\lim_{y \to 0} \int_0^t e^{-\frac{y^2}{4Dz}} \frac{y}{\sqrt{z}} dz = 0.$$

By utilizing these expressions for the last integrals, we obtain the following expression for the flux of substance to the disk surface:

$$j = Dc_0 \left[ \frac{1}{\sqrt{\pi Dt}} + 1{,}02 \frac{\omega^{3/2}}{\nu^{1/2}} t \right]. \tag{62.21}$$

The first term in formula (62.21) evidently is the flow of substance carried by molecular diffusion. It was first obtained in the above-cited study by A. N. Sokolov. The second term in the right side of the formula characterizes the convective flow of substance for a period of time t that has passed since the start of the reaction, and for which the convection flow may still be considered extremely small compared to its molecular flow. It is obvious that the convective flow of substance is small, in comparison to the molecular flow, over time periods from the start of the reaction that are shorter than the characteristic time T, defined by the expression

$$\frac{1}{\sqrt{\pi DT}} \approx \frac{\omega^{3/2}}{\nu^{1/2}} T, \tag{62.22}$$

T being the time during which a stationary regime of convective diffusion becomes established. Obviously, for times $t \geq T$, formula (62.21), which was derived on the assumption that the convective flow of substance is very small compared to the diffusional flow, is inapplicable. Following the elapse of time T, a stationary regime of convective diffusion of substance becomes established. This means that, at $t \approx T$, it is no longer the term $v_y \frac{\partial c}{\partial y}$ which becomes small in the convective diffusion equation (62.1), but the term $\frac{\partial c}{\partial t}$, and the latter can therefore be omitted. Then the distribution and flow of substance on the reaction surface are not functions of time.

## 63. ESTABLISHMENT OF A REGIME AT A GIVEN CURRENT DENSITY

The other case of establishment of a regime, under conditions in actual practice, is the passage through an electrochemical cell of a current of given density [55], which exceeds the limiting density of the diffusion current (at given mixing conditions).

The maintenance of a constant current density is achieved by switching in the current through a very high ohmic resistance, high in comparison with all other resistances in the circuit. The boundary condition on the surface of the electrode in this case is not (62.3) but

$$i = (nF)\, D \left( \frac{\partial c}{\partial y} \right)_{y=0} = \text{const.} \tag{63.1}$$

The initial conditions and those for infinity remain the same as previously. For small values of t, the convective term in equation (62.1) can be omitted.

The concentration distribution is determined here by equation (62.9). The solution of equation (62.9) satisfying the initial condition (62.2) and the boundary condition (62.4) can be written in the form

$$c = \frac{1}{\sqrt{4\pi Dt}} \left\{ c_0 \int_0^\infty e^{-\frac{(\xi - y)^2}{4Dt}} d\xi + \int_0^\infty f(-\xi)\, e^{-\frac{(\xi + y)^2}{4Dt}} d\xi \right\}. \tag{63.2}$$

Boundary condition (63.1) can be satisfied by the establishment of the function f(- ξ), equal to

$$f(-\xi) = c_0 - \frac{2\sqrt{\pi Dt} i}{(nF)\, D}.$$

Substituting this value into equation (63.2), and performing simple transformations, we derive a solution that satisfies the above boundary and initial conditions, in the form

$$c = c_0 - \frac{2i}{nF} \sqrt{\frac{t}{\pi D}}\, e^{-\frac{y^2}{4Dt}} + \frac{2iy}{nFD\sqrt{\pi}} \int_{\frac{y}{2\sqrt{Dt}}}^{\infty} e^{-z^2} dz. \tag{63.3}$$

Passing from an integration along z to an integration along t and integrating by parts, the latter expression can be rewritten in the form

$$c = c_0 - \frac{i}{nF\sqrt{\pi D}} \int_0^t \frac{d\tau}{\sqrt{\tau}} e^{-\frac{y^2}{4D\tau}}. \tag{63.3'}$$

In particular, the concentration on the electrode surface, at y = 0, is

$$c_1 = c_0 - \frac{2i}{(nF)} \sqrt{\frac{t}{\pi D}}. \tag{63.4}$$

The latter equation was derived by Sand [56], but in deriving it the author permitted a certain inaccuracy.

It follows from (63.4) that the concentration at the surface decreases with time and becomes zero after the elapse of a time $T_0$,

$$T_0 = \left(\frac{nF}{2i}\right)^2 \pi D c_0^2. \tag{63.5}$$

The decrease in the concentration potential of the ions being determined is matched by a corresponding increase in the cathode potential $\varphi_C$, following the Nernst formula:

$$\varphi_C = \frac{RT}{(nF)} \ln c_{y=0} + \varphi_0$$

As $t \to T_0$, the cathode potential increases indefinitely. In this case, the regime of the process is disturbed (for example, by the start of a new electrochemical reaction).

It should be noted, however, that equation (63.5) giving the time for establishment of the regime was derived without taking into account the effect of convection. The convective transfer of matter contributes to a slower rate of depletion of the solution at the surface of the electrode, i.e., to an increase in the time $T_0$ of establishment of the regime. To account for that effect, let us introduce the convective term $v_y \frac{\partial c}{\partial y}$, as a small correction, into the equation of transfer (62.1). Thus, let us seek a solution of equation (62.12) in the form

$$c = c_1 + c_2, \tag{63.6}$$

where $c_1$ is the solution of equation (62.9).

Substituting expression (63.6) into equation (62.1), we have, for $c_2$,

$$\frac{\partial c_2}{\partial t} + v_y \frac{\partial c_1}{\partial y} = D \frac{\partial^2 c_2}{\partial y^2}. \tag{63.7}$$

Substituting the value of $\frac{\partial c_1}{\partial y}$ from the solution (63.3), we find:

$$c = c_1 + c_2 = c_0 - \frac{2i}{nF} \sqrt{\frac{t}{\pi D}} e^{-\frac{y^2}{4Dt}} + \frac{2iy}{nFD\sqrt{\pi}} \int\limits_{\frac{y}{2\sqrt{Dt}}}^{\infty} e^{-y^2} dy + 0.17 \frac{\omega^{3/2} j y^2 t}{\sqrt{\nu D}} + 0.08 i t^2 \frac{\omega^{3/2}}{\nu^{1/2}}. \tag{63.8}$$

The same remarks may be made on the subject of formula (63.8) as for expression (62.19). On the surface of the electrode

$$c = c_0 - \frac{2i}{nF}\sqrt{\frac{t}{\pi D}} + 0.08\,\frac{\omega^{3/2} i t^2}{\nu^{1/2}}. \tag{63.9}$$

Formula (63.9) may be used only for short periods of time, when its third term is small compared to the second.

As the current passes through the solution, the surface concentration decreases and becomes zero at $t = T$. In this case, obviously,

$$c_0 - \frac{2i}{nF}\sqrt{\frac{T}{\pi D}} + 0.08\,\frac{\omega^{3/2} T^2 i}{\sqrt{\nu}} = 0. \tag{63.10}$$

Equation (63.10) shows that the time for the surface concentration to become zero is increased due to convection. In the absence of mixing, the curve $i = i\left(\frac{1}{\sqrt{T}}\right)$ becomes a straight line passing through the origin of coordinates. Mixing changes the form of the function $i\left(\frac{1}{\sqrt{T}}\right)$ and i is no longer a linear function of $\frac{1}{\sqrt{T}}$ (Figure 67).

The effect of natural convection on the time for the establishment of a stationary regime was studied by Butler and Armstrong [57], who proposed an empirical formula for expressing the dependence of i on $\frac{1}{T}$:

$$i = i_0 + \frac{a}{T}, \tag{63.11}$$

where $i_0$ and a are constants that are functions of the mixing regime.

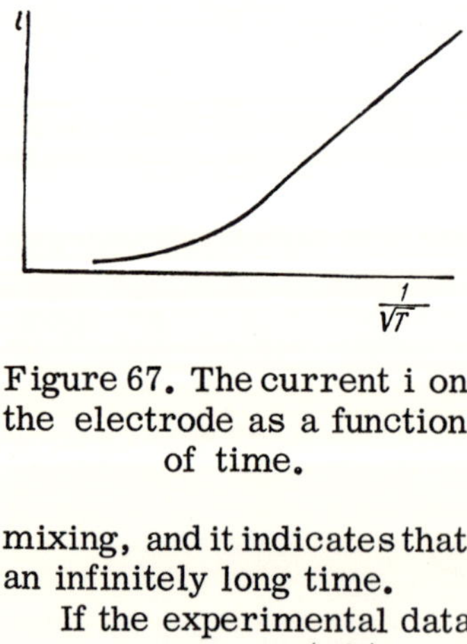

Figure 67. The current i on the electrode as a function of time.

This form of dependence of i on T appears to be very strange: formula (63.11) does not become (63.5) upon decrease in the intensity of mixing, and it indicates that the establishment of the regime requires an infinitely long time.

If the experimental data of Butler and Armstrong are expressed on a curve $i = i\left(\frac{1}{\sqrt{T}}\right)$, a curve is obtained such as that shown in Figure 67. It agrees qualitatively with formula (63.10): at high values of $\frac{1}{\sqrt{T}}$, it becomes a straight line, while, at small $\frac{1}{\sqrt{T}}$, the curve's rise is slower than linear. It may be assumed, therefore, that formula (63.11) is merely an unfortunate correlation of experimental data. Of course, to make a quantitative comparison of the theoretical formula (63.10) with the experimental data of Butler and Armstrong is impossible, due to the quite different mixing conditions.

In conclusion, it should be pointed out that new methods of studying the kinetics of electrochemical processes in recent years have received wide acceptance. They have permitted studies in static solutions. These methods are based on measurements of short duration of the volt-ampere characteristics of the discharge. Such measurements are made, for example, by means of oscillograph instruments. Since the fluid, for all practical purposes, does not have time to start moving during the measurements, the solution may be considered as being completely at rest.

The development of the techniques indicated for experimental studies was accompanied by the development of the theory for the integration of diffusion equations in static media with various boundary conditions which correspond to the various conditions of conducting the experiment (application of a continuous current, sawtooth voltage, a voltage increasing in time in a linear manner, etc.), and different kinetics (reversible processes, reactions of various orders in concentration).

A description of these problems cannot be an objective of this book. We therefore refer to the monograph by Delahay [48], which treats problems in this field, and to studies that have been published since the issuance of this monograph.

Substantial progress in the development of the kinetics of electrode reactions and reactions in solutions has been achieved in recent studies investigating kinetic and catalytic processes. These studies, begun by Czech researchers, have been especially well developed on the basis of the theory of kinetic and catalytic processes in polarography, which was originally developed by J. Koutecký.

In Section 60, for the case of a disk electrode, a mathematical method developed by J. Koutecký for a dropping electrode was used. Applied to the latter, however, the computation is quite unwieldy. Numerous catalytic and kinetic reactions have been studied by an analogous method [59]. The monograph by Delahay contains an extensive bibliography [48].

## BIBLIOGRAPHY

1. Frumkin, A. N., V. S. Bagotskiy, Z. A. Iofa, B. N. Kabanov, Kinetika elektrodnykh protsessov [Kinetics of Electrode Processes], Moscow State Univ. Press, 1952.
2. Eucken, A., Z. phys. Chem. 59, 72 (1907).
3. Agar and Bowden, Proc. Roy. Soc. A169, 217 (1938); Agar, Disc. Far. Soc. 1, 26 (1947).
4. Volmer, M., Zhur. Fiz. Khim. 5, 319 (1934); A. N. Frumkin, Phys. Z. Sowjetunion 4, 239 (1933); Ergebn. exakt. Naturwiss.

7, 235 (1928); V. A. Zarinskiy, Perenapryazheniye vodoroda na rtutnom katode i dzeta-potentsial [Hydrogen Overpotential on a Mercury Cathode and Zeta Potential], State United Publishing House of Science and Technology, 1938; M. Dole, Osnovy elektrokhimii, United Scientific and Technical Publishing Houses, 1937 [see also: Principles of Experimental and Theoretical Electrochemistry, McGraw-Hill, 1935]; M. I. Temkin, Zhur. Fiz. Khim, 15, 296 (1941); A. T. Vagramyan, Elektroosazhdenie metallov [Electrolytic Deposition of Metals], Acad. Sci. USSR Press, 1950; [see also: K. Vetter, Elektrochemische Kinetik, Springer, 1961].

5. Kantorovich, L. V., and V. I. Krylov, Priblizhennyye metody vysshego analiza, Gostekhizdat, 1949 [see also English translation: Approximate Methods of Higher Analysis, Noordhoff, 1958].
6. Levich, V.G., and A.N. Frumkin, Zhur. Fiz. Khim. 15, 748 (1941) [for brief version in English, see: Acta Physicochimica URSS 18, 325 (1943)].
7. Wagner, C., J. Electrochem. Soc. 98, 116 (1951).
8. Wagner, B., Handb. Metallphysik 1, n. 2, 196 (1940); A. N. Frumkin, Zhur. Fiz. Khim. 23, 1477 (1949); V. S. Daniel'-Bekh, Zhur. Fiz. Khim. 22, 697 (1948); T. Hoar and J. Agar, Disc. Far. Soc. 1, 162 (1947).
9. Levich, V. G., Zhur. Fiz. Khim. 18, 335 (1944) [see also Acta Physicochimica 17, 257 (1942); 19, 117 (1944)]; O prokhozhdenii toka cherez rastvory (dissertation) [Concerning the Passage of Current Through Solutions], Kazan', 1943.
10. Nernst, Z. phys. Chem. 47, 52 (1904); Nernst and Merriam, Z. phys. Chem. 53, 235 (1905).
11. Centnerszwer, Z. phys. Chem. A141, 297 (1929); Van Name, Z. phys. Chem. 73, 97 (1910).
12. Muller, Acta physicochimica URSS, 3, 465 (1935).
13. Eucken, A., Z. Elektrochem. 38, 341 (1932).
14. Brunner, Z. phys. Chem. 47, 56 (1904).
15. Van Name, Am. Journ. Sci. ser. 4, 29, 237 (1910).
16. King and Schack, J. Amer. Chem. Soc. 57, 1212 (1935); King and Braverman, J. Amer. Chem. Soc. 54, 1744 (1932).
17. Wildermann, Z. phys. Chem. 66, 445 (1909).
18. Kabanov, B. N., and Yu. G. Sïver, Zhur. Fiz. Khim. 22, 53 (1948).
19. Kabanov, B. N., Zhur. Fiz. Khim. 23, 428 (1949).
20. Hogge, E., and M. B. Kraichman, J. Amer. Chem. Soc. 76, 1431 (1954).
21. Aykazyan, E. A., and A. I. Fedorova, Doklady Adad. Nauk SSSR 86, 1137 (1952).
22. Aksel'rud, G. A., Zhur. Fiz. Khim. 27, 1446 (1953).
23. Trümpler and Zeller, Helv. Chim. Acta 34, 952 (1951).
24. Brenner, A., Proc. Educational Sessions of Twenty-ninth Annual Convention, Amer. Electrochem. Soc., p. 28 (1941).

25. Ibl, N., Helv. Chim. Acta 37, 1149 (1954); N. Ibl, W. Ruegg, and G. Trumpler, Helv. Chim. Acta 36, 1624 (1953).
26. Bagotskaya, I. A., Doklady Akad. Nauk SSSR 85, 1057 (1952).
27. Fedorova, A. I., and G. L. Vidovich, Doklady Akad. Nauk SSSR 109, 135 (1956).
28. Eckert, E., Heat and Mass Transfer, McGraw-Hill, 1950 [or 1959], p. 141; G. A. Inkhailov, Sov. kotloturbostyeniye [Soviet Boiler and Tube Construction], no. 12 (1939); G. N. Kruzhilin and V. A. Shvab, Zhur. Tekh. Fiz. 5, 703 (1935).
29. Kolthoff and Laitinen, J. Phys. Chem. 45, 1079 (1941); Skobets, Zavodskaya Lab. 14, 141 (1948); Chirkov, Zavodskaya Lab. 14, 1300 (1948).
30. Fedorova, A. I., G. L. Vidovich, L. I. Boguslavskiy, and V. D. Yukhtanova, Doklady na IV Soveshchanii po elektrokhimii [Report at the 4th Conference on Electrochemistry].
31. Aykazyan, E.A., and Yu. V. Pleskov, Zhur. Fiz. Khim. 31, 205 (1957).
32. Stackelberg, M., M. Pilgram, and V. Toome, Z. Elektrochem. 57, 342 (1953).
33. Glasstone, S., Vvedenie v elektrokhimiyu, IL, 1951, chapter XIV [see also: An Introduction to Electrochemistry, Van Nostrand, 1942].
34. Akimov, G. V., Teoriya i metody issledovaniya korrozii metallov [Theory and Methods of Studying Metal Corrosion], Acad. Sci. USSR Press, 1945; N. D. Tomashov, Korroziya metallov s kislorodnoy depolyarizatsiyey [Metal Corrosion with Oxygen Depolarization], Acad. Sci. USSR Press, 1947.
35. Kayander, N. N., Zhur. Russ. Fiz.-Khim. Obshch. 13, part 1, (1881).
36. Frumkin, A.N., Trudy Vtoroy konferentsii po korrozii metallov [Transactions of the Second Conference on Metal Corrosion], Acad. Sci. USSR Press, vol. I, 1940.
37. Shultin, A.I., Trudy Vtoroy konferentsii po korrozii metallov [Transactions of the Second Conference on Metal Corrosion], v. I; Zhur. Fiz. Khim. 15, 359, 370 (1941); 18, 61, 69 (1944).
38. Kolotyrkin, Ya. M., and A. N. Frumkin, Comp. Rend. Acad. Sci. URSS, 33, 445, 450 (1941); Zhur. Fiz. Khim. 15, 346 (1941).
39. Bagotskaya, I. A., Zhur. Fiz. Khim. 25, 459 (1951).
40. Durdin, Ya. V., Zhur. Obshch. Khim. 17, 844, 862 (1947).
41. Van Name, Z. phys. Chem. 73, 97 (1910).
42. Durdin, Ya. V., and Z. U. Dukhnyakova, Sb. statei po obshchei khimii [Collection of Articles on General Chemistry], v. I, 1953, p. 157.
43. Bircumshaw, L. L., and A. C. Riddiford, J. Chem. Soc., p. 598 (1951); p. 1490 (1951); p. 698 (1952); p. 701 (1952).
44. Halpern, J., J. Electrochem. Soc. 100, 421 (1953).
45. Kolotyrkin, Ya. M., Zhur. Fiz. Khim. 25, 1248 (1951).

46. Akimov, G. V., Uspekhi Khimii 12, 374 (1943); G. V. Akimov and G. B. Klark, Doklady Adad. Nauk SSSR 58, 819, 1669, 1973 (1947); 59, 57, 399 (1948); V. Pal'mayer, Trudy Mendeleyevskogo s'ezda [Transactions of the Mendeleev Congress] 2 (1937).
47. Brdicka, R., Collection Czechoslov. Chem. Commun., Suppl. II, 19, 41 (1954).
48. Delahay, P., Novyye pribory i metody v elektrokhimii, IL, 1957 [see also New Instrumental Methods in Electrochemistry, Interscience, 1954].
49. Koutecký, J., Doklady na IV soveshchanii po elektrokhimii [Report at the 4th Conference on Electrochemistry], Moscow, 1956.
50. Koutecký, J., Collection Czechoslov. Chem. Commun. 18, 597 (1953).
51. Koutecký, J., and R. Brdicka, Collection Czechoslov. Chem. Commun. 12, 337 (1947).
52. Koutecký, J., Nature, 174, 233 (1954).
53. Koutecký, J., Collection Czechoslov. Chem. Commun. 18, 311 (1953).
54. Sokolov, A. N., Trudy Otdeleniya fizicheskikh nauk Obshchestva lyubiteley yestestvoznaniya [Transactions of the Physics Department of the Society of Amateurs of Natural Sciences], vol. 3, 2nd ed., 1890, p. 22.
55. Levich, V. G., Acta physicochimica URSS 19, 133 (1944).
56. Sand, Phil. Mag. 1, 45 (1901); Karnoglanoff, Z. Elektrochem. 12, 5 (1906).
57. Butler and Armstrong, Proc. Roy. Soc. A139, 406 (1933).
58. Frumkin, A. N., and E. A. Aykazyan, Doklady Akad. Nauk SSSR 100, 315 (1955).
59. Matsuda, H., Bull. Chem. Soc., Japan 26, 342 (1953); I. Kolthoff, J. Jordan, and S. Prager, J. Am. Chem. Soc. 76, 5221 (1954); H. Rubin and F. Collins, J. Phys. Chem. 58, 958 (1954); A. Hickling and J. Maxwell, Trans. Far. Soc. 51, 44 (1955); L. E. Gierst, Z. Elektrochem. 59, 784 (1955); W. Stricks and I. Kolthoff, J. Am. Chem. Soc. 78, 2085 (1956); T. Kambara, Bull. Chem. Soc., Japan 27, 523, 527, 529 (1954); I. Tachi and T. Kambara, Bull. Chem. Soc., Japan 28, 25 (1955); T. Kambara, Z. phys. Chem., N. F. 5, 52 (1955); M. Smutek, Chem. Listy 47, 963 (1953).
60. Dogonadze, R. R., V. G. Levich, and Yu. A. Chizmadzhev, Zhur. Fiz. Khim. 33, 1111 (1959) [see also: Zhur. Fiz. Khim. 34, 2320 (1960) and English translation of latter, Russ. J. Phys. Chem. 34, 1102 (1960)].
61. Coehn and Schnurmann, Z. Physik 46, 354 (1928; C. Reed and W. Schriever, J. Chem. Phys. 17, 935 (1949); W. Schriever and Ferguson, J. Chem. Phys. 19, 609 (1951); J. Faria, Comp. Rendus, 235, 1297 (1952); D. MacInnes, T. Shedlovsky, and L. Longsworth, J. Chem. Phys. 18, 233 (1950).

62. Levich, V. G., Stat'ya v sb. Nekotorye voprosy teoreticheskoy fiziki [article in coll.: Some Questions in Theoretical Physics], Moscow Physics and Engineering Institute, 1958.
63. Dogonadze, R. R., ISFKh 32, 2437 (1958).
64. Friedlander, S. K., A. I. Ch. E. Journ. 3, 43 (1957).
65. Ranz, W. E., and W. R. Marshall, Chem. Eng. Progr. 48, 141, 173 (1952).
66. Kramers, H., Physica 12, 61 (1946).
67. Frumkin, A. N., and L. I. Nekrasov, Doklady Akad. Nauk SSSR 126, 115 (1959 [see also English translation].
68. Ivanov, Yu. B., and V. G. Levich, Stat'ya v sb. Nekotorye voprosy teoreticheskio fiziki [article in coll.: Some Questions in Theoretical Physics], Moscow Physics and Engineering Institute, 1958; Doklady Akad. Nauk SSSR 126, 1029 (1959) [see also English translation].
69. Dukhin, S. S., and B. V. Deryagin, Kolloidnyy Zhurnal 20, 705 (1958); Doklady Akad. Nauk SSSR 121, 503 (1958); Kolloidnyi Zhurnal, 21, 37 (1959) [see also English translations of all three].
70. Koutecký, J., and V. G. Levich, Zhur. Fiz. Khim. 32, 1565 (1958); Doklady Akad. Nauk SSSR 117, 441 (1957) [see also English translation].

## Reviews

Agar, Disc. of Far. Soc., 1947, no. 1, p. 26.
Tobias, J. Electrochem. Soc. 99, 359C (1952).
Vielstich, Z. Elektrochem. 57, 646 (1953).
N. Ibl, Chimica, 9, 135 (1955).

CHAPTER VII

# Motion Induced by Capillarity

## 64. THE INTERFACIAL REGION

The boundary between contiguous bulk phases of matter is known as the interfacial region or, simply, the interface [1].

The interfacial region is that thin layer surrounding a geometric surface of separation, within which the physical properties differ noticeably from those in either of the bulk phases. The thickness of this layer is illdefined because the variation of physical properties across it is continuous. Under conditions well away from the critical point, however, the layer is only a few molecules thick and, in any event, is very small compared to the wavelength of visible light. This is obvious from the reflection of light from the surface of a liquid. Close to the critical point the thickness of the interfacial region increases to become quite appreciable.

Here, we adopt an approximation in which the interface is infinitely thin; i.e., we regard the phase boundary as a geometric surface, and assume that the properties right up to the interface are unchanged from those of the respective bulk phases.

Since the thickness of the interface is of the order of molecular dimensions, such an approximation is fully justified in treating the macroscopic properties of liquids. Because the properties of matter in the interfacial region differ from those in bulk, the interface has free energy per molecule $F^{\Sigma}$ which is peculiar to it.

We consider only the case of one-component systems.

The surface free energy is defined as

$$F^{\Sigma} = F - (F_1 + F_2),$$

where F is the total free energy of the system consisting of two phases separated by an interface, and $F_1 + F_2$ is the sum of the free energies of the two bulk phases, computed on the assumption that both remain homogeneous right up to the surface of separation. This surface free energy is the thermodynamic potential for which temperature T and surface area $\Sigma$ are the independent variables: therefore

$$dF^{\Sigma} = -S^{\Sigma}\,dT + \sigma\,d\Sigma,$$

where $S^{\Sigma}$ is the surface entropy.

The quantity*

$$\sigma = \left(\frac{\partial F^{\Sigma}}{\partial \Sigma}\right)_T$$

is the surface free energy per unit area at constant temperature. It is obvious that the work of reversible isothermal expansion of the surface dA is equal to the decrease in free energy, i.e.,

$$dA = -\sigma\,d\Sigma.$$

The quantity $\sigma$ is known as the interfacial tension or surface tension.

The surface free energy in a one-component system is fully determined by two parameters, for example by the temperature T and surface area $\Sigma$. The surface tension $\sigma$, therefore, is a function only of temperature T.

The surface tension of liquids decreases with temperature. There is, however, no generally accepted or sufficiently well grounded theory describing the dependence of surface tension on temperature. The general character of this dependence becomes apparent only near the critical point.

## 65. CONDITIONS OF EQUILIBRIUM BETWEEN TWO FLUID PHASES AT REST

Let us determine the conditions for thermodynamic equilibrium in a two-phase, gas-liquid system, separated by an interface possessing surface tension $\sigma$ [2].

The first obvious equilibrium condition is the equality of temperature in both phases. Having assumed the temperatures of the two phases to be the same, let us determine the equilibrium conditions at constant temperature.

Thermodynamic equilibrium in an isothermal system requires that the free energy of the system be a minimum; i.e., the following condition must be fulfilled

$$\delta F = 0,$$

where $\delta F$ is an arbitrary and infinitesimal variation of the free energy of the system. Let us imagine two cylinders, one in each of

*This definition is valid only for the one-component systems that are considered here.

the two phases, resting on a surface element $d\Sigma$, and let us relate all quantities to this element. The free energy of the system consists of the free energies of the bulk phase in each of the cylinders $F_1$ and $F_2$ of the surface free energy. Therefore,

$$\delta F = \delta F_1 + \delta F_2 + \delta F^{\Sigma} = 0. \tag{65.1}$$

The change in the free energy of the system at constant temperature can occur only as a result of changes in volume of the two phases and a change in area of the interface. The change in the free energy within the bulk phases at constant temperature is given by

$$\delta F_1 = -p_1 \delta V_1, \tag{65.2}$$

$$\delta F_2 = -p_2 \delta V_2, \tag{65.3}$$

where $p_1$ and $p_2$ are the absolute pressures and $\delta V_1$ and $\delta V_2$ are changes in the volumes of the two cylinders. The change in surface free energy at constant temperature equals

$$\delta F^{\Sigma} = \sigma \, d\Sigma. \tag{65.4}$$

Since the total volume of the system remains constant, an increase in the volume of one phase causes a corresponding decrease in the volume of the other; i.e.,

$$\delta V_1 = -\delta V_2. \tag{65.5}$$

Therefore, the total change in the free energy of system is

$$\delta F = -(p_1 - p_2)\,\delta V_1 + \sigma\,\delta\Sigma. \tag{65.6}$$

The relation between the two quantities $\delta V$ and $\delta\Sigma$ should now be established. Let us first examine the case in which the curvature of the interface is very small. Here, in computing the change in free energy the interface may be considered a plane surface.

First, let us determine the variation in the surface area $\delta\Sigma$, if all of its elements are subjected to some arbitrary and infinitesimal displacement along the normal to an undeformed surface, i.e., the plane $z = 0$.

We consider a surface element $d\Sigma$ on the interface. The sides of the element are dx and dy and therefore $d\Sigma_0 = dxdy$. The deformation of the surface element occurs through extention or contraction, which transform the element of area to

$$d\Sigma = \sqrt{1 + \left(\frac{\partial \zeta}{\partial x}\right)^2}\, d\Sigma_0, \tag{65.7}$$

where $\zeta$ is the vertical displacement over the plane $z = 0$.

The formula takes into account only such deformations of linear elements on the surface (x, y) as result in a change of the surface area of the element $d\Sigma$.

For the sake of simplicity, we have also assumed that the displacements $\zeta$ occur in the plane y = const, producing a surface that is curved in only one direction.

At small displacements $\zeta$, equation (65.7) may be approximated by

$$d\Sigma \approx \left(1+\frac{1}{2}\left(\frac{\partial\zeta}{\partial x}\right)^2\right)dx\,dy. \qquad (65.8)$$

The surface area after deformation is

$$\Sigma = \int\int \left(1+\frac{1}{2}\left(\frac{\partial\zeta}{\partial x}\right)^2\right)dx\,dy.$$

The variation in the surface area is

$$\delta\Sigma = \int\int \frac{\partial\zeta}{\partial x}\frac{\partial\delta\zeta}{\partial x}dx\,dy =$$
$$= \int\left(\frac{\partial\zeta}{\partial x}\delta\zeta\Big|_1^2\right)dy - \int\int\frac{\partial^2\zeta}{\partial x^2}\delta\zeta\,dx\,dy = -\int\int\frac{\partial^2\zeta}{\partial x^2}\delta\zeta\,dx\,dy, \qquad (65.9)$$

since two surface points may be considered fixed, and $\delta\zeta_1 = \delta\zeta_2 = 0$.

The change in the volume $\delta V_1$ of the first phase produced by an infinitesimal displacement of points on the surface is

$$\delta V_1 = \int \delta\zeta\,dx\,dy. \qquad (65.10)$$

Substituting expressions (65.9) and (65.10) into the equation of total change in the free energy of the system (65.6), we get

$$\delta F = \int\int\left[-(p_1-p_2)+\sigma\frac{\partial^2\zeta}{\partial x^2}\right]\delta\zeta\,dx\,dy. \qquad (65.11)$$

In an equilibrium state the free energy is a minimum and $\delta F = 0$. Since $\delta\zeta$ is an arbitrary and an infinitesimal quantity, it follows from equation (65.11) that

$$p_1 - p_2 = -\sigma\frac{\partial^2\zeta}{\partial x^2}. \qquad (65.12)$$

The difference in pressures in the two contiguous phases, in the presence of surface tension, is known as capillary pressure $p_\sigma$. It follows from (65.12) that, with small deformations of a surface curved only in one direction, we have

$$p_\sigma = p_1 - p_2 = -\sigma \frac{\partial^2 \zeta}{\partial x^2}. \tag{65.13}$$

The expression $p_\sigma = p_1 - p_2$ gives the change in mechanical equilibrium of two contiguous phases when the interface is not a plane. Formula (65.13) and analogous expressions that are derived below determine the shape of the free surface of the liquid.

A similar expression for capillary pressure can be derived for the case of a surface that is curved in two directions.

For subsequent discussion, we need to derive expressions for capillary pressure at small deformations of a cylindrical and a spherical surface.

The surface area in cylindrical coordinates $(r, z, \psi)$ is

$$\Sigma = \int\int \sqrt{r^2 + r^2\left(\frac{\partial r}{\partial z}\right)^2 + \left(\frac{\partial r}{\partial \psi}\right)^2}\, d\psi\, dz. \tag{65.14}$$

Assuming that the undeformed cylindrical surface had a radius a, and that the deformed surface has a radius $a + \zeta(\psi, z)$, then the deformed surface for a small displacement $\zeta(\psi, z)$ is described by

$$\Sigma = \int\int (a+\zeta)\sqrt{1 + \left(\frac{\partial \zeta}{\partial z}\right)^2 + \frac{1}{a^2}\left(\frac{\partial \zeta}{\partial \psi}\right)^2}\, dz\, d\psi \approx$$
$$\approx \int\int (a+\zeta)\left[1 + \frac{1}{2}\left(\frac{\partial \zeta}{\partial z}\right)^2 + \frac{1}{2a^2}\left(\frac{\partial \zeta}{\partial \psi}\right)^2\right] dz\, d\psi \approx$$
$$\approx \int\int \left[a + \zeta + \frac{a}{2}\left(\frac{\partial \zeta}{\partial z}\right)^2 + \frac{1}{2a}\left(\frac{\partial \zeta}{\partial \psi}\right)^2\right] dz\, d\psi.$$

The variation of the surface area is

$$\delta\Sigma = \int\int \left[\delta\zeta + a\left(\frac{\partial \zeta}{\partial z}\right)\frac{\partial \delta\zeta}{\partial z} + \frac{1}{a}\left(\frac{\partial \zeta}{\partial \psi}\right)\frac{\partial \delta\zeta}{\partial \psi}\right] dz\, d\psi.$$

Integrating the second and third terms of the last equation by parts with respect to the variables z and $\psi$, we obtain

$$\int\int \frac{\partial \zeta}{\partial z}\frac{\partial \delta\zeta}{\partial z}\, dz\, d\psi = \int \frac{\partial \zeta}{\partial z}\delta\zeta\Big|_0^L d\psi - \int\int \frac{\partial^2 \zeta}{\partial z^2}\delta\zeta\, dz\, d\psi =$$
$$= -\int\int \frac{\partial^2 \zeta}{\partial z^2}\delta\zeta\, dz\, d\psi,$$
$$\int\int \frac{\partial \zeta}{\partial \psi}\frac{\partial \delta\zeta}{\partial \psi}\, dz\, d\psi = \int \frac{\partial \zeta}{\partial \psi}\delta\zeta\Big|_0^{2\pi} dz - \int \frac{\partial^2 \zeta}{\partial \psi^2}\delta\zeta\, dz\, d\psi.$$

Here, L designates the length of the cylinder. At the limits of integration the variation $\delta\zeta$ may be assumed to be zero.

We then find:

$$\delta\Sigma = \int\int \delta\zeta\left[1 - a\frac{\partial^2 \zeta}{\partial z^2} - \frac{1}{a}\frac{\partial^2 \zeta}{\partial \psi^2}\right] d\psi\, dz. \tag{65.15}$$

The variation in the volume for an infinitesimal displacement of the surface may be written in cylindrical coordinates in the form

$$\delta V = \int\int \delta\zeta\,(a+\zeta)\,dz\,d\psi. \tag{65.16}$$

Substituting (65.15) and (65.16) into (65.6), we get

$$\delta F = \int\int -(p_1 - p_2)\,\delta\zeta\,(a+\zeta)\,dz\,d\psi + \\ + \int\int \sigma\left[1 - a\frac{\partial^2\zeta}{\partial z^2} - \frac{1}{a}\frac{\partial^2\zeta}{\partial\psi^2}\right]\delta\zeta\,d\psi\,dz.$$

The condition of minimum free energy gives

$$-(p_1 - p_2) + \frac{\sigma}{a+\zeta}\left[1 - a\frac{\partial^2\zeta}{\partial z^2} - \frac{1}{a}\frac{\partial^2\zeta}{\partial\psi^2}\right] = 0,$$

whence

$$p_\sigma = (p_1 - p_2) \approx \frac{\sigma}{a} - \frac{\sigma}{a^2}\left[\zeta + a^2\frac{\partial^2\zeta}{\partial z^2} + \frac{\partial^2\zeta}{\partial\psi^2}\right].$$

The first term $\frac{\sigma}{a}$ is a constant that is immaterial here and may be omitted. We then have

$$p_\sigma \approx -\frac{\sigma}{a^2}\left[\zeta + a^2\frac{\partial^2\zeta}{\partial z^2} + \frac{\partial^2\zeta}{\partial\psi^2}\right]. \tag{65.17}$$

The capillary pressure in the case of deformation of a spherical surface may be derived in exactly the same manner.

In spherical coordinates $(r, \theta, \psi)$, the surface area is

$$\Sigma = \int\int \sqrt{1 + \frac{1}{r^2}\left(\frac{\partial r}{\partial\theta}\right)^2 + \frac{1}{r^2\sin^2\theta}\left(\frac{\partial r}{\partial\psi}\right)^2}\; r^2\sin\theta\,d\theta\,d\psi.$$

If the undeformed sphere has a radius R, then the radius of the deformed surface is $R + \zeta(\theta, \psi)$. For a small displacement $\zeta$, the deformed surface is given by

$$\Sigma = \int\int \sqrt{1 + \frac{1}{R^2}\left(\frac{\partial\zeta}{\partial\theta}\right)^2 + \frac{1}{R^2\sin^2\theta}\left(\frac{\partial\zeta}{\partial\psi}\right)^2}\,(R+\zeta)^2\sin\theta\,d\theta\,d\psi \approx \\ \approx \int\int\left[1 + \frac{1}{2R^2}\left(\frac{\partial\zeta}{\partial\theta}\right)^2 + \frac{1}{2R^2\sin^2\theta}\left(\frac{\partial\zeta}{\partial\psi}\right)^2\right](R+\zeta)^2\sin\theta\,d\theta\,d\psi.$$

The variation of the surface is then expressed as follows:

$$\delta\Sigma = \int\int\left\{2(R+\zeta)\,\delta\zeta + \frac{\partial\zeta}{\partial\theta}\frac{\partial\delta\zeta}{\partial\theta} + \frac{1}{\sin^2\theta}\frac{\partial\zeta}{\partial\psi}\frac{\partial\delta\zeta}{\partial\psi}\right\}\sin\theta\,d\theta\,d\psi.$$

Integration of the last two terms by parts gives

$$\int \frac{\partial \zeta}{\partial \theta}\frac{\partial \delta\zeta}{\partial \theta}\sin\theta\, d\theta = \left[\left(\frac{\partial \zeta}{\partial \theta}\sin\theta\right)\delta\zeta\Big|_0^\pi\right] - \int \delta\zeta \frac{\partial}{\partial \theta}\left(\sin\theta \frac{\partial \zeta}{\partial \theta}\right) d\theta,$$

$$\int \frac{\partial \zeta}{\partial \psi}\frac{\partial \delta\zeta}{\partial \psi}\, d\psi = \left[\frac{\partial \zeta}{\partial \psi}\delta\zeta\Big|_0^\pi\right] - \int \delta\zeta \frac{\partial^2 \zeta}{\partial \psi^2}\, d\psi.$$

Since the variation $\delta\zeta$ at the limits of integration is zero we find

$$\delta\Sigma = \int\int \left\{2(R+\zeta) - \frac{1}{\sin\theta}\frac{\partial}{\partial\theta}\left(\sin\theta\frac{\partial\zeta}{\partial\theta}\right) - \frac{1}{\sin^2\theta}\frac{\partial^2\zeta}{\partial\psi^2}\right\}\delta\zeta \sin\theta\, d\theta\, d\psi. \quad (65.18)$$

The variation in the volume enclosed by the surface is

$$\delta V = \int \delta\zeta (R+\zeta)^2 \sin\theta\, d\theta\, d\psi. \quad (65.19)$$

Substituting (65.18) and (65.19) into (65.6) we have:

$$\delta F = \int \Big\{ -(p_1 - p_2)(R+\zeta)^2 + 2\sigma(R+\zeta) - \\ - \frac{\sigma}{\sin\theta}\frac{\partial}{\partial\theta}\left(\sin\theta\frac{\partial\zeta}{\partial\theta}\right) + \frac{\sigma}{\sin^2\theta}\frac{\partial^2\zeta}{\partial\psi^2}\Big\}\delta\zeta\sin\theta\, d\theta\, d\psi.$$

The requirement of minimum free energy leads to the expression for capillary pressure

$$p_\sigma = -\frac{2\zeta\sigma}{R^2} - \frac{\sigma}{R^2}\left[\frac{1}{\sin\theta}\frac{\partial}{\partial\theta}\left(\sin\theta\frac{\partial\zeta}{\partial\theta}\right) - \frac{1}{\sin^2\theta}\frac{\partial^2\zeta}{\partial\psi^2}\right]. \quad (65.20)$$

Equations (65.13), (65.17) and (65.20) are often presented in the following form:

$$p_\sigma = \sigma\left(\frac{1}{R_1} + \frac{1}{R_2}\right), \quad (65.21)$$

where $R_1$ and $R_2$ are the principal radii of curvature of the surface. Equation (65.21) is known as the Laplace equation. It may be shown that, at small curvatures (large radii of curvature), the expressions for the sum of the radii of curvature correspond to the formulas given above.

For simple geometric configurations, equation (65.15) may be fully integrated. For example, let a liquid be adjacent to one side of a plane vertical wall (Figure 68). Let us designate the height of the meniscus by y and its thickness by h, and let $\theta$ be the angle of contact between the liquid and the wall. Let the y axis be vertically upward, and the surface of the liquid be at the plane $y = 0$. The gas pressure $p_2$ is constant and may be selected as the reference for pressure measurements. The pressure in the liquid at rest is

$$p_1 = \text{const} - \rho g y. \tag{65.22}$$

One of the radii of curvature, $R_2$, for example, is equal to zero.

Condition (65.21), therefore, gives

$$-\rho g y = \frac{\sigma}{R_1} + \text{const.} \tag{65.23}$$

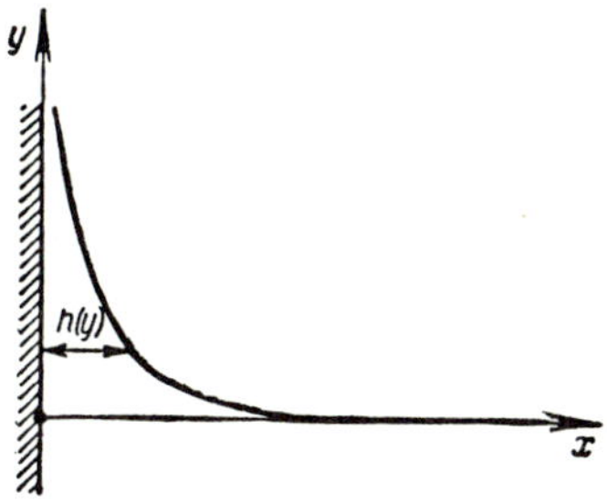

Figure 68. Meniscus at a vertical plate.

For the radius of curvature, we may write

$$\frac{1}{R_1} = -\frac{\frac{d^2h}{dx^2}}{\left(1+\left(\frac{dh}{dx}\right)^2\right)^{3/2}}. \tag{65.24}$$

Assuming that the constant in equation (65.23) is equal to zero, we have

$$\frac{\frac{d^2h}{dx^2}}{\left(1+\left(\frac{dh}{dx}\right)^2\right)^{3/2}} = \frac{\rho g x}{\sigma}. \tag{65.25}$$

Integrating twice equation (65.25) yields an expression which defines the shape of the surface

$$h(x) = -\sqrt{\frac{\sigma}{\rho g}} \operatorname{arg\,ch} 2\sqrt{\frac{\sigma}{\rho g x^2}} + \sqrt{\frac{2\sigma}{\rho g}}\sqrt{2 - \frac{\rho g x^2}{2\sigma}} + \text{const.} \tag{65.26}$$

The constant here is determined by the magnitude of the angle of contact.*

Under given conditions, tangential forces may be exterted on the surface of the liquid, along with normal pressure. If the surface tension of the liquid changes from point to point, a tangential force will be exerted in addition to the pressure normal to the surface. This force is similar to the volume force that appears in a medium with nonuniform pressure, and its magnitude is determined by the surface tension gradient.

---

*Editorial note. In the elegant study of M. M. Kusakov and D. N. Nekrasov [9], the capillary rise in capillaries of arbitrary shapes is examined.

The tangential force per unit area is

$$p_t = \text{grad}\,\sigma. \tag{65.27}$$

The plus sign preceding the gradient indicates that the force $P_t$ tends to move the surface of the liquid in a direction form lower to higher surface tension.

## 66. MOTION INDUCED BY CAPILLARITY

The presence of surface tension on the free surface of a liquid may, in certain cases, considerably affect the regime of motion of the liquid. Those cases of liquid motion in which surface tension plays an important role, are often described by the general term "capillary hydrodynamics."

The presence of an interface between fluid phases may exert an influence on the motion of the liquid in two instances: when the interface has finite curvature, and when the interfacial tension varies from point to point in the liquid surface. In both instances, forces appear in the interfacial region that generally affect the liquid motion. The effect of surface tension on the motion of the liquid is shown by a new system of boundary conditions for the interface between two contiguous liquids.

Let us set up a system of boundary conditions for the interface between two liquid phases.

As pointed out in Section 1, the following conditions obtain at the interface between two liquid phases:

1) The tangential velocity component at the liquid surface must be continuous:

$$v_t^{(1)} = v_t^{(2)}. \tag{66.1}$$

Condition (66.1) indicates that the molecular interactions of particles in both phases occur without slippage.

2) The velocity components normal to the interface of two immiscible phases equal zero:

$$v_n^{(1)} = v_n^{(2)} = 0. \tag{66.2}$$

The existence of interfacial tension has no effect on boundary conditions (66.1) and (66.2). In addition to these kinematic conditions, dynamic conditions must also be fulfilled at the interface.

3) Normal force components per unit area of the interface (normal components of the stress tensor) must be continuous:

$$p_{nn} + p_\sigma = p'_{nn}. \tag{66.3}$$

Condition (66.3) is the direct generalization of condition (65.13) and represents the balance of the normal component of forces exerted on the interface.

4) The tangential force component per unit area must be continuous. This is the condition of continuity of the tangential components of the stress tensor

$$p_{tt} + p_t = p'_{tt}. \qquad (66.4)$$

The above boundary conditions include the quantities $p_\sigma$ and $p_t$, which are functions of surface tension. It is obvious that the effect of surface tension is reflected in conditions (66.3) and (66.4). It should be noted here that while condition (66.3) can be fulfilled on the surface of a static liquid, condition (66.4) cannot be fulfilled under static conditions.

Indeed, the normal stress tensor component in a liquid at rest, according to formula (1.5), reduces to the usual pressure, and equations (66.3) and (65.13) will agree.

Conversely, by definition (1.5), the tangential stress tensor component vanishes for a liquid at rest. Boundary condition (66.4) in such a liquid, therefore, cannot be satisfied since $p_{tt} = p'_{t't'} = 0$, and $p_t \neq 0$. The occurrence of variations in surface tension always sets the liquid into motion.

Another general remark can be made with respect to boundary conditions (66.3) and (66.4): there is no effect of the liquid surface in large scale motions.

The extent of this effect may be evaluated from boundary conditions (66.3) and (66.4). For example, when motion takes place in a field of gravity, it follows from condition (66.3) that capillary pressure may be neglected, provided that

$$p_\sigma \approx \frac{\sigma}{\lambda} \ll \rho g \lambda, \qquad (66.5)$$

where $\lambda$ is the scale of motion.

In other words, if the scale of motion $\lambda$ is large in comparison with the capillary constant $\left(\frac{\sigma}{\rho g}\right)^{1/2}$, the surface forces do not affect the motion. For example, as shown in Sec. 93, surface tension affects the wave process on the surface of a liquid only in the case those waves whose length $\lambda$ is shorter than $\left(\frac{\sigma}{\rho g}\right)^{1/2}$.

In general, the effect of varying surface tension on the motion of a liquid which has a free surface or which is contiguous to another liquid may be viewed as follows. Let us reduce the boundary conditions, for example, condition (66.4), to a dimensionless form. Assuming that the motion of the liquid is viscous, we may write

$$\frac{\mu U}{L}\frac{\partial V_x}{\partial Y} - \frac{1}{l}\frac{\partial \sigma}{\partial X} = \frac{\mu' U'}{L}\frac{\partial V'_x}{\partial Y} \qquad (66.6)$$

or

$$\frac{\partial V_X}{\partial Y} - \left|\frac{\partial \sigma}{\partial X}\right| \frac{1}{\mu U} \frac{L}{l} = \frac{\mu' U'}{\mu U} \frac{\partial V'_X}{\partial Y} \tag{66.7}$$

Assuming the dimensionless derivative $\frac{\partial V_X}{\partial X}$ to be of the order of unity, the surface tension term may be omitted if

$$\frac{L}{l}\left|\frac{\partial \sigma}{\partial X}\right| \frac{1}{\mu U} \ll 1 \tag{66.8}$$

$$\left|\frac{\partial \sigma}{\partial x}\right| \frac{L^2}{\rho \nu^2} \ll \mathrm{Re} \tag{66.9}$$

Since the motion of the liquid was assumed to be viscous, the Reynolds number must be small compared to unity. Therefore, formula (66.9) may now be rewritten

$$\left|\frac{\partial \sigma}{\partial x}\right| \frac{L^2}{\rho \nu^2} \ll 1. \tag{66.10}$$

It follows from (66.10) that the effect of the varying surface tension on the liquid motion may be disregarded for changes in surface tension such that

$$\left|\frac{\partial \sigma}{\partial x}\right| \ll \frac{\rho \nu^2}{L^2}.$$

Thus, when the liquid velocity is low and the flow region is sufficiently thick, the effect of surface forces arising in variations of surface tension from one point to another may be neglected only at very small values of $\frac{\partial \sigma}{\partial x}$.

Of course, the characteristic thickness L must not be too large, if the Reynolds number is to remain small compared to unity.

## 67. RATE OF CAPILLARY RISE

As the simplest example of capillary motion, we examine the rise of liquid in a cylindrical capillary.

Capillary rise of liquids is often encountered in practice and in nature; it plays an important role in liquid motion through porous media.

The surface of a liquid filling a vertical capillary is subjected to a capillary force equal to $p_\sigma \cdot S$, where $p_\sigma$ is the capillary pressure and S is the free surface of the liquid. Capillary pressure is equal to $p_\sigma = \frac{2\sigma}{r}$ and the liquid column of height h exerts a hydrostatic pressure equal to $\rho gh$. The liquid is, therefore, subject to the difference in pressures $\Delta p = p_\sigma - \rho gh$. Under the influence of this pressure difference, the liquid moves at a rate which may be easily computed.

Since the pressure gradient along the capillary is constant, the Poiseuille equation may be used to compute the rate of the rise. We may write

$$v = \frac{dh}{dt} = \frac{r^2}{8\mu}\frac{\Delta p}{h} \tag{67.1}$$

where v is the rate of rise produced by the pressure difference.

Substituting the value of $\Delta p$, we find

$$\frac{dh}{dt} = \frac{r^2}{8\mu h}\left(\frac{2\sigma}{r} - \rho gh\right). \tag{67.2}$$

Integrating this equation, we obtain

$$t = \frac{8\mu}{r^2\rho g}\left(h_0 \ln\frac{h_0}{h_0 - h} - h\right) \tag{67.3}$$

In order to simplify equation (67.3), we have designated the equilibrium height of capillary rise by $h_0$, which is defined by $\frac{2\sigma}{r} - \rho gh_0 = 0$.

Equation (67.3) shows that although infinite time is necessary for a rise of the liquid to its maximum height, the liquid actually reaches a height that differs little from the maximum in a relatively short period of time.

Similar considerations for the case of a horizontal capillary lead to the following expression for the time it takes the liquid to travel a distance $l$ along the capillary:

$$t = \frac{2\mu l^2}{r\sigma}$$

In arriving at formula (67.3), we employed Poiseuille's law (67.1) which is valid for steady flow. In real cases, however, capillary rise is not a strictly steady-state phenomenon.

At large values of t, capillary rise becomes quasi-stationary. Therefore, formula (67.3) should be used only for large values of t, i.e., when $t \gg \frac{\rho r^2}{\mu}$.

Experiments conducted by S. S. Kozlovskii [4] with transformer oil confirmed the applicability of formula (67.3) to the capillary rise of liquids in a glass tube when $t \gg \frac{\rho r^2}{\mu}$.

## 68. THERMOCAPILLARY MOTION

As a second example of capillary motion, we examine the motion of a liquid with varying surface tension. The variation in surface tension from one point to another leads to the appearance of tangential stresses on the liquid surface as described by equation (65.27).

The reason for the variation in surface tension of the liquid is not important here. Later, we present numerous examples of reasons for the variation in surface tension, such as the introduction of surface-active substances with concentrations that vary from point to point, the presence of a varying electric charge on the liquid surface, and so on.

The simplest reason, however, is the variation in surface temperature.

Let us assume that the liquid is placed in a vessel whose sides are at different temperatures. The surface temperature of the liquid will, in that case, vary from point to point. Since the surface tension of the liquid is a function of temperature, it will also vary from point to point across the surface. When tangential forces act on the liquid surface, a motion will be produced. This motion is known as capillary convection. Obviously, the temperature difference in the vessel walls will produce the usual convective motion within the liquid in addition to the capillary convection. In certain cases, however, the usual convection will play a relatively unimportant role. When the surface of the liquid is sufficiently large compared to its volume, e.g., if the liquid is poured into a sufficiently wide shallow pan, the usual thermal convection will produce liquid velocities that are small compared to those caused by capillary convection. Indeed, the surface effects in this case must be large compared to those in the bulk of the liquid, because the surface dimensions are relatively large, and the surface tension forces are very significant compared to the gravitational forces, which are proportional to extremely small changes in liquid density.

Quantitative calculations confirm these qualitative considerations [5].

We examine the motion of a liquid poured into a shallow pan to a depth h. The two walls of the pan are at temperatures $T_1$ and $T_2$, with $T_2 > T_1$. We assume that the usual mixing by thermal convection may be neglected and that the temperature is constant along the vessel. This assumption is justified later.

Let us find the expression for the variation in surface tension of the liquid. We take the x axis in the direction of the temperature gradient, and the y axis in the direction perpendicular to the surface of the liquid. The variation in surface tension is given by

$$\operatorname{grad} \sigma = \frac{\partial \sigma}{\partial T} \operatorname{grad} T. \tag{68.1}$$

The liquid surface is at the plane $y = 0$, and the bottom of the pan is at the plane $y = h$.

The temperature coefficient for the surface tension of all liquids $\left(\frac{\partial\sigma}{\partial T}\right) < 0$; i.e., $\sigma$ decreases with increasing liquid temperature. As a first approximation, we assume that the temperature gradient is constant along the length of the pan

$$\sigma = \sigma(T_1) + \left(\frac{\partial\sigma}{\partial T}\right)\left(\frac{T_2 - T_1}{l}\right)x. \tag{68.2}$$

The surface tension varies from point to point along that surface, attaining a maximum at the colder wall of the pan, and decreasing linearly in the direction of the warmer wall.

According to expression (65.27), the force exerted per 1 cm² of surface is

$$p_t = \frac{\partial\sigma}{\partial x} = \left(\frac{\partial\sigma}{\partial T}\right)\operatorname{grad} T. \tag{68.3}$$

We determine the motion of the liquid produced by this force. Since the depth of the pan is very small compared to its other dimensions, and the assumption has been made that usual convection can be neglected, the hydrodynamic equations can be simplified.

Since there is no temperature gradient across the pan, there will be no motion in that direction. Furthermore, in the absence of convection arising in heating of the liquid, there can be no forces exerted in the direction of the y axis. The velocity component in the y direction therefore is zero. The equation for the velocity component in the x direction, in accordance with expression (1.2), is

$$\frac{\partial p}{\partial x} = \mu\left(\frac{\partial^2 v_x}{\partial x^2} + \frac{\partial^2 v_x}{\partial y^2}\right). \tag{68.4}$$

Since the depth of the pan is very small in comparison to its other dimensions, the derivative $\partial v_x/\partial x$ will be very small in comparison with the derivative $\partial v_x/\partial y$. Hence, we may omit the term $\partial^2 v_x/\partial x^2$ in equation (68.4) as a higher order infinitesimal. We then obtain

$$\frac{\partial p}{\partial x} = \mu\frac{\partial^2 v_x}{\partial y^2}. \tag{68.5}$$

Again, since the pan is shallow, it may be assumed that the pressure is not a function of the distance y. This is clearly evident from the equation for the velocity component in the y direction

$$\frac{\partial p}{\partial y} = 0. \tag{68.6}$$

In equation (68.5), therefore, the pressure p may be considered a function of the distance x only.

The system of equations (68.5) and (68.6) must be completed with the continuity equation. In this case, the continuity equation can be conveniently expressed in terms of the total flow of the liquid in the pan by setting the equation equal to zero. Indeed, the flow of liquid at the surface, set in motion by surface forces, is accompanied by flow in the opposite direction in the remainder of the vessel. The reverse flow is due to the pressure gradient along the pan as given in equation (68.5).

Thus, the continuity equation in this case becomes

$$\int_0^h v_x\,dy = 0. \tag{68.7}$$

We can now write the system of boundary conditions for equations (68.5) to (68.7). At the bottom of the vessel, the liquid velocity is zero, and

$$(v_x)_{y=h} = 0. \tag{68.8}$$

The condition of continuity of the tangential stress component must be fulfilled at the free liquid surface; i.e., the viscous stress and the surface force per unit area must be equal:

$$\mu\left(\frac{\partial v_x}{\partial y}\right)_{y=0} = p_t. \tag{68.9}$$

Using the definition of $p_t$ given above, the boundary condition at the surface is

$$\mu\left(\frac{\partial v_x}{\partial y}\right)_{y=0} = \frac{\partial \sigma}{\partial T}\,\mathrm{grad}\,T. \tag{68.10}$$

Integrating equation (68.5), and remembering that pressure changes only in the x direction, we obtain

$$v_x = a + by + \frac{1}{2}\frac{1}{\mu}\cdot\frac{\partial p}{\partial x}\cdot y^2. \tag{68.11}$$

Boundary conditions (68.8) and (68.10) yield the following values for the constants a and b:

$$b = -\frac{1}{\mu}\frac{\partial \sigma}{\partial x}; \qquad a = \frac{1}{\mu}\frac{\partial \sigma}{\partial x}h - \frac{1}{2\mu}\frac{\partial p}{\partial x}h^2.$$

As an important example of an actual operation in which thermocapillary motion plays the principal part, we now examine the mass transfer through a thin liquid film.

The transfer of oxygen through thin layers of a liquid plays an important role in metal corrosion when the metal is covered with a film of electrolyte. The electrolyte is in contact with the atmosphere, and equilibrium concentration is established on its outer surface.

The corrosion rate strongly depends on the magnitude of the oxygen flux to the metal surface.

The convective diffusion equation in the thin film is

$$v_x \frac{\partial c}{\partial x} = D \frac{\partial^2 c}{\partial y^2},$$

where $v_x$ is determined from equation (68.11). On the assumption that the diffusion resistance is in a region of low velocities (i.e., the thickness of the diffusion layer is small compared to that of the film), the following expression may be written for $v_x$

$$v_x \approx \alpha y',$$

where $y'$ is the vertical distance $y' = y - h$, measured from the bottom, and $\alpha = \frac{1}{2\mu}\frac{\partial \sigma}{\partial x}$. The boundary conditions in this case are

$$c \to c_0 \quad \text{as} \quad y' \to \infty,$$
$$c = 0 \quad \text{at} \quad y' = 0.$$

Thus, we obtain

$$v_x = \frac{1}{\mu}\frac{\partial \sigma}{\partial x}(h - y) - \frac{1}{2\mu}\cdot\frac{\partial p}{\partial x}(h^2 - y^2). \tag{68.11'}$$

Substituting the velocity as given above into the continuity equation, we get

$$\frac{\partial p}{\partial x} = \frac{3}{2h}\frac{\partial \sigma}{\partial x}.$$

Integrating the equation for the pressure gradient, we find the pressure distribution in the liquid

$$p = p_0 + \frac{3}{2h}[\sigma(x) - \sigma(0)]. \tag{68.12}$$

The constant $p_0$ is an undetermined constant pressure.

Using the pressure gradient as given in equation (68.12), we arrive at the final formula for velocity distribution

$$v_x = \frac{1}{4h\mu}\frac{\partial \sigma}{\partial T}(3y^2 - 4hy + h^2)\frac{dT}{dx}. \tag{68.13}$$

The maximum velocity of the liquid at the surface is

$$(v_x)_{y=0} = \frac{1}{4\mu} \cdot \frac{\partial \sigma}{\partial T} \cdot h \cdot \frac{dT}{dx}. \tag{68.14}$$

The velocity increases with the temperature gradient and with the depth of the liquid layer in the pan. This is true only for thin liquid layers. If the depth of the vessel becomes sufficiently great, the above solution will not apply, because it was obtained on the assumption that the following inequality is valid:

$$v_x \frac{\partial v_x}{\partial x} \ll \nu \frac{\partial^2 v_x}{\partial y^2};$$

i.e., it was assumed that the depth $h^2$ of the liquid layer is small compared to $\frac{\nu l}{|v_x|}$. Using expression (68.14), the following inequality is the condition for applicability of our solution,

$$h^3 \ll \frac{4\nu^2 \rho l}{\left|\frac{\partial \sigma}{\partial T}\right| \left|\frac{\partial T}{\partial x}\right|}. \tag{68.15}$$

The velocities arising in the liquid under the action of the surface forces are quite large. For example, for water, at $|\text{grad } T| = 0.1$ degree/cm, in a vessel whose depth is $h = 0.03$ cm, and where $\frac{\partial \sigma}{\partial T} = -0.15$ erg/cm$^2$·degree, we find:

$$(v_x)_{y=0} = 0.1 \text{ cm/sec.}$$

Such velocities are substantially greater than those arising in the liquid due to ordinary convection, which in turn is caused by the density differences in the liquid at the walls.

It should also be borne in mind that equation (68.14) was derived on the assumption that the value of the Reynolds number is low. This equation may be used only when

$$\text{Re} = \frac{(v_x)_{y=0} h}{\nu} = \frac{1}{4\rho\nu^2} \frac{\partial \sigma}{\partial T} \frac{dT}{dx} h^2 \ll 1. \tag{68.16}$$

The above inequality corresponds to the general condition (66.10).

The diffusion problem is found to be identical with that of diffusion at the inlet section of a tube as presented in Sec. 20. Using equation (20.11), the mass flux equation is

$$I_{\text{diff}} \approx 0.6 \frac{c_0 S D^{2/3}}{L^{1/3}} \left(\frac{1}{\mu} \frac{\partial \sigma}{\partial x}\right)^{1/3},$$

where S is the area of the film and L its length (the larger dimension).

The mass flux is independent of the film thickness provided the latter is large compared to the thickness of the diffusion layer; i.e.,

$$\delta' = L^{1/3} D^{1/3} \frac{1}{\left(\frac{1}{\mu}\frac{\partial\sigma}{\partial x}\right)^{1/3}} \ll h.$$

At values for $\delta'$ of the order of h, the flux is given by

$$I \sim \frac{Dc_0 S}{h}.$$

The expressions given above were derived by I. L. Rosenfel'd and K. A. Zhigalova [6].

Measured values of the average current density equal to $\frac{IFn}{S}$, using oxygen, are shown in Figure 69.

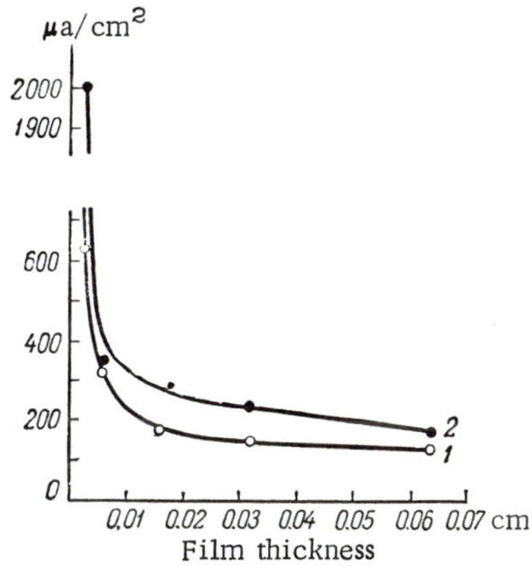


Figure 69. Dependence of the rate of oxygen reduction on the thickness of the electrolyte film. The electrolyte is 0.1 N NaCl; the cathode material is copper. Vapor pressure of water: 1) 17 mm Hg; 2) 1.15 mm Hg.

We see that the theoretical requirement that I be independent of h is found at $h > 3 \cdot 10^{-2}$ cm. In this case $\delta' \approx \frac{4}{100} h$.

An estimate of the value of $\frac{\partial\sigma}{\partial x}$ in these experiments gives $\frac{\partial\sigma}{\partial x} \sim 1$ (at a film length L of a few centimeters). Such variation in surface tension occurred because of nonuniformity of evaporation, change in solution composition during electrolysis, and changes in concentration of the electrolyte in various parts of the film.

Thermocapillary motion causes the formation of various types of films in an unevenly heated liquid.

## 69. EFFECT OF SURFACE-ACTIVE SUBSTANCES ON LIQUID FLOW

Experiments with liquids having clean surfaces are relatively rare in actual practice. Usually, the liquid is covered with surface-active substances to a greater or lesser degree [7].

Such a film modifies the forces exerted on the surface of the fluid. The modification is not simply the reduction of the surface tension. It involves tangential forces arising on the liquid surface. This leads to a change in forces of capillarity as well as to the appearance of additional surface forces which are equal in magnitude to the capillary pressure.

The presence of surface-active substances, therefore, may lead to a significant change in the hydrodynamic regime. Indeed, the motion of the liquid and the change in shape of the surface caused by the motion lead to a change in the distribution of the concentration of the surface-active substance. Thus, the concentration of the latter varies from point to point on the surface

$$\Gamma = \Gamma(x, y, z),$$

where (x, y, z) are related to each other by the equation for the surface. A change in concentration of the surface-active substance results in a variation in surface tension along the surface. Therefore, capillary pressure; i.e., the normal force acting per 1 cm$^2$ of surface area

$$p_n = \frac{2\sigma(\Gamma)}{R} \tag{69.1}$$

will be different at different points on the surface.

When the surface tension of the liquid varies from point to point, then, a tangential force arises on the liquid surface in addition to the capillary pressure. The tangential force is directed from points of greater to points of lesser surface tension. This force per unit of liquid surface area is given by

$$p_t = \operatorname{grad} \sigma = \frac{\partial \sigma}{\partial \Gamma} \operatorname{grad} \Gamma \tag{69.2}$$

and is directed tangentially to the surface.

Thus, the presence of surface-active substances leads to the appearance of surface forces and changes the boundary conditions that must be fulfilled by the liquid velocity in the surface.

The boundary conditions at the interface between two liquids (or liquid-gas) are

$$p_{nn} + p_{\sigma} = p'_{nn}, \tag{69.3}$$

$$p_{tt} + p_t = p'_{tt}, \tag{69.4}$$

where $p_{nn}$ and $p_{tt}$ are the normal and tangential components of the viscous stress tensor, $p_{\sigma}$ is the capillary pressure, and $p_t$ is the tangential force (per 1 $cm^2$ of surface area).

The difference in boundary conditions on a liquid surface covered with a film, as compared to pure liquid, in turn leads to a different flow regime.

The effect of surface-active substances on liquid flow is apparent as long as the condition given in (69.4) is controlling. This condition concerns only the viscous stresses in the liquid. At high Reynolds numbers, therefore, when the viscous forces are small compared to inertia and the liquid viscosity is not important, the tangential velocity of the liquid is not limited in any way. Thus, at high Reynolds numbers, the presence of surface-active substances does not influence the liquid flow.* The meaning of this assertion is simple. At high liquid velocities, the surface-active substances will be entirely swept away ("blown off"). Conversely, in a viscous flow regime, the effect of surface-active substances could be very significant. It should be stressed that, along with the effect of the surface-active substance on liquid flow, the motion of the surface-active substance should also be examined. Indeed, the energy dissipation caused by irreversible processes occurs not only in the liquid, but also in the film formed by the surface-active substance. One can speak of a surface viscosity which manifests itself when there is motion in the film.

However, since films of surface-active substances generally are composed of a single layer of surface-active molecules, the dissipation of energy in this layer is small compared to that in the bulk phase. Later (Sec. 77) we discuss in detail the problem of viscosity of monolayers and the circumstances under which this viscosity is important. In any event, the viscosity of the monolayer can be taken into account in formulating all further equations. The viscosity of

---

*It should be borne in mind, however, at high Reynolds numbers, that surface-active substances may have an indirect effect on the motion of the liquid. S. I. Kosterin and M. N. Rubanovich [8] showed that for liquid flow in a tube, the presence of surface-active substances may cause persistent foaming. The presence of numerous bubbles adhering to the tube walls leads to an increase in the hydrodynamic resistance coefficient (separating the bubbles from the walls requires the expenditure of a fairly considerable amount of energy).

the monolayer may be determined only from experimental data — data on the energy dissipation in liquids covered by a monolayer of surface-active materials. Experiments always measure the sum of the effects, and provide no subdivision of the energy dissipation into contributions from the surface and bulk phases.

For separate determination of the energy dissipated in the liquid phase and in the monolayer, it is necessary to know the flow regime in the presence of a surface-active material. Then, the energy dissipated in the liquid could be computed. The difference between the experimentally measured energy dissipation and the calculated figure (or of two equivalent quantities) then would represent the energy dissipation due to viscous flow in the surface phase.

It is shown later than in many instances — such as, for example, the fall of a droplet through a liquid when the liquid surface is covered by a surface-active substance — viscous motion in the monolayer and the energy dissipation caused by it are so small compared with the energy dissipation in the bulk phase that this portion of the energy dissipated may be disregarded. In this sense the viscosity of the monolayer need not be taken into account. It would be incorrect, however, to regard this conclusion as generally applicable.

In other instances, the energy dissipation in the monolayer may be so important that it must be taken into account in computing the total energy dissipated.

The presence of noticeable viscous energy dissipation in the monolayer can be determined only by experiments and calculations of the liquid flow regime. In these calculations the effect of surface-active substances should be considered.

The calculation must consider boundary conditions (69.3) and (69.4) at the surface of the liquid covered by an absorbed surface-active material.

The normal and tangential forces — $p_\sigma$ and $p_t$, respectively — in expressions (69.3) and (69.4) may be computed, provided the distribution of the surface-active substance on the surface of the liquid, i.e., $\sigma(x, y, z)$, is known.

Assuming that $\sigma = \sigma(\Gamma)$ the distribution of the surface-active material on the moving liquid surface must be known in order to determine the surface forces.

This distribution is affected by several factors. First, the surface-active material on the surface is entrained by liquid flow. As a result, a convective flux of the surface-active substance arises along the liquid surface

$$j_{surf} = \Gamma \mathbf{v}_t, \tag{69.5}$$

where $\mathbf{v}_t$ is the tangential liquid velocity at the surface. Since the concentration of the surface-active material on the surface varies from point to point, a diffusion of molecules of that material appears in the liquid surface. The diffusion proceeds from points of greater to points of lesser concentration. The diffusional mass flux arising on the surface is given by

$$\mathbf{j}_{\text{diff}} = -D_s \operatorname{grad} \Gamma, \tag{69.6}$$

where $D_s$ is the surface diffusion coefficient.

If the surface-active material is soluble in the liquid, it can pass from the surface phase into the bulk phase, or vice versa, at any point on the surface.

Let $j_n$ designate the mass flux per unit area from the surface to the bulk phase, or vice versa. The flux is governed by the slower of the two steps into which this may be subdivided in either direction:

1) the desorption (or adsorption) process, and

2) the passage of dissolved surface-active materials into the bulk of the solution (or their passage to the surface from the bulk of the solution).

The law of conservation of surface-active material may then be written

$$\frac{\partial \Gamma}{\partial t} + \operatorname{div}(\mathbf{j}_{\text{surf}} + \mathbf{j}_{\text{diff}}) + j_n = 0,$$

or

$$\frac{\partial \Gamma}{\partial t} + \operatorname{div}(\Gamma \mathbf{v}_t - D_s \operatorname{grad} \Gamma) + j_n = 0. \tag{69.7}$$

In the case of a plane liquid surface represented in Cartesian coordinates, we get

$$\frac{\partial \Gamma}{\partial t} + j_n + \frac{\partial}{\partial x}\left(\Gamma v_x - D_s \frac{\partial \Gamma}{\partial x}\right) + \frac{\partial}{\partial y}\left(\Gamma v_y - D_s \frac{\partial \Gamma}{\partial y}\right) = 0. \tag{69.8}$$

It follows from equation (69.7) that the distribution of the material in the surface may be found if the velocity distribution in the liquid is known (the flux $j_n$ may be a function of the velocity distribution in the liquid, if the flux is determined by the rate of transfer of the surface-active material).

In turn, to find the velocity distribution, it is necessary to know the forces acting on the liquid surface, i.e., the distribution of the surface-active material.

Thus, in order to solve the complete problem of liquid flow in the presence of surface-active materials, it is necessary to solve simultaneously the equations of viscous liquid motion, with consideration of boundary conditions (69.3) and (69.4), and equation (69.7) for the conservation of surface-active material.

## BIBLIOGRAPHY

1. Adam, N., Fizika i khimiya poverkhnostey, Gostekhizdat, 1947. [N. K. Adam, The Physics and Chemistry of Surfaces, 3rd ed., Oxford, 1941]; E. K. Rideal, Khimiya poverkhnostnykh yavlenii (The Chemistry of Surface Phenomena), United Scientific and Technical Publishing Houses, 1936 [E. K. Rideal, An Introduction to Surface Chemistry, 2nd ed., Cambridge, 1930. Cf. J. T. Davies and E. K. Rideal, Interfacial Phenomena, Academic Press, New York, 1961].
2. Davidov, A. Yu., Teoriya kapillyarnykh yavlenii (Theory of Capillary Phenomena), Moscow, 1851; Bakker, Kapillarität und Oberflächenspannung, Handb. d. Exp. Phys., Bd. 6, Lpz., 1928.
3. Pickett, F., J. Appl. Phys. 15, 623 (1944).
4. Kozlovskiy, S. S., Trudy Grozn. neft. in-ta. (Transactions of the Grozniy Petroleum Institute), No. 7, 1949; Bikerman, Surf. chem. industr. Res., No. 4, 1948, p. 20; [J. J. Bikerman, Surface Chemistry for Industrial Research, Academic Press, New York, 1947. Cf. 2nd ed. (Surface Chemistry: Theory and Application), 1958].
5. Fedosov, A. I., Candidate dissertation, Moscow, 1948 [Cf. M. J. Block, Nature 178, 650 (1956); J. R. A. Pearson, J. Fluid Mech. 4, 489 (1958)].
6. Rozenfel'd, I. L., K. A. Zhigalova, Dokl. Akad. Nauk SSSR, 104, 876 (1955).
7. Levich, V. G., K teorii poverkhnostnykh yavlenii (On the Theory of Surface Phenomena), "Sov. Science," 1941.
8. Kosterin, S. I., M. N. Rubanovich, Izv. Akad. Nauk SSSR, Dept. of Techn. Sci., No. 7, 1085 (1949).
9. Kusakov, M. M., D. N. Nekrassov, Dokl. Akad. Nauk SSSR, 119, 107 (1958).

CHAPTER VIII

# Motion of Drops and Bubbles in Fluid Media

## 70. MOTION OF LIQUID DROPS IN FLUID MEDIA

The motion of liquid drops in a fluid medium has been studied extensively both experimentally and theoretically. The study of this process yields important conclusions concerning the properties of the interface between two liquids, and between a liquid and a gas when the medium in which the drop moves is a gas.

The study of the laws governing drop motion in a fluid medium is essential in the solution of experimental and practical engineering problems. This includes various experiments on the motion of drops in an emulsion, extraction from liquid drops, the subdivision and atomization of drops in fuel injection into internal combustion engines, motion of raindrops, etc.

The motion of liquid mercury and other metallic drops in electrolytic solutions is characterized by the existence of charges on the surfaces of these drops. It will, therefore, be considered separately in Chapter IX.

The motion of liquid drops is closely tied to the size of the diffusional flux to the drop-solution interface. Finding the latter is a matter of great practical and theoretical significance. Problems related to diffusion at the drop-solution interface are discussed in Section 72.

We shall first consider the motion of drops so small that the motion is in the viscous regime ($\mathrm{Re} \ll 1$). We have to deal most frequently with the fall of drops in a gravitational field. In what follows, therefore, we shall assume that gravity is the driving force.

The motion of a solid sphere for $\mathrm{Re} \ll 1$ is determined by the well-known Stokes formula.

Computations by Rybczynski [1] and those undertaken independently by Hadamard [2] have shown that a change in the boundary conditions on the surface of a drop leads to a significant change in the velocity of the falling drop. Since we shall later need to perform computations similar to those made in the derivation of the Rybczyński-Hadamard formula, we present its complete derivation here.

Let us consider the fall of a liquid drop in a liquid medium, the liquids being immiscible and differing in density and viscosity. Let the viscosity of the liquid medium be $\mu$, and its density, $\rho$. The same properties for the drop are $\mu'$ and $\rho'$.

Let us direct the x axis vertically downward, so that the gravitational force acting on the drop is directed along the x axis and is

$$F=(\rho-\rho')g\cdot V,$$

where V is the volume of the drop.

Under the action of the force of gravity, the drop falls in the liquid at a certain velocity U. During its motion, the liquid within the drop attains a velocity $\mathbf{v}'$, while the liquid medium surrounding the drop attains a velocity $\mathbf{v}$. In this case, the drop preserves its spherical shape while falling and is not deformed. Some time after the start of the fall, the motion of the drop in the liquid medium becomes steady.

The Navier-Stokes equations for the steady motion of the liquid medium and for the liquid within the drop have the form

$$\operatorname{grad} p=\mu\,\Delta\mathbf{v}+\mathbf{i}(\rho-\rho')g, \tag{70.1}$$

where p is the pressure and $\mathbf{i}$ is the unit vector in the positive direction of the x axis. Correspondingly

$$\operatorname{grad} p'=\mu'\,\Delta\mathbf{v}', \tag{70.2}$$

where $p'$ is the pressure in the liquid inside the drop. Velocities $\mathbf{v}$ and $\mathbf{v}'$ satisfy the continuity equation

$$\left.\begin{aligned}\operatorname{div}\mathbf{v}&=0,\\ \operatorname{div}\mathbf{v}'&=0.\end{aligned}\right\} \tag{70.3}$$

It is convenient to reduce equation (70.1) to a form similar to (70.2) by introducing an additional effective pressure instead of the external force [3],

$$\pi=(\rho-\rho')gx=\frac{Fx}{V}.$$

It is not difficult to see that when the shear stress which includes this effective pressure is integrated over the surface, the total force exerted on the spherical drop will be precisely equal to the force F.

Equation (70.1) then acquires the form

$$\operatorname{grad}(p-\pi)=\mu\cdot\Delta\mathbf{v}, \tag{70.4}$$

which is similar to (70.2).

The similarity of the equations describing the motion of the inner and outer liquids makes the computations simpler.

To find the distributions of the velocities v and v′ and of the pressures p and p′, the system of equations (70.1) to (70.4) must be solved taking the appropriate boundary conditions into consideration.

The symmetry of the problem suggests using a system of spherical coordinates $(r, \theta, \varphi)$ with the origin placed at the center of the drop. The angle $\theta$ is measured counterclockwise from the point of incidence of the flow.

Since the motion of the drop is symmetrical with respect to the x axis, the velocities of the inner and outer liquids are not functions of the angle $\varphi$; i.e., they have the following components $v_r(r, \theta)$, $v_\theta(r,\theta)$, $v'_r(r,\theta)$, and $v'_\theta(r, \theta)$.

The Navier-Stokes equations in spherical coordinates have the form

$$\frac{\partial p}{\partial r} = \mu\left(\frac{\partial^2 v_r}{\partial r^2} + \frac{1}{r^2}\frac{\partial^2 v_r}{\partial \theta^2} + \frac{2}{r}\frac{\partial v_r}{\partial r} + \frac{\cot\theta}{r^2}\frac{\partial v_r}{\partial \theta} - \right.$$
$$\left. - \frac{2}{r^2}\frac{\partial v_\theta}{\partial \theta} - \frac{2v_r}{r^2} - \frac{2\cot\theta}{r^2} v_\theta\right), \qquad (70.5)$$

$$\frac{1}{r}\frac{\partial p}{\partial \theta} = \mu\left(\frac{\partial^2 v_\theta}{\partial r^2} + \frac{1}{r^2}\frac{\partial^2 v_\theta}{\partial \theta^2} + \frac{2}{r}\frac{\partial v_\theta}{\partial r} + \right.$$
$$\left. + \frac{\operatorname{ctg}\theta}{r^2}\frac{\partial v_\theta}{\partial \theta} + \frac{2}{r^2}\frac{\partial v_r}{\partial \theta} - \frac{v_\theta}{r^2 \sin^2\theta}\right), \qquad (70.6)$$

and the continuity equation has the form

$$\frac{\partial v_r}{\partial r} + \frac{1}{r}\frac{\partial v_\theta}{\partial \theta} + \frac{2v_r}{r} + \frac{v_\theta \cot\theta}{r} = 0. \qquad (70.7)$$

The equations of motion for the inner liquid have similar forms.

In order to formulate the boundary conditions, it is convenient to pass to a system of coordinates which moves with the center of gravity of the falling drop. In such a system, the drop as a whole is considered motionless, while the outer liquid, which, as a whole, moves in the direction opposite to that of the actual motion of the drop, moves at a velocity (—U) (U being the velocity of the drop's motion).

Thus, far from the falling drop, the velocity distribution in the liquid medium has the form

$$\left.\begin{aligned} v_r &= -U\cos\theta, \\ v_\theta &= +U\sin\theta \end{aligned} \quad \text{as} \quad r \to \infty. \right\} \qquad (70.8)$$

The normal component $p_{rr}$ and the tangential component $p_{r\theta}$ of the viscous stress tensor are continuous at the liquid interface. In spherical coordinates the normal and tangential components of the stress tensor have the form

$$\left.\begin{aligned} p_{rr} &= -p + 2\mu\frac{\partial v_r}{\partial r}, \\ p_{r\theta} &= \mu\left(\frac{1}{r}\frac{\partial v_r}{\partial \theta} + \frac{\partial v_\theta}{\partial r} - \frac{v_\theta}{r}\right). \end{aligned}\right\} \tag{70.9}$$

The pressure in the expression for the normal component $p_{rr}$ of the stress tensor is the total pressure in the liquid.

In our notation, the total pressure in the outer liquid is $(p - \pi)$. The boundary conditions for the stresses on the surface of the drop, at $r = a$, can therefore be written in the form

$$-(p-\pi) + 2\mu\left(\frac{\partial v_r}{\partial r}\right)_{r=a} = -p' + 2\mu'\left(\frac{\partial v'_r}{\partial r}\right)_{r=a}, \tag{70.10}$$

$$\mu\left(\frac{1}{r}\frac{\partial v_r}{\partial \theta} + \frac{\partial v_\theta}{\partial r} - \frac{v_\theta}{r}\right)_{r=a} = \mu'\left(\frac{1}{r}\frac{\partial v'_r}{\partial \theta} + \frac{\partial v'_\theta}{\partial r} - \frac{v'_\theta}{r}\right)_{r=a}. \tag{70.11}$$

In addition, the velocities of the inner and outer liquids must satisfy the following conditions on the surface of the drop:

1) The normal components of the inner and outer fluid velocities must be zero

$$v_r = v'_r = 0 \quad \text{at} \quad r = a. \tag{70.12}$$

Indeed, since the drop is not deformed and its surface does not pulsate (at high velocities, this is not so; see Section 79) and since the inner and outer liquids are immiscible, the surface of the liquid drop is the same as the impenetrable surface of a solid particle.

2) The boundary condition for the tangential velocity components on the liquid interface is significantly different from the boundary condition on a solid surface.

The tangential velocity components of the two liquids at the interface between them must be continuous; i.e.,

$$v_\theta = v'_\theta \quad \text{at} \quad r = a. \tag{70.13}$$

The continuity requirement for the tangential velocity component at the interface between the liquids follows directly from the well-known experimental fact that there is no slip between liquids at their interface.

The last condition which must be satisfied by the liquid within the drop is a condition which replaces conditions (70.8) for the velocity of the outer liquid. This condition states that the velocity components $v'_r$ and $v'_\theta$ must remain finite at all points within the drop, particularly at the center of the drop (the origin of the coordinates):

$$v'_r \text{ and } v'_\theta \text{ are finite at } r = 0. \tag{70.14}$$

The solution of equations (70.1) to (70.4) for boundary conditions (70.8) to (70.14) allows us to find the velocity and pressure distributions in the liquids.

The boundary conditions (70.8) at infinity show that a solution of equations (70.1) to (70.4) should be sought in the form

$$\left.\begin{aligned} v_r = f(r)\cos\theta; \quad v_\theta = \varphi(r)\sin\theta, \\ p = \mu\psi(r)\cos\theta \end{aligned}\right\} \tag{70.15}$$

and similarly

$$\begin{aligned} v'_r = f'(r)\cos\theta; \quad v'_\theta = \varphi'(r)\sin\theta, \\ p' = \mu'\psi'(r)\cos\theta. \end{aligned}$$

Substituting these solutions (70.15) into equations (70.5), (70.6), and (70.7), we find:

$$\frac{d^2f}{dr^2} + \frac{2}{r}\frac{df}{dr} - \frac{4(f+\varphi)}{r^2} = \frac{d\psi}{dr}, \tag{70.16}$$

$$\frac{d^2\varphi}{dr^2} + \frac{2}{r}\frac{d\varphi}{dr} - \frac{2(f+\varphi)}{r^2} = -\frac{\psi}{r}, \tag{70.17}$$

$$\frac{df}{dr} + \frac{2(f+\varphi)}{r} = 0. \tag{70.18}$$

A similar system of equations is obtained for the inner liquid. Expressions (70.16) to (70.18) are a system of linear, ordinary differential equations, which may be solved easily. Namely, we obtain from equation (70.18):

$$\varphi = -\frac{r}{2}\frac{df}{dr} - f. \tag{70.19}$$

Substituting the value of $\varphi$ into equation (70.17), we obtain:

$$\frac{\psi}{r} = \frac{d^2}{dr^2}\left(\frac{r}{2}\frac{df}{dr} + f\right) + \frac{2}{r}\frac{d}{dr}\left(\frac{r}{2}\frac{df}{dr} + f\right) - \frac{1}{r}\frac{df}{dr}, \tag{70.20}$$

or

$$\psi = \frac{1}{2}r^2\frac{d^3f}{dr^3} + 3r\frac{d^2f}{dr^2} + 2\frac{df}{dr}.$$

Substituting the value of $\psi$ into equation (70.16), we obtain:

$$r^3\frac{d^4f}{dr^4} + 8r^2\frac{d^3f}{dr^3} + 8r\frac{d^2f}{dr^2} - 8\frac{df}{dr} = 0. \tag{70.21}$$

Expression (70.21) is an Euler equation. In accordance with the general theory of integration of equations of this type, we shall seek a solution in the form

$$f \approx r^n.$$

For n we then have the equation

$$n(n-1)(n-2)(n-3)+8n(n-1)(n-2)+8n(n-1)-8n=0,$$

or

$$n(n-2)(n+1)(n+3)=0.$$

The roots of this are:

$$n_1=0; \quad n_2=+2; \quad n_3=-1; \quad n_4=-3.$$

Therefore, we obtain for f:

$$f=\frac{b_1}{r^3}+\frac{b_2}{r}+b_3+a_1 r^2. \tag{70.22}$$

On substituting f into equations (70.20) and (70.19), we obtain:

$$\varphi=\frac{b_1}{2r^3}-\frac{b_2}{2r}-b_3-2a_1 r^2, \tag{70.23}$$

$$\psi=\frac{b_2}{r^2}+10a_1 r. \tag{70.24}$$

Substituting the above determined values of the functions f, $\varphi$, and $\psi$ from (70.22), (70.23), and (70.24) into (70.15), we obtain:

$$v_r=\left(\frac{b_1}{r^3}+\frac{b_2}{r}+b_3+a_1 r^2\right)\cos\theta, \tag{70.25}$$

$$v_\theta=\left(\frac{b_1}{2r^3}-\frac{b_2}{2r}-b_3-2a_1 r^2\right)\sin\theta, \tag{70.26}$$

$$p=\mu\left(\frac{b_2}{r^2}+10a_1 r\right)\cos\theta. \tag{70.27}$$

The expressions for the velocity components and the pressure (70.25) to (70.27) represent the general solution of equations (70.1) to (70.3). Exactly the same expressions can be written for the velocity and pressure distributions within the drop. The values of the constants should be selected so as to satisfy conditions (70.8) to (70.14). In order for the outer solution to satisfy conditions (70.8), it is necessary that the constant $a_1$ for the outer liquid be equal to zero. Otherwise $v_r$ and $v_\theta \to \infty$ as $r \to \infty$. Thus, the outer solution must have the form

$$v_r = \left(\frac{b_2}{r} + \frac{b_1}{r^3} + b_3\right)\cos\theta, \tag{70.28}$$

$$v_\theta = \left(\frac{b_1}{2r^3} - \frac{b_2}{2r} - b_3\right)\sin\theta, \tag{70.29}$$

$$p = \mu\frac{b_2}{r^2}\cos\theta. \tag{70.30}$$

In accordance with (70.14), the velocity distribution within the drop must be so expressed that the velocity components have a finite value at r = 0. For this, the constants for $\frac{1}{r}$ and $\frac{1}{r^2}$ in the inner solution must be equated to zero. The velocity and pressure distributions within the drop have the form

$$v'_r = (a_1 r^2 + a_2)\cos\theta, \tag{70.31}$$

$$v'_\theta = -(2a_1 r^2 + a_2)\sin\theta, \tag{70.32}$$

$$p' = \mu' a_0 + 10\mu' a_1 r\cos\theta. \tag{70.33}$$

Five unknowns, $b_1$, $b_2$, $b_3$, $a_1$, and $a_2$, are involved in equations (70.28) to (70.33). The boundary conditions (70.8) to (70.13) contain six equations for the determination of these constants and of a sixth, the quantity U. In the course of the calculations, the constant in (70.33) was equated to zero. This arbitrary step is related to the fact that only the pressure difference in the liquid has a definite value. Substituting the expressions (70.28) and (70.29) into boundary condition (70.8), we find:

$$b_3 = -U. \tag{70.34}$$

Substituting the values of the corresponding quantities $v_r$, $v'_r$, $v_\theta$, $v'_\theta$ p, p′, and $\pi$ into conditions (70.10) to (70.13) and following elementary, although somewhat lengthy computations, we obtain

$$U = \frac{2(\rho' - \rho)ga^2}{3\mu}\,\frac{\mu + \mu'}{2\mu + 3\mu'}. \tag{70.35}$$

This formula is known as the Rybczyński-Hadamard formula. It passes into the Stokes equation

$$U_{St} = \frac{2}{9}\,\frac{(\rho' - \rho)ga^2}{\mu}, \tag{70.36}$$

when the viscosity of the interior liquid is large, $\mu' \gg \mu$.

Comparing expressions (70.36) and (70.35), we note that

$$\frac{U}{U_{St}} = 3\,\frac{\mu + \mu'}{2\mu + 3\mu'} > 1. \tag{70.37}$$

The meaning of this result is simple. Due to the mobility of the interface, the velocity gradients present in the liquid are smaller than those in the case of a solid interface. Decreasing the velocity gradients leads to a decrease in the energy dissipated in the liquid.

With the same driving force (gravity) and a smaller energy dissipation, the velocity of steady fall of a liquid drop should be greater than the velocity of fall of a solid sphere.

The liquid on the surface of the drop moves at a velocity

$$(v_\theta)_{r=a} = v_0 \sin\theta, \tag{70.38}$$

where $v_0$ is the magnitude of the velocity at the drop's equator,

$$v_0 = \frac{\mu}{2}\frac{U}{\mu + \mu'}. \tag{70.39}$$

For $\mu' \gg \mu$, the velocity on the surface is small and in the limit becomes zero (the case of a solid surface). For $\mu' \approx \mu$, $v_0 \approx \frac{U}{4}$, and for $\mu \gg \mu'$, $v_0$ has its maximum value: $v_0 \approx \frac{U}{2}$.

## 71. COMPARISON OF THE RYBCZYŃSKI-HADAMARD FORMULA WITH EXPERIMENTAL DATA

Comparisons of the Rybczyński-Hadamard formula with experimental data have been made in many instances. In the extremely careful measurements made by A. A. Lebedev [4], the velocities of fall of mercury drops in castor oil coincided exactly with the velocities for solid spheres. Similar results have been obtained by other investigators [5,6]. On the basis of these data it was generally concluded that the Rybczyński-Hadamard formula is not applicable under actual conditions. As an explanation of this contradiction between theory and experiment, Boussinesq [7] advanced the hypothesis that a thin layer of higher viscosity exists near a liquid interface. This assumption of the existence of increased viscosity in a thin, but macroscopic, layer near the interface has received wide acceptance and will be examined in greater detail later. According to Boussinesq's theory, velocity of fall of the drop is given by the formula

$$U = \frac{2}{3}\frac{\rho - \rho'}{\mu} g a^2 \frac{\mu + \mu' + \frac{2e}{3a}}{2\mu + 3\mu' + \frac{2e}{a}}. \tag{71.1}$$

Here e is the surface viscosity coefficient which expresses the relation between surface tension and the rate at which the surface of the liquid changes.

It can be seen from equation (71.1) that, for large drop radii, surface viscosity may be neglected relative to that of the bulk liquid. However, for small radii, the surface viscosity is very important.

Bond [8] established that drops of sufficiently large dimensions fall at velocities close to those given by the Rybczyński-Hadamard formula. Small drops, on the other hand, fall like solid spheres. Bond attempted to interpret his result as being due to a change in surface tension on the drop which prevents tangential motion of the liquid interface. Such an explanation, however, causes confusion; there is no change in surface tension during the fall of a drop in a pure liquid. The time required to establish equilibrium between the surface and the bulk of a pure liquid is very small. Thus, for all practical purposes, one may always consider that the surface and the bulk of the liquid are in a state of statistical equilibrium. However, the changing composition of the equilibrium phases and the interchange of molecules between them cannot result in a change of the free energy. In particular, the renewal of the molecules in the surface layer of a drop, during the latter's motion, cannot lead to a change in the magnitude of the surface tension. Therefore, the transition of the regime of fall of a drop from that of a solid ball (following Stokes' law) to that of a drop with a mobile surface (following the Rybczyński-Hadamard law) when the drop's radius passes through a critical value $a = a_{cr}$, as was observed in the experiments of Bond, is certainly not the result of a change in surface tension. Later we shall show that the results of these experiments can be fully explained without assuming, as was done by Boussinesq, the existence of a significant surface viscosity.

In contrast to the studies mentioned above, measurements of the velocity of a falling drop of molten lead in fused $Be_2O_3$, conducted by M. P. Volarovich and A. A. Leont'yeva [9] showed that the velocity of the falling drop was one and one-half times that of the Stokes' velocity, which follows from the Rybczynski-Hadamard formula for $\mu \gg \mu'$. A similar result was obtained in a study by A.N. Frumkin and I. A. Bagotskaya [10], who established the fact that drops of mercury in extremely pure glycerine, specially purified to be free of surface-active substances, fall at velocities corresponding exactly to the Rybczyński-Hamard formula.

From what has been said, it follows that measured data for velocities of falling drops have often contradicted the hydrodynamic Rybczyński-Hadamard theory, but, at the same time, because of the extreme variation of the results, could not be used as proof of the Boussinesq hypothesis.

All of this induced the author of this book, jointly with A. N. Frumkin, to undertake a study of the effect of surface-active materials on the motion of drops [11].

In this study, as well as in special investigations conducted for the purpose of verifying the proposed theory, the mechanism of the effect of surface-active materials on drop motion was elucidated. It was found that surface-active materials in certain instances may retard motion on the surface of the drop (when the drop is small), and consequently the drop falls like a solid sphere. Under other conditions, the motion of the drop strictly obeys the Rybczyński-Hadamard law. Special attention will be given later to an analysis of the reasons for this phenomenon. We shall first examine the problem of diffusional flux to a drop. This flux is of the utmost importance in the development of a theory of drop motion in the presence of surface-active materials. In discussing the area of applicability of the expression for the diffusional flux, it should be noted (as will be shown in Section 78) that surface-active materials cannot retard the surface motion of a relatively large drop. Therefore, even in a medium containing such materials, drops of large dimensions fall under conditions of unretarded surface motion.

## 72. DIFFUSIONAL FLUX TO A MOVING DROP

The examination of convective diffusion at the interface between two liquids is based on the same general principles that were used in Chapter II for a fluid-solid body interface.

Computation of the diffusional flux to the moving surface of a drop is very important in a large variety of practical processes. Thus, the industrial process of extraction from a liquid phase is often conducted from drops falling through liquid media. The extracted material in this case passes from the falling drops into the solution. The rate of the extraction process, in most cases, is determined by the magnitude of the convective diffusional flux from the surface of the drop. Calculation of the diffusional flux on the drop is also important in polarography.

The process of convective diffusion at a fluid-fluid interface differs significantly from that at a fluid-solid body interface. This is explained, first of all, by the different hydrodynamic conditions at the phase interface. The velocity of fluid motion at the immediate surface of a solid body is always zero. However, at a liquid phase interface the liquid is still mobile. Only the normal velocity component is zero at such an interface, not the tangential velocity.

Since the fluid motion is somewhat retarded near a fluid-solid body interface, the convective transfer of material by the moving fluid to the surface is considerably less than in the bulk of the solution. Diffusion to a fluid-fluid interface, however, is more favorable since the flow is less retarded. For this reason, the convective diffusion of a substance to a fluid-fluid interface is more intense than to a fluid-solid body interface.

We shall now compute the diffusional flux to the surface of a drop falling in a liquid medium [12]. We assume that the liquids are immiscible and have different densities. To be specific, let the density of the liquid in the drop be greater than that of the medium, so that the drop under consideration falls by gravity. Henceforth we shall consider only the diffusion to a drop of sufficiently small dimensions, so that the Reynolds number is substantially less than unity. Nevertheless, as we have seen in the example of a solid particle (Section 14), the Peclet number may be sufficiently large so that a diffusion boundary layer is formed near the surface of the drop.

The concentration distribution in the boundary layer is described by equation (14.1),

$$v_r \frac{\partial c}{\partial r} + v_\theta \frac{1}{r} \cdot \frac{\partial c}{\partial \theta} = D \frac{\partial^2 c}{\partial r^2}, \tag{72.1}$$

where c is the concentration of the material diffusing from the surface of the drop into the solution. The following serve as boundary conditions:

$$c = c_0 \quad \text{as} \quad r \to \infty,$$
$$c = c_1 \quad \text{at} \quad r = a,$$

where a is the radius of the drop, and $c_0$ and $c_1$ are given concentrations.

In order to integrate equation (72.1), let us introduce the stream function $\psi$, related to $v_r$ and $v_\theta$ by the ratios (14.3),

$$v_\theta = -\frac{1}{r \sin \theta} \frac{\partial \psi}{\partial r}; \quad v_r = \frac{1}{r^2 \sin \theta} \frac{\partial \psi}{\partial \theta}.$$

Let us change to new variables, for which we select the polar angle $\theta$ and the stream function $\psi$ (compare Section 13). Following proper transformation of derivatives, we find:

$$\frac{\partial c}{\partial r} = \frac{\partial c}{\partial \psi} \frac{\partial \psi}{\partial r} = -r \sin \theta v_\theta \left(\frac{\partial c}{\partial \psi}\right)_\theta,$$

$$\left(\frac{\partial c}{\partial \theta}\right)_r = \left(\frac{\partial c}{\partial \theta}\right)_\psi + \frac{\partial c}{\partial \psi} \frac{\partial \psi}{\partial \theta} = \left(\frac{\partial c}{\partial \theta}\right)_\psi + r^2 \sin \theta v_r \left(\frac{\partial c}{\partial \psi}\right)_\theta.$$

Since equation (72.1) is valid in the boundary layer near the surface of the drop, the value of the velocity component $v_\theta$ should be taken at values of r close to a.

We introduce as a new variable the distance y from the surface of the drop,

$$y = r - a,$$

and we shall assume that $y \ll a$.

Let us express r in terms of y in equation (72.1) and let us restrict ourselves to terms of the first order. We then obtain:

$$\frac{\partial c}{\partial \theta} = Da^3 (v_\theta)_{y=0} \sin^2\theta \left(\frac{\partial^2 c}{\partial \psi^2}\right). \tag{72.2}$$

The value of $(v_\theta)_{y=0}$ (i.e., the value of $v_\theta$ on the surface of the drop) is determined by equation (70.38). The value of $\frac{\partial^2 c}{\partial \psi^2}$ is taken at small values for y, where the following expression is valid for $\psi$

$$\psi \approx -v_0 ay \sin^2\theta. \tag{72.3}$$

Substituting the value of $(v_\theta)_{y=V_0}$ into (72.2), we finally obtain:

$$\frac{\partial c}{\partial \theta} = Da^3 v_0 \sin^3\theta \frac{\partial^2 c}{\partial \psi^2}.$$

Introducing a new variable

$$t = Da^3 v_0 \int \sin^3\theta \, d\theta = Da^3 v_0 \left(\frac{\cos^3\theta}{3} - \cos\theta\right) + a_1, \tag{72.4}$$

where $a_1$ is the integration constant, we find

$$\frac{\partial c}{\partial t} = \frac{\partial^2 c}{\partial \psi^2}. \tag{72.5}$$

In this way, as a result of the change of variables, we have arrived at the classical heat conduction equation.

The boundary conditions expressed in the new variables have the form

$$\begin{aligned} c &= c_0 \quad \text{as} \quad \psi \to -\infty, \\ c &= c_1 \quad \text{as} \quad \psi \to 0. \end{aligned} \tag{72.6}$$

At the forward stagnation point (the point $r = a$, $\theta = 0$), the solution is not depleted, and its concentration must be the same as that far from the drop.

The last condition in terms of variables $(t, \psi)$ can be written in the form

$$c = c_0 \quad \text{at} \quad t = t_0 = a_1 - \frac{2}{3} Da^3 v_0.$$

The solution of equation (72.5) satisfying the boundary conditions is

$$c = \frac{2(c_0 - c_1)}{\sqrt{\pi}} \int_0^{\frac{-\psi}{2\sqrt{t - t_0}}} \exp\{-z^2\}\, dz + c_1.$$

Substituting the value of $\psi$ from (73.3) and that of t from (72.4), and returning to the variables $(y, \theta)$, we obtain:

$$c = \frac{2(c_0 - c_1)}{\sqrt{\pi}} \int_0^N e^{-z^2} dz + c_1,$$

where N is

$$N=\frac{av_0 \sin^2 \theta y}{2\sqrt{Dv_0a^3\left(\frac{2}{3}-\cos\theta+\frac{\cos^3\theta}{3}\right)}}.$$

For the density of the diffusional flux to the surface of the drop, we find

$$j=D\left(\frac{\partial c}{\partial y}\right)_{y=0}=\frac{Dv_0a\sin^2\theta\,(c_0-c_1)}{\sqrt{D\pi v_0a^3\left(\frac{2}{3}-\cos\theta+\frac{\cos^3\theta}{3}\right)}}=$$
$$=\left(\frac{Dv_0}{a}\right)^{1/2}\sqrt{\frac{3}{\pi}}\sqrt{\frac{(1+\cos\theta)^2}{2+\cos\theta}}\,(c_0-c_1)=\frac{D\,(c_0-c_1)}{\delta}, \tag{72.7}$$

where $\delta$ is the thickness of the diffusion layer, equal to

$$\delta=\sqrt{\frac{\pi}{3}}\left(\frac{aD}{v_0}\right)^{1/2}\sqrt{\frac{2+\cos\theta}{(1+\cos\theta)^2}}. \tag{72.8}$$

Equation (72.7) shows that the diffusional flux is proportional to the square root of the liquid velocity on the surface of the drop and is inversely proportional to the square root of the radius of the drop, and is also a function of the angle $\theta$. The highest flux (and the thinnest part of the diffusion layer) is at the point of incidence of fresh solution, $\theta = 0$. As $\theta$ increases, the flux gradually decreases and attains a value of zero at the point $\theta = \pi$. Correspondingly, at this point ($\theta = \pi$) the thickness of the diffusion layer is infinite. It should be borne in mind, however, that all computations based on the concept of the existence of a diffusion boundary layer are valid only where the thickness of this layer is small compared to the radius of the drop. Actually, the fact that the thickness $\delta$ of the diffusion layer becomes infinite indicates that the diffusion layer thickness at the downstream part of the drop is comparable to the radius of the drop. Correspondingly, the diffusional flux in the part of the drop, in terms of an order of magnitude, is equal to $\frac{D(c_0 - c_1)}{a}$.

For the drop, the role of the tangential transfer of matter to the reaction surface, as we have already pointed out in Section 18, is manifested with especial clarity. At the mobile interface between the phases, where only the normal velocity component is zero while the tangential component is of the same order of magnitude as in the undisturbed flow, the role of tangential transfer is especially important. In the main, the substance is transferred just along the surface of the drop, and the depletion of the solution in the downstream part of the drop is especially pronounced. The total flow of substance to the surface of the drop is obtained from expression (72.7) by integration over the entire surface of the drop.

$$I = 2\pi a^2 \int_0^\pi j \sin\theta \, d\theta = 2\sqrt{3\pi}\, a^2 \left(\frac{Dv_0}{a}\right)^{1/2} \int_0^\pi \frac{1+\cos\theta}{\sqrt{2+\cos\theta}} \sin\theta \, d\theta \, (c_0 - c_1).$$

Evaluation of the integral gives:

$$I = 8\,(c_0 - c_1)\sqrt{\frac{\pi}{3}}\left(\frac{Dv_0}{a}\right)^{1/2} a^2. \tag{72.9}$$

Let us rewrite formula (72.9) by substituting from (70.39) the velocity U of the fall of the drop for $v_0$, the velocity of liquid motion on the surface.

Expression (72.9) then takes the form

$$I = 8\sqrt{\frac{\pi}{3}}\left(\frac{D}{2a}\cdot\frac{\mu}{\mu+\mu'}\right)^{1/2} a^2 (c_0 - c_1)\sqrt{U}. \tag{72.10}$$

Introducing into this formula the dimensionless quantities $\text{Re} = \frac{Ua}{\nu}$ and Pr, the dimensionless diffusional flux Nu to the surface of the drop can be expressed as

$$\text{Nu} = \frac{Ia}{4\pi a^2 D\,(c_0 - c_1)} = \frac{2}{\sqrt{6\pi}}\,\text{Pr}^{1/2}\text{Re}^{1/2}\left(\frac{\mu}{\mu+\mu'}\right)^{1/2}. \tag{72.11}$$

It is interesting to compare expression (72.9) for I with the expression for the total flow I of material to the surface of a solid sphere, given by equation (14.8). We see that expression (72.9) contains the Prandtl number to a higher power than (14.8). For the same Reynolds numbers, the diffusional flow to the drop is greater than to a solid particle.

The expressions derived for total flow I are valid only at small Reynolds numbers, $\text{Re} \ll 1$. For $\text{Re} > 1$, the velocity distribution near the surface of the drop changes, and at sufficiently high velocities of fall, the drop loses its spherical shape.

The theory of drop motion for $\text{Re} > 1$ has not been developed. Nevertheless, it can be supposed that the general form of the dependence of I on Re and Pr, as given in formula (72.10), is valid at Reynolds numbers substantially greater than unity (but where the motion remains laminar). Indeed, if a certain diffusion boundary layer exists on the surface of a particle, its thickness $\delta$ can be found (in terms of an order of magnitude) from the relation

$$v_0 \frac{c}{a} \approx D \frac{c}{\delta^2}. \tag{72.12}$$

Formula (72.12) is analogous to (10.13) and shows that, at a distance $\delta$ from the surface of the drop, the diffusion term in the convective diffusion equation (72.1) is of the same order of magnitude as the convective term.

No matter what the velocity distribution at the surface of the drop may be for Re > 1, the tangential velocity $v_0$ at the surface is undoubtedly equal to the velocity of motion U of the drop, to an order of magnitude. Therefore we find from formula (72.12):

$$\delta \approx \sqrt{\frac{Da}{U}} \tag{72.13}$$

and, consequently,

$$I \approx \frac{D(c_0 - c_1) \cdot 4\pi a^2}{\sqrt{\frac{Da}{U}}}. \tag{72.14}$$

In dimensionless form, formula (72.14) leads to the same functional dependence of Nu on Re and Pr as (72.11), but with an undetermined numerical coefficient. When a turbulent wake forms downstream from the drop, the situation becomes considerably more complex, and a formula like (72.11) is not valid in the downstream portion.

Experimental verification has confirmed the correctness of formula (72.11) under conditions where the motion on the surface of the drop is not retarded by surface-active materials (see Section 115).

## 73. FALL OF DROPS IN LIQUID MEDIA IN THE PRESENCE OF SURFACE-ACTIVE SUBSTANCES

If dissolved molecules of a surface-active material are present in the liquid medium in which a drop falls, they will be adsorbed at the interface between the two liquids [11].

In accordance with the general considerations regarding the effect of surface-active particles on fluid motion presented in Section 69, it may be assumed that adsorption of these materials on the surface of the drop will have a noticeable effect on its motion. Indeed, the density of molecules of the adsorbed material on the upstream part of the drop is lower than that corresponding to equilibrium with the bulk solution, because of the constant surface renewal. On the other hand, the surface density at the downstream part of the drop exceeds the equilibrium density. The liquid motion carries the molecules of surface-active substances on the surface of the drop toward the rear. The resulting accumulation of surface-active substances lowers the surface tension in the rear part of the drop. A force then appears which is exerted along the surface of the drop and tends to retard its surface motion, thereby inhibiting the further accumulation of surface-active material in the rear portion of the drop.

In order to obtain a quantitative understanding of this process and to find the resulting velocity of fall, we employ the general considerations set forth in Section 69. Here, we assume that the surface active substances are soluble in the liquid medium, but not in the liquid of the drop. As in Section 69, we accept the proposition that surface tension is a function only of the amount $\Gamma$ of adsorbed material present (i.e., we do not take into account the possibility of surface relaxation processes which would result from deformation and which are not a function of $\Gamma$, for example, orientation processes in the adsorbed film).

Let us examine the fall of a drop on whose surface a certain surface-active material has been adsorbed to the equilibrium amount $\Gamma$ per unit area.

The equations of motion for the inner and outer liquids are (70.1) to (70.4).

Boundary conditions (70.8), (70.12), and (70.13) likewise remain unchanged. However, the presence of surface-active substances and the resulting variable surface tension lead to a change in boundary conditions (70.10) and (70.11).

Let us write the latter conditions, restricting ourselves to the viscous fall regime where $\mathrm{Re} \ll 1$:

$$-p+2\mu\frac{\partial v_r}{\partial r}+\pi+p_\sigma=-p'+2\mu'\frac{\partial v'_r}{\partial r}, \tag{73.1}$$

$$\mu\left(\frac{1}{r}\frac{\partial v_r}{\partial\theta}+\frac{\partial v_\theta}{\partial r}-\frac{v_\theta}{r}\right)+p_t=\mu'\left(\frac{1}{r}\frac{\partial v'_r}{\partial\theta}+\frac{\partial v'_\theta}{\partial r}-\frac{v'_\theta}{r}\right), \tag{73.2}$$

where $p_\sigma$ is the capillary pressure and $p_t$ is the tangential force exerted per $\mathrm{cm}^2$ of drop surface unevenly covered by the surface-active material. Measuring the angle $\theta$ counterclockwise, we have

$$p_\sigma=\frac{2\sigma}{a}, \tag{73.3}$$

$$p_t=-\frac{1}{a}\frac{\partial\sigma}{\partial\theta}. \tag{73.4}$$

In the absence of surface-active substances, surface tension $\sigma = \mathrm{const}$, so that $p_t = 0$, and the capillary pressure is reduced to an inconsequential constant since the radius of the drop remains unchanged.

In the presence of surface-active substances, their concentration, as was explained above, varies from point to point on the surface of the drop; i.e., $\Gamma = \Gamma(\theta)$. Since the surface tension, in turn, is a function of the concentration $\Gamma$ of surface-active material, it also varies along the surface; i.e., $\sigma(\Gamma) = \sigma[\Gamma(\theta)]$.

It is convenient to express the surface tension at a given point in terms of the surface tension at the drop's equator ($\theta = \frac{\pi}{2}$) in the following manner:

$$\sigma(\theta)=\sigma_{\frac{\pi}{2}}+\int\limits_{\frac{\pi}{2}}^{\theta}\frac{\partial\sigma}{\partial\Gamma}\frac{d\Gamma}{d\theta}d\theta.$$

Then we can write the capillary pressure and the tangential force as the following functions of the angle $\theta$:

$$p_{\sigma}=\frac{2\sigma(\theta)}{a}=\frac{2\sigma_{\pi/2}}{a}+\frac{2}{a}\int\limits_{\frac{\pi}{2}}^{\theta}\frac{\partial\sigma}{\partial\Gamma}\frac{d\Gamma}{d\theta}d\theta, \tag{73.5}$$

$$p_t=-\frac{1}{a}\frac{\partial\sigma}{\partial\Gamma}\frac{d\Gamma}{d\theta}. \tag{73.6}$$

Thus, in order to find the forces exerted on the surface of the drop, the distribution $\Gamma(\theta)$ of the adsorbed material over its surface must be known.

The amount of material adsorbed per unit area must satisfy the law of conservation of matter, which, in the case of a drop, can be expressed as

$$j_n=\operatorname{div}_s(\Gamma\mathbf{v}_t)-\operatorname{div}(D_s\operatorname{grad}_s\Gamma), \tag{73.7}$$

where $\Gamma v_t$ is the flux of adsorbed material transferred along the surface by the tangential component of the flow velocity, and $\operatorname{div}_s(\Gamma v_t)$ is the surface divergence of that flux. The quantity $j_n$ represents the flux of material removed into the bulk of the medium ($j_n < 0$) or conveyed ($j_n > 0$) to a unit area of the drop from the solution; ($-D_s \operatorname{grad}_s \Gamma$) expresses the diffusional flux of adsorbed material in the surface of the liquid. Here $D_s$ designates the surface diffusion coefficient of the surface-active material. Thus, in the expression for the total flux of surface-active material, the right side expresses the full divergence of the diffusional and convective fluxes of material along the surface of the drop. The left side gives the flux of material to or from the bulk of the solution. The magnitude of the latter is evidently a function of the velocity distribution near the surface of the drop, while the magnitude of the convective flux along the surface of the drop is determined by the velocity on the drop surface.

Thus, to find the velocity of the drop, the distribution of adsorbed material on its surface must be known, while, in order to find the latter, the velocity distribution must be known. The problem can be solved, however, if it is assumed that the shape of the drop, in the presence of adsorbed surface-active materials, does not lose its spherical symmetry.

The magnitude $j_n$ of the flux of surface-active material from the surface of the drop into the bulk of the liquid is determined by the slower of the two processes:

1) adsorption or desorption;

2) transfer of molecules of surface-active material from the bulk of the liquid to the surface of the drop.

If the rate of adsorption-desorption is small compared to the rate at which the material is transferred to the surface of the drop, the rate $j_n$ of surface-active molecules leaving (or arriving at) 1 cm$^2$ of surface per unit time is determined by the total number of desorbed (or adsorbed) molecules. If, on the other hand, the kinetics of adsorption and desorption are rapid compared to the rate of transfer of surface-active molecules from the bulk of the solution to the surface of the drop, the number of surface-active molecules leaving (or arriving at) 1 cm$^2$ of surface per unit time is equal to the diffusional flux in the moving liquid.

Henceforth, we shall assume that the deviation $\Gamma'$ of the adsorbed amount $\Gamma$ from the equilibrium amount is always small, so that we can write

$$\Gamma = \Gamma_0 + \Gamma', \tag{73.8}$$

where $\Gamma' \ll \Gamma_0$.

Substituting the value of $\Gamma$ from (73.8) into (73.7), we have:

$$\begin{aligned} j_n = \Gamma_0 \operatorname{div} \mathbf{v}_t + \operatorname{div}_s (\Gamma' \mathbf{v}_t) - \operatorname{div}_s (D_s \operatorname{grad}_s \Gamma_0) - \\ - \operatorname{div}_s (D_s \operatorname{grad}_s \Gamma') \approx \Gamma_0 \operatorname{div}_s \mathbf{v}_t - \operatorname{div}_s (D_s \operatorname{grad} \Gamma'), \end{aligned} \tag{73.9}$$

since $\Gamma_0$ is constant. The product $\Gamma' v_t$ may be neglected since the motion is assumed to be slow and this quantity is a second order infinitesimal.

Assuming that the drop is spherical, let us write the tangential surface velocity $|v_t|$ in the form $v_0 \sin\theta$, where $v_0$ is the velocity on the drop's equator and is an unknown subject to later determination. Then

$$\operatorname{div}_s \mathbf{v}_t = \frac{1}{a \sin\theta} \frac{\partial}{\partial\theta} (v_0 \sin^2\theta) = \frac{2 v_0 \cos\theta}{a}. \tag{73.10}$$

In addition,

$$\operatorname{div}_s \operatorname{grad}_s \Gamma' = \frac{1}{a \sin\theta} \frac{\partial}{\partial\theta} \left( \sin\theta \frac{d\Gamma'}{d\theta} \right). \tag{73.11}$$

Therefore the final equation for the material balance at the surface can be written in the approximate form

$$j_n = \Gamma_0 \frac{2 v_0 \cos\theta}{a} - \frac{D_s}{a \sin\theta} \frac{\partial}{\partial\theta} \left( \sin\theta \frac{d\Gamma'}{d\theta} \right). \tag{73.12}$$

## 74. FALL OF DROPS IN THE PRESENCE OF A SURFACE-ACTIVE MATERIAL, ADSORPTION BEING THE RATE DETERMINING STEP

Let us first consider the phenomena which occur during the fall of a drop in a liquid when the adsorption-desorption process controls the rate at which the surface-active material is exchanged between the bulk of the liquid and the drop surface. In this case we shall assume that the surface diffusion of surface-active molecules is sufficiently small and can be neglected.

Let us designate by $P(\Gamma)$ the number of molecules desorbed per unit area per unit time and by $Q(c_1, \Gamma)$ the number of molecules adsorbed per unit area per unit time, where $\Gamma$ is the surface density of surface-active molecules and $c_1$ is the concentration in the liquid near the drop.

In accordance with the assumption that transfer of material from the bulk of the solution is many times faster than its adsorption rate the concentration of the material in the solution near the surface of the drop can be considered to be the same as the concentration $c_1$ in the solution.

The net rate of the adsorption-desorption process per unit area is

$$j_n = -P(\Gamma) + Q(c_0, \Gamma) = -P(\Gamma_0 + \Gamma') + Q(c_0, \Gamma_0 + \Gamma') \approx \approx -P(\Gamma_0) + Q(c_0, \Gamma_0) - \frac{\partial P}{\partial \Gamma}\Gamma' + \frac{\partial Q}{\partial \Gamma}\Gamma' = -\alpha\Gamma', \tag{74.1}$$

where

$$\alpha = \frac{\partial P}{\partial \Gamma} - \frac{\partial Q}{\partial \Gamma}. \tag{74.2}$$

For the equilibrium value $\Gamma_0$, the rates of the adsorption and desorption processes are the same, so that

$$P(\Gamma_0) = Q(c_0 \Gamma_0).$$

Substituting for $j_n$ in expression (73.12) its value from equation (74.1), we find:

$$\Gamma' = -\frac{2\Gamma_0 v_0}{\alpha a}\cos\theta. \tag{74.3}$$

Using this value for $\Gamma'$ in formulas (73.5) and (73.6), we find the distribution of tangential and normal stresses caused by the variations in surface tension

$$p_t = -\frac{2\Gamma_0 v_0}{\alpha a^2} \cdot \frac{\partial \sigma}{\partial \Gamma}\sin\theta, \tag{74.4}$$

$$p_\sigma = \frac{2\sigma}{a} = \frac{2\sigma_{\pi/2}}{a} - \frac{4}{a^2} \cdot \frac{\partial \sigma}{\partial \Gamma} \cdot \frac{\Gamma_0 v_0}{\alpha}\cos\theta. \tag{74.5}$$

Since the falling drop has spherical symmetry, the distribution of velocities and pressures inside and outside the drop should be sought in the form of (70.28) to (70.33). To determine the constants $b_1$, $b_2$, $b_3$, $a_1$, $a_2$, and $v_0$, boundary conditions (70.8), (70.12), (70.13), (73.1), and (73.2) should be used. The values of $p_\sigma$ and $p_t$ from (74.4) and (74.5) should be substituted into the last two.

In addition, the boundary conditions contain two unknown constants, U and $v_0$. The latter can be easily expressed in terms of $b_1$, $b_2$, and $b_3$ since $v_0$ is the value of $v_\theta$ at $r = a$ and $\theta = \frac{\pi}{2}$. Then we have left six constants linked by six boundary conditions.

The solution of this system of algebraic equations presents no difficulties.

For U and $v_0$, we derive the values:

$$U = \frac{2}{3}\frac{(\rho' - \rho)\, ga^2}{\mu} \cdot \frac{\mu + \mu' + \dfrac{2\Gamma_0}{3\alpha a} \cdot \dfrac{\partial\sigma}{\partial\Gamma}}{2\mu + 3\mu' + \dfrac{2\Gamma_0}{\alpha a}\dfrac{\partial\sigma}{\partial\Gamma}}, \tag{74.6}$$

$$v_0 = \frac{1}{3}\frac{(\rho' - \rho)\, ga^2}{2\mu + 3\mu' + \dfrac{2\Gamma_0}{\alpha a}\dfrac{\partial\sigma}{\partial\Gamma}}. \tag{74.7}$$

Let us introduce the symbols

$$\gamma_1 = \frac{2\Gamma_0}{3\alpha a} \cdot \frac{\partial\sigma}{\partial\Gamma}, \tag{74.8}$$

$$U_0 = \frac{2}{9}\frac{(\rho - \rho')\, ga^2}{\mu}. \tag{74.9}$$

Then, finally,

$$U = 3U_0 \frac{\mu + \mu' + \gamma_1}{2\mu + 3\mu' + 3\gamma_1}, \tag{74.10}$$

$$v_0 = \frac{3U_0}{2}\frac{\mu}{2\mu + 3\mu' + 3\gamma_1}. \tag{74.11}$$

Formula (74.10) becomes the Stokes formula $U = U_0$ for $\gamma_1 \gg (\mu + \mu')$, and the Rybczyński-Hadamard formula $U = 3U_0 \frac{\mu + \mu'}{2\mu + 3\mu'}$ when the inequality $\gamma_1 \ll (\mu + \mu')$ is satisfied. For $\gamma_1 \gg (u + \mu')$, we have $v_0 \ll U_0$, as should have been anticipated.

Thus, surface-active materials exert a retarding effect on the motion of the drop. This effect is expressed quantitatively by the value of $\gamma_1$. At a sufficiently high value of this retardation coefficient, motion on the surface of the drop becomes fully damped, and the velocity $v_0$ of the liquid becomes zero.

Let us examine the retardation coefficient $\gamma_1$ in more detail. Let us rewrite it in the form

$$\gamma_1 = \frac{2}{3}\frac{\Gamma_0}{\alpha a} \cdot \frac{\partial\sigma}{\partial c_0} \cdot \frac{\partial c_0}{\partial\Gamma_0}. \tag{74.12}$$

Utilizing the Gibbs equation

$$\frac{\partial \sigma}{\partial c_0} = \frac{\Gamma_0 RT}{c_0}, \tag{74.13}$$

this coefficient may be written in the form

$$\gamma_1 = \frac{2}{3} \cdot \frac{\Gamma_0^2 RT}{\alpha a c_0} \cdot \frac{\partial c_0}{\partial \Gamma_0}. \tag{74.14}$$

The value of $\gamma_1$, expressed in terms of $\partial c_0/\partial \Gamma_0$, depends on the mechanism of the adsorption process. Indeed, it is easy to show that the quantity $\partial c_0/\partial \Gamma_0$ depends on the combination of the kinetics of adsorption and desorption.

Let us assume that the equilibrium dependence of $\Gamma_0$ on $c_0$ is expressed by the Langmuir adsorption isotherm. This assumption holds approximately for adsorption at a mercury-solution interface for materials containing 3 to 5 carbon atoms per molecule. Langmuir's adsorption isotherm may be interpreted kinetically in various ways. If, as is usually done in an elementary derivation of the Langmuir formula, it is assumed that the desorption rate is equal to $P = \alpha_0 \Gamma_0$ and the adsorption rate is $Q = \beta_0 c_0 \left(1 - \frac{\Gamma_0}{\Gamma_\infty}\right)$ where $\alpha_0$ and $\beta_0$ are constants, it follows that

$$c_0 = \frac{\alpha_0}{\beta_0} \cdot \frac{\Gamma_0}{1 - \frac{\Gamma_0}{\Gamma_\infty}}. \tag{74.15}$$

Introducing the quantity $k = \frac{\beta_0}{\alpha_0}$, we find the equilibrium adsorption equation

$$\Gamma_0 = \frac{k c_0}{1 + \frac{k c_0}{\Gamma_\infty}}. \tag{74.16}$$

For such equilibrium values of P and Q, the quantity $\alpha$ is

$$\alpha = \frac{\partial P}{\partial \Gamma} - \frac{\partial Q}{\partial \Gamma} = \frac{\alpha_0}{1 - \frac{\Gamma_0}{\Gamma_\infty}}. \tag{74.17}$$

Substituting the values for the appropriate quantities into the expression for $\gamma_1$, we have:

$$\gamma_1 = \frac{2RT}{a} \cdot \frac{\Gamma_0}{3\alpha_0}. \tag{74.18}$$

Substituting into equation (74.18) the value of $\Gamma_0$ from expression (74.16), the retardation coefficient $\gamma_1$ may be presented in the form

$$\gamma_1 = \frac{2RT}{3\beta_0 a} \cdot \frac{k^2 c_0}{1 + \frac{k c_0}{\Gamma_\infty}}. \tag{74.19}$$

Since the quantity $\beta_0$ must be primarily a function of the properties of the polar group, it would appear that $\beta_0$ undergoes little change, and even decreases somewhat, when the chain is lengthened. Thus, when the surface is only partially filled, the retardation effect in a homologous series (in the fatty-acid series, for example) increases as the square of the adsorbability. It follows from equation (74.19) that the quantity $\gamma_1$ increases monotonically with increasing $\Gamma_0$ and tends toward the limit $\frac{2RTk\Gamma_\infty}{3\beta_0 a}$ as $\Gamma_0$ approaches $\Gamma_\infty$, i.e., as $c_0 \to \infty$.

Langmuir proposed [13] a different scheme of adsorption kinetics in which the adsorption rate is assumed to be always $Q = \beta_0 c_0$ and is not a function of the degree to which the surface is filled. If we assume that equilibrium adsorption is expressed, as previously, by the Langmuir isotherm, the desorption rate must be given by the formula

$$P = \frac{\alpha_0 \Gamma_0}{1 - \frac{\Gamma_0}{\Gamma_\infty}}.$$

In such a scheme for the adsorption process, the quantity $\alpha$, introduced by us, is found to be

$$\alpha = \frac{\partial}{\partial \Gamma}(P - Q) = \frac{\alpha_0}{\left(1 - \frac{\Gamma_0}{\Gamma_\infty}\right)^2}. \tag{74.20}$$

Substituting this value into equation (74.14) and substituting for $c_0$ and $\frac{\partial c_0}{\partial \Gamma_0}$ their values from formula (74.15), we find the following expression for the retardation coefficient:

$$\gamma_1 = \frac{2RT\Gamma_0}{3\alpha_0 a}\left(1 - \frac{\Gamma_0}{\Gamma_\infty}\right) = \frac{2RT}{3\beta_0 a} k\Gamma_0\left(1 - \frac{\Gamma_0}{\Gamma_\infty}\right). \tag{74.21}$$

The quantity $\gamma_1$ given by formula (74.21) first rises, then passes through a maximum, and finally decreases as $\Gamma_0 \to \Gamma_\infty$.

For low degrees of surface coverage, equations (74.21) and (74.18) coincide.

## 75. FALL OF DROPS IN THE PRESENCE OF A SURFACE-ACTIVE MATERIAL, THE TRANSFER RATE BEING DETERMINED BY DIFFUSION IN THE BULK OF THE MEDIUM AND BY SURFACE DIFFUSION

Computations are more complex in the case where the over-all rate of the exchange of surface-active materials is determined by their rate of transfer from the bulk of the medium to the surface of the drop, i.e., by the rate of convective diffusion. The kinetics of the adsorption-desorption process, in this case, may be considered to be as rapid as desired, so that the concentration of the solution near the surface has a value $c_1$ which is in equilibrium with $\Gamma$, but differs from the concentration $c_0$ in the bulk of the solution. The quantity $c_0$ evidently corresponds to the equilibrium concentration $\Gamma_0$ on the surface.

As a result, there is a flow of surface-active material between the solution adhering to the drop and the bulk of the solution. The density of this flux can evidently be expressed in the form

$$j_n = D\left(\frac{\partial c}{\partial r}\right)_{r=a}, \tag{75.1}$$

where D is the diffusion coefficient of surface-active molecules and the derivative is evaluated at the surface of the drop.

Therefore in the case of rapid kinetics, $j_n$ in the boundary condition (73.12) has the form of expression (75.1), and the boundary condition itself assumes the form (if surface diffusion is neglected)

$$D\left(\frac{\partial c}{\partial r}\right)_{r=a} = \Gamma_0 \operatorname{div}_s \mathbf{v}_t. \tag{75.2}$$

To find the general solution of the hydrodynamic equations with boundary condition (75.2), it is also necessary to find the concentration distribution in the solution. The latter is determined by the equation of convective diffusion

$$\mathbf{v}\,\operatorname{grad} c = D\,\Delta c. \tag{75.3}$$

The solution of the system of hydrodynamic and convective transfer equations with the appropriate boundary conditions presents very serious mathematical difficulties.

Namely, the system of equations (70.1) to (70.4) and (75.3) does not allow solutions that are dependent on the angle $\theta$ in the form of the first spherical (Legendre) function and its derivative (i.e., as $\cos\theta$ and $\sin\theta$). This indicates that the shape of the drop must necessarily deviate from spherical, and the method of solution developed by us for the case discussed above cannot be strictly applied to the solution of the problem posed here. Henceforth, we limit ourselves to a qualitative explanation.

In order to obtain an approximate expression for the velocity of the drop's motion, let us assume that the variation in the surface tension along the drop is small compared to its absolute value, so that we can neglect the deviation in the shape of the drop from spherical. Let us further assume that we can also neglect the dependence of the diffusional flux on the angle $\theta$ and that we may replace the true value of the flux in expression (75.1) by its mean value over the drop.*

---

*Editor's note: Actually we shall replace the true value of the thickness of the diffusion boundary layer by its mean value over the drop [see (75.4)]. The mean flux is zero.

The mean value $\bar{j}_n$ of the flux can then be presented in the form

$$\bar{j}_n = D\frac{\Delta c}{\delta}, \tag{75.4}$$

where $\Delta c$ is the difference in concentration between the bulk of the solution and a point near the surface and $\delta$ is the average thickness of the diffusion layer.

Assuming the shape of the drop to be constant, let us write the tangential velocity component on the surface in the form $v_0 \sin\theta$ and, utilizing formula (73.10) and the value for $j_n$ from expression (75.4), let us write boundary condition (75.2) in the form

$$\frac{D\,\Delta c}{\delta} = \frac{2v_0\Gamma_0 \cos\theta}{a}. \tag{75.4'}$$

Assuming, further, that

$$\Delta c = \frac{\partial c}{\partial\Gamma}\Delta\Gamma = -\frac{\partial c}{\partial\Gamma}(\Gamma - \Gamma_0) = -\frac{\partial c}{\partial\Gamma}\Gamma',$$

we find the expression for the distribution of the amount of molecules adsorbed along the surface

$$\Gamma' = -\frac{\delta}{D}\frac{2v_0\Gamma_0\cos\theta}{a\left(\frac{\partial c}{\partial\Gamma}\right)}. \tag{75.5}$$

Relation (75.5) is analogous to expression (74.3) for the capillary pressure.

The distribution of surface tension along the drop can be derived from the approximate dependence of $\Gamma'$ on the angle $\theta$:

$$\sigma = \sigma_{\pi/2} + \int_{\pi/2}^{\theta}\frac{\partial\sigma}{\partial\theta}d\theta = \int\frac{\partial\sigma}{\partial\Gamma}\frac{\partial\Gamma}{\partial\theta}d\theta + \sigma_{\pi/2} =$$
$$= \sigma_{\pi/2} - \frac{\partial\sigma}{\partial\Gamma}\frac{2\Gamma_0 v_0\delta}{Da\left(\frac{\partial c}{\partial\Gamma}\right)}\cos\theta = \sigma_{\pi/2} - \frac{2\Gamma_0 v_0\delta}{Da}\left(\frac{\partial\sigma}{\partial c}\right)\cos\theta, \tag{75.6}$$

where $\sigma_{\pi/2}$ is the surface tension on the drop's equator.

Knowing the dependence of the surface tension on the angle $\theta$, it is possible to proceed with the formulation of boundary conditions and to the determination of the constants $b_1, b_2, b_3, a_1$, and $a_2$ in the expressions for the components of the velocity, the velocity $U$ of the fall of the drop, and the constant $v_0$ which we introduced into the expression for the tangential velocity component on the surface of the drop (a total of seven constants).

Computations no different from those preceding lead to expressions for $U$ and $v_0$ that are identical with (74.10) and (74.11), in which the quantity $\gamma_2$ has the value

$$\gamma_2 = \frac{2RT\Gamma_0^2\delta}{3Dac_0}. \tag{75.7}$$

If $\Gamma_0$ is expressed by formula (74.16), we have for $\gamma_2$:

$$\gamma_2 = \frac{2RT\delta}{3Da} \cdot \frac{k^2 c_0}{\left(1 + \frac{kc_0}{\Gamma_\infty}\right)^2} = \frac{2RT\delta}{3Da} \cdot k\Gamma_0 \left(1 - \frac{\Gamma_0}{\Gamma_\infty}\right). \tag{75.8}$$

All quantities in formula (75.8), except the adsorption coefficient k, are either weak functions or are independent of the nature of the surface-active material. The retardation coefficient, therefore, is proportional to the square of the adsorbability $k^2$ of the material. If the dependence of the retardation coefficient $\gamma_2$ on the nature of the material is studied in a homologous hydrocarbon series, then, according to formula (75.8), at small adsorbabilities, the retardation coefficient is proportional to the square of the Traube coefficient.

As the concentration $\Gamma_0$ increases, so does $\gamma_2$, which attains a maximum value at $\Gamma_0 = \frac{\Gamma_\infty}{2}$. At the maximum point,

$$(\gamma_2)_{\max} = \frac{k\Gamma_\infty RT\delta}{3aD}. \tag{75.9}$$

A further increase in the concentration of the solution leads to a decrease in $\gamma_2$.

For the quantitative estimation of $\gamma_2$, we may utilize the values for $\delta$ derived in Sections 14 and 72 for two limiting cases — a liquid drop and a solid sphere.

As may be seen from a comparison of expressions (14.17) and (72.8), the quantity $\delta$ is several times greater in the first case. This reflects the more favorable conditions of convective diffusion in the case of the liquid-liquid interface.

To estimate the maximum value of $\gamma_2$ that may be anticipated in the case of retardation by highly soluble surface-active substances, it is necessary to replace $\delta$ in formula (75.7) by its value from expression (14.17), thereby assuming that the motion of the liquid on the surface of the drop is fully damped by the surface-active substance.

Thus,

$$\gamma_2 = \frac{2RT\Gamma_0^2}{3Dac_0}\left(\frac{Da^2}{U}\right)^{1/3}. \tag{75.10}$$

In a case of weak retardation, for example, at sufficiently small concentrations of surface-active substances, we obtain a more correct value for $\gamma_2$ by replacing $\delta$ in formula (75.8) by its value from expression (72.8),

$$\gamma_2 = \frac{2RT\Gamma_0}{Dac_0}\left(\frac{Da}{U}\right)^{1/2}. \tag{75.11}$$

We see that the retardation decreases with increasing velocity in proportion to $U^n$, where n is between 1/3 and 1/2.

Let us now examine the retardation caused by a surface-active material which dissolves so slowly that it can be considered to remain on the surface of the drop during the entire period of the latter's motion. For the sake of brevity, let us agree to call such materials insoluble. Their flux to the surface $j_n$ is equal to zero, so the only mechanism for equalizing the concentration of the surface-active material along the surface of the drop is surface diffusion.

The distribution of the concentration of surface-active materials over the surface of the drop is determined by the equation of surface convective diffusion

$$\mathrm{div}_s(\Gamma v_t) = D_s \Delta \Gamma,$$

or, in spherical coordinates,

$$\frac{1}{a \sin\theta} \frac{\partial}{\partial\theta} (\sin^2\theta \cdot \Gamma \cdot v_\theta) = \frac{D_s}{a^2} \cdot \frac{1}{\sin\theta} \cdot \frac{\partial}{\partial\theta}\left(\sin\theta \frac{\partial\Gamma}{\partial\theta}\right), \tag{75.12}$$

where $D_s$ is the surface diffusion coefficient of the surface-active material.

Assuming, as before, that the variations in concentration are small and that the velocity of motion of the liquid along the surface is equal to $v_0 \sin\theta$, we find:

$$2av_0\Gamma_0 \cos\theta = \frac{D_s}{\sin\theta} \frac{\partial}{\partial\theta}\left(\sin\theta \frac{\partial\Gamma}{\partial\theta}\right),$$

or

$$\Gamma' = -\frac{2av_0\Gamma_0}{D_s} \cos\theta. \tag{75.13}$$

Thus, the distribution of surface-active material on the surface, in this instance, is quite similar to that characterized by expression (75.5), differing from it only by the value of the constant coefficient. Therefore, without repeating the computations of the preceding section, it is possible to write the following expression for the velocity of fall of the drop:

$$U = 3U_0 \frac{\mu + \mu' + \gamma_3}{2\mu + 3\mu' + 3\gamma_3}, \tag{75.14}$$

where the coefficient is $\gamma_3 = \dfrac{2a\Gamma_0}{3D_s}\left|\dfrac{\partial\sigma}{\partial\Gamma}\right|$ and expresses the retardation of the drop by the surface-active material. For small degrees of surface coverage by the adsorbed surface-active material,

$$\gamma_3 = \frac{2\Gamma_0 a}{3D_s}\left|\frac{\partial\sigma}{\partial\Gamma}\right| \approx \frac{2\Gamma_0 aRT}{3D_s}; \tag{75.15}$$

i.e., the retardation increases in proportion to the amount adsorbed.

## 76. COMPARISON OF DIFFERENT THEORIES

In the study of Boussinesq [7], the following expression was derived for the velocity of drop fall

$$U = \frac{2}{3} \frac{(\rho - \rho') g a^2}{\mu} \cdot \frac{\mu + \mu' + \frac{2e}{3a}}{2\mu + 3\mu' + \frac{2e}{a}}, \tag{76.1}$$

where e is the surface viscosity coefficient.*

Surface viscosity is defined by Boussinesq as the coefficient in the relation

$$\Delta\sigma = e \operatorname{div}_s \mathbf{v}_t,$$

whence

$$e = \frac{\Delta\sigma}{\operatorname{div}_s \mathbf{v}_t}. \tag{76.2}$$

Thus, the quantity e, which has been introduced, does not represent ordinary viscosity, but a second viscosity for the surface motion.

Using the values derived for $\operatorname{div}_s \mathbf{v}_t$ and $\Delta\sigma$, it is possible to find the quantity in our theory which corresponds to Boussinesq's surface viscosity. It is clear that various values can be derived for e, depending on the nature of the slow step that controls the over-all kinetics of the exchange.

Namely, from formulas (76.1) and (74.10) and from a comparison of formula (76.2) with expressions (73.10) and (75.6) for the case where the controlling step of the process is the diffusion of the material in the bulk of the solution, we find:

$$e = \left| \frac{\partial\sigma}{\partial c} \right| \frac{\Gamma_0 \delta}{D} = \frac{3\gamma_2 a}{2}. \tag{76.3}$$

Similarly, in the case of surface diffusion, using the same formulas and (75.13), taking the value of $v_0$ into account, we have:

$$e = \left| \frac{\partial\sigma}{\partial\Gamma} \right| \frac{\Gamma_0 a^2}{D_s} = \frac{3\gamma_3 a}{2}. \tag{76.4}$$

Thus, the assumption that the retardation is controlled by a constant surface viscosity and is a function exclusively of the surface properties of the liquid is not justified in either instance. This is especially important in the case of insoluble films because it is precisely in this case that the existence of a surface viscosity, in the sense of Boussinesq, would appear to be most likely. It is only under conditions of solubility of surface-active materials and slow kinetics of adsorption that the quantity e is in reality determined by surface properties and, as seen from expressions (74.5) and (74.3) and from definitions (76.2) and (73.10), can be formulated as follows:

---

*An incorrect value for the numerical coefficient of e is given in this study by Boussinesq.

$$e = \left| \frac{\partial\sigma}{\partial\Gamma} \right| \frac{\Gamma_0}{\alpha}. \tag{76.5}$$

The fact that, in the case of slow adsorption kinetics, the quantity e is a function neither of the radius nor of the flow regime follows directly from the fact that the equalization of the excess surface concentration depends here only on conditions in the surface layer and on the value of the concentration in the bulk phase. The latter, by assumption, remains constant. Nevertheless, in this case too, the similarity with Boussinesq's conclusion is purely formal since the physical meaning of the coefficient in the theory developed by us is quite different and, in constrast with the Boussinesq surface viscosity concept, cannot be understood without taking into account the existence of the bulk phase. With these reservations, the quantity $e = \frac{\gamma a}{2}$ may be regarded as the measure of a certain dynamic elasticity of the surface layer which determines its ability to resist lasting deformation and which is comparable to the elasticity values of surface layers derived by Gibbs [14].

Gibbs examined the change in the interfacial tension when the extent of the surface varies and introduced the concept of the elasticity E of a thin liquid lamella between two gaseous phases, defining E by the equation

$$2d\sigma = \frac{E\,dS}{S}. \tag{76.6}$$

The factor of 2 in the Gibbs equation indicates the presence of two interfaces. In defining the quantity E, Gibbs accepted the proposition that equilibrium always exists between the surface of the layer and adjacent parts of the body of the layer, because any disturbances of the layer accompanying the changes in S are for all practical purposes immediately negated because of the thinness of the film. On the contrary, the equilization of the concentrations in the thin layer along the surface may virtually be neglected. Under these conditions, the value of E can be easily computed from the quantities which determine the equilibrium state of the system. Let us apply Gibbs' argument to the analogous case that we are examining. Namely, let us assume that the adsorbed material is evenly distributed between the surface and the volume V of a certain phase that is adjacent to a second phase not containing that component, and let us designate by g the over-all amount of material that is being adsorbed. Obviously,

$$E = S\left(\frac{\partial\sigma}{\partial S}\right)_g = S\frac{\partial\sigma}{dc}\left(\frac{\partial c}{\partial S}\right)_g,$$

and, since

$$S\Gamma_0 + Vc = g,$$

then

$$E = -S\frac{d\sigma}{dc}\cdot\frac{\Gamma_0}{V + S\frac{\partial\Gamma_0}{\partial c}}. \qquad (76.7)$$

Equation (76.7) is equivalent to the Gibbs equation (655) in the cited reference. Let us examine separately two limiting cases, when $S\Gamma_0 \ll Vc$, i.e., when the amount of material on the surface is negligible compared to its amount in the volume, and the reverse case when $S\Gamma_0 \gg cV$. In the first instance, it follows from equation (76.7) that

$$E = -\frac{d\sigma}{dc}\cdot\frac{\Gamma_0 S}{V} = \frac{RT\Gamma_0^2}{c}\cdot\frac{S}{V} = \frac{RT\Gamma_0^2}{ch}, \qquad (76.8)$$

where $h = \frac{V}{S}$ is the thickness of the layer. In the second case

$$E = \Gamma_0\left|\frac{\partial\sigma}{\partial\Gamma}\right|, \qquad (76.9)$$

or for small degrees of filling of the surface

$$E = RT\Gamma_0. \qquad (76.10)$$

Equation (76.8) is very similar to the relation which we derived for e in the case of retardation under conditions of slow diffusion in the volume:

$$e = \frac{RT\Gamma_0^2}{c}\cdot\frac{\delta}{D}.$$

The latter expression, however, instead of the reciprocal of the thickness of the layer $1/h$, contains the quantity $\delta/D$, which has the dimensions $cm^{-1}$ sec. This difference in dimensions is quite natural since the equilibrium elasticity E of the film as interpreted by Gibbs expresses the variation in the surface tension for a given deformation. The ratio $e/E$ must, therefore, have time as its dimension. Similarly, for the case of retarded surface diffusion, the quantity $e = \Gamma_0\left|\frac{\partial\sigma}{\partial\Gamma}\right|\frac{a^2}{D_S}$ differs from the expression obtained for the elasticity of an insoluble film by the coefficient $a^2/D_S$, which also has the dimension of time. It is easy to see that the ratio $e/E$ always defines to an order of magnitude the time needed for diffusion, under the given conditions of the system, to equalize the disturbed equilibrium of surface and volume concentrations caused by a single deformation of the surface.

## 77. COMPARISON OF THE DIFFERENT RETARDING MECHANISMS

As before, we shall designate the retardation coefficient by $\gamma$; the cases where adsorption, diffusion in the volume, or diffusion on the surface constitutes the slow step of the process are distinguished by the subscripts 1, 2, and 3, respectively. Let us first clarify the relative importance of the volume and surface diffusions. Since these two processes are parallel, the more rapid will be decisive. On it will depend the rate of establishment of a state of equilibrium concentrations. The more rapid process will evidently lead to the smaller value for $\gamma$.

In accord with equations (75.7) and (75.15),

$$\frac{\gamma_2}{\gamma_3} = \left(\frac{2RT\Gamma_0^2\delta}{3Dac_0}\right) : \left(\frac{2a\Gamma_0}{3D_s}\left|\frac{\partial\sigma}{\partial\Gamma}\right|\right) = \frac{RT\Gamma_0\delta D_s}{Dc_0a^2\left|\frac{\partial\sigma}{\partial\Gamma}\right|}.$$

When the surface is sparsely covered with the material being adsorbed, this expression can be simplified because $\frac{\partial\sigma}{\partial\Gamma} \approx -RT$ and, therefore,

$$\frac{\gamma_2}{\gamma_3} = \frac{\Gamma_0\delta D_s}{Dc_0a^2}. \tag{77.1}$$

When this phenomenon is studied with the usual dropping mercury electrode, a and $\delta$ have the values: $a \approx 5\cdot10^{-2}$ and $\delta \approx 3\cdot10^{-3}$. Taking into account the fact that D and $D_S$ are quantities of the same order,* it follows from equation (77.1) that

$$\frac{\gamma_2}{\gamma_3} \approx \frac{\Gamma_0}{c_0}. \tag{77.2}$$

For surface diffusion to be significant compared with diffusion in the bulk phase, it is necessary for the relation $\frac{\gamma_2}{\gamma_3} \geq 1$ to be satisfied. For a 0.01 M solution of tertiary amyl alcohol in 1N NaCl, according to the data of A. N. Frumkin [15], $\frac{\Gamma_0}{c_0} \approx 10^{-5}$ on a mercury-solution interface. Surface diffusion, therefore, is significantly manifested only for materials whose surface activity exceeds that of the substance which we have chosen as a typical example of a soluble, surface-active compound. Such materials, however, are virtually insoluble. The importance of surface diffusion may increase if the equalization of concentrations in the volume phase is

---

*Apparently there are no experimental data for $D_S$ in the case of liquid interfaces. It can be assumed, however, at least for materials which do not have an excessively long hydrocarbon chain and when the surface is sparsely covered, that $D_S$ on a water surface and on a water-mercury interface do not differ from D in terms of orders of magnitude.

additionally retarded by the slowness of the adsorption step. This retardation, however, would hardly be sufficiently significant to cause an equalization of the large differences between the rates of the surface and the volume phase processes. Thus, if the surface-active material is transferred from the solution to the surface of the drop, the surface diffusion process apparently plays no significant part in the equalization of the concentration differences which appear because of convective transfer. This reasoning is no longer valid for drops of microscopic dimensions, at $a \approx 10^{-4}$ for example. In this case, the surface diffusion rate may become comparable to that in the volume phase.

Let us now compare the rate of diffusion in the volume phase with the rate of the adsorption process. Since these two processes are consecutive, the decisive role in determining the over-all rate of the reaction is exerted by the slower process, which accordingly contributes the larger value to the retardation coefficient $\gamma$. For this computation, we employ only the first kinetic interpretation of the Langmuir isotherm. According to equations (75.7) and (74.18)

$$\frac{\gamma_2}{\gamma_1} = \left(\frac{2RT\Gamma_0^2\delta}{3Dac_0}\right) : \left(\frac{2RT}{a} \cdot \frac{\Gamma_0}{3\alpha_0}\right) \approx \frac{\Gamma_0\delta\alpha_0}{Dc_0} = \frac{\beta_0\delta}{D}\left(1 - \frac{\Gamma_0}{\Gamma_\infty}\right). \tag{77.3}$$

Thus, the ratio of the diffusion retardation to that of adsorption decreases with an increase in the concentration of the solution and a decrease in $\delta$, i.e., with an increase in the velocity of motion.

As shown by elementary computations, when a clean interface comes in contact with a solution of constant concentration $c_0$, the coating of the interface changes with time according to the following relation:

$$\Gamma = \Gamma_0\left[1 - \exp\left(-\frac{\alpha_0 t}{1 - \frac{\Gamma_0}{\Gamma_\infty}}\right)\right]. \tag{77.4}$$

Thus, the quantity $\frac{1}{\alpha_0} = \left(1 - \frac{\Gamma_0}{\Gamma_\infty}\right)$ should be regarded as the time $\tau_A$ required for the establishment of an adsorption equilibrium, and therefore

$$\alpha_0 = \frac{1 - \frac{\Gamma_0}{\Gamma_\infty}}{\tau_A}. \tag{77.5}$$

Substituting the value of $\alpha_0$ from (77.5) into (77.3), we obtain:

$$\frac{\gamma_2}{\gamma_1} = \frac{\Gamma_0\delta\left(1 - \frac{\Gamma_0}{\Gamma_\infty}\right)}{Dc_0\tau_A}. \tag{77.6}$$

It follows from equation (77.6) that, for the condition $\frac{\gamma_2}{\gamma_1} < 1$ to be satisfied, it is necessary that

$$\tau_A > \frac{\Gamma_0 \delta \left(1 - \frac{\Gamma_0}{\Gamma_\infty}\right)}{D c_0} = \frac{k\delta}{D}\left(1 - \frac{\Gamma_0}{\Gamma_\infty}\right)^2. \tag{77.7}$$

In the case of a dilute solution of tertiary amyl alcohol which we examined, $\frac{\Gamma_0}{c_0} \approx 10^{-5}$, $\delta \approx 3 \cdot 10^{-3}$, $D \approx 10^{-5}$, and $1 - \frac{\Gamma_0}{\Gamma_\infty} \approx 1$. Thus, it follows from inequality (77.7) that

$$\tau_A > 3 \cdot 10^{-3} \text{ sec.}$$

In other words, for the retardation to be controlled by adsorption kinetics rather than by diffusion, the time for establishing an adsorption equilibrium must exceed several milliseconds. Such a value of $\tau_A$ is possible. The choice between the adsorption and the diffusion mechanisms, therefore, can be made only on the basis of experimental data and, in particular, on the basis of studies of the effect of stirring, which has different effects on the magnitude of the retardation coefficient $\gamma$ in the two cases.

It follows from formulas (75.7) and (74.18) that the coefficient $\gamma$ for both of these retardation mechanisms increases with a decrease in the radius a. As a result, in the case of drops of colloidal dimensions, it may be anticipated that the presence of even negligible amounts of surface-active impurities in the solution will lead to a loss of surface mobility. Thus, under ordinary conditions such drops should behave as solid spheres. Certain limitations are imposed on this conclusion by the decrease in $\gamma_3$ with a decrease in a.

## 78. LIMITS OF APPLICABILITY OF THE THEORY

The applicability of the above theory for the fall of a drop in a medium containing surface-active substances is limited, first of all, by the requirement that the Reynolds number be small during the drop's motion.

The second condition is the fulfillment of the inequality $|\Gamma'| \ll \Gamma_0$, which indicates the applicability of equation (73.12) or of the equivalent inequality $(\Delta\sigma) \ll \sigma_0 - \sigma$, where $\Delta\sigma$ is the variation of the surface tension caused by convective transfer and $\sigma_0$ is the surface tension of the liquid phase interface when $\Gamma_0 = 0$. Attainment of this condition may be difficult in many cases. We shall use this second inequality for a quantitative study. The maximum value for $|\Delta\sigma|$ according to equation (75.6) is

$$|\Delta\sigma| = \sigma(0) - \sigma\left(\frac{\pi}{2}\right) = \frac{2RT\Gamma_0^2 \delta v_0}{D a c_0} = \gamma v_0, \tag{78.1}$$

where $\sigma(0)$ and $\sigma(\frac{\pi}{2})$ are values of $\sigma$ at $\theta = 0$ and $\theta = \frac{\pi}{2}$.

Thus, the inequality under examination is equivalent to the relation

$$\gamma v_0 \ll |\sigma_0 - \sigma|. \tag{78.2}$$

The physical meaning of this limitation is that the changes in surface concentration become comparable to the initial concentration when these changes are equalized slowly. In the lower part of the drop, in particular, this may lead to complete removal of the adsorption layer from the surface and to an approach of the surface tension toward $\sigma_0$. It is evident that under these conditions the true retardation will lag behind the values computed from the formulas and as derived for $\gamma$.

For a freely falling drop and considerable retardation, $v_0$ is equal to $\frac{\mu U_0}{2\gamma}$, according to equation (74.11). Substituting this value and the expression for $U_0$ given by the Stokes formula into equation (78.1), we obtain:

$$|\Delta\sigma| = \frac{\mu U_0}{2} = \frac{1}{9}(\rho' - \rho)ga^2. \qquad (78.3)$$

From this simple relation, the following criterion for the applicability of the theory can be derived. The quantity $2|\Delta\sigma|$ is equal to the difference in interfacial tensions between the fore and aft ends of the drop. This quantity evidently cannot exceed a certain limit $|\Delta\sigma_{max}|$, equal to the maximum decrease in the initial interfacial tension observed in solutions of the given surface-active material. Applicability of the theory, therefore, requires that

$$\frac{2}{9}(\rho' - \rho)ga^2 \ll |\Delta\sigma|_{max},$$

or

$$a \ll \left[\frac{\frac{9}{2}|\Delta\sigma|_{max}}{(\rho' - \rho)g}\right]^{1/2}. \qquad (78.4)$$

This inequality determines the value of a certain critical radius $a_{cr}$ above which the retardation due to changes in the interfacial tension no longer compensates for the viscous stresses, and the drop moves like a liquid drop. This critical radius was first introduced by Bond (see Section 71) on the basis of dimensional analysis and experimental data.

For $|\sigma_0 - \sigma| < \frac{1}{2}|\Delta\sigma_{max}|$, deviations from the quantitative theory of retardation should begin at smaller values of a, determined by the inequality

$$a \ll \left[\frac{3(\sigma_0 - \sigma)}{(\rho' - \rho)g}\right]^{1/2}. \qquad (78.5)$$

Thus, for example, if adsorption on the mercury-water interface leads to a one unit decrease in $\sigma_0$, the retardation that it causes in the free fall of the mercury drop can becomes apparent only at $a_{cr} \approx 6 \cdot 10^{-2}$ cm.

In conclusion, it is interesting to draw attention to the following. For very large values of the retardation coefficient $\gamma$, while equation (74.11) is still strictly applicable, the quantity $v_0$ is very small. As a passes through $a_{cr}$, $v_0$ begins to rise sharply because retardation cannot arrest the surface motion. This conclusion is in agreement with the results of Bond's measurement of the velocity of the drop's fall. Thus, under these conditions, the surface behaves as though it possesses the property of non-Newtonian viscosity.

Let us now examine the limits of applicability of the theory in the case of retardation of surface motion by a layer of insoluble material. It follows from equation (75.15) that extremely small amounts of insoluble substances must have a strong retarding effect on the tangential motion of the surface. Indeed, noticeable retardation will start if

$$3\gamma_3 = \frac{2RT\Gamma_0 a}{D_s} > 2\mu + 3\mu', \tag{78.6}$$

i.e.,

$$\sigma_0 - \sigma \approx RT\Gamma_0 > \frac{D_s(2\mu + 3\mu')}{2a}.$$

Since, for a water-mercury interface, $2\mu + 3\mu' = 0.068$, assuming $D_s \approx 10^{-5}$ and $a = 5 \cdot 10^{-2}$, we have $\sigma_0 - \sigma > 10^{-5}$, where $\sigma_0 - \sigma$ characterizes the decrease in interfacial tension caused by the presence of an insoluble surface-active material. Since such a "cleansing" of a motionless surface to rid it of all insoluble impurities in fulfillment of the condition $\sigma_0 - \sigma < 10^{-5}$ is probably unattainable in practice, it appears that the conclusion to be drawn from equation (78.4) is that tangential motions on macroscopic surfaces can never in general be observed. At sufficiently small $\Gamma_0$, however, the theory is found to be applicable only in cases of very low velocities of motion. At significant velocities, due to convective transfer, most of the surface is cleansed of the insoluble material, which accumulates in a small area. This fact is well known from the practice of cleaning water surfaces by means of an air jet.

Indeed, it follows from the condition $|\Gamma'| < \Gamma_0$ and equation (75.13) that:

$$v_0 < \frac{D_s}{2a},$$

whence, for the same values for the parameters a and $D_s$, we obtain

$$v_0 \ll 10^{-4} \text{ cm/sec}.$$

If, instead of surface velocity $v_0$, the over-all velocity U is given, the limits of applicability of the theory are determined, as in the case of retardation by a soluble material, by the condition

$$\frac{\mu U}{2} \ll |\sigma_0 - \sigma|. \tag{78.7}$$

or, for a freely falling drop, by the equivalent condition

$$(\sigma_0 - \sigma) \gg \frac{(\rho' - \rho)\, a^2}{3}.$$

At the same time, the condition for strong retardation (78.6) must be fulfilled

$$(\sigma_0 - \sigma) > \frac{D_s\,(2\mu + 3\mu')}{2a},$$

which is possible only if

$$(\sigma_0 - \sigma) > \frac{D_s\,(2\mu + 3\mu')}{2a_0},$$

where $a_0$ is the value of the radius determined from the equation

$$\frac{(\rho' - \rho)\, a_0^2}{3} = \frac{D_s\,(2\mu + 3\mu')}{2a_0}. \qquad (78.8)$$

For the case of a freely falling drop of mercury in water, on the basis of equation (78.8), we find that $a_0 \approx 4 \cdot 10^{-4}$ and, consequently,

$$\sigma_0 - \sigma > 10^{-3}.$$

This value for $\sigma_0 - \sigma$ is not as small as the value derived previously without taking into account the limits of the theory's applicability. Nevertheless, it is still very small. For $a > a_0$, the lower limit of $\sigma_0 - \sigma$, at which retardation is apparent, is defined by equation (78.5) For $a > a_0$, it is defined by equation (78.6). If the inevitable contamination of the surface of the drop is estimated as corresponding to $\sigma_0 - \sigma \simeq 10^{-2}$ it follows from inequalities (78.5) and (78.6) that, in the range of values for a from $5 \cdot 10^{-5}$ to $10^{-3}$, the drops must virtually behave as solid spheres.

## 79. MOTION OF LARGE DROPS

We have examined above the motion of drops of relatively small dimensions, where the regime of motion is viscous. i.e., when the Reynolds numbers are small. In actual practice, however, the motion of relatively large drops is often encountered, where the Reynolds numbers are considerably greater than unity.

Let us first examine drop motion in air. In this case, the density and viscosity of the medium are extremely low compared to those for the drop. Its motion, therefore, is no different from that of a solid body of the same shape and dimensions.

In the case of fall in a gravitational field, the drop's velocity can be derived from the following obvious relation

$$\rho' g V = K_f \rho \frac{U^2}{2} S$$

or

$$U = \left(\frac{2\rho' gV}{K_f \rho S}\right)^{1/2} = \sqrt{\frac{2}{3}\frac{\rho' ga}{K_f \rho}}. \tag{79.1}$$

Here, $\rho'$ is the density of the liquid, $\rho$ that of air, and S is the area of the midsection (the largest crosssection) of the drop. The drag coefficient $K_f$ is of the order of 0.5 for a body of spherical shape.

The measured free fall velocities of individual drops in air, under normal conditions, agree well with equation (79.1) for drops ranging within the following size limits: $0.01 \leq a \leq 0.05$ cm.

Larger drops are deformed and their shape changes greatly from spherical. The drop flattens out during its fall and gradually acquires the shape of a convex bean with unequal sides. This deformation is accompanied by pulsations of relatively small amplitude. The velocity of fall ceases to follow (79.1) but increases with an increase in the radius more slowly than indicated by (79.1).

For drops 0.2 to 0.3 cm in radius, the velocity of fall is a weak function of the radius. Experimental data on this subject are not very accurate, but all authors have noted a limit on the velocity vs. dimension curve.

The limiting velocity, according to data of Khrgian [16], is 917 cm/sec. According to other authors it is somewhat smaller. Humphreys [17] sets the limiting velocity at 800 cm/sec.

Upon further increase in the radius of the falling drop, the large drop breaks up into smaller ones. This phenomenon of fragmentation will be examined in more detail later. A rigorous examination of the motion of a deformed and pulsating drop presents enormous difficulties. The velocity of fall of such drops, nevertheless, can be estimated.

In the air passing by the drop at the velocity U, a pressure difference appears between the fore and aft portions of the drop which, according to the Bernoulli law, is

$$\Delta p = \frac{\rho}{2} U^2.$$

The drop is deformed under the influence of this pressure difference. Capillary pressure prevents further deformation, and as a result of the competing effects of these forces, the drop acquires a certain steady shape related to the given velocity U.

The condition for the balance of the pressure and capillary forces can be written in the form

$$\Delta p S\, \delta h + \sigma\, \delta S = 0, \tag{79.2}$$

where S is the area of the deformed drop and $h$ is the drop's dimension in the direction of the motion.

The first term of (79.2) represents the action of pressure forces expended in the deformation $\delta h$. The second term represents the effect of the forces of surface tension for this deformation.

it is obvious that, for a constant volume of drop, the following condition must be fulfilled

$$S\hbar = V = \frac{4\pi}{3} a^3 = \text{const.} \tag{79.3}$$

From (79.2) we obtain

$$\frac{\Delta p S}{\sigma} = -\frac{\delta S}{\delta \hbar}, \tag{79.4}$$

and from (79.3), we have:

$$\frac{\delta S}{\delta \hbar} = -\frac{V}{\hbar^2} = -\frac{S}{\hbar}. \tag{79.5}$$

Combining (79.4) and (79.5), we find, to an order of magnitude, that the thickness of the drop is

$$\hbar = \frac{\sigma}{\Delta p} \sim \frac{2\sigma}{\rho U^2}. \tag{79.6}$$

Its surface, to an order of magnitude, is

$$S = \frac{V}{\hbar} \approx \frac{\rho U^2 V}{2\sigma}. \tag{79.7}$$

Let us now write the equation of motion for a beanlike drop in a gravitational field.

We have, evidently,

$$K_f \frac{\rho U^2}{2} S = \rho' g V.$$

The drag coefficient $K_f$ has its usual value, which is close to unity.

Substituting the value of the area of the midsection from (79.7) we obtain:

$$K_f \frac{V\rho^2 U^4}{4\sigma} \approx \rho' g V,$$

whence

$$U \approx \sqrt[4]{\frac{4\rho' g\sigma}{K_f \rho^2}}. \tag{79.8}$$

The velocity U is found to be independent of the dimensions of the drop and to be inversely proportional to the square root of the density of the medium and proportional to the 1/4 power of the surface tension.

The meaning of this result is quite obvious: with an increase in velocity, the drop is flattened out, and its area consequently increases, which leads to a greater hydrodynamic resistance. The increased resistance compensates exactly for the increase in the driving force.

The greater the surface tension of the liquid, the less pronounced is the deformation of the drop. The velocity of fall, therefore, is higher for particles which have a greater surface tension and become less flattened.

Substituting the value of g and of the density of air under normal conditions, and assuming that $K_f = 1$, we find:

$$U \sim 740 \text{ cm/sec}.$$

For $K_f = 0.5$, we have $U \sim 860$ cm/sec.

Considering the rough nature of the computations, it should be pointed out that the numerical value of U is unexpectedly close to experimental values. Unfortunately we do not have experimental material at our disposal that would allow the dependence of U on surface tension $\sigma$ to be established. That alone would permit a test of equation (79.8).

In the case of a drop moving in a dense medium—for example, a liquid drop falling in another liquid—the picture becomes more complex. Apparently, here there is also a certain range of dimensions for which a formula of the type of (79.1) is applicable. Since the density of the medium, however, is usually comparable to that of the liquid in the drop, instead of (79.1) we must write

$$U = \sqrt{\frac{2}{3}\frac{(\rho' - \rho)\, ga}{K_f \rho}}. \tag{79.9}$$

The velocity of the fall, therefore, is much slower than in air.

Consequently, the dynamic force which deforms a drop falling in a liquid medium is smaller than in the case of fall in air, in the ratio $\frac{\rho' - \rho}{\rho}$. Deformation and transition to a constant fall velocity should take place for drops of larger dimensions.

It should be pointed out that what is frequently encountered in practice is not the motion of individual drops, but the motion of a whole system or column of drops, moving at small distances from each other. In this case, the conditions of the motion change and the preceding reasoning loses its validity. This interaction between drops plays an important role and, to a large extent, determines the nature of the motion of the system.

## 80. MOTION AND DISSOLUTION OF GAS BUBBLES IN LIQUIDS

The problem of the motion of gas bubbles in liquids has attracted the attention of numerous investigators, for a great variety of reasons. In the first place, the study of the motion of gas bubbles can furnish valuable information on the properties of the simplest phase interface, that of a liquid-gas interface. In the second place, the problem of the motion of gas bubbles is of considerable importance in engineering. It is sufficient to point out that, over such wide industrial fields as basic chemicals, foods, and concentration of ores, there is widespread use of processes and equipment in which the operations are directly related to the laws governing the motion of gas or steam bubbles. We have in mind the bubbling of gases through liquids, a process used in many branches of the chemical and food industries and in the flotation process for the concentration of ores. Thus, for example, many chemical production processes involve the absorption of a great variety of gaseous substances by various liquid reagents in specially constructed absorbers. The gases are bubbled into the liquids. The bubbling of carbon dioxide in saturation boilers in the sugar beet industry is one of the basic processes for sugar production, etc. For these reasons bubble motion has been studied from every aspect in many theoretical and experimental investigations.

The regime of bubble motion varies considerably with the Reynolds number $Re = \frac{Ua}{\nu}$, where U is the velocity of bubble motion, a is the bubble radius, and $\nu$ is the kinematic viscosity of the liquid in which the bubble moves.

In the following, we shall treat systematically bubble motion in different regimes, i.e., at various values of the Reynolds number. Experimentally, bubble motion has been studied in various media and for different bubble diameters [18, 19, 34].

The results of all these studies can be summarized as follows. Bubbles with diameters less than 0.01 cm rise like solid spherical particles (in accordance with Stokes' law). For diameters in excess of 0.01 cm ($Re > 1$), the velocity of rise deviates from that indicated by Stokes' law. The resistance to the rising bubble is larger than in the viscous regime. In the region of Reynolds numbers from $Re \geq 1$ to $Re < 10{,}000$, the law of resistance has the same form as in the case of solid spheres (at the same Reynolds numbers).

For Reynolds numbers above 200 to 300, the law of resistance is usually expressed approximately by the formula

$$F = K_f \frac{\rho U^2 \pi a^2}{2},$$

where the drag coefficient $K_f$ is equal to 0.6.* The shape of the bubble remains almost exactly spherical. On the basis of the resistance law stated above, the velocity of the bubble's rise can be determined from the relation

$$F_g V = W,$$

where $F_g$ is the buoyancy force exerted per 1 cm$^3$ of the bubble's volume and V is its volume. This gives

---

*Actually, in the range of Reynolds numbers from 200 to 1000, this coefficient varies from about 0.65 to about 0.55.

$$\rho g \cdot \frac{4}{3} \pi a^3 = 0.6 \rho \frac{U^2}{2} \pi a^2,$$

whence

$$U \approx 2 \sqrt{\frac{ga}{0.9}}.$$

At Reynolds numbers close to 700, which corresponds to bubble diameters of 0.2 to 0.3 cm, the nature of the motion of the bubbles changes sharply, and they start to be noticeably deformed, acquiring the shape of flattened ellipsoids whose minor axis is in the direction of bubble motion (Figure 70). At the same time, the bubbles begin to vibrate, and they rise by a spiral motion instead of in a straight line. The velocity of motion of bubbles with diameters of 0.2 to 1.5 cm is almost independent of the bubble diameter and amounts to about 28 to 30 cm/sec (the corresponding Reynolds numbers lie within the range of 700 to 4500). Bubbles of larger diameters rise at somewhat greater velocities, namely at about 35 to 40 cm/sec, but are not very stable and tend to subdivide into smaller bubbles. This picture of bubble motion can be interpreted theoretically. Bowden termed the Reynolds number of a bubble moving within the range of Re = 1 to Re = 700 to 800 as moderate, and Reynolds numbers above 800 as large. It follows from what has been said that for motion at moderate Reynolds numbers, the shape of the bubble remains unchanged and can be considered spherical. At large Reynolds numbers, the bubble loses its spherical shape.

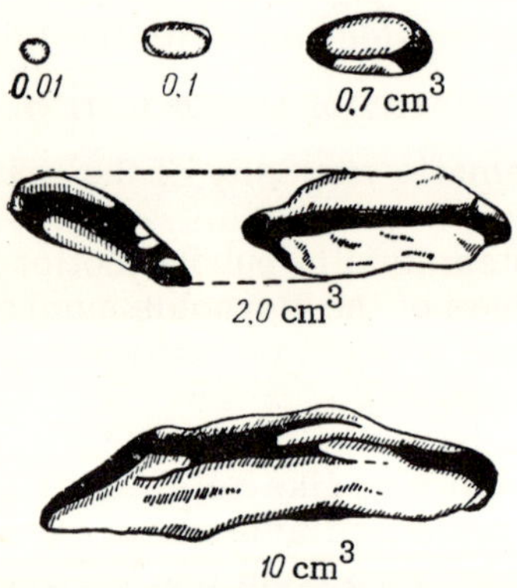


Figure 70. Shapes of air bubbles rising in water (approximately $\frac{1}{2}$ actual size).

## 81. MOTION OF VERY SMALL BUBBLES

The motion of very small bubbles corresponds to Reynolds numbers that are small. This means that the regime of liquid motion near the surface of the bubble is viscous.

In the initial studies devoted to the investigation of the motion of small bubbles, resistance was computed by means of the usual Stokes formula. Rayleigh had already pointed out, however, that the existence of a tangential motion near the gas-liquid interface should change the velocity distribution in the liquid when a gas bubble moves in it, compared to the liquid movement (at the same velocity) for the fall of a solid sphere. The problem of gas bubble motion is closely related to that of the motion of a liquid drop in a liquid medium, a problem which we examined in Section 70. The velocity of rise of a gas bubble in a liquid can be derived from formula (70.35), provided that $\mu'$ and $\rho'$ are assumed to be negligible compared to the viscosity $\mu$ and the density $\rho$ of the medium. This gives:

$$U = -\frac{1}{3}\frac{ga^2}{\nu}, \tag{81.1}$$

where g is the gravitational acceleration, $\nu$ is the kinematic viscosity of the liquid, and a is the radius of the bubble. The minus sign in expression (81.1) indicates that the bubbles rise in the gravitational field.

Formula (81.1) is applicable for small Reynolds numbers, Re ≪ ≪ 1, i.e., when the following inequality holds

$$\frac{ga^3}{3\nu^2} \ll 1. \tag{81.2}$$

For a bubble moving through water, this yields:

$$a \ll 2 \cdot 10^{-2} \text{ cm.}$$

For motion in more viscous liquids, the Rybczyński-Hadamard formula is applicable to larger bubbles.

Comparing the velocity of rise of a bubble calculated from (81.1) with Stokes' law, we see that it is greater than the velocity of a solid sphere of the same density by the ratio $\frac{1}{3} : \frac{2}{9} = \frac{3}{2}$. This is due to the fact that the velocity of the liquid does not become zero at the gas-liquid interface, as it does at a solid surface, but is equal to

$$v_0 = \frac{U_0}{2}$$

according to expression (70.39). This is the reason why the velocity difference between the motionless liquid and the surface of a bubble is smaller than for a solid sphere. Thus, the corresponding energy dissipation is also smaller.

Many investigators have devoted effort to the study of the velocity of rise of small bubbles. Thus, Allen [20] measured the velocity of rising air bubbles in water and in aniline, presenting his data in graphic form showing the drag coefficient as a function of the Reynolds number.

The theoretical value of the drag coefficient for a gas bubble is

$$K_f = \frac{W}{\frac{\rho U^2}{2}\pi a^2} = \frac{8\nu}{Ua} = \frac{8}{\text{Re}},$$

where $Re = \frac{Ua}{\nu}$, i.e., one and one-half times smaller than for a solid sphere.

As noted from Figure 71 (see Section 83), Allen's experimental data completely disagree with the theory and lead to values for the drag coefficient which coincide exactly with the drag on a solid sphere.

The results of measurements made by other students are in good agreement with Allen's data and force us to the general conclusion that, at small Reynolds numbers, gas bubbles move like solid spheres. This fact has usually been interpreted as a confirmation of the Boussinesq theory presented previously (see Section 71).

Actually, the apparent "solidification" of moving bubbles is related to the effect of surface-active materials present in water, in aniline, and in other liquid media. None of the experimenters took any special measures to rid the liquids of surface-active substances when conducting their measurements. The rise velocity of a gas bubble in a liquid medium is expressed by (74.10), where the value of the retardation coefficient is governed by expression (74.8) or by (75.8). Since $\gamma \approx \frac{1}{a}$, an estimate of the values of $\gamma$ for small bubble radii shows that, for $a < 10^{-2}$, the magnitude of $\gamma$ substantially exceeds the viscosity $\mu$, even in the presence of minute traces of surface-active substances. In this way, it is possible to anticipate that gas bubbles of very small dimensions should move in the same manner as solid spheres in all liquid media that have not been specially treated to rid them of surface-active materials. This conclusion has been fully confirmed in experiments conducted by A. V. Gorodetskaya. It will be shown later that all available experimental data on gas bubble motion become understandable if the presence of surface-active materials is taken into account, and that there is no need for such unfounded hypotheses as that of Boussinesq.

## 82. MOTION OF MEDIUM-SIZE BUBBLES

Let us examine bubble motion at moderate Reynolds numbers [21]. In the motion of a solid body at Reynolds numbers $Re \gg 1$, a hydrodynamic boundary layer exists near the surface of the solid body.

The reason for its appearance, as shown in Section 3, is that for $Re \gg 1$, the motion should be viewed as the flow of an ideal fluid, devoid of friction and capable of slipping along the solid surface. Such slippage is impossible in an actual fluid, and the velocity at the surface itself must be zero.

Let us pass to a system in which the bubble is at rest and the liquid moves at a velocity U. At large Reynolds numbers, the liquid can be considered ideal, and the energy dissipated due to viscous friction can be neglected.

If we examine the motion of the liquid near the surface of the bubble, i.e., near the gas-liquid interface, it becomes immediately clear that the conditions of motion near that interface are significantly different from those at a liquid-solid body interface. Since any type of tangential motion of liquid in the liquid-gas interface is possible on the surface of the bubble, the tangential velocity component, in this case, does not become zero. At first glance it might appear that the flow past a gas bubble occurs in the same way as for a body immersed in the flow of an ideal liquid. In other words, it is possible to assume that the existence of viscosity in an actual liquid has no effect on flow past a gas bubble. Actually, however, this is not so. In flow past a bubble, the tangential components of the stress tensor must remain continuous at the phase interface. If the density of the gas and its viscosity are disregarded in comparison with those of the liquid, the following conditions must be fulfilled at the surface of the bubble

$$p = p', \tag{82.1}$$

$$\mu \frac{\partial v_t}{\partial r} = 0. \tag{82.2}$$

The first expresses the equality of pressures in the liquid and gas phases. The second shows that the tangential force must be zero on the surface of the bubble. In addition, the normal velocity component on the bubble's surface must also be zero

$$v_n = 0. \tag{82.3}$$

In the flow of an ideal liquid it is evident that conditions (82.1) and (82.3) are fulfilled. Condition (82.2), however, cannot be satisfied in this case: no limitations can be imposed on the tangential velocity component in the flow of an ideal liquid. It is clear from this that, despite the possibility of liquid flow on the surface of the bubble, viscous effects should become significant in a certain thin layer near the surface. In other words, a peculiar boundary layer appears near the gas-liquid interface, in which a manifestation of viscosity, in the case of an actual liquid, is inevitable. The role of viscous forces in the case of a bubble is not as important as in the flow past a solid body. The viscous forces establish that the tangential velocity at the surface of the bubble will be constant but do not assure that it is zero. The constancy of the tangential velocity affects the derivative of the velocity, but not the velocity of the moving liquid itself. Therefore, the change in the velocity distribution near the bubble's surface and the difference of this distribution from that of an ideal liquid flowing past a sphere are less pronounced than in the case of flow past a solid sphere. This allows us to assume that the velocity distribution, in the case of flow past a gas bubble, is relatively little different from the velocity distribution in an ideal liquid. Let $\mathbf{v}_0$ designate the velocity of an ideal liquid flowing past a sphere whose radius is a. Let us seek the velocity distribution in the liquid flowing past the gas bubble, in the form

$$\mathbf{v}=\mathbf{v}_0+\mathbf{v}', \tag{82.4}$$

where $\mathbf{v}'$ is the deviation of the velocity from that of an ideal liquid. Similarly, we whall seek the pressure distribution in the liquid in the form

$$p=p_0+p', \tag{82.5}$$

where $p_0$ is the pressure in an ideal liquid and $p'$ is the deviation caused by the change in this velocity distribution. We shall consider the quantities $v'$ and $p'$ to be small compared to $v_0$ and $p_0$. Subsequent calculations show the extent to which this assumption may be justified.

The velocity and pressure distributions in an ideal liquid, flowing past a sphere of radius a, are determined from the equations

$$\operatorname{div}\mathbf{v}_0=0, \tag{82.6}$$

$$(\mathbf{v}_0 \operatorname{grad})\,\mathbf{v}_0=-\frac{1}{\rho}\operatorname{grad} p_0 \tag{82.7}$$

with boundary conditions

$$\begin{aligned} \mathbf{v}_0 &\to \mathbf{U} \quad &&\text{as} \quad r\to\infty, \\ (v_0)_r &= 0 \quad &&\text{at} \quad r=a. \end{aligned} \tag{82.8}$$

The first of the above conditions expresses the fact that the velocity of the moving liquid far from the body has a value U. The second shows that the normal velocity is zero on the surface of the sphere.

Since the flow of an ideal liquid past a sphere is irrotational, we can assume that:

$$\mathbf{v}_0=\operatorname{grad}\varphi,$$

where $\varphi$ is the velocity potential satisfying the equation

$$\Delta\varphi=0.$$

The solution of this equation for $\varphi$, satisfying the boundary condition for $r\to\infty$, has the form

$$\varphi=\left(ar+\frac{b}{r^2}\right)\cos\theta.$$

Therefore

$$v_r^{(0)}=\frac{\partial\varphi}{\partial r}=\left(a-\frac{2b}{r^3}\right)\cos\theta,$$

$$v_\theta^{(0)}=\frac{1}{r}\frac{\partial\varphi}{\partial\theta}=-\left(a+\frac{b}{r^3}\right)\sin\theta.$$

Boundary conditions (82.8) yield:

$$v_r^{(0)} = -U\left[1 - \frac{a^3}{r^3}\right]\cos\theta, \tag{82.9}$$

$$v_\theta^{(0)} = U\left[1 + \frac{a^3}{2r^3}\right]\sin\theta. \tag{82.10}$$

The pressure distribution is given by the Bernoulli equation

$$\frac{\rho v_0^2}{2} + p_0 = \text{const.} \tag{82.11}$$

Let us examine the velocity distribution near the bubble's surface at $r \approx a$. Here we may introduce the variable

$$y = r - a$$

and assume that $y \ll a$.

Expanding expressions for $v_r^{(0)}$ and $v_\theta^{(0)}$ in a power series in $y/a$, we obtain:

$$v_r^{(0)} = -3U\frac{y}{a}\cos\theta, \tag{82.12}$$

$$v_\theta^{(0)} = \frac{3}{2}U\left(1 - \frac{y}{a}\right)\sin\theta. \tag{82.13}$$

Knowing the velocity distribution in the flow of an ideal liquid, we may pass to a determination of $\mathbf{v}'$ and $p'$.

Substituting the values of $\mathbf{v}$ and $p$ from (82.4) and (82.5) into the general hydrodynamic equations (1.1) and (1.2), we find:

$$(\mathbf{v}_0 \operatorname{grad})\mathbf{v}_0 + (\mathbf{v}_0 \operatorname{grad})\mathbf{v}' + (\mathbf{v}' \operatorname{grad})\mathbf{v}_0 + (\mathbf{v}' \operatorname{grad})\mathbf{v}' = \tag{82.14}$$

$$= -\frac{\operatorname{grad} p_0}{\rho} - \frac{\operatorname{grad} p'}{\rho} + \nu\Delta\mathbf{v}_0 + \nu\Delta\mathbf{v}',$$

$$\operatorname{div}(\mathbf{v}_0 + \mathbf{v}') = 0. \tag{82.15}$$

Taking equations (82.6) and (82.7) into account and omitting the term $(\mathbf{v}' \operatorname{grad})\mathbf{v}'$, which is a second order infinitesimal, and also bearing in mind that the motion of an ideal, incompressible, irrotational fluid satisfies the condition:

$$\Delta\mathbf{v}_0 = \operatorname{grad}\operatorname{div}\mathbf{v}_0 - \operatorname{rot}\operatorname{rot}\mathbf{v}_0 = 0,$$

we obtain, after simple transformations, the following equations for the quantities $\mathbf{v}'$ and $p'$:

$$(\mathbf{v}_0 \operatorname{grad})\mathbf{v}' + (\mathbf{v}' \operatorname{grad})\mathbf{v}_0 = -\frac{\operatorname{grad} p'}{\rho} + \nu\Delta\mathbf{v}', \tag{82.16}$$

$$\operatorname{div}\mathbf{v}' = 0. \tag{82.17}$$

The quantities $\mathbf{v}'$ and $p'$ are not zero in the very thin boundary layer of liquid immediately adjacent to the surface of the bubble. Since the thickness of this layer can be considered to be very small compared to the bubble radius, equation (82.16) can be simplified further.

Namely, we should take the approximate values given by (82.12) and (82.13) as an expression for the values of the components of velocity $v_0$ in expression (82.16). Further, we can neglect the curvature of the boundary layer and consider it, approximately, as a plane. Then, by introducing rectangular coordinates x and y plotted along the surface of the bubble and along the normal to the surface, and by designating the projections of the velocities on the x and y axes by $u_0$, $u'$ and $v_0$, $v'$, we can rewrite equations (82.16) and (82.17) in terms of the projections

$$u_0\frac{\partial u'}{\partial x}+v_0\frac{\partial u'}{\partial y}+u'\frac{\partial u_0}{\partial x}+v'\frac{\partial u_0}{\partial y}=-\frac{1}{\rho}\frac{\partial p'}{\partial x}+\nu\frac{\partial^2 u'}{\partial y^2}, \tag{82.18}$$

$$u_0\frac{\partial v'}{\partial x}+v_0\frac{\partial v'}{\partial y}+u'\frac{\partial v_0}{\partial x}+v'\frac{\partial v_0}{\partial y}=-\frac{1}{\rho}\frac{\partial p'}{\partial y}+\nu\frac{\partial^2 v'}{\partial y^2}, \tag{82.19}$$

$$\frac{\partial u'}{\partial x}+\frac{\partial v'}{\partial y}=0.^{*} \tag{82.20}$$

The boundary condition on the surface of the bubble will have the form $\frac{\partial u}{\partial y}=0$, or, by substituting $u_0+u'$ in place of u,

$$\frac{\partial u_0}{\partial y}+\frac{\partial u'}{\partial y}=0. \tag{82.21}$$

Boundary condition (82.8) gives:

$$u'\to 0 \qquad \text{as} \quad y\to\infty \tag{82.22}$$

(since, as $y\to\infty, u_0\to U$). Let us estimate the order of magnitude of all terms in (82.18) and (82.19).

It follows from equation (82.20) that

$$v'=-\int\frac{\partial u'}{\partial x}\,dy, \tag{82.23}$$

while expression (82.6) gives:

$$v_0=-\int\frac{\partial u_0}{\partial x}\,dy. \tag{82.24}$$

The term $v'\frac{\partial u_0}{\partial y}$ in equation (82.18) can therefore be omitted, since it is small in comparison with the term $v_0\frac{\partial u'}{\partial y}$.

An inspection of equation (82.19) shows (see Section 3) that all terms are quantities of the second order, so that the equation is reduced to:

$$\frac{\partial p'}{\partial y}=0. \tag{82.25}$$

---

*Editor's note: This continuity equation is incorrect for axisymmetric flow. This introduces a number of errors into the subsequent development. See B.F. Chao, Physics of Fluids 5, 69 (1962).

This indicates that the pressure is constant through the boundary layer under examination (as is true also in the ordinary Prandtl boundary layer). However, since $p' = 0$ outside the boundary layer, we have, instead of equation (82.18),

$$u_0 \frac{\partial u'}{\partial x} + v_0 \frac{\partial u'}{\partial y} + u' \frac{\partial u_0}{\partial x} = \nu \frac{\partial^2 u'}{\partial y^2}, \tag{82.26}$$

or, substituting the value of $v_0$ from expression (82.24), we finally obtain:

$$u' \frac{\partial u_0}{\partial x} + u_0 \frac{\partial u'}{\partial x} - \left(\int \frac{\partial u_0}{\partial x} dy\right) \frac{\partial u'}{\partial y} = \nu \frac{\partial^2 u'}{\partial y^2}. \tag{82.27}$$

Substituting the value for $u_0$ from expression (82.13) into equation (82.27) we obtain:

$$\frac{3U}{2a} u' \cos\theta + \frac{3U}{2a} \sin\theta \frac{\partial u'}{\partial \theta} - \frac{3U\cos\theta}{2a} \cdot \frac{\partial u'}{\partial y} y = \nu \frac{\partial^2 u'}{\partial y^2}. \tag{82.28}$$

This expression permits us to estimate the thickness $\delta_0$ of the boundary layer. Namely, at $y \approx \delta_0$, the viscous and convective terms in expression (82.28) should be quantities of the same order. This estimation yields:

$$\left.\begin{aligned} \nu \frac{\partial^2 u'}{\partial y^2} &\approx \nu \frac{u'}{\delta_0^2}, \\ \nu \frac{u'}{\delta_0^2} &\approx \frac{Uu'}{a}, \end{aligned}\right\} \tag{82.29}$$

$$\delta_0 \approx \left(\frac{\nu a}{U}\right)^{1/2}. \tag{82.30}$$

For $Re \gg 1$, the thickness of the boundary layer is small compared to the bubble's radius.

Upon substitution of the value of $u_0$ from expression (82.13), boundary condition (82.21) takes the form

$$\frac{\partial u'}{\partial y} = \frac{3U}{2a} \sin\theta \qquad \text{at} \quad y = 0. \tag{82.31}$$

In order to obtain a solution for equation (82.28) for boundary conditions (82.31) and (82.22), let us introduce a new unknown function $f = u' \sin\theta$. Equation (82.28) then acquires the form

$$\sin\theta \frac{\partial f}{\partial \theta} - \cos\theta \cdot y \frac{\partial f}{\partial y} = k \frac{\partial^2 f}{\partial y^2}, \tag{82.32}$$

where the following quantity is designated by k

$$k = \frac{2a\nu}{3U}. \tag{82.33}$$

Boundary condition (82.31) becomes condition

$$\left(\frac{\partial f}{\partial y}\right)_{y=0} = \frac{3U}{2a} \sin^2 \theta. \tag{82.34}$$

In order to integrate equation (82.32), we introduce a new variable instead of y

$$\psi = y \sin \theta. \tag{82.35}$$

Hence,

$$\frac{\partial f}{\partial y} = \frac{\partial f}{\partial \psi} \sin \theta; \qquad \left(\frac{\partial f}{\partial \theta}\right)_y = \left(\frac{\partial f}{\partial \theta}\right)_\psi + \frac{\partial f}{\partial \psi}\frac{\partial \psi}{\partial \theta}.$$

We have, therefore:

$$\left(\frac{\partial f}{\partial \theta}\right)_\psi = k \sin \theta \frac{\partial^2 f}{\partial \psi^2}. \tag{82.36}$$

From equation (82.34), we have, accordingly:

$$\left(\frac{\partial f}{\partial \psi}\right)_{\psi=0} = \frac{3U}{2a} \sin \theta. \tag{82.37}$$

It is convenient to introduce a new variable $t = 1 - \cos \theta$ into equation (82.36) instead of $\theta$. We then have finally:

$$\frac{\partial f}{\partial t} = k \frac{\partial^2 f}{\partial \psi^2} \tag{82.38}$$

and

$$\left(\frac{\partial f}{\partial \psi}\right)_{\psi=0} = \frac{3U}{2a} \sqrt{1-(t-1)^2}. \tag{82.39}$$

Condition (82.22), which defines the behavior of $u'$ at infinity, should now be written in the form

$$f \to 0 \quad \text{as} \quad \psi \to \infty. \tag{82.40}$$

Besides, the symmetry of the problem requires that

$$f = 0 \quad \text{at} \quad t = 0. \tag{82.41}$$

This indicates that velocity $u' \neq \infty$ at the point of division of the incident flow, $\theta = 0$.

Expression (82.38) represents the usual type of heat conduction equation. Its general solution has the form

$$f(\psi, \theta) = -\left(\frac{3U\nu}{2\pi a}\right)^{1/2} \int_0^{1-\cos\theta} \frac{[1-(z-1)^2]^{1/2}}{[1-\cos\theta - z]^{1/2}} e^{-\frac{\psi^2}{4k(1-\cos\theta - z)}} dz. \tag{82.42}$$

The correction to the irrotational velocity is

$$u'(\psi, \theta) = \frac{-1}{\sin\theta} \left(\frac{3U\nu}{2\pi a}\right)^{1/2} \int_0^{1-\cos\theta} \frac{[1-(z-1)^2]^{1/2}}{[1-\cos\theta - z]^{1/2}} e^{-\frac{\psi^2}{4k(1-\cos\theta - z)}} dz. \tag{82.43}$$

On the surface of the bubble, in particular at $\psi = 0$,

$$u'(0,\ \theta) = \frac{-1}{\sin\theta}\left(\frac{3U\nu}{2\pi a}\right)^{1/2} \int\limits_0^{1-\cos\theta} \frac{[1-(z-1)^2]^{1/2}}{[1-\cos\theta - z]^{1/2}}\, dz. \tag{82.44}$$

On the strength of the assumption that we have made, expression (82.44) should lead to values of $u' \ll u_0$. Expression (82.44) shows that the velocity $u'(0,\theta)$ has a very pronounced angular dependence:

as $\theta \to 0$

$$u'(0,\ \theta) \to \text{const}\ \frac{U}{\sqrt{\text{Re}}} \ll U; \tag{82.45}$$

as $\theta \to \pi$

$$u'(0,\ \theta) \to \frac{1}{\sin\theta}\int\limits_0^2 \sqrt{z}\, dz \approx \frac{1}{\sin\theta} \to \infty.^{*} \tag{82.46}$$

Thus, in the vicinity of the point of incidence and up to very large angles $\theta$, the velocity correction is very small compared to the velocity of the main flow (the flow of an ideal liquid). At $\theta \approx \pi$, however, the value of the tangential velocity component increases sharply. In conformity with the continuity equation, the normal velocity component in the boundary layer also increases sharply in the same region. There is a liquid flow directed from the surface of the bubble into the bulk of the medium, and the velocity of this flow is high compared to the velocity $u_0$. This indicates that, at $\theta \approx \pi$, a separation phenomenon takes place on the bubble surface. A turbulent flow appears in the region of separation, and a turbulent wake appears behind the bubble.

In order to examine this separation phenomenon, the method proposed by L.D. Landau can be used [22]. Let us note that, at the point where $u'$ increases indefinitely, so does $v'$. This means that the flow is diverted from the interface and is directed into the bulk of the liquid. The usual conditions of separation that occur on the surface of a solid body are not applicable to a liquid surface since the equality $\frac{\partial u}{\partial y} = 0$ is automatically fulfilled over the entire liquid surface. Utilizing the Landau method, the separation conditions for a liquid surface can be written in the form

$$\frac{\partial u}{\partial y} = 0; \qquad \frac{\partial^2 u}{\partial y^2} = 0.$$

Thus, the position of the separation point can be calculated. It is clear, however, that the angle $\theta_1$ at which $u'$ becomes equal in order of magnitude to $U$ (i.e., the dimension of the separation zone) is very

---

*Editor's note: a minus sign should be introduced here, which may alter some of Levich's arguments concerning "separation."

Iyer P. A.



small. This angle can be derived from the condition $u'(0,\theta) \approx U$. The substitution of the value of the velocity $u'$ from formula (82.44) into this condition gives:

$$\theta_1 \approx \pi - \theta \approx \frac{1}{\sqrt{\text{Re}}} \ll 1. \tag{82.47}$$

The overwhelming portion of the bubble area, therefore, is occupied by unseparated flow. Along the surface of the bubble from $\theta = 0$ to $\theta = \pi - \theta_1$, the distribution of liquid flow velocities differs little from the velocity distribution in an ideal liquid flowing past a solid sphere.

Comparing this picture with that set forth in Section 5, we see that the mechanism of the flow past a bubble and, consequently, the law of resistance, are substantially different from what takes place in the flow past a solid sphere. In the latter instance, the separation region occupies almost the entire downstream portion of the sphere at Reynolds numbers of the order of several hundred, and occupies a substantial portion of it at Reynolds numbers that are smaller (but still greater than unity).

In the flow past a bubble, the size of the separation zone does not exceed $2\theta_1$, and, consequently, covers a very narrow area at the bubble's wake. This region is characterized by turbulent liquid motion. The turbulent velocities are small compared to those of the main flow since the area of turbulent motion is very narrow. One can assume that the order of magnitude of the eddy velocities, in a zone approximately $a\theta_1$, is

$$u_{\text{turb}} \approx \frac{Ua\theta_1}{a} \approx U\theta_1 \approx \frac{U}{\sqrt{\text{Re}}}. \tag{82.48}$$

The total friction exerted over the entire surface of the bubble is composed of two parts: resistance that acts on the bubble surface up to the separation point, and forces that act in the turbulent zone. Resistance in the first zone is viscous in nature. It can be derived most simply by computing the energy dissipated.

According to formula (1.15), the dissipated energy $-\frac{dE}{dt}$ is equal to

$$-\frac{dE}{dt} = \mu \int [\text{curl}\, \mathbf{v}]^2\, dV + \mu \int \frac{\partial \mathbf{v}^2}{\partial n}\, ds + 2\mu \int [\mathbf{v}\ \text{curl}\, \mathbf{v}]\, ds, \tag{82.49}$$

where $n$ is the direction of the outer normal at the surface past which the flow is moving and $\mu$ is the viscosity. According to formula (82.4) $v = v_0 + v'$. In the zone of viscous flow past the bubble, the deviation of the velocity from that of an ideal liquid is $v' \ll v_0$, and it can be neglected when substituting into expression (82.49). The velocity of the undisturbed flow is irrotational, so that formula (82.49) can be simplified and becomes the following

$$-\frac{dE}{dt}=\mu\int\left\{\frac{\partial}{\partial r}(v_r^{(0)})^2+\frac{\partial}{\partial r}(v_\theta^{(0)})^2\right\}_{r=a}2\pi a^2\sin\theta\,d\theta=$$

$$=\mu\int\left(2v_r^{(0)}\frac{\partial v_r^{(0)}}{\partial r}+2v_\theta^{(0)}\frac{\partial v_\theta^0}{\partial r}\right)_{r=a}2\pi a^2\sin\theta\,d\theta=$$

$$=\mu\cdot 4\pi a^2\int v_\theta^{(0)}\frac{\partial v_\theta^{(0)}}{\partial y}\sin\theta\,d\theta=-12\pi aU^2\mu. \quad (82.50)$$

In this case we have taken account of the fact that, in view of condition (82.3), $(v_r^{(0)})_{r=a}=0$, and we have substituted in place of $v_\theta^{(0)}$ its value from (82.13). The total dissiptative force F exerted on the bubble is

$$F=-\frac{1}{2}\frac{\partial}{\partial U}\left(\frac{dE}{dt}\right)=12\pi\mu aU. \quad (82.51)$$

We see that, at small Reynolds numbers, this force differs from the one acting on a solid sphere only by the factor 2. As for the resistance that appears due to the action of the turbulent wake, it is, in terms of an order of magnitude, equal to

$$F_{\text{turb}}\approx K_f\rho\frac{U^2}{2}S_1,$$

where $S_1$ is the area covered by the turbulent wake. Therefore, substituting the value of $S_1$ into $F_{\text{turb}}$, we obtain

$$F_{\text{turb}}\approx K_f\rho\frac{U^2}{2}a^2\theta_1^2.$$

Replacing $\theta_1$ with its value from expression (82.47), we obtain

$$F_{\text{turb}}\approx K_f\frac{\rho a^2U^2}{2\,\text{Re}}. \quad (82.52)$$

This force is small compared to that of dissipation, provided only $12\pi \gg K_f$. Thus, the resistance sustained by the bubble at Reynolds numbers $\text{Re}\gg 1$ is basically dissipative in nature and proportional to the liquid viscosity and to the first power of the velocity. Basically, the flow past the bubble occurs without separation. Thus, for example, at Re = 625, the separation zone extends 2° to either side of the line of symmetry.

Turbulent resistance, therefore, has no substantial effect on the over-all resistance during the motion of the bubble. The velocity of the bubble's rise can be derived from the obvious relation

$$F=-\frac{4\pi\rho g a^3}{3}, \quad (82.53)$$

which expresses the equality between the lifting and the resisting forces. This leads to a value for the velocity of rise

$$U = -\frac{1}{9} \cdot \frac{ga^2}{\nu}, \qquad (82.54)$$

which is one half the velocity of the bubble's rise in the range of small Reynolds numbers.

From a hydrodynamic viewpoint, this result may appear somewhat paradoxical. In the motion of any solid body at Re $\gg 1$, the resistance is the sum of the viscous resistance, which is proportional to the $\frac{3}{2}$ power of the velocity, and of form drag (turbulence resistance), which is proportional to the square of the velocity.

In the case of unstreamlined bodies, the latter is many times greater than the resistance due to friction. But for a bubble, due to the mobility of the interface, the separation point is situated so far downstream that the turbulent-wake region covers only a minute portion of the bubble's surface. In this case, it is natural for form drag to be very small. The viscous friction is also different because of the mobility of the interface.

Two assumptions form the basis of the above reasoning:

1) The bubble's shape is strictly spherical;

2) No surface-active materials are present on its surface.

The first assumption is justified when the capillary pressure in the bubble remains high compared to the pressure exerted on the bubble by the flowing liquid. The latter, according to the Bernoulli law (3.3), is of the order of magnitude of

$$p \approx \rho \frac{U^2}{2} \approx \frac{\rho}{2} \cdot \frac{\rho^2 a^4 g^2}{81\mu^2}.$$

Noticeable deformation occurs at $a = a_{cr}$, which is derived from the condition

$$\frac{2\sigma}{a_{cr}} \approx \rho^3 \frac{a_{cr}^4 g^2}{162\mu^2},$$

or

$$a_{cr} \approx \left(\frac{324\mu^2\sigma}{\rho^3 g^2}\right)^{1/5}.$$

The same condition can be written in a different form. Substantial deformation occurs when

$$\frac{2\sigma}{a_{cr}} \approx \frac{F}{S} \approx \frac{4}{3} \frac{\pi\rho g a_{cr}^3}{4\pi a_{cr}^2} \approx \rho g a_{cr},$$

where S is the area of the bubble and the resistance is expressed by the buoyancy force. In this case, we have:

$$a_{cr} \approx \left(\frac{\sigma}{\rho g}\right)^{1/2} = K^{1/2},$$

where K is the capillary constant.

It should be pointed out, however, that for $a \approx \sqrt{K}$ the shape of the bubble differs considerably from spherical. Actually, departure from the spherical shape occurs much sooner, at $a \approx \frac{\sqrt{K}}{3}$, which corresponds to $\mathrm{Re} \approx 800$ to 1000.

We shall show in Section 84 that the law of motion of deformed bubbles differs radically from that of bubbles which are not subject to deformation.

The assumption that there are no surface-active materials in the system is equally important. If such materials are present in the liquid, they will appear on the surface of the bubble. The rapid liquid motion on the surface will carry such molecules to the rear of the bubble, where they form a saturated monolayer. Here, two possibilities arise: either a portion of the bubble, of area $S_0$, remains covered by a nondeforming monolayer during the entire period of the motion, or the liquid motion carries away the monolayer, cleansing the surface completely.

The second case is realized, apparently, only for bubbles which are deformed (see below). If the surface $S_0$ of the bubble is covered by a monolayer, the liquid velocity on this surface is zero, just as on the surface of a solid body. Then, separation occurs in a region $S_0$ and is accompanied by an additional form drag

$$F_0 \approx K_f \rho \frac{U^2}{2} S_0. \tag{82.55}$$

The ratio of viscous drag to form drag is

$$\frac{F}{F_0} \approx 12\pi \frac{a^2}{S_0} \cdot \frac{1}{\mathrm{Re}}.$$

It will be less than unity when

$$12\pi \frac{a^2}{S_0} \cdot \frac{1}{\mathrm{Re}} < 1,$$

or

$$S_0 > \frac{12\pi a^2}{\mathrm{Re}}. \tag{82.56}$$

If $F > F_0$, the resistance law (82.51) is replaced by (82.52). The velocity of rise is determined by the equality

$$K_f \rho \frac{U^2}{2} S_0 = \frac{4\pi}{3} \rho g a^3.$$

Therefore,

$$\mathrm{Re} \approx \frac{Ua}{\nu} \approx (4ga^3)^{1/2} \frac{a}{\nu \sqrt{S_0}}.$$

Substituting Re into expression (82.56), we have:

$$S_0 > \frac{12\pi a^2 \nu \sqrt{S_0}}{(4ga^3)^{1/2} a},$$

$$\sqrt{S_0} > \frac{12\pi\nu}{(4ga)^{1/2}}.$$

The ratio of the portion of the bubble's surface covered by the monolayer to the entire surface area must satisfy the inequality

$$\frac{S_0}{4\pi a^2} > \frac{(12\pi\nu)^2}{4\pi a^2 \cdot 4ga} \approx \frac{28\nu^2}{ga^3}.$$

For the quadratic law of resistance to be manifested, in the case of the motion of a bubble in water, the radius of the bubble being $a \approx 0.05$, it is sufficient for only 2% of the entire surface of the bubble to be covered by a monolayer. Thus, the effect of surface-active materials is very significant, despite the high bubble velocities. At first glance, this circumstance would appear to reduce materially the practical significance of the formulas that have been derived, with regard to their application to the types of bubbles encountered in industrial processes. It should be borne in mind, however, that, under the conditions prevailing in industry, an enormous number of bubbles pass through relatively small liquid volumes. This results in a very extensive gas-liquid interfacial area. In certain instances, therefore, water which has not undergone special treatment to eliminate surface-active substances can attain a very high degree of purity due to the passage of large amounts of gas through it.

## 83. EXPERIMENTAL VERIFICATION OF THE THEORETICAL EXPRESSION FOR THE VELOCITY OF BUBBLE MOTION IN A LIQUID MEDIUM

The theoretical expression for the velocity of bubble motion, derived in the preceding section, is applicable within the range of Reynolds numbers $\mathrm{Re} \approx 50$ to 800, which applies to radii of less than 0.1 cm (or diameters smaller than 0.2 cm).

The indicated relation is in complete disagreement with the data of many authors who studied this problem previously.

In this connection. A. V. Gorodetskaya conducted a special experimental study [23] to measure the velocity of bubbles rising in water and in aqueous solutions. The conditions of purity that are required for the applicability of our theoretical formula were carefully observed.

The measurement of the bubble rise velocity was conducted in a constant temperature vessel (to avoid convection) by the motion picture method. It seems to us that these experiments fully clarify the reason for the disagreement between the theory and the experimental results previously obtained.

The measurements by A. V. Gorodetskaya were conducted in ordinary water, in alcohol solutions, as well as in water that had been boiled with permanganate and then double distilled. The data for bubble motion in water are presented graphically in Figure 71. For purposes of comparison, the theoretical curve (82.54) is also shown, along with the experimental data of Allen, Luchsinger, and Shabalin and the velocity curve for the fall of glass spheres, which were used in the test. The velocities of these glass spheres were adjusted to correspond to the density of the bubbles.

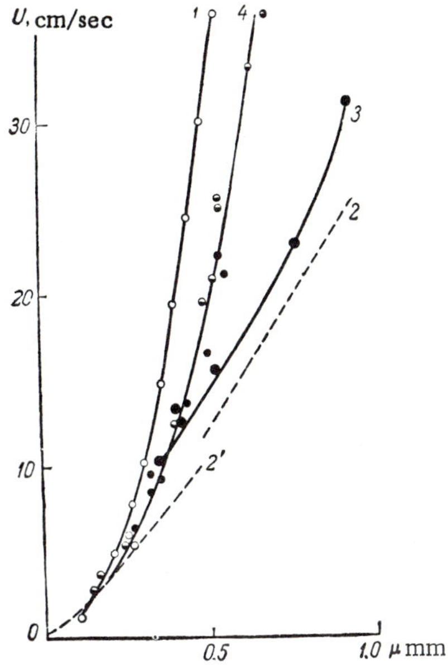


Figure 71. Measurements of bubble velocities in water. 1–theoretical curve by Levich; 2–Allen's experimental curve; 2′ – Shabalin's experimental curve; 3 – Luchsinger's experimental curve; 4 – double distilled water (photographic and visual observations).

In examining the results of A.V. Gorodetskaya's measurements, it becomes quite evident that purification of the water is very important. Curve 2 for ordinary tap water agrees well with Allen's data. Curve 3, obtained by Luchsinger, who used distilled water, is somewhat above Allen's curve. Curve 4 for the special double-distilled water differs markedly from the others and, in general, corresponds to the theoretical curve 1. The discrepancy between the theory and the experimental data does not exceed 30%.* This may be ascribed to experimental errors, to residual surface-active substances, to the effect of form resistance not taken into account in formula (82.54), and to slight deviations in bubble shape from spherical.

The curves in Figure 72 also clearly show the sharply expressed effect of surface-active substances on the motion of the bubbles Alcohols dissolved in water lower the velocity of bubble motion. The maximum retardation of the largest bubbles is 250%. The effect of

---

*We must take into account the fact that an assumption as to the complete absence of bubble deformation was made in the computations. At the same time, there are no arbitrary constants in the formula derived for the velocity of bubble motion.

an alcohol increases with its surface activity. It is most pronounced for hexyl and octyl alcohols, and weakest for butyl alcohol.

The effect of such alcohols cannot be related to changes in surface tension or to the corresponding bubble deformation, since the alcohol concentration in the water was extremely low and the change in surface tension would not exceed 0.5 dyne.

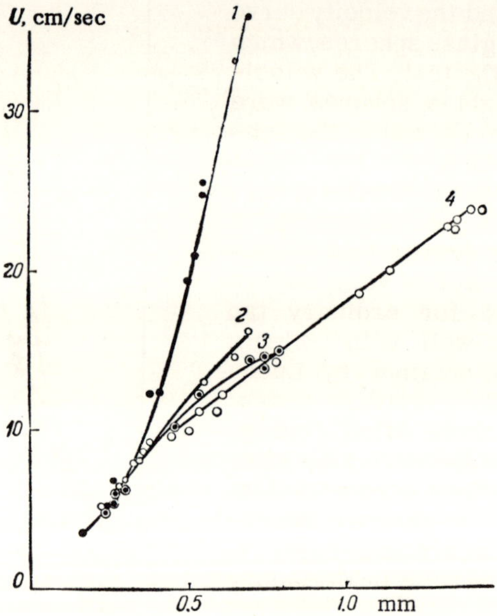


Figure 72. Visual observations.

1 — double-distilled water; 2 — tap water; 3 — n-amyl alcohol, $10^{-4}$ M; 4 — glass balls (adjusted for specific gravity).

From the above data and discussion, we may conclude that the theoretical consideration set forth in the preceding section are in complete agreement with experimental data. Earlier experimental data apply to water that had not been purified, and they should not be used for a comparison with the theory. If follows from A. V. Gorodetskaya's experiments that even minute traces of surface-active substances have an effect on bubble motion.*

---

*Proofreading note. This conclusion was confirmed for hydrocarbon media in a study by M. M. Kusakov et al. [33].

In an investigation by B.K. Kozlov and M.A. Mologin [24], the motion of large bubbles was studied. The authors present Hoefer's empirical formula, which agrees well with formula (82.54) in a range of Reynolds numbers up to several hundred (Figure 73).

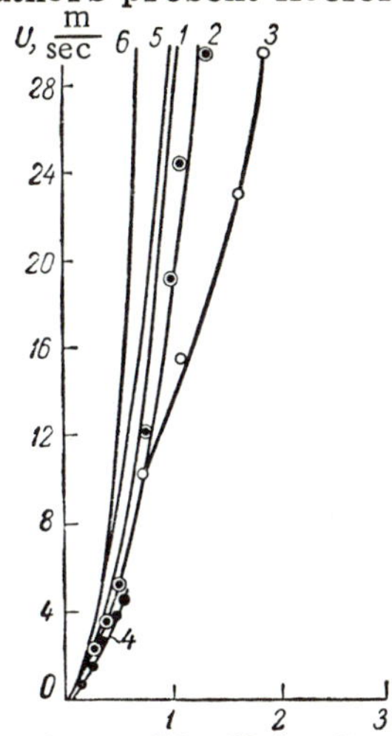


Figure 73. Velocity of rise of air bubbles in motionless water.

Tube diameter 75 mm; water depth 2.0 m; water temperature 18° C. The curves: 1 — according to formula (82.54); 2—Gorodetskaya's experiments; 3-Luchsinger's experiments; 4—Allen's experiments; 5-Hoefer's experimental formula; 6—Stokes' theoretical formula.

## 84. MOTION OF LARGE BUBBLES

We have repeatedly pointed out above that for Re > 800 to 1000 (diameter > 0.2 cm), the bubble becomes deformed. Study of the motion of a deformed bubble presents difficulties, especially in view of the fact that its shape is not strictly fixed and pulsations appear within the bubble, as a result of which its motion ceases to be steady and the bubble no longer moves in a straight line.

The most important peculiarity of the motion of large bubbles, however, is the fact that the velocity of rise is not a function of the bubble's dimensions. All sufficiently large bubbles rise at a velocity close to 30 cm/sec.

Large bubbles acquire the shape of a disk (see Figure 70) whose thickness h is small compared to the transverse dimension $l$. In this case, it is possible to estimate the order of magnitude of the velocity of the bubble's rise, as well as the thickness h. We shall assume that the picture of flow past the bubble remains the same in principle as for moderate Reynolds numbers, so that the distribution of liquid velocities differs little from that of an ideal liquid. The radial velocity component on the surface of the deformed bubbles is zero, and the tangential component is close to the velocity of the external flow (bubble's rising velocity)

$$v_\theta \sim U.$$

Let us seek the velocity of rise of a deformed bubble in the same manner as we sought and derived the velocity for an undeformed bubble. First, we calculate the energy dissipation. According to (82.50) we have:

$$\frac{dE}{dt} = -\mu\left[\frac{\partial}{\partial n}(v_\theta)^2\right] S \qquad (84.1)$$

The derivative along the normal to the surface of the bubble, to an order of magnitude, may be written in the form

$$\frac{\partial}{\partial n} v_0^2 \sim \frac{U^2}{h}, \tag{84.2}$$

where h is the thickness of the flattened bubble.

In order to find the thickness, we can follow the method of reasoning adopted in Section 79, which is very general. Thus, we can express the thickness h of the bubble by formula (79.6).

Inserting the value of the normal derivative into (84.1), we find that the energy dissipated by the motion of the deformed bubble is

$$\frac{dE}{dt} = -\frac{\alpha}{2} \mu \frac{U^2}{h} S,$$

where $\alpha$ is a numerical coefficient.

Here, the drag is

$$F = \alpha \mu U \frac{S}{h}.$$

We can find the velocity of rise of the bubble by equating the drag to the buoyancy force, i.e.,

$$\alpha \mu U \frac{S}{h} = \rho' g S h.$$

Utilizing formula (79.6) for h, we obtain:

$$U \approx \left( \frac{4\sigma^2 g}{\alpha \rho \mu} \frac{\rho'}{\rho} \right)^{1/5}. \tag{84.3}$$

Despite the purely qualitative method of deriving this formula, it accurately portrays the basic peculiarity of the rise of deformed bubbles: velocity, indeed, is independent of the bubble's dimensions. In addition, this formula permits a correct estimate of the order of magnitude of the velocity of rise.

It is natural to assume that the constant $\alpha$ has the same order of magnitude as in the case of undeformed bubbles, i.e., $\alpha \approx 30$. For water ($\sigma \approx 50$, $\rho = 1$), we then compute the velocity of rise to be:

$$U \sim 30 \text{ cm/sec}, \tag{84.4}$$

in agreement with experimental data. It should be stressed, however, that formula (84.3) is not very sensitive to inaccuracies in the values of the quantities entering into it, since it contains a fifth root. This makes its comparison with experiment difficult.

## 85. BREAKUP OF BUBBLES

For the motion of a sufficiently large individual bubble, there is not only deformation, but also fragmentation into bubbles of smaller dimensions.

Several experimental studies have been devoted to this fragmentation problem. It has been established that a rising bubble flattens out and starts to pulsate, forming a thin film at the center. This film breaks suddenly, and the bubble breaks up into a group of smaller bubbles. The radius of bubbles which undergo fragmentation is about 2 to 3 cm; i.e., their volume is approximately 50 $cm^3$.

The fragmentation process is usually discussed in the following manner [25]: if the dynamic pressure $\frac{\rho U^2}{2}$ of the medium exceeds the capillary pressure $\frac{\sigma}{R}$, where R is the radius of curvature, the bubble is deformed and finally breaks up. If we express the radius of curvature as

$$\frac{1}{R} \approx \frac{1}{\sqrt{\frac{V}{h}}},$$

this condition for fragmentation becomes

$$k_f \frac{\rho U^2}{2} > \frac{\sigma}{\sqrt{\frac{V}{h}}}. \tag{85.1}$$

Utilizing formula (79.6) for h, we find that fragmentation should occur for

$$V > \frac{8\sigma^3}{\rho^3 U^6} \approx 2 \cdot 10^{-3} \text{ cm}^3,$$

which is clearly contrary to the facts. If, on the other hand, deformation is not taken into consideration and it is assumed that $\frac{1}{R} = \frac{1}{a}$, the fragmentation condition takes the form

$$a > \frac{2\sigma}{k_f \rho U^2} \approx \frac{4 \cdot 80}{30 \cdot 30} \approx \frac{1}{3} \text{ cm}$$

for $k_f \sim 0.5$, which is also not in accordance with the facts, not to mention the assumption of nondeformation of the bubbles, which cannot be accepted. It is not quite clear, from a physical standpoint, why inequality (85.1) should be interpreted as a fragmentation condition rather than as a condition for bubble deformation. It would seem that by deforming and acquiring the shape of a tablet, the bubble acquires a shape like a torus with a larger radius of curvature and a greater resistance to rupture than the undeformed bubble.

In view of this unsatisfactory state of the problem, it has seemed natural to us to hypothesize the following mechanism for the bubble's destruction: the drag exerted by the liquid sets the gas inside into motion. This motion, being rotational or possibly turbulent in nature, creates a dynamic pressure $\frac{\rho' U'^2}{2}$ within the bubble. This dynamic pressure is directed outward from inside the bubble. If it exceeds the capillary forces holding the bubble together, the bubble must inevitably break up. No deformations can possibly occur in this case which would increase the relative stability of the bubble.

The fragmentation condition can, then, be written in the form

$$k_f' \frac{\rho' U'^2}{2} \geqslant \frac{\sigma h}{l^2} \sim \frac{\sigma \pi h^2}{V}. \tag{85.2}$$

Since the density of the gas is low compared to the surrounding medium, it may be assumed that there is complete entrainment of the gas on the interface, so that the velocity of the gas $U'$ is equal to that of the liquid U. Then, replacing h in formula (85.2) according to formula (79.6), we have:

$$k_f' \frac{\rho' U^2}{2} > \frac{3\sigma^3}{a^3 \rho^2 U^4},$$

whence we derive the critical radius of the bubble, the radius at which breakup begins:

$$a_{\text{cr}} \cong \sqrt[3]{\frac{3}{k_f'}} \frac{\sigma}{U^2 \sqrt[3]{\rho' \rho^2}}. \tag{85.3}$$

Substituting, in the case of water, $\sigma \sim 70$, $\rho = 1$, $\rho' \approx 10^{-3}$ and assuming $k_f' = 0.5$ and the velocity U to be equal to the limiting velocity of rise $U = 30$ cm/sec, we have:

$$a_{\text{cr}} \sim 1.8 \text{ cm}.$$

This value agrees well with experimental data. It is a relatively weak function of the numerical value of the only unknown quantity $k_f'$ which enters into (85.3).

## 86. BREAKUP OF DROPS

If the radius of individual drops falling in a gaseous medium exceeds a critical value of 0.30 to 0.35 cm, they start to break up.

This fragmentation also occurs for fall in a liquid medium, but for drops of larger dimensions.

The process of drop fragmentation is very important in many phenomena, such as in rainfall, in the atomization of fuel in internal combustion engines, etc. The process, therefore, has been studied experimentally under various conditions. Unfortunately, it should be admitted that the complexity of this phenomenon has prevented convincing and sufficiently accurate results to be obtained. The superficial aspects of the process have been studied in sufficient detail. It has been established that a falling drop becomes sharply constricted around its center. Then a thin film appears at the center of the drop, the film bursts, and the drop becomes a torus, which immediately breaks up into a system of small droplets.

We examined above the problem of bubble fragmentation, assuming that this fragmentation was caused by an inner motion resulting from the motion of the bubble relative to the medium. We assumed in this case that the process of subdivision takes place when condition (85.2) is fulfilled. It so happened, on the basis of the hypothesis that was advanced, that it was possible to estimate successfully the critical size of the subdividing bubbles (see Section 85).

Let us now extend the concepts thus developed to the case of subdivision of liquid drops.

To transform criterion (85.2) into a quantitative criterion, the velocity inside the drop must be calculated. This velocity is determined, on the one hand, by the action of the external flow and, on the other hand, by the properties of the liquid in the drop. For the time being, we can assume that the Reynolds numbers $\mathrm{Re} = Ua/\nu$ and $\mathrm{Re}' = U'a/\nu'$ ($U'$ and $\nu'$ apply to the inner liquid) are large compared to unity. The dynamic pressure exerted on the liquid interface from the inside is $k'_f \frac{\rho' U'^2}{2}$, where $k'_f$ is a certain numerical coefficient. If this dynamic pressure exceeds the smallest value of the capillary pressure at any point on the surface of the drop, the latter will burst apart.

The conditions of subdivision can be formulated approximately, if we assume that the deformed drop has the form of an ellipsoid with semiaxes $l$, $l$, $h$ ($h \ll l$). Here $h$ is determined by formula (79.6) and $l$ can be obtained from the relation

$$V = \frac{4}{3}\pi a^3 \approx \pi l^2 h. \tag{86.1}$$

The lowest capillary pressure is at the point of incidence, and at the downstream point of the drop. Here, it is equal to $\frac{\sigma h}{l^2}$.

The condition for rupture can therefore be written in the form

$$k'_f \frac{\rho' U'^2_{cr}}{2} = \frac{\sigma h}{l^2} = \frac{\sigma \pi h^2}{V} = \frac{4\pi\sigma^3}{\frac{4\pi}{3} a^3 \rho^2 U^4_{cr}} = \frac{3\sigma^3}{a^3_{cr}\rho^2 U^4_{cr}},$$

where $a_{cr}$ and $U_{cr}$ are the critical radius and velocity at which subdivision begins.

The velocity $U'$ of the interior motion must be related to that of the air $U$.

Let us rewrite the Navier-Stokes equation in the form

$$(U'\nabla U') = -\frac{1}{\rho'}\nabla p + \nu' \Delta U'.$$

Neglecting viscosity, and estimating the remaining terms to orders of magnitude

$$\frac{U'^2}{L} \sim \frac{p}{\rho' L},$$

where $L$ is a certain characteristic dimension of the drop and $U'$ is the characteristic velocity, we obtain

$$\rho' U'^2 \sim p \sim \rho U^2. \tag{86.2}$$

From the condition of force balance, we obtain:

$$k_f \rho U_{cr}^2 \approx \frac{6\sigma^3}{a_{cr}^3 \rho^2 U_{cr}^4}.$$

The critical radius is

$$a_{cr} \approx \sqrt[3]{\frac{6}{k_f}} \frac{\sigma}{\rho U_{cr}^2}. \tag{86.3}$$

For $k_f \sim 0.5$

$$a_{cr} \approx 2.3 \frac{\sigma}{\rho U_{cr}^2}. \tag{86.4}$$

It should be noted that, in the case of drops, this formula virtually coincides with (85.3). The appearance of the radius in it, however, is now well founded. As has been indicated above in the case of bubble fragmentation, the difference between (85.3) and the formula derived on the basis of the concepts developed above is very material.

Assuming that $\sigma = 72$ and $U_{cr} = 800$ cm/sec, we find that $a_{cr} = 0.26$ cm. Thus, the critical values of the radius are in good agreement with experimental values.

The formulas derived evidently can be applied to the case of drops moving in a liquid of comparable density. The size of the drops is then found to be substantially smaller.

It should be noted that the following important conclusion results from the concepts set forth above: fragmentation does not take place in a Stokes regime of liquid motion, i.e., if the dynamic pressure exerted on the drop by the surrounding medium and the pressure exerted on the drop from the inside are not expressed by the quadratic law $k_f \frac{\rho U^2}{2}$. In other words, the Reynolds number $Re = \frac{Ua}{\nu}$ of the outer liquid must exceed 10, approximately. If this Reynolds number $Re = \frac{Ua}{\nu} < 1$, there will be no fragmentation of the drop. Here $Re' = \frac{U'a}{\nu'}$ is also less than unity. Indeed, $U'$ is always less than U, and the inner dynamic pressure cannot cause fragmentation We still have to examine the case of $Re = \frac{Ua}{\nu} \gg 1$, for $Re' = \frac{U'a}{\nu'} < 1$, which corresponds to a high viscosity of the liquid in the drop.

It is clear that the presence of high viscosity cannot in itself serve as a stabilizing factor. If the viscosity is sufficiently high, however, the time required for deformation and fragmentation of the drop can be very significant.

In terms of an order of magnitude, the time required for the deformation and fragmentation of a drop of an extremely viscous liquid can be estimated on the basis of the Navier-Stokes equation. Omitting its quadratic term and estimating the orders of magnitude of the other terms, we can write:

$$\frac{\nabla p}{\rho} \sim \nu \Delta \mathbf{v},$$

or

$$\frac{\sigma}{\rho a^2} \sim \nu \frac{U}{a^2} \sim \frac{\nu}{a\tau},$$

where $\tau$ is a characteristic time required for the displacement of the liquid particles in the drop. From this we obtain:

$$\tau \sim \frac{\rho a \nu}{\sigma}. \tag{86.5}$$

If each drop is subjected to the influence of the external flow during a period of time that is shorter than $\tau$, its deformation and fragmentation actually may not take place, despite the existence of the required conditions.

## 87. BREAKUP OF DROPS IN TURBULENT LIQUID FLOW

If a certain mass or drop of liquid is placed in a turbulent stream of an immiscible liquid, the drop will break up under the influence of turbulence eddies. This phenomenon was first studied experimentally and theoretically by M. K. Baranayev, Ye. N. Teveroskiy, and E. L. Tregubova [26]. The theory of this phenomenon, based on the theory of homogeneous, isotropic turbulence was developed by A. N. Kolmogorov [27]. The fragmentation phenomenon is related to the fact that the velocity in a turbulent liquid stream varies from one point to another. The velocity of the liquid at the surface of the drop also varies from point to point. Therefore, different dynamic pressures will be exerted at different points on the surface of the drop. Under certain conditions, this will inevitably lead to deformation and breakup of the drop.

Let us first examine drop fragmentation in a field of homogeneous, isotropic turbulence. For reasons that will be explained later, we examine the fragmentation of a drop caused by eddies the scale of which is large compared to the interior scale of turbulence $\lambda_0$.

In addition, we shall assume that the density of the liquid medium is close to the density of the drop. (The case of fragmentation of drops in a gaseous medium will be considered separately.) Then, the difference in the dynamic pressures exerted on opposite sides of a drop with a diameter of 2a is

$$Q = k_f \frac{\rho (v_1^2 - v_2^2)}{2}, \tag{87.1}$$

where $v_1$ and $v_2$ are the velocities of the medium at points separated from each other by a distance 2a. It is clear that large-scale eddies, which vary little over a distance of the order of the drop's dimensions, do not influence the drop and that its deformation and fragmentation are caused by relatively small eddies.

According to (4.9), for a change in the eddy velocity over a region of length 2a, we have $v_\lambda \sim (\epsilon_0 \lambda)^{1/3}$. Therefore,

$$Q = \frac{k_f \rho}{2} \varepsilon_0^{2/3} (2a)^{2/3}, \tag{87.2}$$

where $\epsilon_0 = \frac{\epsilon}{\rho}$ is the energy dissipation per unit mass. From the condition of force balance, and substituting the value of Q, we have:

$$\frac{k_f \rho v^2}{\lambda_0^{2/3}} (2a_{cr})^{2/3} \sim \frac{2\sigma}{a_{cr}},$$

whence, for the radius of the drop that is being formed, we obtain:

$$a_{cr} = \left(\frac{\sigma}{k_f \rho v^2}\right)^{3/5} \lambda_0^{2/3} = L^{2/5} \left(\frac{\sigma}{k_f \rho}\right)^{3/5} \frac{\sqrt{2}}{U^{6/5}}. \tag{87.3}$$

The latter formula was derived by A.N. Kolmogorov [27] (due to a typographical error in his text, however, it was presented in a distorted form). A very similar formula was also proposed in the above-cited study by M.K. Baranayev, Ye.N. Teverovskiy, and E.L. Trugubova.

The radius of the drops being formed decreases with increase in velocity, almost inversely with the velocity, and is also a function of the scale of turbulence.

In view of the assumptions made in its derivation, formula (87.3) is valid for $a > \lambda_0$.

For a given velocity U of the homogeneous, isotropic flow, drops formed in it must be of a single size (87.3). If, however, the drops are not in the flow for a sufficiently long period of time, an equilibrium with respect to drop size may not have time to become established. In addition, it should be borne in mind that the law $v \sim \lambda^{1/3}$, which we used in computing the dynamic pressure, gives only an average value for the difference in velocities at a distance $\lambda$. It is also possible that velocity fluctuations may be present in the turbulent flow, in which large velocity changes may occur over a zone $\lambda$. In that case, drops of smaller dimensions will be formed, smaller than those determined by formula (87.3). The computation of the probability of such eddies cannot be carried out under the present state of development of the theory of local turbulence. It should be stressed that, in view of the circumstances noted at the close of the preceding section, formula (87.3) is found to be valid only for values of a that do not exceed a certain minimum radius $a_{min}$.

Indeed, in order for the difference in dynamic pressure exerted on the drop to cause its deformation and fragmentation, the dynamic thrust acting on the drop must be expressed by the quadratic law;

i.e., the Reynolds number of the liquid flow past the drop must satisfy the condition

$$\mathrm{Re} = \frac{v \cdot 2a_{\min}}{\nu} \sim 2\alpha \sim 10.$$

Therefore $a_{min}$ satisfies the condition

$$\frac{\rho\alpha^2\nu^2}{a_{\min}^2} \sim \frac{\sigma}{a_{\min}}$$

or

$$a_{\min} \sim \frac{\rho\alpha^2\nu^2}{\sigma}. \tag{87.4}$$

Smaller drops cannot be formed by turbulent eddies, regardless of the velocity of the main flow. Therefore we always have

$$a > a_{\min} = \frac{\rho\alpha^2\nu^2}{\sigma}.$$

The latter inequality and formula (87.3) for the radius of the drop that has been formed lead to the inequality

$$\left(\frac{\sigma}{\rho\nu^2}\right)^{3/5} \lambda_0^{8/5} > \frac{\rho\nu^2\alpha^2}{\sigma},$$

or

$$\lambda_0 > \frac{\rho\nu^2}{\sigma}\alpha^{5/4}, \tag{87.5}$$

which limits the value of the inner turbulence scale, for which condition $a > \lambda_0$ is satisfied.

In the zone of scales $\leq \lambda_0$, the motion occurs at Reynolds numbers $< 1$. A flow past a drop at any Reynolds number lower than unity does not cause deformation or fragmentation of the drop. This applies, in particular, to unsteady flow past a drop. Therefore, turbulence eddies with a scale less than $\lambda_0$ cannot cause fragmentation of a drop. In view of this, it is difficult to agree with the method of deriving the formula for the radius of drops given by A.N. Kolmogorov, for drops that are formed as a result of fragmentation by turbulence eddies the scale of which is less than $\lambda_0$.

The formula for the drop radius was derived by A.N. Kolmogorov by substituting expression (32.9) for the change in flow velocity within the zone $\leq \lambda_0$, into (87.1). It has the form

$$a = \left(\frac{\sigma\lambda_0}{\mathrm{const}\, \nu^2\rho}\right)^{1/3} \lambda_0, \qquad a \ll \lambda_0. \tag{87.6}$$

The contradictory nature of this formula is seen, in particular, from the fact that, according to (87.5), we always have $\left(\frac{\sigma\lambda_0}{\nu^2\rho}\right)^{1/3} > 1$, while formula (87.6) is valid, by assumption, for $a \ll \lambda_0$. However, and independently of (87.5), it is easy to note that, in water for example, we have $\nu = 10^{-2}$, $\sigma \geq 20$, and $\left(\frac{\sigma\lambda_0}{\nu^2\rho}\right)^{1/3} < 1$ only at values for $\lambda_0 < 10^{-6}$. Such a value for $\lambda_0$ is quite unrealistic.

A.M. Rozen [32] analyzed and studied very extensive experimental data on the breakup of drops suspended in a flowing, turbulent liquid.

It was found that all data correlated well with an empirical formula for an average drop radius of

$$r \sim \frac{\sigma^{0.5-0.6}}{U},$$

which virtually coincides with the theoretical formula derived above.

## 88. BREAKUP OF DROPS IN THE TURBULENT FLOW OF A LIQUID PAST A WALL

Let us now examine the breakup of drops caused by turbulence eddies in the nonhomogeneous turbulent flow past a solid surface. To be specific, we shall examine fragmentation for flow near the walls of a tube. In this case, the flow velocity and the spatial velocity gradient vary from point to point over the tube's crosssection.

The distribution of the average velocity of the flow over a distance (y) can be written in the form

$$U = \frac{v_0}{\sqrt{a}} \ln \frac{y}{\delta_0}. \tag{88.1}$$

The variation of small-scale eddy velocities, within the scale limits of $y \sim R$, takes place much more rapidly than the change in the average velocity. In terms of an order of magnitude, this variation is determined by formula (4.9). At small distances $y \sim \delta_0$ from the wall, however, the change in the average velocity is so abrupt that eddy velocities here vary in the same way as average velocities. In regions inside the tube that are distant from the walls, therefore, drop fragmentation occurs according to (87.3). Velocity changes near the walls are determined by (88.1).

Let us examine drop fragmentation in the vicinity of the tube walls. The difference in dynamic pressures exerted on the two sides of the drop is

$$Q \approx 6\rho v_0^2 \left\{ \left[\ln\left(\frac{y+2a}{\delta_0}\right)\right]^2 - \left[\ln\frac{y}{\delta_0}\right]^2 \right\}.$$

Assuming that $a \ll y$, we have

$$Q \approx 6\rho v_0^2 \frac{4a}{y} \ln\frac{y}{\delta_0} \approx 25 v_0^2 \rho \frac{a}{y} \ln\frac{y}{\delta_0}.$$

By substituting this expression in the force balance condition we derive:

$$(a_{cr})_{wall} \sim \sqrt{\frac{\sigma}{25\rho v_0^2}} \sqrt{\frac{y}{\ln \frac{y}{\delta_0}}}. \qquad (88.2)$$

The following conclusion may be drawn from formula (88.2): at small y there is, near the wall, a region of rather rapid changes in velocities, in which the drops that develop are smaller than those of (87.3)

Formula (88.2) can be rewritten by replacing the characteristic velocity $v_0$ with the average velocity of the flow U:

$$(a_{cr})_{wall} \sim \frac{1}{U} \sqrt{\frac{\sigma}{\rho}} \sqrt{\frac{y}{\ln \frac{y}{\delta_0}}}. \qquad (88.3)$$

The size of the drop is found to be proportional to $\sim \frac{1}{U}$, and decreases as the wall is approached.

It is clear that it makes sense to use formula (88.3) only at short distances from the solid wall, where the size of the drops determined by formula (88.3) is smaller than the size given by formula (87.3).

The smallest drop size will be found in the wall region at the boundary of the viscous layer $y \sim \delta_0 \sim \frac{\nu}{v_0}$. At shorter distances, the Reynolds numbers become less than unity. To a logarithmic accuracy, we have, evidently,

$$(a_{cr})_{wall} \sim \sqrt{\frac{\sigma \nu}{25\rho v_0^3}}. \qquad (88.4)$$

Here, $(a_{cr})_{wall}$ must be no smaller than $a_{min}$, as determined from formula (87.4).

Let us compare the size of the drops that are formed by fragmentation in the core of the flow with those formed in the wall layer:

$$\frac{a_{vol}}{a_{wall}} \sim \left(\frac{\sigma}{\rho}\right)^{1/10} \frac{L^{2/5} v_0}{U^{6/5}} \left(50 \ln \frac{y}{\delta_0}\right)^{1/2} \cdot \frac{1}{y^{1/2}}. \qquad (88.5)$$

For $y \sim \delta_0 \sim \frac{\nu}{v_0}$, this ratio is equal to

$$\frac{a_{vol}}{a_{wall}} \sim \left(\frac{\sigma}{\rho}\right)^{1/10} \frac{L^{2/5} \sqrt{v_0}}{U^{1/5} \sqrt{\nu}} \left(\frac{v_0}{U}\right). \qquad (88.6)$$

In cases when $\frac{a_{vol}}{a_{wall}} > 1$, drop fragmentation in the wall layer may be the source of drops of varying dimensions and, in particular, of those smaller than the drops formed in the core of the flow.

Thus, in the course of drop fragmentation in a tube, a distribution of drop sizes is formed in the range determined by formulas (88.4) and (87.3).

It is probably impossible to determine the spectrum of this range since, on reaching varying distances from the surface, the drops are subjected to additional fragmentation. The form of the size spectrum, therefore, should vary as the flow proceeds along the tube.

At a given distance from the inlet to the tube, a certain distribution of drops by sizes becomes established, within the limits determined by (88.4) and (87.3). The average drop size at a given distance from the tube inlet may be considered proportional to

$$\bar{a} \sim \frac{1}{U} \sqrt{\frac{\sigma}{\rho}}, \tag{88.7}$$

since the size of the drops near the tube walls is proportional to this quantity, according to (88.3), while the size of drops formed in the core of the flow and accurately determined by formula (87.3) can be well approximated by formula (88.3), because the difference in the exponents for U and $\frac{\sigma}{\rho}$ in the last two equations is relatively small. Exactly the same formulas are obtained in the examination of turbulent flow in the boundary layer past the surface of a plate or of another body.

Let us pass to a comparison of the formulas that have been derived with experimental data, and let us first turn to the experiments conducted by M. K. Baranayev, Ye. N. Teverovskiy, and E. L. Tregubova, which are devoted to a study of the sizes of drops formed in a turbulent fluid, and to a verification of formula (87.3). In these experiments, a thin liquid jet (0.2 cm in diameter) with a viscosity $\nu \approx 10^{-2}$ and low surface tension $\sigma \approx 25$ dynes/cm was introduced into a turbulent stream of water in a tube 1.5 cm in diameter. The diameters of the drops that were formed were measured at a distance 175 cm from the point of injection. The results of the experiments are shown in Table 17, which has been adapted from that study.

Table 17

| $\bar{U}$, cm/sec | Re | $\bar{d}_{calc} \cdot 10^2$ cm | $\bar{d}_{obs} \cdot 10^2$ cm | $d_{min} \cdot 10^4$ cm |
|---|---|---|---|---|
| 41 | 615 | 9.2 | 4.6 | 2.4 |
| 118 | 17,700 | 2.7 | 2.8 | 1.2 |
| 188 | 28,300 | 1.6 | 1.6 | 1.2 |
| 290 | 45,537 | 0.9 | 1.4 | 1.2 |

It is apparent from this table, first, that as a result of fragmentation, a spectrum of drop sizes was formed. The radius of the smallest particles was found to be about 100 times smaller than the average. The average diameter d decreased with velocity somewhat more slowly than $\frac{1}{U}$.

When the diameter becomes smaller than $10^{-4}$, it is no longer a function of the Reynolds number of the stream.

The authors also discovered that the number of drops of a size close to the smallest size increases the longer they remain in the stream and becomes very considerable after about 1 sec. It may be assumed, under the conditions of this experiment, that proper conditions existed for the fragmentation of drops in the wall layer.

In terms of an order of magnitude, according to (4.40), we have $v_0 \sim \frac{0.6}{3\,\mathrm{Re}^{1/8}} U$. For $\mathrm{Re} \sim 10^4$, we have $v_0 \sim 0.1\, U$.

The minimum drop size for drops formed in the wall region is, according to formula (88.4),

$$a_{\text{wall}} \sim \sqrt{\frac{25 \cdot 10^{-2}}{25U^3 10^{-3}}} \sim 6 \cdot 10^{-4} \text{ cm} \quad (\text{for } U \sim 300 \text{ cm/sec}).$$

At the same time, according to (87.3), drops formed in the volume phase have the following dimension

$$a_{\text{vol}} \approx 10^{-2} \text{ cm}.$$

The minimum possible drop size is $a_{min} \approx 4 \cdot 10^{-4}$.

Considering the rough nature of the theory, the hope that it might coincide with experimental data to any degree of accuracy would appear vain.

Qualitatively, however, it gives an accurate picture. The main point that remains unclear undoubtedly is the slower decrease observed in the average size. It should be also noted that the scatter of experimental points is also rather extensive. It would be very important to conduct more accurate measurements of the dependence of drop size distribution on the velocity of the liquid.

The authors give another interpretation of the fact that drops 100 times smaller than the largest were found. They believe that such drops are formed by eddies of a smaller scale than the scale of inner turbulence $\lambda_0$. Since eddies of a scale $\lambda < \lambda_0$ have Reynolds numbers that are smaller than unity, they cannot cause fragmentation.

The study also advances the proposal that the observed drop size distribution is related to the fact that the drops did not remain in the flow for a sufficiently long period of time. The authors believe that there is insufficient time for a subdivision of the drops to take place, considering the period during which they remained in the turbulent stream—since the fragmentation process requires time. If they had remained in the flow for a sufficient length of time, the diameter of the drops would have been of the order of $10^{-4}$ cm.

Although the last assertion appears to be undisputable, its interpretation is subject to criticism. Indeed, if the time necessary for the subdivision of a drop $\sim 10^{-2}$ cm in size is 1 sec, then the velocities of the inner motion, in terms of an order of magnitude, can be estimated to be about $10^{-2}$ cm/sec.

Eddy velocities of the exterior liquid, which cause fragmentation, are of the order of magnitude of

$$v' \sim \Delta U \sim U\left(\frac{a}{L}\right)^{1/3} \sim \frac{U}{10} \sim 30 \text{ cm/sec}.$$

At the same time, the viscosities of the two liquids—the inner and outer liquids—were very close to each other, while the densities were practically the same.

It seems most improbable that the exterior liquid could entrain the interior liquid to so small an extent that the interior liquid's velocity would be several hundred times less than that of the exterior liquid.

Actually it would appear that, over a period of time of the order of 1 sec, all particles under the conditions of the given experiment had the time to pass into the wall zone, in which drop fragmentation down to the smallest dimensions would take place.

## 89. BREAKUP OF DROPS IN TURBULENT GAS FLOW AND BREAKUP OF BUBBLES

The fragmentation of a drop in a turbulent gas stream, i.e., where the density of the medium is low as compared to that of the liquid within the drop, occurs in a somewhat different manner. In this case, inertial effects, examined in Section 33, play a material role in the mechanism of the drop's motion.

A particle of density $\rho$ suspended in a turbulent stream with $\rho_0 \ll \rho$ is only partially entrained by small-scale eddies. Computations made in Section 33 show that the liquid flows past the particle at a velocity determined by formula (33.10). The dynamic pressure Q thereby created on the surface of the particle is

$$Q = \frac{K_f \rho_0}{2}\left(\frac{\rho}{\rho_0}\right)^{2/3} \frac{\varepsilon^{2/3}}{\rho_0^{2/3}} a^{2/3} \sim K_f \rho^{2/3} \rho_0^{-1/3} \varepsilon^{2/3} a^{2/3}. \tag{89.1}$$

Comparing this with (87.2), we see that the dynamic thrust, which appears becuase of inertial effects, is greater than that caused by changes in eddy velocities over the length of the drop by the ratio of $\left(\frac{\rho}{\rho_0}\right)^{2/3}$.

Substituting the value for Q from (89.1) into the force balance condition, we find that the size of the drops that are formed is determined by the relation

$$K_f \rho_0^{-1/3} \rho^{2/3} \frac{\nu^2}{\lambda_0^{8/3}} a_{cr}^{2/3} \sim \frac{2\sigma}{a_{cr}},$$

whence

$$a_{cr} \sim \left(\frac{\sigma}{K_f \rho^{2/3} \rho_0^{1/3} \nu^2}\right)^{3/5} \lambda_0^{8/5} \sim \left(\frac{\sigma}{K_f \rho^{2/3} \rho_0^{1/3}}\right)^{3/5} \frac{L^{2/5}}{U^{6/5}} = \left(\frac{\sigma}{K_f \rho_0}\right)^{3/5} \left(\frac{\rho_0}{\rho}\right)^{2/5} \frac{L^{2/5}}{U^{6/5}}. \tag{89.2}$$

We see that taking the inertial effects into account results in a change in the formula that gives the size of the fragmentation drops. If the computation of their dimensions were to be made on the basis of A.N. Kolmogorov's formula (87.3), the dimensions obtained would be too large by the factor $\left(\frac{\rho}{\rho_0}\right)^{2/5}$. For $\frac{\rho}{\rho_0} \sim 10^3$, this would constitute a difference of an order of magnitude.

Let us now examine the problem of bubble fragmentation in a turbulent stream. The theory of bubble fragmentation in principle is no different from that of drops. The only actual difference is the fact that the critical size of the bubble at which fragmentation occurs is determined by formula (85.3), which differs from the analogous formula (86.4), where the critical size applies to drops. To simplify computations, let us rewrite formula (85.3) in the form

$$a_{\text{cr}} \approx \left(\frac{1}{k_f'}\right)^{1/3} \frac{\sigma}{\frac{\rho U^2}{2}} \left(\frac{\rho}{\rho'}\right)^{1/3}. \tag{89.3}$$

In the case where fragmentation of the bubbles occurs in a zone of homogeneous, isotropic turbulence, we obtain, by making the substitution in the formula (89.3) of the value of the difference in dynamic pressures exerted on various portions of the drop from formula (87.2):

$$a_{\text{cr}} \sim L^{2/3} \left(\frac{\sigma}{k_f' \rho}\right)^{3/5} \frac{1}{U^{6/5}} \left(\frac{\rho}{\rho'}\right)^{1/5}. \tag{89.4}$$

The radius of the fragmented bubbles that form in the turbulent liquid is found to be somewhat greater than the radius of fragmented drops.

## 90. DISSOLUTION OF GAS BUBBLES. ELEMENTARY THEORY OF THE BUBBLING PROCESS

As has been pointed out above, processes of gas absorption in liquids play an important role in chemical technology and, especially, in inorganic chemistry.

In certain fields of modern engineering and, especially, in the production of nitric acid and soda as well as in the heating of liquids by live steam, the so-called process of bubbling is used. During bubbling the liquid and the gas are mixed by the passage of gas bubbles through a layer of liquid. The bubbles are formed by gas jets flowing from special nozzles.

One of the principal hindrances to a widespread use of the bubbling process in gas absorption is the fact that absorption in this process has not been sufficiently studied.

Systematic studies of bubbling in absorbers, conducted in the USSR mainly by I.N. Kuz'minykh [29] and K.N. Shabalin [30], give grounds for an assumption that the future of the application of this process is most promising.

In this connection, problems of the dissolution of gas bubbles acquire special significance. The complexity of the bubbling process is such that the development of a quantitative theory seems doubtful. We have placed a more modest objective before ourselves, namely, to examine the process of gas absorption from an individual bubble. This, in a way, is the elementary act of bubbling. Certain other aspects of the bubbling process are examined later (Section 135).

The experimental study of gas absorption from an individual bubble was conducted by I.N. Shabalin [30]. The process was also studied by Ledig et al. [31].

The course of the process depends significantly on the nature of the gas and that of the liquid. The process of absorption of a certain component of the gas mixture that comprises the bubble proceeds in three steps: the dissolving component must be conveyed to the gas-liquid interface; it must pass through that interface; it must be removed from the surface of the bubble into the bulk of the solution.

The second stage is very often complicated by a chemical reaction in which the dissolving component* participates. The over-all rate of the process is controlled, as usual, by the slowest stage.

Actual practice shows that the absorption rate for various gases is determined by the different steps of the process. In particular, the dissolution rate of easily soluble gases is determined by the first stage—convective diffusion of the component within the bubble. In the opposite limiting case of sparingly soluble gases, the rate of the process is determined by the stage involving convective diffusion in the liquid.

The dissolution of the gas in an individual bubble may be termed the elementary act of bubbling sparingly soluble gases. It should be stressed, however, that mass bubbling is a very complex process and is determined by a variety of factors, the most important of which are: gas dissolution from an unbroken jet, the interaction of bubbles, the foaming process, etc. Dissolution from individual bubbles in these complex and, for the time being, largely unknown processes unquestionably plays an important role, especially if the liquid layer is thick (deep bubbling), and gas jet velocities high.

## 91. CALCULATION OF THE RATE OF GAS BUBBLE DISSOLUTION

Let us first examine the problem of dissolution of slightly soluble gases, the rate of which is controlled by the removal of the dissolving material from the surface of the bubble into the body of the liquid.

---

*For example, during the process of carbonation of the juices resulting from the purification in the sugar beet industry, carbon dioxide breaks down the calcium saccharate, transferring the sugar into the solution and forming a calcium carbonate deposit.

Let us assume that the concentration of the dissolving gas at the surface of the bubble has a constant value, and the dissolution is steady and follows the convective transfer of dissolved molecules. The concentration distribution of the dissolved gas in the liquid can be meaningfully written in the form of an equation for convective diffusion in spherical coordinates

$$v_r \frac{\partial c}{\partial r} + \frac{v_\theta}{r} \frac{\partial c}{\partial \theta} = D \frac{\partial^2 c}{\partial r^2} \tag{91.1}$$

with boundary conditions

$$c = c_0 \quad \text{at} \quad r = a, \tag{91.2}$$

$$c \to 0 \quad \text{as} \quad r \to \infty. \tag{91.3}$$

The nature of the velocity distribution in the liquid will vary, depending on the regime of the motion. At small Reynolds numbers, i.e., for very small bubbles, the computation of the diffusional flow is no different from that of a drop, which was performed in Section 72. The diffusional flux that we are seeking is determined from equation (91.1) and the boundary conditions (91.2) and (91.3) via formulas (72.7) and (72.39)

$$j_{\text{diff}} = D\left(\frac{\partial c}{\partial y}\right)_{y=0} = \sqrt{\frac{3}{\pi}} \left(\frac{DU}{2a}\right)^{1/2} \frac{1 + \cos\theta}{\sqrt{2 + \cos\theta}} c_0. \tag{91.4}$$

In the same way, the total diffusional flow is

$$I = \int j\, dS = 8 \sqrt{\frac{\pi}{3}} \left(\frac{DU}{2a}\right)^{1/2} a^2 c_0. \tag{91.5}$$

The diameter of the bubbles encountered in the bubbling process varies over wide limits, but apparently lies in the range $0.1 < 2a < 2$ cm; i.e., it corresponds to moderate and large Reynolds numbers. Strictly speaking, therefore, (91.5) is not applicable under conditions of bubbling.

At moderate Reynolds numbers, the diffusion of dissolved molecules can be calculated exactly over the entire surface, except for the separation zone.

Substituting $v_r \approx v_r^{(0)}$ and $v_\theta \approx v_\theta^{(0)}$ in the convective diffusion equation and utilizing expressions (82.12) and (82.13) for $v_r^{(0)}$ and $v_\theta^{(0)}$ in the vicinity of $r = a$, we obtain:

$$\frac{3U}{2a} \sin\theta \frac{\partial c}{\partial \theta} - \frac{3U}{a} y \cos\theta \frac{\partial c}{\partial y} = D \frac{\partial^2 c}{\partial y^2}. \tag{91.6}$$

Comparing this expression with equation (72.1) and taking into account the values of $v_r$ and $v_\theta$, we may note the identity of the two equations if we state:

$$\frac{3U}{2} = v_0.$$

The diffusional flux can therefore be derived by employing expression (72.7)

$$j=\sqrt{\frac{3}{\pi}}\left(\frac{3DU}{2a}\right)^{1/2}\frac{1+\cos\theta}{\sqrt{2+\cos\theta}}c_0, \tag{91.7}$$

and the total diffusional flow can be presented in the form

$$I=8\sqrt{\frac{\pi}{2}}\left(\frac{DU}{a}\right)^{1/2}a^2c_0. \tag{91.8}$$

In dimensionless form, the flow of substance becomes

$$\mathrm{Nu}=\frac{Ia}{SDc_0}=\sqrt{\frac{2}{\pi}}\,\mathrm{Re}^{1/2}\mathrm{Pr}^{1/2}. \tag{91.9}$$

In this case, we have disregarded the diffusional flow in the region of separation since it occupies a small area.

Taking expression (82.54) into account, we determine the total diffusional flow

$$I=8\sqrt{\frac{\pi}{2}}\left(\frac{D\rho g}{9\mu}\right)^{1/2}a^{5/2}c_0. \tag{91.10}$$

Let us now examine the dissolution of large (deformed) bubbles. The derivation of quantitative relations, in this case, is difficult. Since it can be assumed, however, that the nature of the flow past a large bubble is not materially different from that past an undeformed bubble, it is possible to assume with the same degree of certainty that the following formula is valid in this case

$$j=\mathrm{const}\left(\frac{DU}{l}\right)^{1/2}c_0, \tag{91.11}$$

where the value of the constant and the angular dependence differ from the corresponding quantities in equation (91.7).

The total diffusional flow has the form

$$I=\mathrm{const}\left(\frac{DU}{l}\right)^{1/2}Sc_0. \tag{91.12}$$

Since U in this case is not a function of the bubble's dimensions

$$I\sim l^{3/2}, \tag{91.13}$$

or

$$I\sim V^{3/4}h^{-3/4}\sim a^{9/4}h^{-3/4}, \tag{91.14}$$

where a is the initial radius of the bubble (prior to deformation). The quantity h is not a function of the volume of the bubble.

The total diffusional flow per unit of bubble surface can be represented by the formula

$$\frac{I}{4\pi a^2}\sim a^{1/4}. \tag{91.15}$$

This formula shows that the diffusional flow on a unit of a large and badly deformed surface is a relatively weak function of the bubble's radius. The latter conclusion agrees with K.N. Shabalin's experiments. His experiments consisted in the measurement of the rate of dissolution of carbon dioxide from gas bubbles in water. The bubbles passed singly through the water and were 0.45 to 1 cm in diameter. This author came to the conclusion that the dissolution rate per unit of bubble surface is not a function of the bubble's dimensions. The accuracy of the measurements was not great and errors amounted to ± 15 %. The variation of the diffusional flow in these experiments, according to formula (91.15), could amount to about 20 %.

In conclusion, let us compute the period of time required for the establishment of a steady convective diffusion process on the bubble. Substituting the value of $\delta_0$ from (82.30) into formula (62.6), we derive:

$$t \sim \frac{\delta_0^{4/3}}{U^{4/3}D^{1/3}} \sim \frac{\nu^{2/3}a^{2/3}}{U^{4/3}D^{1/3}}. \tag{91.16}$$

The time needed for the establishment of a steady bubble regime is very short.

## 92. DISSOLUTION OF A GAS BUBBLE SUSPENDED IN A TURBULENT LIQUID STREAM

The dissolution of bubbles suspended in a stirred liquid is often encountered in actual practice. The liquid is either stirred artificially or, more often, it is stirred by the bubbles themselves as they pass through the liquid.

Vigorous stirring of the liquid leads to the establishment of a regime of developed turbulence in it. In this case, all the reasoning set forth in Section 33 may be repeated, with but one essential difference: for bubbles of not very large size, the force exerted on the bubble by the liquid is determined by formula (82.51). Therefore, the condition, analogous to (33.2), which defines the scale of eddies which cannot entrain a rising bubble can be written in the form

$$\frac{4\pi}{3}\rho a^3 \frac{u}{\tau} = \frac{4}{3}\pi a^3 (\rho_0 - \rho)\sqrt{\frac{\varepsilon}{\rho\tau}} - 12\pi\mu a u, \tag{92.1}$$

where $\rho$ is the gas density and $\rho_0$ is the density of the liquid.

From expression (92.1), we obtain:

$$u = \frac{\rho_0 - \rho}{\rho}\sqrt{\frac{\varepsilon}{\rho}}\,\frac{\sqrt{\tau}}{\frac{9\nu\tau}{a^2}+1}. \tag{92.2}$$

The maximum velocity of the eddies in the liquid that flows past the gas bubble, account being taken of (4.3), is

$$u_{\max} \approx \frac{\rho_0 - \rho}{\rho} \sqrt{\frac{a^2}{\nu}} \cdot \sqrt{\frac{\varepsilon}{\rho}} \approx \frac{\rho_0}{\rho} \frac{U^{3/2} a}{\sqrt{L\nu}}. \tag{92.3}$$

Substituting the value of $u_{\max}$ from expression (92.3) into equation (91.8), we obtain an expression for the flow of matter to an individual bubble,

$$I \approx D^{1/2} \left(\frac{\rho_0}{\rho}\right)^{1/2} \frac{U^{3/4} a^2 c_0}{L^{1/4} \nu^{1/4}}. \tag{92.4}$$

Expression (92.4) can be rewritten in the dimensionless form

$$\mathrm{Nu} = \frac{Ia}{SD} = \mathrm{const} \left(\frac{\rho_0 - \rho}{\rho}\right)^{1/2} \left(\frac{a}{L}\right) \mathrm{Re}^{3/4} \mathrm{Pr}^{1/2}, \tag{92.5}$$

where Re is the Reynolds number of the motion of the entire liquid.

There are, unfortunately, insufficient experimental data to permit verification of formula (92.4). The difficulty of such verification is associated particularly with the fact that in actual practice the total amount of dissolving gas is computed as

$$N = Vm,$$

where m is the number of bubbles per 1 $cm^3$ and V is the volume of the reactor. However, the number of bubbles m in turn is a function of a series of factors (radius of the orifice through which the bubbles issue, the velocity of the gas, surface tension, etc.).

## BIBLIOGRAPHY


1. Rybczyński, Bull. intern. acad. sci. Cracovie (A) 1911, p. 40.
2. Hadamard, Comp. Rend. 154, 1735 (1911).
3. Lamb, H., Gidrodinamika, Gostekhizdat, 1947 [see also: Hydrodynamics, Dover, 1945].
4. Lebedev, A. A., Zhur. Russ. Fiz.-Khim. Obshch., phys. sect. 48, issue 3 (1916).
5. Nordlund, I., Ark. Mat. Astron. och Fysik 9, no. 13 (1913).
6. Silvey, Phys. Rev. 7, 106 (1916).
7. Boussinesq, J., Comp. Rend. 156, 983, 1035, 1124 (1913). Ann. Chim. et Phys. 29, 349, 357, 364, (1913).
8. Bond, W., Phil. Mag. 4, 889 (1927); W. Bond and D. Newton, Phil. Mag., ser. 7, 5, 794 (1928).
9. Volarovich, M.P., and A. A. Leont'yeva, Acta Physicoch. URSS 11, 251 (1939).
10. Frumkin, A. N., and I. A. Bagotskaya, Comp. Rend. Acad. Sci. URSS 55, 131 (1947); I. A. Bagotskaya, Zhur. Fiz. Khim. 24, 3 (1950).
11. Frumkin, A. N., and V. G. Levich, Zhur. Fiz. Khim. 21, 1183 (1947).
12. Levich, V. G., Zhur. Fiz. Khim. 22, 721 (1948).

13. Langmuir, J. Chem. Soc., p. 525 (1940).
14. Gibbs, W., Termodinamicheskiye, raboty, Gostekhizdat, 1950 [see also: Collected Works, v. 1, Longmans, Green, 1928].
15. Frumkin, A. N., Elektrokapillyarnyye yavleniya i elektrodnyye potentsialy [Electrocapillary Phenomena and Electrode Potentials], Odessa, 1919, p. 152.
16. Khrgian, A. Kh., Fizika atmosfery [Physics of the Atmosphere], Gostekhizdat, 1953 [see also 1958 edition].
17. Humphreys, Physics of the Air, p. 279 (1940).
18. Hoefer, VDI Forschungsheft, no. 138 (1913); Pickert, Forschung 3, 308 (1932); Gibson, Phil. Mag. 26, 952 (1913); O'Brien and Gosline, Ind. Eng. Chem. 27, 1436 (1935); Ôyama and Iwase, Sci. Papers Inst. Phys. Chem. Res. 35, 131 (1939); Migagi, Techn. Rept. 4, 1 (1924); Binder, Chem. and Met. Eng. 51 (August, 1944), 104; H. D. Arnold, Phil. Mag., ser. 6, 22, 755 (1911); Luchsinger, Koll. Z. 81, 180 (1937).
19. Shabalin, K. N., Khim. Prom. 1945 (no. 11), 14.
20. Allen, Phil. Mag., ser 5, 50, 323 (1900).
21. Levich, V. G., Zhur. Eksp. i Teoret. Fiz. 19, 18 (1949).
22. Landau, L., and E. Lifshits, Mekhanika sploshnykh sred [Mechanics of Continuous Media], Gostekhizdat, 1953, p. 147 [see also English translation, Fluid Mechanics, Pergamon Press, 1959].
23. Gorodetskaya, A. V., Zhur. Fiz. Khim. 23, 71 (1949).
24. Kozlov, B. K., and M. A. Mologin, Izvestiya Akad. Nauk SSSR, Otd. Tekhn. Nauk, no. 8, 1188 (1951).
25. Prandtl, L., Gidroaeromekhanika, IL, 1951 [see also: Applied Hydro- and Aeromechanics, McGraw-Hill, 1934].
26. Baranayev, M. K., Ye. N. Teverovskiy, and E. L. Tregubova, Doklady Akad. Nauk SSSR 66, 821 (1949).
27. Kolmogorov, A. N., Doklady Akad. Nauk SSSR 66, 825 (1949).
28. Davies, R., and G. Taylor, Proc. Roy. Soc. A200, 375 (1950).
29. Kuz'minykh, I. N., Ye. I. Surkov, et al., Zhur. Priklad. Khim. 10, 1020 (1937).
30. Shabalin, K. N. et al., Zhur. Khim. Prom. 16, 10 (1939).
31. Ledig and Weaver, J. Amer. Chem. Soc. 46, 650 (1924); Ind. Eng. Chem. 15, 931 (1923); Ôyama and Iwase, Sci. Papers Inst. Phys. Chem. Res. 35, 131 (1939).
32. Rozen, A. M., Nauchnyye doklady vysshey shkoly [Scientific Reports of Higher Schools], no. 2 (1958).
33. Koshevnik, A. Yu., M. M. Kusakov, and N. M. Lubman, Zhur. Fiz. Khim. 33, 197 (1959).
34. Kutateladze, S. S., and M. A. Styrakovich, Gidravlika gaso-zhidkostnykh sistem [Hydraulics of Gas-Liquid Systems], Gosenergoizdat, 1958.

# CHAPTER IX

# Motion of Particles in Electrolytic Solutions

## 93. ELECTROKINETIC PHENOMENA

By particles in electrolytic solutions, we mean particles whose dimensions are so large that they may be considered to form a separate phase. Some examples are colloidal objects of various nature and source, emulsified droplets of water in oil and of oil in water, drops of mercury of relatively large dimensions, etc.

As has been repeatedly stressed above, a region of charged solution, known as the electric double layer, appears at the interface between phases when one of the phases is an electrolytic solution. The presence of an electric double layer on the surface of particles causes peculiar electrical, hydrodynamic effects which are manifested in the motion of the particles through electrolytic solutions. These effects are generally known as electrokinetic phenomena and constitute one of the most important branches of colloid chemistry. Certain electrokinetic phenomena also play an important role in electrochemistry.

The study of this field began with the investigations of F. F. Reyss [1], whose first published work involved liquid motion in dispersed systems. Electrokinetic phenomena were later studied in detail both experimentally and theoretically.

It should be noted at this point that the terminology which has become established in physical chemistry is so involved that one may get the impression that a large number of distinct various electrokinetic phenomena can be observed in nature. Actually, the mechanism of all such phenomena is the same and is associated with the relative displacement of different phases. If a solution containing suspended, dispersed particles is subjected to an external electric field, the particles will begin to move. This has been termed electrophoresis. Electrophoresis is characterized by a relative motion, caused by the electric field, between the dispersed and the liquid phases. Along with electrophoresis, a converse effect is observed in which, due to motion of the particles as a result of nonelectric forces (for example, precipitation in a gravitational field), an electric field (the sedimentation potential) is established in the solution.

Thus, the relative motion of two phases brings about a potential difference in the solution.

The same behavior can also be observed when the moving phase is the solution rather than the dispersed particles. In other words, if an electric field is applied to a solid body-solution interface, i.e., when the solid phase is not dispersed but represents an immobile wall, the solution itself will start to move. It is clear then that only the relative motion of the phases is significant, and that the physical nature of both phenomena — particle motion relative to the solution and motion of the solution relative to fixed walls under the action of an external electric field — are the same even though they may be manifested in different ways. The motion of a solution in a field is termed electro-osmosis. Conversely, when an electrolytic solution flows along an immobile wall, an effect known as the streaming potential is observed, which is essentially the same as the sedimentation potential. Thus, as in the case of sedimentation potentials, the motion of the solution relative to the solid phase causes the appearance of an electric field.

It is natural then that a considerable amount of published material should have been devoted to electrokinetic phenomena. In particular, the electrophoretic process has been the subject of especially detailed experimental and theoretical studies which are of considerable practical significance. Since all electrokinetic phenomena are closely related, it is sufficient to examine the theory of electrophoresis in order that all other phenomena can later be related to it.

In a theoretical examination of electrophoresis, the greatest difficulties are encountered in the case of extremely small particles whose dimensions are comparable to the double layer on their surface. Electrophoresis of such particles has an applied value, especially in biological studies. In most cases, however, we must deal with much larger particles whose dimensions are considerably greater than the thickness of the double layer.

We are concerned in what follows with the latter case, especially in view of the fact that the theory of electrophoresis of small particles has not as yet been sufficiently studied, although in the final analysis the difference in the electrophoresis of large and small particles reduces to a change in the coefficient in the electrophoretic law. This permits us to analyze the basic physical phenomena that occur during electrophoresis without going into unwieldy computations.

## 94. ELECTROPHORETIC MOTION NEAR A PLANE SURFACE (ELECTRO-OSMOSIS)

The simplest case of electrophoretic motion is the motion of a solution along a given plane.

Let us assume that the solution is bounded by a dielectric plane at $y = 0$. An electric double layer will be set up on the surface of the dielectric, and near the infinite plane y = 0, a charged region in the volume of the solution will appear whose potential satisfies the Poisson equation

$$\frac{d^2\varphi}{dy^2} = -\frac{4\pi}{D}\rho. \tag{94.1}$$

In the solution let there be an electric field E, whose direction is along the x axis tangential to the solid surface. Then, a force $F_t = \rho E_t$ will be exerted per $cm^3$ of charged solution (of the double layer) tangential to the surface of the solid body and cause the solution to move according to the equation

$$\mu\frac{d^2v}{dy^2} = -F_t, \tag{94.2}$$

where v is the tangential component of velocity.

Replacing $F_t$ in expression (94.2) by $\rho E_t$, we have:

$$\mu\frac{d^2v}{dy^2} = -\rho E_t. \tag{94.3}$$

and, because of equation (94.1),

$$\mu\frac{d^2v}{dy^2} = \frac{DE_t}{4\pi}\frac{d^2\varphi}{dy^2}. \tag{94.4}$$

Integration of equation (94.4) yields:

$$v = \frac{DE_t}{4\pi\mu}\varphi(y) + Ay + B. \tag{94.5}$$

As the distance from the solid wall increases, the velocity v must remain finite, and, therefore, A = 0. Furthermore, at a certain point in the solution near the solid surface, the velocity of the liquid will be zero due to the absence of slippage. Now, although it is most difficult to determine theoretically the distance from the plane y = 0 (the plane of the solid wall) where the velocity becomes zero, experiments show that it is approximately equal to two or three molecular diameters.

Let us then designate by $\varphi_s$ the potential in the region where the velocity becomes zero, so that

$$B = -\frac{DE_t}{4\pi\mu}\varphi_s. \tag{94.6}$$

Far from the wall, the potential $\varphi$ has a constant value, $\varphi_0$, equal to the potential difference between the bulk of the solution and the wall. Here, the velocity is also constant and equal to

$$v_0 = \frac{DE_t}{4\pi\mu}(\varphi_0 - \varphi_s). \tag{94.7}$$

The difference $(\varphi_0 - \varphi_S)$, which represents the drop in potential across the mobile portion of the double layer, has been termed the $\zeta$-potential

$$\zeta = \varphi_0 - \varphi_s;$$

so that equation (94.7) may then be expressed in a form that is commonly encountered in colloid chemistry:

$$v_0 = \frac{DE_t\zeta}{4\pi\mu}.$$

For later use, it is also convenient to introduce, instead of the $\zeta$-potential, the effective charge $\epsilon$ and the thickness d of the double layer as determined by the relation (which is valid for a plane condenser)

$$4\pi\varepsilon d = D\zeta.$$

Then, the velocity of the motion near the solid plane surface can be written in the form

$$v_0 = \frac{\varepsilon d}{\mu} E_t. \tag{94.8}$$

This phenomenon of a liquid motion at a solid surface, when that motion is caused by an electric field applied along the surface, is usually termed electro-osmosis.

## 95. ELECTROPHORESIS OF SOLID DIELECTRIC PARTICLES

From a knowledge of the velocity of electro-osmotic motion in the vicinity of an infinite plane it is now possible to compute without undue difficulty the velocity of electrophoretic motion of colloidal particles of rather large dimensions. If the size of the particles is large compared to the thickness of the double layer, every section of the particle surface can, with sufficient accuracy, be considered to be a plane. Equation (94.8) may then be applied to the separate sections of the surface.

Let us next pass on to a system in which the particle is at rest, while the liquid moves with a velocity U at infinity. We assume that the particle is a sphere of radius a, and since the electrophoretic velocity is small for all fields attainable in practice, we can also suppose that the motion of the liquid past the particle is of the creeping flow type $\left(\mathrm{Re} = \frac{Ua}{\nu} \ll 1\right)$.

The equations for creeping flow past a sphere have the form of equations (70.5), (70.6), and (70.7). To derive their solution, a system of boundary conditions must be formulated.

In accordance with what has been said earlier, the liquid far from the particle will move with a constant velocity U which is parallel to the electric field. If the direction of the electric field is chosen as the x axis, the conditions at infinity are

$$\begin{aligned} v_r &\to -U\cos\theta, \\ & \qquad\qquad r\to\infty. \\ v_\theta &\to +U\sin\theta, \end{aligned} \tag{95.1}$$

The boundary conditions on the surface of the particle may on the other hand be formulated in the following manner: at a distance d from the surface, the liquid has a tangential motion with a velocity $v_0$. Since the particle dimensions are extremely large compared to the thickness of the double layer d, it is possible to assume with sufficient accuracy that the liquid slips past the surface of the particle with a velocity $v_0$. Thus, we require that

$$v_\theta = +v_0\sin\theta, \tag{95.2}$$

$$\text{at} \quad r\approx a.$$

$$v_r = 0. \tag{95.3}$$

Equation (95.2) represents an original way of accounting for the external force acting on the particle. As soon as the liquid motion along the surface is specified, there is no further need to account for any other forces exerted on the surface, and as a last boundary condition we can therefore require that the viscous friction on the particle be zero. Actually, of course, the viscous friction is not zero. Along with it, however, an external driving force is exerted on the particle which, for steady-state motion, develops precisely that velocity $v_0$ and the corresponding viscous friction on the surface of the object which matches the driving force. Thus, instead of equating the driving force with the friction, we may assume that the driving force creates the velocity $v_0$ and that the viscous surface force on the particle equals zero. Now, the total viscous friction exerted on the surface of a spherical particle is determined by the expression

$$\iint (p_{rr}\cos\theta - p_{r\theta}\sin\theta)\,dS = \iint (p_{rr}\cos\theta - p_{r\theta}\sin\theta)\sin\theta\,d\theta\,d\varphi\cdot a^2,$$

so that the last condition which the solution must obey can be written in the form

$$\iint (p_{rr}\cos\theta - p_{r\theta}\sin\theta)\sin\theta\,d\theta\,d\varphi\cdot a^2 = 0. \tag{95.4}$$

Here, in evaluating $p_{rr}$ and $p_{r\theta}$, we should account for the fact that the liquid on the surface of the particle is not motionless, but that it has a velocity given by formula (95.2).

Let us now write down the expressions which characterize the potential distribution in the solution. Outside the double layer the solution is electrically neutral, and the potential satisfies the Laplace equation

$$\Delta\varphi = 0. \tag{95.5}$$

Far from the particle, the potential must approach that of the constant electric field which is parallel to the x axis,

$$\varphi \to Er\cos\theta. \tag{95.6}$$

The electric field in the vicinity of the dielectric particle differs from the uniform field at infinity: on the one hand, because of its dielectric properties and, on the other hand, because of the convective transfer of ions by the moving solution. Indeed, the motion of the solution entrains the ions, so that a convective electric current appears on the particle surface

$$\mathbf{j}_s = \varepsilon\mathbf{v}_t,$$

where $\mathbf{j}_S$ is the density of the surface current and $\mathbf{v}_t$ is the tangential velocity vector at the surface of the particle; $\mathbf{v}_t$ varies from point to point and its surface divergence, therefore, is not zero; $\operatorname{div}_S \mathbf{j}_S = \operatorname{div}_S(\varepsilon\mathbf{v}_t) \neq 0$. The law of conservation of charge requires than the following condition to be fulfilled

$$\varkappa\left(\frac{\partial\varphi}{\partial r}\right)_{r=a} = \operatorname{div}_s\left(\varepsilon\mathbf{v}_t\right) \tag{95.7}$$

where $\varkappa\left(\frac{\partial\varphi}{\partial r}\right)_{r=a}$ is the normal component of the current density at the surface of the particle. The relation (95.7) is the direct generalization of formula (69.7) for the case of the transfer of ions.

The velocity distribution (Figure 74) and the pressure distribution in the liquid can be presented in the form (70.28) to (70.30).

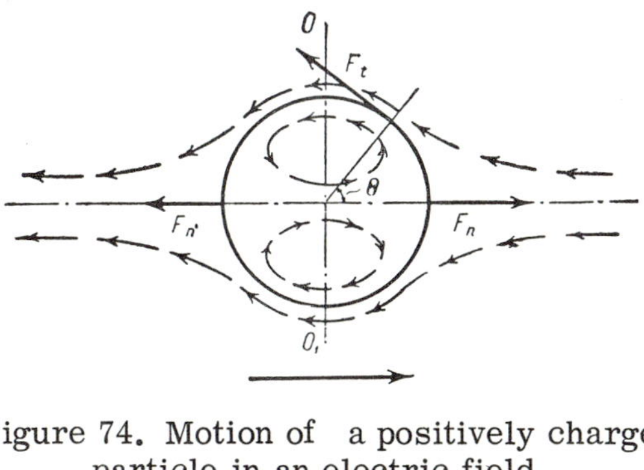


Figure 74. Motion of a positively charged particle in an electric field.

Substitution of the value of $v_t$, derived in a manner analogous to (73.10), into the law of conservation of charge (95.7) yields:

$$\varkappa\left(\frac{\partial\varphi}{\partial r}\right)_{r=a}=\frac{2\varepsilon v_0\cos\theta}{a}. \tag{95.8}$$

The solution of the Laplace equation satisfying conditions (95.6) and (95.8) has the form

$$\varphi=\left[r+\left(\frac{1}{2}-\frac{\varepsilon v_0}{\varkappa Ea}\right)\frac{a^3}{r^2}\right]E\cos\theta, \tag{95.9}$$

which, for $v_0 = 0$, evidently becomes the usual potential distribution near a spherical, nonconducting particle.

Substituting the values for the velocity components $v_r$ and $v_\theta$ and the pressure $p$ from (70.28) to (70.30) into the general expression for the components of the viscous stress tensor and introducing the expressions thereby derived for $p_{rr}$ and $p_{r\theta}$ into (95.4), we obtain:

$$\iint(p_{rr}\cos\theta-p_{r\theta}\sin\theta)\sin\theta\,d\theta\,d\varphi\cdot a^2=-2b_2=0.$$

from which it follows that $b_2=0$. The constants $b_1$ and $b_3$ can on the other hand be determined from boundary conditions (95.1) and (95.3). Elementary calculations yield:

$$U=\frac{2}{3}v_0=\frac{2}{3}\frac{\varepsilon d\,(E_t)_{r=a}}{\mu}. \tag{95.10}$$

Also, from equation (95.9), we find that, for $r = a$:

$$E_t=\left(\frac{3}{2}-\frac{v_0\varepsilon}{\varkappa Ea}\right)E. \tag{95.11}$$

Thus, by substituting this value for $E_t$ into (95.10), we derive the expression

$$v_0=\frac{3}{2}\frac{\varepsilon Ed}{\mu}\cdot\frac{1}{1+\frac{\varepsilon^2 d}{\varkappa\mu a}} \tag{95.12}$$

for the slippage velocity at the surface. The velocity of electrophoretic motion, therefore, can be put into the final form

$$U=\frac{\varepsilon Ed}{\mu+\frac{\varepsilon^2}{\varkappa}\frac{d}{a}}. \tag{95.13}$$

Equation (95.13) determines the velocity of motion of the particle as a function of the magnitude of the applied field E, the charge $\epsilon$ of the mobile portion of the double layer and its effective thickness d, the radius a of the particle, and the liquid properties — viscosity $\mu$ and conductivity $\varkappa$.

It is immediately clear from this formula that one constant, $\zeta = \epsilon d$, is not sufficient to describe the properties of the double layer, for, depending upon the magnitude of $\frac{\epsilon^2}{\varkappa\mu}\frac{d}{a}$, expression (95.13) can be written in either one of two limiting formulas:

$$U = \frac{\varepsilon d E}{\mu}; \quad \frac{\varepsilon^2}{\varkappa\mu} \cdot \frac{d}{a} \ll 1, \tag{95.14}$$

$$U = \frac{a \varkappa E}{\varepsilon}; \quad \frac{\varepsilon^2}{\varkappa\mu} \cdot \frac{d}{a} \gg 1. \tag{95.15}$$

Thus, the velocity of electrophoretic motion, at small $\epsilon$, is proportional to $\epsilon$, while, at large $\epsilon$, it is inversely proportional to it and, as a function of $\epsilon$, passes through a maximum at

$$\varepsilon = \varepsilon_{\max} = \sqrt{\frac{\varkappa a \mu}{d}}. \tag{95.16}$$

Equation (95.14) was first derived by M. von Smoluchowski [3], who also pointed out the need to introduce into the formula corrections for convective conductivity, which becomes significant in cases where electrophoresis occurs in a poorly conducting medium, without, however, making the necessary computations.

It is not difficult to see that (95.14) is obtained as a particular case of (95.13) when the correction for convective conductivity can be disregarded. Indeed, the appearance of the second term in the denominator of this expression relates to the convective transfer of ions along the surface of the drop. Let us characterize surface conductivity by the quantity $\varkappa_s$, determined, by analogy with the usual conductivity, from the expression

$$-\varkappa_s \operatorname{grad}_s \varphi = \mathbf{j}_s = \varepsilon \mathbf{v}_t. \tag{95.17}$$

Taking into account the fact that the magnitude of the potential gradient along the surface is $|\operatorname{grad}_s \varphi| = \frac{1}{a}\frac{\partial\varphi}{\partial\theta}$, and $|v_t| = v_0 \sin\theta$, and utilizing expressions (95.9) and (95.13), we obtain:

$$\frac{1}{a}\frac{\partial\varphi}{\partial\theta} = \frac{3}{2}\frac{\mu E \sin\theta}{\mu + \frac{\varepsilon^2}{\varkappa}\cdot\frac{d}{a}} = \frac{3}{2}\frac{\mu}{\varepsilon d} U \sin\theta,$$

whence

$$\varkappa_s = \frac{|\varepsilon \mathbf{v}_t|}{|\operatorname{grad}_s \varphi|} = \frac{\varepsilon v_0 \sin\theta}{\frac{3}{2} U \frac{\mu}{\varepsilon d} \sin\theta} = \frac{\varepsilon^2 d}{\mu}. \tag{95.18}$$

Therefore

$$\frac{\varepsilon^2 d}{\varkappa\mu a} = \frac{\varepsilon^2 d}{\mu a} \cdot \frac{\partial\varphi}{\partial\theta} a : \varkappa \frac{1}{a} \frac{\partial\varphi}{\partial\theta} \cdot a^2$$

represents, to an order of magnitude, the ratio of the surface convective current, $j_{sur} = \varkappa_s \frac{1}{a} \frac{\partial\varphi}{\partial\theta} a$, to the current in the volume, $j_{vol} = \frac{\varkappa}{a} \frac{\partial\varphi}{\partial\theta} a^2$, of a conductor with a cross section equal to the area of the particle. This ratio is usually small, and the term $\frac{\epsilon^2 d}{\varkappa a}$ in the denominator of (95.13) may be neglected. Conversely, in weakly conducting solutions, the convective current predominates over the current in the volume, and formula (95.13) reduces to (95.15). Thus U, as a function of $\epsilon$, should pass through a maximum, a conclusion which unfortunately cannot be verified from available experimental data.

Electrophoresis is used in actual practice to determine the charge $\epsilon$ of the mobile portion of the double layer or, what amounts to the same thing, the $\zeta$-potential of the particle.

It should also be noted that many authors have developed a theory of surface conductivity, especially Bikerman [4] and Hermans [5]. Bikerman pointed out that, in addition to convective surface conductivity, it is necessary to consider the surface electrical conductivity, which is related to the presence of an excess of ions adsorbed on the particle surface. Thus, if $\epsilon = 0$ but if equal amounts of adsorbed cations and anions are present on the surface, the surface electrical conductivity will evidently be equal to $F(U_c + U_a) \cdot \Gamma$, where $\Gamma$ is the amount of adsorbed material in equivalents per unit area, and $U_c$ and $U_a$ are the mobilities of the cations and anions, respectively, in the surface layer. In the case of motion near a solid wall, this effect should be of the same order of magnitude as the convective effect. Indeed, the ratio $\frac{\epsilon^2 d}{\mu} : F(U_c + U_a) \cdot \Gamma$ is of the order of $\zeta : \frac{e_0}{Dr_0}$, where $e_0$ and $r_0$ are the charge and radius of the ion respectively, D is the dielectric constant of the medium, and $\zeta : \frac{e_0}{Dr_0} \approx 1$. However, it is difficult to accept the method of calculation adopted by Bikerman, who, in computing the total surface conductivity, simply added the convective and conductive terms. The error in this procedure becomes clear as soon as one examines the case of a surface layer containing ions of only one sign, cations, for example. The amount of electricity carried by the charges in the surface layer in a field of unit intensity, according to Bikerman (but with our notation), should then be equal to $\int_0^\infty \epsilon_y (U_c + v_y)\, dy$, where $\epsilon_y$ and $v_y$ are, respectively, the charge density and the velocity of the liquid at a distance y from the wall. It is evident, however, that in reality, when the distances between ions of the same sign are small on the average compared with the distances of the ions from the wall, the liquid medium should move together with the ions as a unit, and that the quantity of electricity transported should equal $\int_0^\infty \epsilon_y v_y dy$, or, for a linear potential distribution, $\frac{\epsilon^2 d}{\mu}$. In the opposite case, i.e., when the distances between the ions are large compared to the distances of the ions from the wall, the amount of electricity transferred should be equal to $\int_0^\infty \epsilon_y U_{cy} dy$, where $U_{cy}$, the mobility of the cation, is slightly changed by the proximity of the wall. In the general case, the surface conductivity should lie between these values, rather than be equal to their sum as Bikerman assumes. The equations set forth in the present study account for only the convective term because these equations are necessary for comparison with the equations of electrocapillary motion, where, as will be shown later, convective conductivity exceeds the electrical conductivity by several orders of magnitude.

In his investigation, Hermans considers polarization of a double layer by an applied electric field with, however, an incorrect boundary condition for the charge density. In other words, he sets the change in the surface charge density of the particle under the action of an external field equal to zero, whereas the actual boundary condition requires that the normal component of the current be equal to zero. As for the charge density, computation shows that its change is by no means equal to zero on the surface, form which it follows that the conclusions of this study are erroneous.

In the preceding, we have confined our discussion to an examination of electrophoretic motion of solid, spherical, nonconducting particles whose dimensions are large compared to the effective thickness of the double layer. In actual practice, however, electrophoresis involves particles which may not satisfy all these conditions. Electrophoresis of metal particles is examined separately in the next section.

Strictly speaking, colloidal particles probably cannot be expected to be exactly spherical in shape. An additional factor of the order of unity which takes into account the deviation of the particle shape from the spherical should therefore be introduced into formula (95.13). Unfortunately, the computation of that factor for the complex geometrical shape of an actual particle appears to be impossible from a practical point of view, and this fact should be borne in mind whenever one evaluates those studies in which the electrophoretic velocity of extremely small particles ($a \ll d$) was determined. Calculations [6] have shown that, for $a \ll d$, the velocity of the motion is less than what is predicted by equation (95.14) by a factor whose limiting value is 2/3, and that the computation of the electrophoretic velocity of small particles is beset with many serious mathematical difficulties that often lead to a series which does not have a well-defined convergence [7].

Considering the fact that all changes in the formula for electrophoretic velocity involve only the numerical coefficient, which is difficult to calculate in most cases, we have decided to omit the detailed analysis.

## 96. ELECTROPHORESIS OF IDEALLY POLARIZABLE METAL PARTICLES. CALCULATION OF FORCES FOR THE CASE OF A DIFFUSE DOUBLE LAYER

So far, we have examined the electrophoresis of nonconducting particles. Let us now pass to the electrophoresis of metal particles and let us first find the forces that act on a spherical, metallic particle in an electrolytic solution in the presence of an electric field. We shall start by assuming that the particle is ideally polarizable, i.e., that the potential differences at the metal-solution interface are such that there occurs neither a discharge of ions of the solution nor the formation of new ions [8]. Under these conditions, the normal component of current at the surface of the particle, and, therefore, the normal component of the field in the solution outside the double layer, becomes zero, which means that the distribution of the lines of force outside the double layer is the same as in the case of a nonconducting particle.

Let us then calculate for the case of a diffuse double layer the magnitude of the force $(F_t)_y$ acting at a distance y from the metal-solution interface. We shall limit ourselves to a system with a plane interface at $y = 0$ (i.e., a sufficiently large particle whose curvature may be neglected), since the extension from a plane to a spherical particle will not add anything conceptually new. For a plane interface, the field intensity in the electrolyte outside the double layer can be considered constant. As before, let us designate by $E_t$ the field strength in a direction parallel to the surface of the metal at a distance which is large compared to the thickness of the double layer. Also we suppose that the positive direction of the y axis is into the solution perpendicular to the surface of the metal, and that $\varphi = 0$ at $x = 0$ and for a sufficiently large y. Then, outside the double layer, we have $\varphi = -E_t x$.

In the presence of an external electric field, the equilibrium in the double layer is destroyed, and an electric current flows through it. On the surface of an ideally polarizable particle, however, the normal component of current is zero. The equation of continuity for the current is

$$\frac{\partial j_x}{\partial x} + \frac{\partial j_y}{\partial y} = 0 \tag{96.1}$$

and, since all gradients parallel to the wall are small compared to the gradients in the perpendicular direction,

$$\frac{\partial j_y}{\partial y} = 0, \quad \text{or} \quad j_y = \text{const} = 0.$$

Therefore, for a sufficiently large particle whose curvature can be neglected, the equilibrium may be considered as undisturbed, which means that the equation for the diffuse double layer (for a monovalent electrolyte) will take the form

$$\frac{D}{4\pi}\frac{\partial^2\varphi}{\partial y^2}=-(Fc_0)\left\{e^{-\frac{(\varphi+E_tx)F}{RT}}-e^{\frac{(\varphi+E_tx)F}{RT}}\right\}, \tag{96.2}$$

where $c_0$ is the concentration of the solution.

As in equation (96.1), we have, on account of their smallness, neglected derivatives parallel to the wall. The electrical force acting in a direction parallel to the surface of the metal on a volume element of the diffuse double layer lying between y and y + dy, and whose base is 1 cm$^2$, equals $\rho(y)\,(E_t)_y dy$, where $(E_t)_y$ is the tangential component of the field and $\rho(y)$ is the charge density at a distance y from the interface, so that

$$(E_t)_y\,\rho(y)\,dy=\frac{\partial\varphi}{\partial x}\cdot\frac{D}{4\pi}\frac{\partial^2\varphi}{\partial y^2}\,dy.$$

In addition, in a polarized, diffuse double layer, a gradient of osmotic pressure also exists (since the ion concentration varies from point to point); this gives a ponderomotive force equal to $-\frac{\partial p(y)}{\partial x}dy$, where p(y) is the osmotic pressure at a distance y from the interface. Thus the resultant ponderomotive force $(F_t)_y dy$ is

$$(F_t)_y\,dy=\left\{\frac{D}{4\pi}\frac{\partial\varphi}{\partial x}\cdot\frac{\partial^2\varphi}{\partial y^2}-\frac{\partial p(y)}{\partial x}\right\}dy. \tag{96.3}$$

Let us designate by $(c_c)_y$ and $(c_a)_y$ the concentration of cations and anions, respectively, at a distance y from the interface. Then

$$p(y)=\{(c_c)_y+(c_a)_y\}\,RT=c_0RT\left\{e^{-\frac{(\varphi+E_tx)F}{RT}}+e^{\frac{(\varphi+E_tx)F}{RT}}\right\}.$$

Integrating equation (96.2) from y to ∞, we obtain:

$$\frac{D}{8\pi}\left(\frac{\partial\varphi}{\partial y}\right)^2=c_0RT\left\{e^{-\frac{(\varphi+E_tx)F}{RT}}+e^{\frac{(\varphi+E_tx)F}{RT}}-2\right\}=p(y)-p_\infty, \tag{96.4}$$

which can be combined with equations (96.2) and (96.3) to yield:

$$\begin{aligned}(F_t)_y=&-c_0F\left\{e^{-\frac{(\varphi+E_tx)F}{RT}}-e^{\frac{(\varphi+E_tx)F}{RT}}\right\}\left(\frac{\partial\varphi}{\partial x}\right)+\\&+c_0F\left\{e^{-\frac{(\varphi+E_tx)F}{RT}}-e^{\frac{(\varphi+E_tx)F}{RT}}\right\}\left(\frac{\partial\varphi}{\partial x}+E_t\right)=E_t\rho(y).\end{aligned} \tag{96.5}$$

Thus, by taking the osmotic pressure gradient into account, we arrive not only at the same expression for the ponderomotive force in the double layer of the metal particle as in the case of a non-conducting particle, but we also obtain the same dependence on distance as determined by the quantity $\rho(y)$.* The difference between the two cases boils down to the fact that the force acting on the inner sheet of the double layer on the surface of the metal is equal to zero, or, more correctly, that the point at which it is applied is displaced by a distance of the same order of magnitude as the particle radius. The possibility of utilizing equation (96.5) for a determination of the ponderomotive force in a double layer of arbitrary structure or in the presence of specific adsorption forces has not yet been clarified. In the most general case, therefore, we are limited to the method of Section 99, which is based on the Lippmann-Helmholtz equation.

We shall now show that the terms $\frac{D}{4\pi}\cdot\frac{\partial^2\varphi}{\partial y^2}\frac{\partial\varphi}{\partial x}dy$ and $\frac{\partial p(y)}{\partial x}dy$ in equation (96.3) are of comparable magnitude. Indeed,

$$\int_0^\infty \frac{D}{4\pi}\frac{\partial^2\varphi}{\partial y^2}\frac{\partial\varphi}{\partial x}dy = \frac{D}{4\pi}\left(\frac{\partial\varphi}{\partial y}\cdot\frac{\partial\varphi}{\partial x}\right)\Bigg|_0^\infty - \frac{D}{8\pi}\int_0^\infty \frac{\partial}{\partial x}\left(\frac{\partial\varphi}{\partial y}\right)^2 dy = -\int_0^\infty \frac{\partial p(y)}{\partial x}dy$$

---

* In an unpolarized, diffuse double layer, the excess lateral pressure on an element of the surface layer of thickness dy, whose integral over the entire surface layer gives the magnitude of the decrease in surface tension, is equal to the sum of the excess osmotic pressure $[p(y) - p_\infty]$ dy of ions in the double layer and the Maxwell pressure $\frac{D}{8\pi}\left(\frac{\partial\varphi}{\partial y}\right)^2 dy$. The terms in this sum are equal to each other. However, if we were to try to find the force $(F_t)_y dy$ from the relation

$$(F_t)_y\,dy = -\frac{\partial}{\partial x}\left\{\frac{D}{8\pi}\left(\frac{\partial\varphi}{\partial y}\right)^2 + [p(y) - p_\infty]\right\}dy, \tag{I}$$

we would arrive at an incorrect result. The reason for this can be traced back to the fact that, in expressing the electric portion of the ponderomotive force in the double layer by means of the components of the Maxwell stress tensor, not only the $T_{xx}$ components, but the $T_{xy}$ components as well must be considered. The latter are absent in an unpolarized double layer. In this case, by neglecting the term that contains $\left(\frac{\partial\varphi}{\partial x}\right)^2$ in the expression for $T_{xx}$, we can arrive at the correct expression for the electric portion of the ponderomotive force, as follows:

$$\left(\frac{\partial T_{xx}}{\partial x} + \frac{\partial T_{xy}}{\partial y}\right) = \frac{\partial}{\partial x}\left\{-\frac{D}{8\pi}\left(\frac{\partial\varphi}{\partial y}\right)^2\right\} + \frac{\partial}{\partial y}\left(\frac{D}{4\pi}\frac{\partial\varphi}{\partial x}\cdot\frac{\partial\varphi}{\partial y}\right) = \frac{D}{4\pi}\frac{\partial^2\varphi}{\partial y^2}\frac{\partial\varphi}{\partial x}. \tag{II}$$

The quantity $T_{xy}$ becomes zero at both boundaries of the double layer, and, as a consequence, integration of the incorrect relation (I) does lead to a correct result.

and, according to equations (96.2) and (96.4),

$$\int_0^\infty \frac{D}{4\pi}\frac{\partial^2\varphi}{\partial y^2}\cdot\frac{\partial\varphi}{\partial x}\,dy = -\int_0^\infty \frac{\partial p\,(y)}{\partial x}\,dy = -\frac{1}{2}E_t\varepsilon,$$

where, as before, $\epsilon$ is the charge on the inner sheet of the double layer. The ratio of the electric component of the ponderomotive force to its osmotic component is then

$$\frac{D}{4\pi}\frac{\partial^2\varphi}{\partial y^2}\cdot\frac{\partial\varphi}{\partial x} : -\frac{\partial p\,(y)}{\partial x} = -\left(\frac{\partial\varphi}{\partial x}\right):\left(\frac{\partial\varphi}{\partial x}+E_t\right). \tag{96.6}$$

At relatively large distances from the particle, this becomes infinite, and the ponderomotive force is consequently determined solely by its electrical component. On the contrary, at sufficiently small distances, this ratio becomes zero, which means that the ponderomotive effect depends entirely on the osmotic term.

## 97. ELECTROPHORESIS OF IDEALLY POLARIZABLE METAL PARTICLES. CALCULATION OF FORCES FOR THE CASE OF A HELMHOLTZ DOUBLE LAYER

Let us assume now that the double layer has a Helmholtz structure, i.e., that the centers of gravity of the ions forming the outer sheet of the double layer lie on the surface of a sphere of radius $a + d$, which is concentric with the particle surface. It is usually postulated that the outer sheet of a Helmholtz double layer is rigidly bound to the surface of the metal. One could suppose, however, that there could exist a layer of adsorbed molecules on the surface of the metal which would act as a dielectric in the double layer, keeping the ions of the outer sheet at a constant distance from the surface of the metal in such a way that these ions would retain their mobility. Regardless of the soundness of such an assumption, we shall use here the simple Helmholtz model of the double layer in order to investigate the electrokinetic behavior of metal particles, since such a layer enables us to illustrate best the relations under discussion.

Under our assumptions, the distribution of the lines of force for a positively charged particle is illustrated in Figure 75. The direction of the x-axis has been taken parallel to the lines of force at a great distance from the particle. Let us designate by E the field intensity far from the particle, by r the distance measured from the center of the sphere, and by $\theta$ the angle between the radius vector and the x-axis ($0 \le \theta \le \pi$). Let us also choose our zero of potential in the solution at $r = a + d$ and $\theta = \frac{\pi}{2}$. Then

for $r \ge a + d$

$$\varphi = -E\left\{r + \frac{1}{2}\frac{(a+d)^3}{r^2}\right\}\cos\theta, \tag{97.1}$$

so that, at $r = a + d$

$$\varphi = -\frac{3}{2} E(a+d)\cos\theta. \quad (97.2)$$

Inside the metal and on its surface, i.e., for $r \leq a$,

$$\varphi = \varphi_0 \qquad (\varphi_0 \text{ is a constant}). \quad (97.3)$$

The solution of the Laplace equation satisfying boundary conditions (97.2) and (97.3) has the form

$$\varphi = \varphi_0\left(\frac{a+d}{r} - 1\right)\frac{a}{d} - \frac{3}{2}\frac{E\cos\theta}{1-\left(\frac{a}{a+d}\right)^3}\left(r - \frac{a^3}{r^2}\right). \quad (97.4)$$

The quantities which characterize the double layer and the forces acting on it can then be determined with the aid of relations (97.1) to (97.4).

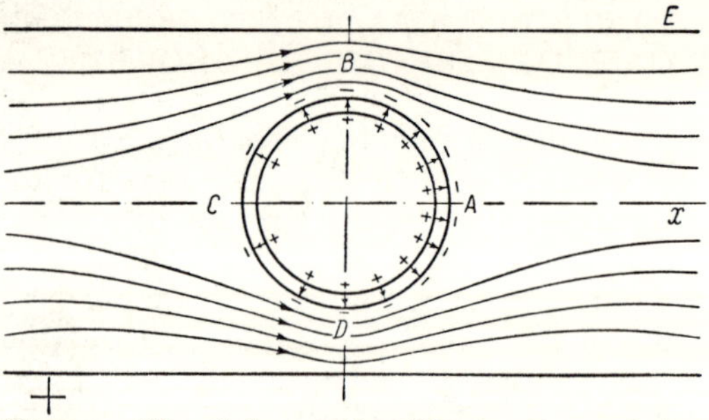


Figure 75. Schematic illustration of the double layer of an ideally polarizable metal particle in an external electric field.
The arrows show the direction of the field lines.

Let us next introduce the following notation: $\Delta\varphi$ as the jump in potential between the metal and the solution in the double layer, $\epsilon$ as the charge per unit area of the inner sheet of the double layer, $\epsilon'$ as the charge per unit area of the outer sheet, $E_t$ as the field intensity tangential to the double layer, considering $E_t$ positive when it forms an acute angle with the x axis, $E_{n_1}$ as the field component normal to the surface at the boundaries of the double layer and directed into it, and $E_{n_2}$ as the same directed outward from the double layer. In order now to compute the forces we consider a section of the double layer consisting of a surface element of the inner sheet, dS, and a surface element of the outer sheet, $\frac{(a+d)^2}{a^2}\,dS$, included in the same solid angle. We designate by $F_t dS$ the tangential force exerted on this element, and by $F_n dS$ the normal force, where the positive direction of $F_t$ is the same as the positive direction of $E_t$, while the positive direction of $F_n$ is from the center of the sphere (see Figure 74).

In determining the value of each of these quantities, we shall assume that $d \ll a$ and also limit all final expressions to the first term of the series expansion in powers of $\frac{d}{a}$. To derive these results however, it becomes necessary to determine the intermediate values of $E_n$ and $\epsilon$ up to an accuracy one order of magnitude higher.

It follows from equations (97.1) to (97.4) that:

$$\Delta\varphi = \varphi_0 + \frac{3}{2} Ea \cos\theta. \tag{97.5}$$

The dependence of $\Delta\varphi$ on $\theta$ corresponds (for $\Delta\varphi > 0$) to the distribution of the lines of force in the double layer (schematically presented in Figure 75), which indicates the polarization of the double layer under the influence of an external field.

Furthermore,

$$(E_{n_1})_{r=a} = -\left(\frac{\partial\varphi}{\partial r}\right)_{r=a} = \varphi_0\left(\frac{1}{d} + \frac{1}{a}\right) + \frac{3}{2} Ea \cos\theta \left(\frac{1}{d} + \frac{2}{a}\right), \tag{97.6}$$

$$(E_{n_1})_{r=a+d} = \left(\frac{\partial\varphi}{\partial r}\right)_{r=a+d} = -\varphi_0\left(\frac{1}{d} - \frac{1}{a}\right) - \frac{3}{2} Ea \cos\theta \cdot \frac{1}{d}, \tag{97.7}$$

where the values of $\frac{\partial\varphi}{\partial r}$ are determined from equation (97.4),

$$(E_{n_2})_{r=a} = (E_{n_2})_{r=a+d} = 0, \tag{97.8}$$

$$(E_t)_{r=a+d} = \frac{1}{a+d} \cdot \left(\frac{\partial\varphi}{\partial\theta}\right)_{r=a+d} = \frac{3}{2} E \sin\theta, \tag{97.9}$$

$$\begin{aligned}\varepsilon &= \frac{D}{4\pi}\left\{(E_{n_1})_{r=a} + (E_{n_2})_{r=a}\right\} = \\ &= \frac{D}{4\pi}\left\{\varphi_0\left(\frac{1}{d} + \frac{1}{a}\right) + \frac{3}{2} Ea\cos\theta\left(\frac{1}{d} + \frac{2}{a}\right)\right\},\end{aligned} \tag{97.10}$$

$$\begin{aligned}\left(\frac{a+d}{a}\right)^2 \varepsilon' &= \frac{D}{4\pi}\left\{(E_{n_1})_{r=a+d} + (E_{n_2})_{r=a+d}\right\}\left(\frac{a+d}{a}\right)^2 = \\ &= -\frac{D}{4\pi}\left\{\varphi_0\left(\frac{1}{d} + \frac{1}{a}\right) + \frac{3}{2} Ea\cos\theta\left(\frac{1}{d} + \frac{2}{a}\right)\right\} = -\varepsilon,\end{aligned} \tag{97.11}$$

$$(E_t)_{r=a} = 0, \tag{97.12}$$

Here D is the dielectric constant of the medium in the double layer.

The quantities $F_t$ and $F_n$ can be easily obtained from equations (97.6), (97.10), and (97.12). Limiting our final expression to only the first term of the series expansion, we find that

$$F_t = (E_t)_{r=a}\,\varepsilon + (E_t)_{r=a+d}\,\varepsilon'\left(\frac{a+d}{a}\right)^2 = -(E_t)_{r=a+d}\,\varepsilon \qquad (97.13)$$

and, consequently,

$$F_t = -\frac{D}{4\pi d}\left\{\varphi_0 + \frac{3}{2}Ea\cos\theta\right\}\frac{3}{2}E\sin\theta = \frac{1}{a}C\Delta\varphi\frac{\partial\varphi}{\partial\theta} = \frac{1}{a}\frac{\partial\Delta\sigma}{\partial\theta}, \qquad (97.14)$$

where $C = \frac{D}{4\pi d}$ is the capacitance of the double layer per unit area for $d \ll a$, and $\Delta\sigma = \frac{C(\Delta\varphi)^2}{2}$. The quantity $\Delta\sigma$, in the case of a Helmholtz double layer, expresses the lowering of the interfacial tension due to the surface charges. The direction of the force $F_t$ is shown in Figure 74.

We have, for the normal component,

$$F_n = -\frac{1}{2}\left\{(E_{n_1})_{r=a} - (E_{n_2})_{r=a}\right\}\varepsilon + \frac{1}{2}\left\{(E_{n_1})_{r=a+d} - (E_{n_2})_{r=a+d}\right\}\cdot\varepsilon'\left(\frac{a+d}{a}\right)^2 = -\frac{C(\Delta\varphi)^2}{a} = -\frac{2\Delta\sigma}{a}. \qquad (97.15)$$

It is seen from equation (97.15) that the absolute value of $F_n$ is greater in the right hemisphere than in the left hemisphere. The directions and the relative values of the force $F_n$ for $\theta = 0$ and $\theta = \pi$ are shown in Figure 74.

The resulting forces acting on the particle as a whole can be derived from equations (97.14) and (97.15). The force $F_t$ has a component parallel to the x axis equal to $F_t \sin\theta$, so that the tangential force acting on the element of the double layer between the angles $\theta$ and $\theta + d\theta$ is

$$2\pi a^2 F_t \sin\theta\sin\theta\, d\theta = 2\pi a\sin^2\theta\frac{\partial\Delta\sigma}{\partial\theta}d\theta = -3\pi a^2 EC\sin^3\theta\left(\varphi_0 + \frac{3}{2}Ea\cos\theta\right)d\theta,$$

which can be integrated over the entire surface to yield:

$$-\int_0^\pi 3\pi a^2 EC\sin^3\theta\left(\varphi_0 + \frac{3}{2}Ea\cos\theta\right)d\theta = -4\pi a^2\cdot C\varphi_0 E = -4\pi a^2\varepsilon_0 E = -eE, \qquad (97.16)$$

where $\epsilon_0 = C\varphi_0$ is the charge density on the surface of the metal in the absence of an external field, and $e = 4\pi a^2\epsilon_0$ is the total charge on the metal surface. Thus, the forces acting on the double layer in a direction tangential to the surface of the metal give rise to a component which is parallel to the field intensity and equal in magnitude to the force acting on a particle with a free charge e, but opposite in sign since this force originates from the charges on the outer sheet of the double layer.

On the other hand, the component of the force $F_n$ parallel to the x axis is $F_n \cos\theta$, which when integrated over the entire surface of the particle reduces to:

$$-\int_0^\pi 2\pi a^2 F_n \cos\theta \sin\theta \, d\theta = 4\pi a \int_0^\pi \Delta\sigma \sin\theta \cos\theta \, d\theta = \\ = 4\pi a^2 C \varphi_0 E = eE. \tag{97.17}$$

Thus, the sum of the tangential and normal forces acting over the entire double layer is equal to zero. This should have been expected. The results are different, however, if instead of the surface as a whole only a portion of the particle, for example the spherical segment which is included within the angles $\theta$ and $\pi - \theta$, is considered. The sum of the forces parallel to the x axis is evidently

$$2\pi a^2 \int_\theta^{\pi-\theta} (F_t \sin^2\theta - F_n \sin\theta \cos\theta)\, d\theta = \\ = 2\pi a \int_\theta^{\pi-\theta} \left(\sin^2\theta \cdot \frac{\partial \Delta\sigma}{\partial\theta} + 2\Delta\sigma \sin\theta \cos\theta\right) d\theta = \\ = 2\pi a \Delta\sigma \sin^2\theta \Big|_\theta^{\pi-\theta} = \frac{3}{2} eE \cos\theta \sin^2\theta. \tag{97.18}$$

The expression vanishes at $\theta = 0$, in accordance with what has been said above, and passes through a maximum $\frac{eE}{\sqrt{3}}$ at $\cos^2\theta = \frac{1}{3}$. Since the force exerted on the surface of the particle, consisting of two spherical layers — one defined by the range of angles from 0 to $\theta$, and the other, by the range from $\pi - \theta$ to $\pi$ — is equal in magnitude and opposite in sign to the derived value, so-called Maxwellian stresses should appear in the solid metal particle under the action of these forces. These stresses are absent in the case of a nonconducting particle. In a liquid metal particle (see below), the same forces, applied at points separated by distances of the order of magnitude of the particle radius, should give rise to a motion far more intense than what is usually observed in electrokinetic phenomena. The derivation set forth here, based directly on a consideration of the electrical forces operating on the charges in the simplest model of a double layer, is somewhat involved and depends in addition on the specific assumptions concerning the structure of the double layer as set forth in the model. Nevertheless, it clearly shows the origin of the forces exerted on a metal particle.

Let us now consider the forces acting on an element of the particle's surface between the angles $\theta$ and $\theta + d\theta$. The resultant of all surface tensions acting on this portion of the surface is evidently directed along the x axis and is equal to

$$-\frac{\partial}{\partial\theta}(2\pi a\sigma\sin^2\theta)\,d\theta=-2\pi a\sin^2\theta\frac{\partial\sigma}{\partial\theta}\,d\theta-\\-2\pi a^2\sin\theta\cos\theta\frac{2\sigma}{a}\,d\theta=(F_t\sin\theta-F_n\cos\theta)\,dS. \tag{97.19}$$

The first term in expression (97.19) gives the magnitude of the tangential forces projected onto the x axis, while the second gives the projection of the normal forces. During the integration over the entire surface of the sphere, the quantity $F_t \sin\theta$ does not change its sign, but the quantity $F_n \cos\theta$ has opposite signs in the two hemispheres separated by the plane $O_1O$ (Figure 74). But, since the value of the surface tension (for $\epsilon > 0$) in the left hemisphere is greater than in the right hemisphere, the resultant of all the normal forces is positive and exactly cancels out the resultant of all the tangential forces, because, according to (97.19), we have that

$$\int_0^\pi F_n\cos\theta\,dS=\int_0^\pi F_t\sin\theta\,dS.$$

In the absence of a field, i.e., for $\sigma$ = const, each of these integrals vanishes and all the forces reduce to a uniform compression $2\sigma_0/a$. Thus, detailed examination of the forces acting on the surface of an ideally polarizable particle shows that, despite its metallic nature, forces are generated in every portion of its surface which set the solution into motion (we assume the particle to be motionless). The magnitude of these forces is no different from the forces exerted on the surface of a nonconducting particle. An ideally polarizable metal particle must therefore pass into the same electrophoretic motion as a dielectric particle, with its velocity given by equation (95.13), which is general for nonconducting and for ideally polarizable, conducting particles. This conclusion is in full agreement with data obtained by N. A. Balashova and A. N. Frumkin [9] in a study of the electrophoresis of metal sols known to be free of oxide films (platinum sols, for example), and other substances.

The problem of the electrophoresis of metal particles has been considered theoretically by Henry [6], who arrived at the erroneous result that this phenomenon cannot be present under such circumstances. This erroneous conclusion came about because the author failed to account for the polarizing properties of metal particles in solutions.

## 98. ELECTROCAPILLARY MOTION OF MERCURY DROPS IN AN ELECTRIC FIELD

The motion of mercury drops under the influence of an electric current was first described in detail by Christiansen [10], who showed that, when a current passes through a drop of mercury placed in an electrolytic solution, the potential difference across the mercury-solution interface changes, inducing changes in the boundary layer. If the mercury in the solution is initially positively charged, the interfacial tension increases at the point where the current enters the drop (the negative pole of the drop) and decreases at the point of exit (the positive pole) thereby causing the mercury to move from the positive to the negative pole along the surface and in the opposite direction within the drop itself. The surface motion drags along the adjacent layers of the solution, and this results in a reactive movement of the drop as a whole along the lines of current. This motion, named electrocapillary motion by Christiansen, can be observed, for example, in the deviation from the vertical of mercury drops falling in an electric field. The velocity of the motion is very great, considerably exceeding that of the ordinary electrokinetic motion, but as the solution is diluted, the velocity of the drop in the field, for the same potential gradient, decreases. Besides the motion of the drop as a whole, Christiansen described the changes in shape which it undergoes. The curvature decreases at the positive pole and increases at the negative pole, so that the drop moves with the blunt end forward.

It follows from the theory of electrocapillarity that the drop will remain motionless in the field if its initial potential corresponds to the maximum on the electrocapillary curve, i.e., to the point of zero charge. This conclusion was confirmed by Frumkin [11] for falling drops, and by Bodforss [12] and Craxford [13] for drops at rest on a glass surface.

The motion of mercury drops in an electric field was first investigated throetically by Craxford, Gatty, and McKay [14], who concluded that the electrophoretic force acting on the drop is equal to the electric field intensity multiplied by the charge of the inner sheet of the double layer, as measured by the usual electrocapillary methods. It is obvious, however, that the sum of the electric forces acting on a system whose total charge is zero cannot itself be different from zero, so that such a conclusion must be incorrect. These authors consider the motion of mercury drops under small potential gradients as a particular and, at the same time, the simplest instance of ordinary electrophoresis.

In addition to observations on the motion of relatively large drops, published data are available on the electrokinetic motion of mercury particles of colloidal dimensions ($10^{-4}$ cm) in mercury sols in the presence of low concentrations of electrolytes. The velocities observed in such instances do not differ in order of magnitude from those common in electrokinetic motion.

N. A. Bakh [15] studied the potential difference occurring in a shower of mercury drops falling in a column of an electrolyte. This is an effect similar to the currents caused by the fall of solid particles in liquids, the theory of which has been given by Smoluchowski

[3]. The potential differences induced by mercury droplets falling in a normal $KNO_3$ solution are, however, about $10^5$ times greater than those possible for solid particles, and the ratio, in the case of $10^{-5}$ N $KNO_3$, is approximately 10. Thus, the observed potential differences, especially in the case of highly conducting solutions, cannot be interpreted as ordinary electrokinetic effects. This phenomenon was explained as a reversal of the Christiansen effect. As the drop falls through the solution, the surface of the mercury, along with the charges of the double layer, is pulled toward the rear of the drop. Thus the charge density of the double layer on the forward part of the drop decreases, while it increases at the rear, which in turn leads to potential differences in the electrolyte between the ends of the drop and, hence, to induced currents.

Interest in electrocapillary motion increased markedly in connection with the problem of the maxima on the current-voltage curves for a mercury electrode (see Chapter X). In a study by A. N. Frumkin and B. P. Bruns [16], it was shown that the high values of the current observed at these maxima were due to the agitation of the solution by the electrocapillary motions of the mercury surface. The electrocapillary motions giving rise to these maxima were later studied in detail, both for the case of a mercury electrode with a constant surface area [17], and for the case of a dropping mercury electrode. The latter system was also investigated, in particular, by Antweiler and Stackelberg [18]. Antweiler considered that the motions of the mercury surface were electrokinetic, and that high velocities were attained because of the absence of strong friction, which exists on the surface of a solid body, and because of the mobility of the charges in the inner sheet of the double layer, which results from the high conductivity of mercury. In this case, and in contrast to the electrokinetic motion at a solid surface, the double layer at the mercury surface moves as a single unit. Antweiler's point of view is that the passage of current through the metal-solution interface is essential for the appearance of motion. This is especially stressed by Stackelberg [19], according to whom the motion described by Christiansen is related to the presence of dissolved oxygen in the solution. Thus, if in the case of an ideally polarizable drop the oxygen is completely removed, then the motion should cease.

T. A. Kryukova and B. N. Kabanov [20] drew attention to the motion which arises in a solution near the surface of the falling mercury drop and which is caused by the very process of dropping. They found such motions to be especially noticeable at relatively high concentrations of the electrolyte (higher than 0.1 N).

The papers cited above do not throw sufficient light on the questions concerning the influence of the electrolyte on the motion of the mercury surface, the hydrodynamic properties of this surface in conducting solutions, and the relationship between electrocapillary and electrokinetic motions. Whereas the majority of authors have treated these two phenomena as completely independent, in the investigations conducted by Craxford et al., both were considered to be identical, despite the enormous differences in the observed velocities of motion. A. N. Frumkin and the author of this book developed a theory for the electrocapillary motion of mercury drops and established the relationship between this phenomenon and the electrokinetic motion of solid particles.

In Section 96 we examined the system of forces exerted on the surface of a metal particle immersed in an electrolytic solution in an electric field. A mercury drop is a particular case of a metal particle. The only difference between it and a solid particle is that the forces acting on the surface of a drop are balanced by viscous rather than by elastic forces.

## 99. MOTION OF LIQUID METAL DROPS IN AN ELECTRIC FIELD

Let us examine the motion of a liquid metal particle in a solution under the influence of an applied electric field. The conductivity of the metal particle may be considered infinitely high compared to that of the solution [21]. Let us also assume that the radius of the particle is large in comparison with the effective thickness of the double layer and suppose at first that the particle is ideally polarizable. This latter assumption implies that, in the given solution at a given potential on the surface of the particle, no discharge or formation of ions can occur and that no current can flow through the particle.

The electric field potential $\varphi$ in the solution is determined by the Poisson equation

$$\Delta\varphi = -\frac{4\pi\rho}{D},$$

where $\rho$ is the volumetric charge density and D is the dielectric constant of the medium.

The potential $\varphi$ at a great distance from the particle coincides with that of the applied field E. It has a constant value, which we shall later select as the zero of potential, on the surface of the metal particle itself.

If the direction of the external field is assumed to be positive in the direction of the x axis, the potential at infinity is

$$\varphi \to Er\cos\theta. \tag{99.1}$$

while, on the surface of the particle, at r = a,

$$\varphi = 0. \tag{99.2}$$

Let us now introduce spherical coordinates with the origin at the center of the particle. The entire solution surrounding the particle may be subdivided into two regions: the region of the double layer where the charge density is not zero, and the region of the electrically neutral solution. If d is the effective thickness of the double layer, in the first region $r < a + d$, and in the second, $r > a + d$. Here, we assume that $a \gg d$.

Let us further suppose that the field gradient at the surface of the metal outside the double layer is small compared to the potential gradient in the double layer along the normal to the surface, and that the radius of the particle is sufficiently large so that the interfacial tension is not a function of the radius. With these assumptions, the state of the surface layer and, consequently, the surface tension $\sigma$ at the metal-solution interface, are uniquely determined, for each point of the metal surface, by the potential difference between the metal and the solution $\Delta\varphi = \varphi_0 - \varphi_S$(where $\varphi_S$ is the value of the potential in the solution at the surface of the metal, but outside the double layer) or, since $\varphi_0$=const, $\sigma$ is determined by $\varphi_S$ itself and is independent of the potential difference at the neighboring points. If $\Delta\varphi$ were constant, $\sigma$ would also have a constant value, and no tangential forces would exist on the surface of the metal. In the case of a polarized particle, however, $\Delta\varphi$ and, hence, $\sigma$ vary from point to point. The tangential force is $F_t = \text{grad}\,\sigma$ and, since $\sigma$ is a function of $\theta$ only and $F_t$ is considered positive if the tangential force is in the direction of decreasing values of $\theta$, then

$$F_t dS = -\frac{1}{a}\frac{\partial\sigma}{\partial\theta}dS. \tag{99.3}$$

It is known now that the surface tension is related to the jump in potential $\Delta\varphi$ at the mercury-solution interface and to the surface charge $\epsilon$ by the Lippmann-Helmholtz relation

$$\frac{\partial\sigma}{\partial\,\Delta\varphi} = -\varepsilon. \tag{99.4}$$

If the jump in potential $\Delta\varphi$ varies from point to point on the particle surface, the surface tension at any point on the surface, characterized by the angle $\theta$, may be expressed in the form

$$\sigma(\theta) = \sigma_{\pi/2} + \int_{\pi/2}^{\theta}\frac{\partial\sigma}{\partial\,\Delta\varphi}\cdot\frac{\partial\,\Delta\varphi}{\partial\theta}d\theta = \sigma_{\pi/2} - \int_{\pi/2}^{\theta}\varepsilon\frac{\partial\,\Delta\varphi}{\partial\theta}d\theta,$$

where $\pi/2$ is the surface tension at the drop's equator.

We assume, in what follows, that the applied field is sufficiently weak to allow us to neglect all second order (quadratic) quantities. It is obvious then that $\epsilon$ will correspond to the density of the surface charge of the drop in the absence of a field — a charge which is not a function of the angle $\theta$ and which can therefore be placed outside the integral sign. Thus,

$$\sigma = \sigma_{\pi/2} - \varepsilon \int\limits_{\pi/2}^{\theta} \frac{\partial \Delta\varphi}{\partial\theta} d\theta = \sigma_{\pi/2} - \varepsilon \Delta\varphi. \tag{99.5}$$

The presence of varying surface tension leads to the appearance of normal $p_{rr}^{(\sigma)} = F_n$ and tangential $p_{r\theta}^{(\sigma)} = -F_t$ stresses, acting on the boundary of the metal-solution interface,

$$F_n = \frac{2\sigma}{a} = \frac{2\sigma_{\pi/2}}{a} - \frac{2\varepsilon \Delta\varphi}{a}, \tag{99.6}$$

$$F_t = \operatorname{grad} \sigma = -\varepsilon \operatorname{grad}_\theta \Delta\varphi = -\frac{\varepsilon}{a} \frac{\partial \Delta\varphi}{\partial\theta}. \tag{99.7}$$

The ratio in (99.6) differs from formula (97.15) only in that it contains the term $\frac{2\sigma_{\pi/2}}{a}$. It is absent in equation (97.15) because the latter represents the part of the force that is electrical in origin.

The derivation of the expressions for the forces exerted on the surface — based on an application of the thermodynamic Lippmann-Helmholtz formula — is more general than the derivations set forth in Sections 96 and 97. It is independent of any model regarding the structure of the double layer, and yet at the same time, it has the disadvantage that it can only lead to expressions for the total forces and not for their spatial distribution.

Since field intensities different from zero are applied to the liquid boundary, the liquid cannot remain at rest and must move. The velocity of this motion will then be such that viscous stresses will match the surface stresses described above.

In order to determine this motion, it is necessary, as may be seen from formulas (99.6) and (99.7), to find how the potential difference $\Delta\varphi$ in the double layer surrounding the particle depends on the polar angle $\theta$, and to accomplish this, we must know the potential distribution in the outer sheet of the double layer (the sheet that faces the solution), since we have taken the value of the potential in the metal as zero. This potential distribution can be derived from a solution of the equation for the potential outside the double layer in the neutral solution

$$\Delta\varphi = 0 \tag{99.8}$$

with appropriate boundary conditions. In this way we do not need to consider specific models of the double layer.

The requirement that the potential must coincide with that of the applied field, i.e., equation (99.1), serves as the boundary condition at infinity. In order to formulate the boundary condition at the neutral solution-double layer interface, it is necessary, as shown below, to take into account the motion of the solution and the corresponding convective transfer of ions.

If, under the action of the applied field, the liquid surrounding the particle moves with a velocity v, the liquid motion in a direction tangential to the surface of the particle will lead to a convective ion flow.

Let $\epsilon$ be the charge per unit area of the mobile portion of the double layer. For liquid metal particles we assume that the entire double layer is mobile, so that $\epsilon$ represents the entire charge per unit area $\left(\epsilon = \int_0^\infty \rho(r)\, dr\right)$. Also let $v_\theta$ be the tangential component of the fluid velocity in the mobile portion of the double layer (more exactly, at the interface $\theta$ between the two liquids). Then the convective surface current flowing across the surface of the particle has a density of $j_s = \epsilon v_0$.

The tangential velocity varies from point to point on the surface of the particle, which means that the surface divergence of the convective current is not zero: $\text{div}_s(\epsilon v_t) \neq 0$. In other words, the convective transfer of ions by the flowing liquid leads to a situation in which the ions leave the double layer in order to enter the region of neutral solution at some points of the surface, while at other points they enter the double layer from the outside (it follows from the assumption that the particle is ideally polarizable that the ions cannot lose their charge on the surface).

The law of conservation of charge requires now that condition (95.7) be fulfilled at the interface between the double layer and the electrically neutral solution. The required boundary condition is therefore

$$\varkappa \frac{\partial \varphi}{\partial r} = \text{div}_s(\varepsilon \mathbf{v}_t). \tag{99.9}$$

Thus, it may be seen from (99.9) that, the hydrodynamic problem must first be solved, before the electrical problem can be tackled and that these two problems are inseparable. In particular, even though problems of this type cannot usually be solved exactly, the high degree of geometric symmetry in this case (a spherical drop a uniform external field) permits the construction of a solution under certain assumptions.

In formulating the hydrodynamic equations, we shall assume that the particle motion occurs at small Reynolds numbers $Re = \frac{Ua}{\nu}$, where U is the velocity of the drop and $\nu$ is the kinematic viscosity of the solution. It is convenient, in what follows, to use a system of coordinates related to the drop in such a way that the drop is considered at rest, while the liquid is supposed to move relative to it with a velocity which is equal in value but opposite in direction to the actual motion of the drop proper. Taking the symmetry of the problem into account, let us then use the Navier-Stokes and continuity equations in spherical coordinates — equations (70.5) to (70.7).

The boundary conditions for $v_r$ and $v_\theta$ at infinity are

$$v_r = -U\cos\theta, \quad v_\theta = +U\sin\theta, \qquad \text{as } r \to \infty, \tag{99.10}$$

where U is the velocity of liquid motion at infinity (velocity of the drop relative to the solution). The velocity of the inner liquid must remain finite at the center of the drop (the origin of the coordinates), while the normal velocity components of the exterior and interior liquids on the surface of the drop must vanish, i.e.,

$$v_r = 0, \quad v'_r = 0, \qquad \text{at } r = a, \tag{99.11}$$

Lastly, the tangential velocity component of both liquids must be continuous

$$v_\theta = v'_\theta \qquad \text{at } r = a. \tag{99.12}$$

In addition, however, the condition of continuity for the normal and tangential components of the stress tensor must be fulfilled at the interface between the liquids, which means that if $p_{rr}^{(0)}$ and $p_{r\theta}^{(0)}$ are the normal and tangential components of the viscous stress tensor, we must require that at the interface

$$p_{rr}^{(0)} + p_{rr}^{(\sigma)} = p'_{rr}, \tag{99.13}$$

$$p_{r\theta}^{(0)} + p_{r\theta}^{(\sigma)} = p'_{r\theta}, \qquad \text{at } r = a. \tag{99.14}$$

Now, in order to solve this system of hydrodynamic and electrodynamic equations with the indicated boundary conditions, we should also note that the electric field and the velocity of fluid flow far from the particle must be parallel to the x axis; i.e., conditions (99.1) and (99.10) must be obeyed. In a similar manner we shall assume that the field potential near the drop is a function of the external field and of the angle in such a manner that

$$\varphi = \alpha E r \cos\theta, \tag{99.15}$$

where $\alpha$ is an unknown constant whose value will be derived later in the course of the solution.

Let us also assume that the drop retains its spherical symmetry during its motion, so that the expressions for the radial and tangential components of velocity and the pressure, satisfying the hydrodynamic equations and condition (99.10) at infinity, may be written in the form of the first terms of an expansion in spherical harmonic functions, i.e., in the form of formulas (70.28) to (70.33). The five unknown constants — $a_1$, $a_2$, $b_1$, $b_2$, and $b_3$ — contained in these expressions must then be determined from boundary conditions. The latter include the value of $\Delta\varphi$ as obtained from the solution of the electrical part of the problem.

We shall show next that, for a consistent choice of the constants, all the boundary conditions of the electrical and hydrodynamic problems can be satisfied and that in this fashion the complete solution can be found. This in turn will justify our original assumption about the form of the solution, which should be sought in the same fashion as was done for the similar problem of the motion of a drop in the presence of surface-active agents.

For the tangential component of the velocity on the surface of the drop, let us first write the expression

$$(v_\theta)_{r=a} = v_0 \sin\theta, \tag{99.16}$$

where $v_0$ is a new unknown constant which has to be determined. Then, for the surface divergence of the velocity entering into boundary condition (99.9), we have an expression analogous to (73.10)

$$\operatorname{div}_s(\varepsilon \mathbf{v}_t) = \frac{\varepsilon}{a\sin\theta}\left(\frac{\partial}{\partial\theta}(v_\theta \sin\theta)\right)_{r=a} = \frac{2\varepsilon v_0\cos\theta}{a}.$$

If we substitute the above for $\operatorname{div}_s(\epsilon \mathbf{v}_t)$ into expression (99.9), we see that

$$\varkappa\left(\frac{\partial\varphi}{\partial r}\right)_{r=a} = \frac{2\varepsilon v_0\cos\theta}{a}. \tag{99.17}$$

where we suppose that $\epsilon$ is constant along the surface and that the value of $\partial\varphi/\partial r$ is taken at $r = a$ and not at $r = a + d$. The first assumption follows from the postulate that all perturbations are small, and the second is due to the inequality $a \gg d$.

Solution of equation (99.8) for the potential satisfying boundary condition (99.17) gives

$$\varphi = \left[r + \left(\frac{1}{2} - \frac{\varepsilon v_0}{\varkappa Ea}\right)\frac{a^3}{r^2}\right]E\cos\theta. \tag{99.18}$$

and therefore the value of the constant $\alpha$ in (99.15) is $\left(\frac{3}{2} - \frac{\epsilon v_0}{\varkappa Ea}\right)$.

Thus, from a knowledge of the potential outside the double layer, we can obtain the expression

$$\Delta\varphi = -\varphi_{\text{met}} + \varphi_{r=a+d} = +\varphi_{r=a} = \left(\frac{3}{2} - \frac{v_0\varepsilon}{\varkappa Ea}\right)Ea\cos\theta \tag{99.19}$$

for the potential jump at the metal-solution interface.

Consequently, because of (99.6) and (99.7),

$$p_{rr}^{(\sigma)} = \frac{2\sigma_{\pi/2}}{a} - 2\varepsilon\left(\frac{3}{2} - \frac{v_0\varepsilon}{\varkappa Ea}\right)E\cos\theta,$$

$$p_{r\theta}^{(\sigma)} = \left(\frac{3}{2} - \frac{v_0\varepsilon}{\varkappa Ea}\right)\varepsilon E\sin\theta.$$

Similarly, from the stress distribution on the surface of the drop we can determine the constants with the aid of the boundary conditions for the hydrodynamic problem. By substituting the values of $p_{rr}^{(0)}$, $p_{r\theta}^{(0)}$, $p'_{rr}$, and $p'_{r\theta}$ from expressions (70.9) into the boundary conditions (99.13) and (99.14), we have:

$$\cos\theta\left[-\left(\frac{3b_2\mu}{a^2}+\frac{6\mu b_1}{a^4}\right)+2\varepsilon E\left(\frac{3}{2}-\frac{v_0\varepsilon}{\varkappa Ea}\right)+6\mu' a_1 a\right]=0,$$

$$-\frac{3\mu b_1}{a^4}+\varepsilon E\left(\frac{3}{2}-\frac{v_0\varepsilon}{\varkappa Ea}\right)=-3\mu' a_1 a\ ,$$

and by substituting the values of the velocities from expressions (70. 28) to (70.33) into the boundary conditions (99.11) and (99.12), we obtain

$$\frac{b_2}{a}+\frac{b_1}{a^3}+b_3=0, \qquad a_1 a^2+a_2=0,$$

$$v_0=\frac{b_1}{2a^3}-\frac{b_2}{2a}-b_3, \qquad -2a_1 a^2-a_2=v_0.$$

Finally, from equation (99.10), we find that

$$b_3=-U.$$

The unknown quantities $b_1$, $b_2$, $a_1$, $a_2$, $v_0$, and U can be found from this system of equations, which may be solved by elementary transformations

$$b_2=0, \qquad \frac{b_1}{a^3}=U,$$

$$a_1 a^2=-\frac{3}{2}U, \qquad a_2=\frac{3U}{2}, \tag{99.20}$$

$$v_0=\frac{3}{2}U=\frac{3}{2}\frac{\varepsilon Ea}{2\mu+3\mu'+\frac{\varepsilon^2}{\varkappa}},$$

$$U=\frac{\varepsilon Ea}{2\mu+3\mu'+\frac{\varepsilon^2}{\varkappa}}. \tag{99.21}$$

Thus, the velocity distribution in the flowing liquid has the form (in a coordinate system related to the drop)

$$v_r=U\left(\frac{a^3}{r^3}-1\right)\cos\theta, \tag{99.22}$$

$$v_\theta=U\left(\frac{a^3}{2r^3}+1\right)\sin\theta, \tag{99.23}$$

$$v'_r=\frac{3}{2}U\left(1-\frac{r^2}{a^2}\right)\cos\theta, \tag{99.24}$$

$$v'_\theta=\frac{3}{2}U\left(\frac{2r^2}{a^2}-1\right)\sin\theta, \tag{99.25}$$

and the potential distribution is given by the relation

$$\varphi = E\left[r + \left(\frac{1}{2} - \frac{3}{2}\frac{\varepsilon U}{\varkappa E a}\right)\frac{a^3}{r^2}\right]\cos\theta. \tag{99.26}$$

The velocity of the particle in the electric field is evidently determined by formula (99.21). For $\frac{2\mu+3\mu'}{\epsilon^2/\varkappa} \gg 1$, i.e., at high values of viscosity and electrical conductivity and low surface charge, the drop velocity is

$$U = \frac{\varepsilon E a}{2\mu + 3\mu'}, \tag{99.27}$$

while for the reverse inequality $\frac{2\mu+3\mu'}{\epsilon^2/\varkappa} \ll 1$, i.e., at low values of viscosity and electrical conductivity and for a high surface charge, the drop velocity is

$$U = \frac{\varkappa E a}{\varepsilon}. \tag{99.28}$$

The physical meaning of both limiting values of U is simple. The first limit corresponds to the case where the convective current around the drop, which tends to equalize the potential in the solution near the surface of the drop, is so small that it can be neglected. Then, it may be assumed that the potential difference in the solution near the drop is not equalized by the conductivity of the solution. The forces exerted on the surface of the drop can be balanced only by viscous stresses in the moving liquid. Thus U, as given by formula (99.27), is the velocity of the drop corresponding to such a balance of forces. In particular, the drop velocity is very low for the case of a very high viscosity $\mu'$ of the interior liquid. In the transition to a solid particle, i.e., with $\mu'$ tending toward infinity, the interior motion cannot be maintained and the drop will cease to move. The system of surface forces is then matched by a system of stresses in the solid particle. In reality, however, the particle can still pass into the usual electrophoretic motion.

In the second instance, for a low electrical conductivity of the medium, the convective current along the surface of the drop is so large that it completely equalizes the potential difference along the drop. The potential in the solution near the drop surface is then constant along the drop, which, therefore, behaves as a completely unpolarizable metal particle with a current flowing through it.

In order that the potential jump at the drop-solution interface may remain constant at all times, it is necessary that the convective current of the ions of the outer sheet of the double layer, carried by the liquid moving along the surface of the drop, be equal to this electric current. The convective current of the charges in the inner sheet is then short-circuited by a conduction current flowing within the mercury drop, and, since the electrical conductivity of mercury is high compared with that of the solution, the potential difference thus established can be ignored. This condition is fulfilled when the velocity of the drop's surface motion is $v_0 = \frac{3}{2}U$, where U is determined by formula (99.28).

It is convenient next to introduce a quantity Z expressing the mobility of the particle, i.e., the ratio of the velocity U to the strength of the external field,

$$Z = \frac{U}{E} = \frac{\varepsilon a}{2\mu + 3\mu' + \frac{\varepsilon^2}{\varkappa}}, \tag{99.29}$$

or a quantity z called the specific mobility, i.e., the mobility of the particle with a radius equal to unity,

$$z = \frac{\varepsilon}{2\mu + 3\mu' + \frac{\varepsilon^2}{\varkappa}}. \tag{99.30}$$

The dual role of the surface charge density is revealed with particular clarity in formulas (99.29) and (99.30). The charge, on the one hand, is the source of the motion of the drop and, on the other, causes a self-retarding action, which, for large $\epsilon$, leads to a damping of the motion. The mobility reaches its maximum value at a charge density $\epsilon_{max}$,

$$\varepsilon_{max} = \sqrt{\varkappa(2\mu + 3\mu')}, \tag{99.31}$$

and is equal, at the maximum, to

$$z_{max} = \frac{1}{2}\sqrt{\frac{\varkappa}{2\mu + 3\mu'}}. \tag{99.32}$$

It follows from formula (99.26) that the quantity 3Ea is equal to the maximum potential difference between two points on the surface in the absence of motion. Let us denote this quantity by $\Delta\Phi_0$. Equation (99.21) may then be written in the form:

$$U = \frac{1}{3}\frac{\varepsilon(\Delta\Phi_0)}{2\mu + 3\mu' + \frac{\varepsilon^2}{\varkappa}} = \frac{1}{3}z(\Delta\Phi_0). \tag{99.33}$$

It is also interesting to observe that equation (99.21) for the velocity of a metal drop is very similar in form to the corresponding expression (95.13) for a solid (metal or dielectric) particle. As a matter of fact, if we ignore the difference in the numerical coefficients of the $\mu$ quantities, equation (99.21) can be derived from formula (95.13) by substituting a for d in the coefficients for $\epsilon$ and

$\frac{\epsilon^2}{\varkappa}$. This result may be interpreted in the following manner. In the case of a solid particle, the viscous motion of the liquid due to the field should decay within the limits of the double layer, i.e., over distances of the order of d. For a liquid metal particle, however, it should decay within the particle itself, i.e., over distances of the order of its radius a. Hence, the viscous stresses in the second case are smaller, while the corresponding mobility is greater than in the first by the ratio $\frac{a}{d}$. Similarly, in the transition from a solid to a liquid particle, the convective conductivity of the surface and, consequently, the retardation of the motion due to the electric field of the charges that are carried by the convective current on the particle surface also increase in the same ratio.

Let us now consider the mobility of mercury particles which are moving in an extremely viscous medium, and let the first regime of motion apply where the following inequalities are satisfied

$$\frac{\varepsilon^2}{\varkappa} \ll \mu \text{ and } \mu' \ll \mu.$$

For the velocity of the drop, we have

$$U = \frac{\varepsilon E a}{2\mu}. \tag{99.34}$$

A simple computation shows that the mobility of a mercury drop in an extremely viscous and highly conducting medium is equal to three quarters of the mobility that would be possessed, in a nonconducting medium, by a solid particle carrying a free charge equal to that of the inner sheet of the double layer of the mercury drop (and to one half the mobility of the same particle if it were liquid). Since the charge density in the double layer can be many times greater than the possible density of a free charge, mercury drops in a viscous, highly conducting solution have especially high mobilities.

It is also of interest to compute the surface conductivity of a metal drop and to compare it with that of a solid particle in electrophoretic motion. Because of equation (95.17) and on the basis of expressions (99.17), (99.19), (99.20), and (99.21), we have

$$\dot{\varkappa}_s = \frac{|\varepsilon v_t|}{|\operatorname{grad}_s \varphi|} = \frac{\varepsilon^2 a}{\mu}. \tag{99.35}$$

A comparison of this value for $\varkappa_s$ with expression (95.18) will show then that the surface electrical conductivity of a liquid metal drop exceeds that of a solid particle by the ratio a/d. Since the computed surface conductivity evidently represents the convective conductivity, which is related to the transfer of ions by the motion of the solution, high velocities of motion on the surface of the drop correspond to very high values of $\varkappa_S$.

In addition to convective surface conductivity, the drop also possesses the usual ohmic conductivity. The latter is somewhat higher on the surface than in the bulk of the solution because the ion concentration in the double layer is greater than in the main stream. It is evident, however, that, in the case of particles of macroscopic dimensions ($a \gg d$), ohmic surface conduction can be neglected in comparison to convective conduction.

In deriving formula (99.21) it was assumed that in electrocapillary motions the surface tension at all points on the surface of the drop should depend only on the potential. This condition will not be fulfilled if, for example, surface-active materials are present in the solution and if the establishment of an equilibrium between the surface and the bulk of the solution requires time due to the slowness of the process of adsorption or diffusion of the adsorbed material. The surface tension in this case, even for constant $\Delta\varphi$, is greater on those parts of the drop where the mercury surface is distended (in the right portion of the drop in Figure 75), and less where the surface is contracted (the left portion of the drop). This increases the drag on the drop and reduces its mobility Z.

It should be possible to observe similar phenomena even in the absence of specifically adsorbed materials, provided that the establishment of equilibrium between the boundary layer and the bulk of the solution at a given potential is not instantaneous. Thus, for example, if the mercury surface is negatively charged and if both monovalent and polyvalent cations are present in the solution, there will be a relative excess of polyvalent cations in the surface layer, and, when the surface contracts, the adjacent layers of the solution will become richer in these polyvalent cations. The result is that the interfacial tension is different at various parts of the surface even though they are at the same potential. It is clear then that the current flowing through the solution does not equalize the changes which occur in the composition of the solution. This is possible only by diffusion, i.e., in a relatively slow manner, so that the retarding effects that take place can be very considerable.

The applicability of the theory developed above is also restricted to low velocities and small potential gradients. Thus, according to this theory, if the change in the interfacial tension over the surface is small compared to its initial value and if the motion of the drop falls within the region of small Reynolds numbers, the shape of the drop remains spherical throughout the process. Numerical estimates show, however, that, for large values of $\epsilon$ and sufficiently large drops, the drop's velocities U, as calculated by equation (99.21), are so great that the Reynolds number $Re = \frac{Ua}{\nu}$ is by no means small compared to unity. Under these conditions, then, equation (99.21) is not valid and must accordingly be modified.

The motion of liquid bodies for Reynolds numbers greater than unity has not been studied in hydrodynamics in sufficient detail either theoretically or experimentally. Under these conditions a drop loses its spherical shape and becomes egg shaped, with the blunt end pointing forward. Furthermore, it follows from general considerations that the viscous drag is then small compared with the dynamic pressure on the drop and can therefore be neglected.

The dynamic pressure per cm² of drop surface can be presented in the form

$$\frac{F}{S}=\frac{K_f \rho U^2}{2},$$

where $K_f$ is the drag coefficient, which is a complicated and as yet unknown function of the Reynolds number for the case of the liquid drop which concerns us. It may be assumed, however, that it is not very different from the analogous function for a solid, egg-shaped body and that the drag coefficient first decreases rather rapidly with a rise in Re and then remains almost constant over a wide range of Reynolds numbers.

The velocity of the drop may be estimated very roughly for values $Re \gg 1$ by equating the driving force and the dynamic pressure. Then

$$\varepsilon E \approx K_f \rho U^2,$$

i.e.,

$$U \approx \sqrt{\frac{\varepsilon E}{\rho K_f}}. \tag{99.36}$$

Since $K_f$ here is an unknown function which, only in the roughest approximation, may be considered constant, the nature of formula (99.36) is qualitative. It shows that at large Reynolds numbers the velocity of the drop should increase with the charge more slowly than would follow from formula (99.21). Also, it should be a weak function of the viscosity of the mercury and of the solution.

It may also be assumed that certain changes in the shape of the drop, described by Christiansen [10], are precisely related to this effect (simple estimates show that the value of Re in his experiments was considerably greater than unity) and not to the change in interfacial tension of the drop, as was postulated by Christiansen. This is also apparent from the fact that if the change in the shape of the drop were caused by the variation in interfacial tension, the drop would become egg shaped with the pointed end forward, instead of backward as is actually observed. Indeed, the interfacial tension in the forward section of the drop would be lowered, and, if the deviation from a spherical shape were caused by changes in surface tension, it is this part which would jut forward, forming a pointed end on the drop.

Up to this point we have assumed in the course of the computations that the maximum variation in the potential along the drop is small compared to the initial potential jump

$$\varphi_0 = \frac{\varepsilon}{C},$$

where C is the specific capacitance of the double layer. However, the maximum change of the potential jump in the double layer is, in accordance with expression (99.26),

$$\frac{2\mu + 3\mu'}{2\mu + 3\mu' + \frac{\varepsilon^2}{\varkappa}} \frac{3Ea}{2}.$$

and hence, for the theory to be applicable, the following inequality must hold

$$\frac{2\mu + 3\mu'}{2\mu + 3\mu' + \frac{\varepsilon^2}{\varkappa}} Ea \ll \frac{\varepsilon}{C}. \qquad (99.37)$$

This will always be satisfied for $Ea \ll \varphi_0$. Conversely, if inequality (99.37) is not met, the variation in charge density over the surface of the drop can no longer be neglected in the computations. In fact, the variation may then be so great that one side of the drop is positively charged and the other negatively. In such a case, motions of the type described here should arise at both of the oppositely charged ends of the drop, but these motions will be in opposite directions. The over-all velocity of the drop should then obviously decrease sharply, and the drop itself might even break up. This is what Christiansen observed.

Let us also note that the motion of emulsified drops in an electric field can be examined in the very same manner as for the similar motion of mercury drops. The appropriate computations were made by A. I. Fedosov [26].

An emulsion drop differs from a mercury drop in that the electrical conductivity of the interior phase is then comparable to that of the exterior phase. In view of this, an electric field exists inside the drop, and condition (99.9) must be satisfied for both the interior and the exterior liquids.

Computations lead to the following expression for the velocity of electrocapillary motion of emulsion drops in an electric field:

$$U = \frac{\varepsilon Ea}{2\mu + 3\mu' + \varepsilon^2\left(\frac{1}{\varkappa} + \frac{2}{\varkappa'}\right)}, \qquad (99.38)$$

where $\varkappa'$ is the conductivity of the interior phase.

For emulsions of water in oil $\varkappa' \gg \varkappa$, and equation (99.38) becomes (99.21). Here, it should be borne in mind, however, that water emulsion drops in oil will in fact not move at high velocities due, on the one hand, to the small value of $\epsilon$ for water drops and, on the other, to the low value of E in oil.

In the case of emulsions of oil in water, equation (99.38) acquires, in view of the smallness of $\varkappa'$, the form

$$U = \frac{\varepsilon E a}{2\mu + 3\mu' + \frac{2\varepsilon^2}{\varkappa'}} \approx \frac{\varkappa' E a}{2\varepsilon}, \tag{99.39}$$

so that the velocity of the electrocapillary motion of oil drops is proportional to $\varkappa'$ and inversely proportional to $\epsilon$. In particular, in view of the smallness of $\varkappa'$, this velocity is always very low.

## 100. MOTION OF NONIDEALLY POLARIZABLE DROPS IN AN ELECTRIC FIELD

Up to this point we have examined only the motion of ideally polarizable liquid particles.

Let us now pass [22] to an examination of a partially polarizable particle. Since we assume that the change of the potential jump at the metal-solution interface, upon the passage of a current, is small, it follows that for a drop to be incompletely polarizable — i.e., for ions to be formed or lose their charge on its surface — there must be, prior to the application of a current, an interchange of ions between the metal and the solution, as is the case, for example, with a drop of mercury in a solution containing $Hg^{++}$ ions, or with a drop of zinc amalgam in a solution containing $Zn^{++}$ ions.

The balance of the amount of electricity at the double layer-electrically neutral solution interface can, in the case of a nonideally polarizable particle, be written in the form

$$\varkappa \left(\frac{\partial \varphi}{\partial r}\right)_{r=a} = 2\,\frac{\varepsilon v_0}{a} \cos\theta + i, \tag{100.1}$$

where i is the current density flowing through the surface of the drop. We assume this current to be sufficiently weak, so that the departure from equilibrium in the double layer can be neglected.

The relation between i and the potential difference $\varphi$ between the metal and the solution depends on the particle's polarization mechanism.

In the simplest case of chemical polarization (either a hindered ion discharge or a particularly slow step in any of the other stages of the electrochemical reaction on the surface of the particle), this relation can be represented, for weak currents, by the linear law

$$i = k_0 \varphi. \tag{100.2}$$

where $k_0$ may be termed the conductance of the metal-solution interface, and $\frac{1}{k_0} = \omega$, its resistance.

Equation (100.1) can then be rewritten in the form

$$\varkappa\left(\frac{\partial \varphi}{\partial r}\right)_{r=a} = 2\,\frac{\varepsilon v_0}{a}\cos\theta + \frac{\varphi}{w}. \tag{100.3}$$

Now, the solution of equation (99.8) for the potential satisfying boundary conditions (99.9) and (100.3) has the form

$$\varphi = E\left(r + \frac{\left(\frac{1}{2} - \frac{\varepsilon v_0}{\varkappa E a} - \frac{a}{2w\varkappa}\right)}{1 + \frac{a}{2w\varkappa}}\,\frac{a^3}{r^2}\right)\cos\theta, \tag{100.4}$$

and therefore the quantity $\alpha$ in equation (99.15) is, for a nonideally polarizable particle,

$$\alpha = 1 + \frac{\frac{1}{2} - \frac{\varepsilon v_0}{\varkappa E a} - \frac{a}{2w\varkappa}}{1 + \frac{a}{2w\varkappa}}.$$

By repeating computations similar to those presented previously we can derive the following expression for the velocity of the drop:

$$U = \frac{\varepsilon E a}{(2\mu + 3\mu')\left(1 + \frac{a}{2w\varkappa}\right) + \frac{\varepsilon^2}{\varkappa}}. \tag{100.5}$$

As should have been expected, the resulting conductance at the metal-solution interface leads to a decrease in the velocity of the particle (as compared to an ideally polarizable object). This reduction in the velocity of a nonpolarizable particle is related to the fact that the ponderomotive forces of such a system are smaller than those in an ideally polarizable particle, with the ratio of the first to the second being $1 : \left(1 + \frac{a}{2\omega\varkappa}\right)$. This applies both to the forces that are a function of the external field and to those associated with the self-retarding effect. Equation (100.5) should, therefore, be written in the form

$$U = \frac{\varepsilon E a\left(1 + \frac{a}{2w\varkappa}\right)^{-1}}{2\mu + 3\mu' + \frac{\varepsilon^2}{\varkappa}\left(1 + \frac{a}{2w\varkappa}\right)^{-1}}. \tag{100.6}$$

Also, since according to equation (100.4) we have for a nonideally polarizable drop

$$\Delta\Phi_0 = 3Ea\left(1+\frac{a}{2\varkappa w}\right)^{-1}, \tag{100.7}$$

we can in this case replace equation (100.6) by:

$$U = \frac{1}{3}\frac{\varepsilon(\Delta\Phi_0)}{2\mu+3\mu'+\frac{\varepsilon^2}{\varkappa}\left(1+\frac{a}{2w\varkappa}\right)^{-1}}. \tag{100.8}$$

According to this result then, a decrease in $\omega$ produces a decrease in the velocity of the moving drop. The significance of the current's origin disappears if $\omega$ is sufficiently large compared to $\frac{a}{\varkappa}$. It is also unimportant if the drop is in the second regime of motion since the term $\left(1+\frac{a}{2\omega\varkappa}\right)$ will cancel out in equation (100.6). Indeed, if the external field is fully equalized by the convective current caused by motion of the double layer, the fact that a current can flow through the drop will have no effect on the velocity of the motion. As has been pointed out above, however, the derivation of equation (100.8) assumes that $\omega$ is not a function of $\theta$. This condition is not fulfilled when the agitation of the solution is determined by the motion of the drop.

Let us consider qualitatively the phenomena which should be observable in this situation, taking as an example a positively charged drop of mercury in a mercury-salt solution. As the current flows through the drop (see Figure 75), mercury ions form on its right side, and the ion concentration in the solution increases. On the left side of its surface, however, the ions are discharged, and the solution becomes depleted. However, the changes in concentration which occur under these conditions differ in magnitude. Because of capillary motion, a stream of fresh solution containing the free stream concentration flows against the right side of the drop, whereas the left side is in contact with that part of the solution which has been enriched during its flow past the right side. Any excess ion concentration is deposited from the solution onto the mercury, so that the stream flowing from the drop in the direction of negative values of x has an average ion concentration equal to that of the bulk. Since, during the motion of the stream, excess mercury ions have time to diffuse to some extent from the surface of the drop into the bulk of the solution, a layer that is somewhat depleted compared to the free stream will exist immediately adjacent to the left side of the drop, while farther from the drop the solution will retain a higher concentration for a while. At a sufficiently great distance from the drop, all these variations in composition should disappear, and the concentrations at all points should return to their initial value. With such a distribution in composition, it is evident that deviations from the initial concentration, and, consequently, the change in potential as well, are greater near the right side of the drop than near the left side. In other words, as already pointed out above, the quantity $\omega$ is a function of the angle $\theta$ and decreases as $\theta$ varies from 0 to $\pi$. The flow of current, under these conditions, should likewise cause a certain shift in the average value of the potential jump in the double layer, tending to increase the positive charge of the drop.

On the other hand, if the drop is negatively charged, the picture is reversed. The stream of liquid moves from left to right, with respect to the drop, and the deposition of ions is from a solution containing the initial concentration of metal ions. Thus depleted, the solution arrives at the right side of the surface, where its concentration increases to its initial value. The shift in potential, under these conditions, is larger in absolute value on the left side than on the right, so that the average value of the potential of the drop is shifted in a direction of an increased negative charge of the metal surface.

It is clear that, in the presence of such a complicated dependence of $\omega$ on the angle $\vartheta$, it is not possible to compute accurately the effect of incomplete polarization on the velocity. An approximate estimate can, however, be given, if equation (100.8) is retained with a certain average value of $\omega$, which may be obtained from the expression for the limiting diffusion current $i_{lim}$, and from the relation $\omega = \frac{RT}{zFi_{lim}}$. We found, in Section 72 and Section 14, that the average density of the limiting diffusion current is determined from the expressions:

$$i_{lim} = zFj = \frac{zF\overline{I}}{4\pi a^2} = 0.6\left(\frac{v_0 D}{a}\right)^{1/2} zFc_m \qquad (100.9)$$

and for a solid particle

$$i_{lim} = zFj = \frac{zF\overline{I}}{4\pi a^2} = 0.6\left(\frac{UD^2}{a^2}\right)^{1/3} zFc_m. \qquad (100.10)$$

Here, I is the total diffusional flow to the drop; z and $c_m$ are respectively the metal ion valence and the initial concentration of the ions of the drop metal (in gram equivalents per $cm^3$).

As an example, let us determine the magnitude of the coefficient $k = 1 + \frac{a}{2\omega\varkappa}$ with $a = 10^{-2}$ in the case of the deviation in an electric field of a mercury drop, falling by gravity in a solution containing a neutral electrolyte of concentration c and mercury ions of concentration $c_m$. It is assumed in this example that the hydrodynamic conditions of drop motion are preserved $\left(\frac{\epsilon^2}{\varkappa(2\mu + 3\mu')} \ll 1\right)$. Since $U \approx 30$ cm/sec,

$$D \approx 0.8 \cdot 10^{-5}; \quad \varkappa \approx 1.2 \cdot 10^2 c; \qquad z = 2,$$

$$\frac{a}{2w\varkappa} = \frac{0.6\, a^{1/2} U^{1/2} D^{1/2} F c_m}{\varkappa \frac{RT}{F} \cdot \frac{1}{z}} \approx 10^2 \frac{c_m}{c}. \qquad (100.11)$$

Thus, the quantity $k = 1 + \frac{a}{2\omega\varkappa}$, which may be termed the depolarization coefficient, is a function of the ratio of the concentration of the metal ions in the drop to the over-all concentration of the solution. If $\frac{c_m}{c} \approx 10^{-2}$, this coefficient is noticeably different from unity, and for $c_m = c$ (i.e., for a mercury drop in a solution containing only a mercury salt) the deviation from the vertical of a falling drop in an electric field is about 100 times less than in the case of an ideally polarizable drop. Of major importance is the fact that, contrary to some published statements, the incomplete polarizability of a drop always leads to a decrease in its mobility as compared to that of an ideally polarizable drop.

## 101. COMPARISON OF THEORY AND EXPERIMENT

An experimental verification of the theory was conducted [24b] by means of an instrument (Figure 76) which was used to determine the deviation from the vertical of falling drops of mercury in an electric field. The mercury, under the pressure of a column of the metal, flowed from a capillary tube in drops 0.036 cm in radius. They fell through a solution of KBr in glycerine $C_3H_5(OH)_3$. The drops were spaced in time about 0.9 to 1 sec apart. The viscosity of the solutions which were employed was about 2.6 to 3.3 poise, at 21 to 22°. The velocity of fall of the drop due to gravity was about 1.2 to 1.3 cm/sec, and the Reynolds number was, therefore, of the order of 0.04. The falling drops formed a column of mercury droplets along the axis of the instrument, and were separated from each other by a distance of approximately 1.3 cm.

As the drops grew on the end of the capillary they formed a circuit with an auxiliary electrode (Figure 76) and were charged by an applied emf that directed a charging current over the growing drop. The charge density on the drop was calculated by measuring this current and dividing it over the area of the drop.

The intensity of the electric field was varied from 0.25 to 1.8 v/cm, so that the potential difference in the solution did not exceed 0.1 v.

The mobility of the mercury drops in the electric field was determined by their deviation from the vertical path, which is the one followed by the drops when the electric field is absent.

The dependence of mobility on the charge per unit area was determined by varying the value of the charging current on the drop.

The points in Figure 77 illustrate the experimentally determined mobility of mercury drops in an electric field as a function of $\epsilon$, while the solid curves represent mobilities as computed by means of equation (99.30). Curves a refer to the most concentrated solution, and curves b to the most dilute.

The general conclusion may be reached on the basis of the curves in Figure 77 that the theory presented above is in good agreement with the experimental results. Thus, in accordance with the theory, the experimentally observed mobility of mercury drops first increased with an increase in the absolute charge on the drop, and then began to decrease. In the transition from concentrated to dilute solutions, the maximum value of the mobility of the mercury drops decreased on both the positive and the negative branches of the curve. The maximum, which is entirely absent in curves for concentrated solutions, is shown especially well in this case.

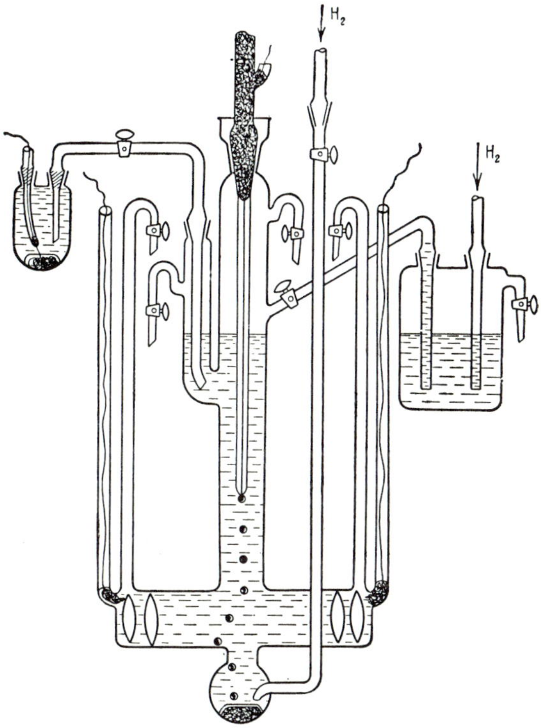


Figure 76. Instrument for determining the mobility of mercury drops in a horizontal electric field.

For the solutions which were tested, however, the values for the observed mobility, as compared to the theoretical predictions, were shifted in the direction of negative charges. Thus, the mobility of the mercury drops passed through zero, not at a charge equal to zero as expected theoretically, but at a slightly negative charge equal to approximately 3 to $6 \cdot 10^{-6}$ coul/cm$^2$. The mobilities observed in dilute solutions were somewhat less than those computed, especially for $\epsilon < 0$. This is probably related to a certain drop in the potential of the applied field across the droplet, which could have led to a decrease in the drop's velocity in the region of a rapidly charging $U(\epsilon)$ function (steep curve). Despite these small differences between the computed and the observed mobilities, it is justified, considering the complexity of the phenomenon and the absence of any arbitrary constants in the theoretical formulas, to assert that a satisfactory quantitative agreement between theory and experiment has been established.

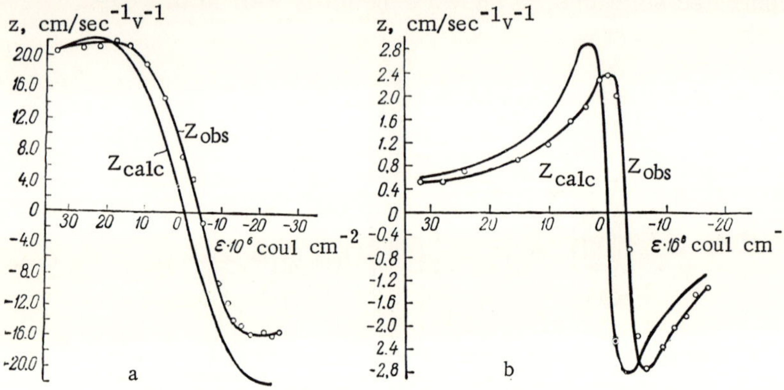


Figure 77. Mobility of mercury drops in an electric field: a — 1.5 N KBr, $\varkappa = 7 \cdot 10^{-3}$; b — 0.02 N KBr, $\varkappa = 1.9 \cdot 10^{-5}$.

The observed deviations apparently are related to variations in the charge during the fall of the drop, which could have been caused by the action of remnants of dissolved oxygen that could not be entirely eliminated from the solution.

## 102. FALL OF MERCURY AND EMULSIFIED DROPS IN A GRAVITATIONAL FIELD

The presence of a charge on the surface of a mercury drop is of very great significance in determining the motion of the drop not only in an electric, but also in a gravitational field [22]. Let us then examine the fall of a mercury drop in an electrolytic solution. The resulting motion of the solution on the surface of the drop causes an entrainment of the ions in the outer sheet of the double layer, which are thus carried to the rear of the drop. A convective current of ions at its rear generates an electric field in the solution near the drop, even though no external electric field is present.

The electric field that thus appears in the solution tends to equalize the ion concentration on the drop surface by moving them in a direction opposite to that of the fluid convection. However, in addition to the equalization of potential brought about in this manner, the potential difference is also equalized by electric currents in the bulk of the solution, which are passing through the solution from the back to the front of the drop.

To obtain the velocity of fall of a drop of mercury in an electrolytic solution, we must first determine the potential distribution in the solution and the magnitude of the retarding forces that are generated. Since there are no charges in the bulk of the solution, the potential distribution in this region is found from the Laplace equation

$$\Delta\varphi = 0, \tag{102.1}$$

together with condition (99.9), i.e.,

$$\varkappa \frac{\partial\varphi}{\partial r} = \operatorname{div}_s (\varepsilon \mathbf{v}_t), \tag{102.2}$$

which evidently serves as the boundary condition on the surface of the drop. Also, since there is no external field at infinity, we must require that far from the drop

$$\varphi \to 0. \tag{102.3}$$

For $v_t$ different from zero, it is obvious that the potential distribution around the drop satisfying equation (102.2) cannot be spherically symmetric. Along the drop, therefore, a certain change in potential takes place which in turn causes a variation of surface tension and the appearance of the corresponding surface forces. The system of hydrodynamic equations which determines the velocity distribution in the solution and in the drop must be solved by taking these forces into account. Let us restrict our further analysis to the case of small Reynolds numbers, for which equations (70.28) to (70.33) can be written for the velocity distribution inside and outside the drop.

Equations (99.11) to (99.14) now serve as boundary conditions on the surface of the drop. Here, as was done in Section 70, an effective pressure $\pi$ must be introduced into equation (99.13) for the normal component in order to account for the volumetric effect of gravity on the drop. Also, the surface forces $F_t$ and $F_n$ are given, as previously, by equations (99.6) and (99.7), while the velocity at infinity must satisfy equation (99.10).

If we designate the velocity of the liquid at the mercury-solution interface by $v_0$, we can set:

$$(v_\theta)_{r=a} = v_0 \sin\theta, \tag{102.4}$$

which when introduced into boundary condition (102.2) yields:

$$\left(\frac{\partial\varphi}{\partial r}\right)_{r=a} = \frac{2}{\varkappa}\frac{v_0\varepsilon}{a}\cos\theta. \tag{102.5}$$

The solution for the potential satisfying equation (102.1) with boundary conditions (102.5) and (102.3) is:

$$\varphi = -\frac{\varepsilon v_0 a^2}{\varkappa r^2}\cos\theta. \qquad (102.6)$$

Thus, it follows from equations (99.6), (99.7), and (102.6) that the forces exerted on the drop's surface are:

$$F_n = \frac{2\sigma_{\pi/2}}{a} - \frac{2\varepsilon\,\Delta\varphi}{a} = \frac{2\sigma_{\pi/2}}{a} + \frac{2\varepsilon^2 v_0}{\varkappa a}\cos\theta, \qquad (102.7)$$

$$F_t = \mathrm{grad}_\theta\,\sigma = -\varepsilon\,\mathrm{grad}_\theta\,\Delta\varphi = -\frac{\varepsilon}{a}\frac{\partial\,\Delta\varphi}{\partial\theta} = \frac{\varepsilon^2 v_0}{\varkappa a}\sin\theta. \qquad (102.8)$$

Introducing expressions (102.7) and (102.8) into equations (99.13) and (99.14), we have, at r = a:

$$-p + 2\mu\frac{\partial v_r}{\partial r} + \pi = -p' + 2\mu'\frac{\partial v_r'}{\partial r} - \frac{2\sigma_{\pi/2}}{a} + \frac{2\varepsilon^2 v_0}{\varkappa a}\cos\theta, \qquad (102.9)$$

$$\mu\left(\frac{1}{r}\frac{\partial v_r}{\partial\theta} + \frac{\partial v_\theta}{\partial r} - \frac{v_\theta}{r}\right) = \mu'\left(\frac{1}{r}\frac{\partial v_r'}{\partial\theta} + \frac{\partial v_\theta'}{\partial r} - \frac{v_\theta'}{r}\right) + \frac{\varepsilon^2 v_0}{\varkappa a}\sin\theta. \qquad (102.10)$$

Equations (102.9) and (102.10) reduce to:

$$-\mu\left(\frac{3b_2}{a^2} + \frac{6b_1}{a^4}\right) = -6\mu' a_1 a - (\rho - \rho')ga + \frac{2\varepsilon^2 v_0}{\varkappa a}, \qquad (102.11)$$

$$-\frac{3\mu b_1}{a^4} = -3\mu' a_1 a + \frac{\varepsilon^2 v_0}{\varkappa a}, \qquad (102.12)$$

by substitution of the appropriate expression for the velocities and the pressure. The boundary conditions (99.11) and (99.12), together with the conditions at infinity, yield

$$\frac{b_2}{a} + \frac{b_1}{a^3} - U = 0, \qquad (102.13)$$

$$\frac{b_1}{2a^3} + U - \frac{b_2}{2a} = v_0, \qquad (102.14)$$

$$a_1 a^2 + a_2 = 0, \qquad (102.15)$$

$$-2a^2 a_1 - a_2 = v_0. \qquad (102.16)$$

Following elementary transformations, we finally obtain $v_0$ and U:

$$v_0 = \frac{(\rho - \rho')ga^2}{3\left(2\mu + 3\mu' + \frac{\varepsilon^2}{\varkappa}\right)}, \qquad (102.17)$$

$$U = \frac{2(\rho - \rho')ga^2}{3\mu}\,\frac{\mu + \mu' + \frac{\varepsilon^2}{3\varkappa}}{2\mu + 3\mu' + \frac{\varepsilon^2}{\varkappa}}. \qquad (102.18)$$

For the sake of brevity, let us write down the velocity distributions inside and outside the drop without giving an explicit solution for the other constants:

$$v_r = \frac{1}{\mu}\left\{2\mu\left(\frac{a}{r} - 1\right) + \left(\frac{a}{r} - \frac{a^3}{3r^3} - \frac{2}{3}\right)\left(3\mu' + \frac{\varepsilon^2}{\varkappa}\right)\right\} v_0 \cos\theta, \qquad (102.19)$$

$$v_\theta = \frac{1}{\mu}\left\{\mu\left(2 - \frac{a}{r}\right) - \left(\frac{a^3}{2r^3} + \frac{3}{2}\frac{a}{r} - 2\right)\left(\mu' + \frac{\varepsilon^2}{3\varkappa}\right)\right\} v_0 \sin\theta, \qquad (102.20)$$

$$v'_r = \left(1 - \frac{r^2}{a^2}\right) v_0 \cos\theta, \qquad (102.21)$$

$$v'_\theta = \left(2\frac{r^2}{a^2} - 1\right) v_0 \sin\theta. \qquad (102.22)$$

We may now present the final expressions for the distribution of the potential and for the density of the electric current on the surface of the drop

$$\varphi = \frac{\varepsilon(\rho - \rho')ga}{3\left(2\mu + 3\mu' + \frac{\varepsilon^2}{\varkappa}\right)\varkappa} \cdot \frac{a^3}{r^2}\cos\theta, \qquad (102.23)$$

$$i = -\varkappa\left(\frac{\partial\varphi}{\partial r}\right)_{r=a} = \frac{2}{3} \cdot \frac{(\rho - \rho')ga\varepsilon}{\left(2\mu + 3\mu' + \frac{\varepsilon^2}{\varkappa}\right)}\cos\theta. \qquad (102.24)$$

Equation (102.18) shows that, as in the case of an electric field, two limiting cases of motion in a gravitational field exist.

1) $2\mu + 3\mu' \gg \frac{\epsilon^2}{\varkappa}$. Here the velocity U for the fall of the drop is equal to that of an uncharged drop, as given by the Hadamard-Rybczyński formula (70.35). This result may be given the following physical interpretation. The potential differences is evened up through the solution so rapidly that the convective current of ions along the surface is compensated by currents through the adjacent layers of the solution without any noticeable accumulation of charges at the rear end of the drop. Thus, in this instance, the convective transfer of ions along the surface is unhindered, and the electric field that is generated does not retard the motion of the liquid along the surface. The velocity of fall of the drop is, therefore, determined by the Hadamard-Rybczyński formula in the same way as in the absence of an electrolyte.

2) $2\mu + 3\mu' \ll \frac{\epsilon^2}{\varkappa}$. Here the velocity of the drop's fall coincides with that of a solid sphere, as given by the Stokes formula

$$U_{\text{St}} = \frac{2}{9}\frac{(\rho - \rho')ga^2}{\mu}. \qquad (102.25)$$

This is caused by the fact that no time is available for the equalization of the potential, so that the ions accumulate continuously at the rear end of the drop until the generated electric field fully retards the convective transfer of ions along the surface. In other words, the electric field generated in this instance arrests the motion of mercury along the surface of the drop, and causes it to fall like a solid sphere. The velocity of fall of such a "solidified" drop is fully described by Stokes' law [25].

Let us next examine the ratio of the velocity of the drop's fall to that of a solid sphere, i.e., the ratio of equation (102.18) to (102.25),

$$\frac{U}{U_{St}} = \frac{3\mu + 3\mu' + \frac{\varepsilon^2}{\varkappa}}{2\mu + 3\mu' + \frac{\varepsilon^2}{\varkappa}}. \tag{102.26}$$

It is clear from (102.26) that this ratio attains its highest value when the viscosity $\mu$ of the medium is considerably greater than the viscosity $\mu'$ of mercury. Under such conditions

$$\frac{U}{U_{St}} = \frac{3\mu + \frac{\varepsilon^2}{\varkappa}}{2\mu + \frac{\varepsilon^2}{\varkappa}}. \tag{102.27}$$

The dependence of the ratio $\frac{U}{U_{St}}$ on the charge $\epsilon$ is shown by the solid line in Figures 78 and 79. At the point $\epsilon = 0$, the ratio $\frac{U}{U_{St}}$ attains its maximum value of 1.5.

Let us now clarify the region of validity of equation (102.18). To assure its applicability, the following conditions must be fulfilled:

1) the Reynolds number must be extremely small compared to unity;

2) surface conductivity must be low compared to that of the bulk of the solution;

3) the maximum potential variation along the drop must be small compared to the initial jump in potential in the double layer;

4) the maximum variation in the surface tension must not exceed a certain limit $\Delta\sigma_{max}$;

5) surface tension must depend only in the charge per cm$^2$ of the drop's surface.

The value of $\Delta\sigma_{max}$ is determined by the condition that the charge accumulation at the rear end of the drop must, at some moment, generate an electrolysis between the two ends of the drop, which could, for example, be related to the passage of mercury ions from the double layer into the solution (if the drop's charge is positive). The meaning of the first condition is obvious. The second has been discussed in Section 99, so let us examine conditions 3) and 4).

The requirement that the potential difference along the drop be small compared to that in the double layer can be represented by the expression

$$(\Delta\Phi)_{\max} = \frac{2}{3} \frac{\varepsilon(\rho - \rho')ga^2}{\varkappa\left(2\mu + 3\mu' + \frac{\varepsilon^2}{\varkappa}\right)} < \frac{\varepsilon}{C}, \tag{102.28}$$

where C is the capacitance of the double layer. For the case of a viscous solution, and for $\frac{\epsilon^2}{\varkappa} \ll 2\mu + 3\mu'$, equation (102.28) can be written in the form

$$\frac{3U_{\text{St}}}{2\varkappa} < \frac{1}{C},$$

or

$$\varkappa > \frac{3}{2} CU_{\text{St}}. \tag{102.29}$$

Thus, equation (102.28) is not applicable to very dilute solutions.

The fourth condition places a limitation on the size of the falling drops. Indeed, since

$$\Delta\sigma = \varepsilon(\Delta\Phi)_{\max} = \frac{2}{3}(\rho - \rho') \cdot \frac{ga^2\varepsilon^2}{\varkappa} \frac{1}{2\mu + 3\mu' + \frac{\varepsilon^2}{\varkappa}} < \Delta\sigma_{\max},$$

it follows that, for high charges,

$$\frac{2}{3}(\rho - \rho')ga^2 < (\Delta\sigma_{\max}),$$

or

$$a < \sqrt{\frac{3}{2}} \frac{(\Delta\sigma_{\max})^{1/2}}{\sqrt{(\rho - \rho')g}} = a_{\text{cr}}. \tag{102.30}$$

The further accumulation of charges at the rear of the drop and an increase in their retarding effect on it are impossible for values of a exceeding $a_{cr}$. This is so because, for large drops, the retarding forces can no longer compete with the liquid's inertial forces, and drops greater than $a_{cr}$ must fall in a "liquid" regime. It may normally be assumed that $\Delta\sigma_{max}$ is 20 to 30% of the total surface tension. It should also be remarked that equation (102.30) coincides with (78.4), which was derived for the case of the retarding action of surface-active materials.

The last limitation, the fifth, implies that surface-active materials are absent from the solution.

An experimental verification of equation (102.18) was made by A. N. Frumkin and I. A. Bagotskaya [24a]. The fall of the drops was studied, under conditions most favorable for a verification of the theory, in solutions of KBr and $Na_2SO_4$ in glycerine with a viscosity of 5 to 7 poises (which is considerably in excess of that of mercury, which is 0.01 poise). The fall of drops 0.056 cm in diameter took place, with intervals of 3 to 5 cm between drops. The solution was carefully cleansed of oxygen, although traces of oxygen remained and amounted to a 2 to $3 \cdot 10^{-6}$ molar solution. By diffusing to the surface of the drop, the oxygen caused a gradual reduction in its charge during the fall. The authors introduced a correction for this effect by computing the number of oxygen molecules arriving at the surface according to equation (72.10). The ratio $\frac{U}{U_{St}}$ which was computed in this manner is presented by the broken curves in Figures 78 and 79, while the circles in the same figures indicate the measured values for this ratio.

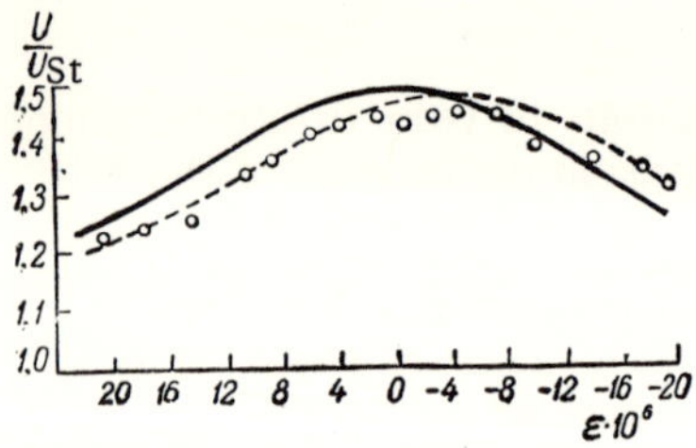


Figure 78. Velocity of fall of a mercury drop in 1 N KBr in glycerine, as a function of the surface charge density, $\varkappa = 5 \cdot 10^{-4}$, $\mu = 4.8$, a = 0.434 mm. Solid line represents $U_{calc}$ according to (102.26); broken curve shows $U_{calc}$ according to (102.26) taking into account (72.10) and assuming $\epsilon - \epsilon_0 = 8 \cdot 10^{-6}$; ooo gives $U_{obs}$.

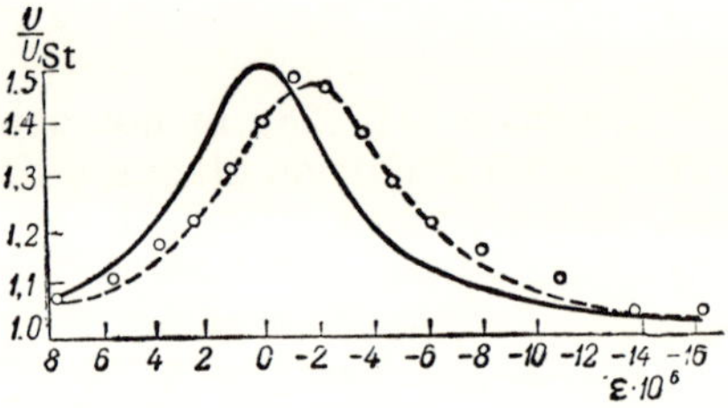


Figure 79. Velocity of fall of a mercury drop in saturated $Na_2SO_4$ in glycerine, $\varkappa = 8.1 \cdot 10^{-6}$, 7.2, a = 0.442 mm. Solid line represents $U_{calc}$ according to (102.26); dotted line shows $U_{calc}$ according to (102.26); taking account of (72.10) and assuming $\epsilon - \epsilon_0 = 3.4 \cdot 10^{-6}$; ooo gives $U_{obs}$.

The agreement of the theoretical formulas with the experimental data is essentially complete.

I. A. Bagotskaya pointed out that the introduction of mercury ions into the solution brings about a considerable reduction of the retarding action. This is in qualitative agreement with the conclusion in Section 100 for an incompletely polarizable drop.

Equation (102.18) for the velocity of fall may now easily be generalized for the case of falling drops of an emulsion. Here, however, a phenomenon should be noted which has no analogue in the case of mercury drops falling in a conducting medium. Namely, where the motion takes place in a medium of low conductivity, the effects noted in Section 100 may appear.

The transfer of ions in a dilute binary solution may lead to their unequal distribution on the drop. This distribution will not have time to be equalized by conduction (migration) and diffusion, and as a result, an electrical convective-diffusion field will appear. The total current in the drop will, in this case, be zero, and in order to estimate the field thus generated, equation (51.24) may be utilized. This formula may be conveniently rewritten with the help of formula (75.4′), so that

$$E \sim \frac{D_2 - D_1}{(z_1 D_1 + z_2 D_2)} \frac{RT}{F} \frac{v_0 \Gamma_0}{D_{\text{eff}}} \frac{\delta}{a c_0} \sim \frac{D_2 - D_1}{D_1 D_2} \cdot \frac{RT v_0 \Gamma_0}{Fa} \frac{\delta}{c_0}, \qquad (102.31)$$

where only the ion concentration in the bulk of the solution and the surface concentration appear.*

## 103. POTENTIALS OF FALLING DROPS

We saw in the preceding section that a potential difference $\Delta\Phi$, given by equation (102.28), appears across the ends of a mercury drop falling in an electrolytic solution. If a series of drops fall in succession in the solution, a certain potential difference will be established along the column of drops. This is termed the falling-drop potential or sedimentation potential [25].

In order to compute this potential difference, let us assume that the number of drops per unit volume of the solution is sufficiently small, so that the distance between the drops is large compared to their own dimensions. We can then consider that the drops fall independently of one another and that the electric fields of the drops are additive. The sedimentation potential can in this instance be calculated exactly.

Let us first examine in more detail the case of a single drop falling through a column of liquid, and let us find the average value $\overline{\varphi}_u$ of the potential in a plane S, which is situated above the drop at a small distance compared with the radius of the column, but in such a manner that it does not intersect the drop. The cross-sectional dimensions of the liquid column in this case are large compared to the dimensions of the drop. Let us place the origin of the coordinates in the center of the drop. The area of that element of the plane situated between the angles $\theta$ and $\theta + d\theta$ is $\frac{2\pi r^2 \sin\theta}{\cos\theta} d\theta$. With the help of equation (102.23) $\overline{\varphi}_u$ may therefore be expressed as

---

*Proofreading note. This conclusion was arrived at independently by S. S. Dukhin and B. V. Deryagin (see p. 286).

$$\overline{\varphi}_u = \frac{1}{S} \int \frac{2\pi r^2 \sin\theta}{\cos\theta} \varphi \, d\theta =$$

$$= \frac{2\pi}{S} \frac{\varepsilon(\rho - \rho') g a^4}{3\varkappa\left(2\mu + 3\mu' + \frac{\varepsilon^2}{\varkappa}\right)} \int\limits_{\pi/2}^{\pi} \sin\theta \, d\theta = -\frac{2\pi}{S} \frac{\varepsilon(\rho - \rho') g a^4}{3\varkappa\left(2\mu + 3\mu' + \frac{\varepsilon^2}{\varkappa}\right)}.$$

In the same manner, we obtain for the portion of the plane which is situated below the drop:

$$\overline{\varphi}_d = \frac{2\pi}{S} \frac{\varepsilon(\rho - \rho') g a^4}{3\varkappa\left(2\mu + 3\mu' + \frac{\varepsilon^2}{\varkappa}\right)},$$

so that the sedimentation potential caused by the drop falling in the liquid column is

$$\overline{\varphi}_d - \overline{\varphi}_u = \frac{4\pi}{S} \frac{\varepsilon(\rho - \rho') a^4 g}{3\varkappa\left(2\mu + 3\mu' + \frac{\varepsilon^2}{\varkappa}\right)}. \tag{103.1}$$

Let us also designate by n the average number of drops per unit volume. The sedimentation potential E in a liquid column 1 cm long in which there is a shower of falling drops is, obviously,

$$E = nS(\overline{\varphi}_d - \overline{\varphi}_u) = \frac{4\pi n\varepsilon(\rho - \rho') g a^4}{3\varkappa\left(2\mu + 3\mu' + \frac{\varepsilon^2}{\varkappa}\right)} = \frac{2\pi a^2 \mu\varepsilon n U}{\varkappa\left(\mu + \mu' + \frac{\varepsilon^2}{3\varkappa}\right)}, \tag{103.2}$$

where S is the cross section of the tube in which the drops are falling. The magnitude of the potential E of the falling drops is proportional to $\dfrac{\epsilon}{2\mu + 3\mu' + \dfrac{\epsilon^2}{\varkappa}}$, to the number n of drops per $cm^3$ of solution, and to the radius of the drops. As with the mobility of a drop in an external field according to (99.30), this quantity E passes through a maximum at $\epsilon_{max} = \sqrt{\varkappa(2\mu + 3\mu')}$.

When we complete the circuit formed by the liquid column of length L and cross section S with an external resistance W, the potential difference in the external portion of the circuit balances, so that a current I flows through the entire liquid column and the external resistance. This current is superimposed on the system of local currents at the falling drops. Since the emf in this circuit is EL, we have

$$I = \frac{EL}{\frac{L}{S\varkappa} + W} = \frac{I_0 \frac{L}{S\varkappa}}{\frac{L}{S\varkappa} + W}, \tag{103.3}$$

where $I_0$ is the current which flows when W = 0, i.e., when the column is short-circuited. According to equations (103.2) and (102.18)

$$I_0 = ES\varkappa = \frac{4\pi nS\varepsilon\,(\rho - \rho')\,ga^4}{3\left(2\mu + 3\mu' + \frac{\varepsilon^2}{\varkappa}\right)} = \frac{2\pi a^2 nSU\varepsilon\mu}{\mu + \mu' + \frac{\varepsilon^2}{3\varkappa}}. \tag{103.4}$$

In the case of positively charged drops, the current in the external circuit flows upward, while in the column itself it flows from the top toward the bottom, i.e., in a direction opposite to that of the local currents near each drop. Indeed, short-circuiting the ends of the liquid column is equivalent to applying across it an external potential difference equal in magnitude and opposite in sign to the total sedimentation potential EL. In this case, the potential determined by equation (102.23) decreases. Let us find the magnitude of this correction for a case which is especially important, namely, when the drop is in a "liquid" motion regime where $\frac{\epsilon^2}{\varkappa} \ll 2\mu + 3\mu'$. Then, according to equation (99.26), the potential in the liquid column near the drop, caused by the flow of current $I_0$, is

$$-E\left\{r + \frac{1}{2}\frac{a^3}{r^2}\right\}\cos\theta,$$

which leads to the following expression for the potential distribution and electric current density on the surface of the drop, in place of (102.23):

$$\varphi = \frac{\varepsilon\,(\rho - \rho')\,ga\cos\theta}{3\varkappa\,(2\mu + 3\mu')}\left\{\frac{a^3}{r^2} - 4\pi na^3\left(r + \frac{1}{2}\frac{a^3}{r^2}\right)\right\}. \tag{103.5}$$

At the surface of the drop, i.e., at r = a, the ratio of the second term in equation (103.5) to the first is small and equals $6\pi na^3$ when the distance between the drops is large compared to their radius. In other words, the completion of the external circuit has no significant effect on the potential distribution in the immediate vicinity of each drop, thus justifying our use of (103.2) in equation (103.4).

For the case of an open external circuit, the total current through each horizontal cross section of the liquid column is equal to zero, which means that the amounts of electricity carried by the convective current caused by the motion of the outer sheets of the double layer and by the conduction current are equal and opposite in sign. One might be tempted to conclude that the value of the conduction current is equal to the "current of the falling drops" which flows in the external circuit and in the liquid column when the latter is short-circuited. As may be easily proved by the following argument, however, this is not in accordance with the facts. Let us draw a horizontal plane through the drop, and let $\theta_a$ be the angle between the vertical and the radius vector a, drawn from the center of the drop to points where the surface of the drop intersects this plane. The current emerging from the lower portion of the drop and entering the upper portion, and passing, therefore, through the indicated cross section of the solution, is

$$I' = -\varkappa \int_0^{\theta_a} 2\pi a^2 \left(\frac{\partial \varphi}{\partial r}\right) \sin\theta \, d\theta = \frac{2\pi\varepsilon (\rho - \rho') g a^3 \sin^2\theta_a}{3\left(2\mu + 3\mu' + \frac{\varepsilon^2}{\varkappa}\right)}. \tag{103.6}$$

The number of drops in the tube for $\theta_a$ between $\theta$ and $\theta + d\theta$ is $nSa \sin\theta \, d\theta$, so that the total current is

$$I'' = \int_0^{\pi} I' nSa \sin\theta \, d\theta = \frac{8}{9} \cdot \frac{\pi\varepsilon (\rho - \rho') g a^4 nS}{2\mu + 3\mu' + \frac{\varepsilon^2}{\varkappa}} = \frac{2}{3} I_0. \tag{103.7}$$

The reason for the difference between $I''$ and $I_0$ becomes apparent if it is recalled that these two currents, the total current flowing through the liquid when the column is short-circuited and the current in the open circuit, are related to the corresponding potential differences at the ends of the column. In terms of absolute values, these differences are equal. However, the solution acts like a uniform conductor for the current $I_0$, while the current lines of $I''$, which is the sum of the local currents of individual drops, curve around the drop surfaces and deviate from the shortest route.

It follows from equation (103.4) for the limiting "liquid" flow regime that:

$$I_0 = 2\pi n \varepsilon a^2 S \mu \frac{U}{\mu + \mu'}. \tag{103.8}$$

Since the quantity $4\pi\epsilon a^2 nSU$ expresses the total charge on the inner sheet of the double layer around the drops passing through a cross section of the tube in unit time, the numerical value of the current generated by the falling drops differs, in this instance, from that which would result from the fall of spheres with a free charge equal to the total charge on the inner sheet of the double layer, only by the coefficient

$$\frac{\mu}{2} \cdot \frac{1}{\mu + \mu'} \approx 0.2.^*$$

*At first glance, such a result appears paradoxical. It is simple to understand, however, that in the first limiting case the convective current transfers a quantity of electricity of the order of $2\pi a \epsilon U$ from the lower to the upper part of the drop, i.e., through a distance a. The product of these two quantities, $2\pi a^2 \epsilon U$, is the same order of magnitude as the product of the velocity of fall and the total free charge, e.g., $4\pi a^2 \epsilon U$ in the case of a charged sphere.

This reasoning loses validity in a "dense" shower of mercury drops, when the distance between the drops is comparable to their radius, i.e., $\sqrt[3]{n} \approx a$. In this case, which, incidentally, is only of theoretical interest because of the difficulty of achieving it in actual practice, the electric fields near each drop can no longer be viewed independently from one another, and the local deviations from equipotentiality are averaged out. This is akin to an increase in the conductivity of the medium surrounding the drop, as a result of which the retarding action of the charge on the motion of the drop's surface is decreased. In addition, the completion of the external circuit has, under such conditions, a significant effect on the potential distribution near each drop. For the limiting case of a very "dense" shower, the retarding action of the charge should disappear on short-circuiting the external circuit. This would occur if the liquid were pressed through a solid diaphragm. The closing of the external circuit in a "dense" shower would also have an effect on the velocity of fall of the drops.

Equation (103.4) can be applied only if all the conditions which were formulated in the preceding section are strictly fulfilled.

All the formulas presented here relate to the case of an ideally polarizable drop. In the presence of mercury ions, the variation of the potential jump across the double layer, due to convective currents, is partially equalized by a discharge on the surface of the mercury. This leads to a reduction in the sedimentation potential. At the same time, the retarding effect on the motion is also reduced. However, just as for the similar case examined in Section 100, a rigorous computation is impossible because of the complex nature of the relation between the magnitude of the limiting diffusion current and the angle $\theta$. If the limiting diffusion current were not a function of $\theta$, then, for incompletely polarizable drops, equation (103.2) would have to be replaced by:

$$E = \frac{4\pi}{3} \cdot \frac{n\varepsilon(\rho - \rho')\, g a^4 k^{-1}}{\varkappa(2\mu + 3\mu') + \varepsilon^2 k^{-1}} = \frac{4\pi n\varepsilon(\rho - \rho')\, g a^4}{3[\varkappa k(2\mu + 3\mu') + \varepsilon^2]}, \tag{103.9}$$

where

$$k = 1 + \frac{a}{2\varkappa w}.$$

In any event, this expression allows one to estimate approximately the reduction in the sedimentation potential which is brought about by the incomplete polarizability of the drops.

The fall of mercury drops in a liquid containing mercury ions should cause a change in the mercury ion concentration in addition to a separation of the electric charges.

For positively charged drops, the concentration will increase in the upper portion of the column and decrease in the lower portion. This should result in the appearance of a definite potential difference between mercury electrodes of the first kind in contact with the solution, the sign of which is opposite to that of the sedimentation potential. In contrast to the latter, this potential difference does not vanish when the fall of drops through the solution ceases, but remains until such time as the concentration difference in the column is equalized by diffusion. If the mercury enters the upper part of the column in the form of drops and these drops combine into a single mass at the lower end, concentration differences that are well known in the theory of the dropping electrode also appear. In this case, the mercury ion concentration decreases in the upper part of the column and increases in the lower part; i.e., the sign of the change is opposite to that caused by the process of fall. It is possible that this may explain some of the phenomena observed by Billitzer in his experiments with dropping electrodes.

An experimental verification of equation (103.2) was made by A. N. Frumkin and I. A. Bagotskaya [23]. To assure the condition Re ≪ 1, the mercury drops were allowed to fall in a solution of KBr

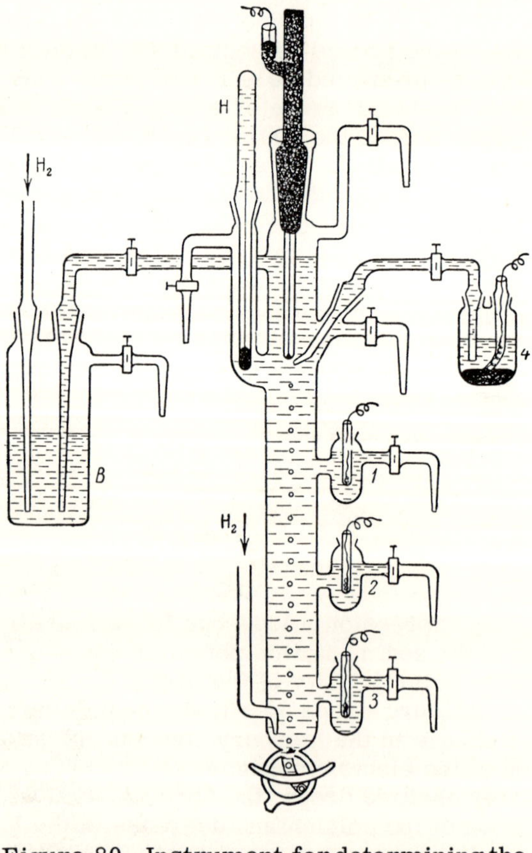


Figure 80. Instrument for determining the potential of falling drops.

in glycerine (with a viscosity of about 10 poises). The instrument for measuring the potential of the falling drops is shown in Figure 80. These potentials were measured by means of electrodes 1, 2, and 3, situated at distances of 4 and 5.7 cm from each other. The drops were charged by means of the current from an external emf through the auxiliary electrode 4. Drops 0.06 cm in radius fell at a velocity of 1 cm/sec. The distance between them in the column was about 0.4 cm, so that the shower would not be considered dense. The potentials created by the drops may consequently be assumed to be additive when forming the total potential E. In the case of a viscous solution, the viscosity of mercury in formula (103.2) can be neglected, and we can write:

$$E = \frac{\varepsilon n a}{2\mu\varkappa + \varepsilon^2} \frac{4\pi a^3}{3} g(\rho - \rho'). \tag{103.10}$$

Figure 81 depicts the curves which characterize the decrease in the potential E over a column of unit length as a function of $\epsilon$ in 0.03 N KBr in glycerine. The two curves which are presented compare a theoretical one, computed according to equation (103.10), and one measured experimentally. It is seen that the agreement between them is entirely satisfactory.

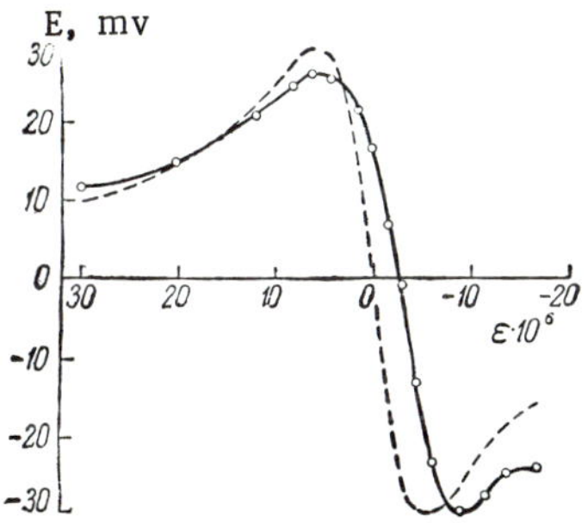


Figure 81. Dependence of the potential E of falling drops on the charge density $\epsilon$ in 0.03 N KBr in glycerine: ---- theoretical curve; $\mu = 9.0, \varkappa = 1.1 \cdot 10^{-5}$; ooo observed values for E.

The observed slight shift between the experimental curve relative to the theoretical one is apparently caused by discharge of the drops due to residual traces of oxygen in the solution. This is indicated directly by a sharp increase in the effect with an increase in the amount of dissolved oxygen (when the solution is cleaned less thoroughly). It should also be pointed out that the observed potentials of the falling drops exceeded those of a column of falling solid particles by a factor of $10^5$ to $10^6$.

## 104. EFFECT OF A MAGNETIC FIELD ON THE FALL OF DROPS OF MERCURY [27]

The effect of a magnetic field on the processes that occur in electrolysis has been given little study to date. It would be of interest, therefore, to clarify the question of whether a magnetic field can have an effect on the motion of mercury drops.

We shall conduct the examination in a system in which the drop is at rest. Let us assume that an external magnetic field of strength H is applied to a drop falling in a gravitational field. Let the magnetic field be in a direction perpendicular to the vertical axis and let us direct the y axis along the magnetic field and the z axis vertically downward.

The total current flowing through the system (the falling drop) is zero. We have seen, however, that there exists a distribution of surface and volume currents in the system, which compensate each other only in terms of the resulting net current. The potential distribution outside the double layer is determined by equation (102.23), so there is a current in the bulk of the solution around the drop, with a density j

$$\mathbf{j}^{\text{soln}} = -\varkappa \operatorname{grad} \varphi.$$

When a volume current flows, the Lorentz force per unit volume of solution is

$$\mathbf{F}^{\text{soln}} = \frac{1}{c}\left[\mathbf{j}^{\text{soln}}\mathbf{H}\right]. \qquad (104.1)$$

In addition to the volume current outside the drop, there is also a current within the drop. Indeed, the surface current formed by the motion of the charges in the inner sheet of the double layer forms a volume current inside the drop as required by the law of conservation of charge, which states that

$$\operatorname{div}(\mathbf{j}_{\text{sur}}) + j_{\text{n}}^{\text{mer}} = 0, \qquad (104.2)$$

where $\mathbf{j}_{\text{sur}}$ is the surface current, equal to

$$\mathbf{j}_{\text{sur}} = -\varepsilon \mathbf{v}_t,$$

and $j_{\text{n}}^{\text{mer}}$ is the component of the current in the drop normal to its surface. By substituting the value for $v_t$, we obtain:

$$\operatorname{div} \mathbf{j}_{\text{sur}} = -\frac{2\varepsilon v_0 \cos\theta}{a}, \qquad (104.3)$$

so that, with the aid of (104.2) and (104.3), the current inside the drop is:

$$\mathbf{j}^{\text{mer}} = -\frac{2v_0\varepsilon}{a}\mathbf{k},$$

where k is the unit vector in the direction of the z axis. Then the force exerted on a unit volume of the mercury drop is

$$\mathbf{F}^{\text{mer}} = \frac{1}{c}[\mathbf{j}^{\text{mer}}\mathbf{H}] = \frac{2v_0\varepsilon H}{ac}\mathbf{i}, \tag{104.4}$$

where i is the unit vector in the direction of the x axis.

The force exerted by the magnetic field on the double layer is equal to zero since the sheets of the double layer have opposite charges and move in the same direction.

Now that the system of volume forces exerted on the solution and on the liquid inside the drop has been determined, we may write the equations of motion for both liquids. Namely, in the system of coordinates in which the drop is at rest, we have for the exterior liquid (the solution),

$$\left.\begin{aligned} \operatorname{grad}\frac{p}{\rho} &= \nu\,\Delta\mathbf{v} + \frac{V(\rho-\rho')}{\rho}g\mathbf{k} + \mathbf{F}^{\text{soln}}, \\ \operatorname{div}\mathbf{v} &= 0. \end{aligned}\right\} \tag{104.5}$$

Similarly, for the interior liquid (the mercury),

$$\left.\begin{aligned} \operatorname{grad}\frac{p'}{\rho'} &= \nu'\,\Delta\mathbf{v}' + \mathbf{F}^{\text{mer}} \\ \operatorname{div}\mathbf{v}' &= 0. \end{aligned}\right\} \tag{104.6}$$

The following system of boundary conditions must be fulfilled at the interface between the two liquids

$$\left.\begin{aligned} p_{rr} &= p'_{rr} + p_\sigma, \\ p_{r\theta} &= p'_{r\theta}, \\ p_{r\varphi} &= p'_{r\varphi}, \\ v_r &= v'_r = 0, \\ v_\theta &= v'_\theta, \\ v_\varphi &= v'_\varphi \end{aligned}\right\} \quad \text{at} \quad r = a. \tag{104.7}$$

In addition, the velocity of the interior liquid must remain finite at r = 0.

The difference between these equations and boundary conditions and those for a drop falling by gravity in the absence of a magnetic field is due to the fact that in the presence of a magnetic field the velocity component $v_\varphi$ and the corresponding components of the stress tensor cannot equal zero.

We seek a solution of the above system of equations and boundary conditions by assuming that the velocity acquired by the drop under the action of the magnetic field is small compared to that which it develops in the field of gravity. This assumption, which is fulfilled for all reasonable values of magnetic field strength, permits a solution of the problem by the method of successive approximations.

Let us write the velocities and pressures for the interior and exterior liquids in the form

$$\mathbf{v} = \mathbf{v}_0 + \mathbf{v}_1,$$
$$p = p_0 + p_1,$$
$$\mathbf{v}' = \mathbf{v}'_0 + \mathbf{v}'_1,$$
$$p' = p'_0 + p'_1,$$

where the subscript zero designates the values of the quantities in the absence of a magnetic field, and the quantities with the subscript 1 satisfy the following equations

$$\left.\begin{aligned} \operatorname{grad}\frac{p_1}{\rho} &= \nu\,\Delta\mathbf{v}_1 + \mathbf{F}^{\text{soln}},\\ \operatorname{div}\mathbf{v}_1 &= 0, \end{aligned}\right\} \tag{104.8}$$

$$\left.\begin{aligned} \operatorname{grad}\frac{p'_1}{\rho'} &= \nu'\,\Delta\mathbf{v}'_1 + \mathbf{F}^{\text{mer}}\\ \operatorname{div}\mathbf{v}'_1 &= 0. \end{aligned}\right\} \tag{104.9}$$

We seek a solution of equations (104.8) and (104.9) in the form

$$\left.\begin{aligned} v_r &= f(r)\sin\theta\cos\varphi,\\ v_\theta &= \Psi(r)\cos\theta\cos\varphi,\\ v_\varphi &= \sin\varphi\,[\alpha(r) + \beta(r)\sin^2\theta],\\ p &= \mu\gamma(r)\sin\theta\cos\varphi, \end{aligned}\right\} \tag{104.10}$$

where $f(r)$, $\Psi(r)$, $\alpha(r)$, and $\beta(r)$ are unknown functions of the radius, and where $\theta$ and $\varphi$ are the appropriate angles in a system of spherical coordinates.

If equation (104.8) is expressed in spherical coordinates, it can be shown that the above radial functions must satisfy the system of equations

$$\left.\begin{aligned} \alpha(r) &= -\Psi(r),\\ f' - \frac{2\Psi}{r} + \frac{2f}{r} + \frac{\beta}{r} &= 0,\\ -\gamma' + \frac{4f'}{r} + f'' - \frac{\varepsilon v_0 a^2 H}{c\mu r^3} &= 0,\\ \frac{\gamma}{r} + \alpha'' + \frac{2\alpha'}{r} + \frac{2\beta}{r^2} - \frac{2f}{r^2} + \frac{2\Psi}{r^2} - \frac{2\varepsilon v_0 a^2 H}{c\mu r^3} &= 0,\\ \beta'' + \frac{2\beta'}{r} - \frac{6\beta}{r^2} + \frac{3\varepsilon v_0 a^2 H}{c\mu r^3} &= 0. \end{aligned}\right\} \tag{104.11}$$

Here the prime marks above the functions indicate differentiation with respect to r.

The liquid motion inside the drop is determined by means of similar equations.

The solution of this system can be effected by standard methods and results in the following expressions for the velocities and pressure inside the drop

$$\left.\begin{aligned} v'_r &= (F + Xr^2)\sin\theta\cos\varphi, \\ v'_\theta &= \left(F + \frac{B}{2}r^2 + 2Xr^2\right)\cos\theta\cos\varphi, \\ v'_\varphi &= \sin\varphi\left[\left(-F - \frac{B}{2}r^2 - 2Xr^2\right) + Br^2\sin^2\theta\right], \\ p' &= 10X\mu' r\sin\theta\cos\varphi + \frac{2v_0\varepsilon H}{ac}r\sin\theta\cos\varphi, \end{aligned}\right\} \quad (104.12)$$

where the quantities F, X, and B are integration constants. These velocities remain finite at the origin of the coordinates.

Also, we define

$$n = \frac{v_0\varepsilon H}{c}; \quad (104.13)$$

then the solution of a system of equations similar to (104.11) yields for the velocities and the pressure outside the drop:

$$\begin{aligned} v_r &= \left(\frac{D_1}{r^3} + \frac{E_1}{r} + U\right)\sin\theta\cos\varphi, \\ v_\theta &= \left(\frac{A_1 - D_1}{2r^3} + \frac{na^2}{4\mu r} + \frac{E_1}{2r} + U\right)\cos\theta\cos\varphi, \\ v_\varphi &= \sin\varphi\left[\left(\frac{D_1 - A_1}{2r^3} - \frac{na^2}{4\mu r} - \frac{E_1}{2r} - U\right) + \left(\frac{A_1}{r^3} + \frac{na^2}{2\mu r}\right)\sin^2\theta\right], \\ p &= \mu\left(\frac{na^2}{2\mu r^2} + \frac{E_1}{r^2}\right)\sin\theta\cos\varphi. \end{aligned}$$

These velocities remain finite as $r \to \infty$.

The constants $F_1$, $D_1$, $E_1$, $A_1$, etc., appearing in the velocities must be determined from boundary conditions (104.7), which yield the following system of equations:

$$\begin{aligned} F + Xa^2 &= 0, \\ \frac{D_1}{a^3} + \frac{E_1}{a} + U &= 0, \\ F + \frac{Ba^2}{2} + 2Xa^2 &= \frac{A_1}{2a^3} - \frac{D_1}{2a^3} + \frac{na}{4\mu} + \frac{E_1}{2a} + U, \\ Ba^2 &= \frac{A_1}{a^3} + \frac{na}{2\mu}, \\ \mu' Ba &= \left(-\frac{4A_1}{a^4} - \frac{n}{\mu}\right)\mu, \\ -6Xa\mu' - 2n &= -\frac{n}{2} - \frac{3\mu E_1}{a^2} - \frac{6D_1\mu}{a^4}, \\ \mu'\left(3Xa + \frac{Ba}{2}\right) &= \mu\left(\frac{3D_1}{a^4} - \frac{n}{2\mu} - \frac{2A_1}{a^4}\right). \end{aligned}$$

The solution of this somewhat involved system of equations yields the velocity of the liquid motion at infinity relative to the drop. It is found to be directed along the x axis, i.e., in a direction perpendicular to the magnetic field, and equal to

$$U = -\frac{a}{\mu}\frac{v_0 \varepsilon H}{c}\frac{\mu+\mu'}{3\mu'+2\mu}, \tag{104.14}$$

where $v_0$ is the velocity on the surface. The velocity of the drop in the direction of the x axis is finally found to be

$$U = \frac{ga^3}{3\mu}\frac{\varepsilon H}{c}\frac{(\mu+\mu')(\rho-\rho_1)}{(3\mu'+2\mu)\left(2\mu+3\mu'+\frac{\varepsilon^2}{\varkappa}\right)}. \tag{104.15}$$

Equation (104.15) shows that, as a result of the action of the magnetic field on the system of volume currents, the falling mercury drop acquires a velocity that increases with increasing falling speed and with increasing intensity of the magnetic field. The absolute value of this velocity is not great, since the speed of light c enters into the denominator of (104.15), but numerical estimates show that it has an observable magnitude in fields with an intensity of the order of $10^5$ gauss.

## BIBLIOGRAPHY


1. Reyss, F.F., Soc. Nat. Met. Moscou 2, 327 (1809).
2. Levich, V. G., Zhur. Fiz. Khim. 21, 689 (1947).
3. Smoluchowski, M. V., in Graetz, Handbuch der Elektrizitat und des Magnetismus, Barth. Leipzig, v. II, p. 366, 1921.
4. Bikerman, Z. physik. Chem. A163, 378 (1933); 171, 209 (1934); Trans. Far. Soc. 36, 154 (1940); A. N. Frumkin and V. G. Levich, Zhur. Fiz. Khim. 21, 399 (1947).
5. Hermans, Phil. Mag. 26, 650 (1938).
6. Henry, D., Proc. Roy. Soc. A133, 106 (1931).
7. Booth, F., Proc. Roy. Soc. A203, 514 (1950); J. Chem. Phys. 18, 1361 (1950).
8. Frumkin, A. N. and V. G. Levich, Zhur. Fiz. Khim. 19, 573 (1945) [for English version, see Acta physicochimica URSS 20, 769 (1945)].
9. Balashova, N. A. and A. N. Frumkin, Doklady Akad. Nauk SSSR 20, 449 (1938) [for German version, see Comp. Rend. Acad. Sci. URSS 20, 449 (1938)].
10. Christiansen, Ann. Phys., ser. 4, 12, 1072 (1903).
11. Frumkin, A. N., Zhur. Russ. Fiz.-Khim. Obshch. 49, 207 (1917).
12. Bodforss, Z. Elektrochem. 29, 121 (1923).
13. Craxford, Phil. Mag. 16, 268 (1933).
14. Craxford, Gatty and McKay, Phil. Mag. 23, 1079 (1937).
15. Bakh, N. A., Acta physicochimica URSS 1, 27 (1934).

16. Frumkin, A. N. and B. P. Bruns, Acta physicochimica URSS 1, 232 (1934).
17. Bruns, B. P., A. N. Frumkin, L. V. Vanyukova, E. Yu. Zolotarevskaya, and Z. A. Iofa, Acta physicochimica URSS 9, 359 (1938); Z. A. Iofa and A. N. Frumkin, Doklady Akad. Nauk SSSR 20, 293 (1938) [for English version, see Comp. Rend. Acad. Sci. URSS 20, 293 (1938)]; Z. A. Iofa, V. V. Losev, and I. I. Tret'yakov, Zhur. Fiz. Khim. 19, 358 (1945).
18. Stackelberg, Antweiler, and Kieselbach, Z. Elektrochem. 44, 663 (1938); Antweiler, Z. Elektrochem. 44, 719, 831, 888 (1938).
19. Stackelberg, Z. Electrochem. 45, 466 (1939).
20. Kryukova, T. A. and B. N. Kabanov, Zhur. Fiz. Khim. 13, 1454 (1939); 15, 475 (1941).
21. Levich, V. G., Zhur. Fiz. Khim. 21, 689 (1947); A. N. Frumkin and V. G. Levich, Zhur. Fiz. Khim. 19, 573 (1945).
22. Levich, V. G., Zhur. Fiz. Khim. 21, 689 (1947); A. N. Frumkin and V. G. Levich, Zhur. Fiz. Khim. 21, 953 (1947) [for English version, see Acta physicochimica URSS 21, 193 (1946)].
23. Bagotskaya, I. A. and A. N. Frumkin, Zhur. Fiz. Khim. 21, 1033 (1947) [for English version, see J. Phys. and Colloid Chem. 52, 1 (1948)].
24a. Bagotskaya, I. A., Zhur. Fiz. Khim. 24, 3 (1950) [see also Comp. Rend. Acad. Sci. URSS 55, 131 (1947)].
24b. Bagotskaya, I. A., Zhur. Fiz. Khim. 23, 1231 (1949).
25. Frumkin, A. N. and V. G. Levich, Zhur. Fiz. Khim. 21, 953 (1947).
26. Fedosov, A. I., Zhur. Fiz. Khim. 29, 822 (1955).
27. Levich, V. G. and V. A. Myamlin, Zhur. Fiz. Khim. 31, 2453 (1957) [see also V. A. Myamlin, V. A. Kibardin, and Yu. Ya. Gurevich, Proc. Acad. Sci. USSR, phys. chem. section (in translation), 137, 357 (1961)].

CHAPTER X

# Theory of the Polarographic Method

## 105. THE POLAROGRAPHIC METHOD

In recent years, increased use has been made of the polarographic method for chemical analysis of solutions. This method is based on interpretation of the current-voltage curves obtained during reduction or oxidation of materials on a dropping mercury electrode.

Many studies, articles and monographs have been devoted to the theory and practice of the polarographic method.

The first polarograph was constructed by Heyrovsky in 1925. This was an electrolysis cell in which the cathode was a drop of mercury that issued from a small capillary (Figure 82). The relationship between voltage and current (the current-voltage curve) was automatically recorded on a photographic film. Such diagrams are termed polarograms, and the current-voltage curves are also termed polarographic curves. The polarograms obtained with a dropping mercury electrode are no different in principle from the current-voltage curves obtained with ordinary solid electrodes (see Figure 46). Before the potential reaches a sufficiently high value to initiate the electrochemical reaction, an extremely low charging current flows through the polarographic cell. When the applied potential difference attains the necessary magnitude, the current in the cell rises sharply. As the applied potential is increased further, a potential region is reached in which the current through the polarographic cell becomes constant because of the establishment of complete concentration polarization at the surface of the dropping mercury electrode. This current represents the limiting diffusion current.

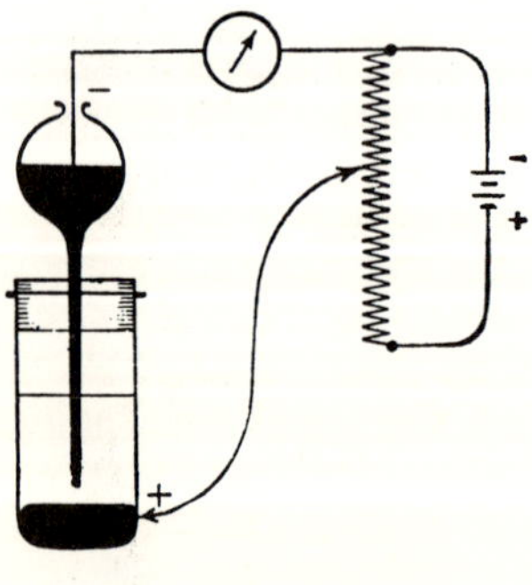

Figure 82. Polarographic circuit.

The magnitude of this current at both dropping mercury and solid electrodes, as shown later, is proportional to the concentration of the material undergoing electrolysis at the electrode. This serves as the basis for the polarographic method. If the solution contains several substances that may be electrolyzed at different values of the applied emf, a jump or step on the polarographic curve, at the appropriate emf value, indicates the start of the process for each of these substances. Such steps are termed polarographic waves. The appearance of a polarographic wave at a certain voltage is characteristic for a given substance and may be taken as evidence of its presence in a solution. Figure 45 shows the polarogram of a solution containing several different ions. Qualitative analysis of the solution is based on the position of the different waves in a polarogram related to a reference electrode.* Quantitative analysis is based on the change in the height of the waves, i.e., on the magnitude of the limiting diffusion current.

Since diffusion currents are proportional to ion concentration, and the diffusion coefficient is known (see below), the unknown concentration can be obtained directly from the height of the waves.

The above reasoning is equally valid for all solid electrodes.

The features of a mercury electrode that have resulted in its widespread use in actual practice are the following:

1. The surface of a dropping mercury electrode is periodically and rather frequently renewed. As a result the electrochemical reaction always occurs at a fresh metal surface, which is free of impurities, reaction products, bubbles generated during the reaction, etc.

2. The periodic fall of the mercury drop causes a considerable and easily reproducible convection in the solution.

3. The theory of a dropping mercury electrode was developed some time ago by Ilkovic, who computed the magnitude of the limiting diffusion current on a dropping mercury electrode.

Among the above-listed features of a dropping mercury electrode, only the first is significant at present, because the attainment of consistently reproducible agitation on a solid electrode does not present special difficulties. Moreover, the theory of a solid electrode has been developed as completely as that of the dropping mercury electrode.

The renewal of the surface, however, is of such great practical importance that, despite the complexity and fragile nature of a polarographic apparatus, dropping mercury electrodes are more frequently used at present in analytical chemistry than are solid electrodes. A dropping mercury electrode usually serves as the cathode of the cell, although it may sometimes be used as the anode.

---

*It should be pointed out that in analytical practice, what is determined is not the decomposition potential, which is a function of the concentration of the solution, but the so-called half-wave potential (see below).

Its use as an anode is limited to the range of potentials below the oxidation potential of mercury. The second electrode is a mercury electrode of a relatively large area with a practically constant potential which is not related to applied emf. We can, therefore, further examine the theory of a dropping mercury electrode without reference to polarization phenomena at the second electrode.

In many instances, polarograms do not have normal shapes (see Figure 89), but manifest more or less sharp peaks, known as polarographic maxima. The theory of this phenomena is also of importance in polarography, and is discussed later.

The polarographic method is used in the analysis of solutions containing ions of all metals, oxygen, hydrogen and a large number of organic substances. It is applicable to the determination of extremely low concentrations (of the order of $10^{-2}$ to $10^{-6}$ mole per liter, and in some instances, even lower concentrations). The accuracy of polarographic data is ±5 percent for the lowest concentrations and ±2 percent for concentrations of the order of $10^{-2}$ to $10^{-4}$ mole per liter [2]. The use of the polarograph is not limited to analytical measurements. It is a very convenient instrument, as shown below, for studying the kinetics of electrochemical reactions.

## 106. LIQUID MOTION IN A HANGING MERCURY DROP

Of basic significance in formulating a theory for the polarographic method is the distribution of velocities within the growing drop of mercury and in the region of the solution contiguous to the drop. The motion of the liquid inside the drop was elucidated by direct visual observations [3] of a small jet of colored water issuing from a wax-coated capillary tube into a drop of clear water hanging at its end.

Visual observations of motion inside a drop were made by B.N. Kabanov and T. A. Kryukova [4]. Appropriately treated carbon powder was introduced into the solution. By observing the motion of the carbon particles, T.A. Kryukova determined the velocities in different zones of the solution with considerable accuracy (of the order of 10 to 15 percent). The data thus obtained showed that the velocity distribution both within and outside the drop are fairly complex and vary considerably with several factors: the velocity at which the mercury issues from the capillary tube, the potential of the drop, and the concentration of the solution.

Two limiting regimes of liquid motion can be described.

The first regime is manifested most clearly when the potential of the dropping mercury electrode is equal to the maximum potential of the electrocapillary curve. The liquid jet penetrates to the bottom of the drop, where it divides tangentially (Figure 83, a). A system of symmetric eddies is generated within the drop. These eddies are similar to those that exist in a freely moving drop. The surface motion of a mercury drop entrains the adjacent layers of the solution. The resulting velocity distribution in the solution near the free bottom surface of the drop is akin to that which exists near the forward surface of a freely moving drop. At the upper portion of the drop, due to the masking effect of the aperture of the capillary tube, the velocity distribtuion, of course, is quite different from that in the vicinity of the rear end of a freely moving drop. If the velocity at which the mercury issues from the capillary is maintained constant, but the potential on the surface of the drop is varied by shifting it to one or the other side of the maximum of the electrocapillary curve, then the regime of the motion also varies (Figure 83, b). The tangential motion on most of the drop surface becomes

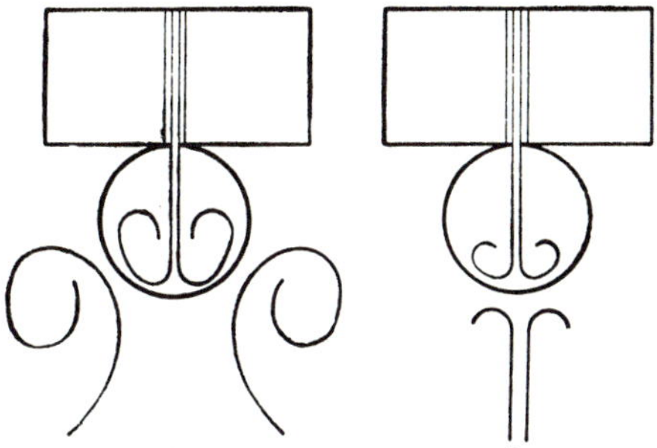

a b

Figure 83. Motion in the solution near a mercury drop and within the drop.
a—in 0.1 N KCl at the maximum of the electrocapillary curve; b—in same solution, at 1.6 volts.

retarded. The portion of the drop surface that is in motion gradually becomes smaller, and the rest of the surface moves only in a radial direction. When the potential is sufficiently far from the electrocapillary maximum, the motion within the drop ceases in a tangential direction and becomes exclusively radial. The drop grows uniformly in a radial direction, in a manner similar to a rubber balloon that is being inflated.

The reasons for the retardation of the surface motion of the mercury were examined in detail in Section 99. Here, we merely point out that the retarding mechanism is similar to that of freely moving drops, examined in Section 99. The range of potentials within which tangential eddy motion occurs and the area on the drop surface where the motion in not retarded depend strongly on the velocity at which the mercury issues from the capillary and on the concentration of the supporting electrolyte. As the issuing velocity increases, the zone in which the dynamic thrust is sufficient to overcome the retarding forces on the surface increases. Owing to this, the potential range in which tangential motion is observed also increases, along with the fraction of the area of tangential motion on the drop surface.

Retardation on the surface of tangential motion increases rapidly with decrease in concentration of the supporting electrolyte. Therefore, in transition to low concentrations of the supporting electrolyte, the range of potentials in which tangential surface motion is possible becomes quite narrow. The fraction of tangential motion on the surface decreases at the same time. All of the above is illustrated schematically in Figures 84 and 85. The arrows indicate the

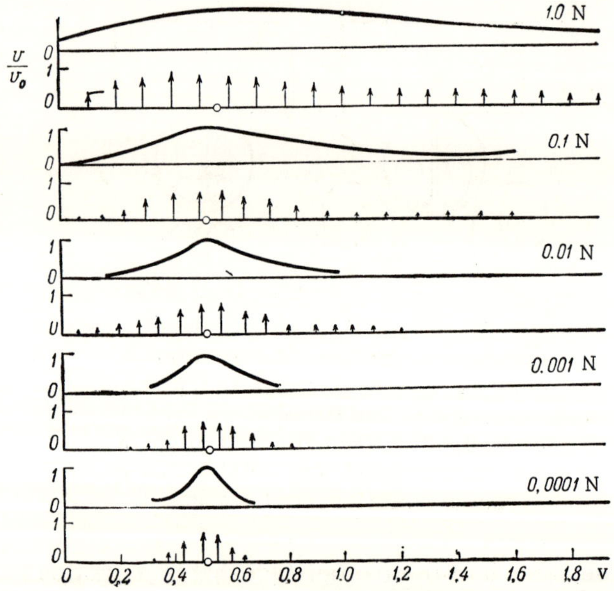


Figure 84. Relative tangential liquid velocity in the vicinity of a growing drop of KCl solutions at various drop potentials and concentrations. (The arrows indicate approximate experimental values according to data by T. A. Kryukova. The solid-line curves represent theoretical calculations.)

tangential velocities on the surface of the mercury found by T. A. Kryukova [4] at different surface potentials of the drop, at different flow rates of the mercury in the capillary, and at several concentrations of the supporting electrolyte. Surface velocities were determined from the motion of carbon particles. Maximum length arrows represent velocities of 5 to 6 mm/sec. Velocity measurements were conducted under conditions of an ideally polarizable electrode with no electrochemical reactions at its surface. Tangential velocity measurements were made at a mercury flow rate of 43 mm/sec at the capillary tip. It is from Figure 85 that at this exit velocity and at a concentration below 0.1 N KCl, tangential motion is virtually nonexistent. On the contrary, at exit velocity approximately twice as great, tangential motion is observed even in relatively dilute solutions, and the potential range associated with this motion widens with increasing solution concentration.

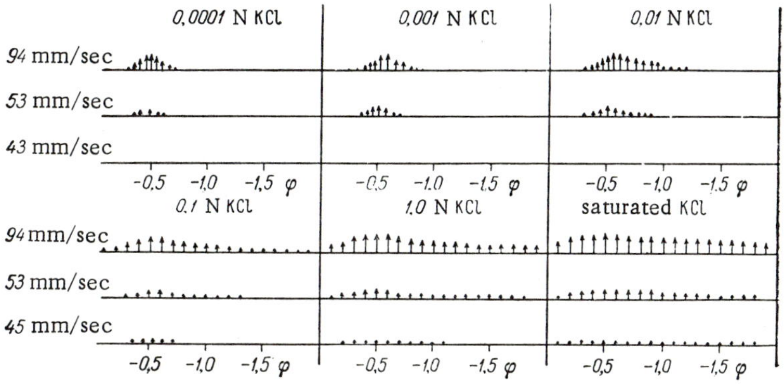


Figure 85. Variation of the velocity of a solution in the vicinity of a mercury drop as a function of the potential, at varying linear mercury flow velocities in the capillary and potassium chloride concentrations (according to data by T.A. Kryukova).

Attempts are usually made to conduct a polarographic analysis when only radial motion of the liquid exists during the growth of the drop. We call such drop growth a radial regime. When eddy motion within the drop adds to the radial growth, the condition is termed an eddy regime. The following sections consider the polarographic theory with a radial regime of drop growth.

## 107. DIFFUSION CURRENT AT A DROPPING MERCURY ELECTRODE IN THE PRESENCE OF SUPPORTING ELECTROLYTES. THE CASE OF REVERSIBLE REACTIONS

It is usual in polarographic practice to employ solutions containing considerable amounts of supporting electrolytes. The supporting electrolyte is used to prevent an undesired decrease in the solution potential which is usually accompanied by certain phenomena causing the appearance of a maximum on the polarographic curves.

We examine the phenomena taking place at a dropping mercury

electrode when a large amount of supporting electrolyte is present in the solution; hence its concentration is high compared with the concentration of the substance undergoing reductions. As pointed out in Section 53, the ion migration can be disregarded in this case and the diffusion of charged and uncharged particles is governed by the same diffusion equations.

Let us consider the surface of a mercury drop issuing from the end of a capillary.

Since the volume of the drop grows with time (prior to its disengagement), the surface of the drop moves in a radial direction. We assume that the drop growth is not accompanied by tangential motion of its surface. Under these conditons, the growth proceeds in a manner similar to a rubber balloon being blown up. The radial velocity of the surface motion may be expressed easily in terms of the velocity of mercury flow in the capillary, if we assume that the drop is spherical (i.e., if distortion in the shape of the drop at the end of the capillary is disregarded). Let the linear velocity of mercury in the capillary be $U_0$.* Then the volume of mercury flowing through a cross section in the capillary is $\frac{\pi d^2 U_0}{4}$, where d is the diameter of the capillary. This quantity represents the increase in drop volume per second:

$$\Delta V = \frac{\pi d^2 U_0}{4}.$$

The volume of a drop that started to grow at time t = 0 is given at time t by

$$V(t) = \Delta V \cdot t = \frac{\pi d^2 U_0 t}{4}.$$

The drop radius at time t, therefore, is

$$a(t) = \left[\frac{3}{4\pi} V(t)\right]^{1/3} = \left(\frac{3d^2 U_0}{16}\right)^{1/3} t^{1/3}.$$

The radial velocity of the drop surface is

$$v_a = \frac{da}{dt} = \frac{1}{3}\left(\frac{3}{16} d^2 U_0\right)^{1/3} \frac{1}{t^{2/3}}.$$

For subsequent calculations it is convenient to introduce the following variable:

$$\gamma = \left(\frac{3d^2 U_0}{16}\right)^{1/3}.$$

Hence, the radial velocity can be given by

$$a(t) = \gamma t^{1/3}. \qquad (107.1)$$

---

*$U_0$ is, of course, the average flow velocity over the cross section of the capillary.

$$v_a = \frac{1}{3}\frac{\gamma}{t^{2/3}}. \tag{107.2}$$

The growth of the drop produces a radial motion in the solution.

We introduce a spherical system of coordinates with the origin at the center of the drop. For the drop growing in a radial direction, the liquid velocity $v_r$ is a function only of the distance r. The continuity equation gives

$$v_r \cdot 4\pi r^2 = \text{const}$$

or, considering that on the drop surface $r = a$ and $v_r = v_a$,

$$v_r = v_a \frac{a^2}{r^2}. \tag{107.3}$$

The radial growth of the drop is important in the diffusion of particles from the solution to the mercury surface. It is manifested in two ways: the solution is set into motion in a direction opposite to the direction of diffusion. This reverse flow complicates the diffusion. As the drop grows, the surface area for diffusion form the solution also increases.

In this respect, the extension of the drop in a radial direction differs from a linear displacement of an infinite plane. It is readily seen that the motion of an infinite plane in an incompressible fluid exerts a purely kinematic effect and does not influence the course of the diffusion.

In spherical coordinates, the equation of convection diffusion to the surface of a growing drop is

$$\frac{\partial c}{\partial t} + v_r \frac{\partial c}{\partial r} = D\left(\frac{\partial^2 c}{\partial r^2} + \frac{2}{r}\frac{\partial c}{\partial r}\right). \tag{107.4}$$

Equation (107.4) shows that the concentration of the diffusing substance is a function only of the radius vector $r$, and not of the spherical angles $\theta$ and $\varphi$. The radial velocity $v_r$ is determined by formula (107.3).

The solution of equation (107.4) must be obtained with the following boundary conditions: at the initial time $t = 0$, the concentration of the solution is constant and

$$c = c_0. \tag{107.5}$$

When an emf is applied at $t = 0$, the dropping mercury electrode potential remains constant. If a reversible reaction occurs on the surface of the mercury electrode and the concentration of the reacting substance is related to the potential by formula (44.3), then a constant potential corresponds to a constant concentration near the electrode surface. Proceeding from the boundary condition for reversible processes we see that

$$c = c(\varphi) \quad \text{at} \quad r = a, \tag{107.6}$$

where $\varphi$ is the potential of the dropping mercury electrode, measured with respect to the anode of the cell. This potential can be expressed in terms of the concentration by

$$c(\varphi) = c_0 A e^{-\frac{nF\varphi}{RT}},$$

where the constant A is proportional to the ionic activity coefficient. This constant is also a function of the nature of the electrochemical process.

The concentration of the reacting substance far from the capillary electrode remains constant in time:

$$c = c_0 \quad \text{as} \quad r \to \infty. \tag{107.7}$$

This condition indicates that we disregard the general depletion of the solution during electrolysis. A solution of equation (107.4) with boundary conditions (107.5) to (107.7) is very difficult to obtain. The equation, however, may be simplified considerably, if we assume that the convective diffusion of the solute takes place in a very thin layer of the solution contiguous to the surface of the dropping electrode. On the basis of previous considerations (see Section 61), we can assume that the decrease in concentration of the reacting substance, within the range from $c_0$ to $c(\varphi)$, occurs in a boundary layer whose thickness is very small compared to the drop radius. Obviously, this assumption does not apply to the initial period of the drop growth. However, we will show later that the current through the drop during this initial period is small compared to the total current through the drop during its lifetime. If the thickness of the diffusion boundary layer $\delta'$ is small compared to the drop radius a, the diffusion layer is almost planar and the liquid velocity in the boundary layer is approximately equal to the velocity on the mercury surface.

Since diffusion occurs in a boundary layer with a small curvature, the distance r in equation (107.4) may be given by

$$r \approx a(t) + y, \tag{107.8}$$

where $y \ll a(t)$.

It is obvious in this case that

$$\frac{1}{r}\frac{\partial c}{\partial r} \approx \frac{1}{a}\frac{\partial c}{\partial r} = \frac{1}{a}\frac{\partial c}{\partial y},$$

$$\frac{\partial^2 c}{\partial r^2} \approx \frac{\partial^2 c}{\partial y^2}.$$

In equation (107.4), the term $\frac{2}{r}\frac{\partial c}{\partial y}$ may be omitted as being small in comparison with $\frac{\partial^2 c}{\partial y^2}$, because

$$\frac{\partial^2 c}{\partial y^2} \approx \frac{c}{\delta'^2} \gg \frac{1}{a}\frac{\partial c}{\partial y} \approx \frac{1}{a}\frac{c}{\delta'}.$$

Further, by substituting the variables in accordance with formula (107.8), we get

$$\left(\frac{\partial c}{\partial t}\right)_r = \left(\frac{\partial c}{\partial t}\right)_y + \left(\frac{\partial c}{\partial y}\right)\frac{\partial y}{\partial t} = \left(\frac{\partial c}{\partial t}\right)_y - \left(\frac{\partial c}{\partial y}\right)\frac{\partial a}{\partial t} = \left(\frac{\partial c}{\partial t}\right)_y - \frac{\gamma}{3t^{2/3}}\frac{\partial c}{\partial y}. \quad (107.9)$$

The liquid velocity $v_r$ during radial growth of the drop is given by

$$v_r = v_a\frac{a^2}{r^2} \approx v_a\frac{1}{\left(1+2\frac{y}{a}\right)} \approx v_a - 2v_a \cdot \frac{y}{a} = \frac{\gamma}{3t^{2/3}} - \frac{2}{3}\frac{y}{t}. \quad (107.10)$$

Substituting $\left(\frac{\partial c}{\partial t}\right)_r$ from equation (107.9) and $v_r$ from formula (107.10) into equation (107.4), and omitting the term of higher order of infinitesimal, $\frac{2}{r}\frac{\partial c}{\partial r}$, equation (107.4) can be rewritten as follows:

$$\left(\frac{\partial c}{\partial t}\right)_y - \frac{\gamma}{3t^{2/3}}\frac{\partial c}{\partial y} + \left(\frac{\gamma}{3t^{2/3}} - \frac{2}{3}\frac{y}{t}\right)\frac{\partial c}{\partial y} = D\frac{\partial^2 c}{\partial y^2},$$

or

$$\frac{\partial c}{\partial t} - \frac{2}{3}\frac{y}{t}\frac{\partial c}{\partial y} = D\frac{\partial^2 c}{\partial y^2}. \quad (107.11)$$

We will attempt to reduce this equation to the usual equation of diffusion in a stationary medium by introducing a new variable

$$z = f(t) \cdot y, \quad (107.12)$$

where f(t) is an unknown function of time, selected in such a manner that the first derivative term in equation (107.11) drops out.

Clearly,

$$\left(\frac{\partial c}{\partial t}\right)_y = \left(\frac{\partial c}{\partial t}\right)_z + \left(\frac{\partial c}{\partial z}\right)\frac{\partial z}{\partial t} = \left(\frac{\partial c}{\partial t}\right)_z + \frac{\partial c}{\partial z}\cdot yf'(t),$$

$$\left(\frac{\partial c}{\partial y}\right)_t = \left(\frac{\partial c}{\partial z}\right)_t\frac{\partial z}{\partial y} = f(t)\frac{\partial c}{\partial z}.$$

Hence, instead of (107.11), we get

$$\frac{\partial c}{\partial t} + \left[yf'(t) - \frac{2}{3}\cdot\frac{y}{t}f(t)\right]\frac{\partial c}{\partial z} = D[f(t)]^2\frac{\partial^2 c}{\partial z^2}. \quad (107.13)$$

By selecting

$$f(t) = t^{2/3},$$

the expression in brackets before $\frac{\partial c}{\partial z}$ vanishes. We then have

$$\frac{\partial c}{\partial t} = Dt^{4/3}\frac{\partial^2 c}{\partial z^2}. \quad (107.14)$$

Introducing a new variable

$$\tau = \frac{3D}{7} t^{7/3}, \tag{107.15}$$

we finally get

$$\frac{\partial c}{\partial \tau} = \frac{\partial^2 c}{\partial z^2}. \tag{107.16}$$

The solution of equation (107.16) that satisfies the initial and boundary conditions

$$c = c_0 \quad \text{at} \quad \tau = 0, \tag{107.17}$$
$$c = c(\varphi) \quad \text{at} \quad z = 0, \quad \tau > 0, \tag{107.18}$$
$$c = c_0 \quad \text{as} \quad z \to \infty \tag{107.19}$$

takes the form

$$c(z, \tau) = \frac{2[c_0 - c(\varphi)]}{\sqrt{\pi}} \int_0^{\frac{z}{2\sqrt{\tau}}} e^{-x^2} dx + c(\varphi). \tag{107.20}$$

In polarography we are interested in the instantaneous diffusion current through the drop. To find it, we must compute the current density i, i.e.,

$$i = (Fn) D\left(\frac{\partial c}{\partial y}\right)_{y=0} = (Fn) D\left(\frac{\partial c}{\partial z}\right)_{z=0} \cdot \frac{\partial z}{\partial y} = (nF) D \frac{c_0 - c(\varphi)}{\sqrt{\pi\tau}} \cdot f(t) =$$
$$= \frac{(nF) D [c_0 - c(\varphi)] t^{2/3}}{\sqrt{\frac{3}{7} \pi D t^{7/3}}} = \left(\frac{7}{3\pi}\right)^{1/2} (Fn) D^{1/2} t^{-1/2} [c_0 - c(\varphi)]. \tag{107.21}$$

The instantaneous current through the drop is

$$I_{\text{Ilk}} = 4\pi a^2 i = 4\left(\frac{7\pi}{3}\right)^{1/2} \gamma^2 D^{1/2} (nF) t^{1/6} [c_0 - c(\varphi)]. \tag{107.22}$$

Equation (107.22) is the fundamental relation of polarographic analysis and was first derived by Ilkovic [5]; hence it is known as the Ilkovic equation.

In practice the quantity $\gamma$ is usually expressed in terms of the weight of the issuing mercury by

$$\gamma' = \left(\frac{3}{16} d^2 U_0\right)^{1/3} = \left(\frac{3}{4\pi} \frac{m}{\rho}\right)^{1/3},$$

where $\rho$ is the density of mercury and m is the mass of mercury issuing per second. Substituting nemerical values, we get

$$I_{\text{Ilk}} = 0.732 \cdot n \cdot D^{1/2} m^{2/3} t^{1/6} [c_0 - c(\varphi)]. \tag{107.23}$$

Here, c is expressed in moles per milliliter. In practical units, and at a concentration c in millimoles per liter

$$I_{\text{Ilk}} = 706 F D^{1/2} m^{2/3} t^{1/6} [c_0 - c(\varphi)] \quad \mu\text{a}. \tag{107.24}$$

The Ilkovic equation shows that the current through a dropping mercury electrode is proportional to $t^{1/6}$, to the concentration of the substance reacting at the electrode, and to the 2/3 power of the mass of mercury issuing from the capillary in unit time (t is the lifetime of the drop).

The increase in the instantaneous current at the dropping mercury electrode with time is explained quite clearly by the following reasoning. The effective thickness $\delta'$ of the diffusion layer, as shown by formula (107.21), is

$$\delta' = \frac{FnD\,[c_0 - c(\varphi)]}{i} = \left(\frac{3\pi}{7}\right)^{1/2} \sqrt{Dt}, \qquad (107.25)$$

and increases with time as the $\sqrt{t}$, as is also the case for a stationary solution (see Section 61).

The surface of the drop increases as the square of its radius $a^2$, i.e., as $t^{2/3}$ [see equation (107.1)]. The increase in the instantaneous current is caused by the more rapid growth of the drop surface in comparison with the increase in the diffusion layer thickness $\delta'$. However, we should remember that the period of drop formation usually is very short. In practice, therefore, several drops usually succeed one another while the measurements are taken.

In actual measurements an average galvanometer reading is taken which is proportional to the average current through the drop during a time interval T. This average current is defined as

$$\bar{I}_{\text{Ilk}} = \frac{1}{T} \int_0^T I(t)\,dt = 605 n D^{1/2} m^{2/3} T^{1/6} [c_0 - c(\varphi)] \qquad (107.26)$$

(the current is measured in microamperes).

The constancy of the average current in the polarograph, despite the succession of the drops, is due to the good reproducibility of the hydrodynamic conditions for each drop.

During the short lifetime of the drop, the diffusion layer lacks sufficient time to penetrate the bulk of the solution, and the depleted solution zone is relatively small. Moreover, the stirring of the solution by the drops separating from the capillary assures satisfactory reproductibility of the current through the drop.

Equation (107.26) has been thoroughly verified by experiments.

In order to obtain the true value of $I_{\text{Ilk}}$, the measured current should be corrected for the residual current value (charging current). This is accomplished by subtracting the extrapolated value of the charging current from the total current (see Figure 48).

It has been shown that the true value of the diffusion current derived in this manner is proportional to the concentration of ions discharging at the mercury electrode with an accuracy of $\pm 0.5\%$ [1].

The linear relation between the diffusion current and the product $m^{2/3}T^{1/6}$ holds with an accuracy of 3% for various capillaries and at various mercury flow rates, if the interval between drops is between 3 and 6 seconds.

The polarographic currents as measured and as computed by the Ilkovic equation (197.26) agree within the same degree of accuracy. Equation (107.26) represents the theoretical basis for polarographic analysis, and in practice is employed in the quantitative analysis of solutions.

Thus, the impression is created that the Ilkovic equation is confirmed by experimental data to a high degree of accuracy. This, in turn, indicates the validity of the initial assumptions regarding the velocity distribution in the growing drop.

The real situation, however, is considerably more complex. The very careful measurements undertaken by T. A. Kryukova showed that the area of applicability of the Ilkovic formula in pure solutions is, in fact, very restricted. As explained in Section 112, the "radial" growth regime of a drop occurs only under exceptional circumstances.

The difficulty in achieving a purely radial regime is related to the fact that, along with the above described mechanism for the appearance of tangential motion on the drop surface, another mechanism exists causing such motion. The other mechanism is related to the appearance of electrocapillary motion on the surface of the mercury drop in an electrical field.

We have established the fact (see below) that the presence of tangential motion has a significant effect on the amount of current flowing through a mercury electrode, and that under such conditions, the Ilkovic equation is inapplicable.

The appearance of tangential motion due to the effluence from the capillary becomes evident at low concentrations of the supporting electrolyte, while the tangential motion due to the electrocapillary effects becomes apparent at high concentrations of the supporting electrolyte. Therefore, the range of validity of the Ilkovic equation is very narrow. The very close quantitative agreement between the measured values of current through a dropping mercury electrode and those computed by the Ilkovic equation may be explained by the presence of impurities and additions in the solution that damp the tangential motion, and by random compensation of errors arising in the derivation of the equation.

## 108. CORRECTIONS TO THE EQUATION FOR THE DIFFUSION CURRENT AT A DROPPING MERCURY ELECTRODE ALLOWING FOR SURFACE CURVATURE AND NON-STEADY EFFLUX OF THE LIQUID

In order to increase the accuracy of polarographic measurements closer approximations must be made in the formula for the diffusion current at a dropping mercury electrode.

In deriving formula (107.26) through series expansion in powers of $\left(\frac{y}{a}\right)$, we used only the first term in the expansion. This implies that the curvature of the diffusion layer was disregarded. With the long drop times that are used in modern polarography, the resulting error is found to be significant.

The accuracy of the equation for diffusion current at a dropping mercury electrode can be increased by correcting for the curvature of the diffusion layer. This is accomplished by using the higher-order terms in the series expansion in powers of $\left(\frac{y}{a}\right)$.

Thus, instead of (107.10), we get

$$v_r = v_a \frac{a^2}{r^2} \approx v_a - 2v_a \frac{y}{a} + 3v_a \frac{y^2}{a^2} = v_a - \frac{2}{3}\frac{y}{t} + \frac{y^2}{\gamma t^{4/3}}. \quad (108.1)$$

In this approximation, the derivative and the Laplace operator are

$$\left(\frac{\partial c}{\partial t}\right)_r = \left(\frac{\partial c}{\partial t}\right)_y + \left(\frac{\partial c}{\partial y}\right)\frac{\partial y}{\partial t} \approx \left(\frac{\partial c}{\partial t}\right)_y - \frac{\partial c}{\partial y}\frac{da}{dt} = \left(\frac{\partial c}{\partial t}\right)_y - \frac{\gamma}{3t^{2/3}}\frac{\partial c}{\partial y};$$

$$\Delta c = \frac{d^2c}{dr^2} + \frac{2}{r}\frac{dc}{dr} \approx \frac{\partial^2 c}{\partial y^2} + \frac{2}{a}\frac{\partial c}{\partial y} = \frac{\partial^2 c}{\partial y^2} + \frac{2}{\gamma t^{1/3}}\frac{\partial c}{\partial y}.$$

In deriving the Ilkovic formula we disregarded the last term in the expression for the Laplace operator.

Introducing the above series into the convective diffusion equation for a dropping mercury electrode (107.4), we get

$$\frac{\partial c}{\partial t} - \left(\frac{2}{3}\frac{y}{t} - \frac{y^2}{\gamma t^{4/3}}\right)\frac{\partial c}{\partial y} = D\left(\frac{\partial^2 c}{\partial y^2} + \frac{2}{\gamma t^{1/3}}\frac{\partial c}{\partial y}\right). \quad (108.2)$$

Since the two terms in equation (108.2) that correspond to the quadratic terms in the power expansion of $\left(\frac{y}{a}\right)^2$ are small, the solution of equation (108.2) may be obtained by the successive approximation method.

We denote the concentration c as

$$c = c_1 + c_2, \quad (108.3)$$

where $c_1$ is the solution of the first approximation equation (107.14), given by (107.20); and $c_2$ is the assumed small correction to $c_1$, related to the curvature of the diffusion layer. It is obvious that $c_2$ satisfies the equation

$$\frac{\partial c_2}{\partial t} - \frac{2}{3}\frac{y}{t}\frac{\partial c_2}{\partial y} - D\frac{\partial^2 c_2}{\partial y^2} = \frac{2D}{\gamma t^{1/3}}\frac{\partial c_1}{\partial y} - \frac{y^2}{\gamma t^{4/3}}\frac{\partial c_1}{\partial y}. \quad (108.4)$$

The solution of equation (108.4) can be obtained by replacing the variables as explained in Section 107.

Introducing the variables defined by equations (107.12) and (107.15), and substituting the value of $c_1$ from (107.20), we get

$$\frac{\partial c_2}{\partial \tau}-\frac{\partial^2 c_2}{\partial z^2}=\frac{2}{\sqrt{\pi}}\frac{[c_0-c(\varphi)]}{\gamma\left(\frac{7}{3D}\right)^{3/7}\tau^{13/14}}e^{-\frac{z^2}{4\tau}}-$$

$$-\frac{[c_0-c(\varphi)]}{\sqrt{\pi}D\left(\frac{7}{3D}\right)^{10/7}}\frac{z^2e^{-\frac{z^2}{4\tau}}}{\gamma\tau^{27/14}}. \tag{108.5}$$

The boundary conditions for the unknown function $c_2$ are

$$\left.\begin{aligned} c_2&=0 \quad \text{at} \quad \tau=0,\ z>0,\\ c_2&=0 \quad \text{as} \quad z\to\infty,\\ c_2&=0 \quad \text{at} \quad z=0,\ \tau>0.\end{aligned}\right\} \tag{108.6}$$

The solution of equation (108.5) with boundary conditions (108.6) is obtained by considering the following equation*

$$\frac{\partial f}{\partial \tau}-\frac{\partial^2 f}{\partial z^2}=\frac{\alpha z^n}{\tau^{k+\nu}}e^{-\frac{z^2}{4\tau}}, \tag{108.7}$$

where f is an unknown function of the variables $\tau$ and z, n and k are arbitrary positive numbers, and $\nu$ is a known number.

The general solution of equation (108.7) is given by

$$f=\sum_k\sum_n c_{kn}\frac{z^n}{\tau^{k+\nu}}e^{-\frac{z^2}{4\tau}}. \tag{108.8}$$

Substitution of (108.8) into (108.7) leads to the following recurrence relation for determination of the coefficients of $c_{kn}$:

$$c_{k-1,n}\left(-k-\nu+n+\frac{3}{2}\right)-c_{k,n+2}(n+2)(n+1)=\delta_{kk'}\delta_{nn'}\alpha, \tag{108.9}$$

where $\delta_{kk'}$ and $\delta_{nn'}$ are Kronecker deltas

$$\delta_{ik}=\begin{cases}0, & i\neq k,\\ 1, & i=k.\end{cases}$$

Using the recurrence relation (108.9), we obtain the particular solution of equation (108.5)

$$c'=-\frac{7}{11\sqrt{\pi}}\frac{[c_0-c(\varphi)]}{\gamma\left(\frac{7}{3D}\right)^{10/7}}\frac{z^2e^{-\frac{z^2}{4\tau}}}{\tau^{13/14}}+\frac{28}{11\sqrt{\pi}}\frac{[c_0-c(\varphi)]}{\gamma\left(\frac{7}{3D}\right)^{3/7}}\tau^{1/14}e^{-\frac{z^2}{4\tau}}. \tag{108.10}$$

---

*At our request, the solution was obtained by I. G. Ivanter.

At the zone boundary $z = 0$, and at the time $\tau = 0$, $c'$ takes the following form:

$$\left.\begin{aligned} c' &= \frac{28}{11\sqrt{\pi}}\frac{[c_0 - c(\varphi)]}{\gamma\left(\frac{7}{3D}\right)^{3/7}}\tau^{1/14} && \text{at } z = 0, \\ c' &= 0 && \text{at } \tau = 0, \quad z > 0, \\ c' &\to 0 && \text{as } z \to \infty, \ \tau > 0. \end{aligned}\right\} \qquad (108.11)$$

We seek a solution of equation (108.5) that satisfies both the initial and the boundary conditions (108.6) in the following form:

$$c_2 = c' + c'', \qquad (108.12)$$

where $c'$ is a particular solution of (108.10) and $c''$ is a solution of the homogeneous diffusion equation (108.5) that satisfies the initial and boundary conditions

$$c'' = -c' \quad \text{for} \quad \begin{cases} z = 0, \ \tau > 0, \\ \tau = 0, \ z > 0, \\ z \to \infty, \ \tau > 0. \end{cases} \qquad (108.13)$$

Using the well-known general solution for the diffusion equation, we get

$$c'' = -\frac{2}{\sqrt{\pi}}\cdot\frac{28}{11\sqrt{\pi}}\frac{c_0 - c(\varphi)}{\left(\frac{7}{3D}\right)^{3/7}}\int\limits_{\frac{z}{2\sqrt{\tau}}}^{\infty}\left(\tau - \frac{z^2}{4x^2}\right)^{1/14} e^{-x^2}\,dx. \qquad (108.14)$$

Thus,

$$\begin{aligned} c = c(\varphi) &+ \frac{2[c_0 - c(\varphi)]}{\sqrt{\pi}}\int\limits_0^{\frac{z}{2\sqrt{\tau}}} e^{-x^2}\,dx + \frac{28}{11\sqrt{\pi}}\frac{[c_0 - c(\varphi)]}{\gamma\left(\frac{7}{3D}\right)^{3/7}}\tau^{1/14}e^{-\frac{z^2}{4\tau}} - \\ &- \frac{7}{11\sqrt{\pi}}\frac{[c_0 - c(\varphi)]}{\gamma\left(\frac{7}{3D}\right)^{10/7}}\frac{z^2 e^{-\frac{z^2}{4\tau}}}{\tau^{13/14}} - \frac{2}{\sqrt{\pi}}\times \qquad (108.15) \\ &\times\frac{28}{11\sqrt{\pi}}\frac{[c_0 - c(\varphi)]}{\gamma\left(\frac{7}{3D}\right)^{3/7}}\int\limits_{\frac{z}{2\sqrt{\tau}}}^{\infty}\left(\tau - \frac{z^2}{4x^2}\right)^{1/14} e^{-x^2}\,dx. \end{aligned}$$

The current density at the drop surface is

$$\begin{aligned} i = FnD\left(\frac{\partial c}{\partial y}\right)_{y=0} &= FnD\left(\frac{\partial c_1}{\partial y} + \frac{\partial c_2}{\partial y}\right)_{y=0} = \\ &= i_1 + FnD\left(\frac{\partial c_2}{\partial y}\right)_{y=0} = i_1 + i_2, \end{aligned}$$

where $i_1$ is the current density given by Ilkovic and $i_2$ is the computed correction

$$i_2 = FnD\left(\frac{\partial c_2}{\partial y}\right)_{y=0} =$$
$$= -\lim_{z\to 0}\frac{\partial}{\partial z}\frac{2}{\sqrt{\pi}}\frac{FnDf(t)}{\gamma}\frac{28}{11\sqrt{\pi}}\frac{c_0 - c(\varphi)}{\left(\frac{7}{3D}\right)^{3/7}}\int\limits_{\frac{z}{2\sqrt{\tau}}}^{\infty}\left(\tau - \frac{z^2}{4\mu^2}\right)^{1/14} e^{-\mu^2}\,d\mu. \quad (108.16)$$

Differentiation and substitution of the variables yields

$$\eta = \frac{\mu^2\tau}{z^2} - \frac{1}{4},$$

which can be used for rewriting the integral as

$$i_2 = FnD\frac{c_0 - c(\varphi)}{11\pi\gamma\tau^{3/7}\left(\frac{7}{3D}\right)^{3/7}}\lim_{z\to 0}\int\limits_0^{\infty}\frac{d\eta e^{-\frac{z^2}{\tau}\left(\eta+\frac{1}{4}\right)}}{\eta^{13/14}\left(\eta+\frac{1}{4}\right)^{4/7}} =$$
$$= FnD\frac{28\,[c_0 - c(\varphi)]}{11\pi\gamma\tau^{3/7}\left(\frac{7}{3D}\right)^{3/7}}\int\limits_0^{\infty}\frac{d\eta}{28\eta^{13/14}\left(\eta+\frac{1}{4}\right)^{4/7}}. \quad (108.17)$$

Numerical integration yields

$$\frac{1}{28}\int\limits_0^{\infty}\frac{d\eta}{\eta^{13/14}\left(\eta+\frac{1}{4}\right)^{4/7}} \approx 1.05 \pm 0.03.$$

Therefore,

$$i_2 \approx FnD\frac{28\,[c_0 - c(\varphi)]}{11\pi\gamma\tau^{3/7}\left(\frac{7}{3D}\right)^{3/7}}\cdot 1.05. \quad (108.18)$$

From the current density we may obtain the total current at the surface of the drop. Averaging over one drop lifetime, we easily obtain the final equation for the average current on a polarograph drop

$$\bar{I} = I_{\mathrm{Ilk}}\left(1 + 3{,}39\frac{D^{1/2}T^{1/6}}{m^{1/3}}\right), \quad (108.19)$$

where $I_{\mathrm{Ilk}}$ is given by equation (107.23) and m is the volume of mercury in $cm^3$ issuing per second.

Dimensional analysis shows that $\bar{I}$ can be expressed as $I = F\left(\frac{D^{1/2}T^{1/6}}{m^{1/3}}\right)$, where F is a certain function. Two terms of the expansion of this function in powers of its argument have been computed above.

Equation (108.19) was first derived by Koutecky [7], who integrated the convective diffusion equation using a series expansion. This derivation is, however, quite cumbersome.

Equation (108.19) was found to be in good agreement with the experimental data of Stackelberg [8, 9].

## 109. DIFFUSION CURRENT AT A DROPPING MERCURY ELECTRODE IN THE PRESENCE OF SUPPORTING ELECTROLYTE. IRREVERSIBLE REACTIONS

As mentioned above, a polarograph is a very convenient instrument for studying the kinetics of electrochemical processes. The renewal of the mercury surface at a dropping mercury electrode is an important feature in studying the kinetics of electrode reactions — reactions that are sensitive to the presence of impurities.

One of the first reactions studied at a dropping mercury electrode was the discharge of hydrogen ions. Equation (54.5) shows that this most important reaction, in the presence of an excess of ions of a supporting electrolyte, is a first order reaction in concentration. In this case, the boundary condition at the surface of the dropping mercury electrode is

$$\beta (nF) \left(\frac{\partial c}{\partial r}\right)_{r=a} = c_{r=a}, \tag{109.1}$$

where $\beta = \frac{1}{k}$ is defined by equation (54.5).

The change in the boundary condition at the surface of a dropping mercury electrode is due to the change in diffusion kinetics from a reversible to an irreversible reaction.

Therefore, the Ilkovic equation is not directly applicable to reactions whose rate is comparable to the rate of transfer of the reacting substances. We present here a derivation of the expressions for current density and for total current flowing through a dropping mercury electrode, in a reaction involving the discharge of hydrogen ions. This derivation was originally obtained by Meyman [10]. The presentation here is made in accordance with the R. R. Dogonadze method.

We have demonstrated in the preceding sections that substitution of variables in accordance with equations (107.12) and (107.15) greatly simplifies the convective diffusion equation.

In order to find the current through a dropping mercury electrode on which a first order reaction is taking place, we must first express the boundary condition (109.1) in terms of the variables z and $\tau$. Simple substitutions yield

$$D (nF) \frac{\partial c}{\partial r} = k\tau^{2/7} \frac{\partial c}{\partial z},$$

where $k = (nF)\, D\left(\frac{7}{3D}\right)^{2/7}$.

The initial condition at $\tau = 0$ and the boundary condition far from the electrode are not determined by kinetics of the electrode reaction and remain the same as in Section 107.

The concentration distribution and the current through the electrode for a first order reaction can be found by solving the diffusion equation

$$\frac{\partial c}{\partial \tau} = \frac{\partial^2 c}{\partial z^2} \tag{109.2}$$

at boundary conditions

$$\beta k \tau^{2/7} \frac{\partial c}{\partial z} = c \qquad \text{at} \quad z = 0, \tag{109.3}$$

$$c \to c_0 \qquad \text{as} \quad z \to \infty \tag{109.4}$$

and with initial condition

$$c = c_0 \qquad \text{at} \quad \tau = 0. \tag{109.5}$$

Solution of this boundary value problem is much more complex than the solution of the comparable problem for a reversible reaction, because the mixed boundary condition (109.3) is more complex than condition (107.6).

To obtain the solution we must reduce the partial differential equation (109.2) to an ordinary differential equation.

We seek the solution of (109.2) in the form

$$c = c_0 + \tau^{\nu} \varphi_{\nu}(\eta),$$

where $\varphi_{\nu}(\eta)$ denotes a function of the new variable $\eta$, defined by

$$\eta = \frac{z}{\sqrt{\tau}}. \tag{109.6}$$

The exponent $\nu$ is not yet defined.

We assume that the general solution of (109.2) is expressed by the series

$$c = c_0 + \sum \tau^{\nu} \varphi_{\nu}(\eta). \tag{109.7}$$

Substituting the series (109.7) into the initial equation (109.2), we see that the unknown functions $\varphi_{\nu}(\eta)$ must satisfy the following ordinary differential equation:

$$\frac{d^2 \varphi_{\nu}}{d\eta^2} + \frac{\eta}{2} \frac{d\varphi_{\nu}}{d\eta} - \nu \varphi_{\nu} = 0. \tag{109.8}$$

Equation (109.8) represents a simple relationship between the functions and the Hermite functions of an imaginary argument.

The latter can be expressed in terms of Hermite functions of a real argument. With the help of appropriate formulas, or by direct comparison, we see that the functions $\varphi_\nu(\eta)$ which satisfies the initial condition (109.5) and the boundary condition (109.3) can be presented as

$$\varphi_\nu(\eta) = A_\nu H_{-2\nu-1}\left(\frac{\eta}{2}\right) e^{-\frac{\eta^2}{4}}, \tag{109.9}$$

where $H_l\left(\frac{\eta}{2}\right)$ is a Hermite function of the first type of argument $\frac{\eta}{2}$ and parameter $l = -(2\nu + 1)$.

Indeed, by the simple substitution of (109.9) into (109.8), it is seen that (109.3) is satisfied if $H l$ satisfies the equation

$$H_l'' - 2\eta H_l' + 2lH_l = 0,$$

which is the equation for Hermite functions.

The boundary condition, at $\eta = 0$, and the initial condition for $\varphi_\nu(\eta)$ obviously are

$$\varphi \to 0 \quad \text{as} \quad \eta \to \infty.$$

Since at high values of the argument $H l \ \frac{\eta}{2}$ we have the asymptotic expression

$$H_l\left(\frac{\eta}{2}\right) \sim \eta^l \left[\sum_{i=0}^{n} \frac{(-1)^i \Gamma(2i-l)}{i!\,\Gamma(-l)\,\eta^{2i}} + O(\eta^{-2n-2})\right]$$

as $\eta \to \infty, \varphi_\nu$ defined by equation (109.9) approaches zero.

We seek the general solution of the boundary value problem in the form

$$c = c_0 + \sum_\nu \tau^\nu A_\nu H_{-2\nu-1}\left(\frac{\eta}{2}\right) e^{-\frac{\eta^2}{4}}. \tag{109.10}$$

Substituting the series (109.10) into the boundary condition (109.3), we obtain the following equation for the determination of the constants $A_\nu$:

$$c_0 + \sum_\nu A_\nu \tau^\nu H_{-2\nu-1}(0) = \frac{\beta k}{2} \sum_\nu A_\nu H'_{-2\nu-1}(0)\, \tau^{\nu - \frac{3}{14}}. \tag{109.11}$$

In order to satisfy (109.11) at all values of $\tau$, $\nu$ must pass through the series of values

$$\nu = \frac{3}{14} n, \quad \text{where } n = \begin{cases} 1, 2, 3, \ldots \\ 0, -1, -2, \ldots \end{cases}$$

Both sequences of the integer n lead to a correct result. We have, therefore, two equivalent forms of the solution of the boundary value problem: one in the form of a series of sets of positive $\nu$, and the other in the form of a series of sets of negative $\nu$. Let us derive both of them.

Form 1. $\nu > 0$. At n = 1, equation (109.11) yields

$$c_0 = \frac{\beta k}{2} A_{\frac{3}{14}} H'_{-\frac{3}{14}\cdot 2-1}(0). \qquad (109.12)$$

For n ≠ 1, taking into account (109.12), we obtain a recurrence relation among the coefficients $A_\nu$

$$A_\nu H_{-2\nu-1}(0) = \frac{\beta k}{2} A_{\nu+\frac{3}{14}} H'_{-2\left(\nu+\frac{3}{14}\right)-1}(0). \qquad (109.13)$$

Using the following equations:

$$H_l(0) = \frac{2^l \Gamma\left(\frac{1}{2}\right)}{\Gamma\left(\frac{1-l}{2}\right)}, \quad H'_l(0) = \frac{2^l \Gamma\left(-\frac{1}{2}\right)}{\Gamma\left(-\frac{l}{2}\right)},$$

where $\Gamma$ is a gamma function, we get from (109.13) and (109.12)

$$A_\nu = A_{\frac{3}{14}n} = -c_0(-1)^{n-1} \frac{2^{\frac{3n}{7}+1} \Gamma\left(\frac{3}{14}+\frac{1}{2}\right)}{(\beta k)^n \Gamma\left(\frac{1}{2}\right)} \prod_{i=1}^{n-1} \frac{\Gamma\left(\frac{3i}{14}+\frac{1}{2}\right)}{\Gamma\left(\frac{3i}{14}+1\right)}. \qquad (109.14)$$

The concentration distribution is obtained by substituting (109.12) and (109.14) into (109.9), and followed by substitution into (109.7),

$$c = c_0 \left[ 1 - \sum_{n=1}^{\infty} (-1)^{n-1} \left( \prod_{i=1}^{n-1} \frac{\Gamma\left(\frac{3i}{14}+\frac{1}{2}\right)}{\Gamma\left(\frac{3i}{14}+1\right)} \right) \times \right.$$
$$\left. \times \frac{2^{\frac{3n}{7}+1} \Gamma\left(\frac{3n}{14}+\frac{1}{2}\right)}{\Gamma\left(\frac{1}{2}\right)} H_{-2\nu-1}\left(\frac{\eta}{2}\right) \mu^n e^{-\frac{\eta^2}{4}} \right], \qquad (109.15)$$

where

$$\mu = \frac{\tau^{3/14}}{\beta k}. \qquad (109.16)$$

On the electrode surface, at $\eta = 0$, considering the expression for $Hl(0)$, we have

$$c(0,\tau)=c_0\left[1+\sum(-1)^n\mu^n\prod_{i=1}^{n}\frac{\Gamma\left(\frac{3i}{14}+\frac{1}{2}\right)}{\Gamma\left(\frac{3i}{14}+1\right)}\right]. \tag{109.17}$$

Similarly, for the current density at the electrode surface, we get

$$i=nFD\left(\frac{\partial c}{\partial r}\right)_{r=a}=\frac{1}{\beta}c_{r=a}=\frac{1}{\beta}c(0,\tau)=$$
$$=\frac{c_0}{\beta}\left[1+\sum(-1)^n\mu^n\prod_{i=1}^{n}\frac{\Gamma\left(\frac{3i}{14}+\frac{1}{2}\right)}{\Gamma\left(\frac{3i}{14}+1\right)}\right]. \tag{109.18}$$

If the reaction rate is very low, and hence $\beta\to\infty$, we see from (109.17), as expected, that $c\to c_0$ and $i\to 0$. The series in (109.18) converges rapidly and can be conveniently used in computations for $\mu<1$, i.e., for times earlier in a drop life.

For $\mu>1$, the series (109.18) converges slowly, and it is more convenient to use the other form of solution.

Form 2. Letting $\nu\leq 0$. From (109.11), at $\nu=-\frac{3}{14}|n|$ we find the recurrence relation for the coefficients $A_\nu$

$$A_\nu=-\frac{\beta k}{2^{\frac{3}{7}}}A_{\nu-\frac{3}{14}}\frac{\Gamma\left(1-\frac{3|n|}{14}\right)}{\Gamma\left(\frac{1}{2}-\frac{3(|n|-1)}{14}\right)}.$$

Proceeding as we did in the case of $\nu>0$, we get

$$c=c_0\left[1+\sum_{|n|=0}^{\infty}(-1)^{|n|+1}2^{\frac{3|n|}{7}+1}\mu^{-|n|}\left(\prod_{i=0}^{|n|-1}\frac{\Gamma\left(1-\frac{3|n|}{14}\right)}{\Gamma\left(\frac{1}{2}-\frac{3|n|}{14}\right)}\right)\times\right.$$
$$\left.\times\frac{\Gamma\left(1-\frac{3|n|}{14}\right)}{\Gamma\left(\frac{1}{2}\right)}H_{\frac{3}{14}(2|n|-1)}\left(\frac{\eta}{2}\right)e^{-\frac{\eta^2}{4}}\right]. \tag{109.19}$$

Here, the surface current density at a dropping mercury electrode is given by

$$i=\frac{1}{\beta}c(0,\tau)=\frac{c_0}{\beta}\left[1+\sum_{|n|=1}^{\infty}(-1)^{|n|+1}\left(\prod_{i=0}^{|n|-1}\frac{\Gamma\left(1-\frac{3i}{14}\right)}{\Gamma\left(\frac{1}{2}-\frac{3i}{14}\right)}\right)\mu^{-|n|}\right]. \tag{109.20}$$

It should be stressed that both series (109.18) and (109.19) are different forms of the same function. The series in (109.20) converges rapidly at $\mu>1$, and slowly at $\mu<1$.

It is evident from equation (109.19) that, for a high reaction rate $\beta \to 0$, i.e., $\mu \to \infty$, $c(\tau, 0) \sim \frac{c_0}{\mu} \to 0$, and the Ilkovic equation must be used to determine i.

The numerical values of i, for $\mu < 1.2$, can be conveniently found from equation (109.20), and for $\mu > 2.5$ from equation (109.18). In the intermediate zone, i.e., for $1.2 < \mu < 2.5$, the values of $c(0, \tau)$ are obtained by graphical interpolation.

The relationship between $\frac{c(0, \tau)}{c_0}$ and $\mu$ is given in Figure 86.

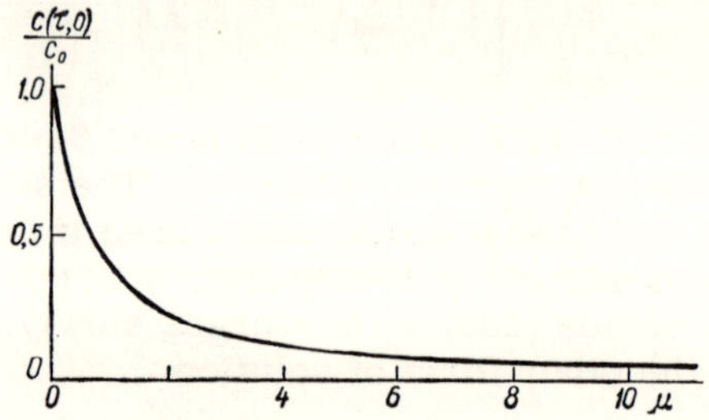


Figure 86. Dependence of the relative concentration on $\mu$.

Let us now consider the following. The products in (109.19) contain factors of the form

$$\frac{\Gamma\left(1 - \frac{3}{14} \cdot 7 \cdot 2k\right)}{\Gamma\left(\frac{1}{2} - \frac{3}{14} \cdot 7\,(2k - 1)\right)},$$

which are indeterminant when the argument of the $\Gamma$-functions is an integer. Expansion of the numerator and denominator of this indeterminate form shows that the ratio is equal to $\frac{1 - 2k}{2k(3k - 1)}$.

The total current flowing to the drop is

$$I = 4\pi a^2 i = \frac{4\pi\gamma^2}{\beta} c(0, \tau)\, t^{2/3}. \qquad (109.21)$$

The average current over the drop time is

$$\overline{I} = \frac{1}{T} \int_0^T I(t)\, dt.$$

Introducing

$$J_M = \frac{\overline{I}}{\overline{I}_{Ilk}} = \frac{\overline{I}}{\frac{6}{7} \frac{4\pi\gamma^2}{\sqrt{\pi}} \left(\frac{7}{3}\right)^{1/3} n D^{1/2} T^{1/6} c_0}, \qquad (109.22)$$

the following values for $J_M$ can be easily found,

at $\mu_0 < 1.2$

$$J_M = 1.241\mu_0 - 1.332\mu_0^2 + 1.278\mu_0^3 - \ldots, \tag{109.23}$$

at $\mu_0 > 10$

$$J_M = 1 - 0.656\mu_0^{-1} + 0.303\mu_0^{-2}, \tag{109.24}$$

where $\mu_0$ denotes the value of $\mu$ at $t = T$. Obviously, $\mu_0$ is inversely proportional to $\beta$.

In reactions involving the discharge of hydrogen ions, $\beta$ is an exponential function of the electrode potential. Therefore, $\ln \mu_0$ is a linear function of the electrode potential $\varphi$.

The current $\bar{I}$ as a function of the electrode potential is given in Figures 87 and 88. The curves were calculated using equations (109.20) and (109.21) for various hydrogen ion concentrations. The curves agree with the values computed by the Ilkovic equation (107.26) at $\mu \gg 1$, but a considerable difference between them exists at $\mu \ll 1$. The values of J determined by Meyman were experimentally checked and verified by V. S. Bagotskiy [11].

V. S. Bagotskiy measured the hydrogen discharge current in $3.9 \cdot 10^{-4}$ N and $0.48 \cdot 10^{-4}$ N HCl solutions in the presence of 0.33 N KCl. The experimental points are shown as circles on Figures 87 and 88.

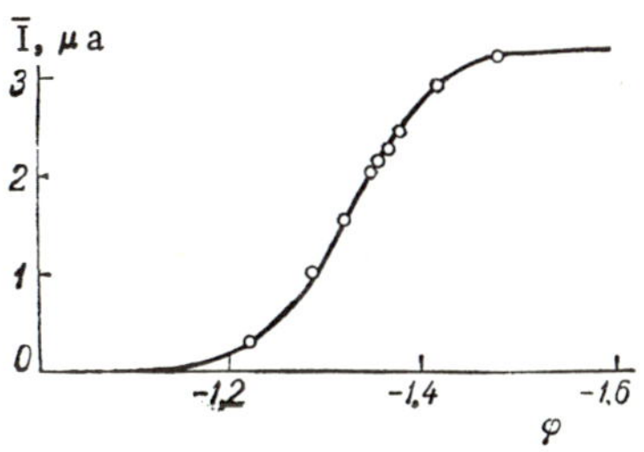


Figure 87. Concentration polarization curve in a solution of $3.9 \cdot 10^{-4}$ N HCl + 0.33 N KCl.

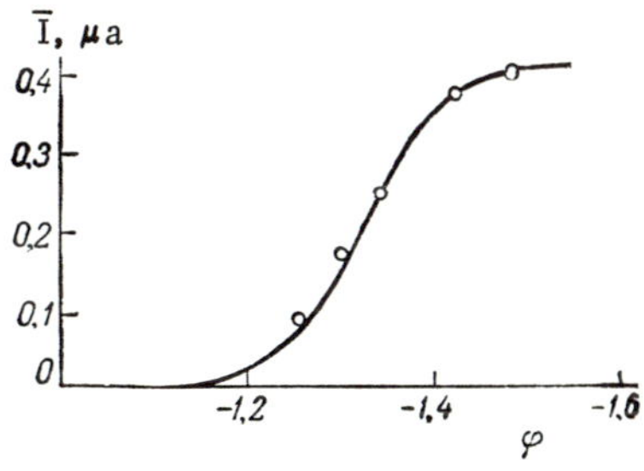


Figure 88. Concentration polarization curve in a solution of $0.48 \cdot 10^{-4}$ N HCl + 0.33 N KCl.

The agreement between Meyman's theory and experimental data is excellent. We can therefore draw two important conclusions: equations (109.20) and (109.21) for the average current at a dropping mercury electrode may be satisfactorily used in the case of a first order, irreversible cathodic reaction. Also, the agreement between the measured and the computed values of the current indicates that the equations expressing the dependence of the rate of hydrogen discharge on the electrode potential, derived for considerably higher concentrations (10,000 times), are also applicable to discharge reactions at low hydrogen ion concentrations.

In addition, the agreement between the computed and the measured values of the hydrogen discharge current at a dropping mercury electrode provides a sufficient basis for rejecting the hypothesis proposed by Heyrovsky regarding the difference between the mechanism of hydrogen ion discharge on fixed and moving electrodes. This hypothesis, advanced by Heyrovsky [12] and others [13], was based on the lack of agreement between data for a stationary (fixed) and a dropping mercury electrode. The dependence of the electrode potential on the current density showed good agreement with ion discharge data on a stationary electrode, while no such confirmation was obtained with a dropping mercury electrode. The source of this discrepancy does not reside in the difference between the discharge kinetics of these two cases, but in the use of an ordinary polarogram with irreversible reactions.

## 110. CURRENT AT A DROPPING MERCURY ELECTRODE IN A BINARY SALT SOLUTION

It is sometimes desirable to conduct a polarographic analysis without a supporting electrolyte. If we disregard, for the time being, the complexities arising from possible polarographic maxima, the difference in the current flow at a dropping mercury electrode caused by the presence or the absence of supporting electrolyte can be related to the migration of ions in an electrical field. Such migration takes place in a binary salt solution.

We examine here the theory of a dropping mercury electrode in a binary salt solution. We assume that the dropping mercury electrode serves as the cathode in the polarographic cell. The diffusion and migration of ions were examined in Section 51 in a general form. It has been shown that migration could be accounted for by introducing an effective diffusion coefficient.

In this case, the equation for concentration distribution in the solution is

$$\frac{\partial c}{\partial t}+v_r\frac{\partial c}{\partial r}=D_{\text{eff}}\left(\frac{\partial^2 c}{\partial r^2}+\frac{2}{r}\frac{\partial c}{\partial r}\right), \qquad (110.1)$$

where c is the molar concentration of the solution and $D_{\text{eff}}$ is the effective diffusion coefficient.

The general solution of equation (110.1) can be obtained in exactly the same manner as that of (107.4).

The current density at the dropping mercury electrode as a function of the applied potential difference is found by the following method: the deviation of the field in the vicinity of the drop from spherical symmetry is disregarded, and a purely radial field distribution is assumed. This assumption is based on the fact that most of the change in concentration and in the potential difference occur in a liquid layer surrounding the dropping mercury electrode surface. The thickness of this layer $\delta'$ is given by equation (107.25). The potential difference between the dropping mercury cathode and the anode of the cell is given by equation (51.22). In this case, we integrate with respect to the radius vector and equation (51.22) becomes

$$\varphi = \frac{RTi}{F^2 (z_1 D_1 + z_2 D_2) z_1 z_2} \int \frac{dr}{c(r, t)} - \\ - \frac{RT (D_1 - D_2)}{F (z_1 D_1 + z_2 D_2)} \int \frac{dc}{dr} \frac{dr}{c} + \frac{RT}{Fz_1} \ln \frac{c_1}{c_l}. \qquad (110.2)$$

The boundary condition on the surface of the dropping mercury electrode is that the radial component of current density for the current carried by anions is zero:

$$-\frac{\partial c}{\partial r} + \frac{Fz_2 c}{RT} \frac{\partial \varphi}{\partial r} = 0. \qquad (110.3)$$

The boundary conditions far from the electrode may be given approximately as follows: let us draw a sphere at a certain distance $\rho$ from the electrode, a distance that is substantially greater than the thickness of the effective diffusion layer, and let us assume that the field within the limits of this sphere is solely radial.

Let $\varphi_\rho$ be the potential on the surface of the sphere. Since the concentration of the solution beyond the sphere of radius $\rho$ is constant, we can assume that

$$\varphi_\rho = \varphi_a - wi, \qquad (110.4)$$

$$c_\rho = c_l, \qquad (110.5)$$

where w is the appropriate ohmic resistance over the region between the anode and the sphere of radius $\rho$, and $c_l$ is the molar concentration of the solution.

The approximate nature of this analysis is justified later by the fact that the current-voltage curve is actually a weak function of the magnitude chosen for $\rho$.

The general solution of equation (110.1) is

$$c(y, \tau) = a_1' \int_{\rho}^{\frac{y}{2\sqrt{D_{\text{eff}}\tau}}} \exp(-x^2)\, dx + a_2. \qquad (110.6)$$

Boundary condition (110.5) yields

$$a_2 = c_l. \qquad (110.7)$$

The constant $a_1$ is determined by substituting solution (110.6) and equation (110.2) into equation (110.3)

$$a_1 = \frac{2i}{z_1 F D_1 (z_1 + z_2)} \cdot \sqrt{\frac{3}{7} D_{\text{eff}} t}. \tag{110.8}$$

The potential difference between the mercury cathode and the sphere of radius $\rho$ is given by

$$V_l = \varphi - \eta_{\text{conc}} = -\frac{RT(D_1 - D_2)}{F(z_1 D_1 + z_2 D_2)} \ln \frac{c_1}{c_l} + \frac{RT}{F z_1} \ln \frac{c_1}{c_l} + \\ + \frac{RTi}{F^2 (z_1 D_1 + z_2 D_2) z_1 z_2} \int\limits_{\rho}^{0} \frac{dr}{\frac{2i\sqrt{3 D_{\text{eff}} t}}{\sqrt{7} z_1 F D_1 (z_1 + z_2)} \int\limits_{\rho}^{\frac{r-a}{2\sqrt{\frac{3}{7} D_{\text{eff}} t}}} \exp[-x^2]\, dx + c_l} \tag{110.9}$$

The integral in equation (110.9) can be evaluated by subdivision into convenient domains

$$I = \int\limits_{\rho}^{\delta'} \frac{dy}{a_1 \int\limits_{\rho}^{y/2\sqrt{\frac{3}{7} D_{\text{eff}} t}} \exp[-x^2]\, dx + c_l} + \int\limits_{\delta'}^{0} \frac{dy}{a_1 \int\limits_{\rho}^{y/2\sqrt{\frac{3}{7} D_{\text{eff}} t}} \exp[-x^2]\, dx + c_l}.$$

The first integral is taken outside the diffusion layer in the constant concentration region of the solution

$$\int\limits_{\rho}^{\delta'} \frac{dy}{a_1 \int\limits_{\rho}^{y/2\sqrt{\frac{3}{7} D_{\text{eff}} t}} \exp[-x^2]\, dx + c_l} \approx \int\limits_{\rho}^{\delta'} \frac{dy}{c_l} \approx -\frac{\rho}{c_l}.$$

For the second integral, in view of the rapid convergence of the integrand at $y < \delta'$, we get

$$\int\limits_{\delta'}^{0} \frac{dy}{a_1 \int\limits_{\rho}^{y/2\sqrt{\frac{3}{7} D_{\text{eff}} t}} \exp[-x^2]\, dx + c_l} \approx \int\limits_{\delta'}^{0} \frac{dy}{a_1 \int\limits_{\infty}^{0} \exp[-x^2]\, dx + c_l} = \\ = -\frac{\delta'}{c_l - \frac{\sqrt{\pi} a}{2}}$$

Finally,

$$I = -\frac{\rho}{c_l} - \frac{\delta'}{c_l - \frac{\sqrt{\pi} \cdot a_1}{2}}. \tag{110.10}$$

Substituting this value of I into equation (110.9), we get

$$V_l = \frac{D_2(z_1+z_2)RT}{Fz_1(z_1D_1+z_2D_2)}\ln\frac{c_1}{c_l} - w_l i - \frac{RT\delta' l}{z_1z_2F^2(z_1D_1+z_2D_2)c_l}\cdot\frac{1}{1-\frac{\sqrt{\pi}a_1}{2c_l}}, \quad (110.11)$$

where

$$w_l = \frac{RTl}{z_1z_2F^2(z_1D_1+z_2D_2)c_l} \quad (110.12)$$

is the ohmic resistance of a layer of thickness $l$.

We note that, according to equation (110.6), concentration on the mercury surface is

$$c_1 = c_l - \frac{\sqrt{\pi}a_1}{2}. \quad (110.13)$$

Therefore, the equation for $V_l$ can be given as

$$V_l = \frac{D_2RT(z_1+z_2)}{Fz_1(z_1D_1+z_2D_2)}\ln\frac{c_1}{c_l} - wi - \frac{RT\delta' l}{z_1z_2F^2(z_1D_1+z_2D_2)}\cdot\frac{1}{c_1}. \quad (110.14)$$

By including in the ohmic resistance the resistance over the region between the anode and the sphere of radius $\rho$, we can assume that $V_l$ is the total potential difference between the anode and the mercury cathode.

Equation (110.14) can be simplified by examining two limiting cases: $c_1 \approx c_l$, which corresponds to a small potential difference in the circuit, and $c_1 \to 0$, which corresponds to a sizeable potential difference and to a limiting current region.

In the first case, we have:

$$-\frac{a_1}{c_l}\cdot\frac{\sqrt{\pi}}{2} \approx \ln\left(1-\frac{\sqrt{\pi}a_1}{2c_l}\right) = \ln\frac{c_1}{c_l},$$

and, further,

$$\frac{RT\delta' l}{z_1z_2F^2(z_1D_1+z_2D_2)}\frac{1}{c_l} \approx -\frac{RT\delta' l}{z_1z_2F^2(z_1D_1+z_2D_2)}\cdot\frac{1}{\frac{\sqrt{\pi}a_1}{2}\left(1-\frac{2c_l}{\sqrt{\pi}a_1}\right)} \approx$$

$$\approx -\frac{RTD_1(z_1+z_2)}{z_2F(z_1D_1+z_2D_2)}\ln\frac{c_1}{c_l}.$$

Therefore,

$$V_l = wi + \frac{RT}{F}\frac{(z_1+z_2)}{z_1\cdot z_2}\ln\frac{c_1}{c_l} = V_{\text{ohm}} + \frac{RT}{F}\frac{(z_1+z_2)}{z_1\cdot z_2}\ln\left(1-\frac{i}{i_{\text{lim}}}\right), \quad (110.15)$$

where

$$i_{\lim} = \frac{FD_1 z_1 (z_1 + z_2) c_l}{\delta'}. \tag{110.16}$$

As $c_1 \to 0$, equation (110.14) becomes meaningless, and the current flow is directly determined by the condition $c_1 \to 0$. Equations (110.8) and (110.13) yield

$$\lim (i)_{c_1 \to 0} = i_{\lim}$$

The current-voltage curve, therefore, can be plotted using an interpolation equation similar in form to (110.15).

Equation (110.15) differs from the Ilkovic equation for a polarographic wave in the presence of a supporting electrolyte, by yielding a different value for the limiting current $i_{\lim}$.

Substituting the thickness $\delta'$ of the diffusion boundary layer into equation (110.16), we obtain the limiting current density

$$i_{\lim} = \frac{F z_1 D_1 (z_1 + z_2) c_l}{\sqrt{\frac{3\pi}{7} D_{\text{eff}} t}}. \tag{110.17}$$

The difference between the limiting currents on a dropping mercury electrode in the absence and presence of a supporting electrolyte is the same as for a solid electrode (although the expressions for limiting current on liquid and on solid electrodes are, of course, different).

The ratio of limiting currents in solutions of a binary salt and in a solution with supporting electrolyte is

$$\frac{i_{\lim}}{i_{\text{Ilk}}} = \frac{c_l}{c_0}(z_1 + z_2) \frac{\sqrt{D_1}}{\sqrt{\frac{D_1 D_2 (z_1 + z_2)}{z_1 D_1 + z_2 D_2}}} = \left(1 + \frac{z_2}{z_1}\right) \sqrt{\frac{(z_1 D_1 + z_2 D_2)}{D_2 (z_1 + z_2)}}. \tag{110.18}$$

Equation (110.18) was later derived by Stackelberg [9], who expressed the current in terms of the number of cations transferred

$$T_1 = \frac{u_1}{u_1 + u_2} = \frac{D_1 z_1}{D_1 z_1 + D_2 z_2}.$$

The Stackelberg equation

$$\frac{i_{\lim}}{i_{\text{Ilk}}} = \left(\frac{D_{\text{eff}}}{D_1}\right)^{1/2} \left(\frac{1}{1 - T_1}\right) = \frac{\sqrt{\frac{D_1 D_2 (z_1 + z_2)}{z_1 D_1 + z_2 D_2}}}{\sqrt{D_1}} \left(\frac{1}{1 - \frac{D_1 z_1}{z_1 D_1 + z_2 D_2}}\right) =$$

$$= \sqrt{\frac{D_2 (z_1 + z_2)}{z_1 D_1 + z_2 D_2}} \frac{z_1 D_1 + z_2 D_2}{z_2 D_2} = \left(1 + \frac{z_1}{z_2}\right) \sqrt{\frac{z_1 D_1 + z_2 D_2}{D_2 (z_1 + z_2)}}$$

is identical with equation (110.18).

## 111. POLAROGRAPHIC MAXIMA

We mentioned in Section 105 that in many instances the current-voltage curve sharply deviates from the Ilkovic equation. Waves appear on such curves that do not correspond to any reduction process that could possibly take place in a given solution. In such cases, the linear relationship predicted by the Ilkovic equation for the amplitude of the wave (the current) and the concentration of the electroactive substance is often violated.

Typical "anomalous" polarograms [14] are shown in Figure 89. The current voltage curve shows a current increase up to a certain voltage, followed in some cases by a sharp decrease and in others by a smooth drop in the circuit current to values predicted by the Ilkovic equation. This phenomenon has been designated as a polarographic maximum (on the current-voltage curve).

Experimental studies of these maxima have shown that the mechanism leading to the appearance of the maxima illustrated in Figure 89 (curve a) which are followed by a sharp drop in the current differs greatly from the mechanism which causes the appearance of maxima followed by a smooth decrease in the current (curve b).

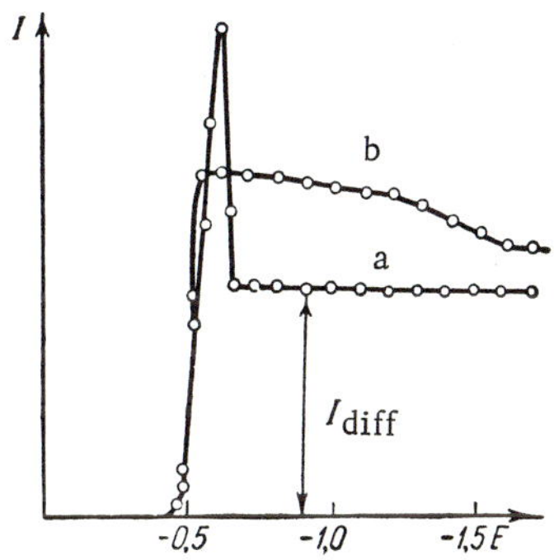


Figure 89. Different forms of maxima for lead.

We will consider the maxima of the first kind (curve a) and of the second kind (curve b).

Polarographic maxima have long been the subject of numerous experimental studies. It has been found that this phenomenon is always related to the appearance and subsequent disappearance of tangential motion on the surface of a liquid mercury cathode [15, 16].

The appearance of tangential motion in the case of maxima of the first kind is caused by differences in surface tension at different points of the cathode, and is not related to the nature of the dropping mercury electrode. Maxima of the first kind are also observed at nondropping mercury electrodes. A distinguishing feature of maxima of the first kind is that a decrease in the maximum current results from an increase in the concentration of the supporting electrolyte in the solution. This phenomenon has been termed repression of the maximum.

Polarographic maxima of the second kind, discovered and studied by B. N. Kabanov and T. A. Kryukova [4], are not repressed by the addition of a supporting electrolyte. They are characteristic of a droping mercury electrode.

Study of maxima of the second kind has revealed that they can be attributed to tangential motion of the drop surface which, in turn, is caused by mercury flow from the capillary.

Study of polarographic maxima has both theoretical and practical importance. The reasons for the appearance of tangential motion and the related increase in the current at a dropping mercury electrode were suggested by A. N. Frumkin et al. [15, 16]. Until recently, however, no one succeeded in elucidating the mechanism that leads to decreasing current beyond the maximum point.

Attempts [17] to relate the decrease in current on the drop to a rearrangement of the double layer on the mercury surface were disproved by experiments [18], which showed that the polarographic maxima disappear completely when the solution is vigorously stirred. Elucidation of the theory of maxima is of interest in the computation methods used in polarography. In many instances the maxima form false waves that may be erroneously taken for discharge reactions of cations which are actually not present in the solution. As an example, we may cite false waves that resemble the reduction of $Fe^{++}$ and $Ni^{++}$. In many cases polarographic maxima can be used for analytical purposes.

Before the discussion of the theory of maxima, we note that in the current terminology a maximum is termed positive if it occurs at potentials corresponding to the positive branch of the electrocapillary curve, and negative in the reverse case.

## 112. POLAROGRAPHIC MAXIMA OF THE SECOND KIND

We begin the examination of the theory of polarographic maxima with maxima of the second kind.

We have shown in Section 106 that at certain conditions the radial growth of the drop is replaced by an eddy regime. This occurs when the dropping mercury electrode potential approaches that at the maximum of the electrocapillary curve, and when there is large flow of mercury and high concentration of the supporting electrolyte. In the eddy regime, the liquid surface acquires a tangential velocity component, and the assumptions made in deriving the Ilkovic equation do not hold.

The regime of motion in the lower portion of the drop is found to be close to that of a freely falling drop. Thus, the convective flow of ions from the solution to the drop surface caused by radial expansion of the drop is accompanied by convective flow caused by the transfer of ions by the tangential velocity component of the solution.

If the region of tangential motion extended over the entire surface of the drop, the instantaneous diffusion flow caused by tangential motion can be computed from equation (72.10).

In reality, the motion of the liquid in the upper portion of the drop is distorted by the presence of the capillary. Nevertheless, as noted in Section 99, the general considerations concerning the appearance and retardation of the liquid metal-solution interface apply to any shape of the interface. Therefore, equations derived in Section 72 can be used for qualitative estimates.

The velocity at any point on the surface of the mercury drop can be determined by an equation similar to equation (100.6). The numerical coefficient in this formula, however, is a function of the shape of the surface of a freely falling drop, a shape that is different in a hanging drop. In computing the diffusion current at a dropping mercury electrode, therefore, it is preferable to employ measured values of the surface velocity. It is important to note, however, that the unknown numerical coefficient at the ratio $\frac{v_\theta}{v_{max}}$ (where $v_0$ is the tangential velocity at a given point on the drop at any surface potential, and $v_{max}$ is the velocity at the same point at a maximum potential) can be cancelled, and using (100.6), we get

$$\frac{v_0}{v_{max}} = \frac{2\mu + 3\mu'}{2\mu + 3\mu' + \frac{\varepsilon^2}{\varkappa}\left(1 + \frac{a}{2\varkappa\omega}\right)^{-1}}. \tag{112.1}$$

In an ideally polarizable surface, $\omega \to \infty$ and

$$\frac{v_0}{v_{max}} = \frac{2\mu + 3\mu'}{2\mu + 3\mu' + \frac{\varepsilon^2}{\varkappa}} = \frac{1}{1 + \frac{\varepsilon^2/\varkappa}{2\mu + 3\mu'}} = \zeta. \tag{112.2}$$

These two equations show that the tangential velocity of the drop at a given surface potential (i.e., at a given charge $\epsilon$ per unit of surface area) is lower than at zero charge in comparison to the factor $\zeta$. As the surface charge $\epsilon$ increases, the motion is quickly retarded. The mechanism of this retarding action was examined in detail in Section 102. It is important to note that the tangential velocity must pass through a maximum at the point of zero surface charge. The surface charge is equal to zero when the potential reached corresponds to the maximum point on the electrocapillary curve. Hence, at this point the tangential velocity for any rate of effluence of mercury from the capillary and for any concentration of the supporting solutions and with a high supporting electrolyte concentration, the conductance $\varkappa$ of the solution is high, hence the retarding factor $\gamma = \frac{\epsilon^2/\varkappa}{2\mu + 3\mu'}$ is relatively small. In these solutions, the retarding factor $\gamma$ becomes large only at such potentials which are significantly different from the zero point potential, where the surface charge becomes relatively high. In other words, the tangential motion of the surface, in the case of concentrated solutions, occurs within a relatively wide range of potentials around the zero point. On the contrary, the conductance is low in dilute solutions and the retarding factor is large even at relatively low surface charges and at potentials which are close to the zero point potential. In dilute solutions, therefore, tangential motion can occur only with a very narrow range.

Figure 84 shows the ratio $\frac{v}{v_0}$, computed from equation (112.2) for an ideally polarizable drop as a function of the potential in KCl solutions of various concentrations (solid curves) [19]. The values of the mercury cathode potential in Figure 84 are relative to a calomel reference electrode; hence the point of zero charge lies at potentials of: 0.56 v for 1 N KCl, at 0.82 v for 0.1 to 0.01 N KCl, and at 0.51 v for 0.0001 N KCl. The values of $\epsilon$ have been computed from the capacitance of the mercury surface measured in a KCl solution [20]. Arrows on the same figure show the experimental values of the ratio $\frac{v_0}{v_{max}}$. These values have been visually determined by T. A. Kryukova [4], in a series of experiments designed to verify the theory.

The experimental conditions were such that the drop could be considered ideally polarizable. It is evident that the theoretical curves reproduce well the tangential velocity profile in the solution. The presence of tangential motion in the solution results in a considerable increase in the diffusion current to the surface of the drop. The diffusional flux to a freely moving drop is computed in Section 72. Equation (72.7), however, can be applied only to the lower portion of a hanging drop at the end of a capillary. The velocity profile in the upper portion of the drop is distorted by the capillary. However, according to equation (72.7), the diffusion flux to the front half of the drop exceeds that to the rear half. We can therefore assume that this equation also gives the order of magnitude of the current to a drop in a polarograph. Combining equations (72.7) and (112.1), we get the current in the case of tangential motion of the drop:

$$I_t \approx (nF)\, D^{1/2} \bar{a}^{3/2} \bar{v}_0^{1/2} c_0 \zeta^{1/2}; \tag{112.3}$$

where $\bar{v}_0$ is the average value of the maximum tangential velocity over the drop period.

Formula (112.3) shows that a change in tangential velocity (change in factor $\zeta$) must be accompanied by a corresponding change in the current $I_t$ flowing to the drop. The usual polarographic current, determined by the Ilkovic equation (107.22), flows to the drop along with the current $I_t$. Since both currents are given by similar equations, and since the numerical values of the coefficients are close, the total current to the drop can be conveniently expressed by

$$I^2 = k_1 \bar{v}_r + k_2 \bar{v}_0, \tag{112.4}$$

where $k_1$ and $k_2$ are constants.

Substitution of numerical values into equation (112.4) yields the total current, which is comparable to experimental values (see Section 113).

Thus, when the electrode potential passes through the point of zero charge on the surface, then the tangential velocity and the current on the drop must simultaneously pass through a maximum. This is clearly illustrated by curve 1 in Figure 90.

The maximum value of the current to the drop is at the point of zero charge. At the same point the maximum tangential velocity is observed in the solution near the drop. Substituting the numerical values of coefficients $k_1$ and $k_2$ into equation (112.4), and assuming that coefficient $k_2$ is the same as for a freely falling drop, a numerical estimate of the current for a maximum of the second kind can be obtained. The tangential velocity data were obtained by T. A. Kryukova [4].

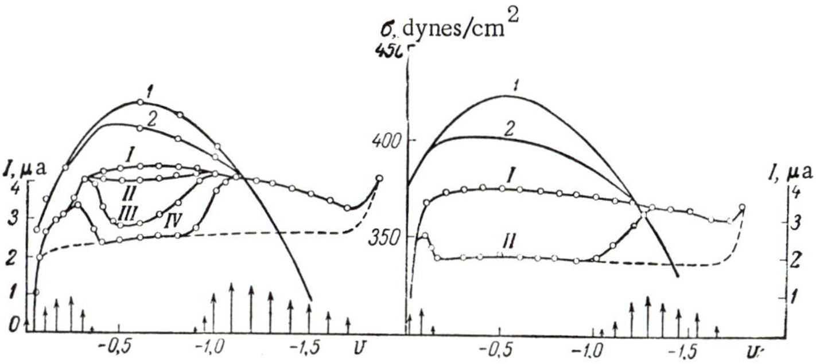


Figure 90. Comparison of polarization and electrocapillary curves:
I — solution of 0.0003 N $HgCl_2$ + satd. KCl without amyl alcohol; II — same solution with amyl alcohol at a concentration of $1 \cdot 10^{-4}$ M; III — same, $1 \cdot 10^{-3}$ M; IV — same, $1 \cdot 10^{-2}$ M*; 1 — electrocapillary curve, 1 N NaCl; 2 — same solution with amyl alcohol $1 \cdot 10^{-2}$ M. The right side illustrates the same for n-butyl alcohol. Curves 1 and I — pure solution 1N $Na_2SO_4$. Curves 2 and II — same solution with 0.1 M butyl alcohol. Arrows indicate surface velocity on mercury for curves of greatest alcohol concentration.

Table 18 gives the calculated and observed values of the difference $I - I_{IIk}$. Steps on the current-voltage curve correspond to the start of the discharge process for appropriate ions. They differ from the steps on a usual polarogram only by the absence of the horizontal portions.

It is clear from Table 18 that the observed current values at the polarographic maxima were higher than the calculated values in each case. This should have been expected, since the motion of the liquid

---

*Ed. Note: The original Russian text gives $10^{-3}$, which is probably erroneous.

Table 18

| Composition of Solution | $a \times 10^{-2}$ | v | $D \times 10^{-2}$ | $\bar{I}_{calc} \times 10^{-6}$ | $\bar{I}_{exp} \times 10^{-6}$ |
|---|---|---|---|---|---|
| 0.0006 N $O_2$* + 0.001 N KCl | 5 | 1.7 | 1.86 | 23.4 | 18 |
| 0.00057 N $O_2$ + N KCl | 5 | 0.93 | 1.86 | 16.5 | 11.2 |
| 0.001 N $CdCl_2$ + N KCl | 7 | 0.4 | 0.72 | 19.5 | 15.8 |
| 0.001 N $CdCl_2$ + N KCl | 7 | 0.3 | 0.72 | 16.8 | 9.3 |
| 0.001 N $CdCl_2$ + N KCl | 4 | 0.28 | 0.72 | 7.0 | 3.4 |

*Current measurements were relative to the first stage of oxygen reduction, so that, for a molecule of $O_2$, n = 2.

in the upper portion of the drop is strongly retarded by the aperture of the capillary. Nevertheless, the order of magnitude and the general configuration of the curve that represents the difference $(I - I_{IIk})$, computed from equation (112.4), agree well with those observed experimentally.

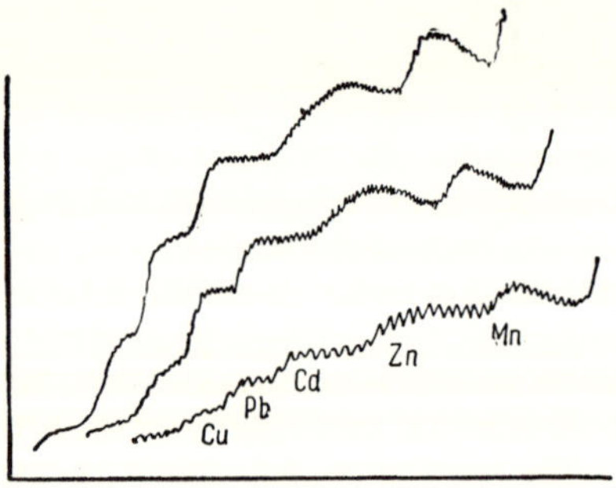


Figure 91. Polarograms of a 0.001 N solution containing Cu, Pb, Cd, Zn and Mn ions in a supporting electrolyte of 2 N KCl, obtained at different mercury pressures.

Thus, the theory of the maximum of the second kind presented here can be considered semiquantitative. Due to the complicated effect of the capillary on liquid flow, the numerical value of $k_2$ in equation (112.4) cannot be determined with any greater accuracy than indicated above. Formula (112.3) leads to an important practical conclusion: in a second-kind maximum, the current at a drop is directly proportional to the concentration of current-carrying ions. As a result of this, proportionality between current and concentration, the second-kind maxima can be used in polarographic analysis (T. A. Kryukova [28]).

Analysis in the region of a maximum of the second kind offers several advantages, since short drop times (of the order of 1 sec) are frequently encountered in modern polarographic practice. As shown in Section 115, the undesirable effects of impurities that may be present in the solution are reduced with short drop times. The latter is obtained by increasing the rate of flow of mercury through the capillary. Thus, the polarograph operates in a region corresponding to a maximum of the second kind. To assure normal operation of the polarograph under these conditions, a sufficiently large concentration of the supporting electrolyte solution is required and a constant rate of flow of mercury must be maintained.

Polarographic waves under conditions permitting maxima of the second kind are shown in Figure 91.

## 113. POLAROGRAPHIC MAXIMA OF THE FIRST KIND

Development of a theory for maxima of the first kind is considerably more difficult than for maxima of the second kind. The relationship between the appearance of electrocapillary motion on the mercury surface and the appearance of polarographic maxima has been established in 1939 [16]. However, it has been very difficult to explain the sharp drop in the current beyond the maximum point.

Attempted explanations, which assumed a rearrangement of the double layer on the mercury-solution interface [17], were disproved by experiments. The experiments [18] showed that the maxima disappears when the solution is vigorously stirred. In other words, it has been shown that all aspects of the phenomenon are governed by hydrodynamic rather than molecular processes of some kind.

We examine the behavior of a hanging mercury drop [21] at the end of a capillary. The drop serves as the cathode of a polarograph. Since an electric field is present in the instrument, electrocapillary motion (examined in Section 100) appears in the drop. The drop attains a surface velocity which is governed, on the one hand, by the potential difference along the drop at a given field configuration and, on the other, by hydrodynamic factors and the physicochemical properties of the liquid.

The results obtained in Section 100 applied to a freely falling drop and, therefore, are not applicable to a dropping mercury electrode. However, as we mentioned above, it was shown that the velocity distribution in the bottom and middle portions of the drop is identical with the velocity distribution in a freely falling drop. We can therefore make a basic theoretical assumption regarding polarographic maxima: the surface motion of a dropping mercury electrode in an applied electrical field obeys the same laws as in a free drop. This assumption, of course, is not valid for the upper portion of the drop, where both the liquid motion and the field distribution are distorted by the presence of the capillary. The tangential motion, however, and the additional stirring of the solution that accompanies it occur mostly in the bottom and middle portions of the drop, where the effect of the capillary can be disregarded. Thus, it is impossible to develop a semiquantitative theory of polarographic maxima of the first kind.

The surface velocity relative to the center of the mercury drop can be given by equation (100.8), which, taking into account (99.20), can be conveniently presented as

$$v_0 = \frac{1}{2} \cdot U(\Delta\Phi_0), \tag{113.1}$$

where $(\Delta\Phi_0)$ is the potential difference along the drop at a given field with stirring of the solution, and with $\epsilon = 0$; U is the drop velocity at $\Delta\Phi = 1$. According to (100.7), the potential difference along the drop is

$$(\Delta\Phi_0) = 3aE\left(1 + \frac{a}{2\varkappa w}\right)^{-1}.$$

The expression for $v_0$ must also account for the difference between a dropping mercury electrode and a free drop. This difference consists in that the total instantaneous current flowing on a free drop is zero, while it is not zero at a dropping mercury electrode. To accomplish this, the potential difference $(\Delta\Phi_0)$ and the effective resistance w must be expressed in terms of the current density i. In a first approximation, we assumed that the potential difference was uniform, which permits the total difference to be given by

$$(\Delta\Phi_0) = 3k\left(\frac{ia}{\varkappa}\right)\left(1 + \frac{a}{2\varkappa w}\right)^{-1}, \tag{113.2}$$

where the proportionality coefficient k is determined by the geometric configuration of the field. For the time being, we consider the field to be uniform. At the same time, we assumed that the potential difference over a distance of 1 cm (field at the surface of a drop that has a zero charge) is according to Ohm's law equal to $\frac{i}{\varkappa}$. Using these simplifying assumptions, the surface velocity of the mercury drop may be written as

$$v_0 = \frac{1}{2} \cdot Ui\frac{3ka}{\varkappa}\left(1 + \frac{a}{2\varkappa w}\right)^{-1}. \tag{113.2'}$$

The effective resistance w in equation (113.1) and (113.2) is, by definition, equal to

$$w = -\frac{\partial\varphi}{\partial i}, \tag{113.3}$$

where $\varphi$ is the potential difference between the mercury and the solution. The minus sign indicates that this difference is taken from the mercury toward the solution.

For ions with a valence n discharging at the cathode (and forming an amalgam with the mercury), the potential difference between the mercury and a distant point in the solution can be expressed in terms of the half-wave potential

$$\Delta\varphi = \frac{RT}{nF}\ln\left(\frac{i_{\lim}}{i} - 1\right) + \varphi_{1/2}, \tag{113.4}$$

where $i_{\lim}$ is the density of the limiting diffusion current and $\varphi_{1/2}$ is the half-wave potential [1].

Substituting $\Delta\varphi$ from (113.4) into equation (113.3), we find

$$w = \frac{RT}{nF} \cdot \frac{1}{i} \cdot \frac{i_{\lim}}{i_{\lim} - i}. \tag{113.5}$$

Substituting (113.2), (100.8) and (113.5) into the general equation for the velocity of a charged surface in an electrical field, (113.1), we get

$$v_0 = \frac{3}{2\varkappa} \frac{\varepsilon k a i \left(1 + \frac{anFi}{2\varkappa RT} \cdot \frac{i_{\lim} - i}{i_{\lim}}\right)^{-1}}{\left[(2\mu + 3\mu') + \frac{\varepsilon^2}{\varkappa}\left(1 + \frac{anFi}{2\varkappa RT} \cdot \frac{i_{\lim} - i}{i_{\lim}}\right)^{-1}\right]}. \tag{113.6}$$

The limiting diffusion current $i_{\lim}$ in equation (113.6) is the sum of the current that appears during radial growth of the drop $i_r$, and the current caused by the tangential motion of the surface $i_t$. The latter is a function of the velocity $v_0$. However, before high tangential velocities are attained, the second component of the diffusion current is relatively small, and equation (113.6) can be approximated by

$$v_0 \approx \text{const} \frac{\frac{ai\varepsilon}{\varkappa}\left(1 + \frac{nFai}{2\varkappa RT} \cdot \frac{i_r - i}{i_r}\right)^{-1}}{2\mu + 3\mu' + \frac{\varepsilon^2}{\varkappa}\left(1 + \frac{anFi}{2\varkappa RT} \cdot \frac{i_r - i}{i_r}\right)^{-1}}. \tag{113.7}$$

Formula (113.7) determines the dependence of the velocity $v_0$ on the current i. It is of greater interest, however, to derive the dependence of the velocity $v_0$ on the step-change in the potential difference $\Delta\varphi$ at the mercury-solution interface. This step-change, according to equation (113.3), causes a change in the current density i. Moreover, the step-change in the potential difference produces a change in the specific surface charge $\epsilon$.

It is evident from equation (113.4) that the change in the current density i with $\Delta\varphi$ is exponential. The change in $v_0$ with $\Delta\varphi$, therefore, is primarily determined by the factor i in the numerator of formula (113.7). We can therefore approximate formula (113.7) by

$$v_0 \approx \text{const}\, i \approx \text{const} \frac{1}{1 + \exp\left(\frac{nF\Delta\varphi}{RT}\right)}. \tag{113.8}$$

This formula shows that an increase in the negative step-change in the potential difference $\Delta\varphi$ causes a sharp increase in both the current density and in the tangential surface velocity. At current densities close to the limiting value, formula (113.7) can be simplified again. In this case the effective resistance becomes very large (since $\frac{i_r - i}{i_r} \to 0$), and formula (113.7) can be rewritten as

$$v_0 = \text{const}\,\frac{\varepsilon}{2\mu + 3\mu' + \varepsilon^2/\varkappa}, \tag{113.9}$$

where all quantities that are not functions of the potential difference $\Delta\varphi$ are combined in the constant (including the limiting diffusion current). According to formula (113.9), the surface velocity at $i \approx i_r$ is determined by the magnitude of the specific charge on the surface. It passes through a maximum at the point $\epsilon = \epsilon_{max} = [(2\mu + 3\mu')\varkappa]^{1/2}$ (see Section 99), and then decreases with increasing surface charge. Since the variation of $\epsilon$ with the change in the potential difference is dissimilar for positively and negatively charged surfaces, the change in $v_0$ with $\Delta\varphi$ must be examined separately for different signs of surface charge.

Figure 92 shows the change in surface velocity $v_0$ with the value of $\Delta\varphi$.

We will now consider a positively charged surface.

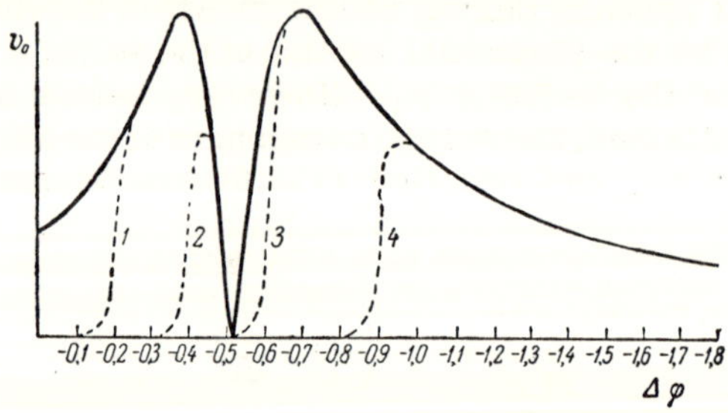


Figure 92. Theoretical dependence of the mercury surface velocity on the potential with passage of current for the case of reduction of a bivalent cation $\frac{c_k}{c} = 0.1$.

The broken curves 1 to 4 correspond to half-wave potentials of: - 0.2, - 0.4, - 0.6 and - 0.9 v (relative to a calomel reference electrode). Upon the appearance of concentration polarization, these curves become the solid-line curve expressed by general equation (113.9). Velocity values are given disregarding the direction of motion, which changes when passing through the point of zero charge, at $\varphi = -0.52$ v.

The curves are drawn for 0.01 N KCl for the reduction of a bivalent cation (n = 2) at a concentration of 0.001 N. The different curves correspond to different values of the half-wave potential $\varphi_{1/2}$ relative to the point of zero charge. All potentials in Figure 92 are relative to a calomel reference electrode, and therefore the zero point lies at $\Delta\varphi = -0.52$ v. The point $\epsilon = \epsilon_{max}$ lies in the 0.01 N KCl solution at a potential of 0.38 v. Three typical cases have been selected. Curve 1 is for a half-wave potential of $\varphi_{1/2} = -0.1$. This indicates that the electrochemical process starts to develop far from the point of zero charge under conditions of strong concentration polarization. Here, the dependence of $v_0$ on $\Delta\varphi$ is given by formula (113.9). In the second limiting case, $\varphi_{1/2} = -0.4$ v (curve 2), the electrochemical process starts to develop very close to the point of zero charge and there is insufficient time for the concentration polarization to become established at the start of the process. According to the general formula (113.7), the velocity $v_0$ is proportional to the product $(i\epsilon)$. The current i rises rapidly with an increase in the negative $\Delta\varphi$. On the contrary, the specific charge of the positively charged surface decreases with an increase in $\Delta\varphi$.

At the start of the process, the increase in i predominates, following which the increase in $v_0$ with i is noticeably reduced by the decrease in the surface charge $\epsilon$. At a high current density $i \approx i_r$, the current ceases to increase and the velocity passes through a maximum and becomes zero at the point of zero charge. The second case is the most typical one. Here, the change in $\epsilon$ with the potential at the start of the process is small, since it begins far from the point $\epsilon = \epsilon_{max}$. The velocity $v_0$ increases rapidly due to the increase in i, until i reaches a value close to the limiting current. Further change in $v_0$ then proceeds in accordance with formula (113.9).

In the case of a negative surface charge, three typical cases should be considered (Figure 92). If the process takes place under conditions of concentration polarization (curve 3), the velocity varies in accordance with formula (113.8): passing through a maximum at $\epsilon = -\epsilon_{max}$ (at a potential of 0.66 v) and accompanied by an increase in the negative potential of the mercury surface. In cases where the electrochemical process develops in the immediate vicinity of the zero point, to the left of the point $\epsilon = -\epsilon_{max}$ the velocity $v_0$ increases with an increase in the product $(i\epsilon)$. In the case of a negative charge, both factors in the product and the velocity $v_0$ increase in accordance with formula (113.7) until the current i attains the limiting value. Further changes in $v_0$ obey equation (113.8). Most commonly, this process develops at more negative potentials than the potential $\epsilon = \epsilon_{max}$. The velocity increases with the factor i and decreases with the factor $\frac{\epsilon}{2\mu + 3\mu' + \epsilon^2/\varkappa}$. As soon as current i attains the limiting value, the velocity $v_i$ starts to decrease with the factor $\frac{\epsilon}{2\mu + 3\mu' + \epsilon^2/\varkappa}$, since the transition to a condition of concentration polarization occurs after the point $\epsilon = -\epsilon_{max}$ (curve 4).

The total current flowing through the drop is composed of the usual current associated with the radial growth of the drop, and an additional current that arises from the additional stirring caused by the tangential motion of the drop surface. According to (72.10), the latter current is proportional to $v_0^{1/2}$. The total current flowing on the drop, therefore, is related to the mercury cathode potential by

$$I = I_{Ilk} + \text{const}\sqrt{v_0(\Delta\varphi)}, \qquad (113.10)$$

and passes through a maximum simultaneously with the velocity $v_0(\Delta\varphi)$. The dependence of I on $\Delta\varphi$ is represented by curve 1 shown in Figure 93.

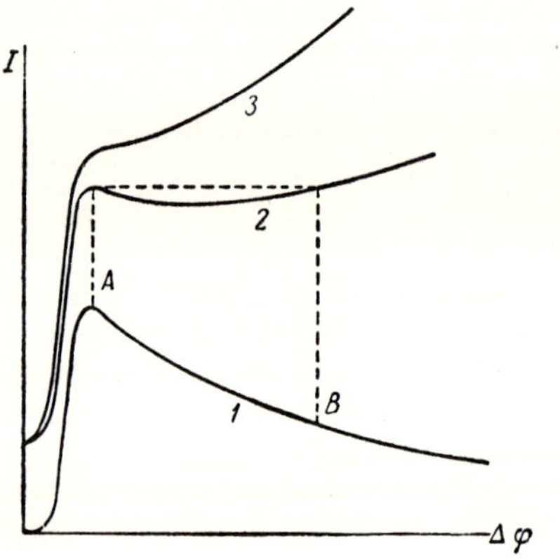


Figure 93. Relationship between current density, applied voltage and solution-cathode potential. 1 — current density; 2 — applied voltage $\frac{E_0}{w} = I - \frac{\varphi}{w}$; 3 — applied voltage at a lower value of w. The position of curves 2 and 3 on the ordinate is arbitrary, since it depends on the choice of the anode for the cell.

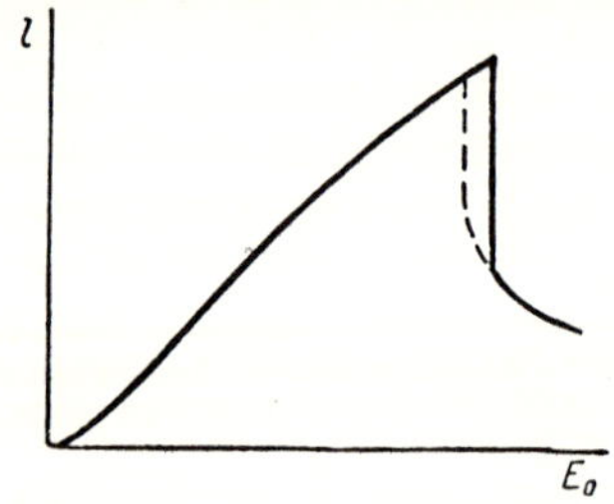


Figure 94. Relationship between the current density and applied voltage. The solid line curve gives the direct, and the broken line curve the reverse current.

At some value of the potential, the current starts to rise sharply. Upon further increase in the negative potential of the drop, the current reaches a maximum and then decreases smoothly. This variation of the current on a polarograph, as a function of the dropping mercury electrode potential $\Delta\varphi$, agrees with the experimental curve of the current I plotted as a function of the applied voltage E. Indeed, if the curve $I(\Delta\varphi)$ has the form of curve 1 in Figure 93, the quantity $\frac{E}{w_0} = -\frac{\Delta\varphi}{w_0} + I(\Delta\varphi)$, where $w_0$ is the ohmic resistance of the circuit, has the form shown in Figure 94. Curves 2 and 3 in Figure 93 represent $E/w_0$, rather than E alone (to preserve the scale). Curves 2 and 3 are obtained from curve 1 by subtracting the quantity $\frac{\Delta\varphi}{w_0}$. Curve 2 represents the case of moderately low values of $w_0$. At very low values of resistance $w_0$ (in concentrated solutions), the relationship between $\frac{E}{w_0}$ and $\Delta\varphi$ is given by curve 3.

The case of moderately low resistances, given by curve 2 in Figure 93, is of interest in discussing the theory of polarographic maxima.

When the voltage E reaches its maximum value at the point A, any further continuous rise in the voltage can occur only by a step-change in $\Delta\varphi$ to point B on the second segment of curve 2 with a positive slope. The corresponding change in the electrode potential is shown by a broken line in Figure 93. In this case, the current flowing through the drop, decreases sharply by a step-change from $I_A$ to $I_B$. The characteristic maximum, accompanied by a sharp decrease in the current on the current-voltage curve, corresponds to the smooth maximum on the current-electrode potential curve. This point has been verified experimentally (Figure 94).

Thus, using the theory of electrocapillary motion, and accounting for the effect of tangential surface motion of the mercury cathode, a semiquantitative theory for polarographic maxima of the first type can be constructed. In this case, there is no need to advance any arbitrary and unfounded hypotheses regarding either the nature of adsorption forces or the kinetics of formation of the double layer. Such hypotheses have been advanced by Antweiler and Stackelberg [17] and other authors, in their unsuccessful attempts to explain the appearance of polarographic maxima.

The following general conclusion of the theory presented above can now be given: when the potential of the mercury cathode varies over the range $(\Delta\varphi) > (\Delta\varphi_{max})$ to $(\Delta\varphi) > (-\Delta\varphi)_{max}$, i.e., when the drop surface charge varies from a positive value of $\epsilon < \epsilon_{max}$ to a negative value $(-\epsilon) > (-\epsilon_{max})$, two maxima should appear on the current-potential curve. One of the maxima must occur at a surface potential which corresponds to the maximum surface charge $\epsilon_{max}$. The other maximum will occur at a potential which corresponds to $-\epsilon_{max}$. A certain difference, although not a basic one, exists between the maxima on the positive and negative branches of the electrocapillary curve. The shape of the curve $I(\Delta\varphi)$, for $\epsilon > 0$, differs somewhat from its shape at $\epsilon < 0$. The most typical current-potential curve is shown in Figure 95. We assume, of course, that the ions in the solution shown in Figure 95 discharge at positive potentials of the electrocapillary curve. The current in Figure 95 has been calculated by formulas (113.6) and (72.10). Characteristic polarographic maxima on the current-voltage curve correspond to the smooth maxima on the current-potential curve.

The derivation given here is semiquantitative. The complexity of geometric conditions precludes an accurate computation of the numerical coefficient in the formula for the current through the drop.

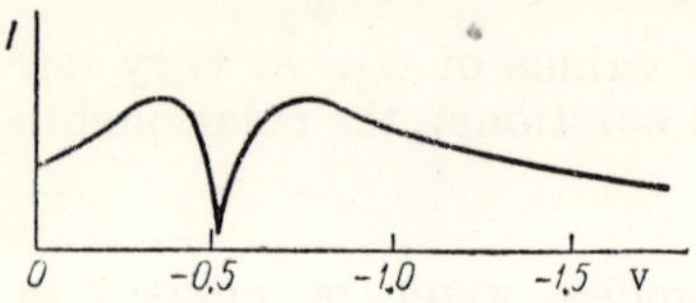

Figure 95. Theoretical relationship between the maximum current and potential in a 0.01 N KCl solution.

A very important assumption was made in the derivation of the above theory: we considered that the decrease in potential along the drop is given by equation (113.2). Here, the relationship between $v_0$ and the potential (113.2′) is determined solely by the dependence on the potential in the product (Ui). We will show later that the assumption of a uniform potential decrease along the drop is very important in explaining the appearance of the two maxima on either side of the zero point.

It is shown in Section 114 that this assumption does not apply to the conditions prevailing during the operation of a polarograph. Therefore, our basic conclusion regarding the existence of two maxima is not valid for the usual operating conditions.

As shown in greater detail below, the two maxima are usually not observed in practice. As a rule, only one of them is attained at positive potentials of the electrocapillary curve. However, the absence of the second maximum on the $I(\Delta\varphi)$ curve cannot yet be considered as invalidating the theory presented above. Therefore, the absence of the two theoretically predicted and almost symmetrical maxima on the $I(\Delta\varphi)$ curve does not per se contradict that theory.

To verify the theory, the experimental conditions must positively assure a uniform potential decrease along the drop.

Such conditions were attained in specially designed experiments [22]. Tangential motion of mercury was caused by an external electrical field applied to a dropping mercury electrode (Figure 96). The solid arrows show the direction of the applied field between the electrodes A and C. The current flowing between the electrodes A and C is not related to the electrical reduction occurring at the mercury cathode. The broken lines represent the lines of the current flowing to the dropping mercury electrode, where the discharge of mercury ions occurs.

To maintain a sufficiently slow mercury flow from the capillary (in order to prevent the appearance of maxima of the second kind), the drop time was set at 5 to 6 seconds. Experiments were conducted in a very low concentration of the substance being reduced — mercury bichloride $HgCl_2$ and mercury perchlorate $Hg(ClO_4)_2$ at $0.14 \cdot 10^{-4}$ to $6 \cdot 10^{-4}$ N — and in the presence of a supporting electrolyte. 3 N calcium chloride and calcium perchlorate were used as the supporting electrolyte. Under these conditions, the first kind maxima were not present and the current to the dropping mercury electrode (in the absence of an applied external field) was exactly equal to that computed by the Ilkovic equation.

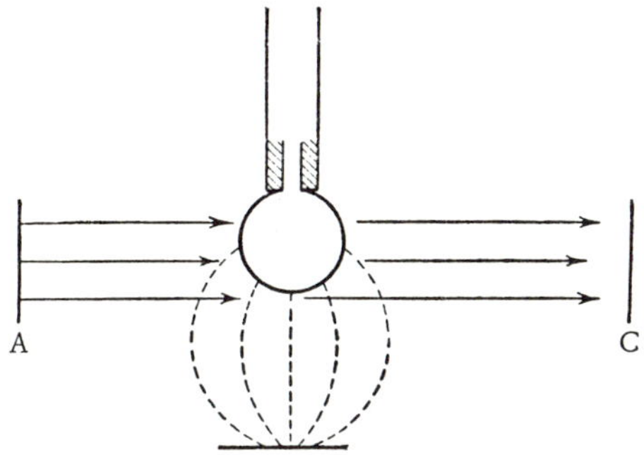


Figure 96. Experimental setup for studying the motion of a mercury drop in an electrical field.

When an external field with a voltage of E = 0.2 to 0.7 v/cm was applied, a tangential motion appeared on the mercury surface. The direction of the motion depended, of course, on the sign of the surface charge.

The motion was in opposite directions for positively and negatively charged mercury surface (the sign of the charge could be changed by adjusting the potential of the drop). The tangential velocity was calculated from equation (113.1), in which the decrease in the potential ($\Delta\varphi_0$) along the drop is equal to 3Ea, where E is the applied field voltage. Under these conditions the decrease in the potential along the drop was very symmetrical, and the distorting effect of the capillary was reduced to a minimum. The above formulas can therefore be considered applicable to the case of tangential motion of the mercury drop surface. The tangential mercury surface motion led to improved stirring conditions and to an increase in the flux of the substance being reduced to the dropping mercury electrode.

The relationship between the current flowing to the drop and the potential of the drop was determined in this experiment. It was found that, in the presence of an external field, two maxima appeared on the $I(\Delta\varphi)$ curve. The maxima were located on the positive and negative segments of the electrocapillary curve. In the interval between them, at the electrocapillary zero, the current decreased to a value equal to the current in the absence of an external field. The latter value coincided with that computed by the Ilkovic equation. The value of the current at the maxima exceeded the value of the current in the absence of a field by several times; it increased with an increase in the applied field. The general path of the $I(\Delta\varphi)$ curves was found to coincide with that theoretically predicted (Figure 97).

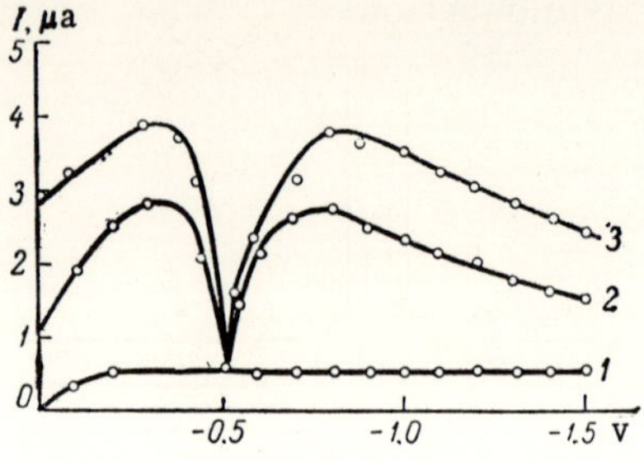


Figure 97. Polarization curves in $0.78 \cdot 10^{-4}$ N $Hg_2(ClO_4)_2$ + 0.01 N KCl.
1 — in the absence of external field; 2 — with an applied field voltage of E = 0.21 v/cm; 3 — with an applied field voltage of E = 0.42 v/cm; $\varphi$ is computed relative to the 0.01 N calomel reference electrode.

Quantitative verification of the theory was made by comparing the experimental and predicted positions of three characteristic points: the two points of maximum current ($\epsilon_{max}$) and ($-\epsilon_{max}$), and the point at which the current dropped to a minimum value (which, according to the theory, must correspond to the zero point on the electrocapillary curve).

Table 19 shows the calculated and experimentally measured positions of these characteristic points.

Table 19

| Potassium perchlorate concentration | Electrocapillary zero potential, derived from electrocapillary curves | Potential at current decrease obtained from polarographic curves | Calculated potentials corresponding to maximum slopes | Observed potentials corresponding to maximum current values |
|---|---|---|---|---|
| 0.01 | - 0.62 | - 0.55 | {- 0.47<br>- 0.82 | - 0.4 — 0.45<br>- 0.80 — 0.85 |
| 0.002 | - 0.62 | - 0.6 | {- 0.52<br>- 0.72 | - 0.5<br>- 0.7 |

Table 19 shows that the observed positions of the three characteristic points are somewhat shifted, in comparison with theoretical values, in the direction of the negative potentials (all potentials are given relative to 0.01 N calomel reference electrode). Nevertheless, we note not only a confirmation of the theoretical prediction of the existence of two maxima, but also the relatively satisfactory agreement between the experimental data and the theory.

In spite of the semiquantitative nature of the computed results, the theory correctly reflects the mechanism of the phenomena, and can be used to determine the position of the characteristic points on the $I(\Delta\varphi)$ curve with sufficient accuracy. It should be noted that in the study of T. I. Popova and T. A. Kryukova [22], the symmetrical shape of the maxima was distorted at high mercury ion concentrations, and the maxima shifted considerably in the direction of more negative potentials. Apparently, complex tangential motion caused by the external field was superposed on the usual tangential motion of the mercury drop itself.

## 114. MAXIMA ON THE POSITIVE AND NEGATIVE SEGMENTS OF THE ELECTROCAPILLARY CURVE

We have stated above that the theoretical predictions regarding the existence of two symmetrical polarographic maxima are not confirmed under normal operating conditions. A survey of numerous experimental studies shows that investigators have observed only a single maxima, located either on the positive or on the negative segment of the electrocapillary curve. Even before the cited experimental verification of the theory was conducted, the same authors advanced the possibility that this discrepancy between theory and practice could be explained by the failure to maintain the uniform decrease in the potential along the drop. In other words, the authors have supposed that it is not sufficient to analyze only the factor (Ui) of equation (113.1) in order to formulate a workable theory of polarographic maxima that could be applicable to normal operating conditions. An additional analysis of the dependence of $(\Delta\Phi_0)$ in equation (113.2) on the electrode geometry on the type of flow and on the polarity of the surface charge is required. It is obvious that a change in the charge is a function of the polarity of the charge on the mercury surface. This was first indicated by Z. A. Iofa, V. V. Losev et al. (Chapter IX [17]). In determining the potential difference between various points on the electrode, only the difference between the current paths is usually considered. If the ohmic resistance $w_0$ in the path to any part of the electrode is higher, then, of course, the current density and, consequently, the polarization of that part of the surface are lower. In considering the conditions for electrode polarization in a liquid stream, the profile of the variation in the potential on the surface is also a strong function of the rate and type of delivery of the substance being reduced to the various points on the surface. The rate and type of flow of this substance affect the value of w, which, in turn, determines the polarizability of the electrode [see (113.3)]. If a fresh stream of the solution is delivered to a portion of the electrode surface, the quantity w and therefore the shift in the potential at a given current density become smaller. On the other hand, we can now examine the situation where the solution being delivered to the surface has already made a previous contact with the electrode during its course of flow. As a result of frictional velocity loss (and the impoverishment of the solution) the rate of delivery of the substance to the surface is lowered. Thus, at this portion of the electrode surface, the value of w and, consequently, the potential shift become higher. In the general case, both the current distribution along the electrode surface and the quantity ($\Delta\Phi_0$) are functions of the electrode geometry, which determines the magnitude of w, and of the type of liquid flow. The behavior of positively and negatively charged surfaces, from this standpoint, must be very different, as is easily shown using a dropping mercury electrode as an example. The geometrical conditions in this electrode are such that the lower surface of the drop is more easily polarized because the upper portion is somewhat farther from the anode and is shielded by the capillary. Thus, the current flow to the mercury becomes more difficult.

We will now examine the two cases individually.

When the mercury surface is positively charged, the higher polarization of the lower portion causes a relative increase in surface tension. The mercury surface and the adjacent layers of the solution move downward (Figure 98). The stream of the fresh solution, in this case, first comes into contact with the upper portion of the drop. As a result, the quantity w in the upper portion becomes smaller than in the lower portion, and the shift in potential (which has already become lower because of the unfavorable geometrical conditions) decreases further compared to the potential shift in the lower portion. In the case of a positively charged drop, therefore, the effects associated with the difference between w and $w_0$ are additive and strengthen each other.

In the case of a negatively charged surface, the flow of current reduces surface tension. The reduction is more pronounced in the lower portion of the drop, and the motion of both the surface and of surrounding solution is upward. Here, the fresh solution is delivered to the lower portion of the drop, and consequently the quantity w is smaller than in the upper portion. The higher current passing through the lower portion of the drop, therefore, causes a relatively small shift in the potential. The effects related to the difference between w and $w_0$ in the upper and lower portions of the drop are not additive. In this case they are subtractive and mutually weakening. As a result, all other conditions remaining equal, the intensity of motion at negative maxima is lower than at positive maxima.

The complexity of geometric and dynamic conditions applying to a drop issuing from a capillary has prevented the formulation of a quantitative theory of this phenomenon up to date.

We are, therefore, forced to limit our analysis to a very approximate evaluation, that can only augment the quantitative method. As a model of a non-uniformly polarized dropping mercury electrode, we examine the combination of two sections of an electrode, with different current lines and different values of the limiting current. The first of these sections corresponds to the lower portion, and the second to the upper portion of the drop. We designate by the subscript 1 the values of w, $w_0$, i and $i_{lim}$ in the first section, and by the subscript 2, the values in the second section of the electrode. Symbols without subscripts indicate the average values of the same quantities. Let $\Delta w$, $\Delta w_0$, $\Delta i$ and $\Delta i_{lim}$ indicate the differences $w_1 - w_2$, $w_{01} - w_{02}$, etc. We assume the quantities $\frac{\Delta w}{w}$, $\frac{\Delta w_0}{w_0}$, $\frac{\Delta i}{i}$ and $\frac{\Delta i_{lim}}{i_{lim}}$ to be small. We first assume that the potential differences at both sections of the electrode are a certain function of the ratio $\frac{i}{i_{lim}}$, where i is the actual current flowing to the section, i.e., $\varphi = f\left(\frac{i}{i_{lim}}\right)$. We make no specific assumptions regarding the form of this function, but assume temporarily that it is the same at both sections of the electrode; hence $\varphi_1 = f\left(\frac{i_1}{i_{1lim}}\right)$ and $\varphi_2 = f\left(\frac{i_2}{i_{2lim}}\right)$, where $\varphi_1$ and $\varphi_2$ are the metal-solution potential differences at both sections of the electrode.

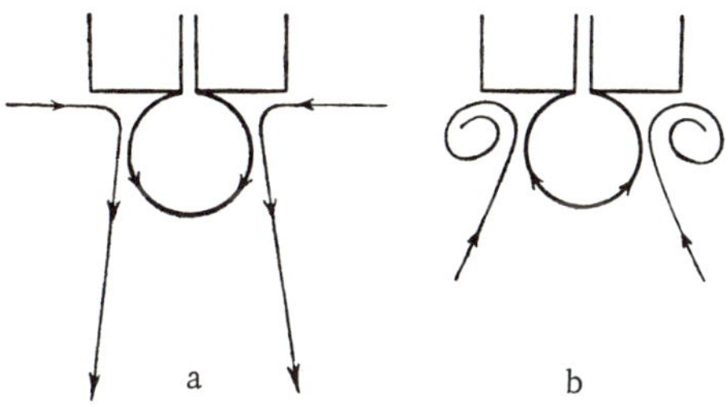

a b

Figure 98. Motion of the mercury surface and adjacent layers of the solution caused by nonuniform polarization.
a — positively charged surface;
b — negatively charged surface.

Our problem here is to compute the potential difference ($\Delta\Phi_0$) between points in the solution and sections 2 and 1 of the electrode.

By definition,

$$(\Delta\Phi_0) = -(\varphi_2 - \varphi_1) = f\left(\frac{i_1}{i_{1\lim}}\right) - f\left(\frac{i_2}{i_{2\lim}}\right) = \frac{\partial f}{\partial i}\left(\frac{\Delta i}{i_{\lim}} - \frac{i\,\Delta i_{\lim}}{i_{\lim}^2}\right),$$

and since

$$w = -\frac{1}{i_{\lim}}\frac{\partial f}{\partial i},$$

then

$$(\Delta\Phi_0) = -w\left(\Delta i - \frac{i}{i_{\lim}}\Delta i_{\lim}\right). \qquad (114.1)$$

Let a potential difference E be applied between the electrode under consideration and some auxiliary nonpolarizable electrode. Then, obviously,

$$E = -w_{1,0}i_1 + \varphi_1 = -w_{1,0}i_1 + f\left(\frac{i_1}{i_{1\lim}}\right), \qquad (114.2)$$

$$E = -w_{2,0}i_2 + \varphi_2 = -w_{2,0}i_2 + f\left(\frac{i_2}{i_{2\lim}}\right). \qquad (114.3)$$

From equations (114.2) and (114.3), with an accuracy to the first order of infinitesimals, we get

$$-\Delta w_0 i - w_0\Delta i - w\left(\Delta i - \frac{i}{i_{\lim}}\Delta i_{\lim}\right) = 0,$$

whence,

$$\Delta i = \frac{-\Delta w_0 + \frac{w}{i_{\lim}}\cdot\Delta i_{\lim}}{w_0 + w}\,i,$$

and, consequently, from equation (114.1) and to the same degree of accuracy, we get

$$(\Delta\Phi_0) = \frac{i_{\lim}\Delta w_0 + w_0\,\Delta i_{\lim}}{i_{\lim}\left(1 + \frac{w_0}{w}\right)}\,i = \frac{w_{1,0}i_{1\lim} - w_{2,0}i_{2\lim}}{i_{\lim}\left(1 + \frac{w_0}{w}\right)}\,i. \qquad (114.4)$$

Equation (114.4) shows that the magnitude and the sign of the potential difference across the double layer are determined both by the changes in $w_0$ and in $i_{\lim}$ along the surface of the electrode. If the stirring is caused by the motion of the nonuniformly polarized electrode surface, then a substantial difference exists between the behavior of the positively and negatively charged mercury surface.

To be specific we assume that $w_{1,0} < w_{2,0}$; i.e., the current path to section 1 is shorter than to section 2. Section 1 of the model then corresponds to the lower, more easily polarizable portion of the capillary electrode, and section 2 corresponds to its upper portion. In this case $i_{1lim} = i_{2lim}$ $(\Delta\Phi_0) < 0$ and, therefore, $v_0 < 0$; i.e., the motion is in the direction from 2 to 1 (downward) at $\epsilon > 0$, and from 1 to 2 (upward) at $\epsilon < 0$. If the stirring is significantly dependent on the electrocapillary motion of the electrode, then, in the first instance, the fresh stream of the solution is delivered to section 2 (the upper part) of the electrode, and, consequently, $i_{1lim} > i_{2lim}$. Then, from equation (114.4) the absolute values of $(\Delta\Phi_0)$ and consequently of $v_t$ increase compared to their values at $i_{1lim} = i_{2lim}$.

In the second instance, the fresh solution reaches section 2, i.e., the lower portion of the electrode, and, therefore, $i_{1lim} > i_{2lim}$. This, in accordance with equation (114.4), must lead to a decrease in the absolute values of both $(\Delta\Phi_0)$ and $v_0$. At relatively high values of the ratio $\frac{i_{1lim}}{i_{2lim}}$, the quantity $(\Delta\Phi_0)$, and therefore $v_0$, may even become zero, or may reverse their sign with respect to their polarity at $w_{1,0} < w_{2,0}$ and $i_{1lim} = i_{2lim}$. In this sense, it can be stated that the conditions for the appearance of electrocapillary motion and polarographic maxima are less favorable and more complex in the case of a negatively charged mercury surface than in a positively charged surface. The appearance of the electrocapillary motion and of the maxima in a negatively charged surface are governed by an additional condition, which states that the quantity $\frac{i_{1lim}}{i_{2lim}}$ must not exceed certain limits.

In reality, this difference goes farther than could be concluded on the basis of equation (114.4). The effect of the direction of the stream on polarization distribution must increase with an increase in the value of the current flowing through the drop. While this current is small compared to the limiting diffusion current, there is no significant depletion of the solution that contacts the mercury. Thus, the concentration of the solution that arrives at the upper portion of the drop differs little from the initial concentration. When considerable concentration polarization takes place, the solution loses almost all of the substance being reduced when it comes into contact with the mercury surface. As a result a highly depleted solution reaches the upper portion of the drop. Since the velocity distribution is such that the delivery of the substance to the upper portion of the drop is hindered, only small currents can pass through this portion of the drop. Here, even a low current will produce a substantial shift in the potential. In other words, in a negatively charged surface an increase in concentration polarization causes an increase in the polarizability of the upper portion with respect to the lower portion of the drop. This argument is fully confirmed by the general relaxation theory of diffusion presented in Section 18.

## 115. SUPPRESSION OF POLAROGRAPHIC MAXIMA AND SOME PRACTICAL APPLICATIONS

It has been clarified above that tangential motion of the surface of a dropping mercury electrode is caused by two different mechanisms: by the flow of mercury from the capillary and by electrocapillary motion in an electrical field. In section 112 we have established that the tangential velocity at any drop charge is proportional to the velocity $v_{max}$ at zero charge. The latter velocity in turn increases with the velocity of mercury flowing from the capillary. The tangential velocity $v_0$ is given by equation (112.2). This equation shows that $v_0$ decreases with increase in the retardation factor $\frac{\epsilon^2/\varkappa}{2\mu + 3\mu'}$. If the conductivity of the solution decreases (i.e., at lower $\varkappa$), the retardation factor increases. At the same time, the tangential motion of the drop may occur with an ever decreasing range of values of the surface charge. Figure 84 shows that the quantity $v_0$ becomes practically zero at low concentrations of the solution. (The concentration decreases with increasing exit velocity from the capillary and with increasing departure of electrode potential from the point of zero potential.) Thus, suppression of tangential motion, and the second kind maxima on the current-voltage curves, can be obtained by lowering both the concentration of the solution and the exit velocity of mercury from the capillary. In practice, the drop time must be reduced to 5 to 6 sec, which causes many operating difficulties and inconveniences.

As is well known from experience, in order to suppress the maxima of the first kind, the conductivity of the solution must be increased substantially by adding considerable amounts of supporting electrolyte. This conclusion also follows from the theory of the maxima of the first kind. From the general equation (113.2′), the tangential velocity is proportional to the product $U \cdot \frac{1}{\varkappa}$. At the maximum point $U = U_{max}$, which is determined from $\epsilon = \epsilon_{max}$, the tangential velocity is

$$(v_0)_{max} \sim \frac{1}{\varkappa}. \tag{115.1}$$

According to equation (72.10), the maximum current is proportional to

$$I_{max} \sim \frac{1}{\sqrt{\varkappa}}. \tag{115.2}$$

Formulas (115.1) and (115.2) show that the tangential motion and the first kind of polarographic maximum can be suppressed by increasing the conductivity of the solution. Thus, the conditions for suppression of maxima of the first and second kind are opposite in relation to the solution.

In polarographs with short drop times ($\tau \approx 1$ to 2 sec) it is difficult to attain a concentration of the supporting electrolyte that would supress both the first and the second kind of maximum. In practice, therefore, the damping effect of surface-active agents is widely used to suppress polarographic maxima. Such agents as gelatin, dyes, alcohols and other surfactants have been used for this purpose for a relatively long time [23]. We will show below, however, that such a completely empirical approach may, and actually did, lead to serious errors.

The following discussion is based on the theory of retardation of surface motion by surface-active agents as given in Sections 73 to 75. In these sections, we studied the mechanism by which the surface motion of a freely falling drop is retarded. It was shown that the surface velocity of the drop is given by

$$v_0 = v_{max} \frac{1}{1 + \frac{\gamma}{2\mu + 3\mu'}} = v_{max}\zeta, \tag{115.3}$$

where $v_{max}$ is the tangential velocity on the surface of the drop in the absence of a surface-active agent and $\gamma$ is the magnitude of retardation of the motion by a surface-active agent. In form, equation (115.3) coincides with equation (99.20), in which an entirely different retardation mechanism is considered. Owing to the similarity of the two equations, we can express the tangential velocity in the general case as

$$v_0 = v_{max} \frac{1}{1 + \frac{\gamma + \varepsilon^2/\varkappa}{2\mu + 3\mu'}}. \tag{115.4}$$

The value of the total retardation coefficient $\left(\gamma + \frac{\epsilon^2}{\varkappa}\right)$ can be determined experimentally be measuring currents flowing to a drop. Indeed, from equations (115.4) and (112.4), we get

$$\frac{v_0}{v_{max}} = \frac{I^2 - I_{Ilk}^2}{I_{max}^2 - I_{Ilk}^2}, \tag{115.5}$$

where I is the current at a tangential velocity $v_0$, $I_{max}$ is the current at the maximum tangential velocity in the absence of retardation, and $I_{Ilk}$ is the current correction for radial motion of the drop. From (115.3) and (115.2), we find

$$\zeta = \frac{2\mu + 3\mu' + \gamma + \varepsilon^2/\varkappa}{2\mu + 3\mu'} = \frac{I_{max}^2 - I_{Ilk}^2}{I^2 - I_{Ilk}^2}. \tag{115.6}$$

The value of $\gamma$ can be determined from equation (115.6) by substituting the experimentally measured values of I and $I_{max}$.

The value of $\gamma$ that characterizes surface retardation has been examined in detail in Sections 73 to 75. We showed that, depending on the nature of the process that determines the adsorption equilibrium, the theory leads to various values for $\gamma$. Simple estimates show that for any significant concentration of the solution, the electrical retardation $\epsilon^2/\varkappa$ is small compared to the adsorption retardation $\gamma$. It may be assumed, therefore, that the only significant retardation mechanism is caused by the adsorbed surface-active agents. We then have the following conditions in accordance with Sections 73 to 75:

1) Retardation of the charged surface should occur only in that range of potentials where noticeable adsorption of surface-active agents takes place.

2) If convective diffusion controls the rate of delivery of the substance, then $\gamma$ may be computed by equation (75.8). In that case: a) $\gamma$ increases as the square of the specific adsorption of materials in a homologous series; b) $\gamma$ increases with increasing concentration of the surface-active agent $c_0$ in the solution up to a value of $c_0 \leq \frac{\Gamma_\infty}{k}$, and then decreases with further increase in $c_0$.

3) If the adsorption-desorption process controls the rate of delivery of the substance, $\gamma$ is determined by equation (74.21). In that case: a) $\gamma$ is directly proportional to the specific adsorption in a homologous series; b) $\gamma$ increases monotonically with concentration $c_0$ until the latter attains a certain constant maximum as given by equation (74.18).

If the delivery mechanism is given by equation (75.2), then the theoretical value of I can be found from equations (75.7) and (115.6).

Combining these two equations, we get

$$1 + \frac{2RT\Gamma_0^2\delta}{3(2\mu + 3\mu')Dac_0} = \frac{I_{\max}^2 - I_{\text{Ilk}}^2}{I^2 - I_{\text{Ilk}}^2}. \tag{115.7}$$

The thickness of the diffusion layer $\delta$ may be expressed in terms of the current flowing to the drop,

$$\delta = \frac{(nF)D_{\text{Hg}}c_{\text{Hg}}S}{I}, \tag{115.8}$$

where $c_{\text{Hg}}$ and $D_{\text{Hg}}$ are the concentration and the diffusion coefficient of mercury ions, respectively, and S is the surface area of the drop. Thus, we obtain the following equation for the current I:

$$\frac{I_{\max}^2 - I_{\text{Ilk}}^2}{I^2 - I_{\text{Ilk}}^2} = \frac{2RT\Gamma_0^2}{3(2\mu + 3\mu')D}\,\frac{4\pi a(nF)D_{\text{Hg}}c_{\text{Hg}}}{c_0 \cdot I} + 1. \tag{115.9}$$

The radius of the drop a in equation (115.9) can be substituted by its approximate value at the instant of separation. The value of $I_{\max}$ is determined from equation (72.10).

The value of $\Gamma_0$ as a function of the mercury surface potential can be obtained from the electrocapillary curve for mercury [24].

The theory was verified in the experiments of T. A. Kryukova [25].

The measured values of the current through the drop, the velocity and the surface tension of mercury are given in Figure 99.

The latter shows that the characteristic "cutoff" on the electrocapillary curve caused by adsorption of surface-active substance extends precisely to that region of potential difference in which the current through the drop decreases to values that are limited solely by the radial growth of the drop. Arrows indicate the visually determined tangential surface velocities.

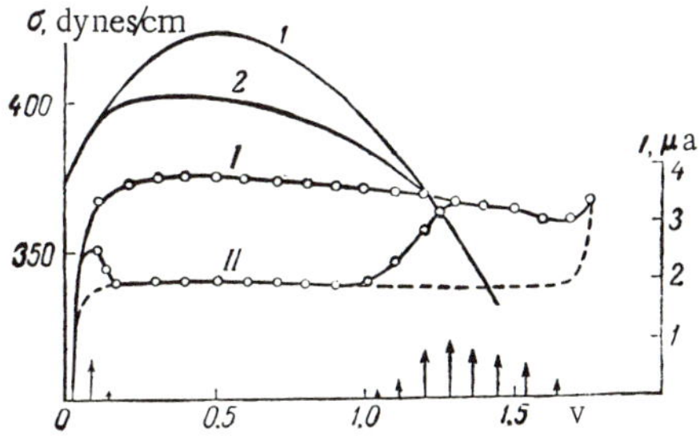


Figure 99. Effect of adsorption on the dependence of current on surface tension in the presence of tangential surface motion.

Curve I shows the dependence of the current I on voltage for 1 N $Na_2SO_4$ + $3.10^{-4}$ N $Hg_2$ . . .; II — same, in the presence of 0.1 M butyl alcohol; broken line curve — same, following complete damping of motion by the addition of gelatin; 1 — dependence of the interfacial tension $\sigma$ on the potential in 1 N $Na_2SO_4$; 2 — same in the presence of 0.1 M butyl alcohol. Arrows show the mercury surface velocity.

Indeed, there is no tangential motion in the region where the current through the drop is reduced.

Considering that the curves were computed without any empirical or theoretical constants, the agreement between theory and experiment must be acknowledged as satisfactory.

Figures 100 and 101 give the values of current I to a dropping mercury electrode at distinctly different rates of flow from the capillary (different values of U). Noticeable features of the curves

$I(\Delta\varphi)$ are the saddlelike depressions of the curves in the electrocapillary zero region and the changes in the shape of the curves with concentration. The former indicates the sharp damping of the tangential motion by the surface-active agent, leading to a decrease in the current to values determined by the Ilkovic equation (right line 7). When the mercury surface charge is large, adsorption decreases sharply and tangential motion of the surface becomes possible. Accordingly, the current to the drop increases to its value in a pure solution. The slope of the curves in the presence of a surface-active substance reflects the characteristic of equation (75.8): the retardation first increases with the concentration (curves 2, 3, 4) and then starts to decrease (curves 5 and 6).

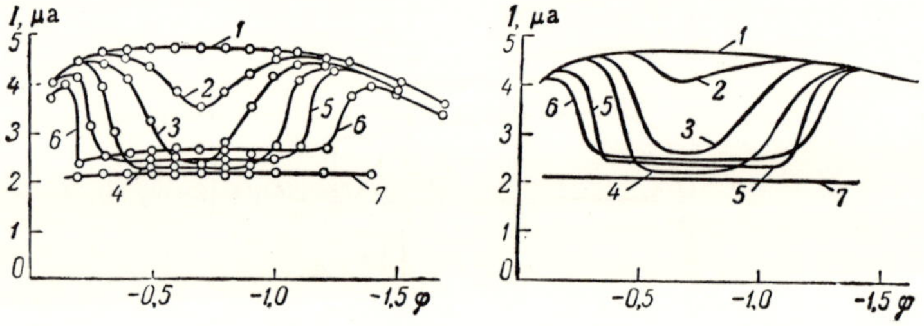


Figure 100. Experimental (left) and theoretical (right) polarization curves in 3 N KCl $+3\cdot10^{-4}$N $HgCl_2$ in the presence of n-butyl alcohol.
1 — pure solution; 2 — same $+6\cdot10^{-3}$ M $C_4H_9OH$; 3 — + 0.01 M $C_4H_9OH$; 4 — + 0.03 M $C_4H_9OH$; 5 — + 0.1 M $C_4H_9OH$; 6 — + 0.3 M $C_4H_9OH$; 7 — polarization curve with complete damping of tangential surface motion by 0.01 percent solution of gelatin. Pressure — 270 mm of mercury.

The maximum retarding effect occurs with a 0.03 M concentration of n-butyl alcohol (curve 4). The existence of maximum retardation at some concentration of the surface-active agent predicted by equation (75.8) shows that the diffusion of molecules of surface-active agents is the slow stage in the establishment of an adsorption equilibrium on the surface of a drop in a solution of n-butyl alcohol.

The existence of retardation maxima at relatively low concentration of surface-active agents is indicative of the care that should be exercised in the use of surfactants for the suppression of polarographic maxima. Additions of excessive amounts of surface-active agents could lead to a decrease and even to disappearance of the effect.

The theory of the motion-damping effect of surfactants provides the explanation for the appearance of so-called false polarographic waves, and offers methods for preventing them.

False polarographic waves are maxima that appear on a current-voltage curve not as a result of some new reduction or oxidation process, but because of cessation of the retarding action on the surface of the drop.

When a polarographic analysis is conducted in the presence of a surface-active agent, the surface motion on a drop becomes damped and the current flowing through it is given by the Ilkovic equation.

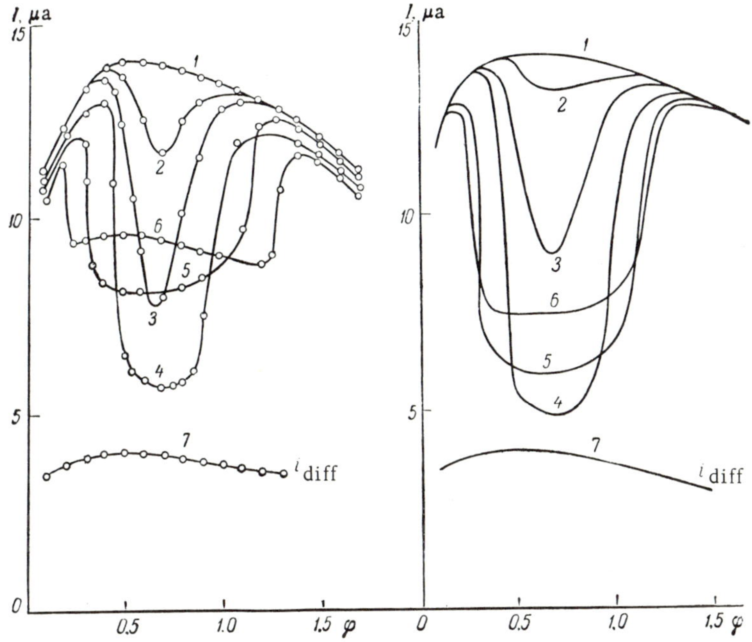


Figure 101. Experimental (left) and computed (right) polarization curves in 3 N KCl $+3 \cdot 10^{-4}$N $HgCl_2$ in the presence of n-butyl alcohol.

1 — pure solution; 2 — same $+6 \cdot 10^{-3}$ M $C_4H_9OH$; 3 — same + 0.01 M $C_4H_9OH$; 4 — + 0.03 M $C_4H_9OH$; 5 — + 0.1 M $C_4H_9OH$; 6 — + 0.3 M $C_4H_9OH$; 7 — polarization curve with complete damping of tangential surface motion by 0.01 percent solution of gelatin. Pressure — 850 mm of mercury.

If the applied emf is increased, however, the cathode potential can attain a value at which the surface-active agents become desorbed from the surface of the drop. The drop surface thus freed of the adsorbed substances begins to move and a new wave appears on the current-voltage curve. This new wave simulates the reduction process on the cathode.

T. A. Kryukova [26] showed that such a false maximum was erroneously interpreted by Orleman and Kolthoff [27] as the process of total reduction of water. In order to explain this process, the authors have developed a special theory which is, however, in contradiction with the principles of thermodynamics.

The elimination of false waves can be accomplished by the following means: operation with very pure reagents, free of even the most minute traces of surface-active agents; operation in the presence of very considerable amounts of surface-active agents which can be adsorbed at all potentials; operation at very high dropping rates when the mercury surface changes too rapidly to permit the adsorption of any significant amount of surfactants.

Agreement between values calculated from the theoretical equation (115.9) and the experimental data can be considered an indirect confirmation of equation (72.10) for the mass flow to the surface of a drop.

The retardation coefficient $\gamma_1$, given by equation (74.19), is proportional to the square of the adsorption coefficient k. If surface-active agents of one homologous series are taken as the adsorbing molecules, then, in accordance with the well-known empirical rule of Traube, the adsorption coefficient k increases with an increase in the chain length by one CH group. Therefore, the retarding action of the consecutive terms of a homologous series must become nine times stronger.

If the maximum suppressor used is a highly adsorbable substance that is soluble with difficulty (n-hexyl alcohol, for example), the retarding effect does not pass through a maximum. In this case the ratio $\Gamma_0/c_0$ is always high.

Understanding of the mechanism by which surface-active agents affect polarographic maxima has contributed to the development of new methods of polarographic analysis. In the studies by T. A. Kryukova [28], the effect of retardation of motion on a mercury drop surface was used to develop a new method of polarographic determination of small quantities of surface-active agents. Since even the smallest traces of such substances are revealed in the solution through the damping of the surface motion of mercury, the polarograph can be used for a quantitative analysis of their concentration. It has been found that the sensitivity of the second kind of polarographic maxima in quantitative analysis is considerably greater than that of maxima of the first kind. This is associated with the relatively small adsorption of surface-active substances at potentials that are far removed from the zero point. The suppression of maxima of the second kind by surface-active agents can be used for quantitative analysis of solutions containing extremely minute ($10^{-7}$ molar) quantities of such substances.

The high sensitivity of the method is assured by the rapid rise in the retardation coefficient $\gamma$ with an increase in the number of adsorbed molecules. This method has been used in the analysis of water contamination by surface-active substances. The use of highly purified water is required in some analyses, such as the study of crystallization, interphase potentials, electrocrystallization, in the development of electroplating procedures, etc. In addition, the quality control of water is necessary in many industrial processes, to detect oils in steam condensers, in the analysis of water for adsorbents, for sanitary and hygienic purposes, etc.

## BIBLIOGRAPHY


1. Kolthoff, I. M. and J. J. Lingane, Polyarografiya [Polarography], Gostekhizdat, 1948; G. Semerano, Bibliografia Polarografica, Spoleto, Artigraf. Panetto s. Petrelli, 1949; T. A. Kryukova, Zav. Laboratoriya 14, Nos. 5, 6, 7 (1948); 16, No. 2 (1950).
2. Kolthoff and J. J. Lingane, Polyarografiya [Polarography], Gostekhizdat, 1948, p. 24.
3. Antweiler, H., Zt. f. Elektrochem. 43 (1937) 596; 44, 719, 831, 888 (1938).
4. Kryukova, T. A. and B. N. Kabanov, Zh. Fiz. Khimii 20, 1179 (1946); 21, 365 (1947); Zav. Laboratoriya No. 5, 511 (1948).
5. Ilkovic, D., Collec. Czechoslov. Commun. 6, 498 (1934); J. chim. phys. 35, 129 (1939); D. Mac Gillavry, E. Rideal, Rec. Trav. chim. 56, 1013 (1937).
6. Maas, J., Coll. Czech. Comm. 10, 42 (1938).
7. Koutecký, J., Czech. J. of Phys. 2, 50 (1953).
8. Stackelberg, Zt. f. Elektrochem. 54, 51 (1950).
9. Idem, ibid. 55, 244 (1951).
10. Meyman, N. N., Zh. Fiz. Khimii 22, 1454 (1948).
11. Bagotskiy, V. S., Zh. Fiz. Khimii 22, 1466 (1948).
12. Heyrovsky, I., Coll. Trav. Chim. 9, 150 (1937).
13. Tomes, J., Coll. Trav. Chim. 9, 150 (1937).
14. Kryukova, T. A., Zav. Laboratoriya 14, No. 5 (1948).
15. Frumkin, A. and B. Bruns, Acta physicochimica URSS 1, 232 (1934).
16. Bruns, B., A. Frumkin, Z. Iofa, L. Vanyukova and S. Zolotarevskaya, Zh. Fiz. Khimii 13, 785 (1939).
17. Antweiler, H. — see 3; M. Stackelberg, H. Antweiler, I. Kieselbach, Zt. f. Elektrochem. 44, 663 (1938).
18. Iofa, Z. and A. Frumkin, Dokl. Ak. Nauk SSSR 20, 293 (1938).
19. Frumkin, A. and V. Levich, Zh. Fiz. Khimii 20, 953 (1947).
20. Borisova, T. I. and M. A. Proskurin, Zh. Fiz. Khimii 21, 463 (1947); M. A. Vorsina and A. N. Frumkin, Dokl. Ak. Nauk SSSR 24, 918 (1939).
21. Frumkin, A. N. and V. G. Levich, Zh. Fiz. Khimii 21, 1335 (1947).
22. Popova, T. I. and T. A. Kryukova, Zh. Fiz. Khimii 5, 283 (1951).
23. Heyrovsky, I., Polarography, in the book by W. Boettger, Die physikalischen Methoden der chemischen Analyse, Vol. II, p. 260, Leipzig, 1936; H. Hohn, Chemische Analysen mit den

Polarografen, Berlin, 1937; Cit. from book by Kolthoff and Lingane.

24. Frumkin, A., Trudy of the Karpov Institute of Chemistry 5, 3 (1936); Zt. d. phys. 35, 792 (1928).
25. Kryukova, T. A., Zh. Fiz. Khimii 20, 1179 (1946); Zav. Laboratoriya 639 (1948); T. A. Kryukova and A. N. Frumkin, Zh. Fiz. Khimii 23, 819 (1949).
26. Kryukova, T. A., Zav. Laboratoriya 767 (1948).
27. Orleman, F. and J. Kolthoff, Journ. Amer. Chem. Soc. 64, 833 (1942).
28. Kryukova, T. A., Zav. Laboratoriya 767 (1948); 134 (1950).

CHAPTER XI

# Waves on a Liquid Surface

## 116. WAVES ON THE SURFACE OF AN IDEAL LIQUID

Surface (capillary) properties of the liquid play a vital role in wave motion — an important type of liquid motion.

The theory of wave motion of liquids is one of the best developed areas of hydrodynamics. However, we cannot present here the complete theory of wave motion in liquids, and will confine our analysis to certain interesting situations in physicochemical hydromechanics.

When the particles in a liquid surface, as a result of random disturbances, are subjected to infinitesimal displacement, and if the liquid surface becomes deformed and shifts from the state of equilibrium, then forces tending to return the surface to the state of equilibrium will appear in the liquid. The increase in free surface energy associated with such deformation leads to the appearance of capillary forces that tend to reduce the over-all surface area of the liquid to its equilibrium shape. Furthermore, if the liquid is in a gravitational field and its free surface is a plane, the disturbance gives rise to the appearance of forces that tend to return the surface to its original shape. Also, the liquid particles will tend to return to their original equilibrium position. Owing to inertia, however, the particles overshoot the equilibrium position and are again affected by forces tending to restore the equilibrium, etc., etc. Waves appear on a liquid surface which has been subjected to a random disturbance. When, however, surface tension is the basic cause for the appearance of the waves, the resulting system is known as capillary waves. When the acting forces are gravity forces we speak of gravity waves.

We first examine the theory of surface waves in an ideal liquid in which there is no dissipation of energy due to viscosity. We assume that the waves appear on the surface of a liquid of substantial bulk (depth). As shown below, wave motion is damped in the bulk of the liquid and penetrates only to a depth of the order of one wavelength. We analyze situations where the wavelength is small in comparison with the depth of the liquid. Such waves are known as surface waves (to distinguish them from tidal waves, whose wavelength is considerably greater than the depth of the liquid).

The hydrodynamic equations for non-steady-state motion are

$$\frac{\partial \mathbf{v}}{\partial t} + (\mathbf{v}\,\mathrm{grad})\,\mathbf{v} = -\frac{1}{\rho}\,\mathrm{grad}\,p + \nu\Delta\mathbf{v} + \mathbf{g}, \qquad (116.1)$$

$$\mathrm{div}\,\mathbf{v} = 0. \qquad (116.2)$$

These equations are extremely complex, but in application to wave motion they may be simplified considerably.

In an ideal liquid, the viscosity term can be omitted if the following inequality applies:

$$\frac{\partial \mathbf{v}}{\partial t} \gg \nu\Delta\mathbf{v}. \qquad (116.3)$$

Let $\lambda$ be the wavelength and $\omega$ the frequency. The wavelength $\lambda$ is a characteristic dimension of the wave motion region. Then, approximately

$$\Delta\mathbf{v} \approx \frac{v}{\lambda^2}$$

and, similarly,

$$\frac{\partial \mathbf{v}}{\partial t} \approx v\omega.$$

Therefore, if the inequality (116.3) is to be valid, we must have

$$v\omega \gg \nu\frac{v}{\lambda^2}$$

or

$$\frac{\omega\lambda^2}{\nu} \gg 1. \qquad (116.4)$$

We show later that $\omega$ decreases with the wavelength. Condition (116.4), therefore, shows that viscosity may be disregarded in the case of sufficiently long waves. In the next section we examine the case for which condition (116.4) is not valid.

Even if the viscosity term is omitted from the equation it still remains nonlinear. For the time being, we consider wave motion only at small Reynolds numbers.

Let a be the wave amplitude. Then, approximately

$$\mathbf{v} \approx a\omega.$$

The Reynolds number is given by $\mathrm{Re} = \frac{\mathrm{v}\lambda}{\nu} \approx \frac{\mathrm{a}\lambda\omega}{\nu}$.

At low Reynolds numbers the following inequality applies:

$$\frac{a}{\lambda}\,\frac{\lambda^2\omega}{\nu} \ll 1. \qquad (116.5)$$

For simultaneous validity of conditions (116.4) and (116.5), the wave amplitude must be very small compared to the wavelength.

Let us estimate the order of magnitude of the term (v grad) v in equation (116.1). Obviously,

$$(\mathbf{v}\,\mathrm{grad})\,\mathbf{v} \sim \frac{v^2}{\lambda} \sim \frac{a^2\omega^2}{\lambda}.$$

This term is small in comparison with $\frac{\partial \mathbf{v}}{\partial t}$, if

$$v\omega \gg \frac{a^2\omega^2}{\lambda},$$

or

$$a\omega^2 \gg \frac{a^2}{\lambda}\,\omega^2,$$

or

$$\frac{a}{\lambda} \ll 1. \tag{116.6}$$

Thus, if the wave amplitude is very small (as compared to the wavelength $\lambda$), the nonlinear term in equation (116.1) may also be neglected.

In that case [1], expression (116.1) takes the form

$$\frac{\partial \mathbf{v}}{\partial t} = -\frac{1}{\rho}\,\mathrm{grad}\,p + \mathbf{g}. \tag{116.7}$$

Equation (116.7) shows that a low frequency motion is potential flow, and we may write

$$\mathbf{v} = \mathrm{grad}\,\varphi. \tag{116.8}$$

The function $\varphi$ (velocity potential) satisfies the equation

$$\Delta\varphi = 0, \tag{116.9}$$

which is obtained by substituting (116.8) into (116.2).

Further, we analyze two-dimensional waves on a plane liquid surface. In two-dimensional waves, the motion of the liquid is restricted to the horizontal, along the x axis, and to the vertical, along the z axis. The waves represent a series of parallel crests and hollows, running along the x axis and extending infinitely along the y axis. In this case all the wave characteristics are functions of only the distances along the x and z axes. Along the y axis the motion is homogeneous and $\varphi$ and v are independent of y. The velocity potential, in the case of two-dimensional waves, satisfies the equation

$$\frac{\partial^2\varphi}{\partial x^2} + \frac{\partial^2\varphi}{\partial z^2} = 0.$$

The pressure, from equation (116.7), is

$$\operatorname{grad}\frac{\partial\varphi}{\partial t}=-\frac{1}{\rho}\operatorname{grad}p+\mathbf{g}. \tag{116.10}$$

Whence, recalling that the vector g has only the component along the z axis, we find a simple expression relating the pressure to the velocity potential

$$p=-\rho\frac{\partial\varphi}{\partial t}-\rho gz+F(t), \tag{116.11}$$

where F(t) is an arbitrary function of time. The latter may always be set equal to zero. Indeed, if F(t) is not zero we may always introduce a new velocity potential

$$\varphi'=\varphi+\frac{1}{\rho}\int F(t)\,dt$$

to eliminate F(t) from equation (116.11). The new velocity potential will always result in the same values for the velocity components as before.

We can present $\varphi$ in the form of a periodic function of the distance x and of time

$$\varphi=f(z)\cos(kx-\omega t). \tag{116.12}$$

The quantity $T=\frac{2\pi}{\omega}$ is the wave period, $\omega$ is the angular wave frequency (hereafter simply termed frequency), and k is the wave number, related to the wavelength by

$$k=\frac{2\pi}{\lambda}.$$

Substituting (116.12) into (116.9) the following equation for f(z) is obtained:

$$\frac{d^2f}{dz^2}-k^2f=0,$$

whence

$$f=Ae^{kz}+Be^{-kz}.$$

Finally, the value of $\varphi$ is

$$\varphi=(Ae^{kz}+Be^{-kz})\cos(kx-\omega t). \tag{116.13}$$

And, correspondingly,

$$v_x=\frac{\partial\varphi}{\partial x}=-(Ae^{kz}+Be^{-kz})\,k\sin(kx-\omega t) \tag{116.14}$$

$$v_z=\frac{\partial\varphi}{\partial z}=(Ae^{kz}-Be^{-kz})\,k\cos(kx-\omega t), \tag{116.15}$$

$$p=-\rho gz-\rho\omega(Ae^{kz}+Be^{-kz})\sin(kx-\omega t). \tag{116.16}$$

The constants A and B, as well as the relation between frequency and the wave number, must be derived from boundary conditions which we formulate below. To determine the boundary conditions we must know the displacement of the liquid particles on the surface. Let us assume the plane $z = 0$ to be the equilibrium position of the liquid surface. The displacement of the liquid particles on the surface from the position of equilibrium along the z axis is denoted by $\zeta$. The vertical displacement $\zeta$ is a function of the distance x and time t. The curve $\zeta$ (x) is, at a given time, the wave profile of the liquid surface. The function $\zeta$(t) at a given x describes the time dependence of the displacement of the particles. The vertical velocity component $v_z$ is related to the vertical displacement by

$$v_z = \frac{\partial\zeta}{\partial t} + \frac{\partial\zeta}{\partial x}\frac{dx}{dt} \approx \frac{\partial\zeta}{\partial t}. \qquad (116.17)$$

Because at low amplitudes the quantity $\frac{\partial\zeta}{\partial x}\frac{dx}{dt} = \frac{\partial\zeta}{\partial x}v_x$ is a second order infinitesimal compared to $\frac{\partial\zeta}{\partial t}$, the expression for $\zeta$ can be obtained using the relations (116.8) and (116.17),

$$\frac{\partial\zeta}{\partial t} = \frac{\partial\varphi}{\partial z}. \qquad (116.18)$$

Therefore,

$$\zeta = \int \frac{\partial\varphi(x, 0, t)\,dt}{\partial z} = -(A - B)\frac{k}{\omega}\sin(kx - \omega t).$$

The quantity $(A - B)\frac{k}{\omega}$ represents the wave amplitude on the liquid surface. Henceforth the wave amplitude on the surface is designated by a, and the vertical displacement on the surface is

$$\zeta = a\sin(kx - \omega t). \qquad (116.19)$$

We can now set up the boundary conditions on the surfaces confining the liquid.

Since we are examining only surface waves, the motion is confined to a region, shallow compared with the depth of the liquid. Thus, the liquid may be considered arbitrarily deep. In that case, wave motion velocity in the bulk of the liquid becomes zero at the bottom of the liquid. This is our first boundary condition. Selecting the upward direction of the z axis as positive, the liquid occupies the half-space $z < 0$, and

$$\mathbf{v} \to 0 \quad \text{as} \quad z \to -\infty. \qquad (116.20)$$

The condition (116.20) is satisfied when, in equations (116.14) and (116.15), the constant B is set equal to zero (then, $A = -\frac{\omega}{k}a$).

The following condition must be fulfilled on the free surface of the liquid:

$$p + p_\sigma = p_0 \quad \text{at} \quad z = \zeta, \tag{116.21}$$

where $p_\sigma$ is the capillary pressure, and $p_0$ is the pressure of the gas phase above the liquid.

Substituting the values of p from (116.11) and of $p_\sigma$ from (65.13) into condition (116.21), we obtain:

$$-\rho \frac{\partial \varphi}{\partial t} - \rho g \zeta + \sigma \frac{\partial^2 \zeta}{\partial x^2} = p_0 \quad \text{at} \quad z = \zeta. \tag{116.22}$$

Since the displacement $\zeta$ on the surface is arbitrarily small, the boundary condition (116.22) must be fulfilled on the surface $z = 0$ with an error that is a second-order infinitesimal.

Differentiation of (116.22) with respect to t yields

$$-\rho \frac{\partial^2 \varphi}{\partial t^2} - \rho g \frac{\partial \zeta}{\partial t} + \sigma \frac{\partial^3 \zeta}{\partial t\, \partial x^2} = 0 \tag{116.23}$$

which is exact to a second order infinitesimal. The values of the quantities $\frac{\partial^2 \varphi}{\partial t^2}$ and $\frac{\partial^3 \zeta}{\partial t \partial x^2}$ in condition (116.23) at $z = 0$ may be taken from equations (116.13) and (116.19).

Moreover, replacing $\frac{\partial \zeta}{\partial t}$ with $\frac{\partial \varphi}{\partial z}$, as given in (116.18), and substituting the value of $\varphi$ from equation (116.13), we find

$$\rho\omega^2 - (\rho g k + \sigma k^3) = 0,$$

whence

$$\omega^2 = \frac{\sigma k^3}{\rho} + g k, \tag{116.24}$$

or

$$\omega^2 = \frac{8\pi^3 \sigma}{\rho \lambda^3} + \frac{2\pi g}{\lambda}. \tag{116.25}$$

Equations (116.24) and (116.25) establish the relation between the frequency and the wave number, and/or the wavelength. We can therefore conclude that traveling waves can exist on a liquid surface, and they propagate over that surface with a phase velocity given by

$$c = \frac{\omega}{k} = \sqrt{\frac{\sigma k}{\rho} + \frac{g}{k}}. \tag{116.26}$$

The velocity potential in these waves is

$$\varphi = \frac{-a\omega e^{kz}}{k} \cos(kx - \omega t); \tag{116.27}$$

the distributions of velocities and pressures are

$$v_x = a e^{kz} \omega \sin(kx - \omega t), \tag{116.28}$$

$$v_z = -a e^{kz} \omega \cos(kx - \omega t), \tag{116.29}$$

$$p = -\rho \frac{\partial \varphi}{\partial t} - \rho g z = \frac{a\omega^2}{k} e^{kz} \sin(kx - \omega t) - \rho g z; \tag{116.30}$$

and the wave profile on the surface of the liquid is

$$\zeta = a \sin (kx - \omega t). \qquad (116.31)$$

Equations (116.28) and (116.29) show that the horizontal and vertical displacement velocities of the wave are of the same order of magnitude. In the proximity to the surface they are both of the order of $a\omega$. The velocity decreases exponentially with the depth of the liquid, according to $\exp(kz) = \exp\frac{2\pi z}{\lambda}$. At a depth $(-z)$ of the order of $\lambda$, the velocities are $e^{2\pi} = 535$ times lower than at the surface of the liquid. Frequency is related to wave number by the relation given by (116.24). The amplitude in these equations is an arbitrary constant, the limits of which are determined by condition (116.6).

Let us now analyze the trajectories of moving liquid particles in a wave. We remember that the displacement amplitudes of the liquid particles are small. Therefore, on the right side of equations (116.28) and (116.29) we may replace x and z by their equilibrium values $x_0$ and $z_0$, respectively

$$\frac{dx}{dt} = ae^{kz_0}\omega \sin (kx_0 - \omega t), \qquad (116.32)$$

$$\frac{dz}{dt} = -\omega ae^{kz_0} \cos (kx_0 - \omega t). \qquad (116.33)$$

Integrating these two equations, we obtain:

$$x = x_0 + ae^{kz_0} \cos (kx_0 - \omega t),$$
$$z = z_0 + ae^{kz_0} \sin (kx_0 - \omega t).$$

Eliminating the time, we find the trajectory of a liquid particle

$$(x - x_0)^2 + (z - z_0)^2 = a^2 e^{2kz_0}. \qquad (116.34)$$

Each liquid particle moves along a circumference of a circle with an angular velocity $\omega$. The radius of the circle is $ae^{kz_0}$ and it decreases exponentially with depth.

The frequency $\omega$, according to formula (116.25), decreases with increase in $\lambda$ and depends on g and the capillary constant T. For wavelengths to which the inequalities

$$\frac{g}{\lambda} \gg \frac{4\pi^2\sigma}{\rho\lambda^3}$$

or

$$\lambda \gg 2\pi \sqrt{\frac{\sigma}{\rho g}} = 2\pi \sqrt{T} \qquad (116.35)$$

apply, where T denotes the capillary constant, it is evident that the frequency $\omega$ depends chiefly on the term $g/\lambda$ and is equal to

$$\omega \approx \sqrt{\frac{2\pi g}{\lambda}} \approx \sqrt{gk}. \qquad (116.36)$$

For long waves, the basic effect is that of gravitation. Surface tension and capillary phenomena do not play any significant part. Such waves are known as gravity waves. The propagation velocity of gravity waves is

$$c=\sqrt{\frac{g}{k}}\,. \tag{116.37}$$

On the contrary, for short waves, when the inequality

$$\frac{4\pi^2\sigma}{\rho\lambda^3}\gg\frac{g}{\lambda}$$

is valid, or

$$\lambda\ll 2\pi\sqrt{T},$$

surface tension plays the major role in wave propagation. Such waves are known as capillary waves or (using Faraday's terminology) ripples. Wave frequency and propagation velocity are described by

$$\omega\approx\sqrt{\frac{8\pi^3\sigma}{\rho\lambda^3}}=\sqrt{\frac{\sigma k^3}{\rho}}\,, \tag{116.38}$$

$$c=\sqrt{\frac{2\pi\sigma}{\rho\lambda}}=\sqrt{\frac{\sigma k}{\rho}}\,. \tag{116.39}$$

Comparison of equations (116.37) and (116.39) shows that the propagation velocity of gravity waves increases, whereas that of capillary waves decreases with the wavelength. Therefore, the propagation velocity of surfaces waves c exhibits a minimum at a certain wavelength. The values of $\lambda_{min}$ and of $c_{min}$ are found from the following condition

$$\frac{dc}{d\lambda}=0.$$

Substitution of c as given by (116.26) yields

$$\lambda_{min}=2\pi\sqrt{\frac{\sigma}{\rho g}}\,, \tag{116.40}$$

$$c_{min}=\sqrt[4]{\frac{4\sigma g}{\rho}}\,. \tag{116.41}$$

For pure water ($\sigma$=74 dynes/cm) substitution of numerical values into expressions (116.41) and (116.40) gives $\lambda_{min}$ = 1.78 and $c_{min}$ = =23.5 cm/sec.

Table 20 gives the values for wavelengths, frequency and propagation velocity in the vicinity of the minimum velocity.

If waves of varying lengths are produced on a liquid surface, the resulting motion is determined by the individual types of waves described above. In this case each type of wave plays the role of a separate component in a Fourier series.

Table 20

| Wavelength, cm ...... | 0.5 | 1.0 | 1.7 | 2.5 | 3.0 | 5.0 |
|---|---|---|---|---|---|---|
| Frequency .......... | 63.0 | 24.7 | 13.6 | 9.6 | 8.3 | 4.9 |
| Velocity c, cm/sec .... | 31.5 | 24.7 | 23.1 | 23.9 | 24.9 | 29.5 |

The proposed theory and equation (116.26) for the propagation velocity of surface waves [2] have been verified experimentally for a wide range of wavelengths, down to $\lambda = 0.04$ cm. The waves were produced by touching a static surface with plates connected to tuning forks vibrating at predetermined frequencies. The liquids used were water, mercury, alcohol, ether and carbon disulfide.

There was good agreement between the theoretical values computed from equation (116.26) and the experimental data.

The experiments led to the development of a method for measuring surface tension of liquids based on measurement of the propagation velocity of capillary waves [3].

Standing waves with a predetermined frequency were produced on the surface of a liquid contained in a vessel. The plate used to generate the waves was immersed and withdrawn from the liquid at a certain frequency. The motion of the plate activated a light source which illuminated the surface of the liquid by light flashes. The frequency of the flashes was equal to the frequency of the generated waves. Using this method of lighting and special means of observation, it is possible to determine accurately the position of the crests of waves with extremely small amplitudes.

Thus, the dependence of the wavelength $\lambda$ of standing waves on the generating frequency $\omega$ can be measured. Equation (116.25) can then be used to calculate the surface tension of the liquid.

This method of determining surface tension appears to be one of the most accurate ones.

## 117. WAVES ON THE SURFACE OF A VISCOUS LIQUID

We now assume that the viscosity term in the equation for liquid motion is not small compared to the quantity $\frac{\partial v}{\partial t}$, and that condition (116.4) no longer applies. In other words, we are analyzing wave motion on the surface of a viscous liquid. As before, we assume that the inequality (116.6) holds; i.e., the wave amplitude is assumed to be extremely small compared to its length. The equations of motion are then

$$\frac{\partial \mathbf{v}}{\partial t} = -\frac{1}{\rho}\operatorname{grad} p + \nu\Delta\mathbf{v} + \mathbf{g}; \quad \operatorname{div}\mathbf{v} = 0,$$

or, in component form

$$\frac{\partial v_x}{\partial t} = -\frac{1}{\rho}\frac{\partial p}{\partial x} + \nu\left(\frac{\partial^2 v_x}{\partial x^2} + \frac{\partial^2 v_x}{\partial z^2}\right), \qquad (117.1)$$

$$\frac{\partial v_z}{\partial t} = -\frac{1}{\rho}\frac{\partial p}{\partial z} + \nu\left(\frac{\partial^2 v_z}{\partial x^2} + \frac{\partial^2 v_z}{\partial z^2}\right) + g, \qquad (117.2)$$

$$\frac{\partial v_x}{\partial x} + \frac{\partial v_z}{\partial z} = 0. \qquad (117.3)$$

As before, we are examining a system of two-dimensional surface waves on the surface of a liquid of infinite depth (compared with the wavelength) [1].

Expressions (117.1) and (117.2), in contrast to (116.7) for wave motion of an ideal liquid, are equations of second order in the spatial coordinates. It is logical to seek a solution of these equations in the form

$$v_x = v_x^{(0)} + U_1, \qquad (117.4)$$

$$v_z = v_z^{(0)} + U_2, \qquad (117.5)$$

$$p = p^{(0)}, \qquad (117.6)$$

where the quantities with the superscript 0 relate to an ideal liquid and are of the same form as expressions (116.14) to (116.16). Indeed, $v_x^{(0)}$ and $v_z^{(0)}$ satisfy the continuity equation, and it is clear from physical considerations that the viscosity changes the value of the frequency but not of the pressure distribution.

It is more convenient to write the expression for $v_x^{(0)}$, $v_z^{(0)}$ and $p^{(0)}$ in exponential form

$$\varphi = Ae^{ikx}e^{kz}e^{\alpha t}, \qquad (117.7)$$

$$v_x^{(0)} = \frac{\partial \varphi}{\partial x} = ikAe^{ikx}e^{kz}e^{\alpha t},$$

$$v_z^{(0)} = \frac{\partial \varphi}{\partial z} = kAe^{kz}e^{ikx}e^{\alpha t}, \qquad (117.8)$$

$$p^{(0)} = -\rho\frac{\partial \varphi}{\partial t} - \rho g z = -\rho\alpha Ae^{kz}e^{ikx}e^{\alpha t} - \rho g z. \qquad (117.9)$$

Bearing in mind that the frequency will be a complex number, it is more convenient to write the coefficient of t in a real form.

Substituting the solutions (117.4) and (117.5) into (117.3), we find the relation between $U_1$ and $U_2$.

$$\frac{\partial U_1}{\partial x} + \frac{\partial U_2}{\partial z} = 0. \qquad (117.10)$$

This equation shows that the functions $U_1$ and $U_2$ may be presented as

$$U_1 = -\frac{\partial \psi}{\partial z}, \qquad (117.11)$$

$$U_2 = \frac{\partial \psi}{\partial x}. \tag{117.12}$$

Substituting the values of $v_x$ and $v_z$ from (117.4) and (117.5) into equations (117.1) and (117.2) and considering equations (117.11) and (117.12) we obtain

$$\frac{\partial v_x^{(0)}}{\partial t} - \frac{\partial}{\partial z}\frac{\partial \psi}{\partial t} = -\frac{1}{\rho}\frac{\partial p^{(0)}}{\partial x} + \nu\left(\frac{\partial^2 v_x^{(0)}}{\partial x^2} + \frac{\partial^2 v_x^{(0)}}{\partial z^2}\right) - \nu\frac{\partial}{\partial z}\left(\frac{\partial^2 \psi}{\partial x^2} + \frac{\partial^2 \psi}{\partial z^2}\right),$$

$$\frac{\partial v_z^{(0)}}{\partial t} + \frac{\partial}{\partial x}\frac{\partial \psi}{\partial t} = -\frac{1}{\rho}\frac{\partial p^{(0)}}{\partial z} + \nu\left(\frac{\partial^2 v_z^{(0)}}{\partial x^2} + \frac{\partial^2 v_z^{(0)}}{\partial z^2}\right) + \nu\frac{\partial}{\partial x}\left(\frac{\partial^2 \psi}{\partial x^2} + \frac{\partial^2 \psi}{\partial z^2}\right).$$

By definition both $v_x^{(0)}$ and $v_z^{(0)}$ satisfy the equations of motion of an ideal liquid. In addition, the following equality holds

$$\Delta v_x^{(0)} = \Delta \frac{\partial \varphi}{\partial x} = \frac{\partial}{\partial x}\Delta\varphi = 0,$$

and similarly

$$\Delta v_z^{(0)} = \frac{\partial}{\partial z}\Delta\varphi = 0,$$

since $\Delta\varphi = 0$.

The function $\psi$, therefore, must satisfy the equations

$$\frac{\partial}{\partial z}\left(\frac{\partial \psi}{\partial t} - \nu\Delta\psi\right) = 0, \tag{117.13}$$

$$\frac{\partial}{\partial x}\left(\frac{\partial \psi}{\partial t} - \nu\Delta\psi\right) = 0. \tag{117.14}$$

The last two equations show that the following equality must be valid:

$$\frac{\partial \psi}{\partial t} = \nu\Delta\psi. \tag{117.15}$$

Yet another constant could be introduced into expression (117.15). Such a constant, however, has no meaning, since we are interested in velocities which are related to $\psi$ by equations (117.11) and (117.12), and whose values do not change on adding a constant to $\psi$.

We are seeking a solution of equations (117.15) in the form of a periodic function of time and the horizontal coordinate and an exponential function of the depth

$$\psi = Ce^{lz}e^{ikx}e^{\alpha t}, \tag{117.16}$$

where k is the same wave number as in the solution for an ideal liquid.

By substituting (117.16) into (117.15), we get

$$l^2 = k^2 + \frac{\alpha}{\nu}. \tag{117.17}$$

Then, the final solution of equations (117.1) to (117.3), account being taken of (117.7) to (117.9), (117.11), (117.12) and (117.16), is

$$v_x = (ikAe^{kz} - lCe^{lz})\, e^{ikx+\alpha t}, \qquad (117.18)$$

$$v_z = (kAe^{kz} + ikCe^{lz})\, e^{ikx+\alpha t}, \qquad (117.19)$$

$$p = -\rho\alpha Ae^{kz}e^{ikx+\alpha t} - \rho gz. \qquad (117.20)$$

Extraneous solutions that increase exponentially with liquid depth could have been added to solutions (117.18) to (117.20). We rejected them, however, for the same reasons as presented in the preceding section.

We now seek the vertical and horizontal displacements on the free surface of the liquid. For the former, we get

$$\frac{\partial\zeta}{\partial t} \approx v_z \quad \text{at} \quad z = 0; \qquad (117.21)$$

whence, using expression (117.19), we obtain

$$\zeta = \frac{kA + ikC}{\alpha}\, e^{ikx+\alpha t}. \qquad (117.22)$$

Similarly, for the horizontal displacement $\xi$, with the same approximation, we may write:

$$\frac{\partial\xi}{\partial t} \approx v_x \quad \text{at} \quad z = 0. \qquad (117.23)$$

This gives

$$\xi = \frac{ikA - lC}{\alpha}\, e^{ikx+\alpha t}. \qquad (117.24)$$

Solutions (117.18) to (117.20) contain three unknown quantities A, C and $\omega$. The remaining boundary conditions on the surface of the liquid permit the formulation of two equations that relate these quantities to each other. (The boundary conditions as $z \to -\infty$ have already been used in selecting the solutions.) The wave amplitude remains arbitrary.

The boundary conditions on the free surface of a viscous liquid are

$$\left.\begin{aligned} p_{zz} &= p_\sigma + p_0, \\ p_{zx} &= 0 \end{aligned}\right\} \quad \text{at} \quad z = \zeta, \qquad (117.25)$$

where $p_{zz}$ and $p_{zx}$ are the normal and tangential components of the viscous stress tensors, $p_\sigma$ is the capillary pressure, and $p_0$ is pressure in the gas phase. The latter is selected as the reference level for pressure, and in the following computations we will consider $p_0 = 0$.

The normal component of stress at $z = \zeta$ can, according to expression (1.5), be written in the form

$$p_{zz} = -p + 2\mu\frac{\partial v_z}{\partial z} = \rho\alpha A e^{ikx+\alpha t} + \rho g\zeta + \\ + 2\mu k(kA + ilC)\, e^{ikx+\alpha t} = \left\{\rho\alpha A + \rho g\frac{kA + ikC}{\alpha} + \right. \\ \left. + 2\mu k(kA + ilC)\right\} e^{ikx+\alpha t}. \qquad (117.26)$$

Similarly, the tangential component is given by

$$p_{zx} = \mu\left(\frac{\partial v_x}{\partial z} + \frac{\partial v_z}{\partial x}\right) = \mu\,[ik^2A - l^2C + ik^2A - k^2C]\, e^{ikx+\alpha t} = \\ = \mu\,[2ik^2A - (l^2 + k^2)\,C]\, e^{ikx+\alpha t}. \qquad (117.27)$$

Finally, the capillary pressure is

$$p_\sigma = \sigma\frac{\partial^2\zeta}{\partial x^2} = -\frac{\sigma(k^3A + ik^3C)}{\alpha}\, e^{ikx+\alpha t}. \qquad (117.28)$$

Substituting these values for $p_{zz}$, $p_{zx}$ and $p_\sigma$ into the boundary conditions (117.25) we get

$$(\alpha^2 + 2\nu k^2\alpha + \omega_0^2)\, A + i\,(\omega_0^2 + 2\nu kl\alpha)\, C = 0, \qquad (117.29)$$

$$2i\nu k^2 A - (\alpha + 2\nu k^2)\, C = 0, \qquad (117.30)$$

where $\omega_0 = \sqrt{\frac{\sigma k^3}{\rho} + gk}$ is the frequency in an ideal liquid.

Equations (117.29) and (117.30), with boundary condition (117.17), are used in determining the frequency $\omega$ and the ratio A/C.

Eliminating the ratio C/A from (117.29) and (117.30), we obtain the equation for the determination of $\omega$

$$(\alpha + 2\nu k^2)^2 + \omega_0^2 = 4\nu^2 k^4\sqrt{\frac{\alpha}{\nu k^2} + 1}. \qquad (117.31)$$

## 118. WAVE MOTION ON THE SURFACE OF A LOW-VISCOSITY LIQUID

Before turning to the solution of the equations given above, it should be noted that the expression A/C is the ratio of the irrotational and rotational parts of the velocity vector. Indeed, A represents the amplitude of $v^{(0)}$, which is the velocity of potential flow in an ideal liquid.

Direct calculation gives

$$\mathrm{rot}_y\,\mathbf{v} = -\left(\frac{\partial v_z}{\partial x} - \frac{\partial v_x}{\partial z}\right) = (k^2 - l^2)\, Ce^{lz}e^{ikx+\alpha t}.$$

Thus, in cases where the viscosity of the liquid is relatively low, the ratio A/C must be large, and the motion should be very similar to potential flow in an ideal liquid.

Condition (117.30) shows that taking into account (117.17) we get

$$\frac{A}{C}=\frac{l^2+k^2}{2ik^2}=-i\left(1+\frac{\alpha}{2\nu k^2}\right). \tag{118.1}$$

In order that the ratio A/C be large, either the following inequality

$$\frac{\alpha}{2\nu k^2}\gg 1 \tag{118.2}$$

or an equivalent inequality

$$l^2\gg k^2 \tag{118.3}$$

must be satisfied.

In this case we deal with a low viscosity liquid and assume that the viscosity does not exert any appreciable effect on either the liquid motion or on the wave frequency $\omega$.

Assuming that for waves in an ideal liquid the frequency $\omega$ differs little from the frequency $\omega_0$, let us estimate the range of applicability of inequalities (118.2) and (118.3). Substituting the frequency $\omega_0$, computed in section 116, for the frequency $\omega$, the following inequality can be written:

$$\frac{\omega_0}{\nu k^2}=\frac{c}{\nu k}>\frac{c_{\min}}{\nu k}.$$

Substituting the value of $c_{\min}$ from (116.41), we get

$$\frac{\omega_0}{\nu k^2}>\frac{\lambda}{2\pi}\sqrt[4]{\frac{4\sigma g}{\rho\nu^4}}.$$

For water this yields

$$\frac{\omega_0}{\nu k^2}>4\cdot 10^2\lambda\ (\text{cm}). \tag{118.4}$$

Thus, the inequality is satisfied if the wavelength

$$\lambda\gg 2-3\cdot 10^{-2}\ \text{cm}.$$

Assuming that inequality (118.4) holds, we can now derive the approximate solutions of equation (117.31). Since the motion, in this case, is very similar to that of an ideal liquid, we seek a solution in the form*

$$\alpha=-i\omega_0+\beta, \tag{118.5}$$

where $\beta\ll\omega_0$.

---

*It should be noted that in this approximate solution we lose two real roots of equation (117.31), which relate to nonperiodic motion and are of no interest, since nonperiodic motion is rapidly damped.

Substituting the value of $\alpha$ from (118.5) into equation (117.31), and disregarding small quantities of the order of $\beta^2$, we get

$$2i\beta\omega_0 + 4i\nu k^2\omega_0 = 0,$$

whence

$$\beta = -2\nu k^2. \tag{118.6}$$

Thus, the velocity distribution in the liquid is

$$v_x = \operatorname{Re} ikAe^{kz} e^{i(kx-\omega_0 t)} e^{-2\nu k^2 t} =$$
$$= -Ae^{-2\nu k^2 t} k \sin(kx - \omega_0 t)\, e^{kz} = a\omega e^{-2\nu k^2 t} \sin(kx - \omega_0 t)\, e^{kz}, \tag{118.7}$$
$$v_z = Ae^{-2\nu k^2 t} k \cos(kx - \omega_0 t)\, e^{kz} =$$
$$= -a\omega e^{-2\nu k^2 t} \cos(kx - \omega_0 t)\, e^{kz}. \tag{118.8}$$

Comparing equations (118.7) and (118.8) with (116.28) and (116.29) we see that the velocity distribution for wave motion in a low viscosity liquid differs from that in an ideal liquid only by the factor $e^{-2\nu k^2 t}$. The wave amplitude in a viscous liquid decreases with time. The damping coefficient $\beta$, according to (118.6) is proportional to the kinematic viscosity $\nu$ and inversely proportional to the square of the wavelength.

The above reasoning shows that the concept of a low viscosity liquid is relative. The conditions actually governing the slow damping of waves are the inequalities (118.2) and (118.4), which involve both the wavelength and the viscosity. Waves in the short-wave range of the spectrum are damped most rapidly. On the contrary, for relatively long waves the damping coefficient can be relatively small even in the case of a highly viscous liquid.

Now, let us examine the energy dissipation for a low viscosity liquid.

The general formula for energy dissipation per unit time (1.15) may be greatly simplified for wave motion in a low viscosity liquid. The energy dissipated throughout the body of the liquid per unit time is determined by

$$-\frac{dE}{dt} = 2F =$$
$$= \mu \int \left\{ 2\left(\frac{\partial v_x}{\partial x}\right)^2 + 2\left(\frac{\partial v_z}{\partial z}\right)^2 + \left(\frac{\partial v_x}{\partial z} + \frac{\partial v_z}{\partial x}\right)^2 \right\} dv. \tag{118.9}$$

This formula gives the instantaneous value of the power dissipated. However, we are interested in the average value of $\frac{dE}{dt}$ for a period $T = \frac{2\pi}{\omega}$. The latter quantity is clearly constant in time

$$2\overline{F} = \frac{1}{T} \int_0^T \frac{dE}{dt}\, dt =$$
$$= -\frac{\mu}{T} \int dv \int \left[ 2\left(\frac{\partial v_x}{\partial x}\right)^2 + 2\left(\frac{\partial v_z}{\partial z}\right)^2 + \left(\frac{\partial v_x}{\partial z} + \frac{\partial v_z}{\partial x}\right)^2 \right] dt.$$

Let us solve the integrals with time as the variable

$$\frac{1}{T}\int_0^T\left(\frac{\partial v_x}{\partial x}\right)^2 dt=\frac{k^2\omega^2a^2}{T}\,e^{2kz}\int_0^T e^{-4\nu k^2t}\sin^2(kx-\omega t)\,dt.$$

Because, owing to inequality (118.2), $\nu k^2 T \ll 1$, the exponential time function may be replaced by unity, and

$$\frac{1}{T}\int_0^T\left(\frac{\partial v_x}{\partial x}\right)^2 dt=\frac{k^2\omega^2a^2}{T}\,e^{2kz}\int_0^T\sin^2(kx-\omega t)\,dt=e^{2kz}\,\frac{k^2\omega^2a^2}{2}.$$

It is clear that

$$\frac{1}{T}\int_0^T\left(\frac{\partial v_z}{\partial z}\right)^2 dt=\frac{a^2\omega^2k^2}{2}\,e^{2kz},$$

$$\frac{1}{T}\int_0^T\left(\frac{\partial v_x}{\partial z}+\frac{\partial v_z}{\partial x}\right)^2 dt=$$

$$=\frac{a^2\omega^2e^{2kz}k^2}{T}\int_0^T[\sin^2(kx-\omega t)+\cos^2(kx-\omega t)]\,dt=a^2\omega^2e^{2kz}k^2.$$

Therefore,

$$2\overline{F}=-4\mu a^2\omega^2k^2\int_0^{-\infty}e^{2kz}\,dz\cdot S=-2\mu a^2\omega^2kS=-2\mu a^2k^3c^2S, \qquad (118.10)$$

where S is the surface area of the liquid.

The dissipated energy is found to be proportional to the square of the wave amplitude and increases with $\omega^2 k$.

For long gravity waves

$$2\overline{F}=-2\mu a^2gk^2S=-2\mu\left(\frac{2\pi}{\lambda}\right)^2ga^2S, \qquad (118.11)$$

whereas for capillary waves, the energy dissipation is

$$2\overline{F}=-2\mu\,\frac{\sigma k^4}{\rho}\,a^2S=-2\mu\,\frac{\sigma}{\rho}\left(\frac{2\pi}{\lambda}\right)^4a^2S. \qquad (118.12)$$

We can now compute the energy E of the liquid wave motion. By assuming the viscosity low enough to be disregarded, we have

$$\overline{E}=\overline{T}+\overline{U}=2\overline{T}, \qquad (118.13)$$

where T is the kinetic and U the potential energy of the waves. Here, we take advantage of the well-known property of periodic motion with small amplitude, for which the average kinetic energy is always equal to the average potential energy.

Substituting the kinetic energy into (118.13),

$$T=\int\frac{\rho v^2}{2}\,dv$$

and the values of $v_x$ and $v_z$ from (118.7) and (118.8), we get

$$\overline{E}=\rho\int\overline{v^2}\,dv=\rho\int(\overline{v_x^2}+\overline{v_z^2})\,dv=$$

$$=\rho a^2\omega^2 e^{-4\nu k^2 t}\int\limits_0^{-\infty}e^{2kz}\,dz\cdot S=\frac{\rho a^2\omega^2 S}{2k}. \qquad (118.14)^*$$

At a value of the damping coefficient $\beta = -2\nu k^2$, the values of $\overline{E}$ and $2F$ satisfy the law of conservation of energy: $-\left(\frac{\overline{dE}}{dt}\right) = 2F$. Conversely, using the energy equation, the damping coefficient $\beta$ can be found. To find the damping coefficient it is necessary to substitute the expressions for $\frac{\overline{dE}}{dt}$ and $2F$ into the equation of conservation of energy, writing the time dependence for the latter expression in the form $e^{-2\beta t}$, where $\beta$ is the unknown coefficient.

## 119. WAVE MOTION ON THE SURFACE OF A HIGHLY VISCOUS LIQUID

We define a highly viscous liquid as one for which the following inequality holds

$$\frac{\alpha}{\nu k^2}\ll 1, \qquad (119.1)$$

and which is the inverse of inequality (118.2). In this case we can make the following approximation:

$$l=\sqrt{k^2+\frac{\alpha}{\nu}}=k\sqrt{1+\frac{\alpha}{\nu k^2}}\approx k\left(1+\frac{\alpha}{2\nu k^2}\right).$$

Equation (117.31) may then be rewritten in the form

$$[(\alpha+2\nu k^2)^2+\omega_0^2]^2=\frac{16\,(\nu k^2)^3\,\alpha}{\omega_0^2}.$$

We have approximately [1]

$$\alpha_1\approx-\nu k^2+\nu k^2\left(1-\frac{\omega_0^2}{2\nu^2 k^4}\right)\approx-\frac{\omega_0^2}{2\nu k^2}, \qquad (119.2)$$

$$\alpha_2\approx-0.9\nu k^2. \qquad (119.3)$$

---

*Editor's note: There appears to be an error in equation (118.14). It should read $= -\frac{\rho a^2\omega^2 S}{2k}\,e^{-4\nu k^2 t}$.

Both values of $\alpha$ are found to be real and pertain to a nonperiodic motion. The second root is of no interest, since it is considerably greater than the first, and the motion that it covers is rapidly damped. The first root also pertains to exponential damping of the motion. However, the damping occurs much more slowly.

Since $\omega_0^2 = \frac{\sigma k^3}{\rho} + gk$, the damping of the motion proceeds according to $e^{-\left(\frac{\sigma k}{2\nu\rho} + \frac{g}{2\nu k}\right)t}$. We see that very short-wave disturbances in the liquid generate a periodic motion that is damped with time.

From expression (117.30) we find, for $\frac{\alpha}{\nu k^2} \ll 1$,

$$\frac{A}{C} = -i\left(1 + \frac{\alpha}{2\nu k^2}\right) = -i\left(1 - \frac{\omega_0^2}{4\nu^2 k^4}\right). \tag{119.4}$$

The motion of the liquid is rotational because $|A| \sim |C|$.

Substituting (119.4) into (117.22), we find the displacement of the liquid surface

$$\zeta = \frac{kA + ikC}{\alpha} e^{ikx} e^{\alpha t} = \frac{A}{2\nu k} e^{ikx} e^{-\frac{\omega_0^2 t}{2\nu k^2}}.$$

In the intermediate cases, between those examined in this and in the preceding section, there is a smooth transition of $\alpha$ from values represented by (118.5) to those given by (119.2).

## 120. EFFECT OF SURFACE-ACTIVE SUBSTANCES ON THE WAVE MOTION OF A LIQUID

The effect of a surface-active substance on wave motion was known to the ancient Greeks. Pliny the Elder mentions calming the fury of waves by means of oil poured on the surface of the sea.

Ancient books on seafaring contain a number of suggestions and give advice to the sailors. In critical cases it is recommended that barrels of oil be poured onto the surface of the sea. Preference is given, incidentally, to animal and vegetable as against mineral oils.

The damping effect of surface-active materials on wave motion is widely known and is mentioned in all specialized handbooks.

In his classic work "Oceanography," Yu. M. Shokal'skiy notes that the damping of waves is currently of considerable practical interest.

Similar ideas are given by V. V. Shuleykin [5]. Modern maritime almanacs give practical instructions to seamen regarding the methods by which surface-active materials are to be spread over the surface of the sea, the amounts that should be used, and so on.

Since the most ancient times many experimental and theoretical studies have been devoted to the damping effect of surface-active materials on waves [6]. The study of surface phenomena actually started with observations of the damping effect of surface-active materials on ocean waves. Most of these observations, however, were made by sailors under rather trying conditions, and as a result the information made available was purely descriptive. It has not been possible, therefore, to determine the mechanism of the process on the basis of these descriptions. Until very recently, no quantitative analyses of the damping effect have been made.

At present, several different theories of the damping of waves exist. The first of these amounts to the assertion that the film of surface-active material covering the surface of the liquid reduces friction between the air and the water. It is also asserted that supposedly as a result of the decrease in friction the energy transferred to the liquid by the wind also decreases.

Experimental studies [7], however, have shown that the mechanism by which energy is transferred from air to the liquid is determined by processes completely unrelated to friction at the air-water interface.

Another theory of the damping of capillary waves by surface-active materials ascribes the effect of the latter to a reduction of surface tension in the liquid [8]. As Shuleykin correctly points out, however, the lowering of surface tension cannot result in an increase of the damping coefficient, because the latter is independent of surface tension.

V. V. Shuleykin [5] developed a damping theory based on the viscosity of the film of surface-active material. The deformation of the film, in his opinion, leads to energy dissipation within that film.

The viscosity of adsorbed surface-active layers was examined in Section 76. It was shown that it is impossible, in the general case, to consider viscosity as a constant in relation to energy dissipation in a film. It was also shown that consideration of the elastic properties of the film was equivalent to assuming the existence of a second viscosity of the film.

We advanced a theory of wave damping by surface-active materials, in which the damping effect is related to the effects very generally considered in Section 73. We refer to the effect of surface-active materials on liquid motion in terms of changes in the boundary conditions on the liquid surface.

In Section 121 we analyze the damping coefficient for waves on the surface of a viscous liquid covered by a surface-active substance.

## 121. DAMPING OF CAPILLARY WAVES BY SURFACE-ACTIVE SUBSTANCES

In contrast to the previously advanced theories of the damping of waves in which the damping was explained by some special hypotheses, we developed a theory that may be termed hydrodynamic [9].

This theory considers the presence of the surface-active film and relates it to the wave motion. As in the case of steady-state motion examined in Chapter VII, the presence of the surface-active film produces surface forces that change the regime of motion. In the case of wave motion, these forces result from expansion of the surface-active film on the crest of the wave and its contraction at the trough of the wave. Forces that do not generally exist on the surface of a clean liquid surface exist in those areas of the interface where the density concentration of the surface-active agent in the film varies. In analyzing the steady-state motion of a drop in the presence of surface-active substances, we have seen that the additional forces exerted at the free surface are accounted for in changed boundary conditions there. The cange in boundary conditions, in turn, leads to a change in the velocity distribution in the liquid. According to hydrodynamic theory, the damping effect of the surface-active film on wave motion is related to such change in the velocity distribution. The hydrodynamic theory of wave damping does not involve any special assumptions regarding the damping nature of the film.

It should be stressed that the consideration of phenomena arising in expansion and contraction of the film is equivalent to assigning a certain effective surface viscosity to the film, as was done in the case of falling drops (Section 76).

We will next find the velocity distribution and calculate the wave damping coefficient for a liquid surface covered by a surface-active film.

Let us examine the system of capillary-gravity waves on the surface of a liquid covered by a surface-active film. We assume that the surface-active agent is insoluble, and then generalize the results for the case of a soluble material.

In Section 117 we derived relations for capillary-gravity waves on the clean surface of a viscous liquid.

The presence of the surface-active substance changes the boundary conditions on the liquid surface.

Boundary conditions (117.25), which are valid for a liquid surface, must not be replaced by boundary conditions (69.3) and (69.4).

A general solution of the hydrodynamic equations for waves of small amplitude, as in the case of a clean liquid surface, has the form

$$v_x = (ikAe^{kz} - lCe^{lz})\, e^{ikx+\alpha t}, \tag{121.1}$$

$$v_z = (kAe^{kz} + ikCe^{lz})\, e^{ikx+\alpha t}, \tag{121.2}$$

$$p = -\rho\alpha Ae^{kz}e^{ikx+\alpha t} - \rho gz. \tag{121.3}$$

The values of the quantity $\alpha$ and of the ratio of the amplitudes A/C, however, differ from those computed in Sections 117 to 119.

In Section 117, these constants were determined by substitution of the expressions for the velocity and pressure components into boundary conditions (117.25). In the present case, however, the constants must be found from a set of new boundary conditions.

We first write these boundary conditions in expanded form, and substitute the values of $p_{nn}$ and $p_t$. From the definition of these quantities (cf. Section 1), we obviously have

$$-p + 2\mu \frac{\partial v_z}{\partial z} = p_n, \tag{121.4}$$

$$\mu\left(\frac{\partial v_x}{\partial z} + \frac{\partial v_z}{\partial x}\right) = p_t. \tag{121.5}$$

The general expressions for $v_x$, $v_y$ and p were given above, and we have to find only the expressions for the normal and tangential components of the forces $F_n$ and $F_t$ (per unit of liquid surface).

The normal force $F_n$ exerted on the liquid surface when the surface is slightly deformed is, as in the case of a clean surface, equal to

$$p_\sigma = p_n = \sigma(\Gamma)\frac{\partial^2 \zeta}{\partial x^2}, \tag{121.6}$$

where $\zeta$ is the vertical displacement of the surface, equal to $\zeta = \int v_z dt$. Before substituting the value of (121.6) into condition (69.4), however, we must still account for the dependence of surface tension on the concentration $\Gamma$ of the surface-active agent. Surface tension, therefore, varies from point to point along the surface. The tangential force $p_t$ in expression (121.5), according to (69.2), is

$$p_t = -\frac{\partial \sigma}{\partial \Gamma}\operatorname{grad}\Gamma. \tag{121.7}$$

In order to write an explicit expression for the forces, we must express the concentration of the surface-active agent in terms of the liquid velocity as was done in Section 73 for the case of steady-state motion of a drop. For this purpose we use the equation (69.8) for the conservation of matter. The latter, in the case of mass flow on a surface which is almost planar, becomes

$$\frac{\partial \Gamma}{\partial t} + \frac{\partial}{\partial x}(\Gamma v_x) = D_s\frac{\partial^2 \Gamma}{\partial x^2}, \tag{121.8}$$

where the value for $v_x$ is taken at $z = 0$ on the liquid surface.

We assume, as we did in Section 73, that the concentration $\Gamma$ of the surface-active agent is

$$\Gamma = \Gamma_0 + \Gamma', \tag{121.9}$$

where $\Gamma' \ll \Gamma_0$ and $\Gamma_0$ is the constant concentration on the undeformed surface. Then, assuming that the liquid velocity is low (small-amplitude waves) and disregarding the quantity $\Gamma' \frac{\partial v_x}{\partial x}$, which is a second-order infinitesimal, we can write the following in place of expression (121.8):

$$\frac{\partial \Gamma'}{\partial t} + \Gamma_0 \frac{\partial v_x}{\partial x} = D_s \frac{\partial^2 \Gamma'}{\partial x^2}. \quad (121.10)$$

Substituting the value of $v_x$ from expression (121.1), we seek a solution of equation (121.10) in the form

$$\Gamma' = \gamma e^{ikx+\alpha t},$$

where $\gamma$ is a constant. From equation (121.10), we have:

$$\gamma\alpha + \Gamma_0 (ikA - lC)\, ik = - D_s k^2 \gamma,$$

whence

$$\gamma = -\frac{ik\Gamma_0 (ikA - lC)}{\alpha + D_s k^2}. \quad (121.11)$$

We show later that the following inequality is satisfied for all wave frequencies

$$|\alpha| \gg D_s k^2. \quad (121.12)$$

This inequality signifies that, owing to the small magnitude of the diffusion coefficient to the surface, the influence of the diffusion on the concentration distribution of the surface-active agents may be disregarded. Thus, we can assume that the surface-active agent is fully swept along with the liquid.

Assuming inequality (121.12) to hold, the expression (121.11) for $\gamma$ may be simplified by neglecting the small term $D_s k^2$. Then,

$$\gamma \approx -\frac{ik\Gamma_0 (ikA - lC)}{\alpha}, \quad (121.13)$$

$$\Gamma = \Gamma_0 - \frac{ik\Gamma_0}{\alpha} (ikA - lC)\, e^{ikx+\alpha t}. \quad (121.14)$$

Knowing the distribution of the material on the surface, we can also write the expression for the forces. With accuracy to second order infinitesimals, we have

$$p_\sigma = \sigma(\Gamma) \frac{\partial^2 \zeta}{\partial x^2} \approx \sigma(\Gamma_0) \frac{\partial^2 \zeta}{\partial x^2} = -\frac{\sigma_0 (kA + ikC)\, k^2}{\alpha} e^{ikx+\alpha t}, \quad (121.15)$$

$$\begin{aligned} p_t &= -\frac{\partial \sigma}{\partial \Gamma} \frac{\partial \Gamma}{\partial x} = -k^2 \frac{\Gamma_0}{\alpha} \left(\frac{\partial \sigma}{\partial \Gamma}\right) (ikA - lC)\, e^{ikx+\alpha t} = \\ &= -\varepsilon \left(\frac{ik^3}{\alpha} A - \frac{k^2 l}{\alpha} C\right) e^{ikx+\alpha t} = -\frac{\varepsilon k^2}{\alpha} (ikA - lC)\, e^{ikx+\alpha t}, \end{aligned} \quad (121.16)$$

where $\epsilon$ denotes the quantity

$$\varepsilon = \frac{\partial \sigma}{\partial \Gamma} \Gamma_0. \qquad (121.17)$$

Let us also note that the tangential force $p_t$ can be expressed as

$$p_t = \frac{\varepsilon}{\alpha} \left( \frac{\partial^2 v_x}{\partial x^2} \right)_{z=0}.$$

The latter equation shows that $\epsilon$ plays the part of a coefficient of elasticity in compression of the film.

The normal force $p_\sigma$ in this approximation is identical with the normal force exerted on the free surface of a clean liquid.

By substituting the values from (121.15) and (121.16) into (121.4) and (121.5), and performing some elementary, but somewhat tedious calculations, we obtain the following boundary conditions for the determination of $\alpha$ and of the ratio A/C:

$$\left(\alpha^2 + 2\nu k^2 \alpha + \omega_0^2\right) A + i \left(\omega_0^2 + 2\nu k l \alpha\right) C = 0, \qquad (121.18)$$

$$i (\varepsilon k^3 + 2\nu k^2 \alpha) A - (\alpha^2 + 2\nu k^2 \alpha - \varepsilon k^2 l) C = 0, \qquad (121.19)$$

where the following notation is used:

$$\omega_0^2 = \frac{\sigma(\Gamma_0) k^3}{\rho} + gk. \qquad (121.20)$$

Equations (121.18) and (121.19) determine the complex frequency in exactly the same manner as in Section 118 for a clean liquid surface.

Comparing these equations with the analogous expressions (117.29) and (117.30), we see that equation (121.18), which expresses the dynamic balance of forces, is identical with equation (117.29). Such an identity occurs only with the approximation that the surface tension $\sigma(\Gamma)$ in the expression for $p_\sigma$ can be replaced, according to formula (121.15), by a constant surface tension.

To find the complex quantity $\alpha$ from equations (121.18) and (121.19), it is useful to note that the ratio $\nu k^2/\omega_0$ is small compared to unity at all wavelengths usually encountered in practice. Indeed,

$$\frac{\nu k^2}{\omega_0} < \frac{\nu k}{c_{\min}}, \qquad (121.21)$$

where $c_{\min}$ is the minimum propagation velocity of capillary-gravity waves on an interface having surface tension [see equation (116.41)]

$$c_{\min} = \sqrt[4]{4 g \sigma(\Gamma_0)}. \qquad (121.22)$$

Therefore

$$\frac{\nu k^2}{\omega_0} < \frac{2\pi\nu}{\sqrt[4]{4g\sigma(\Gamma_0)}} \cdot \frac{1}{\lambda}.$$

With $\sigma(\Gamma_0) = 40$ dynes/cm, we have

$$\frac{\nu k^2}{\omega_0} < \frac{6 \cdot 10^{-2}}{\sqrt[4]{16 \cdot 10^4}} \frac{1}{\lambda} \approx \frac{3 \cdot 10^{-3}}{\lambda}.$$

Thus, if the wavelength is large compared to $3 \cdot 10^{-3}$ cm, we can assume that

$$\frac{\nu k^2}{\omega_0} \ll 1. \tag{121.23}$$

Our subsequent examination is limited to waves for which inequality (121.23) holds.

The quantity $\nu k^2$ characterizes the damping coefficient for waves on the surface of a pure liquid. Inequality (121.23) shows, therefore, that we are examining waves that are only slightly damped over a time interval equal to one period.*

We can rewrite equations (121.18) and (121.19) in dimensionless form by dividing them by $\omega_0^2$, and by introducing the quantity $m = \dfrac{\nu k^2}{\omega_0}$. Then

$$\left(\frac{\alpha^2}{\omega_0^2} + 2m\frac{\alpha}{\omega_0} + 1\right)A + i\left(1 + 2m\frac{l\alpha}{k\omega_0}\right)C = 0, \tag{121.24}$$

$$i\left(\frac{\varepsilon k^3}{\omega_0^2} + 2m\frac{\alpha}{\omega_0}\right)A - \left(\frac{\alpha^2}{\omega_0^2} + 2m\frac{\alpha}{\omega_0} - \frac{\varepsilon k^2}{\omega_0}\frac{l}{\omega_0}\right)C = 0. \tag{121.25}$$

By eliminating the constants A and C from equations (121.24) and (121.25), the frequency $\alpha$ may be determined. However, since $\alpha$ enters into the expression for $l$ [see equation (117.17)], it will be determined, in a general case, by a higher order equation. To determine $\alpha$, we can use the method of approximation. We assume that the change in wave frequency compared to the frequency in an ideal liquid is small. Thus, we can write

$$\alpha = i\omega_0 + \beta, \tag{121.26}$$

where $\beta$ is a complex number, but by definition $|\beta| \ll \omega_0$.

---

*It would be incorrect to conclude from the above that waves with $\lambda \ll 3 \cdot 10^{-3}$ cm are absent from the liquid surface. Although such short waves are characterized by a considerable damping coefficient, they are maintained by thermal motion.

The real part of $\beta$ represents the damping coefficient, and the imaginary part, the frequency correction.

It is convenient to seek separate solutions for equations (121.24) and (121.25) for different wavelength regions.

Let us first examine the case of pure capillary waves of frequency

$$\omega_0 = \sqrt{\frac{\sigma k^3}{\rho}}. \tag{121.27}$$

Substituting the value of $\alpha$ from (121.26) into (121.24) and (121.25), and disregarding quantities of the order of magnitude of $\left(\frac{\beta}{\omega_0}\right)^2$, we obtain

$$\left(\frac{2i\beta}{\omega_0} + 2im\right)A + i\left(1 + 2im\frac{l}{k}\right)C = 0, \tag{121.28}$$

$$i\left(\frac{\varepsilon}{\sigma} + 2im\right)A - \left(-1 + 2i\frac{\beta}{\omega_0} + 2im - \frac{\varepsilon}{\sigma}\frac{l}{k}\right)C = 0. \tag{121.29}$$

In equations (121.28) and (121.29) $m \ll 1$. In addition,

$$\frac{l}{k} \approx \sqrt{1 + \frac{i}{m}} \approx \sqrt{\frac{i}{m}} = \sqrt{\frac{i\omega_0}{\nu k^2}}.$$

By eliminating the amplitudes A and C from equations (121.28) and (121.29), we find the real part of the damping coefficient

$$\beta = -\frac{\omega_0}{2\sqrt{2}} \cdot \frac{\left(\frac{\varepsilon}{\sigma}\right)^2\left(\frac{\omega_0}{\nu k^2}\right)^{1/2}}{\left(\frac{\varepsilon}{\sigma}\right)^2\frac{\omega_0}{\nu k^2} - \sqrt{2}\frac{\varepsilon}{\sigma}\sqrt{\frac{\omega_0}{\nu k^2}} + 1} + \frac{2\nu k^2\left[\frac{\varepsilon}{\sigma}\sqrt{\frac{\omega_0}{2\nu k^2}} - 1\right]}{\left(\frac{\varepsilon}{\sigma}\right)^2\frac{\omega_0}{\nu k^2} - \sqrt{2}\frac{\varepsilon}{\sigma}\sqrt{\frac{\omega_0}{\nu k^2}} + 1}. \tag{121.30}$$

The imaginary part of $\beta$, which represents a small frequency correction, has been omitted.

The quantity $\epsilon/\sigma$ in expression (121.30) characterizes the dependence of damping on the physical properties of the surface-active film.

We analyze here two limiting cases: $\frac{\epsilon}{\sigma} \to 0$ and $\frac{\epsilon}{\sigma} \to 1$. In the first case, which corresponds to a clean surface,

$$\beta \to \beta_0 = -2\nu k^2, \tag{121.31}$$

as should have been expected [see equation (118.6)].

In the second case, which corresponds to a surface covered by a surface-active film with a high coefficient of elasticity of the film, we have, from equation (121.30):

$$\beta \to \beta_{\text{lim}} = -\frac{1}{2\sqrt{2}} \nu^{1/2} \omega_0^{1/2} k = -\frac{1}{2\sqrt{2}} \nu^{1/2} \sigma^{1/4} k^{7/4} \rho^{-1/4} \tag{121.32}$$

To simplify equation (121.30), we considered the inequality (121.23).

In this limiting case, the damping coefficient for waves on a surface covered by a surface-active agent is found to be significantly higher than for a liquid surface. Indeed, owing to the inequality (121.23), the ratio $\beta_{\text{lim}}$ to $\beta_0$ becomes

$$\frac{\beta_{\text{lim}}}{\beta_0} = \frac{1}{4\sqrt{2}} \left(\frac{\omega_0}{\nu k^2}\right)^{1/2} \gg 1.$$

Thus, the presence of a surface-active agent causes strong damping of capillary waves on the surface of the liquid. In the limiting case where the values of the coefficient of elasticity $\epsilon$ are very high, the damping coefficient is not a function of the physical properties of the film and is determined solely by the wavelength, surface tension and by the viscosity of the liquid.

In order to elucidate this conclusion, we turn to equation (121.28), and simplify it by assuming that $\frac{\epsilon}{\sigma} \approx 1$.

Since $\frac{l}{k} \approx \sqrt{\frac{i\omega_0}{\nu k^2}} \gg 1$, we have, from equation (121.28)

$$kA - ilC = 0. \tag{121.33}$$

Comparing this relation with expression (121.1), we see that for $\frac{\epsilon}{\sigma} \approx 1$, the following condition is fulfilled

$$(v_x)_{z=0} = 0. \tag{121.34}$$

The above condition signifies that, for a sufficiently high value of the coefficient of elasticity $\epsilon \approx \sigma$, the film on the liquid surface behaves like an incompressible solid plate. As the surface deforms, it merely bends, and neither contracts at the trough of the wave nor expands at its crest. It is obvious, since the film is quite incompressible, that the nature of the liquid motion cannot be a function of film properties. In order to elucidate the dissipation mechanism in the case of an incompressible film, we formulate an equation for the vorticity of liquid suffering wave motion in the presence of an incompressible film.

Direct computation gives

$$\text{rot}_y \mathbf{v} = \frac{\omega_0}{\nu} C e^{lz} e^{i(kx - \omega_0 t)}.$$

Thus, the ratio of the irrotational portion of the motion to the rotational portion is characterized, as previously (cf. Section 118), by the ratio A/C. In the case of a pure liquid, this ratio, according to (118.1), is

$$\frac{A}{C} \approx \frac{\omega_0}{\nu k^2}. \tag{121.35}$$

In the case of a surface covered by an incompressible film, this ratio, according to (121.33), is given by

$$\frac{A}{C} \approx \frac{l}{k} \approx \sqrt{\frac{\omega_0}{\nu k^2}}. \tag{121.36}$$

Comparing (121.36) and (121.35), we see that the ratio A/C in the latter case is much smaller than in the first. This signifies that the rotational part of the motion is considerably increased by the presence of an incompressible film, because when the tangential velocity component becomes zero on the surface, the change in velocity across the surface layer is much sharper than in the case of a clean surface.

The increase in the rotational part of the motion, according to expression (1.15), leads to additional dissipation and damping of capillary waves.

The damping coefficient generally lies between $\beta_0$ and $\beta_{\text{lim}}$, and is strongly dependent on the properties of the film. Condensed films with densely packed molecular chains and with low compressibility exhibit the best damping properties. Such materials are fatty acids, triglycerides and other similar substances. Such insoluble surface-active agents as acetamides, phenols, nitriles and other materials with chains at angles to one another are materials with the least damping properties.

Let us turn to the analysis of damping of longer waves with a frequency

$$\omega_0 = \sqrt{gk}.$$

Substituting this value for $\omega_0$ into boundary conditions (121.24) and (121.25), disregarding the quantities $\sqrt{\frac{\nu k^2}{\omega_0}}$ and $\frac{\nu k^2}{\omega_0}$ that are small compared to unity, and substituting the value of $l/k$ from expression (117.17), we get the boundary conditions

$$\left(\frac{2\beta}{\sqrt{gk}} + \frac{2\nu k^{3/2}}{\sqrt{g}}\right) A - C = 0, \tag{121.37}$$

$$\left(\frac{\varepsilon k^2}{g} + \frac{2i\nu k^{3/2}}{\sqrt{g}}\right) A + i\left(1 - \frac{\sqrt{i}\,\varepsilon k^{5/4}}{g^{3/4}\nu^{1/2}}\right) C = 0. \tag{121.38}$$

It is evident from (121.37) and (121.38) that two regions of gravity waves exist: a region where the wave vector $k \approx k_1$, where $k_1$ can be approximated by

$$\frac{\varepsilon k_1^{5/4}}{g^{3/4}\nu^{1/2}} \approx 1$$

or by

$$k_1 \approx \frac{g^{3/5}\nu^{2/5}}{\varepsilon^{4/5}}.$$

The other region is defined by the condition $k \approx k_2$, where $k_2$ is computed from the following:

$$\frac{\varepsilon k_2}{g} \approx \frac{2\nu k^{3/2}}{\sqrt{g}}.$$

The latter gives

$$k_2 \approx \frac{\nu^2 g}{\varepsilon^2}.$$

A comparison of $k_1$ and $k_2$ shows that $k_1 > k_2$.

At representative values for $\nu$ and $\epsilon$, $k_1 \approx \frac{1}{3}\,\text{cm}^{-1}$, so that $\lambda_1 = \frac{2\pi}{k_1} \approx 15$ cm, $k_2 \approx 10^{-3}\,\text{cm}^{-1}$ and $\lambda = \frac{2\pi}{k_2} \approx 10^3$ cm. In the region of $k \approx k_1$, i.e., in the region of centimeter wavelengths, $\frac{\epsilon k^2}{g} \approx \sqrt{\frac{2\nu k^{3/2}}{\sqrt{g}}}$. Thus, the second term $\frac{2\nu k^{3/2}}{\sqrt{g}}$ in boundary condition (121.38) may be disregarded compared to the first term $\frac{\epsilon k^2}{g}$.

From conditions (121.37) and (121.38), we then find the expression for the real part of the damping coefficient

$$\beta = \frac{1}{2\sqrt{2}} \cdot \frac{g^{1/4}\varepsilon^2\nu^{1/2}k^{15/4}}{\varepsilon^2 k^{5/2} + \nu g^{3/2} - \sqrt{2}\,\varepsilon\nu^{1/2}g^{3/4}k^{5/4}}. \tag{121.39}$$

An approximation of the real part gives

$$\beta \approx \sqrt{\nu g^{1/2} k^{5/2}}. \tag{121.40}$$

The ratio of $\beta$ to the damping coefficient $\beta_0$ in a pure liquid, therefore, has the form

$$\frac{\beta}{\beta_0} \approx \sqrt{\frac{g^{1/2}}{\nu k^{3/2}}};$$

i.e., the ratio is large compared to unity. Thus, the damping effect of the film remains fairly effective for waves of such lengths.

In the region $k \approx k_2$, i.e., in the region of very long waves, in condition (121.38)

$$\frac{\varepsilon k^{5/4}}{g^{3/4}\nu^{1/2}} \ll 1,$$

and therefore this term can also be disregarded.

From conditions (121.37) and (121.38) we find

$$\beta = -2\nu k^2;$$

i.e., this region is characterized by natural damping. The effect of the film is merely a small change of frequency which is of no interest. For even longer waves the latter change disappears.

Thus, the damping action of surface-active agents according to the hydrodynamic theory of damping of capillary waves can be explained without resorting to any special hypotheses regarding special properties of these materials. It is found that the damping effect occurs throughout the wave spectrum, starting from the shortest capillary waves up to centimeter-long gravity waves.

In the limiting case the film behaves as a thin incompressible membrane that covers the surface of the liquid. This should not lead to the conclusion, however, that the damping properties of surface-active agents are independent of their nature. The damping coefficient, as seen from equations (121.30) and (121.39), is determined by the value of the coefficient of elasticity [see equation (121.17)]. The damping coefficient increases with increase in $\epsilon$, and becomes independent of $\epsilon$ only in the limiting case. The damping effect of insoluble surface-active agents increases with the concentration $\Gamma$. That surface-active agent for which the value of $\partial\sigma/\partial\Gamma$ is the highest has the greatest damping effect. However, only in the limiting cases, involving very few of the surface-active agents, does the damping effect cease to be a function of the nature of the material.

The dependence of the damping effect on the nature of the surface-active agent is even more pronounced in the case of soluble materials – a case examined in the next section.

We now turn to the anlysis of viscosity in surface-active agents. We showed in Section 77 that the effect of surface-active substances on the regime of liquid motion in steady-state is equivalent to the introduction of a certain new viscosity e. This viscosity is not only a function of the properties, but also of the nature of the processes which control the rate at which equilibrium between surface and substrate is approached.

We introduce the surface viscosity coefficient for the case of a non-steady-state process. By definition (cf. Section 77), the surface viscosity coefficient is

$$e = \frac{\partial \Delta\sigma}{\partial \operatorname{div}_s \mathbf{v}} = \frac{\partial\sigma}{\partial\Gamma}\frac{\partial\Delta\Gamma}{\partial \operatorname{div}_s \mathbf{v}} = \frac{\partial\sigma}{\partial\Gamma}\frac{\partial\Delta\Gamma}{\partial\left(\frac{\partial v_x}{\partial x}\right)} = \frac{\partial\sigma}{\partial\Gamma}\cdot\frac{1}{ik}\frac{\partial\Gamma}{\partial v_x}.$$

But, according to equations (98.14) and (98.1),

$$\Delta\Gamma = -\frac{ik\Gamma_0}{\alpha} v_x;$$

therefore,

$$e = -\frac{\Gamma_0}{\alpha} \cdot \frac{\partial\sigma}{\partial\Gamma} = -\frac{\varepsilon}{\alpha} \approx -\frac{\varepsilon}{\omega_0}.$$

We see that surface viscosity cannot be viewed as a constant that characterizes the properties of the liquid, because it depends on the wave frequency $\omega_0$. Nevertheless, with this exception e may be regarded as a dynamic characteristic of the surface-active film and one which characterizes the relaxation processes in the material. The quantity $\epsilon$, introduced by us, is the constant that characterizes the properties of the material. Thus, our previous examination, based on the introduction of the constant $\epsilon$, adequately accounts for the surface viscosity. On the contrary, the introduction of surface viscosity as a constant to characterize the properties of the film is inadmissible.

In the light of the preceding discussion and bearing in mind the reasoning set forth in Section 77, it is clear that our analysis of the elastic properties of the film and our disregard of the viscous properties is purely a matter of terminology. We could have equally introduced the concept of surface viscosity and could have argued in terms of the viscosity of the film without mentioning its elastic properties.

The terminology adopted seems more appropriate to us because of the variable nature of the viscosity e, which cannot be a constant of the material. Non-steady-state processes involving surface-active agents, in this respect, are found to be similar to steady-state processes.

## 122. DAMPING OF WAVES BY SOLUBLE SURFACE-ACTIVE AGENTS

Surface-active agents, especially inorganic materials, are usually measurably soluble. This property affects their damping action. Let us consider the damping effect of soluble surfactants.

When a state of equilibrium exists between the surface and the bulk of the liquid, an equilibrium distribution of the surface-active agent is established. When the surface is deformed by waves of frequency $\omega_0$ or of period $1/\omega_0$, two limiting cases are possible: 1) the relaxation time for equilibrium to become established between the surface and the body of the liquid (more exactly, the portion of the liquid body contiguous to the surface) is large compared to the wave period $1/\omega_0$, and 2) this time interval is small compared to $1/\omega_0$.

The deformation in the first instance occurs so rapidly that an equilibrium between the surface and the bulk of the liquid is not established, and the molecules do not have enough time to pass from the body of the solution to the surface and back again. Here, the presence of the bulk of the solution does not affect the process and all conclusions of the preceding section remain valid.

In the second case, we may assume that the surface and bulk of the solution are continually in a state of statistical equilibrium. This is the case of our further examination.

As before, let $\Gamma_0$ denote the surface concentration of an undeformed surface, and let $\Gamma'$ denote the total change in concentration — resulting from surface deformation.

Similarly, let c and $c'$ be the concentrations in the bulk of the solution and its change resulting from diffusion of the dissolved substance from the bulk of the solution to the surface when the latter is deformed.

Then, the equilibrium condition between the bulk and surface solutions is given by

$$\mu_n(\Gamma_0+\Gamma')=\mu_p(c+c'), \tag{122.1}$$

where $\mu_n$ is the chemical potential of the solution on the surface and $\mu_p$ is the chemical potential in the bulk of the solution.

The value of the chemical potential in the bulk of the solution is taken near the surface, i.e., at $z = 0$.

Since the surface deformation and, therefore, the change in concentration are infinitely small, the chemical potentials may be expanded into a power series of which only the leading terms need be retained

$$\Gamma'=\frac{\left(\frac{\partial\mu_p}{\partial c}\right)}{\left(\frac{\partial\mu_n}{\partial\Gamma_0}\right)}c', \tag{122.2}$$

and, since the solution in the bulk can usually be considered weak, we have

$$\mu_p=RT\ln c_0+\psi(p,\ T),$$

where T is the absolute temperature and p is pressure. Whence,

$$\Gamma'=\frac{RT}{c_0}\frac{1}{\left(\frac{\partial\mu_n}{\partial\Gamma_0}\right)}c'. \tag{122.3}$$

The value of $c'$ is taken near the surface, i.e., at $z = 0$.

The change in surface concentration is related to two factors: surface deformation and diffusional flow from the body of the solution, which reduces this change. Therefore, considering the equation of conservation of matter for $\Gamma'$, and without taking surface diffusion into consideration, we have with an accuracy to first order infinitesimals:

$$\frac{\partial\Gamma'}{\partial t}+\Gamma_0\frac{\partial v_x}{\partial x}=D\left(\frac{\partial c'}{\partial z}\right)_{z=0}. \qquad (122.4)$$

The right side of equation (122.4) is the diffusional flow from the bulk and D is the diffusion coefficient.

In the absence of diffusional flow, expression (122.4) is similar to (121.8).

To find the diffusional flow from the bulk of the solution, the diffusion equation for $c'$ must be solved considering boundary conditions (122.2) and (122.4). However, since the boundary conditions include terms determined by the fluid flow equations, the diffusion equation with boundary conditions (122.2) and (122.4) must be solved simultaneously with the fluid flow equation and appropriate boundary conditions, which are

$$p_\sigma=\sigma\frac{\partial^2\zeta}{\partial x^2}, \qquad (122.5)$$

$$p_t=-\frac{\varepsilon}{\Gamma}\frac{\partial\Gamma'}{\partial x}. \qquad (122.6)$$

Condition (122.5) is equivalent to (121.6), and condition (122.6), although similar to (121.7), in this case has a different meaning for $\frac{\partial\Gamma'}{\partial x}$.

Substituting (122.3) into (122.4) we obtain, finally,

$$\frac{RT}{c_0}\frac{1}{\left(\frac{\partial\mu_n}{\partial\Gamma_0}\right)}\left(\frac{\partial c'}{\partial t}\right)_{t=0}=-\Gamma_0\frac{\partial v_x}{\partial x}-D\left(\frac{\partial c'}{\partial z}\right)_{z=0}. \qquad (122.7)$$

The concentration distribution in the bulk of the solution is given by the diffusion equation

$$\frac{\partial c}{\partial t}=D\left(\frac{\partial^2 c}{\partial x^2}+\frac{\partial^2 c}{\partial z^2}\right). \qquad (122.8)$$

The solution of the hydrodynamic equations is

$$v_x=(ikAe^{kz}-lCe^{lz})\,e^{ikx+\alpha t}, \qquad (122.9)$$

$$v_z=(kAe^{kz}+ikCe^{lz})\,e^{ikx+\alpha t}. \qquad (122.10)$$

Therefore, a solution of (122.8) that satisfies boundary condition (122.7) must be sought in the following form:

$$c'=Bf(z)\,e^{ikx+\alpha t}.$$

Substituting this expression for $c'$ into (122.8), we get

$$c'=Be^{\sqrt{\frac{\alpha+Dk^2}{D}}\,z}\,e^{ikx+\alpha t}. \qquad (122.11)$$

The amplitude of diffusion waves in equation (122.11) may be found from (122.7) by substituting the values of $c'$ and $v_x$ from equations (122.11) and (122.9), respectively. Then

$$B = \frac{\Gamma_0 ik\,(ikA - lC)}{\dfrac{RT\alpha}{c_0\left(\dfrac{\partial\mu_n}{\partial\Gamma_0}\right)} + \sqrt{D}\,\sqrt{\alpha + Dk^2}},$$

whence

$$c' = \frac{\Gamma_0 ik\,(ikA - lC)}{\dfrac{RT\alpha}{c_0\left(\dfrac{\partial\mu_n}{\partial\Gamma_0}\right)} + \sqrt{D}\,\sqrt{\alpha + Dk^2}}\, e^{\sqrt{\frac{\alpha + Dk^2}{D}}}\, e^{ikx + \alpha t},$$

$$\Gamma' = \frac{RT}{c_0\left(\dfrac{\partial\mu_n}{\partial\Gamma_0}\right)}\, c' = \frac{\Gamma_0 ik\,(ikA - lC)}{\alpha + \dfrac{\sqrt{D}\,c_0}{RT}\left(\dfrac{\partial\mu_n}{\partial\Gamma_0}\right)\sqrt{\alpha + Dk^2}}\, e^{ikx + \alpha t} = \tag{122.12}$$

$$= \frac{\Gamma_0 \dfrac{\partial v_x}{\partial x}}{\alpha + \dfrac{\sqrt{D}\,c_0}{RT}\left(\dfrac{\partial\mu_n}{\partial\Gamma_0}\right)\sqrt{\alpha + Dk^2}}.$$

Boundary condition (69.5), therefore, may be written as

$$p_t = \frac{\dfrac{\varepsilon}{\alpha}}{1 + \dfrac{\sqrt{D}\,c_0}{\alpha RT}\left(\dfrac{\partial\mu_n}{\partial\Gamma_0}\right)\sqrt{\alpha + Dk^2}}\, \frac{\partial^2 v_x}{\partial x^2}. \tag{122.13}$$

Condition (122.13) shows that, in the case of a soluble substance, the following parameter plays the role of the elastic constant (elasticity coefficient)

$$\varepsilon' = \frac{\varepsilon}{1 + \dfrac{\sqrt{D}\,c_0}{\alpha RT}\left(\dfrac{\partial\mu_n}{\partial\Gamma_0}\right)\sqrt{\alpha + Dk^2}}.$$

Since the diffusion coefficient in the bulk of the solution is of the same order of magnitude on the surface, we may again assume [see (121.12)] that $\alpha \gg Dk^2$, and as

$$\varepsilon' = \frac{\varepsilon}{1 + \dfrac{\sqrt{D}\,c_0}{RT\sqrt{\alpha}}\left(\dfrac{\partial\mu_n}{\partial\Gamma_0}\right)}. \tag{122.14}$$

This parameter is found to be a function not only of the elastic properties of the film, but also of the diffusion coefficient, of the bulk concentration and of the frequency.

It is not difficult to see that, as the bulk concentration decreases, i.e., as $c_0 \to 0$, this effective elastic constant (coefficient of elasticity) becomes equal to $\epsilon$. On the contrary, when $c_0$ increases, the effective elastic constant decreases. In the limit as $c_0 \to \infty$ it becomes zero. Although this expression of course falls far short of high concentrations, because it depends on dilute solution approximations for chemical potentials, it nevertheless indicates the general trend of the elasticity coefficient.

By setting $\alpha = i\omega_0 + \beta$ and substituting the values (122.9) and (122.10) into boundary conditions (121.4) and (122.13), we obtain the equations for determining $\beta$. In this case we will limit our discussion to the most interesting case of capillary waves and, as previously, assume that $\frac{\nu k^2}{\omega_0} \ll 1$ and $\frac{\epsilon}{\sigma} \approx 1$. We now discuss the case when

$$\frac{\sqrt{D}c_0\left(\frac{\partial\mu_n}{\partial\Gamma_0}\right)}{RT\sqrt{\omega_0}} \gg 1.$$

Indeed, it is easily observed that if

$$\frac{\sqrt{D}c_0\left(\frac{\partial\mu_n}{\partial\Gamma_0}\right)}{RT\sqrt{\omega_0}} \ll 1,$$

then the above term may be omitted, and we revert to the case of an insoluble material. If, however,

$$\frac{\sqrt{D}c_0\left(\frac{\partial\mu_n}{\partial\Gamma_0}\right)}{RT\sqrt{\omega_0}} \approx 1,$$

then simple computations show that we obtain the damping coefficient given by (121.32). If

$$\frac{\sqrt{D}c_0\left(\frac{\partial\mu_n}{\partial\Gamma_0}\right)}{RT\sqrt{\omega_0}} \gg 1,$$

then, disregarding the small terms, we get

$$\left[i\frac{2\nu k^2}{\omega_0} + \frac{\varepsilon k^3}{\omega_0^2}\frac{RT\sqrt{i\omega}}{c_0\left(\frac{\partial\mu_n}{\partial\Gamma_0}\right)\sqrt{D}}\right]A - \\ -i\left[-1 + \frac{\varepsilon k^2}{\omega_0^2}\sqrt{\frac{i\omega_0}{\nu k^2}}\frac{RT}{c_0\left(\frac{\partial\mu_n}{\partial\Gamma_0}\right)\sqrt{D}} - \frac{2\beta}{\omega_0}\right]C = 0, \tag{122.15}$$

$$\left(2\frac{\beta}{\omega_0} + \frac{2\nu k^2}{\omega_0}\right)A - C = 0. \tag{122.16}$$

Equation (122.15) shows that there are two regions:

$$\frac{RT\sqrt{\omega_0}}{c_0\left(\frac{\partial\mu_n}{\partial\Gamma_0}\right)\sqrt{D}} \approx \sqrt{\frac{\nu k^2}{\omega_0}}$$

or

$$\left.\begin{aligned} k^{1/2} &\approx \frac{\nu^{1/2}\sqrt{D}\,c_0}{\sigma^{3/4}RT}\left(\frac{\partial\mu_n}{\partial\Gamma_0}\right) \\ \frac{RT\sqrt{\omega_0}}{c_0\left(\frac{\partial\mu_n}{\partial\Gamma_0}\right)\sqrt{D}} &\approx \frac{\nu k^2}{\omega_0}. \end{aligned}\right\} \qquad (122.17)$$

and

or

$$k^{1/4} \approx \frac{\nu\sqrt{D}\,c_0}{\sigma^{3/4}RT}\left(\frac{\partial\mu_n}{\partial\Gamma_0}\right) \qquad (122.18)$$

Omitting the small terms from equation (122.15), we obtain the following damping coefficients.

In the first region

$$\beta_1 = -\frac{1}{2\sqrt{2}}\,\frac{\varepsilon RT\,k^{9/4}}{\sqrt{D}\,\sigma^{1/4}c_0\left(\frac{\partial\mu_n}{\partial\Gamma_0}\right)}\,\frac{1+\frac{\varepsilon RTk^{1/2}}{\sqrt{D}\,\sigma^{1/2}\nu^{1/2}}}{1+\frac{\varepsilon^2(RT)^2k}{D\sigma\nu}}. \qquad (122.19)$$

The damping coefficient in the first region is approximately equal to $\beta_1 \approx \nu^{1/2}\sigma^{1/4}k^{7/4}$. Although the damping coefficient in this region is somewhat lowered, owing to influx of molecules from the bulk of the solution, it nevertheless in order of magnitude agrees with the damping coefficient of an insoluble film.

In the region that is intermediate between the first and the second, i.e., at

$$\sqrt{\frac{\nu k^2}{\omega_0}} \gg \frac{RT\sqrt{\omega}}{\sqrt{D}\,c_0\left(\frac{\partial\mu_n}{\partial\Gamma_0}\right)} \gg \frac{\nu k^2}{\omega_0},$$

the damping coefficient is

$$\beta_{1,2} = -\frac{1}{2\sqrt{2}}\,\frac{\varepsilon RT\,k^{9/4}}{\sqrt{D}\,\sigma^{1/4}c_0\left(\frac{\partial\mu_n}{\partial\Gamma_0}\right)}. \qquad (122.20)$$

Finally, in the second region

$$\beta_2 = -2\nu k_2.$$

Thus, in the second region, where the wavelengths are long or the concentration in the bulk of the solution is very high, the damping coefficient is equal to that in a pure liquid. This result could have been predicted, because, at a sufficiently high bulk concentration, the solution becomes homogeneous and only its surface tension differs from that in a pure liquid.

We see, therefore, that the solubility of a surface-active agent lessens its damping effect by reducing the constant $\epsilon$. The decrease is the greater the smaller the frequency of the waves and the higher the equilibrium concentration (solubility) of the agent in the bulk of the solution.

Thus, a reduction in the damping effect primarily affects the long waves. This is in accordance with available experimental data on the damping effect of surfactants. The use of organic oils as wave-damping agents was recommended in early maritime handbooks. Substances with the best damping effects are seal oil and fish oils which are rich in fatty acids. The least effective are petroleum and kerosene.

These properties of surfactants are related to their solubility and to the value of the elastic constant. The fatty acids are least soluble and the mineral oils are most soluble. Unfortunately all available data on the damping of waves, including the data of R. N. Ivanov (although his experiments were conducted after the publication of the hydrodynamic theory, Ivanov did not refer to it — see [15]), are only qualitative and cannot be used for definitive verification of the theory.

Let us now consider application of the results presented here to the purely practical problem of damping of waves under actual conditions at sea.

The damping effect, as we have seen, influences short waves only. Long gravity waves are not damped by surface-active agents. Surfactants can generally affect only that liquid motion in which viscosity effects are present. In other words, the effect of surface-active agents can be manifested only at small Reynolds numbers or in the boundary layer.

In sea waves with high amplitudes, the motion is turbulent, rather than viscous. Therefore, surface-active agents exert no direct effect on such wave motion. They damp the ripples and small waves that cover the surface of the large ones. This influences the nature of the wave motion in the sense that it reduces the interaction between the wind and the waves. A smooth wave surface devoid of ripples, etc., reduces the transfer of energy from the wind to the wave. This results in reduced asymmetry of the wave and eliminates the deformation of waves which presents a dangerous condition for ships. In other words, the usual wave action is replaced by so-called swells.

## 123. BREAK-UP OF A LIQUID JET AT LOW FLOW VELOCITIES. THE CASE OF SYMMETRICAL DEFORMATIONS

One of the most important problems in capillary hydrodynamics, which has attracted the attention of such outstanding scientists as

Rayleigh [10] and Bohr [11], is the break-up of a liquid jet emanating from a nozzle or a spout into droplets. Interest in this problem has not diminished because of fuel injection systems in internal combustion engines, of emulsification and of many other problems. To the contrary, the problem of break-up of jets under the influence of dynamic air action — air in which the jet motion occurs — has become important along with the study of break-up of a jet caused by capillary forces, the problem that occupied Rayleigh and Bohr. We begin our analysis of the process of jet break-up with the case of a jet at low flow velocities.

At low jet velocities with reference to gas velocities, break-up occurs owing to the action of capillary forces. The surface energy of a cylindrical liquid jet is not a minimum.

As the jet breaks up into droplets, the surface energy decreases, since the liquid approaches its optimum, spherical shape.

Let us consider a cylindrical jet of radius a, issuing from a nozzle at a velocity $u_0$. As it issues from the nozzle, the jet surface is always subjected to disturbances. No matter how well the surface of the nozzle may be polished, a certain roughness of the surface is inevitable. Vibrations of the nozzle, motion of the air surrounding the jet, presence of turbulence in the liquid moving within the nozzle — all may serve as the source of initial disturbances. Later we examine the motion of a liquid surface that has been subjected to an arbitrarily small disturbance.

The basic characteristics of jet break-up are the duration of the interval it remains unbroken and the size of the drops that are formed. The time interval prior to break-up determines the distance of propagation of the jet and the nature of its break-up.

Owing to the symmetry of the problem, the hydrodynamic equations must be written in a cylindrical coordinate system. For the time being, we examine only those disturbances in which there is no motion of the liquid around its axis of symmetry, i.e., disturbances for which we may assume the velocity components $v_\theta = 0$. We term such waves "symmetrical." The jet cross section, in the case of symmetrical waves, remains circular and either contracts or expands.

Equations of liquid motion in a jet with symmetrical waves are conveniently given in cylindrical coordinates

$$\frac{\partial v_r}{\partial t} = -\frac{1}{\rho}\frac{\partial p}{\partial r} + \nu\left[\frac{\partial^2 v_r}{\partial z^2} + \frac{\partial}{\partial r}\left(\frac{1}{r}\frac{\partial}{\partial r}(r \cdot v_r)\right)\right], \qquad (123.1)$$

$$\frac{\partial v_z}{\partial t} = -\frac{1}{\rho}\frac{\partial p}{\partial z} + \nu\left[\frac{\partial^2 v_z}{\partial z^2} + \frac{1}{r}\frac{\partial}{\partial r}\left(r \cdot \frac{\partial v_z}{\partial r}\right)\right], \qquad (123.2)$$

where $v_r$ is the radial velocity component, and $v_z$ is the velocity component along the jet axis. Since the deformations are small, the second-order terms have been omitted.

The continuity equation is given as

$$\frac{\partial v_z}{\partial z}+\frac{1}{r}\frac{\partial}{\partial r}(r\cdot v_r)=0. \qquad (123.3)$$

Boundary conditions for the jet surface must be known. The equation for a jet surface disturbed by some wave $\zeta(r, z, t)$, where $\zeta$ is the displacement of a point on the surface, is

$$r=a+\zeta(z, t).$$

Assuming that $\zeta$ is very small compared to a, we may write the boundary conditions in the form (117.5) as

$$\left.\begin{array}{l} p_{rr}=p_{\sigma}, \\ p_{rz}=0 \end{array}\right\} \text{ at } r\approx a. \qquad (123.4)$$

Moreover, the velocity components along the jet axis, i.e., at r = 0, must be finite.

The solution of the problem is approached similarly as in the case of wave motion on a plane surface of a viscous liquid. We denote the velocity in a viscous liquid by [12]

$$v_r=v_r^{(0)}+u_1, \qquad (123.5)$$

$$v_z=v_z^{(0)}+u_2, \qquad (123.6)$$

where quantities with the superscript zero are identical with the corresponding quantities in an ideal liquid.

The pressure should be the same as in an ideal liquid, because it is evident that the presence of viscosity affects the wave frequency but not the pressure within the liquid.

In an ideal liquid, and using cylindrical coordinates, the velocity is related to the velocity potential $\varphi$ by

$$v_r^{(0)}=\frac{\partial\varphi}{\partial r}, \quad v_z^{(0)}=\frac{\partial\varphi}{\partial z} \qquad (123.7)$$

and

$$\Delta\varphi=\frac{1}{r}\frac{\partial}{\partial r}\left(r\frac{\partial\varphi}{\partial r}\right)+\frac{\partial^2\varphi}{\partial z^2}=0. \qquad (123.8)$$

The pressure is given by

$$p=-\rho\frac{\partial\varphi}{\partial t}. \qquad (123.9)$$

The continuity equation shows that $u_1$ and $u_2$ are related by

$$\frac{\partial u_2}{\partial z}+\frac{1}{r}\frac{\partial}{\partial r}(ru_1)=0.$$

Thus, $u_1$ and $u_2$ may be given by

$$\left.\begin{aligned} u_1 &= -\frac{1}{r}\frac{\partial\psi}{\partial z}, \\ u_2 &= \frac{1}{r}\frac{\partial\psi}{\partial r}, \end{aligned}\right\} \tag{123.10}$$

where $\psi$ is as yet an unspecified function of the coordinates and of time.

Substituting the expressions for velocity components into (123.1) and (123.2), and considering the equations for $u_1$ and $u_2$, we obtain

$$\frac{\partial v_r^{(0)}}{\partial t} - \frac{1}{r}\frac{\partial}{\partial z}\frac{\partial\psi}{\partial t} =$$
$$= -\frac{1}{\rho}\frac{\partial p^{(0)}}{\partial r} + \nu\Delta v_r^{(0)} - \nu\frac{1}{r}\frac{\partial}{\partial z}\left[\frac{\partial^2\psi}{\partial z^2} + r\frac{\partial}{\partial r}\left(\frac{1}{r}\frac{\partial}{\partial r}\psi\right)\right],$$
$$\frac{\partial v_z^{(0)}}{\partial t} + \frac{1}{r}\frac{\partial}{\partial r}\frac{\partial\psi}{\partial t} =$$
$$= -\frac{1}{\rho}\frac{\partial p^{(0)}}{\partial z} + \nu\Delta v_z^{(0)} + \nu\frac{1}{r}\frac{\partial}{\partial r}\left[\frac{\partial^2\psi}{\partial z^2} + r\frac{\partial}{\partial r}\left(\frac{1}{r}\left(\frac{\partial\psi}{\partial r}\right)\right)\right].$$

Since

$$\Delta v_r^{(0)} = \frac{\partial}{\partial r}\Delta\varphi = 0,$$
$$\Delta v_z^{(0)} = \frac{\partial}{\partial z}\Delta\varphi = 0,$$
$$\frac{\partial v_z^{(0)}}{\partial t} = \frac{\partial^2\varphi}{\partial z\,\partial t} = -\frac{1}{\rho}\frac{\partial p^{(0)}}{\partial z},$$

we find that function $\psi$ must satisfy the following equation:

$$\frac{\partial^2\psi}{\partial r^2} - \frac{1}{r}\frac{\partial\psi}{\partial r} + \frac{\partial^2\psi}{\partial z^2} = \frac{1}{\nu}\frac{\partial\psi}{\partial t}. \tag{123.11}$$

We are interested in wave motion in the liquid. We therefore seek a solution for $\varphi$ and $\psi$ as periodic functions in z and exponential functions in t:

$$\varphi = \Phi(r)\, e^{ikz+\alpha t}, \tag{123.12}$$

$$\psi = \Psi(r)\, e^{ikz+\alpha t}. \tag{123.13}$$

Substituting (123.12) and (123.13) into (123.8) and (123.11), we get

$$\frac{d^2\Phi}{dr^2} + \frac{1}{r}\frac{d\Phi}{dr} - k^2\Phi(r) = 0, \tag{123.14}$$

$$\frac{d^2\Psi}{dr^2} - \frac{1}{r}\frac{d\Psi}{dr} - \left(k^2 + \frac{\alpha}{\nu}\right)\Psi = 0. \tag{123.15}$$

The solution of equation (123.14) that retains a finite value along the axis of the jet [13] is

$$\Phi = C_1 I_0(kr), \tag{123.16}$$

where $I_0(kr)$ is the Bessel function of zero order with imaginary argument

$$I_0(kr) = J_0(ikr).$$

We seek a solution of equation (123.15) in the form

$$\Psi(r) = ry(r).$$

Then, for $y(r)$, we have

$$y'' + \frac{y'}{r} - \left(l^2 + \frac{1}{r^2}\right) y = 0, \tag{123.17}$$

where $l$, as in Section 117, is defined by

$$l^2 = k^2 + \frac{\alpha}{\nu}. \tag{123.18}$$

The solution of equation (123.17) that remains finite along the jet axis is

$$y = C_2 I_1(lr),$$

where $I_1(lr)$ is the Bessel function of first order with imaginary argument

$$I_1(lr) = \frac{1}{i} J_1(ilr).$$

Finally, we obtain

$$\varphi = C_1 I_0(kr)\, e^{ikz+\alpha t}, \tag{123.19}$$

$$\psi = C_2 r I_1(lr)\, e^{ikz+\alpha t}. \tag{123.20}$$

The velocity and pressure profiles are found from $\varphi$ and $\psi$ using equations (123.7) and (123.10)

$$v_r = -ik\left\{iC_1 I_0'(kr) + C_2 I_1(lr)\right\} e^{ikz+\alpha t}, \tag{123.21}$$

$$v_z = k\left\{iC_1 I_0(kr) + C_2\left[\frac{I_1(lr)}{kr} + \frac{l}{k} I_1'(lr)\right]\right\} e^{ikz+\alpha t}, \tag{123.22}$$

$$p = -\rho\alpha C_1 \cdot I_0(kr)\, e^{ikz+\alpha t}. \tag{123.23}$$

Equations (123.21) to (123.23) contain three unknown quantities: the amplitudes $C_1$ and $C_2$ and the complex frequency $\alpha$. The second boundary condition (123.4) allows one of the amplitudes to be expressed in terms of the other. The other boundary condition can then be used to derive the equation for $\alpha$. We first compute the tangential component of the stress tensor at the liquid surface. Obviously, we have

$$p_{rz} = \mu\left(\frac{\partial v_z}{\partial r} + \frac{\partial v_r}{\partial z}\right)_{r=a}. \tag{123.24}$$

Setting this equation equal to 0, we find the relation between $C_1$ and $C_2$:

$$C_1 = C_2 \frac{I_1(la)\left[\frac{\alpha}{\nu} + 2k^2\right]}{2ik^2 I_1(ka)} = C_2 \frac{I_1(la)(l^2 + k^2)}{2ik^2 I_1(ka)}. \tag{123.25}$$

The first boundary condition (123.4) contains $p_{rr}$ and $p_\sigma$. The former can be computed directly

$$p_{rr} = 2\rho\nu k e^{ikz+\alpha t}\left\{-kC_1 I_1'(kr) - ilC_2 I_1'(lr) + \frac{\alpha C_1}{2\nu k} I_0(kr)\right\}. \tag{123.26}$$

Capillary pressure, in cylindrical coordinates, is obtained from equation (65.17)

$$p_\sigma = \sigma\left(\frac{1}{R_1} + \frac{1}{R_2}\right) = \frac{\sigma}{a} - \frac{\sigma}{a^2}\left(\zeta + a^2 \frac{\partial^2 \zeta}{\partial z^2}\right), \tag{123.27}$$

where $\zeta$ is the displacement of the jet surface. It is obviously not a function of the azimuthal angle $\theta$ and, therefore, the term involving the derivative $\frac{\partial^2 \zeta}{\partial \theta^2}$ was omitted in (123.27). The value of the displacement of the surface for small displacements is related to the radial velocity component on the surface by

$$\zeta = \int (v_r)_{r=a}\, dt = \frac{ik}{\alpha}\left[ilC_1 I_1(ka) - C_2 I_1(la)\right] e^{ikz+\alpha t}.$$

Substituting this value for $\zeta$ into (123.27), we get

$$p_\sigma = -\frac{\sigma}{a^2}(1 - k^2 a^2)\zeta.$$

(The reference pressure is $p = \frac{\sigma}{a}$.) Substituting the values found for $p_{rr}$ and $p_\sigma$ into boundary condition (123.4) and, eliminating the constants $C_1$ and $C_2$ by using equation (123.25), we get the equation for $\alpha$,

$$\alpha^2 + \frac{2\nu k^2}{I_0(ka)}\left[I_1'(ka) - \frac{2kl}{k^2 + l^2} \cdot \frac{I_1(ka)}{I_1(la)} I_1'(la)\right]\alpha = \\ = \frac{\sigma k}{\rho a^2}(1 - k^2 a^2)\frac{I_1(ka)}{I_0(ka)} \cdot \frac{l^2 - k^2}{l^2 + k^2}. \tag{123.28}$$

In the general case this equation is highly complex [since, according to (123.18), $l$ is a function of $\alpha$] and cannot be solved analytically. We confine our analysis here to the two limiting cases: liquids with very low viscosity and liquids with very high viscosity. If the viscosity is low, and the following inequality holds in the wavelength region of interest*:

$$\frac{\alpha}{\nu} \gg k^2, \tag{123.29}$$

then we may write that

$$l \gg k.$$

It is easy to demonstrate (see below) that for $l \gg k$ the linear term in equation (123.28) is small and can be omitted.

Equation (123.28) then acquires a simple form

$$\alpha^2 = \frac{\sigma k}{\rho a^2}(1 - k^2a^2)\frac{I_1(ka)}{I_0(ka)}. \tag{123.30}$$

In the region where the wavelengths are small compared to the jet radius a the value of $\alpha$ is imaginary and

$$ka > 1.$$

For $ka \gg 1$, we have, in accordance with the well-known formula of the theory of Bessel functions [13],

$$I_n(x) \sim \frac{e^x}{\sqrt{2\pi x}}, \tag{123.31}$$

$$I_1(ka) \approx I_0(ka) \approx \frac{e^{ka}}{\sqrt{2\pi ka}}.$$

Setting the frequency $\omega = -i\alpha$, we get

$$\omega = \sqrt{\frac{\sigma k^3}{\rho}};$$

which corresponds to the frequency of capillary waves on the surface of a low-viscosity liquid.

Of greater interest is the case of long waves, for which the following inequality is satisfied

$$ka = \frac{2\pi a}{\lambda} < 1. \tag{123.32}$$

The quantity $\alpha$ for such waves has a positive real value. Positive values of $\alpha$ correspond to an unlimited rate of growth of wave

*It will be shown below that in this region the wavelength is considerably larger than the jet radius a.

amplitude. It is shown in the theory of turbulence that an exponential growth of wave amplitude signifies the appearance of continuous turbulence eddies in the liquid. The scale of these eddies is of the order of one wavelength. The presence of turbulence eddies in a liquid with a free surface leads to disruption of the surface and ejection of liquid. In the case of a cylindrical liquid jet, the exponential growth of wave amplitude with time leads to instability of its surface and break-up of the jet into drops. The jet surface is unstable for all wavelengths that satisfy the inequality (123.32). Equation (123.30), however, exhibits a maximum at $ka < 1$ for a given wavelength. The position of this maximum is determined by

$$\frac{\partial \alpha^2}{\partial k} = 0 = \frac{I_1'(ka)\, I_0(ka) - I_0'(ka) I_1(ka)}{I_0^2(ka)}(1 - k^2a^2) - \frac{I_1(ka)}{I_0(ka)}\, 2ak.$$

A numerical solution of this equation shows that the maximum corresponds to a value of the wave number k equal to

$$k_{\max} = \frac{2\pi}{9.02a} = 0.70 \cdot \frac{1}{a}, \qquad (123.33)$$

or to a wavelength

$$\lambda_{\max} = 9.02a. \qquad (123.34)$$

This equation was first derived by Rayleigh. The value of $\alpha$ for this maximum can be easily found by substituting the numerical values of the Bessel functions:

$$\alpha_{\max} = 0.12 \sqrt{\frac{\sigma}{\rho a^3}}. \qquad (123.35)$$

The wave number $k = k_{max}$ corresponds to the greatest instability in waves. With $k = k_{max}$ the instability is much greater than in waves with wave numbers satisfying the inequality (123.32). The maximum of the function $\alpha^2(k)$ is sharply defined. Since the amplitude grows exponentially in time, any increase in wave amplitude with a length $\lambda$ that corresponds to the maximum $\alpha$ is decisive in the break-up of the jet.

During the time interval

$$\tau = \frac{1}{\alpha_{\max}} = 8.46 \sqrt{\frac{\rho a^3}{\sigma}} \qquad (123.36)$$

the amplitude increases e times. Break-up of the jet produces drops whose size is of the order of $\lambda_{max}$, i.e., about nine times greater than the radius a of the jet. The length of the jet prior to break-up is of basic interest. If this distance is small and the change in the jet velocity can be neglected, then the length is given by

$$L \approx u_0 \tau = 8.46 u_0 \sqrt{\frac{\rho a^3}{\sigma}}. \tag{123.37}$$

This formula shows that the intact length of the jet is proportional to the 3/2 power of its radius and inversely proportional to the square root of the dynamic surface tension $\sigma/\rho$. Curiously enough, the intact length of the jet is very small, or at least much smaller than is usually assumed on the basis of visual observations. Thus, for example, for $u_0 \approx 100$ cm/sec, $\sigma/\rho \approx 80$ and $a \approx 1$ cm, we obtain $L \approx 10$ cm.

Visual observations fail to give correct values for the length of the unbroken jet due to an optical illusion — the apparent merging of the moving drops into a solid stream. Figure 102 shows the break-up of a jet recorded by flash photography [14].

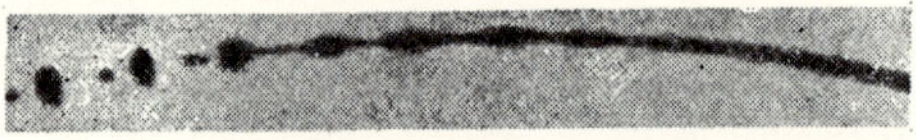

Figure 102. Break-up of horizontal water jet into drops.

We will now define the limits of applicability of equation (123.27). Substituting equations (123.35) and (123.33) into (123.29), we get

$$\sqrt{\frac{a\sigma}{\rho\nu^2}} \gg 1. \tag{123.38}$$

It is evident that inequality (123.38) is also a condition determining the smallness of the linear term in (123.28) compared to the constant.

For water, condition (123.38) is satisfied in almost every case. In the case of a liquid like glycerine ($\nu \sim 10$, $\sigma/\rho \sim 30$), condition (123.38) holds for a jet radius $a > 3$ cm. At even higher viscosities equation (123.37) holds for even larger radii. It is simple to analyze another limiting case, where an inequality opposite to (123.29) is satisfied for a highly viscous liquid. For this case equation (123.28) can be again greatly simplified. It is more meaningful, however, to analyze a more generalized case. This is presented in the next section.

## 124. BREAK-UP OF A LIQUID JET AT LOW FLOW VELOCITIES. THE CASE OF ARBITRARY DEFORMATIONS

In the preceding section we examined the break-up of a jet resulting from the appearance of symmetrical waves on its surface. It was found that break-up of the jet is caused by waves with a length approximately 10 times the radius of the jet. Accounting for viscosity, as shown below, displaces the most unstable deformations into a region of longer waves. There is good reason, therefore, to study the general case of deformations with wavelengths that are large compared to the radius of the jet. This permits the problem to be simplified considerably. The order of magnitude of the velocity components $v_z$, $v_r$ and $v_\theta$ is estimated from the continuity equation

$$\frac{\partial v_z}{\partial z}+\frac{1}{r}\frac{\partial}{\partial r}(rv_r)+\frac{1}{r}\frac{\partial v_\theta}{\partial \theta}=0. \tag{124.1}$$

We have, obviously,

$$\frac{\partial v_z}{\partial z}\approx\frac{\partial v_r}{\partial r}\approx\frac{1}{a}\frac{\partial v_\theta}{\partial \theta}. \tag{124.2}$$

On the other hand, for wave motions in which the wavelength $\lambda \gg a$, we have

$$\frac{\partial v_z}{\partial z}\sim\frac{v_z}{\lambda};\quad \frac{\partial v_r}{\partial r}\sim\frac{v_r}{a};\quad \frac{1}{a}\frac{\partial v_\theta}{\partial \theta}\sim\frac{v_\theta}{a}.$$

It follows from (124.1) that

$$v_z\sim\frac{\lambda}{a}v_r\sim\frac{\lambda}{a}v_\theta. \tag{124.3}$$

Whence we get the inequality

$$v_z\gg v_r,\ v_\theta. \tag{124.4}$$

Since the jet has a free surface and the wavelength is large compared to the jet radius, we may, as the first approximation, ignore the fact that $v_z$ is a function of the radial coordinate r, and examine only the change in $v_z$ along the jet and with time. This is done by expansion of Bessel functions in equation (123.28) into a series in which only the first term is retained.

With $v_r \approx 0$ and $\frac{\partial v_z}{\partial r} \approx 0$, equations (123.1) and (123.2) can be simplified

$$\frac{\partial v_z}{\partial t}=\frac{1}{\rho}\frac{\partial}{\partial z}\left(p_{zz}-\mu\frac{\partial v_z}{\partial z}\right), \tag{124.5}$$

$$\frac{\partial p}{\partial r}=0. \tag{124.6}$$

The second equation shows that pressure in the jet and the velocity $v_z$ may be considered constant in the radial direction. This pressure is

$$p\approx-\frac{1}{3}(p_{zz}+p_{rr}+p_{\theta\theta})_{r=a}=-\frac{1}{3}(p_{zz}+2p_\sigma).$$

In equation (124.5) only the component $p_{zz}$ of the stress tensor is retained since we disregard the fact that all quantities are functions of the radial coordinate within the jet. For $p_{zz}$ we have

$$p_{zz} = -p + 2\mu\frac{\partial v_z}{\partial z} = p_\sigma + 3\mu\frac{\partial v_z}{\partial z}. \tag{124.7}$$

Substituting (124.7) into (124.5), we obtain the equation for $v_z$:

$$\frac{\partial v_z}{\partial t} = \frac{1}{\rho}\frac{\partial p_\sigma}{\partial z} + 2\nu\frac{\partial^2 v_z}{\partial z^2}. \tag{124.8}$$

Since, according to (123.27), $p_\sigma$ is determined by the displacement $\zeta$ of the surface of the liquid, $v_z$ should also be expressed in terms of $\zeta$.

To do this for long waves, it is convenient to employ the continuity equation, which may be conveniently written in our approximation as

$$\frac{\partial S}{\partial t} + \frac{\partial}{\partial z}(Sv_z) = 0, \tag{124.9}$$

where S is the cross-sectional area of the jet:

$$S = \pi(a+\zeta)^2. \tag{124.10}$$

Substituting this expression into (124.9), we have

$$\frac{\partial v_z}{\partial z} \approx -\frac{2}{a}\frac{\partial \zeta}{\partial t}. \tag{124.11}$$

Differentiating (124.8) with respect to z and substituting (124.11) and the value of $p_\sigma$ from (123.27), we obtain the equation that defines $\zeta(z, t, \theta)$

$$\frac{\partial^2\zeta}{\partial t^2} - 2\nu\frac{\partial^3\zeta}{\partial z^2\,\partial t} + \frac{\sigma}{2\rho a}\frac{\partial^2}{\partial z^2}\left(\zeta + \frac{\partial^2\zeta}{\partial\theta^2} + a^2\frac{\partial^2\zeta}{\partial z^2}\right) = 0. \tag{124.12}$$

We seek a solution of equation (124.12) in the form of a series

$$\zeta = \sum_s A_s e^{ikz+\alpha t}\cos(s\theta), \tag{124.13}$$

where $A_s$ is the s-th amplitude, and summing up is from s = 0 to s = ∞.

Substituting (124.13) into (124.12) we obtain the equation defining the complex frequency $\alpha$:

$$\alpha^2 + 2\nu k^2\alpha = \frac{\sigma k^2}{2\rho a}[1 - s^2 - (ka)^2] \approx \frac{\sigma k^2}{2\rho a}(1 - s^2). \tag{124.14}$$

Let us note that equation (124.14) for symmetrical waves s = 0 may be derived directly from (123.28) if the Bessel functions are expanded and the formulas

$$I_0(ka) \approx 1, \quad I_1(ka) \approx \frac{ka}{2}, \quad I_1(la) \approx \frac{la}{2},$$

$$I_1'(ka) = I_1'(la) \approx \frac{1}{2}$$

which hold for ka $\ll 1$ are employed, and if we take account of the relation between k and $l$ from formula (123.18).

Symmetrical waves (s = 0) on the surface of the jet correspond to successive contractions and expansions. Waves s = 1 for which displacements $\zeta$ are proportional to cos $\theta$ correspond to jet deformations in which the cross section, while remaining circular, is displaced relative to the jet axis. Here, the jet acquires a snakelike or wavelike profile (Figure 103).

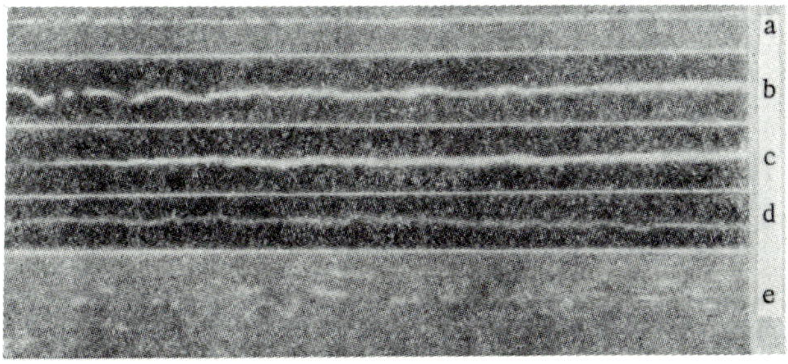


Figure 103. Break-up of thin liquid jets (the jet flows from right to left).

Waves s = 2, with elliptical cross sections, rise and fall. This means that an ellipse with a large horizontal semiaxis is transformed into an ellipse with a long vertical semiaxis. It follows directly from equation (124.14) that symmetrical waves (s = 0) are unstable ($\alpha$ is real). Other types of waves whose length is large compared to the radius of the jet are stable.

In the preceding section, our analysis was limited to symmetrical waves. Using the general equation (124.14), the effect of viscosity on the unbroken length of the jet may be studied. Assuming s = 0, and solving equation (124.14), we find

$$\alpha = -\nu k^2 \pm \sqrt{(\nu k^2)^2 + \frac{\sigma k^2}{2\rho a}(1 - k^2a^2)}. \qquad (124.15)$$

The value of $\alpha$ reaches a maximum at

$$k_{max} = \frac{1}{\left(\nu a\sqrt{\frac{\rho a}{2\sigma}} + 2a^2\right)^{1/2}}. \qquad (124.16)$$

This maximum value is given by

$$\alpha_{max} = \frac{\sqrt{\frac{\sigma}{2\rho a^3}}}{2 + 3\nu\sqrt{\frac{2\rho}{a\sigma}}}. \qquad (124.17)$$

For low viscosity, when inequality (123.38) holds, the values of $k_{max}$ and $\alpha_{max}$ correspond to those computed in the preceding section. If viscosity is high, and the inequality (123.38) is reversed, we find for $k_{max}$ and $\alpha_{max}$, respectively,

$$k_{max} \approx \sqrt{\frac{1}{6\nu a\sqrt{\frac{\rho a}{2\sigma}}}} \approx \sqrt[4]{\frac{\sigma}{18\nu^2\rho a^3}} = 0.5\sqrt[4]{\frac{\sigma}{\nu^2\rho a^3}}, \qquad (124.18)$$

$$\alpha_{max} \approx \frac{\sigma}{6a\nu\rho} = 0.2\frac{\sigma}{a\nu\rho}. \qquad (124.19)$$

The time interval for break-up is

$$\tau \sim \frac{1}{\alpha_{max}} \approx 5\frac{a\nu\rho}{\sigma}. \qquad (124.20)$$

The dimension of the drops formed is

$$\lambda \sim \frac{2\pi}{k_{max}} \sim 2\pi\sqrt[4]{\frac{18\nu^2\rho a^3}{\sigma}} = 13\sqrt[4]{\frac{\nu^2\rho a^3}{\sigma}}. \qquad (124.21)$$

This dimension is proportional to the square root of the liquid viscosity and to the 3/4 power of the jet radius; it is also a weak function of surface tension (decreases with $\sqrt[4]{\sigma}$).

The intact length of the jet [by analogy with (123.37)] is

$$L = u_0\frac{1}{\alpha_{max}} \approx 5\frac{u_0 a\nu\rho}{\sigma}. \qquad (124.22)$$

This length increases linearly with the dynamic viscosity of the liquid $\mu$, and is inversely proportional to surface tension. It should be noted, however, that the assumption regarding the constant velocity of the jet — an assumption that holds for low-viscosity liquids, where the length of the intact portion of the jet is small — may not be valid for highly viscous jets.

Formula (124.22) may be derived as a limiting case of relations established in the study of Timotika [12], who examined the more general problem of the break-up of a jet moving in a liquid medium.

Timotika assumed the densities of both liquids to be comparable, while in all our studies we have assumed the density of the external medium to be very low.

## 125. BREAK-UP OF A JET AT HIGH VELOCITIES. ATOMIZATION

So far we have examined the break-up of a jet moving at such velocities that the dynamic action of the outside medium, air, in particular, played no part in the formation of drops.

Break-up of the jet occurred solely as a result of capillary forces, and was accompanied by a reduction in the surface energy of the system. At high velocities (the meaning of a high velocity will be clarified later) the break-up of the jet into drops presents a different picture. Break-up into a few, relatively large drops gives way to atomization into a cloud of small droplets. The changed picture of the process is illustrated in Figure 104. The number of drops becomes so great that the over-all surface area of the atomized jet is incomparably greater than the initial surface of the intact jet.

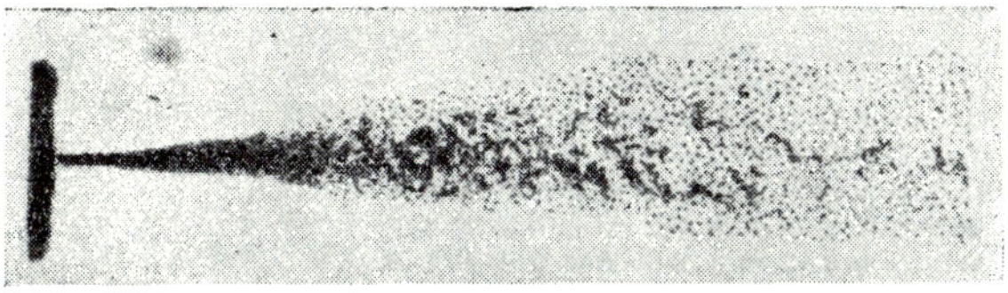

Figure 104. Atomization of a jet of fuel.

The formation of additional surface area is caused by the dynamic effect of surrounding medium on the surface of the jet. We assume that the liquid jet, density $\rho$, is moving in a gas medium with a density $\rho^g \ll \rho$. Quantities that relate to the gas are identified by the superscript g. Owing to the low density of the gas we can ignore, in the first approximation, its viscosity and consider the gas as an ideal fluid. The qualitative picture of the dynamic effect of the gas on the jet surface can be presented as follows: let us assume some disturbance (even infinitesimal) on the surface of the jet. Gas pressure over the crest of a protuberance is less than the average pressure, while at the base of the protuberance the pressure is higher than the average. Let us draw a streamline far from the jet surface, so that the streamline is not disturbed by the protuberance. Then in the zone between the streamline and the jet surface, we find that the gas velocity over the crest of the protuberance is higher than at the base. Pressure, according to the Bernoulli equation, must be lower at points where velocity is higher. Gas pressure at the crest of the protuberance, therefore, must be lower than at its base.

Therefore, the protuberance that has formed somehow on the liquid surface tends to increase. The higher the relative velocity between the liquid and the gas, the more pronounced is this effect.

An increase in the size of the protuberance on the liquid surface results in detachment of a drop from the surface.

For a quantitative description of this phenomenon, we consider the dynamic effect of the gas on wave formation.

We begin by determining the velocity profile in the gas. For the sake of convenience, we assume that the jet is at rest and that the gas far from the jet moves in the positive direction of the z axis at a velocity $u^g$. The problem relating the velocities $u^g$ and $u_0$ to each other, where the latter is the exit velocity of the jet from the nozzle, is examined later.

Ignoring the viscosity of the air in the first approximation, and assuming an ideal fluid, we may write equation (3.5) for the velocity potential

$$\Delta\varphi^g = \frac{\partial^2\varphi^g}{\partial z^2} + \frac{1}{r^2}\frac{\partial^2\varphi^g}{\partial\theta^2} + \frac{1}{r}\frac{\partial}{\partial r}\left(r\frac{\partial\varphi^g}{\partial r}\right) = 0. \qquad (125.1)$$

The boundary conditions at infinity that must be satisfied by the gas velocity components are

$$\left.\begin{array}{l} v_z^g \to u^g, \\ v_r^g \to 0 \end{array}\right\} \quad \text{as} \quad r \to \infty. \qquad (125.2)$$

The following boundary conditions, obviously, must be fulfilled on the jet surface:

1) the gas and liquid velocities on the surface of the jet must be equal

$$\left.\begin{array}{l} v_r^g = v_r, \\ v_z^g = v_z \end{array}\right\} \quad \text{at} \quad r = a; \qquad (125.3)$$

2) the normal stresses on the surface of the liquid must balance; i.e., instead of (123.4), the following equality must hold

$$p_{zz} = p_\sigma + p^g; \qquad (125.4)$$

3) in actual gas flow, tangential stresses are exerted on the surface. Tangential forces play a very important part in many aspects that are examined below, e.g., the transfer of energy by gas to waves and entrainment of liquid by wind. However, the tangential stress is proportional to the square of the wave amplitude and can (in our approximation for waves of small amplitude) be disregarded. We can assume, therefore, that the tangential component of the stress tensor, as in the absence of gas flow, becomes zero,

$$p_{zr} = 0. \qquad (125.5)$$

In studying the dynamic effect of air on the moving jet, we examine two cases: that of waves that are symmetrical relative to the axis of the jet, and the case of relatively long waves of arbitrary symmetry.

We consider first the case of symmetrical waves. If symmetrical deformations occur on the jet surface the gas velocity around the jet obviously is not a function of the angle $\theta$. We are, therefore, seeking a solution of equation (125.1) in the form

$$\varphi^g = -u^g z + \varphi(r, \theta) e^{ikz+\alpha t}. \quad (125.6)$$

In this case, the first boundary condition is automatically fulfilled (for this, it is necessary that $\varphi$ vanish at infinity). Substituting (125.6) into (125.1), we derive the equation for $\varphi(r, \theta)$

$$\frac{1}{r}\frac{\partial}{\partial r}\left(r\frac{\partial \varphi}{\partial r}\right) - k^2\varphi = 0. \quad (125.7)$$

To satisfy the second boundary condition (125.2), a solution for equation (125.7) that vanishes at infinity must be selected.

The following serves as such a solution [13]

$$\varphi(r) = bK_0(kr),$$

where the function

$$K_0(kr) = \frac{i\pi}{2} H_0^{(1)}(ikr) e^{\frac{i\pi n}{2}}$$

is a Bessel function of the second kind with an imaginary argument (often referred to as the MacDonald function) that vanishes at infinity. Thus, we have for $\varphi$:

$$\varphi^g = -u^r z + bK_0(kr) e^{ikz+\alpha t}. \quad (125.8)$$

From the velocity potential (125.8), we find the velocity and pressure distribution in the gas:

$$v_r^g = -\frac{\partial \varphi}{\partial r} = -bkK_0'(kr) e^{ikz+\alpha t}, \quad (125.9)$$

$$v_z^g = -\frac{\partial \varphi}{\partial z} = u^g - ibkK_0(kr) e^{ikz+\alpha t}, \quad (125.10)$$

$$p^g = \rho^g \frac{\partial \varphi}{\partial t} - \frac{1}{2}\rho^g (v^g)^2 = \rho^g \frac{\partial \varphi}{\partial t} - \frac{1}{2}\rho^g \left[(v_r^g)^2 + (v_z^g)^2\right] \approx$$
$$\approx \rho^g bK_0(kr)\alpha e^{ikz+\alpha t} + \rho^g ikbK_0(kr) u^g e^{ikz+\alpha t}. \quad (125.11)$$

Here, the constant $\frac{\rho^g (u^g)^2}{2}$ has been omitted, since the gas pressure can be determined only within the accuracy of an arbitrary constant. A second-order term of the amplitude b has also been omitted.

Boundary condition (125.4) includes pressure $p_\sigma$ expressed in terms of the displacement $\zeta$ on the surface. The gas pressure $p^g$

should also be expressed in terms of $\zeta$. To do this, we employ the relation

$$(v_r)_{r=a} = \frac{\partial \zeta}{\partial t} + \frac{\partial \zeta}{\partial z}\frac{dz}{dt} = \frac{\partial \zeta}{\partial t} + \frac{\partial \zeta}{\partial z} v_z. \tag{125.12}$$

By substituting $v_r$ and $v_z$ from (125.9) and (125.10) and by expressing $\zeta$ as

$$\zeta = \zeta_0 e^{ikz+\alpha t}, \tag{125.13}$$

we obtain, with an accuracy of second order infinitesimals,

$$b = -\frac{\alpha + iku^{\text{г}}}{k} \frac{1}{K_0'(ka)} \zeta_0, \tag{125.14}$$

whence

$$p^g = -\zeta_0 \frac{[\alpha + iku^g]^2}{k} \frac{K_0(kr)}{K_0'(kr)} e^{ikz+\alpha t} \rho^g. \tag{125.15}$$

We can now discuss the actual meaning of $u^g$.

In examining the motion of a real gas past a jet surface with protuberances at high velocities, it becomes evident that a turbulent boundary layer forms in the gas at the jet surface. The amplitude of the waves (protuberances) $\zeta_0$ on the jet surface and the gas velocity $u_0$ at infinity (equal to the jet outlet velocity from the nozzle) are related to the gas velocity $u^g$ at the surface by

$$u_0 = u^{\text{г}} \ln \frac{a}{\zeta_0},$$

which has logarithmic accuracy.*

In certain instances, the difference between $u_0$ and $u^g$ must be taken into account.

The quantities $v_r$, $v_z$ and p for the jet are given, as previously, by equations (123.21) to (123.23).

The constants $C_1$, $C_2$, $\zeta_0$ and the quantity $\alpha$ are determined from boundary conditions (125.2) to (125.5).

The latter agrees with (123.24) and leads to equation (123.25), which relates $C_1$ and $C_2$.

---

*Translator's note: In Russian usage, whenever an equation is given in terms of logarithms, the accuracy is referred to as "logarithmic accuracy." This means that the logarithms of a quanity, rather than the quantity itself, should be used in determining the accuracy of the expression.

Expressing the amplitude b, in (125.9) and (125.10), in terms of $\zeta_0$ by using equation (123.14), and substituting the gas and liquid velocities as well as pressures p, $p^g$ and $p_\sigma$ into equations (125.3) and (125.4), we get, after some tedious algebraic manipulations (similar to those presented in Section 123), the expression for $\alpha$:

$$\alpha^2 + \frac{2\nu k^2}{I_0(ka)}\left[I_1'(ka) - \frac{2kl}{l^2+k^2}\frac{I_1(ka)}{I_1(la)}I_1'(la)\right]\alpha =$$

$$= \frac{\sigma k}{\rho a^2}(1-k^2a^2)\frac{I_1(ka)}{I_0(ka)}\frac{l^2-k^2}{l^2+k^2} + \tag{125.16}$$

$$+ \frac{\rho^g k^2 (u^g)^2}{\rho}\frac{K_0(ka)\,I_1(ka)}{K_1(ka)\,I_0(ka)}\frac{l^2-k^2}{l^2+k^2}.$$

In the calculations we have omitted the terms $\rho^g\alpha^2$ and $\rho^g iku^g\alpha$ (small compared to the terms $\rho\alpha^2$ and $\rho iku^g\alpha$ that were retained). Since $l$ is expressed in terms of $\alpha$, equation (125.16) is very complex. We will, therefore, analyze the limiting cases of short ($\lambda \ll a$) and long ($\lambda \gg a$) waves.

In the case of short waves, the Bessel function arguments (ka) and ($l$a) are large compared to unity, and we can use the asymptotic expressions [13]

$$K_0(x) = -K_0'(x) \approx \sqrt{\frac{\pi}{2x}}\,e^{-x}.$$

Simplifying, we easily get

$$\alpha^2 + 2\nu k^2\alpha\left[1 - \frac{2kl}{l^2+k^2}\right] = -\frac{\sigma k^3}{\rho}\frac{l^2-k^2}{l^2+k^2} + \frac{\rho^g k^2 (u^g)^2}{\rho}\frac{l^2-k^2}{l^2+k^2}. \tag{125.17}$$

Further simplifications are possible for cases of high and low liquid viscosities. Inequality (123.29) holds in the second case.* Here, as shown in Section 123, the linear term in $\alpha$ is small compared to the square term. Moreover, $l \gg$ k. Neglecting the linear term in (125.17) and simplifying, we get,

$$\alpha = \sqrt{\frac{\rho^g k^2 (u^g)^2 - \sigma k^3}{\rho}}. \tag{125.18}$$

Equation (125.18) can be easily derived by considering waves on a plane liquid surface (see, e.g., [1]). The equation shows that at a sufficiently high gas velocity the quantity $\alpha$ may be real, even for short waves. Thus, $\alpha$ is real for wave numbers

$$k \leqslant \frac{\rho^g (u^g)^2}{\sigma}. \tag{125.19}$$

*Editor's note: The original Russian version states: "first case."

The corresponding waves are unstable. This is the difference between jets of high and low velocities. In the latter case, it was found that only very long waves, i.e., $\lambda \gg a$, are unstable. At high velocities, the dynamic effect of the gas stream causes an increase in the amplitudes and produces instability for much shorter waves on the liquid surface.

The quantity $\alpha$ as a function of the wave number k exhibits a broad maximum at

$$k = k_{\max} = \frac{2}{3} \frac{\rho^g (u^g)^2}{\sigma}. \tag{125.20}$$

In this case, the maximum value of $\alpha$ is

$$\alpha_{\max} = 0.4 \frac{(u^g)^3}{\sigma} \sqrt{\frac{(\rho^g)^3}{\rho}}. \tag{125.21}$$

During a time interval

$$\tau = \frac{1}{\alpha_{\max}} \approx 2.6 \frac{\sigma}{(u^g)^3} \sqrt{\frac{\rho}{(\rho^g)^3}}, \tag{125.22}$$

the wave amplitude increases e times.

In the final analysis we can state that an increase in wave amplitude leads to destruction of the free surface of the liquid.

For short waves, this destruction consists in the separation of drops whose size is approximately determined by the length of the unstable wave

$$L_0 \sim \frac{1}{k_{\max}} \sim \frac{\sigma}{\rho^g (u^g)^2}. \tag{125.23}$$

It should be stressed in addition to drops with a dimension determined by formula (125.20), drops of other sizes are formed, because all waves with numbers that satisfy inequality (125.19) are unstable.

We now examine the conditions for which expression (125.23) can be applied. Inequality (123.29) holds if

$$\begin{gathered} l^2 = k^2_{\max} + \frac{\alpha_{\max}}{\nu} \approx \frac{\alpha_{\max}}{\nu} \approx \frac{(u^g)^3}{\nu\sigma} \frac{(\rho^g)^{3/2}}{\rho^{1/2}} \gg k^2_{\max} \approx \left(\frac{\rho^g (u^g)^2}{\sigma}\right)^2, \\ \frac{\mu u^g}{\sigma} \sqrt{\frac{\rho^g}{\rho}} \ll 1. \end{gathered} \tag{125.24}$$

This condition is fulfilled at low viscosity values and moderate jet velocities. The condition is always fulfilled for liquids similar to water, as shown by substitution of numerical values

$$4 \cdot 10^{-6} u^g \ll 1.$$

The break-up of a jet accompanied by the formation of small droplets is known as atomization.

The distribution of the droplets formed can be computed by assuming atomization to be a random process. Atomization continues until the entire jet is broken up into drops.

Let us estimate the time necessary to complete atomization. By knowing the atomization period, the length of the intact portion of the jet may be determined. It is obvious that the atomization period can be calculated only on the basis of some simplifying assumptions. We assume that all the droplets are of the same size $L_0$ as given in (125.23). Other simplifying assumptions are given later.

The law of conservation of energy may be used to compute the atomization period. The energy transferred from the surrounding gas to the liquid jet is partially dissipated and partially used up in the formation of new surface.

The energy transferred by a gas stream at a velocity $u^g$ to a plane liquid surface is computed in Section 129.

The energy transferred to the liquid surface in time T is

$$E = \rho^g \left( u^g - \frac{\omega}{k} \right)^2 \omega k \zeta_0^2 S T,$$

where $\left(u^g - \frac{\omega}{k}\right)$ is the relative gas velocity, $\frac{\omega}{k}$ is the rate of wave propagation for waves of frequency $\omega$, and S is the liquid surface area.

The energy consummed for the formation of the drops is given by

$$E_\sigma = \sigma 4\pi L_0^2 N = \sigma 4\pi L_0^2 \frac{V}{\frac{4\pi}{3} L_0^3} \approx \sigma \frac{V}{L_0}.$$

Finally, the dissipated energy is given by

$$E_\nu = \rho \nu \omega^2 k \zeta_0^2 S T.$$

An energy balance yields

$$\sigma \frac{V}{L_0} + \rho \nu \omega^2 k \zeta_0^2 S T \approx \rho^g (u^g)^2 \omega k \zeta_0^2 S T,$$

whence

$$T = \frac{\sigma V}{L_0 S \zeta_0^2 k \left( -\rho \nu \omega^2 + \rho^g (u^g)^2 \omega \right)}. \qquad (125.25)$$

It is evident that at $\omega \sim \alpha_{max}$, the inequality

$$\rho^g (u^g)^2 \omega \gg \rho \nu \omega^2 \qquad (125.26)$$

is equivalent to inequality (123.29). Therefore,

$$T \approx \frac{\sigma}{L_0 \zeta_0^2 \rho^g (u^g)^2} \frac{1}{\omega k} \frac{V}{S} \approx \frac{\sigma}{L_0 \zeta_0^2 \rho^g (u^r)^2} a \frac{1}{\alpha_{max} k_{max}} \approx \tau \frac{a}{L_0} \approx \sqrt{\frac{\rho}{\rho^g}} \frac{a}{u^g}. \qquad (125.27)$$

This formula shows that the period of atomization is substantially larger than the period of separation of an individual drop, and is inversely proportional to the velocity of the jet.

Since it was assumed that the atomization period is very short and that the jet velocity, within that period, remains constant, the following approximation can be set up for the intact length of the jet:

$$L \approx Tu^g \approx \sqrt{\frac{\rho}{\rho^g}}\, a \approx 30\, a. \qquad (125.28)$$

This length is found to be independent of the jet velocity and, under normal conditions in air, is equal to approximately 30 jet radii.

Above we neglected the decrease in both the jet radius and velocity during atomization. Therefore, the order of magnitude of the atomization period can be only approximated by formula (125.27).

We now consider the jet break-up when inequality (125.26) ceases to hold. In other words, we examine the case of jet break-up when the liquid is highly viscous.

The quadratic term in equation (125.27) is small compared to the linear term. Moreover, $l \approx k$. Simple computations yield, approximately,

$$\alpha \approx \frac{\rho^g (u^g)^2 - \sigma k}{\rho \nu}. \qquad (125.29)$$

When $\rho^g (u^g)^2 > \sigma k$, the value of $\alpha$ is positive and the waves are unstable. It is clear, however, that waves with a low value of k, i.e., long waves, are the least stable. Thus, we conclude that for a highly viscous liquid the region of instability is shifted in the direction of long waves.

It is clear from equation (125.25) when

$$\frac{\mu u^g}{\sigma}\sqrt{\frac{\rho^g}{\rho}} \approx 1,$$

and the viscosity increases, that both terms in the denominator approach each other and the atomization period increases to infinity.

This increase in the atomization period is related to the energy dissipated in the liquid. The unlimited increase in the atomization period indicates that no atomization occurs with a highly viscous liquid and break-up of the jet results in the formation of relatively large drops.

Thus, the atomization phenomenon is characteristic of the break-up of liquid jets with low viscosity.

## 126. BREAK-UP OF HIGH VELOCITY JETS. THE CASE OF LONG WAVES

The intact portion of the jet in long waves can be computed by the same technique as used in the case of low velocity jets.

The presence of a dynamic effect caused by the gas may be taken into account at high velocities by replacing the equality $p = p_\sigma$ with $p = p_\sigma + p^g$.

Then, instead of (124.7) we have

$$p_{zz} = -(p_\sigma + p^g) + 2\mu \frac{\partial v_z}{\partial z}.$$

Equation (124.8) is rewritten as

$$\frac{\partial^2 \zeta}{\partial t^2} = 2\nu \frac{\partial^3 \zeta}{\partial z^2 \partial t} + \frac{1}{\rho} \frac{\partial^2 (p_\sigma + p^g)}{\partial z^2}. \tag{126.1}$$

Taking (123.27) and (125.15) into account, we get

$$\frac{\partial^2 \zeta}{\partial t^2} - 2\nu \frac{\partial^3 \zeta}{\partial z^2 \partial t} + \frac{\sigma}{2\rho a} \frac{\partial^2}{\partial z^2}\left(\zeta + \frac{\partial^2 \zeta}{\partial \theta^2} + a^2 \frac{\partial^2 \zeta}{\partial z^2}\right) + \\ + \frac{\rho^g a\,(iku^g + \alpha)^2}{2\rho k} \frac{\partial^2 \zeta}{\partial z^2} \frac{K_s(ka)}{K_s'(ka)} = 0. \tag{126.2}$$

Here, for the sake of generalization, we replaced the MacDonald function of zeroth order in (126.2) by the MacDonald function of s-th order. We must employ the limiting value of this function, which corresponds to low values of the argument $ka \ll 1$.

At low values of the argument, the ratio $\frac{K_s(x)}{K_s'(x)}$ takes the form [13]

$$\frac{K_s(x)}{K_s'(x)} \approx -\frac{x}{s}, \qquad \frac{K_0(x)}{K_0'(x)} \approx x \ln \frac{x}{2}.$$

Let us seek the displacement $\zeta$ of the surface in the form

$$\zeta = \sum \zeta_s \cos(s\theta)\, e^{ikz + \alpha t}. \tag{126.3}$$

Substituting (126.3) into (126.2), and considering equation (125.15), we find

$$\alpha^2 + 2\nu k^2 \alpha = \frac{\sigma k^2}{2\rho a}(1 - s^2 - (ka)^2) - \frac{\rho^g a^2 k^4 (u^g)^2}{2\rho}\left(\ln \frac{ka}{2}\right). \tag{126.4}$$

This equation differs from (124.14) by the presence of a term which accounts for the dynamic effect of the air. With the help of this equation the value of $\alpha$ can be found. In order to avoid writing cumbersome equations, however, we note that in the case of low-viscosity liquids, for which inequality (125.24) holds, the linear term on the left side of (126.4) is small compared to the quadratic term and can be omitted. Then

$$\alpha = \sqrt{-\frac{\rho^g k^4 a^2 (u^g)^2 \left(\ln \frac{ka}{2}\right)}{2\rho} + \frac{\sigma k^2}{2\rho a}(1 - s^2 - k^2 a^2)}. \tag{126.5}$$

In this equation, as in the case of low velocities, we see that symmetrical waves (s = 0) correspond to the highest positive value of $\alpha$.

We now turn to the analysis of sufficiently high velocities and assume that

$$\rho^g (u^g)^2 (ka)^4 \gg \frac{\sigma}{a} (ka)^2. \qquad (126.6)$$

Here, by omitting the surface tension term under the radical sign we obtain

$$\alpha = \sqrt{\frac{\rho^g}{2\rho} \frac{u^g}{a} (ka)^2 \cdot \left(-\ln \frac{ka}{2}\right)^{1/2}}. \qquad (126.7)$$

For long waves ka < 1 and $\alpha > 0$. The quantity $\alpha$ has a broad maximum at a point defined by

$$4 \ln \frac{ka}{2} + 1 = 0. \qquad (126.8)$$

The root of equation (126.8) is

$$k_{\max} \approx \frac{0.75}{a},$$

with a corresponding

$$\lambda_{\max} \approx 8a. \qquad (126.9)$$

For this wavelength, we obtain

$$\alpha_{\max} \approx \sqrt{\frac{\rho^g}{\rho} \frac{u^g}{a}}. \qquad (126.10)$$

The break-up period of the jet into large drops, caused by the dynamic effect of the air, is given, as usual, by

$$T \sim \frac{1}{\alpha_{\max}} \sim \sqrt{\frac{\rho}{\rho^g} \frac{a}{u^g}}. \qquad (126.11)$$

Correspondingly, the length of the intact length of the jet is

$$L \approx \sqrt{\frac{\rho}{\rho^g}}\, a. \qquad (126.12)$$

Since the maximum is broad, determination of the numerical coefficient in (126.12) has no strong significance. If we ignore the nonessential numerical coefficient, comparison of (126.12) and (125.28) shows that the period of atomization and the period of fragmentation into large-size drops are both given by identical formulas.

Moreover, when inequality (126.6) holds, long waves that are not symmetrical relative to the jet axis will undergo fragmentation within the same time period as symmetrical waves. Thus, the over-all picture of fragmentation of a low-viscosity liquid jet issuing at a sufficiently high velocity from a nozzle appears to be as follows: under the effect of the infinitely small disturbances that are always present, both long and short waves (in comparison with the radius) appear on the jet surface. Owing to the dynamic effect of air, both long and short waves are unstable and their amplitude grows rapidly in time. The growth in amplitude of short waves leads to the separation of small droplets and to atomization of the liquid.

The instability of long waves leads to the appearance of symmetrical waves, of waves of the first order with a snakelike profile, and to the breaking up of the jet into fragments. The dimensions of the fragments exceed several times the jet radius. The times necessary for atomization of the entire mass of the jet and for break-up of the jet into large drops are of the same order of magnitude.

It is clear that the determination of the numerical coefficient for both (126.12) and (125.28) has no significance, because simplifying assumptions were made in the derivation of these equations. Specifically, we assumed that the velocity of the jet relative to the air does not change during break-up. Nor was the reduction in the jet thickness during fragmentation and atomization taken into consideration.

We now consider the case of high viscosity, for which inequality (125.26) is reversed. Here, the second-order term in equation (126.4) can be omitted, and

$$\alpha \approx -\frac{\rho_{\mathfrak{g}}(u_{\mathfrak{g}})^2 (ka)^2}{4\mu}\left(\ln\frac{ka}{2}\right). \tag{126.13}$$

The maximum for $\alpha$ is attained at

$$k_{\max} \approx \frac{1}{a}$$

and at a wavelength

$$\lambda_{\max} = 6a. \tag{126.14}$$

The value of $\alpha$ is given by

$$\alpha_{\max} = 0.2\,\frac{\rho_{\mathfrak{g}}(u_{\mathfrak{g}})^2}{\mu}, \tag{126.15}$$

and the break-up period by

$$T = \frac{1}{\alpha_{\max}} = 5\,\frac{\mu}{\rho_{\mathfrak{g}}(u_{\mathfrak{g}})^2}. \tag{126.16}$$

The intact length of the jet is

$$L = \frac{\mu}{\rho \mathfrak{g}(u\mathfrak{g})}. \tag{126.17}$$

Thus, for a highly viscous liquid, jet fragmentation results in the formation of large drops only. The drop size is not a function of the viscosity. The intact length of the jet increases with the viscosity in accordance with equation (126.17).

It is clear, however, that if the length of the intact portion of the jet is very large, it is no longer possible to ignore the retardation due to the break-up of the jet.

The retarding effect may be estimated, if we assume that the resistance to the jet motion through the air is equal to the resistance to motion of a solid rod having a rough surface. The scale of roughness may be assumed to be equal to $\lambda_{max}$, where $\lambda_{max}$ is the size of the drops that separate. The scale of roughness enters the resistance equation in a logarithmic form, and therefore any inaccuracy in its value does not produce any significant errors.

## 127. BREAK-UP OF A LIQUID MASS OF ARBITRARY SHAPE. CONCLUDING REMARKS

Above, we confined our examination to the break-up of cylindrical jets. By approximating, it is not difficult to show that, by ignoring the numerical coefficients, the formulas derived above may be extended to the case of fragmentation of a liquid mass of arbitrary shape.

Let us consider a liquid mass of arbitrary shape with a surface that is not a minimum (i.e., which is different from spherical).

Owing to random disturbances waves will appear on the liquid surface. Upon development, these waves lead to break-up of the liquid mass similarly as for a cylindrical jet.

The equation of wave motion with small amplitude is

$$\frac{\partial \mathbf{v}}{\partial t} = -\frac{\operatorname{grad} p}{\rho} + \nu \Delta \mathbf{v}. \tag{127.1}$$

Let us estimate the order of magnitude of the quantities in equation (127.1), for the cases of low-viscosity and for high-viscosity liquids.

For the first case, we omit the small viscosity term contained in equation (127.1). Further,

$$\frac{\partial v}{\partial t} \sim \frac{v}{\tau} \sim \frac{a}{\tau^2}; \quad \frac{\operatorname{grad} p}{\rho} \sim \frac{p}{a\rho},$$

where a and $\tau$ are the characteristic dimension and period of the motion respectively.

Pressure p is equal to the capillary pressure

$$p_\sigma \sim \frac{\sigma}{a}.$$

Therefore,

$$\frac{\partial v}{\partial t} \sim \frac{a}{\tau^2} \sim \frac{p_\sigma}{\rho a} \sim \frac{\sigma}{\rho a^2}, \tag{127.2}$$

whence

$$\tau \sim \sqrt{\frac{\rho a^3}{\sigma}}, \tag{127.3}$$

which corresponds to equation (123.36).

In the case of a highly viscous liquid, the motion may be considered to be quasi-stationary. This means that the term $\frac{\partial v}{\partial t}$ is small compared to the other terms in (127.1). In addition, we have,

$$\nu \Delta v \sim \frac{\nu v}{a^2} \sim \frac{\nu}{a\tau}.$$

Comparing this with the pressure gradient term, we get

$$\frac{\nu v}{a^2} \sim \frac{\nu}{a\tau} \sim \frac{p_\sigma}{\rho a} \sim \frac{\sigma}{\rho a^2}, \tag{127.4}$$

whence

$$\tau \sim \frac{\rho \nu a}{\sigma}, \tag{127.5}$$

which corresponds to (124.20).

Let us now examine a liquid mass moving at a sufficiently high velocity. In this case the air pressure $p^g$ is substantially higher than the capillary pressure $p_\sigma$. Here, we must substitute $p_\sigma$ in formulas (127.2) and (127.4) with $p^g \approx \rho^g (u^g)^2$.

In the case of a low-viscosity liquid, this gives

$$\frac{\partial v}{\partial t} \sim \frac{a}{\tau^2} \sim \frac{p^g}{\rho a} \sim \frac{\rho^g (u^g)^2}{\rho a}$$

and

$$\tau \sim \frac{a}{u^g} \sqrt{\frac{\rho}{\rho^g}}. \tag{127.6}$$

In the case of a highly viscous liquid

$$\frac{\nu}{a\tau} \sim \frac{\rho^g (u^g)^2}{\rho a},$$

whence

$$\tau \sim \frac{\nu\rho}{\rho\vartheta(u\vartheta)^2}. \tag{127.7}$$

Formulas (127.6) and (127.7) are identical with formulas (125.7) and (126.16) within the accuracy of the numerical coefficient.

Thus, the limiting formulas derived above for the break-up and atomization periods of jets are very general.

The derived formulas are based on the assumption that infinitesimal disturbances are present on the liquid surface. In real cases a jet surface is often subjected not to infinitesimal, but rather to finite and, sometimes, to extremely large disturbances. These lead to destruction of the jet in a shorter period of time than we have calculated. Moreover, the break-up period is a function of the specific nature of the surface disturbances.

Such circumstances vitiate attempts to study the break-up of jets theoretically under the influence of finite disturbances. Such an attempt was made by Weber [14], who assumed that an analogy could be drawn between a jet and a beam that was deflected by air resistance.

Such an analogy, however, appears to have little justification. Weber assumed arbitrarily that the break-up of a jet at high velocities is related to waves of the first order that develop a snakelike profile. We concluded above that waves of the first order are no different from other waves, and are less effective at moderate velocities than waves of zeroth order, which are symmetrical relative to the jet axis.

Weber assumed that formulas for the bending of a beam by concentrated forces may be used to determine the shape of the jet. This is unwarranted, of course, since a dynamic air pressure is a distributed, not a concentrated load.

The expressions derived above for the break-up period and the intact length of a jet represent the highest possible values of these quantities. They apply to jets issuing from a nozzle with well polished walls and with a smooth cross section that does not disturb the jet.

In conclusion, let us briefly mention the effect of surface-active agents on the break-up of a jet.

At a first glance it might appear that the damping of waves by surfactants could be used to prolong the break-up period of the jet.

However, this is actually not true.

In the case of atomization which is not related to capillary forces, no effect can be exerted by phenomena related to the variation in surface tension.

The fact that surfactants do not affect the break-up of a low-velocity jet is less evident. In the case of a low-viscosity liquid (see Section 69), those surface-active agents that change the boundary conditions on the liquid surface affect the break-up only by lowering the surface tension. In the case of a highly viscous liquid, the effect of a surface-active agent on break-up is again weak, because the long waves are instrumental in causing the break-up of a jet.

In Section 121, however, we saw that surfactants have a weak effect on long waves. The damping of waves by surface-active agents is significant only in the case of short waves.

This reasoning is confirmed by direct computations.

By assuming that a surface-active agent has the maximum possible damping effect, and by replacing the second boundary condition (123.4) by

$$v_z = 0,$$

we can derive the equation for $\alpha$. In the limiting cases of low and high viscosity, and for high jet velocities, this equation corresponds to equations (123.28) and (125.16).

## 128. CAPILLARY WAVES ON THE SURFACE OF A DROP

Drops formed in the breaking up of a jet undergo capillary oscillations about a spherical shape. If we assume that the amplitude of the deviations from the spherical is small, so that the surface displacement $\zeta$ is small compared to the drop radius, the period of natural oscillations can be determined.

We assume that the liquid viscosity is low and that we are not concerned with the damping of the natural oscillations. Then, assuming that the liquid is ideal and that the motion is potential, we can further assume that the velocity potential satisfies the Laplace equation

$$\Delta\varphi = \frac{1}{r^2 \sin\theta}\frac{\partial}{\partial\theta}\left(\sin\theta\frac{\partial\varphi}{\partial\theta}\right) + \frac{1}{r^2\sin^2\theta}\frac{\partial^2\varphi}{\partial\psi^2} + \frac{1}{r^2}\frac{\partial}{\partial r}\left(r^2\frac{\partial\varphi}{\partial r}\right) = 0. \quad (128.1)$$

Here, the liquid pressure is

$$p = -\rho\frac{\partial\varphi}{\partial t}.$$

The boundary condition on the drop surface is

$$p = p_\sigma \quad \text{at} \quad r = R,$$

or, in accordance with (65.20),

$$\rho\frac{\partial\varphi}{\partial t} = \frac{2\sigma\zeta}{R^2} + \frac{\sigma}{R^2}\left[\frac{1}{\sin\theta}\frac{\partial}{\partial\theta}\left(\sin\theta\frac{\partial\zeta}{\partial\theta}\right) + \frac{1}{\sin^2\theta}\frac{\partial^2\zeta}{\partial\psi^2}\right]. \quad (128.2)$$

Expressing the velocity potential $\varphi$ in terms of the displacement by the formula

$$v_r = \frac{\partial \varphi}{\partial r} = \frac{\partial \zeta}{\partial t}$$

and differentiating boundary condition (128.2) with respect to time, we get

$$\rho \frac{\partial^2 \varphi}{\partial t^2} = \frac{2\sigma}{R^2} \frac{\partial \varphi}{\partial r} + \frac{\sigma}{R^2} \left[ \frac{1}{\sin\theta} \frac{\partial}{\partial \theta} \left( \sin\theta \frac{\partial^2 \varphi}{\partial \theta\, \partial r} \right) + \frac{1}{\sin^2\theta} \frac{\partial^3 \varphi}{\partial \psi^2\, \partial t} \right] \quad \text{at} \quad r = R. \tag{128.3}$$

We seek a solution for (128.1) in the form of standing waves

$$\varphi = e^{i\omega t} f(r, \theta, \psi),$$

where $f(r, \theta, \psi)$ obviously satisfies the Laplace equation

$$\Delta f(r, \theta, \varphi) = \frac{1}{r^2 \sin\theta} \frac{\partial}{\partial \theta} \left( \sin\theta \frac{\partial f}{\partial \theta} \right) + \frac{1}{r^2} \frac{\partial}{\partial r} \left( r^2 \frac{\partial f}{\partial r} \right) + \frac{1}{r^2 \sin^2\theta} \frac{\partial^2 f}{\partial \varphi^2} = 0. \tag{128.4}$$

It is known that the following function is a partial solution of equation (128.4):

$$f_{lm} = r^l Y_{lm}(\theta, \psi) = r^l e^{im\psi} P_l^m(\cos\theta), \tag{128.5}$$

where

$$P_l^m(\cos\theta) = \sin^m \theta \frac{d^m P_l(\cos\theta)}{d(\cos\theta)^m},$$

$p_l(\cos\theta)$ is the Legendre polynomial of order $l$.

For various partial solutions, the integer $l$ may run through a series of positive whole numbers $l = 0, 1, 2, \ldots$, while m passes through the values $m = \pm l, \pm(l-1), \pm(l-2), \pm 1, 0$.

The function $Y_{lm}$ satisfies the equation

$$\frac{1}{\sin\theta} \frac{\partial}{\partial \theta} \left( \sin\theta \frac{\partial Y_{lm}}{\partial \theta} \right) + \frac{1}{\sin^2\theta} \frac{\partial^2 Y_{lm}}{\partial \psi^2} + l(l+1) Y_{lm} = 0. \tag{128.6}$$

Each value of the numbers $l$ and m corresponds to a certain natural capillary oscillation of the drop.

Substituting (128.5) into boundary condition (128.3) and taking (128.6) into account, we obtain, after elimination of $Y_{lm}$:

$$-R^l \rho \omega^2 = \frac{2\sigma l}{R^2} R^{l-1} - \frac{\sigma l}{R^2} R^{l-1} l(l+1),$$

whence

$$\omega_l = \sqrt{\frac{\sigma}{\rho R^3} l(l-1)(l+2)}\,. \tag{128.7}$$

Formula (128.7) was first derived by Rayleigh [10].

It is evident that the frequency of natural capillary oscillations is a function only of the number $l$. This indicates that natural oscillations with $(2l + 1)$ various values for m have the same frequency and are, therefore, $(2l + 1)$ times degenerate.

Further, it follows from (128.7) that the values $l = 2$ correspond to the minimum possible natural oscillation frequency. This result is quite obvious. For $l = 0$, the surface displacements of the drop would be spherically symmetrical and independent of the angles $\theta$ and $\varphi$. Since the liquid is incompressible, such oscillations are impossible.

For $l = 1$, the drop would advance only as a whole unit. In the absence of external forces acting on the drop, the nature of such motion could not be oscillatory.

For $l = 2$, we have

$$\omega_2 = \sqrt{\frac{8\sigma}{\rho R^3}}\,.$$

The frequency of natural oscillations decreases fairly rapidly with increase in the drop radius R.

## 129. WAVE GENERATION ON A FLAT LIQUID SURFACE

When we examined the interaction between a gas stream and a liquid, we saw that the gas stream contributes to the generation of waves on the liquid surface. Here, we analyze that aspect in greater detail, because of its importance from several points of view. In particular, it is important to establish the gas stream velocity which generates sustained waves (that will not be damped) on the liquid surface. This velocity is important in such problems as the generation of waves by wind action, the entrainment of a liquid film in scrubber installations, etc.

Let us examine the plane surface of an infinitely deep liquid, i.e., which fills a semispace $z < 0$. Let a gas stream with a density $\rho'$ pass above the surface of this liquid, the gas occupying the entire semispace $z > 0$. The gas far from the liquid surface moves at a constant velocity $U_0$. Find the velocity and pressure distribution in the gas.

This problem was solved by assuming an ideal liquid by Kelvin [15] and Helmholtz [16], in connection with studies of wave generation by wind action.

The basic qualitative conclusion of these studies is that the flow of gas above the liquid tends to sustain any wave motion that is present on the surface of the liquid. At sufficiently high air velocities the amplitude of the interfacial waves on the surface grows exponentially in time and the surface itself is unstable.

This effect is illustrated by Figure 105, which shows the streamlines of particles of the upper fluid. It is clear that particle velocities are lower where the streamlines are sparser, i.e., at the troughs of the waves, and higher at the wave crests. Therefore, using the Bernoulli equation, we may assume that the gas pressure on the interface is higher in the troughs and lower at the crests of the waves. In other words, the upper fluid (gas) acts on the interface in such a manner as to increase the amplitude of the waves.

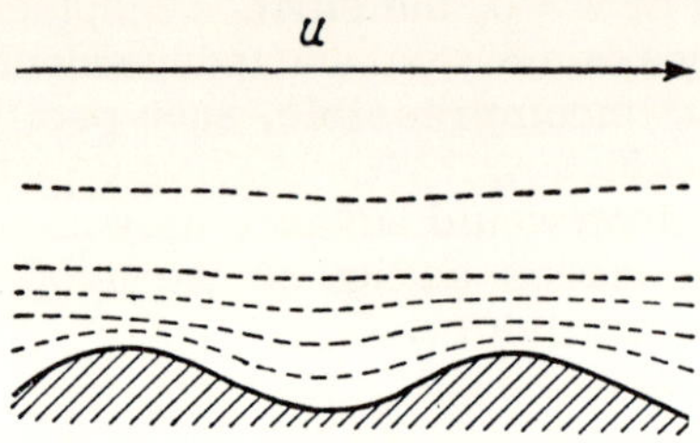


Figure 105. Wave generation by wind action.

The action of the gas stream on the liquid surface results in some work done on the liquid. On the other hand, in a viscous liquid that is in a wave motion there is a constant dissipation of energy. If the energy W transferred by the gas to the liquid per unit time is equal to the energy dissipated ($-\dot{E}$) in the liquid, then waves that will not be damped may exist on the liquid surface. Thus, in order for waves that do not become damped to appear on the surface of the liquid, the following condition must be satisfied

$$W = -(\dot{E}). \tag{129.1}$$

The assumption of an ideal liquid is, however, inadequate to compute the amount of energy transmitted by the gas stream, as shown by P. L. Kapitsa [17].

Of basic significance in the interaction of the gas stream and the liquid surface is the separation of flow that occurs on the slopes of the waves. We analyze the gas flow around one of the waves on the surface of the liquid. Since the liquid density compared to that of the gas is large, the gas flow over the liquid surface is like flow across a solid surface. Thus, the wave may be likened to a protuberance, around which gas flows.

The change in velocity across the upstream portion of the wave is continuous. Across the downstream portion, separation occurs.

The expression for pressure change upstream of the separation is equivalent to the expression for pressure change in an ideal fluid flowing past a protuberance.

Over the distance from the trough of the wave to its crest, the pressure decrease is continuous (since the velocity of an ideal fluid increases from trough to crest).

The pressure in an ideal fluid over the downstream portion of the wave increases steadily. Actually, however, separation occurs at some point on the downstream slope of the wave, and the continuous decrease in velocity becomes a constant average velocity in the separation zone.

In the zone of separation the pressure is also constant. With such a distribution of forces acting on the liquid surface, under no conditions may the liquid surface be assumed to have a truly sinusoidal shape. At small amplitudes, however ($a \ll \lambda$), a turbulent airstream flowing past a wave is akin to flow past a streamlined body. The separation point is located very far downstream (Figure 106,a), i.e., close to the trough of the wave. The separation zone includes a relatively small portion of the wave near the trough itself. The velocity distribution on the rising portion of the wave again reverts to a normal distribution for an incident flow. A similar process is repeated for every wave. At sufficiently small amplitudes, the pressure distribution is closely related to the flow of an ideal fluid over almost the entire length of the wave. Basically, the action of wind on a wave amounts to exertion of normal pressure. The tangential force, which is absent in the case of an ideal fluid, and which is not zero here due to separation, is small when the zone of separation is restricted. In that case, the shape of the wave may be considered to be approximately sinusoidal.

As the wind velocity increases, the point of separation shifts upstream in the direction of flow; i.e., it approaches the crest of the wave. The pressure distribution departs increasingly from the sinusoidal form that corresponds to flow past the wave of an ideal fluid, and the shape of the wave changes.

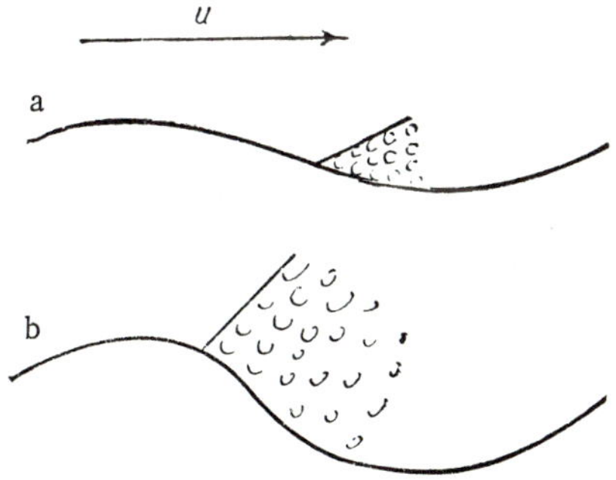


Figure 106. Appearance of separation.
a — small-amplitude waves, b — large-amplitude waves.

The wave becomes flatter on the upwind side and steeper on the downwind side (Figure 106,b).

We now turn to a quantitative estimate of the forces acting on a wave.

The force acting on a bluff object is given by

$$F = k_f \frac{\rho' U_{\text{rel}}^2}{2} \Sigma,$$

where $U_{rel}$ is the relative velocity and $\Sigma$ is the cross-sectional area of the object. In flow past a wave having a phase velocity c the relative velocity of the gas is

$$U_{\text{rel}} = U_1 - c,$$

where $U_1$ is the velocity of the gas at the surface of the liquid. This velocity, in the first approximation, is considered equal to the average velocity $U_0$ of the gas stream, and we may state that

$$U_{\text{rel}} = U_0 - c.$$

Regarding the area of the protuberance, we estimate that $\Sigma$ is of the order of ad, where d is the width of the wave and a is its amplitude (height).

Then, the force acting on a single protuberance (a wave) may be approximated by

$$F \sim k_f \frac{\rho' (U_0 - c)^2 a}{2} d. \tag{129.2}$$

A gas flowing across a liquid surface, of length L, flows over $L/\lambda$ protuberances, and the total force acting on the liquid is

$$F_{\text{tot}} \sim k_f \frac{\rho' (U_0 - c)^2}{2} \frac{a}{\lambda} S, \tag{129.3}$$

where S is the area of the liquid.

The work done by force F per unit time, i.e., the amount of energy transmitted to the liquid, may be approximated by

$$W = F_{\text{tot}} v = \frac{k_f \rho' (U_0 - c)^2 a^2 \omega}{2 \lambda} S = k_f \frac{\rho' (U_0 - c)^2 a^2 k^2 c}{4\pi} S,$$

or

$$W = \gamma_f \rho' (U_0 - c)^2 a^2 k^2 c S, \tag{129.4}$$

where $\gamma_f$ is an unknown proportionality coefficient. Formula (129.4) was first derived by P. L. Kapitsa [17].

P. L. Kapitsa analyzed the flow and showed that an approximation using the assumption of an ideal fluid, as developed in the studies by Jeffreys [18], has no physical significance.

Substituting the value of W given by (129.4) into the energy balance and using formula (118.10) for $(-\dot{E})$, we obtain:

$$\gamma_f \rho' (U_0 - c)^2 a^2 k^2 c S = 2\mu a^2 k^3 c^2 S,$$

$$(U_0 - c)^2 = \frac{2\mu}{\gamma_f \rho'} kc = \frac{4\pi}{\gamma_f} \frac{\mu c}{\rho' \lambda}. \qquad (129.5)$$

Formula (129.5) shows that for every wavelength, there is a particular gas flow velocity at which the wave must propagate over the liquid surface without being damped.

Let us consider the case of long (gravity) waves. For such waves $c = \sqrt{\frac{g\lambda}{2\pi}}$, and equation (129.6) becomes

$$\left(U_0 - \sqrt{\frac{g\lambda}{2\pi}}\right)^2 \sqrt{\frac{g\lambda}{2\pi}} = \frac{\rho}{\rho'} \cdot \nu \frac{2g}{\gamma_f}. \qquad (129.6)$$

It is clear that (129.6) cannot be satisfied by all values of $U_0$, but only at a velocity that exceeds a certain minimum velocity $U_0^{(\min)}$. The latter may be found from the condition

$$\frac{\partial}{\partial c}(U_0 - c)^2 c = 0.$$

Differentiation yields

$$U_0^{(\min)} = 3c^{\min},$$

where

$$c^{\min} = \sqrt[3]{\frac{\nu \rho g}{2\rho' \gamma_f}} = \sqrt{\frac{g\lambda_{\min}}{2\pi}}.$$

Here $c^{\min}$ designates the phase velocity for propagation waves raised by a wind of velocity $U_0^{\min}$ and $\lambda_{\min}$ is the corresponding wavelength.

Jeffreys concluded that there exists a minimum wind velocity that can produce sustained waves that are not damped. This author also conducted experimental measurements of the minimum velocity for generation of waves.

At a wind velocity of $U_0 < U_0^{\min}$, the balance of energy cannot be maintained whatever the wavelength.

Comparison of theory and experiment has been rendered difficult by the fact that the velocity $U_0$, which is present in all formulas, represents the velocity of the air in the immediate vicinity of the liquid surface, while experimental measurements have employed the average wind velocity at a height which is substantially greater than the wave amplitude.

## 130. GENERATION BY WIND OF LARGE-AMPLITUDE DEEP-WATER WAVES AND THEIR DAMPING BY TURBULENT FRICTION

In the preceding section we examined the critical wind velocity at which waves appeared on a liquid surface.

We now examine the limiting situation in which the velocity of the wind above the surface is relatively high.

As the wind velocity increases, the amplitude of the waves generated on the liquid surface and the velocity of the liquid particles also increase until the laminar motion is finally replaced by turbulent liquid motion.

Turbulent motion of a liquid with a free surface may be considered periodic only on the over-all, average basis.

In the turbulent regime, a free liquid surface is covered by waves of varying lengths and amplitudes. The liquid in these waves undergoes an average periodic motion at a frequency expressed by

$$\omega = \sqrt{gk}.$$

The gravitational field produces a certain stability in the motion of a liquid with a free surface. This stabilizing effect of gravity is analogous, to some extent, to the stabilizing effect of gravity on the motion of a liquid divided into stable layers — a problem that was studied by Richardson [20] and later by several other investigators.

Examination of the transfer of wind energy to the liquid and of the turbulent motion of a liquid with a free surface presents great difficulties. In our previous discussions, which followed the work of P. L. Kapitsa, we considered that each wave subject to a gas stream flowing past a liquid surface may be treated independently.

This assumption is justified for a single wave, or several waves, where the distance between the crests is relatively large.

In the latter case, the disturbance to the gas stream due to a turbulent separation from the crest of the wave can be fully damped within the distance up to the next crest.

However, if the number of waves on the liquid surface is large, an accumulation of the perturbing action of crests seems to occur, and the picture of flow past a liquid surface that is covered by waves approaches that of a turbulent flow past an uneven surface.

Wave amplitude plays the part of the height of protuberances and the wavelength corresponds to the distance between them. In this case, in contrast to protuberances on a solid wall, the "protuberances" on the liquid surface move with a velocity $c(\lambda)$.

Turbulent fluid flow in the proximity of an uneven surface has not been studied fully to date. It has only been established that the drag coefficient is a weak function of the ratio $\left(\frac{a}{\lambda}\right)$, and is not a function of the number of protuberances per unit length. The resistance to flow along an uneven surface that we examined in Section 30 must be altered for the case of waves in order to account for the motion of the "protuberances."

If all of the protuberances had the same dimensions, wavelength $\lambda$ and amplitude a, and if all moved at a velocity $c(\lambda)$ relative to the air, then the force acting on a unit surface area could be given by

$$F = \gamma_f \frac{\rho' (U - c)^2}{2} S,$$

where $\gamma_f \sim 10^{-2}$.

Accordingly, the work done by the wind would be

$$W = \gamma_f \frac{\rho' (U - c)^2}{2} Sv = \frac{k\rho' (U - c)^2 a\omega S}{2}. \tag{130.1}$$

Under real conditions, however, the waves on the surface have different amplitudes and lengths, and move at different velocities.

It is not possible, with the present knowledge of the theory of turbulence, to derive an exact expression for the force acting on an uneven surface, where the dimensions of the individual protuberances are different. We are therefore forced to confine our treatment to a rather rough computation. Thus, we assume that the work done by the wind on waves with given a and $\lambda$ is independent of its work done on other waves and is given by formula (130.1).

The work transmitted to a wave is dissipated by the motion of the liquid within the wave, and is not transmitted to other waves. Although these assumptions make no pretense at being complete, they nevertheless permit us to elucidate certain features of wave motion caused by wind.

Steady-state turbulent motion of a liquid results in the formation of the entire wave spectrum on the free liquid surface.

The energy dissipated in the liquid, in this case, is compensated by the energy brought to the liquid by the air stream.

Let us write the energy balance for the dissipated and acquired energy

$$W = -\dot{E},$$

where W is the energy transmitted to the entire surface of the liquid per unit time, and $\dot{E}$ is the total energy dissipated per unit time in the entire liquid volume.

The transmitted energy is given by equation (130.1). Let us find the energy dissipated in turbulent wave motion. The average energy dissipated in the liquid per unit time may be represented by

$$-\dot{E} \sim \rho \overline{\int \nu_{\text{turb}} \left(\frac{\partial v}{\partial l}\right)^2 dV}: \tag{130.2}$$

$\nu_{\text{turb}}$ in this case represents the turbulence viscosity and $\frac{\partial v}{\partial l}$ represents the derivative of the average velocity in a certain direction.

The horizontal line indicates averaging with respect to time (over one period). Here, we have disregarded the energy dissipated directly by molecular viscosity as being very small compared to (130.2).

Since all the velocity components in the wave are of the same order of magnitude, we employ $\left(\frac{\partial v_y}{\partial z}\right)$ to be specific, instead of $\left(\frac{\partial v}{\partial l}\right)$. In this case z is directed downward from the surface.

Then

$$\dot{E} \sim -\rho \overline{\int \nu_{turb}\left(\frac{\partial v_y}{\partial z}\right)^2 dV}.$$

The turbulence viscosity $\nu_{turb}$ may be expressed as

$$\nu_{turb} \sim l^2 \left|\frac{\partial v_y}{\partial z}\right|,$$

where $l$ is the scale of turbulence eddies.

Then, finally

$$\dot{E} \sim -\rho \overline{\int l^2 \left(\frac{\partial v_y}{\partial z}\right)^2 \left|\frac{\partial v_y}{\partial z}\right| dV}. \qquad (130.3)$$

In order to compute the power dissipated, an assumption must be made regarding the magnitude of the unknown scale of turbulence eddies $l$.

A single valued assumption with respect to the length of the mixing path $l$ in turbulent wave motion, and based on considerations of dimensionality, cannot be made as it was in the case of turbulent motion near a solid wall.

In the case under consideration, at least one of the following quantities which have the dimension of length could be chosen as the scale of the motion:

1) the wavelength $\lambda$;
2) the distance $z + a$ to the free surface;
3) the wave amplitude a.

The assumption has been made in a published paper [21] that the scale of the motion is $\lambda$. Two objections may be raised, however, against this proposal: in the first place, $\lambda$ is a kinematic rather than dynamic characteristic of wave motion, and it seems natural for the scale of turbulence eddies to be a function of the dynamic characteristics of the process. In the second place, if we accept the assumption that $l \sim \lambda$, then the expression for wave amplitudes, derived from (130.3) and (130.1), is physically very unlikely (see below).

Adoption of the second assumption, on the other hand, would appear to be quite acceptable. The liquid particles in laminar wave motion and at a low value of the ratio $\frac{a}{\lambda}$ describe circular trajectories, the radius of which is equal to the amplitude a at the surface, and decreases rapidly with depth.

Average velocities, in the case of turbulent motion, also must decrease with the distance from the free surface.

Since viscosity has no direct effect on large-scale turbulent motion, it is natural to assume that the average velocity has the same form as in the case of wave motion in an ideal liquid:

$$v_y \sim a\omega e^{-kz} \cos(ky - \omega t). \tag{130.4}$$

Since this formula has been derived for small-amplitude waves, it may give the order of magnitude of $v_Z$. The liquid particles that participate in the turbulent motion move in accordance with a law of the type of (130.4) only on the average. Random turbulence eddies are superimposed upon this average motion.

As they approach the free surface of the liquid, the paths traveled by turbulence eddies must be approximately equal to the paths traveled by the liquid particles that are in an average motion; i.e., we have $l \sim a$ as $z \to 0$.

It follows from dimensional analysis that $l$ must be a linear function of distance from the surface. Both requirements may be satisfied by [19]

$$l \sim z + a. \tag{130.5}$$

Formula (130.5) is similar to the formula for the scale of turbulence eddies on an uneven surface (see Section 30). The difference lies in that the "size of the protuberance" a is not given, but is determined by a possible eddy scale on the liquid surface.

The length $l$ represents the distance from a given point within the bulk of the liquid to the free surface.

Formula (130.5) becomes very clear if we turn to a system of coordinates in which the air is motionless and the liquid is in motion. The interface is then an uneven surface having a scale of roughness of the order of the wave amplitudes. Turbulence eddies of higher frequencies than that of the average motion flow past such protuberances.

An increase in the scale of turbulence eddies is not contradictory to the decrease in eddy velocities with depth. Indeed, we have

$$v' \sim l\frac{\partial v}{\partial z} \sim (z + a)\, a\omega k e^{-kz} \cos(ky - \omega t);$$

i.e., $v'$ at $z \gg \lambda$ decreases almost exponentially.

Altough assumption (130.5) appears to be quite logical, it should be stressed that subsequent conclusions are not highly dependent on it. Thus, if, instead of (130.5), we had adopted $l \sim a$, the results would be exactly the same.

The dissipated energy may be computed easily with the help of formulas (130.5) and (130.4). Substituting the values of $l$ and $v_y$ into (130.3), we obviously obtain for the average energy dissipated in a wave of length $\lambda$

$$\dot{E} = -\frac{8}{81\pi}\beta\rho a^3\omega^3 S \approx -\frac{\beta}{30}\rho a^3\frac{g^{3/2}}{\lambda\!\!\!^{-}{}^{3/2}} S, \tag{130.6}$$

where $\beta$ is an unknown constant and $\lambda\!\!\!^{-} = \frac{\lambda}{2\pi}$. Here we have omitted those terms that are small compared to the ratio $\frac{a}{\lambda}$.

Equating the power dissipated to that furnished by the wind, we have

$$a = \left(\frac{\gamma_f \rho'}{\beta\rho g}\right)^{1/2}(U_0 - \sqrt{g\lambda\!\!\!^{-}})\,\lambda\!\!\!^{-}{}^{1/2}. \tag{130.7}$$

This equation shows that a power balance may be attained when a certain ratio between the length of the wave and its amplitude exists. The transferred energy may be dissipated at various wavelengths and amplitudes given by equation (130.7).

Thus, we reach the conclusion that in turbulent motion and at a given wind velocity, an entire spectrum of wavelengths may exist on the liquid surface within the range of $0 \le \lambda \le \lambda_{max}$, rather than a wave of definite length and amplitude. The maximum wavelength is determined by (130.7):

$$\lambda_{max} = \frac{U_0^2}{g}.$$

Figure 107 illustrates the dependence of a on $\lambda$, as given by formula (130.7). We see that the amplitude attains a maximum at a definite wavelength.

The maximum amplitude $a_{max}$ and the corresponding wavelength $\lambda a_{max}$ may be determined from formula (130.7) by simple operations

$$a_{max} \approx \left(\frac{\gamma_f}{\beta}\frac{\rho'}{\rho}\right)^{1/2}\frac{U_0^2}{g}, \tag{130.8}$$

$$\lambda\!\!\!^{-}_{a_{max}} = \frac{1}{4}\frac{U_0^2}{g}. \tag{130.9}$$

Formulas (130.8) and (130.9) show that both the length of the wave with maximum amplitude and the maximum amplitude itself increase as the square of the wind velocity near the liquid surface, and their ratio

$$\frac{a_{max}}{\lambda\!\!\!^{-}_{a_{max}}} \approx \left(\frac{\gamma_f \rho'}{\beta\rho}\right)^{1/2} \tag{130.10}$$

is not a function of wind velocity.

The rate of propagation of waves with maximum amplitude obviously is

$$c_{a_{max}} = \frac{1}{2} U_0. \tag{130.11}$$

The investigation of an unsteady-state regime — damping of waves in the absence of supporting wind — is also of considerable interest.

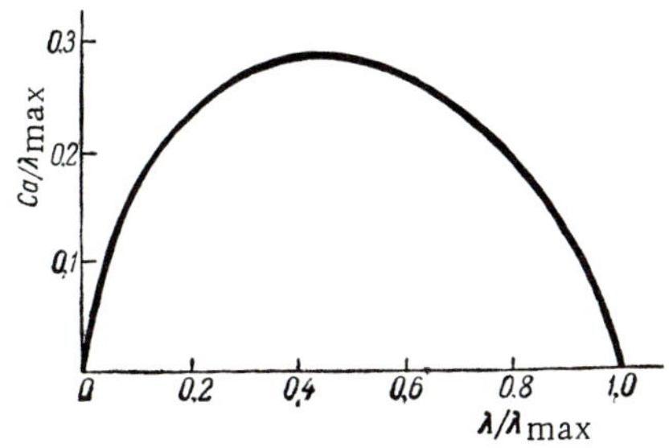


Figure 107. Dependence of $Ca/\lambda_{max}$ on $\lambda/\lambda_{max}$, where C designates the expression $\left(\frac{\beta \cdot \rho}{\gamma\gamma\rho'}\right)^{1/3}$.

To find the expression for damping, let us rewrite the energy balance. The average energy may be given by

$$E = \frac{\rho c^2 k a^2 S}{4}.$$

Therefore,

$$\frac{1}{S}\frac{dE}{dt} = \frac{d}{dt}\frac{\rho c^2 k a^2}{4} = -\beta\rho a^3 \frac{g^{3/2}}{\lambda^{3/2}}. \tag{130.12}$$

If we assume that the length of the wave that is being damped remains unchanged (which is justified in the case of weak damping), then from (130.12), we find

$$\frac{1}{a}\frac{da}{dt} = -\beta g^{1/2} \frac{1}{\lambda^{3/2}},$$

whence

$$\frac{1}{a} = \frac{1}{a_0} + \frac{\beta g^{1/2} t}{\lambda^{3/2}}, \tag{130.13}$$

where $a_0$ is the amplitude at the initial time.

Applying the last formula to a wave with maximum amplitude, we have

$$a_{\max}(x)=\frac{a_{\max}(0)}{1+\frac{\beta a_{\max}(0)x}{\lambda^2_{a_{\max}}}}, \tag{130.14}$$

where $x = ct = \sqrt{g\lambda_{max}}t$ is the path traveled, and $a_{max}(0)$ and $a_{max}(x)$ are the initial amplitude and the amplitude after the wave has travelled a distance x, respectively. Equation (130.14) contains only one unknown constant $\beta$.

Using (130.10), we derive an expression for damping of waves of greatest amplitude in a somewhat different form

$$a_{\max}(x)=\frac{a_{\max}(0)}{1+3\cdot 10^{-3}\beta^{1/2}\frac{x}{\lambda_{a_{\max}}}}. \tag{130.15}$$

The foregoing sketchy computation indicates a wave spectrum that is in qualitative agreement with experimental data, as was shown in the study by Kitaygorodskiy [26], based on very general dimensionality considerations.

The significance of a comparison between theory and experiment can only be very general, owing to the fact that the theory is sketchy and to the nature of the experimental data, which usually are derived from unsteady state regimes.

Theoretical formulas are distinguished by the absence of arbitrary constants [see (130.9)], or they contain a single unknown constant whose value must be obtained by experiment [formulas (130.8), (130.10) and (130.12)].

The experimental data in the published literature are somewhat contradictory, and the experimental points are rather widely scattered.

Nevertheless, the existence of a spectrum of waves of varying amplitude on the surface of a liquid, at a given wind velocity, does not appear to be subject to doubt.

The dependence of the maximum wave amplitude on wind velocity above the surface of the water is given, according to data by Rossby and Montgomery [22], by the empirical formula

$$a_{\max}=\frac{0.3U^2}{g},$$

while the data given by Sverdrup and Munk [23] fit the following formula somewhat better:

$$a_{\max}=\frac{0.2U^2}{g}.$$

It should be noted, however, that the scatter of the experimental points is very large.

Turning to a comparison of formula (130.11) with experimental data, we note that formula (130.11) contains no arbitrary constants. Empirical formulas, as follows, were reported by Cornish [24]

$$c_{a_{\max}} = 0.8U$$

and by Shoot [25]

$$c_{a_{\max}} = 0.76U.$$

It can be stated that there is complete agreement between formula (130.11) and the empirical relations shown above.

Sverdrup and Munk observed waves with a velocity exceeding that of the wind. It may be assumed that they were not steady waves with a maximum amplitude. The existence of waves moving at a velocity greater than that of the wind is in fact possible, since they may originate in a zone where the wind velocity is greater than at the point of observation. Such waves, however, must become damped. We may also estimate the rate at which the waves are damped, by substituting the value of $\gamma_f$, obtained by means of independent measurements, into formula (130.14).

At $\beta \sim 1$, with the help of (130.15), we have

$$a_{\max}(x) = \frac{a(0)}{1 + 10^{-3}3 \cdot \frac{x}{\lambda}}. \tag{130.16}$$

Let us compare formula (130.16) with the rule given by Sverdrup and Munk [23]: "the wave amplitude decreases by 1/3 after it has traveled a distance of x = 6000 λ."

By setting $a_{\max}(x) = \frac{2}{3}a(0)$, we find, from (130.16), that $x \sim 10^3\lambda$. It should be borne in mind that in addition to turbulence viscosity, for wave motion there are energy losses in the liquid due to the turbulence viscosity of the air. In the absence of wind, the waves produce motion of the air, and a significant amount of energy is dissipated in the air.

## BIBLIOGRAPHY


1. Lamb, H., Hydrodynamics, Gostekhizdat, 1948; N. E. Kochin, I. A. Kibel and N. V. Roze, Theoretical Hydromechanics, vol. I, Gostekhizdat, 1948; H. Lamb, Hydrodynamics, 6th edn., Dover reprint, New York, 1945 .
2. Matthiessen, Wied. Ann. 38, 118 (1889).
3. Rayleigh, Phil. Mag. 30, 385 (1890); R. Brown, Proc. Roy. Soc. (A) 48, 312 (1936); L. Tyler, Phil. Mag. (7) 31, 202 (1941).
4. Landau, L. D., and E. M. Lifshits, Mechanics of Continuous Media, Gostekhizdat, 1944; M. M. Kusakov, Acta Physicochimica URSS 19, 286 (1944); L. D. Landau and E. M. Lifshits, Fluid Mechanics, Pergamon Press, London, 1959 .
5. Shuleykin, V. V., Fizika morya Physics of the Sea , AN SSSR Press, 1941; idem, Ocherki po fizike morya Outlines of the

Physics of the Sea), Akad. Nauk SSSR, 1949 cf. U. S. Naval Research Lab. Transl. No. 797 (1960), "Surface Active Films on the Sea," by V. V. Shuleykin.

6. Aitken, Proc. Roy. Soc., Edin. (V) 12, 56 (1883).
7. Ivanov, R. N., Izv. AN SSSR, series geophys., No. 3 (1937).
8. Koppen, W., Ann. Hydrograph. u. Mar. Mat., April 1893.
9. Levich, V. G., Zhurn. Eksp. Teor. Fiz. 10, 1296 (1940); ibid. 11, 340 (1941); V. G. Levich, K teoriipoverkhnostnykh yavleniy On the Theory of Surface Phenomena, Sov. Nauka, 1941.
10. Rayleigh, Teoriya Zvuka (Theory of Sound) vol. II, Gostekhizdat, 1944 Rayleigh, Theory of Sound, 2nd edn., Dover reprint, New York, 1945.
11. Bohr, N., Trans. Roy. Soc. 204 (A), 281 (1909).
12. Timotika, H., Proc. Roy. Soc. 150, 322 (1935); 153, 302 (1936).
13. Watson, G., Teoriya besselevykh funktsiy Theory of Bessel Functions, Gostekhizdat, 1954 G. N. Watson, Treatise on the Theory of Bessel Functions, 2nd edn. Cambridge, 1944.
14. Weber, K., Coll.: Internal Combustion Engines, Vol. I, ONTI, 1936.
15. Lord Kelvin, Phyl. Mag. 4, ser. 42 (1871).
16. Helmholtz, Wissenschaftliche Abhandlungen, Leipzig, 1909.
17. Kapitsa, Dokl. AN SSSR 64, 513 (1949).
18. Jeffreys, H., Proc. Roy. Soc. 107, 197 (1925).
19. Levich, V. G., Dokl. AN SSSR 101, 615 (1955); the calculation error is corrected in the paper by I. S. Brovikov, Trans. of the State Ocanographic Institute, Issue 38 (1957).
20. Richardson, Proc. Roy. Soc. A 97, 354 (1920).
21. Krylov, Yu., Trudy Gosud. Okeanograf. Inst., No. 26, 1954.
22. Rossby, C. G., R. B. Montgomery, Papers in Phys. Oceanogr. a. Metorol. 3, No. 3, 1935.
23. Sverdrup and Munk, Sumposium. Forecasting of Wind Waves, Swell and Surf, IL, 1951 H. U. Sverdrup and W. Munk, Wind, Sea, and Swell: Theory of Relations for Forecasting, U. S. Hydrogr. Office Publ. No. 601 (1947).
24. Cornish, V., Ocean Waves and Kindred Geophysical Phenomena, Cambridge, 1934.
25. Shott, G., Patermans Mitteil Erganzungsheft 109, 82 (1933).
26. Kitaygorodskiy, S. A., Izv. AN SSSR (ser. geoph.), No. 4 (1959).

CHAPTER XII

# Motion and Diffusion in Thin Liquid Films

## 131. FLOW OF THIN LIQUID FILMS

The flow of thin liquid films is a special case of flow in which both surface tension and viscosity play important parts.

The flow of a thin liquid film is of special interest in studying dissolution of gases in liquids and in many other engineering problems.

Most experimental studies of such flow have been made with liquid flowing by gravity down an inclined or vertical wall. Three different flow regimes have been observed experimentally:

1) At Reynolds numbers $Re = \frac{\bar{v}h}{\nu}$ that do not exceed 20 to 30 (where $\bar{v}$ is the average velocity over the cross section of the film and h is its thickness), there exists the usual viscous flow regime and the film thickness is constant;

2) at Reynolds numbers $Re > 30$ to 50, a so-called wave regime appears, in which wave motion is superposed on the forward motion of the film;

3) at $Re \approx 1500$, the laminar regime is replaced by turbulent motion.

In the case of very thin films, the break-up of the film into separate drops is caused by the action of capillary forces. The disruption depends on wetting conditions.

We first consider a film in laminar motion.

The equations of liquid motion in a thin film can be simplified substantially. Because the film thickness is small, all velocity derivatives across the film are large compared to those along the film. In other words, the rate of change of velocity across the film is large compared to the change in velocity along its length. If the motion of the liquid in the film is quasi-unidimensional in nature, then, by orienting the x axis along the direction of the motion and the y axis perpendicularly to the film surface, the Navier-Stokes equation may be written in the form of boundary layer equation (3.8)

$$\frac{\partial v_x}{\partial t} + v_x \frac{\partial v_x}{\partial x} + v_y \frac{\partial v_x}{\partial y} = -\frac{1}{\rho} \frac{\partial p}{\partial x} + \nu \frac{\partial^2 v_x}{\partial y^2} + f, \tag{131.1}$$

where f is the body force per unit mass, in the x direction.

The second boundary layer equation (3.9) is

$$\frac{\partial p}{\partial y} = 0, \tag{131.2}$$

and the continuity equation is

$$\frac{\partial v_x}{\partial x} + \frac{\partial v_y}{\partial y} = 0. \tag{131.3}$$

At $y = h(x)$, the usual conditions at the free surface of a liquid must be fulfilled:

$$\left.\begin{aligned} p &= p_\sigma, && (131.4)\\ \mu \frac{\partial v_x}{\partial y} &= 0, && (131.5) \end{aligned}\right\} \quad y = h(x).$$

At the surface of the solid body we have:

$$v_x = v_y = 0. \tag{131.6}$$

The second boundary layer equation, denoting a constant pressure in the thin film, and boundary condition (131.4), showing that the liquid and capillary pressures are equal, can be combined to eliminate the pressure term from equation (131.2).

Replacing p in equation (131.1) by its value at the boundary, we obtain:

$$\frac{\partial v_x}{\partial t} + v_x \frac{\partial v_x}{\partial x} + v_y \frac{\partial v_x}{\partial y} = -\frac{1}{\rho} \frac{\partial p_\sigma}{\partial x} + \nu \frac{\partial^2 v_x}{\partial y^2} + f. \tag{131.7}$$

Since the film is thin the curvature of its surface cannot be large; therefore, $p_\sigma$ may be approximated from (65.13). Then

$$\frac{\partial v_x}{\partial t} + v_x \frac{\partial v_x}{\partial x} + v_y \frac{\partial v_x}{\partial y} = \frac{\sigma}{\rho} \frac{d^3 h}{dx^3} + \nu \frac{\partial^2 v_x}{\partial y^2} + f, \tag{131.8}$$

where h is the surface coordinate, related to $v_x$ and $v_y$ by

$$v_y = \frac{\partial h}{\partial t} + \frac{\partial h}{\partial x}\frac{dx}{dt} = \frac{\partial h}{\partial t} + v_x \frac{\partial h}{\partial x} \approx \frac{\partial h}{\partial t}, \tag{131.9}$$

since the term $v_x \dfrac{\partial h}{\partial x}$ in this case is very small.

The velocity component $v_y$ can be eliminated by using the continuity equation

$$v_y = -\int \frac{\partial v_x}{\partial x} dy. \tag{131.10}$$

Thus, we get the final equation for $v_x$

$$\frac{\partial v_x}{\partial t} + v_x \frac{\partial v_x}{\partial x} - \left(\int \frac{\partial v_x}{\partial x} dy\right) \frac{\partial v_x}{\partial y} = \frac{\sigma}{\rho} \frac{d^3 h}{dx^3} + \nu \frac{\partial^2 v_x}{\partial y^2} + f. \tag{131.11}$$

The surface coordinate h is related to $v_x$ by

$$\frac{\partial h}{\partial t} = -\frac{\partial}{\partial x}\left(\int v_x \, dy\right). \tag{131.12}$$

Equations (131.11) and (131.12) are the general equations of the motion of liquid in a thin film.

Equations (131.5) and (131.6) serve as boundary conditions in the simplest case of slow steady-state flow down a plate by gravity, and equations (131.11) and (131.12) acquire a simple form, because the liquid film surface remains parallel to the surface of the plate. In this case, the film thickness remains h = const and capillary surface forces are not present.

The equation for slow steady-state motion of a film in a gravitational field is

$$\nu \frac{\partial^2 v_x}{\partial y^2} + g = 0. \tag{131.13}$$

Integration with boundary conditions (131.5) and (131.6) gives

$$v_x = \frac{g}{\nu} y \left(h - \frac{y}{2}\right).$$

Expressing the constant $\frac{g}{\nu}$ in terms of the average velocity over the cross section, defined as

$$\bar{v} = \frac{1}{h} \int_0^h v_x \, dy = \frac{gh^2}{3\nu}, \tag{131.14}$$

we get

$$v_x = 3\bar{v} \cdot \frac{y}{h}\left(1 - \frac{y}{2h}\right). \tag{131.15}$$

We now consider the case when a constant tangential stress F acts on the surface of a film and the body force is absent. The boundary condition (131.5) in that case is replaced by

$$\frac{\partial v_x}{\partial y} = \frac{F}{\mu}, \tag{131.16}$$

and the equation of motion takes the form

$$\nu \frac{\partial^2 v_x}{\partial y^2} = 0. \tag{131.17}$$

A solution of (131.17) that satisfies both equations (131.6) and (131.15) is

$$v_x = \frac{F}{\mu} \frac{y}{h}. \tag{131.18}$$

## 132. THE METHOD OF "BLOWING-OFF" THIN LIQUID FILMS

An important instance is the flow of an extremely thin liquid film under the action of an airstream moving over its surface [2]. The flow of such films has been used to determine the hydrodynamic characteristics of a liquid [3]. The method was proposed and developed by B. V. Deryagin and his collaborators.

As explained below, this method, termed the "blow-off" method, provides a full characterization of the hydrodynamic properties of liquids, with both normal and anomalous (non-Newtonian) viscosity. We examine the motion of a liquid film dragged by air, when the air is blown through a wedgelike slit formed by two planes at an angle $2\alpha$ (Figure 108) and having a radius R. The liquid film lies on the horizontal plane.

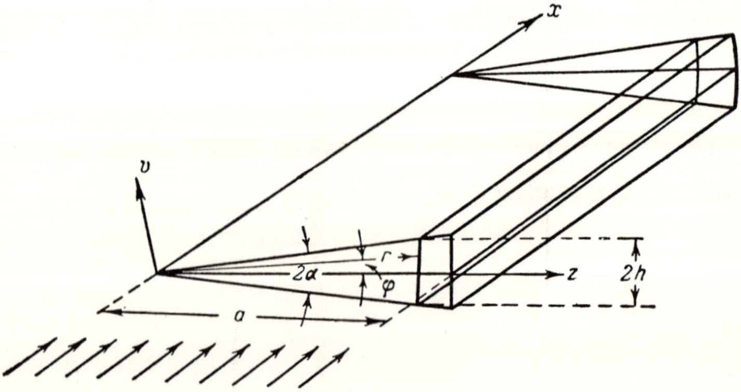


Figure 108. Wedgelike slit.

We assume that the air moves at such velocity that the Reynolds number is small and the flow remains laminar. Since the thickness of the liquid layer is extremely small, it may be disregarded in examining the air motion. Here, the air represents a viscous fluid

flowing through a capillary sector of radius R and angle $2\alpha$. If the velocity distribution in the air flowing through the slit is known, the viscous stress on the liquid film surface can be determined easily. This stress obviously varies with distance from the vertex of the slit. Therefore, the film moves under different stresses in different zones. This variation in the stresses is used later to obtain a complete hydrodynamic characterization of the film.

Air flow in the slit is rectilinear along the x axis with a velocity u.

The Navier-Stokes equation in cylindrical coordinates is

$$\left(\frac{\partial^2 u}{\partial r^2}+\frac{1}{r^2}\frac{\partial^2 u}{\partial \varphi^2}+\frac{1}{r}\frac{\partial u}{\partial r}\right)=\frac{1}{\nu}\frac{\partial p}{\partial x}. \tag{132.1}$$

The pressure gradient is constant throughout the slit, i.e., $\frac{\partial p}{\partial x}=$ const.

To simplify the boundary conditions it is convenient to introduce a new function

$$f=u+\left(\frac{\cos 2\varphi}{\cos 2\alpha}-1\right)\frac{r^2}{\nu}\frac{\partial p}{\partial x}. \tag{132.2}$$

The function f satisfies the equation

$$\frac{\partial^2 f}{\partial r^2}+\frac{1}{r^2}\frac{\partial^2 f}{\partial \varphi^2}+\frac{1}{r}\frac{\partial f}{\partial r}=0 \tag{132.3}$$

and boundary conditions

$$f=0 \qquad \text{at} \quad r=0, \tag{132.4}$$

$$f=0 \qquad \text{at} \quad \varphi=\pm\alpha \quad (0\leqslant r\leqslant R), \tag{132.5}$$

$$f=-R^2\left(\frac{\cos 2\varphi}{\cos 2\alpha}-1\right)\frac{1}{\nu}\frac{\partial p}{\partial x} \quad \text{at} \quad r=R\,(-\alpha\leqslant\varphi\leqslant\alpha); \tag{132.6}$$

i.e., along the perimeter of the slit, the air velocity is zero.

The function f satisfying equation (132.3) and boundary conditions (132.4) to (132.6) can be found from the general solution of equation (132.3)

$$f=\sum (A_k\cos k\varphi+B_k\sin k\varphi)(C_k r^k+D_k r^{-k})$$

by appropriately selecting k and the constants $A_k$, $B_k$, $C_k$ and $D_k$. This function has the form

$$f=\frac{R^2}{\nu}\left(\frac{\partial p}{\partial x}\right)\sum_{n=0}^{\infty}(-1)^n\left(\frac{1}{(2n+1)\pi-4\alpha}+\frac{1}{(2n+1)\pi+4\alpha}-\right.$$
$$\left.-\frac{2}{(2n+1)\pi}\cos\left[(2n+1)\frac{\pi\varphi}{2\alpha}\right]\right)\left(\frac{r}{R}\right)^{\frac{(2n+1)\pi}{2\alpha}}. \tag{132.7}$$

By substituting the value of f into expression (132.2), the air-velocity distribution in the wedgelike slit may be found.

The shear stress exerted on the liquid film is given by

$$F=\mu\frac{1}{r}\left(\frac{\partial u}{\partial \varphi}\right)=$$

$$=\frac{r}{2}\cdot\frac{\partial p}{\partial x}\operatorname{tg}2\alpha+\frac{R}{2}\frac{\partial p}{\partial x}\frac{4}{\alpha}+\sum_{n=0}^{\infty}(-1)^{n+1}\frac{1}{\left[(2n+1)\frac{\pi}{2\alpha}\right]^{2}-4}\left(\frac{r}{R}\right)^{\frac{(2n+1)\pi}{2\alpha}-1} \quad (132.8)$$

The shear stress becomes zero at r = 0 and at r = R.

For small angles of the sector the sum of the series is small compared to the first term on the right, and

$$F\approx\frac{r}{2}\left(\frac{\partial p}{\partial x}\right)\operatorname{tg}2\alpha. \quad (132.9)$$

Under the action of the stress F a film of thickness h is set into motion with a velocity v given by

$$\mu_{\text{liq}}\frac{v}{h}=F. \quad (132.10)$$

In practice, the thickness h and velocity v are measured with an interferometer. The first is measured directly, while the second is measured by the rate of displacement of the interference bands during the blow-off process.

Thus, using equation (132.10), the liquid viscosity can be determined as a function of the shear stress from a single experiment. In the case of normal liquids, this function is a straight line that passes through the origin. At the same time, the blow-off method can be used to determine the thickness of a moving liquid layer. In other words, interferometric measurements can be used to determine that distance from the solid surface at which the liquid particles can undergo displacement relative to the wall.

According to experimental measurements, this distance does not exceed $5\cdot10^{-7}$ cm. This should be considered as the most accurate value for the minimum distance from a solid surface at which a liquid retains its mobility.

The blow-off method and its modifications are widely used in the study of hydrodynamic properties of liquids [4].

## 133. THICKNESS OF FILM REMAINING ON THE SURFACE OF A SOLID WITHDRAWN FROM A QUIESCENT LIQUID

As a typical example of steady-state motion of a liquid film where capillary forces play an important part, we examine a solid withdrawn from a quiescent liquid [5]. A thin film of liquid remains on the surface when the solid is withdrawn. The film thickness is of interest from many practical standpoints, for example, that of coating of solids with a layer of a solute. When the solid object is withdrawn from a solution at a constant speed, its surface remains

covered by a layer of solution which is of uniform thickness. As the solvent evaporates, a thin layer of the solute remains on the solid surface. It is interesting to determine the dependence of the thickness of the layer on the withdrawal rate and the effect of the physicochemical properties of the solution.

Another important example is the migration of petroleum in earth strata. When such migration occurs in the presence of water and gas, a portion of the petroleum is retained in the pores of the strata. It is of interest to determine the factors that govern the thickness of the petroleum layer retained on the pore surface of petroliferous strata.

The thickness of a liquid film remaining on the surface of a solid body is of great importance in accurate handling of chemical solutions. The walls of vessels, when emptied, retain a layer of the liquid solution which constitutes a source of error in accurate measurement of liquid volumes. The presence of liquid on the walls must be considered in the quantitative analysis of chemical solutions, in the use of capillary viscometers for absolute measurements of viscosity, and other cases where the "emptying error" is one of the principal inaccuracies of the method.

At the same time, determination of the film thickness can serve as a method for measuring liquid viscosity.

In view of the great practical significance of this problem, much published material has been devoted to it. In an official review from the General Electric Company [6], a summary of the research in this field was presented. Later, experimental and theoretical studies on the determination of the thickness of a liquid layer were published [7]. In theoretical studies by Jeffreys [8], the thickness of the liquid layer was determined from the balance between gravitational and viscous stresses developed in the film, but without considering capillary forces.

We show below that in the most important case met in practice, that of slow withdrawal of a solid from a liquid, capillary forces play a basic role and disregarding them leads to completely inaccurate results. Equally erroneous were the computations by Couger and Ward, presented in the review from the General Electric Company mentioned above. In studying the flow of a film the authors not only disregarded capillary forces, but also employed the wrong boundary conditions.

On the basis of certain simplifying assumptions we examine the thickness of the liquid layer remaining on the surface of a solid withdrawn from a liquid. We assume that the curvature of the solid surface is very small compared to the thickness of the film remaining on the surface. In this manner we can consider the body as an infinite plane and the film as a closely adhering parallel liquid layer. We also assume that the vessel containing the liquid is relatively large, so that we may disregard the wall effect on the meniscus in the vicinity of the solid. We limit our analysis to the withdrawal of a plate at a constant, low velocity $v_0$. The significance of the low velocity is explained later.

Let us assume that the solid is being withdrawn vertically upward [9]. The liquid far from the plate is at rest, and its surface is horizontal. We take this surface as the plane $y = 0$. We direct the $y$ axis in a direction perpendicular to the plate, and the $x$ axis along the plate.

The liquid that wets the surface of the solid forms a meniscus in the vicinity of the plate. The shape of a meniscus near a moving plate differs substantially from that of a meniscus near a stationary plate, the case examined in Section 65.

In the vicinity of a moving plate the liquid is entrained and set into motion behind the plate. This motion is due to the following: 1) transfer of a certain momentum from the plate to the viscous liquid (the liquid particles on the surface of the plate are fully entrained by the latter); 2) the effect of gravity, which causes the liquid to flow down the plate. In addition, capillary pressure acts at the surface of the liquid, producing a meniscus near the plate.

It is evident that the thickness $h$ of the liquid layer that remains on the plate must be a function of the velocity of withdrawal $v_0$, of the liquid viscosity $\mu$, of the product $\rho g$ and of the surface tension $\sigma$. It varies, of course, with the height along the plate, and

$$h = h(x).$$

At a sufficient height above the liquid surface, the thickness of the entrained film is extremely small and the motion of the liquid is almost rectilinear. On the other hand, far from the plate, the shape of the meniscus remains virtually undistorted by the motion of the plate.

Thus, the liquid surface may be subdivided into two regions, one in which the liquid is directly entrained by the motion of the plate, and another region where the meniscus is static. The liquid in the first region moves almost parallel to the surface of the plate. In the second region, the liquid is almost motionless. The quantitative considerations given below define accurately the dimensions of these two regions. A separate solution of the hydrodynamic equations must be obtained for each region, and then a smooth matching of the two solutions is required. We begin by analyzing the static meniscus region.

The equation for the shape of a liquid surface with a static meniscus was derived in Section 65. In the case of a plate, this equation is

$$\frac{\frac{d^2h}{dx^2}}{\left[1+\left(\frac{dh}{dx}\right)^2\right]^{3/2}} = \frac{\rho g x}{\sigma}. \tag{133.1}$$

The integral of equation (133.1) that satisfies the condition requiring the liquid surface far from the plate to be a horizontal surface is

$$\frac{\frac{dh}{dx}}{\left[1+\left(\frac{dh}{dx}\right)^2\right]^{1/2}} = \frac{\rho g x^2}{2\sigma} - 1. \qquad (133.2)$$

For a moving plate equation (133.2) is valid only far from the plate, where the thickness of the liquid layer is large and its height x is small.

At large x and small h, the entrainment of the liquid cannot be disregarded. A low value of the film thickness in equation (133.2) corresponds to a transition from the static to the entrainment region. In this case, the liquid film is almost vertical and parallel to the plane y = 0. When the derivative $\frac{dh}{dx} = 0$, the film is in a vertical position. Therefore, the transition to the entrainment region occurs as $\frac{dh}{dx} \to 0$. Equation (133.2) shows that $\frac{dh}{dx} \to 0$ as $x \to \sqrt{2}\left(\frac{\sigma}{\rho g}\right)^{1/2}$. Also, equation (133.1) shows that

$$\frac{d^2h}{dx^2} \to \sqrt{2}\left(\frac{\rho g}{\sigma}\right)^{1/2}.$$

Thus, the solution of the equation for the static meniscus region becomes also the solution for the entrainment region when the following conditions are fulfilled simultaneously:

$$h \to 0, \qquad (133.3)$$

$$\frac{dh}{dx} \to 0, \qquad (133.4)$$

$$x \to \sqrt{2}\left(\frac{\sigma}{\rho g}\right)^{1/2}, \qquad (133.5)$$

$$\frac{d^2h}{dx^2} \to \sqrt{2}\left(\frac{\rho g}{\sigma}\right)^{1/2}. \qquad (133.6)$$

Conditions (133.3) to (133.6) must be satisfied by the function h(x) at the boundary of the static meniscus region. The function h(x) over the entire region can be obtained from a solution for the entrainment region that can be matched with the solution of the static surface equation at the boundary between the two regions. These matching conditions are given by (133.3) to (133.6).

We now turn to the analysis of a liquid film on the plate in the entrainment region.

Here, the liquid motion can be considered to be almost vertical. In the entrainment region the vertical velocity component $v_x$, which is parallel to the plate, is very large compared to the horizontal component $v_y$. We examined this type of motion in Section 131.

In slow steady-state motion, the quadratic terms in equation (131.11) may be omitted and, with the body force per unit denoted by g, this equation can be rewritten as

$$\frac{\sigma}{\rho}\frac{d^3h}{dx^3}+\nu\frac{\partial^2 v_x}{\partial y^2}+g=0. \tag{133.7}$$

The boundary conditions here give the entrainment of the film by the rising plate

$$v_x=v_0 \quad \text{at} \quad y=0. \tag{133.8}$$

and

$$\mu\frac{\partial v_x}{\partial y}=0 \quad \text{at} \quad y=h(x). \tag{133.9}$$

Integrating equation (133.7) with respect to y and considering conditions (133.8) and (133.9), we obtain the velocity distribution across the film

$$v_x=v_0-\left(\frac{\rho g}{\mu}+\frac{\sigma}{\mu}\cdot\frac{d^3h}{dx^3}\right)\left(\frac{y^2}{2}-hy\right). \tag{133.10}$$

We introduce the so-called "liquid consumption rate" which represents the liquid flow in the film

$$Q=\int_0^{h(x)} v_x\,dy=\text{const.} \tag{133.11}$$

Substituting $v_x$ from (133.10) into (133.11), we get

$$Q=v_0h+\left(\rho g+\sigma\frac{d^3h}{dx^3}\right)\frac{h^3}{3\mu}. \tag{133.12}$$

Equation (133.12) determines the film thickness h(x) in terms of the given quantities $v_0$, $\rho$, g, $\sigma$ and $\mu$. It is a third-order differential equation. When integrated, three constants appear in the expression for h. In addition, another unknown constant enters equation (133.12), namely, the liquid flow Q. We therefore require four conditions to determine all the constants.

These are the four boundary conditions that must be satisfied by a solution of equation (133.12).

Before turning to a formulation of these conditions, we rewrite (133.12) in dimensionless form.

We introduce a new dimensionless coordinate $\lambda$, defined by

$$\lambda=\left(\frac{3\mu}{\sigma}\right)^{1/3}\frac{v_0^{4/3}}{Q}x, \tag{133.13}$$

and a dimensionless liquid film thickness

$$L=\frac{v_0 h(\lambda)}{Q}. \tag{133.14}$$

Introducing $\lambda$ and L into (133.12), we get

$$\frac{d^3L}{d\lambda^3}-\frac{1-L}{L^3}+\frac{\rho g Q^2}{3\mu v_0^3}=0. \tag{133.15}$$

Now, the boundary conditions that must be satisfied by a solution of (133.15) can be formulated.

At a considerable height above the liquid surface the film must have a constant limiting thickness, and its surface must be parallel to the plate. The limiting thickness $h_0$ is

$$h_0=\frac{Q}{v_0}.$$

The limiting value of film thickness is obtained at a very high value of x. We may assume with sufficient accuracy that

$$h\to h_0 \quad \text{as} \quad x\to\infty,$$

or

$$L\to 1 \quad \text{as} \quad \lambda\to\infty. \tag{133.16}$$

The condition for the verticality of the liquid surface is

$$\frac{dL}{d\lambda}\to 0 \quad \text{as} \quad \lambda\to\infty. \tag{133.17}$$

Since the surface must be planar, its curvature is zero. This gives

$$\frac{d^2L}{d\lambda^2}\to 0 \quad \text{as} \quad \lambda\to\infty. \tag{133.18}$$

The fourth condition is derived from the matching conditions of the solution of equation (133.15) and the static meniscus equation (133.1). To find the matching conditions we first simplify equation (133.15) on the assumption that its last term is small and may be omitted. This assumption is valid when the flow Q is proportional to the plate velocity $v_0$ to a power $n>\frac{2}{3}$ and $v_0$ is sufficiently small.

The computations below show that $Q \approx v_0^{5/3}$ and thus the last term is actually proportional to $v_0^{1/3}$. For a sufficiently low $v_0$, it is therefore small compared to unity. The other terms of equation (133.15) are dimensionless and of the order of unity. We may therefore write

$$\frac{d^3L}{d\lambda^3} - \frac{1-L}{L^3} = 0. \qquad (133.19)$$

Since no dimensional quantities enter equation (133.19), its numerical solution is considerably simplified. Moreover, an important conclusion regarding the fourth boundary condition can be reached. Matching of the solutions of equations (133.1) and (133.19) must occur within a range of film thicknesses that is very large compared to the limiting thickness $h_0$ (the dimensionless thickness $L = 1$), but at the same time, small compared to the film thickness in the static meniscus region. As shown by (133.5) the region of small meniscus thicknesses corresponds to a definite value of x and to a constant but small value of the surface curvature (as a reminder — with small curvature, the surface is given by $\frac{d^2h}{dx^2}$). The matching condition, therefore, requires a constant curvature of the liquid surface in the matching region. More exactly, it is required that the surface curvature found from the static solution as $h \to 0$ be equal to the curvature of the entrained film as $h \to \infty$. The matching condition in dimensionless form is

$$\left(\frac{d^2L}{d\lambda^2}\right)^{\text{stat}}_{L\to 0} = \left(\frac{d^2L}{d\lambda^2}\right)^{\text{entr}}_{L\to\infty}, \qquad (133.20)$$

where $\left(\frac{d^2L}{d\lambda^2}\right)^{\text{stat}}_{L\to 0}$ is the dimensionless static meniscus curvature for small film thicknesses, and $\left(\frac{d^2L}{d\lambda^2}\right)^{\text{entr}}_{L\to 0}$ is the same for the entrained film in a zone of large film thicknesses. Since no dimensional quantities enter equation (133.19), we have

$$\left(\frac{d^2L}{d\lambda^2}\right)^{\text{entr}}_{L\to\infty} = \alpha, \qquad (133.21)$$

where $\alpha$ is an unknown constant. Its value may be found from a numerical solution of equation (133.19). Thus, owing to the fact that equation (133.12) could be transformed into a dimensionless equation (133.19), the matching condition can be greatly simplified by using $\left(\frac{d^2L}{d\lambda^2}\right)^{\text{entr}}_{L\to\infty}$ from the simple relation (133.21).

The value of $\left(\frac{d^2L}{d\lambda^2}\right)^{\text{stat}}_{L\to 0}$ is computed easily from (133.6) and from the definition of L and $\lambda$ given by (133.13) and (133.14), respectively. A simple calculation gives

$$\left(\frac{d^2L}{d\lambda^2}\right)^{\text{stat}}_{L\to 0} \to \sqrt{2}\left(\frac{\rho g}{\sigma}\right)^{1/2} \frac{Q}{v_0^{5/3}}\left(\frac{\sigma}{3\mu}\right)^{2/3}. \tag{133.22}$$

The boundary condition (133.20) can now be rewritten as

$$Q = \frac{\alpha}{\sqrt{2}} \frac{v_0^{5/3}(3\mu)^{2/3}}{\sigma^{1/6}(\rho g)^{1/2}}. \tag{133.23}$$

Numerical integration of (133.19) gives

$$\alpha = 0.63. \tag{133.24}$$

Then the liquid flow through one linear centimeter of film breadth is

$$Q = 0.93\,\frac{v^{5/3}\mu^{2/3}}{\sigma^{1/6}(\rho g)^{1/2}}. \tag{133.25}$$

The limiting thickness of the liquid layer entrained by the plate is given by [5]

$$h_0 = \frac{Q}{v_0} = 0.93\,\frac{(\mu v)^{2/3}}{\sigma^{1/6}(\rho g)^{1/2}}. \tag{133.26}$$

Equation (133.26) shows that the limiting thickness of the entrained film is proportional to the withdrawal velocity of the solid and to the liquid viscosity to the 2/3 power. The thickness is but a weak function, however, of the liquid density and surface tension. The very weak dependence of the thickness on surface tension indicates that surface-active agents should have little effect.

The thickness of a pure water film ($\sigma = 72$, $\mu = 10^{-2}$, $\rho = 1$) entrained by the solid is, according to (133.26),

$$h_0 \approx 8 \cdot 10^{-3} v_0^{2/3}.$$

In deriving equation (133.26), or more exactly, in deriving (133.19) from (133.15), we have neglected $\frac{\rho g Q^2}{3\mu v_0^3}$ as small compared with unity.

Substitution of Q from (133.25) shows that this procedure is valid, provided the following inequality holds

$$v_0 \ll \frac{\sigma}{\mu}. \tag{133.27}$$

For pure water, this inequality gives $v_0 \ll 7 \cdot 10^3$ cm/sec; for glycerine we obtain $v_0 \ll 40$ cm/sec, etc. Although the above solution is valid in principle only at low velocities, it is always valid for withdrawal velocities encountered in practice with low-viscosity liquids.

For high velocities of withdrawal an expression for $h_0$ may be derived by dimensional analysis: the thickness of the film remaining on the solid at high withdrawal velocities cannot depend on the nature of the static liquid meniscus. In other words, $h_0$ is no longer a function of surface tension and is determined exclusively by the quantities g, $\rho$, $\mu$ and $v_0$. The only combination with the dimension of length that may be constructed of these quantities is the expression $\left(\frac{\mu v_0}{\rho g}\right)^{1/2}$.

Therefore, at $v_0 \gg \frac{\sigma}{\mu}$, we have

$$h_0 = A\left(\frac{\mu v_0}{\rho g}\right)^{1/2}. \qquad (133.28)$$

where A is an undetermined constant. It was evaluated by B. V. Deryagin [10] by integrating equation (133.15) and was found to be equal to unity.

In the general case of an arbitrary withdrawal velocity of the plate, we get

$$h_0 = \left(\frac{\mu v_0}{\rho g}\right)^{1/2} f\left(\frac{\mu v}{\sigma}\right), \qquad (133.29)$$

where the function $f\left(\frac{\mu v_0}{\sigma}\right)$ takes the form

$$f\left(\frac{\mu v_0}{\sigma}\right) \approx 0.93\left(\frac{\mu v_0}{\sigma}\right)^{1/6} \quad \text{for} \quad \frac{\mu v_0}{\sigma} \ll 1, \qquad (133.30)$$

$$f\left(\frac{\mu v_0}{\sigma}\right) \approx 1 \quad \text{for} \quad \frac{\mu v_0}{\sigma} \gg 1. \qquad (133.31)$$

Equation (133.29) has been carefully verified in the experiments of B. V. Deryagin and A. S. Titiyevskaya [11] over a wide range of withdrawal velocities.

Figure 109 gives a comparison between the theoretical (solid and broken curves) and measured (circles) values of film thickness.

It is evident that agreement between the theory and experimental data is quite satisfactory. The differences between the computed and the measured thicknesses lie within the range of the experimental error.*

Equation (133.29) may be used to estimate the thickness of the liquid film on a solid of any shape, provided its radius of curvature is extremely large in comparison to the capillary constant $\left(\frac{\sigma}{\rho g}\right)^{1/2}$.

---

*The study by B. V. Deryagin and A. S. Titiyevskaya [11] mentions a discrepancy between the experimental values for $h_0$ and those computed by L. D. Landau and the author. This is based on a misunderstanding. The original article by Landau and Levich contained a serious and obvious typographical error in the substitution of numerical values into equations (133.23) and (133.25); instead of $0.63\frac{\sqrt[3]{9}}{\sqrt{2}} = 0.93$, the text shows = 2.29.

An important experimental verification of the above theory was made by M. M. Kusakov [12]. If some object — a thread, for example — is placed at a certain distance (of the order of the capillary constant) from the plate being withdrawn from the liquid, that object has a significant effect on the film thickness. The object's "shadow" is discovered in interferometric measurements of the film thickness; within the "shadow" the film thickness is less than it is over the balance of the surface of the plate. Introducing the object alters the meniscus, but apparently cannot directly affect the film that remains on the solid. M. M. Kusakov's study clearly demonstrated the role of the meniscus. A change in the meniscus affects the matching conditions, and this, in turn, is reflected in the thickness of the film. Entrainment of liquid by a thin thread ($R < h$)

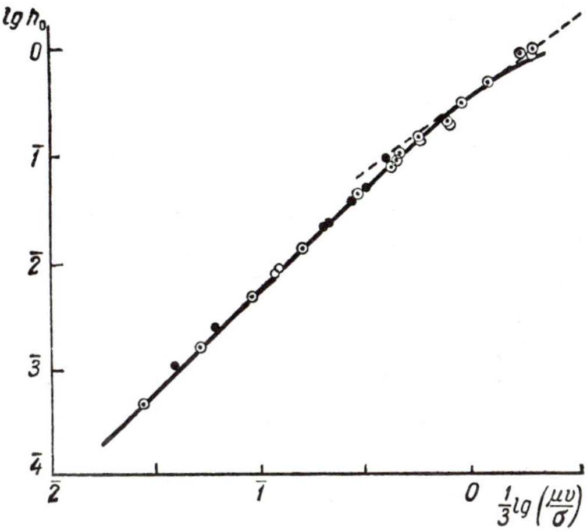


Figure 109. Entrained film thickness $h_0$ as a function of the dimensionless group $\mu v/\sigma$.

and liquid flow down the thread were theoretically examined by B. V. Deryagin [9] and experimentally verified by V. S. Bondarenko [13]. These studies established that capillary waves appear on the liquid surface film. The importance of surface tension during flow of the film at a large height above the meniscus has thus been established. The development of such capillary waves with time results in break-up of the cylindrical layer into a series of large and small drops that succeed one another at regular intervals. The shape of the drops depends in large measure on the contact angle.

## 134. WAVE MOTION IN THIN LIQUID LAYERS

As indicated above, the strictly laminar gravity flow down a plate is superseded by a wave regime at Reynolds numbers in excess of approximately 30 [1, 14]. In the wave regime the surface of the film flowing down a plate is covered with capillary waves. This regime was studied theoretically by P. L. Kapitsa [14].

The main difference between laminar and wave flow is that capillary forces have a principal role in establishing the distribution in the wave regime. These forces appear when the film surface is deformed and they become important when their magnitude is comparable to that of the other active forces — gravity and viscosity. We examine wave motion in a film using the general equation for laminar non-steady-state motion of a thin film (131.11), which here takes the form

$$\frac{\partial v_x}{\partial t}+v_x\frac{\partial v_x}{\partial x}-\left(\int\frac{\partial v_x}{\partial x}\,dy\right)\frac{\partial v_x}{\partial y}=\frac{\sigma}{\rho}\frac{d^3h}{dx^3}+\nu\frac{\partial^2 v_x}{\partial y^2}+g. \tag{134.1}$$

The continuity equation (131.12) states

$$\frac{\partial h}{\partial t}=-\frac{\partial}{\partial x}\left(\int v_x\,dy\right). \tag{134.2}$$

The velocity $v_x$ that we are seeking is a function of the coordinates x and y.

In the following, we are interested only in those surface waves having a length $\lambda$ large compared to the film thickness h.

Equations (134.1) and (134.2) can be greatly simplified in the case of long waves. They may be averaged over the thickness of the film by replacing $v_x(x, y)$ with the average velocity $v(x)$.

The boundary conditions (131.4) and (131.5) are fulfilled when the velocity is given in terms of y by an equation similar to (131.14)

$$v_x=\frac{3\bar{v}(x,t)}{h}\left(y-\frac{y^2}{2h}\right), \tag{134.3}$$

where $\bar{v}(x)$ is the average velocity over the cross section and is a function of the position along the film and of time.

Substituting equation (134.3) into (134.1), we obtain

$$\begin{gathered}3\left(\frac{y}{h}-\frac{y^2}{2h^2}\right)\frac{\partial\bar{v}}{\partial t}+9\left(\frac{y}{h}-\frac{y^2}{2h^2}\right)^2\bar{v}\,\frac{\partial\bar{v}}{\partial x}-\\ -9\bar{v}\,\frac{\partial\bar{v}}{\partial x}\cdot\frac{1}{2h^2}\left(y^2-\frac{y^3}{3h}\right)\left(1-\frac{y}{h}\right)=\frac{\sigma}{\rho}\frac{d^3h}{dx^3}-\nu\frac{3\bar{v}}{h^2}+g.\end{gathered} \tag{134.4}$$

Let us average equation (134.4) over y. To do this, we integrate the equation with respect to y and divide by h; then*

---

*In the original study by P. L. Kapitsa, the term $v_y\frac{\partial v_x}{\partial y}$ was omitted. A somewhat different value of the numerical coefficient for $\bar{v}\frac{\partial\bar{v}}{\partial x}$ resulted. Let us note that, for the regime under consideration, the quadratic terms are not small because of the presence

$$\frac{\partial \bar{v}}{\partial t} + \frac{9}{10}\bar{v}\frac{\partial \bar{v}}{\partial x} = \frac{\sigma}{\rho}\frac{d^3 h}{dx^3} - \frac{3\nu\bar{v}}{h^2} + g. \qquad (134.5)$$

Substituting the value of $v_x$ into the continuity equation (134.2), we obtain:

$$\frac{\partial h}{\partial t} = -\frac{\partial(\bar{v}h)}{\partial x}. \qquad (134.6)$$

We now examine small-amplitude waves (wavelength ≫ amplitude). Accordingly, the thickness of the layer h is written as

$$h = h_0 + h_0\varphi = h_0(1+\varphi), \qquad (134.7)$$

where $h_0$ is the average film thickness (Figure 110) and $h_0\varphi$ is the deviation of the surface from that average thickness.

If we assume that the capillary waves on the film surface are not damped (this is explained below), then all the quantities in equations (134.5) and (134.6) — film thickness and average liquid velocity — are functions of the argument (x - ct), where c is the phase velocity of the waves.

Therefore,

$$\frac{\partial h}{\partial t} = -h_0 c\frac{\partial \varphi}{\partial x}, \qquad (134.8)$$

$$\frac{\partial \bar{v}}{\partial t} = -c\frac{\partial \bar{v}}{\partial x}. \qquad (134.9)$$

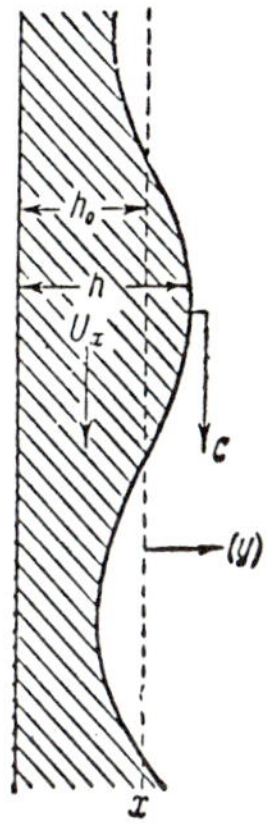

Figure 110. Wave motion in a film flowing down a plate.

Introducing expressions (134.7) to (134.9) into equations (134.5) and (134.6), we obtain

$$-c\frac{\partial \bar{v}}{\partial x} + \frac{9}{10}\bar{v}\frac{\partial \bar{v}}{\partial x} = \frac{\sigma h_0}{\rho}\frac{d^3\varphi}{dx^3} - \frac{3\nu\bar{v}}{h_0^2(1+\varphi)^2} + g, \qquad (134.10)$$

$$\frac{\partial}{\partial x}[(c-\bar{v})h_0(1+\varphi)] = 0. \qquad (134.11)$$

Integrating equation (134.11), we get

$$h_0(c-\bar{v})(1+\varphi) = \text{const} = h_0(c - v_0), \qquad (134.12)$$

where $v_0$ is the velocity at the average stream cross section $h_0$.

---

of an average film motion that does not have a wave character. The coefficient in equation (134.5) was worked out by V. K. Bushmanov at the request of the author.

The relation between $\overline{v}$ and $\varphi$ in the surface wave is determined by equation (134.12)

$$\overline{v}=c-\frac{c-v_0}{1+\varphi}. \tag{134.13}$$

Since the wave amplitude $h_0\varphi$ is small compared to the film thickness $h_0$, we have $\varphi \ll 1$. Therefore, to an accuracy of quantities of the third order of infinitesimals, we have

$$\overline{v}=v_0+(c-v_0)\varphi-(c-v_0)\varphi^2, \tag{134.14}$$

$$\frac{\partial\overline{v}}{\partial x}=(c-v_0)(1-2\varphi)\frac{d\varphi}{dx}. \tag{134.15}$$

Substituting expressions (134.14) and (134.15) into (134.10), we derive the equation governing $\varphi$. In the first approximation, we have

$$\frac{\sigma h_0}{\rho}\frac{d^3\varphi}{dx^3}-(c-v_0)\left(\frac{9v_0}{10}-c\right)\frac{d\varphi}{dx}-\\-\frac{3\nu}{h^2}(c-3v_0)\varphi+g-\frac{3\nu v_0}{h_0^2}=0. \tag{134.16}$$

Equation (134.16) is a linear third-order differential equation. For it to have undamped periodic solution, both the free term and the coefficient of $\varphi$ must equal zero.

The first of these requirements is

$$g-\frac{3\nu v_0}{h_0^2}=0. \tag{134.17}$$

Equation (134.17) signifies that, in a first approximation, the thickness $h_0$ is close to the thickness of the layer in laminar flow down a plate [cf. formula (131.14)].

The second requirement takes the form

$$c-3v_0=0 \tag{134.18}$$

and defines the phase velocity of the wave in the first approximation.

Equation (134.16) becomes

$$\frac{\sigma h_0}{\rho}\frac{d^3\varphi}{dx^3}-(c-v_0)\left(\frac{9v_0}{10}-c\right)\frac{d\varphi}{dx}=0, \tag{134.19}$$

or, after replacing c by $3v_0$, in the first approximation,

$$\frac{\sigma h_0}{\rho}\frac{d^3\varphi}{dx^3}+4,2v_0^2\frac{d\varphi}{dx}=0. \tag{134.20}$$

If k is the wave number, $\varphi$ may be written

$$\varphi=\alpha\sin(kx-\omega t), \tag{134.21}$$

where k is defined by

$$k=\left(\frac{4.2v_0^2\rho}{\sigma h_0}\right)^{1/2}. \tag{134.22}$$

The frequency $\omega$ accordingly is

$$\omega = ck = 3v_0^2 \left(\frac{4{,}2\rho}{\sigma h_0}\right)^{1/2}. \tag{134.23}$$

An important feature of equation (134.23) is that $\omega \approx v_0^2$ and that both approach zero simultaneously. This is the basic difference between surface waves on a film flowing down a plate and ordinary capillary waves. The undamped nature of waves on a film of viscous liquid flowing down a plate also differentiates them from the usual capillary waves. Waves on the film surface are constantly maintained by the force of gravity, which moves the film as a whole. Up to now we have regarded the wave amplitude ($h_0\alpha$) as arbitrary. In fact, however, in the undamped regime the amplitude cannot remain arbitrary.

To maintain an undamped regime, the energy dissipated by wave motion must be balanced exactly by the work done by gravity.

According to (1.15), the energy dissipated in the flow of a thin film is

$$-\frac{dE}{dt} = \mu \int_0^h \left(\frac{\partial v_x}{\partial y}\right)^2 dy.$$

Here, $\frac{dE}{dt}$ applies to a unit length of film. The derivatives $\frac{dv_x}{dx}$ and $\frac{dv_y}{dy}$ are omitted as being small for a thin film. Computation gives

$$-\frac{dE}{dt} = 3\mu \frac{\bar{v}^2}{h}.$$

Averaging this equation over a wavelength, we derive the average energy loss per unit length of film cross section

$$\overline{\frac{dE}{dt}}^x = -3\mu \frac{1}{\lambda} \int_0^\lambda \frac{\bar{v}^2}{h} dx = -3\mu \frac{\overline{\bar{v}^2}^x}{h}.$$

The average work done by gravity per unit of length is

$$\overline{W} = \rho g \overline{\bar{v}h}^x = \rho g Q = \rho g h_0 v_0,$$

where Q is the liquid flow. To maintain a steady-state regime, therefore, the following equality must hold

$$\rho g Q = \frac{d\overline{E}^x}{dt} = 3\mu \left(\frac{\overline{\bar{v}^2}^x}{h}\right). \tag{134.24}$$

Substituting the value of v from (134.13) and that of h from (134.7) into equation (134.24), we obtain:

$$\rho g Q = 3\mu \frac{v_0^2}{h_0} \Phi = 3\mu \frac{Q^2}{h_0^3} \Phi, \tag{134.25}$$

where $\Phi$ designates the quantity

$$\Phi = \frac{1}{\lambda} \int_0^\lambda \frac{\left(1 + \frac{c}{v_0}\varphi\right)^2}{(1+\varphi)^3}\, dx.$$

The average film thickness at a given liquid flow Q, therefore, is determined by

$$h_0^3 = \frac{3\nu Q}{g}\,\Phi. \tag{134.26}$$

Equation (134.26) is used to determine the amplitude of undamped waves. A complication arises in the determination of this amplitude because $\varphi$ is a function of two variables, $\alpha$ and the ratio $c/v_0$. The latter, according to condition (134.18), is equal to 3 only in the first approximation when $\alpha$ is extremely small and the average layer thickness $h_0$ is not a function of $\alpha$. In the second approximation, $h_0$ is found to be a function of $\alpha$.

To find $\alpha$ from equation (134.26) we may follow the Kapitsa method and employ qualitative considerations regarding wave stability. Namely, the quantity $\Phi$ must take on the minimum value possible. Minimum $\Phi$ corresponds to the smallest film thickness with which the balance between energy dissipated and work done by gravity can be maintained. (The film thickness $h_0$ is proportional to $\Phi^{1/3}$.) The minimum value of the average thickness corresponds to the minimum potential energy of the film in the gravitational field and to the most stable flow (at a given liquid flow rate Q).

In the second approximation $h_0$ is a function of $\alpha$; computations based on equation (134.26) and the requirement that $\Phi$ be a minimum have yielded the following values for $\alpha$, c and $\Phi$:

$$\alpha = 0.21, \tag{134.27}$$

$$c = 2.4 v_0, \tag{134.28}$$

$$\Phi = 0.8. \tag{134.29}$$

The wave number is determined from equation (134.19), by taking (134.26), (134.28) and (134.29) into account. We obtain*:

$$k = \left(0.9\,\frac{gQ}{\rho\nu}\right)^{1/2}. \tag{134.30}$$

It has been shown that the above formulas are applicable when

$$\lambda > 14 h_0.$$

---

*For reasons mentioned above, formula (134.30) in the study by P. L. Kapitsa contains the coefficient 0.7.

According to expressions (134.3), (134.14), (134.21) and (134.27), retaining terms up to the second order, the velocity component $v_x$ is

$$\begin{aligned} v_x &= 3(v_0 + 1.4v_0\varphi - 1.4v_0\varphi^2)\left(\frac{y}{h} - \frac{y^2}{2h^2}\right) = \\ &= 3v_0[1 + 0.6\sin(kx - \omega t) - 0.3\sin^2(kx - \omega t)]\left(\frac{y}{h} - \frac{y^2}{2h^2}\right). \end{aligned} \tag{134.31}$$

Accordingly,

$$v_y = -1.8v_0 k\cos(kx - \omega t)\left[1 - \sin(kx - \omega t)\right]\left(\frac{y^2}{2h} - \frac{y^3}{6h^2}\right). \tag{134.32}$$

The results of this theoretical analysis of the wave regime in the flow of film down a plate were experimentally verified in the thorough study of P. L. Kapitsa and S. P. Kapitsa [15].

Direct measurements of wave profile, made by the shadowgraph optical technique, confirmed the theoretical predictions, particularly that of the amplitude.

Agreement between theory and experiment was good. It was found that wave motion of the film readily gives way to turbulent motion when there are external disturbances to the liquid flow.

## 135. TURBULENT MOTION IN A FILM

Turbulent motion in a film appears at Reynolds numbers of the order of 1500. It appears earlier when large disturbances are present. Turbulent motion in a film is a special case of turbulent motion in a liquid with a free surface.

Turbulent motion at a free surface has certain features that distinguish it from turbulent motion in the bulk of a liquid or at a solid wall.

We now examine the flow of a liquid bounded on one side by a solid wall and having a free surface on the other. The average velocity in the film near the solid wall follows the logarithmic law:

$$U = \frac{v_0}{\sqrt{a}} \ln \frac{v_0(h - y)}{a\nu}, \tag{135.1}$$

where the coordinate y is measured from the liquid surface. A turbulent boundary layer appears in the liquid, starting at the wall and gradually embracing the entire thickness of the film. At a fairly large distance from the source of the film (of the order of 50 to 100 h), the entire film may be considered to be in turbulent flow. The average velocity then follows the logarithmic law (135.1) throughout the entire film. The characteristic eddy velocity $v_0$ corresponds to the average velocity U in the film. This characteristic velocity may be expressed in terms of the shear stress by the relation $\tau = \rho v_0^2$.

On the other hand, $\tau$, in the case of steady-state flow, is equal to the motive force acting on the liquid in the film lying alongside 1 $cm^2$ of wall surface. Therefore,

$$v_0 = \sqrt{gh}. \tag{135.2}$$

In the immediate vicinity of the surface the nature of the velocity distribution cannot remain unaltered. Indeed, the tangential stress at the free surface of the liquid must be zero. For the tangential stress to vanish, the derivative $\frac{\partial U}{\partial y}$ must be zero since the tangential stress is always proportional to this derivative. But $\frac{\partial U}{\partial y}$ is never zero in a logarithmic velocity distribution; hence the turbulent regime must be altered in the vicinity of a free surface. This is also evident in the difference between liquid motion near the surface and in the bulk.

The existence of surface tension assures the stability of the surface and changes the velocity distribution at the surface. The forces that oppose deformation of the surface have a retarding effect on the turbulent motion. This effect may be compared to that of a solid surface. As noted above in several examples, this analogy cannot be considered complete: both the eddy and the average velocities must vanish at a solid wall, while the tangential stress is not equal to zero. The average velocity at the free surface does not vanish, and near the surface it must be nearly constant with depth.

Certain assumptions must be made as to the nature of the eddy velocity near the free surface.

If we assume that the retarding action of the surface (caused by surface tension) assures the stability of the surface, it seems reasonable to assume that only in a zone of depth $\lambda$ near the free surface does the flow regime differ from that in the bulk of the liquid.

In the following, we designate y as the distance *to* the free, although deformable, surface of the liquid. We designate x as the coordinate along the surface. It is then possible to state that, starting at some value $\lambda$ of distance *to* the surface, the x component of velocity ceases to increase with increasing y and remains constant.

Since the velocity components must satisfy the continuity equation, and since the x component of velocity is not a function of y, the y component must be proportional to the distance *to* the surface,

$$v_y \sim y. \tag{135.3}$$

The normal velocity component must vanish at the free surface. In contrast with a solid wall, however, where the normal component falls off as the square of the distance *to* the wall (cf. Section 4), near a free surface the normal component decreases only as the first power of the distance *to* the wall. Formula (135.3) may now be rewritten as

$$v_y = v_0 \frac{y}{\lambda}. \qquad (135.4)$$

The scale of the turbulence eddies must also decrease with the distance to the free surface. Dimensional analysis shows that the scale of the motion must be proportional to the distance to the free surface, since no other quantities with the dimension of length enter into its definition.

Therefore,

$$l \approx y. \qquad (135.5)$$

The effective turbulence viscosity for motion near the surface, therefore, is

$$\nu_{\text{turb}} \approx v_y l \approx v_0 \frac{y^2}{\lambda}. \qquad (135.6)$$

If surface tension is the cause of turbulence damping, then the depth $\lambda$ of the zone of damped turbulence near the free surface must be proportional to $\sigma$. In addition, $\lambda$ can only be a function of eddy velocity $v_0$ and of liquid density $\rho$. Here, dimensional analysis shows that

$$\lambda \approx \frac{\sigma}{\rho v_0^2}. \qquad (135.7)$$

Formula (135.7) simply expresses the condition for surface stability. The quantity $\lambda$ represents the order of magnitude of the thickness of the zone which is deformed by turbulence eddies near the liquid surface. The capillary pressure $\sigma/\lambda$ which arises from the deformation compensates for the dynamic thrust $\rho v_0^2$ and prevents the turbulence eddies from splashing the liquid beyond the limits of the liquid surface. The capillary pressure, which opposes the dynamic thrust, reduces the scale and causes damping of the turbulent motion [see equation (135.5)]. The dimension of the zone in which capillary forces appear is $\lambda$.

Substituting the value of $\lambda$ into expressions (135.4) and (135.6), we get

$$v_y \approx \frac{\rho v_0^3}{\sigma} y \qquad (135.8)$$

and

$$\nu_{\text{turb}} \approx \frac{\rho v_0^3 y^2}{\sigma}. \qquad (135.9)$$

Turbulence viscosity decreases with the distance to the free surface; thus at a depth $\delta_0$ near the surface, $\nu_{turb} \approx \nu$. The thickness of the viscous sublayer obviously is

$$\delta_0 \approx \left(\frac{\sigma\nu}{\rho v_0^3}\right)^{1/2}. \quad (135.10)$$

The decrease of the eddy velocity component inside the viscous sublayer $y = \delta_0$ is given by

$$v_y \approx v_y(\delta_0)\frac{y}{\delta_0},$$

where $v_y(\delta_0)$ is the value of $v_y$ at $y = \delta_0$. The change of $v_y$ within the viscous sublayer (135.4) is governed by the same equation which holds beyond that layer. The periodicity of eddies within the viscous sublayer is not a function of the distance to the surface (see Section 4), and is equal to the period that applies at $y = \delta_0$:

$$\tau \approx \frac{\delta_0}{v_y(\delta_0)} \approx \frac{\lambda}{v_0}. \quad (135.11)$$

## 136. DISSOLUTION OF GAS AT THE LIQUID-GAS INTERFACE OF A FLOWING FILM: THE ELEMENTARY SCRUBBING PROCESS

Absorption into a downflowing liquid film is one of the most important and widely used methods for dissolving gases [16]. Absorbers are usually constructed as packed towers or spray chambers and are used in a wide variety of technological processes that involve dissolution. Absorbers packed with saddles (scrubbers) and spray chambers are used in obtaining aqueous gas solutions (e.g., absorption of HCl vapors by water), in the separation of gas mixtures (e.g., absorption of benzene in the production of chemical coke), in the purification of gases (e.g., isolation of $H_2S$ from coke oven gases), and in collecting one of the components of a gaseous mixture (e.g., recovery of volatile solvents). In some cases (highly soluble gases), the over-all rate of the process is controlled by the rate of delivery of the "solute" to the liquid surface through the gaseous phase. In other cases (slightly soluble gases), the over-all rate of the process is controlled by the rate of transport of the dissolved substance (solute) from the surface into the bulk of the liquid film. This is the case which we will analyze below. Intermediate cases, in which the over-all rate is controlled by adsorption or chemisorption, or the case where the diffusion rates in the gas and liquid film are almost equal, all have considerable practical significance, but are too complex for theoretical analysis. Thus, we limit our treatment to calculation of the transfer rate of the dissolved substance from the surface of the liquid film into the bulk of the film. In treating mass transfer the three flow regimes must be distinguished: laminar, wave and turbulent.

We first examine the case of laminar flow. When the velocity distribution over the film cross section is governed by (131.15), the concentration distribution is given by

$$3\bar{v}\left(\frac{y}{h}-\frac{y^2}{2h^2}\right)\frac{\partial c}{\partial x}=D\left(\frac{\partial^2 c}{\partial x^2}+\frac{\partial^2 c}{\partial y^2}\right), \tag{136.1}$$

with the boundary condition

$$c=c_0 \quad \text{at} \quad y=h,$$

where $c_0$ is the equilibrium concentration of the dissolved substance and $c \to 0$ in the bulk of the film ($y \to 0$).

It is generally considered that in effect a liquid diffusion layer must be formed near the free surface. Most of the variation in the concentration of the solute occurs in this layer. The variation of concentration is much less along the film than in the direction perpendicular to the film; hence, in place of equation (136.1), we may write

$$3\bar{v}\left(\frac{y}{h}-\frac{y^2}{2h^2}\right)\frac{\partial c}{\partial x}=D\frac{\partial^2 c}{\partial y^2}. \tag{136.2}$$

In solving equation (136.2), we use the fact that the diffusion layer thickness is very small compared to the film thickness h. The extent to which this is actually the case is established later.

If the indicated condition is satisfied, diffusion occurs to small depths near the surface of the liquid. Consequently, we may suppose that in equation (136.2), $y \sim h$, and we may rewrite this equation as [17]

$$\frac{3}{2}\bar{v}\frac{\partial c}{\partial x}=D\frac{\partial^2 c}{\partial y^2}. \tag{136.3}$$

Expression (136.3) is in the form of the usual equation of heat conduction.

In solving this equation it is convenient to revert to the convention of measuring y from the liquid surface. With the origin in the surface the boundary conditions become

$$c=c_0 \quad \text{at} \quad y=0, \tag{136.4}$$

$$c=0 \quad \text{as} \quad y\to\infty. \tag{136.5}$$

The solution of equation (136.3) which satisfies boundary conditions (136.4) and (136.5) is

$$c=\frac{2c_0}{\sqrt{\pi}}\int_0^{\frac{y\sqrt{\bar{v}}}{2\sqrt{\frac{2}{3}Dx}}}\exp\{-z^2\}\,dz. \tag{136.6}$$

For the diffusion flux at the surface we have

$$j = D\left(\frac{\partial c}{\partial y}\right)_{y=0} = \sqrt{\frac{3D\bar{v}}{2\pi x}} \cdot c_0. \tag{136.7}$$

The effective thickness of the diffusion layer is

$$\delta = \frac{Dc_0}{j} = \sqrt{\frac{2\pi Dx}{3\bar{v}}}. \tag{136.8}$$

The total flow to the surface of the liquid film of width b and length $l$ is

$$I = c_0 b \sqrt{\frac{6D\bar{v}l}{\pi}}. \tag{136.9}$$

Equation (136.9) may be rewritten in dimensionless form

$$\mathrm{Nu} = \frac{Id}{Dc_0 S} = \sqrt{\frac{6}{\pi}}\, \mathrm{Re}^{1/2} \mathrm{Pr}^{1/2}, \tag{136.10}$$

where the Reynolds number is based on the film thickness, i.e., $\mathrm{Re} = \frac{\bar{v}d}{\nu}$.

Equations (136.9) and (136.10) hold under the following assumptions.

1) Re < 50 to 100,
2) $\delta \ll h$,
3) a fully developed flow exists in the film, and the velocity profile is given by formula (131.15).

Equation (136.8) shows that the thickness of the diffusion layer becomes comparable to h at a distance x from the edge of the film given by

$$x \approx 50 h\, \mathrm{Re}. \tag{136.11}$$

At Re $\approx$ 20, we obtain x$\approx$1000 h. It may be assumed, therefore, that in practice, when a liquid film flows down the packing or the walls of a tower, condition 2) is fulfilled.

The situation here is quite analogous to the diffusion toward the inner surface of a tube which was examined in Section 20; owing to the low value of the diffusion coefficient the diffusion layer does not embrace the entire thickness of the film.

Equation (136.2) has been solved by several authors [18]. All of these authors started from the opposite limiting case, in which $\delta \approx h$, i.e., in which the thickness of the diffusion layer is comparable to that of the film. In this case equation (136.2) cannot be further simplified and its solution is a series which is difficult to represent. Its meaning, however, is very simple: in the limit $\delta \approx h$ and we have $j \approx \frac{Dc_0}{h}$, or Nu = constant. It is clear that such a solution, applicable for x > h (50 Re), has no practical value. Equation (136.9) is of course derivable from a rigorous series solution by taking the appropriate limiting case. This computation was made by V. V. Vyazovov [19].

Equation (136.10) agrees well with the experimental data of M. D. Kuznetsov [20], where the Reynolds number was varied within the range from 27 to 110.

Let us examine the wave regime in the film. In this case, the equation for the concentration is

$$\frac{\partial c}{\partial t}+\frac{3}{2}v_0[1+0.6\sin(kx-\omega t)-0.3\sin^2(kx-\omega t)]\frac{\partial c}{\partial x}- \\ -1.8v_0k\cos(kx-\omega t)(1-\sin(kx-\omega t))\frac{h}{3}\frac{\partial c}{\partial y}=D\frac{\partial^2 c}{\partial y^2} \tag{136.12}$$

with the usual boundary conditions. In the expressions for velocity components we have retained only the first terms of series expansions in powers of wave amplitude $h_0\alpha$, and we have adopted the corresponding expressions for small values of $y = h$.

We solve equation (136.12) by the method of successive approximation, assuming that

$$c=c_1+c_2, \tag{136.13}$$

where $c_1$ is the solution for $\alpha = 0$, given equation (136.6), and $c_2$ is a correction term due to wave motion. Since $c_1$ satisfies the boundary condition (136.4), $c_2$ must satisfy the condition

$$c_2=0 \quad \text{at} \quad y=0. \tag{136.14}$$

Since we are interested in a quasi-steady-state distribution of concentration and quasi-steady-state diffusional flow, we take the average of equation (136.12) over one period. We then obtain

$$\frac{\overline{\partial c_2}^t}{\partial t}+\frac{3}{2}v_0\overline{[1+0.6\sin(kx-\omega t)-0.3\sin^2(kx-\omega t)]}^t\frac{\partial c_1}{\partial x}+\frac{3v_0}{2}\frac{\overline{\partial c_2}^t}{\partial x}- \\ -0.6v_0kh\overline{\cos(kx-\omega t)(1-\sin(kx-\omega t))}^t\frac{\partial c_1}{\partial y}=D\frac{\overline{\partial^2 c_2}^t}{\partial y^2}. \tag{136.15}$$

In virtue of equation (136.5) and the following equalities

$$\frac{\overline{\partial c_2}^t}{\partial t}=0;\quad \overline{\sin(kx-\omega l)}^t=\overline{\cos(kx-\omega t)}^t= \\ =\overline{\sin(kx-\omega t)\cos(kx-\omega t)}^t=0;\quad \overline{\sin^2(kx-\omega t)}^t=\frac{1}{2},$$

equation (136.15) can be rewritten as

$$\frac{3}{2}v_0\frac{\overline{\partial c_2}}{\partial x}-\frac{0.9}{2}v_0\frac{\partial c_1}{\partial x}=D\frac{\overline{\partial^2 c_2}}{\partial y^2},$$

or, by substituting the value of $\frac{\partial c_1}{\partial x}$ at $y = 0$ from (136.6) and rearranging, we get

$$\frac{\partial^2 \bar{c}_2}{\partial y^2} - \frac{3v_0}{2D}\frac{\partial \bar{c}_2}{\partial x} = \frac{0.9\sqrt{3}}{16\sqrt{2\pi}} v_0^{3/2} \frac{y}{x^{3/2}} c_0. \tag{136.16}$$

This equation is invariant with respect to the substitution $y \to \lambda y'$, $x \to \lambda^2 x'$. We therefore seek its solution as a function of the variable $\eta = \sqrt{\frac{3}{2}\frac{v_0}{D}}\frac{y}{\sqrt{x}}$. Obviously, we get

$$\bar{c}_2 = c_0 f\left(\frac{y}{\sqrt{x}}\right). \tag{136.17}$$

The complementary solution of the homogeneous equation (136.16) is identical with a solution of the equation for $\bar{c}_1$. Since $c_2$ must satisfy the zero boundary conditions it need not be given. Therefore, $\bar{c}_2$ is given by a particular solution of the inhomogeneous equation (136.16), namely,

$$\bar{c}_2 = \frac{0.15}{\sqrt{\pi}}\sqrt{\frac{3v_0}{2D}}\frac{y}{\sqrt{x}} c_0. \tag{136.18}$$

Substituting these values of $c_1$ and $\bar{c}_2$ into (136.13), we obtain

$$c = \frac{0.3c_0}{2\sqrt{\pi}}\eta + c_0\int_0^{\eta} e^{-\frac{z^2}{2}}\,dz. \tag{136.19}$$

Because we are interested in the solution near to $y = 0$, we have evaluated all functions, and $\frac{\partial c_1}{\partial y}$ in particular, at $y = 0$. Therefore, the solution (136.19) does not satisfy the condition at infinity. However, if we replaced $\frac{\partial c_1}{\partial y}$ by its complete definition, we could obtain an expression for $c_2$ that would converge rapidly at $y = \infty$ and which, at $y = 0$, would take the form of equation (136.19). Appropriate computations confirm these conclusions. Therefore, as $y \to 0$, we obtain

$$c = c_1 + \bar{c}_2 \approx \frac{c_0}{\sqrt{\pi}}\frac{y}{\sqrt{x}}\frac{\sqrt{v_0}}{\sqrt{\frac{2}{3}D}} + \frac{0.15}{\sqrt{\pi}}\sqrt{\frac{3v_0}{2D}}\frac{y}{\sqrt{x}}c_0, \tag{136.20}$$

where the value of $c_1$ is taken at small values of $y$. Accordingly, the time average of the diffusional flux to the surface is

$$j_{\text{wave}} = j(1 + 0.15), \tag{136.21}$$

where $j$ is the flux defined by equation (136.7). We see that the wave motion in the film increases the diffusional flux although its distribution across the film remains the same.

According to equation (136.21), the correction to the flux is approximately 15 percent. P. L. Kapitsa [21]* obtained a similar result using an entirely different method.

We now analyze turbulent motion in the film.

The diffusional flux in turbulent flow can be approximately determined by considering the nature of turbulent motion near the surface. At Pr ≫ 1, we can assume that the major diffusional resistance is offered by the diffusion sublayer, which is thinner than the viscous sublayer. Then, the mass flux in the viscous sublayer is

$$j = D_{\text{turb}} \frac{\partial c}{\partial y}, \tag{136.22}$$

where $D_{\text{turb}}$ is the turblent diffusion coefficient in the viscous sublayer. This equation can be rewritten as

$$D_{\text{turb}} \approx v_y^2 \cdot \tau,$$

where the velocity of turbulent mixing and the period are determined by equations (135.8) and (135.11). Substituting their values, we obtain

$$D_{\text{turb}} \approx v_0 \frac{y^2}{\lambda}. \tag{136.23}$$

The turbulent diffusion coefficient decreases as the surface is approached and, at y = δ, it is equal to the molecular diffusion coefficient D

$$\frac{v_0 \delta^2}{\lambda} \approx D,$$

whence,

$$\delta \approx \left(\frac{D\lambda}{v_0}\right)^{1/2} \approx \left(\frac{D\sigma}{\rho v_0^3}\right)^{1/2}. \tag{136.24}$$

The total diffusional flux is

$$j = \frac{Dc_0}{\delta} = \frac{D^{1/2} \rho^{1/2} v_0^{3/2}}{\sigma^{1/2}} c_0. \tag{136.25}$$

By expressing $v_0$ in terms of the flow velocity U, according to equations (135.1) and (135.2), we finally obtain,

$$j = \left(\frac{D\rho}{\sigma}\right)^{1/2} \frac{U^{3/2} c_0}{\left(\ln \frac{h}{\nu} \sqrt{gh}\right)^{3/2}}. \tag{136.26}$$

---

*The theoretically determined value of the diffusional flux for a wave regime has been experimentally confirmed by B. I. Konobeyev, N. M. Zhavoronkov and V. A. Malyusov [24].

Equation (136.26) can be given in a dimensionless form

$$\mathrm{Nu} = \mathrm{Pr}^{1/2}\mathrm{Re}^{3/2}\left(\frac{\rho\nu}{2h\sigma}\right)^{1/2}\frac{1}{\left(\ln\frac{h}{\nu}\sqrt{gh}\right)^{3/2}}.$$

Thus, Nu is found to be proportional to $\mathrm{Re}^{3/2}$.

In the actual practice of gas absorption in packed scrubbers, the downflow of the film is usually in the wave or turbulent regime. The state of experimentation in this field is, however, such that quantitative comparison of the derived formulas with experimental data would, for the time being, be difficult.

The general conclusions to be drawn from equations (136.21) and (136.26) can be summarized as follows: 1) in wave motion of a liquid film, the rate of dissolution is proportional to $\sqrt{v}$ (v is the linear flow velocity of the film); 2) the magnitude of the diffusional flux in wave motion is greater than in a laminar regime by approximately 15 percent; 3) transition to turbulent flow at the liquid-gas interface should denote a substantial increase of the convective diffusion velocity, which increases proportionally to $v^{3/2}$ in the liquid film.

It can be presumed that these qualitative conclusions agree with experimental data.

N. Kuz'minykh [23] measured the rate of absorption of gases as a function of the linear flow velocity when diffusion in the liquid film was controlling. The absorption rate was found to be proportional to $v^{0.3}$. It should be remembered, however, that the velocity measured in these experiments was the average flow velocity calculated with respect to the effective surface of the packing. In actuality, the entirety of the true liquid surface is not active in absorption; rather only a portion of it is. The figure cited, therefore, does not contradict formula (136.21). The same is true of the work by M. D. Kuznetsov, who studied absorption in a film flowing down a cylindrical wall. The flow regime, however, was not specified.

A. N. Ternovskaya and A. P. Belopol'skiy [22] added a surface-active agent to the liquid phase. They discovered that in cases when the diffusion in the liquid film was controlling, the rate of absorption of gases was noticeably lowered.

On the basis of the concepts developed above, this effect may be interpreted in the following manner: the surface-active agent damps the capillary waves on the surface of the film, causing the wave flow to become laminar and, accordingly, the diffusional flux of the solute in the film is reduced.

Unfortunately, there are no really accurate data on gas absorption in turbulent film flow. The studies of the mechanism of gas dissolution, therefore, cannot be considered complete either from an experimental or from a theoretical standpoint.

In conclusion, we would like to express the hope that this modest effort will be of some benefit to investigators and students in the fields of physical chemistry and chemical engineering, and that it will promote further strengthening of the creative collaboration among physicists, chemists and engineers.

## BIBLIOGRAPHY


1. Friedman, S., and C. Mayer, Ind. Eng. Chem., p. 885 (1941); C. Kirkbride, Trans. Am. Inst. Chem. Eng. 30, 170 (1933); Fallah, Hunter and Nash, J. Soc. Chem. Ind. 53, 369 (1944); P. Semenov, Zhurnal Tekh. Fiz. 14, 427 (1944).
2. Kusakov, M. M., Zh. Eksp. Teoret. Fiz. 16, 451 (1946).
3. Deryagin, B., G. Strakhovskii and D. Malysheva, Zh. Eksp. Teoret. Fiz. 16, 171 (1946); B. Deryagin, M. Kusakov and K. Krym. ibid. 16, 179 (1946).
4. Kusakov, M. M., and K. S. Krym, Zh. Eksp. Teoret. Fiz. 16, 266 (1946); M. M. Kusakov, Dokl. Akad. Nauk SSSR 54, 145 (1946); M. M. Kusakov and E. A. Razumovskaya, Second All-Union Conference on Friction and Wear in Machinery, Reports, Vol. 1, 1947.
5. Landau, L. D., and V. G. Levich, Acta Physicochimica URSS 17, 42 (1942).
6. Research Staff of General Electric Co., Ltd., London, Phil. Mag. 44, 1002 (1922).
7. Bindham and Young, Journ. Ind. Eng. Chem. 14, 1130 (1922); Hopf., Ann. d. Phys. 32, 777 (1910); Redwood, Petroleum and Products, II, London, 1906; Walley, Analyst 44, 288 (1919); Drummond, J. Soc. Chem. Ind. 43, 323 (1924); Stott, Turner and Sloman, Proc. Roy. Soc. (A) 112, 499 (1926); Satterly and Collingwood, Trans. Roy. Soc. Can. 25, 215 (1931); Satterly and Stucky, Trans. Roy. Soc. Can. 26, 131 (1932); Satterly and Givens, Trans. Roy. Soc. Can. 27, 147 (1933); F. Morey, J. of Res. Nat. Bureau of Standards 25, 285 (1940); G. Barr, Viskozimetriya [Viscosimetry], GONTI, 1938 [G. Barr, A Monograph of Viscometry, Oxford University Press, London, 1931].
8. Jeffreys, H., Proc. Cambr. Phil. Soc. 26, 204 (1930).
9. Landau, L. D., and V. G. Levich, Acta Physicochimica URSS 17, 42 (1942); B. V. Deryagin, Dokl. Akad. Nauk SSSR 39, 11 (1943).
10. Deryagin, B. V., Acta Physicochimica URSS 20, 6 (1945).
11. Deryagin, B. V., and A. S. Titiyevskaya, Dokl. Akad. Nauk SSSR 50, 307 (1945).
12. Kusakov, M. M., and K. S. Krym, Izv. Akad. Nauk SSSR, Dept. Techn. Sci. (1940).
13. Bondarenko, V. S., Trans. of the Groznyy Petrol. Inst., 1948; Dissertation, Moscow, 1948.
14. Kapitsa, P. L., Zh. Ekper. Teoret. Fiz. 18, 3 (1948).
15. Kapitsa, P. L., and S. P. Kapitsa, ibid. 19, 105 (1949).

16. Kasatkin, A. G., Osnovnyye protsessy i apparaty khimicheskoy tekhnologii [Basic Processes and Instruments in Chemical Engineering], Goskhimizdat, 1948.
17. Levich, V. G., Zh. Fiz. Khim. 22, 721 (1948).
18. Pozin, M. E., Zh. Priklad. Khim. 20, 205 (1947).
19. Vyazovov, V., Zh. Tekhn. Fiz. 10, 1519 (1940); Zh. Prikl. Khim. 17, 14 (1940).
20. Kuznetsov, M., Zh. Prikl. Khim. 21, 48 (1948); Van Krevelen and Hoftijzer, Rec. trav. chim. 68, 221 (1949).
21. Kapitsa, P. L., Zh. Eksp. Teoret. Fiz. 21, 964 (1951).
22. Ternovskaya, A., and A. Belopol'skiy, Zh. Fiz. Khim. 24, 43, 981 (1950).
23. Kuz'minykh, I. N., Zh. Prikl. Khim. 11, 38 (1934).
24. Konobeyev, B. I., N. M. Zhavoronkov and V. A. Malyusov, Dokl. Akad. Nauk SSSR 117, 671 (1957).